Immunogenomics and Human Disease

Immunogenomics and Human Disease

András Falus

Semmelweis University, Budapest, Hungary

John Wiley & Sons, Ltd

Telephone (+44) 1243 779777

Email (for orders and customer service enquiries): cs-books@wiley.co.uk
Visit our Home Page on www.wileyeurope.com or www.wiley.com

Other Wiley Editorial Offices

John Wiley & Sons Inc., 111 River Street, Hoboken, NJ 07030, USA

Jossey-Bass, 989 Market Street, San Francisco, CA 94103-1741, USA

Wiley-VCH Verlag GmbH, Boschstr. 12, D-69469 Weinheim, Germany

John Wiley & Sons Australia Ltd, 42 McDougall Street, Milton, Queensland 4064, Australia

John Wiley & Sons (Asia) Pte Ltd, 2 Clementi Loop #02-01, Jin Xing Distripark, Singapore 129809

John Wiley & Sons Canada Ltd, 22 Worcester Road, Etobicoke, Ontario, Canada M9W 1L1

Wiley also publishes its books in a variety of electronic formats. Some content that appears in print may not be available in electronic books.

Library of Congress Cataloging-in-Publication Data

Immunogenomics and human disease/[edited by] András Falus.
p. ; cm.
Includes bibliographical references and index.
ISBN-13 978-0-470-01530-6 (HB : alk.paper)
ISBN-10 0-470-01530-6 (HB : alk.paper)
1. Immunogenetics. I. Falus, András.
[DNLM: 1. Immunogenetics–methods. 2. Genetic Techniques.
3. Genomics–methods. 4. Immunity–genetics. QW 541 I332 2006]
QR184.I46 2006
616.07′96–dc22

2005027905

British Library Cataloguing in Publication Data

A catalogue record for this book is available from the British Library

ISBN-13 978-0-470-01530-6 (HB)
ISBN-10 0-470-01530-6 (HB)

Typeset in 10.5/12.5pt Times by Thomson Press (India) Limited, New Delhi
Printed and bound in Great Britain by Antony Rowe Ltd Chippenham, Wilts
This book is printed on acid-free paper responsibly manufactured from sustainable forestry in which at least two trees are planted for each one used for paper production.

Contents

Preface

Immunology–Immunogenomics–Immunomics

One of the major differences that distinguishes vertebrates from nonvertebrates is the presence of a complex immune system characterized by a highly complex web of cellular and humoral interactions. Over the past 500 million years, many novel immune genes and gene families have emerged and their products form sophisticated pathways providing protection against most pathogens, both extrinsic (microbes) and intrinsic (tumorous tissue). In recent years, the Human Genome Project has laid the foundation for the study of these genes and pathways in unprecedented detail.

Genomics or, in other words, genome-based biology offers an entirely new perspective on strategies applicable to the study of distinct physio-pathological conditions. Indeed, analysis of genomic variation at the DNA level and functional genomics has been used extensively by bioscientists to study various disease states and this trend has spread more recently to applications in basic and clinical immunology. This paradigm shift in the study of biology and in immunology is particularly apt for the understanding of immune regulation in sickness and in health. The extreme versatility of the immune system in its responses to environmental changes, such as pathogen invasion or the presence of malignant or allogeneic tissue, and in some cases toward normal autologous tissues, makes it a very complex system. The study of single immunological parameters has so far failed to answer several questions related to the immune-system complexity. The time has come for a global, systematic approach. New methods have been developed that allow a comprehensive vision of genetic processes taking place in parallel at various levels, encompassing genetic variation (single-nucleotide polymorphism analysis), epigenetic changes (methylation-detection arrays or comparative genomic hybridization) and global transcription analysis (cDNA- or oligonucleotide-based microarrays like the lymphochip or the peptide–MHC microarrays). When combined with rapidly developing bioinformatics tools, these methods provide a new way of approaching the description of complex immunological phenomena.

It is likely that database mining will supplement classical experimentally driven scientific thinking with a more interactive *in silico* processing of information integrated by software programs that link material in the literature with extensive databases from different laboratories. The resulting increased data pool can then be used to generate new hypotheses. The novel word 'immunogenomics' describes the switch from hypothesis-driven immunological research to a more interactive and flexible relationship between classical research and a discovery-driven approach. I also believe that immunogenomics will be particularly useful in clinical immunology for the simple reason that genetic variation of patients and their diseases is so much greater in humans than in inbred animal models.

As several of the authors in this book emphasize, immunogenomics is not in competition with traditional hypothesis-driven science. I hope that the 'fishing' in large data sets will complement traditional approaches by revealing new processes or validating known concepts that best fit the reality of human disease.

At least four major aspects of immunogenomics can be identified. Immunogenomics covers:

1. The convergence of studies on distinct elements of the immune system into a network of information regarding the gene structure, the expression profile and the product localization of various components of the innate and the acquired immune responses.

2. The genetic regulation of physiological immune functions by inherited or epigenetic processes such as immunoglobulin or T cell receptor gene rearrangement, somatic hypermutation, immune selection in primary and peripheral immune tissues, antigen processing and presentation by major histocompatibility molecules, and cytotoxic interactions.

3. The genetic changes in immune function in pathological conditions such as oncohaematological diseases, allergy, immune deficiencies, infections, chronic inflammation, autoimmunity and cancer. Genetic linkage analysis of multicase families has recently identified new major susceptibility loci for a few immunologically determined common diseases. However, the greatest potential lies in genome-wide searches for susceptibility genes that individually might have quite modest effects but cumulatively have a large effect on individual risk. This new era of immunogenomics promises to provide key insights into disease pathogenesis and to identify multiple molecular targets for intervention strategies.

4. Personalized approaches to immune therapy – immunogenomics should help to predict which treatments will be successful or have deleterious side effects in certain populations and, therefore, help select therapies appropriate for individual patients. This new era will start when such data can be easily collected during clinical trials through inexpensive high-throughput methods for detection of genomic variation or for expression profiling in large patient populations. With this strategy, it will be possible to immune-phenotype individuals according to

their genetic make-up and the epigenetic adaptations of their immune system. Hopefully, the kinetics of individual immune responses, the network in which they operate and their vulnerability in sickness or in health, as well as adverse drug effects, may be predicted for each individual.

The global immunogenomics approach stands on three foundations: (i) the revolutionary expansion of genome knowledge that is now available in giant computer databanks; (ii) robust nanotechnology such as microarray chips and similar tools that allow real-time measurement of gene variants and gene expression; and (iii) the availability of improved techniques in immune bioinformatics ('immunomics') that could generate data-mining tools for efficient interpretation of otherwise unmanageable biological information.

Given the complexity of the immune system network and the multidimensionality in clinical situations, such as tumour–host interactions, autoimmune and allergic disfunctions, the comprehension of immunology should benefit greatly from high-throughput DNA array analysis, which can portray the molecular kinetics of an immune response on a genome-wide scale. This will accelerate the accumulation of knowledge and ultimately catalyse the development of new hypotheses in cell biology. Although in its infancy, the implementation of DNA array technology in basic and clinical immunology studies has already provided investigators with novel data and intriguing hypotheses concerning the cascade of molecular events that leads to an effective immune response in diseased tissue.

Recent trends in immunology focus on immunological databases, antigen processing and presentation, immunogenomics, host–pathogen interactions and mathematical modelling of the immune system. Immunogenomics represents one of the first stages in the systems biology era in immunology. Computational analysis is already becoming an essential element of immunology research, particularly in the management and analysis of immunological data.

On the basis of the individual chapters in this book, I foresee the emergence of immunomics not only as a collective endeavour by researchers to decipher the sequences of T cell receptors, immunoglobulins and other immune receptors, but also to functionally annotate the capacity of the immune system to interact with the whole array of self and non-self entities, including genome-to-genome interactions.

András Falus
Budapest, June, 2005

List of Contributors

Adarichev, Vyacheslav A. Section of Molecular Medicine, Department of Orthopedic Surgery, Rush University Medical Center, Cohn Research Building, Room 708, 1735 W. Harrison Str., Chicago, IL 60612, USA

Arima, Kazuhiko Division of Medical Biochemistry, Department of Biomolecular Sciences, Saga Medical School, Saga 849-8501, Japan

Batheja, Ameesha Bioinformatics and Functional Genomics, Johnson & Johnson Pharmaceutical Research and Development, Room B238, 1000 Route 202 S, Raritan, NJ 08869, USA

Beretta, Ottavio Department of Biotechnology and Bioscience, University of Milano-Bicocca, Piazza della Scienzia 2, 20126 Milan, Italy

Berinstein, Neil Cancer Vaccine Program, Sanofi Pasteur, Toronto, Ontario, Canada

Buzas, Edit Department of Genetics, Cell- and Immunobiology, Semmelweis University, 1089 Nagyvárad tér 4, 1445 Budapest, Hungary

Capuano, Giusy Department of Biotechnology and Bioscience, University of Milano-Bicocca, Piazza della Scienzia 2, 20126 Milan, Italy

Co, Dominic O. Department of Pathology and Laboratory Medicine, Room 5580 Medical Science Center, 1300 University Avenue, Madison, WI 53706, USA

Darvas, Ferenc ComGenex Inc., Záhony u. 7, H-1031 Budapest, Hungary

Darvas, Susan Department of Genetics, Cell- and Immunobiology, Semmelweis University, 1089 Nagyvárad tér 4, 1445 Budapest, Hungary

Dormán, György ComGenex Inc., Záhony u. 7, H-1031 Budapest, Hungary

Dunford, Paul J. Johnson & Johnson Pharmaceutical Research and Development, LLC, 3210 Merryfield Row, San Diego, CA 92121, USA

Durandy, Anne INSERM U429 Hopital Necker-Enfants Malades, 149 rue de Sevres, 75015 Paris, France

Falus, András Department of Genetics, Cell- and Immunobiology, Semmelweis University, Immunogenomics Research Group, Hungarian Academy of Sciences, H-1089 Budapest, Hungary

Foti, Maria Department of Biotechnology and Bioscience, University of Milano-Bicocca, Piazza della Scienzia 2, 20126 Milan, Italy

Gabius, Hans-Joachim Faculty of Veterinary Medicine, Institute of Physiology, Physiological Chemistry and Animal Nutrition, Ludwig-Maximilians-University Munich, Veterinärstraße 13, 80539 Munich, Germany

Gay, Renate E. Center of Experimental Rheumatology and WHO Collaborating Center for Molecular Biology and Novel Therapeutic Strategies for Rheumatic Diseases, University Hospital Zurich, Glorastr. 23, 8091 Zurich, Switzerland

Gay, Steffan Center of Experimental Rheumatology and WHO Collaborating Center for Molecular Biology and Novel Therapeutic Strategies for Rheumatic Diseases, University Hospital Zurich, Glorastr. 23, 8091 Zurich, Switzerland

Gish, Kurt Protein Design Laboratories, Fremont, California, USA

Glant, Tibor T. Section of Molecular Medicine, Department of Orthopedic Surgery, Rush University Medical Center, Cohn Research Building, Room 708, 1735 W. Harrison Str., Chicago, IL 60612, USA

Gogolak, Peter Institue of Immunology, Medical and Health Science Center, Faculty of Medicine, University of Debrecen, 98 Nagyerdei Blv., H-4012 Debreceb, Hungary

Granucci, Francesca Department of Biotechnology and Bioscience, University of Milano-Bicocca, Piazza della Scienzia 2, 20126 Milan, Italy

Grapow, Martin Division of Cardio-Thoracic Surgery and Department of Research, University Clinics Basel, Veselanium, Vesalgasse 1, CH–4051 Basel, Switzerland

Grussenmeyer, Thomas Division of Cardio-Thoracic Surgery and Department of Research, University Clinics Basel, Veselanium, Vesalgasse 1, CH–4051 Basel, Switzerland

Gut, Ivo Glynne Centre National de Génotypage, Bâtiment G2, 2 rue Gaston, Crémieux, CP 5721, 91057 Evry Cedex, France

Hatanaka, Yasumaru Laboratory of Biorecognition Chemistry, Faculty of Pharmaceutical Sciences, Toyama Medical and Pharmaceutical University, Sugitany 2630, Toyama 930-0194, Japan

Hegyesi, Hargita Department of Genetics, Cell- and Immunobiology, Semmelweis University, 1089 Nagyvárad tér 4, 1445 Budapest, Hungary

Hennessy, Bryan Department of Molecular Therapeutics, University of Texas, MD Anderson Cancer Center, Box 904, 1515 Holcombe Blvd, Houston, TX 77030-4009, USA

Ho, George Bioinformatics and Functional Genomics, Johnson & Johnson Pharmaceutical Research and Development, Room B238, 1000 Route 202 S, Raritan, NJ 08869, USA

Hogan, Laura H. Department of Pathology and Laboratory Medicine, Room 5580 Medical Science Center, 1300 University Avenue, Madison, WI 53706, USA

Homey, Bernhard Department of Dermatology, Heinrich-Heine University, Düsseldorf, Germany

Huber, Lars C. Center of Experimental Rheumatology and WHO Collaborating Center for Molecular Biology and Novel Therapeutic Strategies for Rheumatic Diseases, University Hospital Zurich, Glorastr. 23, 8091 Zurich, Switzerland

Igaz, Peter Department of Medicine II, Semmelweis University, 1089 Nagyváradtér 4, 1445 Budapest, Hungary

Izuhara, Kenji Division of Medical Biochemistry and Division of Medical Research, Department of Biomolecular Sciences, Saga Medical School, Saga 849-8501, Japan

Kanaji, Sachiko Division of Medical Biochemistry, Department of Biomolecular Sciences, Saga Medical School, Saga 849-8501, Japan

Kanaji, Taisuke Division of Medical Biochemistry, Department of Biomolecular Sciences, Saga Medical School, Saga 849-8501, Japan

Krishnan, Sandeep Department of Cellular Injury, Walter Reed Army Institute of Research, Silver Spring, MD 20910, USA

Kuchzak, Ramón Centre National de Génotypage, Bâtiment G2, 2 rue Gaston, Crémieux, CP 5721, 91057 Evry Cedex, France

Lefkovits, Ivan Division of Cardio-Thoracic Surgery and Department of Research, University Clinics Basel, Veselanium, Vesalgasse 1, CH–4051 Basel, Switzerland

Lefkovits, Michael Clinic Barmelweid, Switzerland

Marincola, Francesco Immunogenetics Section, Department of Transfusion Medicine, Clinical Center, National Institutes of Health, Bethesda, MD20892, USA

Masumoto, Kiyonari Division of Medical Biochemistry, Department of Biomolecular Sciences, Saga Medical School, Saga 849-8501, Japan

Matt, Peter Division of Cardio-Thoracic Surgery and Department of Research, University Clinics Basel, Veselanium, Vesalgasse 1, CH–4051 Basel, Switzerland

Mills, Gordon Department of Molecular Therapeutics, University of Texas, MD Anderson Cancer Center, Box 904, 1515 Holcombe Blvd, Houston, TX 77030-4009, USA

Mingozzi, Francesca Department of Biotechnology and Bioscience, University of Milano-Bicocca, Piazza della Scienzia 2, 20126 Milan, Italy

Mueller, Anja Department of Radiation Oncology, Heinrich-Heine University, Düsseldorf, Germany

O'Donnell, Niall Johnson & Johnson Pharmaceutical Research and Development, LLC, 3210 Merryfield Row, San Diego, CA 92121, USA

Pavelka, Norman Department of Biotechnology and Bioscience, University of Milano-Bicocca, Piazza della Scienzia 2, 20126 Milan, Italy

Pelizzola, Mattia Department of Biotechnology and Bioscience, University of Milano-Bicocca, Piazza della Scienzia 2, 20126 Milan, Italy

Pivarcsi, Andor Department of Dermatology, Heinrich-Heine University, Düsseldorf, Germany

Pongor, Sándor Protein Structure and Bioinformatics Group, International Centre for Genetic Engineering and Biotechnology, Area Science Park, I-34012 Trieste, Italy and Bioinformatics Group, Biological Research Center, Hungarian Academy of Sciences, Temesvári krt. 62, H-6726 Szeged, Hungary

Pos, Zoltan Department of Genetics, Cell- and Immunobiology, Semmelweis University, 1089 Nagyvárad tér 4, 1445 Budapest, Hungary

Prohászka, Zoltán Kútvölgyi út 4, H-1125 Budapest, Hungary

Radvanyi, Laszlo G. Department of Melanoma Medical Oncology, University of Texas, MD Anderson Cancer Center, Box 904, 1515 Holcombe Blvd, Houston, TX 77030-4009, USA and Cancer Vaccine Program, Sanofi Pasteur, Toronto, Ontario, Canada

Rajnavölgyi, Eva Institute of Immunology, Medical and Health Science Center, Faculty of Medicine, University of Debrecen, 98 Nagyerdei Blv., H-4012 Debreceb, Hungary

Ricciardi-Castagnoli, Paola Department of Biotechnology and Bioscience, University of Milano-Bicocca, Piazza della Scienzia 2, 20126 Milan, Italy

Sandor, Matyas Department of Pathology and Laboratory Medicine, Room 5580 Medical Science Center, 1300 University Avenue, Madison, WI 53706, USA

Szalai, Csaba Department of Genetics, Cell and Immunobiology, Semmelweis University, 1089 Nagyvárad tér 4, 1445 Budapest, Hungary

Takacs, Laszlo Biosystems International, 93 rue Henri Rochefort, 91000 Evry, France

Thurmond, Robin L. Johnson & Johnson Pharmaceutical Research and Development, LLC, 3210 Merryfield Row, San Diego, CA 92121, USA

Tomohiro, Takenori Laboratory of Biorecognition Chemistry, Faculty of Pharmaceutical Sciences, Toyama Medical and Pharmaceutical University, Sugitany 2630, Toyama 930-0194, Japan

Tsokos, George C. Department of Cellular Injury, Walter Reed Army Institute of Research, Silver Spring, MD 20910, USA

Uhlinger, David Bioinformatics and Functional Genomics, Johnson & Johnson Pharmaceutical Research and Development, Room B238, 1000 Route 202 S, Raritan, NJ 08869, USA

Urbano, Matteo Department of Biotechnology and Bioscience, University of Milano-Bicocca, Piazza della Scienzia 2, 20126 Milan, Italy

Vihinen, Mauno Institute of Medical Technology, University of Tampere and Research Unit, Tampere University Hospital, FI-33014 Tampere, Finland

Vizzardelli, Caterina Department of Biotechnology and Bioscience, University of Milano-Bicocca, Piazza della Scienzia 2, 20126 Milan, Italy

Wang, Ena Immunogenetics Section, Department of Transfusion Medicine, Clinical Center, National Institutes of Health, Bethesda, MD20892, USA

Wang, Xiwei Bioinformatics and Functional Genomics, Johnson & Johnson Pharmaceutical Research and Development, Room B238, 1000 Route 202 S, Raritan, NJ 08869, USA

Xiao, Xiaoyao Bioinformatics and Functional Genomics, Johnson & Johnson Pharmaceutical Research and Development, Room B238, 1000 Route 202 S, Raritan, NJ 08869, USA

Zanoni, Ivan Department of Biotechnology and Bioscience, University of Milano-Bicocca, Piazza della Scienzia 2, 20126 Milan, Italy

Zerkowski, Hans-Reinhard Division of Cardio-Thoracic Surgery and Department of Research, University Clinics Basel, Veselanium, Vesalgasse 1, CH–4051 Basel, Switzerland

1

Genotyping Methods and Disease Gene Identification

Ramón Kucharzak and **Ivo Glynne Gut**

Abstract

Technical approaches are described that are used for the identification of genes that might be associated with disease. As many diseases show association with the major histocompatibility complex (MHC), methods for the analysis of DNA variability in this region of the genome are described and particularities highlighted. A general overview of single-nideotide polymorphism (SNP) genotyping methods is given, starting from very early developments to methods currently used for high-throughput studies. Their optimal window of use is shown in terms of applicability to the study of a limited number of SNPs in a huge cohort to the study of many different SNPs in a limited number of individuals. Further, methods for more effective resequencing that are currently emerging are discussed, as well as an outlook given on the potential future of genome variation studies.

1.1 Introduction

Genotyping – the measurement of genetic variation – has many applications: disease gene localization and identification of disease-causing variants of genes, quantitative trait loci (QTL) mapping, pharmacogenetics and identity testing based on genetic fingerprinting, just to mention the major ones. Genotyping is used in genetic studies, in humans, animals, plants and other species. Classical strategies for identifying disease genes rely on localizing a region harbouring a disease-associated gene by genome scan. Two approaches can be taken: the linkage study, where related individuals are analysed, and the association study where individuals are solely selected on the basis of being affected by a phenotype or not. For a linkage studies, families with affected and nonaffected members are genotyped with usually on the order of 400 microsatellite markers that are fairly evenly spaced throughout the

Immunogenomics and Human Disease Edited by András Falus

genome (this can be done with a commercially available panel of microsatellite markers from Applied Biosystems; www.appliedbiosystems.com). The alleles of the different microsatellites are analysed and associations of alleles with the phenotype sought. A genomic region where affected individuals have the same genotype which is different from that of nonaffected individuals has an increased probability of harbouring a disease-causing variant of a gene. Owing to the usually large number of alleles that each microsatellite can have, the informativity is quite good even if the spacing of markers is 10 cM. However, this also means that many genes fall into the interval between any two neighbouring markers. To home in on the right gene, the region has to be saturated with additional markers. Increasing the mapping density with microsatellites is effective to a certain degree, but limited, as the number of polymorphic microsatellites is not very high. If, for example, an association to the major histocompatibility complex (MHC) is identified, the number of applicable microstellites does not help resolve the problem because too many genes are in the interval and few useful microsatellites exist for fine mapping the MHC. The overall approach taken nowadays is to identify the most promising candidate genes in the interval and either sequence them or genotype the SNPs to increase resolution. This strategy has successfully been applied for the identification of genes responsible for many monogenic disorders. In the case of an association with the MHC human leukocyte antigen (HLA) typing can be carried out to achieve further resolution. Associating functional candidate genes in the MHC, in general, is very difficult owing to the gene density of this locus and the, in many cases, limited understanding of gene function. For pathologies with potentially many genes throughout the genome implicated, such as common diseases like diabetes and cardiovascular disease, this strategy loses power. One way to increase power is to increase the number of individuals entered into a study as well as the density of markers that are tested. Excessively large families affected by a single complex pathology are extremely rare. This forces researchers to move from a linkage to an association study (Cardon and Bell, 2001).

Microsatellite markers (also called short tandem repeats – STRs, or a frequently used subgroup, CA repeats; Figure 1.1) are stretches of a repetitive sequence motif of a few bases. The standard microsatellite markers for linkage studies have a two-base CA sequence repeat motif. Alleles of a microsatellite carry a different number of repetitions of the sequence motif and thus a large number of different alleles is possible. There are some technical challenges associated with the analysis of microsatellites. They are analysed by sizing polymerase chain reaction (PCR) products by electrophoretic separation, usually on automated sequencers. PCR of microsatellites results in stutter products, owing to slippage of the template strand, which gives rise to parasite peaks in the electropherograms. Particularly in cases where two alleles are close together, automated interpretation is very difficult owing

ATATTGAGCTGATCCT(CA)$_N$CCTGGATATTCGATC

Figure 1.1 A microsatellite polymorphism, also known as short tandem repeat (STR), or CA repeat.

to problems of deconvolution. Consequently, substantial manual labour is involved in the interpretation of microsatellite genotyping results. It is necessary to use families so that the transmission of alleles can be used to compensate for the lack of absolute calibration of product peaks in the electropherogrammes. The fact that microsatellites are largely intergenic in the best case puts them into complete linkage disequilibrium (LD) with a disease-causing variant of a gene – never is an allele of a microsatellite the causative variant. SNPs, in contrast, can be intragenic and can thus represent disease-causing variants of genes. However, microsatellite markers still are largely valued for statistical analysis owing to the large number of alleles they can have and the power provided by this.

1.2 Genotyping of Single-Nucleotide Polymorphisms

SNPs (Figure 1.2) are changes in a single base at a specific position in the genome, in most cases with only two alleles (Brookes, 1999). SNPs are found at a frequency of about one in every 1000 bases in human (Kruglyak, 1997). By definition, the rarer allele should be more abundant than 1% in the general population. The relative simplicity, number of methods for SNP genotyping and the abundance of SNPs in the human genome have made them a very popular tool in recent years. First projects using SNPs as markers for disease gene identification have been shown (Ozaki *et al.*, 2002). Yet, there still is quite some debate about the usefulness of SNP markers compared with microsatellite markers for linkage studies and how many SNP markers will have to be analysed for meaningful association studies (Kruglyak, 1999; Weiss and Terwilliger, 2000). It has been reported that the human genome is structured in blocks of complete LD that coincide with events of ancient ancestral recombination (Reich *et al.*, 2001; Patil *et al.*, 2001). Using high-frequency SNPs, usually only a very limited number of haplotypes is detected and, depending on the region of the genome, LD can extend more than 100 kb. The HapMap project has the objective of providing a selection of SNP markers that tag haplotype blocks in order to reduce the number of genotypes that have to be measured for a genome-wide association study (Couzin, 2002; Weiner and Hudson, 2002). This project is nearing completion and first products harvesting the results are hitting the market. Both Affymetrix and Illumina offer array-based solutions for genotyping 100 000 SNP distributed across the human genome.

An approach that is still widely used and requires genotyping of only a limited set of SNPs is the candidate gene study. For this, genes are selected based on criteria such

```
ATATTGAGCTGATCCTGGATATTCGATC
TATAACTCGACTAGGACCTATAAGCTAC

ATATTGAGCTGATCTTGGATATTCGATC
TATAACTCGACTAGAACCTATAAGCTAC
```

Figure 1.2 A single-nucleotide polymorphism.

as likely association based on literature searches or presence of a gene in a likely pathway of action. SNPs in the candidate genes are selected and genotyped. The number of SNPs can be reduced by examining the haplotypes and removing SNPs that are redundant for the capture of haplotypes.

Nevertheless the flurry of activity in the arena of SNP genotyping technology development has still not come to a halt. SNP genotyping methods are constantly being improved and integrated and new methods are still emerging to satisfy the needs of genomics and epidemiology. The reason for this is that each application has to be well suited to the study that is being carried out. By applying the most suitable method, efficiency can be achieved often without using the cheapest method.

Many SNP genotyping technologies have reached maturity in last five years and have been integrated into large-scale genotyping operations. No one SNP genotyping method fulfils the requirements of every study that might be undertaken. The choice of method depends on the scale of a study and the scientific question a project is trying to answer. A project might require genotyping of a limited number of SNP markers in a large population or the analysis of a large number of SNP markers in one individual. Flexibility in choice of SNP markers and DNA to be genotyped or the possibility and precisely quantifying an allele frequency in pooled DNA samples might be issues. Studies might also use combinations of typing methods for different stages.

SNP genotyping methods are very diverse (Syvänen, 2001). Broadly, each method can be separated into two elements, the first of which is a method for interrogating a SNP. This is a sequence of molecular biological, physical and chemical procedures for the distinction of the alleles of a SNP (Gut, 2001). The second element is the actual analysis or measurement of the allele-specific products, which can be an array reader, a mass spectrometer, a plate reader, a gel separator/reader system, etc. Often very different methods share elements, for example, reading out a fluorescent tag in a plate reader (SNP genotyping methods with fluorescent detection were reviewed by Landegren *et al.*, 1998), or the method of generating allele-specific products (e.g. primer extension, reviewed by Syvänen, 1999), which can be analysed in many different analysis formats.

1.3 Methods for Interrogating SNPs

Only a few examples of SNP genotyping methods are cited in the text. The most common ones and the ones most pertinent for genotyping in the MHC are highlighted. A more complete list of SNP typing methods is given in Table 1.1.

Hybridization

Alleles can be distinguished by hybridizing complementary oligonucleotide sequences to a target sequence. The stringency of hybridization is a physically controlled process. As the two alleles of a SNP are very similar in sequence,

Table 1.1 A selection of SNP genotyping methods and their features

Method	Allele distinction	Strength	Reference
		Microarray	
GeneChip	Hybridization	Sophisticated software	Wang *et al.* (1998), www.affymetrix.com
Tag array	Primer extension	Detector for a high degree of multiplexing	www.affymetrix.com
APEX	Primer extension	Data quality	Shumaker *et al.* (1996); Pastinen *et al.* (1996B), 1997, 2000)
OLA	Ligation	High multiplexing	Khanna *et al.* (1999)
EF microarray	Hybridization	Flexibility, stringency	Edman *et al.* (1997); Sosnowski *et al.* (1997)
		Mass spectrometry	
PNA	Hybridization	Simple principle, rapid data accumulation	Ross, Lee and Belgrader (1997); Griffin, Tang and Smith, (1997)
Masscode	Hybridization, mass tag	Very high throughput	www.qiagen.com, Kokoris *et al.* (2000)
Mass tags	Hybridization, mass tag	High throughput	Shchepinov *et al.* (1999)
PROBE	Primer extension	Accurate, rapid data accumulation	Braun, Little and Köster, (1997); Little *et al.* (1997)
MassArray	Primer extension	Accurate, high-throughput, complete system	www.sequenom.com
PinPoint	Primer extension	High-quality data, rapid data accumulation	Haff and Smirnov (1997); Ross *et al.* (1998)
GOOD	Primer extension	Accurate, easy handling, no purification	Sauer *et al.* (2000a, b)
VSET	Primer extension	Accurate	Li *et al.* (1999)
Invader	Cleavage	No PCR, isothermal	Griffin *et al.* (1999)
		Gel	
ARMS	Primer extension	Easy access	Newton *et al.* (1989)
RFLP	Cleavage	Easy access, often used as reference	Botstein *et al.* (1980)
MADGE	Cleavage	Easy access, low set-up cost	Day and Humphries (1994)
SNaPshot™	Primer extension	Multiplexing	www.appliedbiosystems.com
OLA	Ligation	High plex factor for allele generation	Baron *et al.* (1996)
Padlock	Ligation	No PCR	Nilsson *et al.* (1994); Landegren *et al.* (1996)

(*Continued*)

Table 1.1 (*Continued*)

Method	Allele distinction	Strength	Reference
		Plate reader	
Invader	Cleavage	End-point, no PCR, isothermal	DeFrancesco (1998); Mein *et al.* (2000)
SNP-IT	Primer extension	End-point, integrated system for SNPstream UHT	www.orchidbio.com
OLA	Ligation	End-point, easy format	Barany, (1991); Samiotaki *et al.* (1994)
FP-TDI	Primer extension, fluorescence polarization	End-point, homogeneous assay	Chen *et al.* (1997, 1999)
5′-Nuclease (TaqMan™)	Hybridization	Real-time or end-point, homogeneous assay	Holland *et al.* (1991); Livak *et al.* (1995) www.appliedbiosystems.com
Molecular beacons	Hybridization	Real-time, homogeneous assay	Tyagi and Kramer, 1996; Tyagi, Bratu and Kramer (1998)
Kinetic PCR	Allele-specific primer extension	Real-time, homogeneous assay	Germer, Holland and Higuchi, (2000)
Amplifluor™	Allele-specific primer extension	Real-time or end-point, homogeneous assay	Myakishev *et al.* (2001) www.serologicals.com
Scorpions	Hybridization	Real-time or end-point, homogeneous assay	Whitcomb *et al.* (1999)
DASH/DASH-2	Hybridization	Low cost	Howell *et al.* (1999, 2002); Jobs *et al.* (2003)
		Flow cytometer	
Coded spheres	Ligation	Flexibility, high-throughput	Iannone *et al.* (2000)
Coded spheres	Primer extension	Flexibility, high-throughput	Cai *et al.* (2000); Chen *et al.* (2000)
		Sequencing	
Sequencing	Primer extension	Complete sequence information	www.appliedbiosystems.com
Pyrosequencing	Primer extension	Quantitation	Ronaghi *et al.* (1996, 1999); Ronaghi (2001)
		Current generation	
Illumina	Primer extension and ligation	High degree of multiplexing	www.illumina.com
Parallele	Primer extension and ligation	High degree of multiplexing, detection on tag array	www.parallelebio.com
SNPlex	Ligation	50–100-plex reactions	www.appliedbiosystems.com www.serologicals.comtems.com
Perlegen	Hybridization	Whole genome representation	www.affymetrix.com

significant cross-talk can occur. Allele-specific hybridization with two differently labelled oligonucleotides, one for each of the alleles, is generally used in conjunction with stringent washing to remove probes that are not fully complementary. Alternatively the probes can be immobilized and labelled probes hybridized to them. Hybridization generally suffers from a lot of cross-talk of the two alleles. A way to circumvent this difficulty is dynamic allele-specific hybridization (DASH). It relies on the different thermal stability of fully complementary probes compared with probes with one mismatch. Templates are immobilized and probes hybridized. The separation of the duplex is monitored in real time (Howell *et al.*, 1999; Howell, Jobs and Brookes, 2002; Jobs *et al.*, 2003).

HLA typing by sequence-specific oligonucleotide probe hybridization (SSOP; Saiki *et al.*, 1986) and the reverse line blot (Saiki *et al.*, 1989) use direct hybridization.

The 5′-nuclease assay, more commonly known as the TaqMan™ assay (www.appliedbiosystems.com) is a hybridization-based assay, similar to molecular beacons, except that the probe is degraded enzymatically during PCR. A labelled oligonucleotide probe complementary to an internal sequence of a target DNA is added to a PCR (Holland *et al.*, 1991; Livak Marmaro and Todd, 1995; Kalinina *et al.*, 1997). The nucleotide probe carries a fluorescent dye and a fluorescence quencher molecule. Successful hybridization of the oligonucleotide probe due to matching with one allele of the SNP results in its degradation by the 5′- to 3′-nuclease activity of the employed DNA polymerase whereby the fluorescent dye and quencher are separated, which promotes fluorescence. Since the inclusion of minor groove binders into the probes in 2001, the TaqMan™ assay has become a very reliable SNP genotyping method which is easy to integrate and run at high throughput. TaqMan™ was implemented for class I HLA typing by Gelsthorpe *et al.* (1999).

Strictly speaking, TaqMan™ is not a pure hybridization approach for genotyping, but uses a combination of hybridization and nuclease activity for allele distinction. Adding an enzymatic step to distinguish alleles in general helps increase fidelity. A hybridization event is followed by the intervention of an enzyme. Many different enzymes, like DNA polymerases, DNA ligases, sequence- or structure-specific nucleases can be applied. Most enzymatic methods allow the generation of both allele products of an SNP in a single reaction.

Primer extension

Primer extension is a stable and reliable way of distinguishing alleles of an SNP. For primer extension an oligonucleotide is hybridized next to an SNP. Nucleotides are added by a DNA polymerase generating allele-specific products (Syvänen, 1999). Either only terminating nucleotides (ddNTPs result in single base primer extension) or a mix of non-terminating nucleotides (dNTPs) and terminating bases (ddNTPs) are provided. For fluorescence detection a fluorescent molecule can be attached either to the primer (when used in conjunction with fragment sizing) or to the terminating nucleotide (in conjunction with fragment sizing or direct detection). In most instances it is necessary to provide a substantial amount of template material and a reduction of complexity prior to the primer extension reaction. This is effectively achieved by a PCR.

The formation of a product in a PCR is dependent on complementary primers. The concept of ARMS (amplification refractory mutation system; Newton *et al.*, 1989) is used in commercial HLA typing systems (www.dynal.no). Primers are chosen to be completely complementary to certain alleles, giving rise to a PCR product of a specific size only in the case of the template DNA carrying that allele. Multiple primer systems are combined in a kit using different fluorescent dye-labelled primers. Analysis of PCR products is done by gel electrophoresis. An ARMS system was published by Tonks *et al.* (1999).

DNA sequencing is a viable method for assigning alleles of polymorphisms. Automated fluorescence Sanger sequencing is currently not competitive for the kind of throughput that is being targeted for association studies. However, in highly polymorphic regions of the genome such as the MHC, when many SNPs can be captured with a few sequencing reactions, it does become very competitive (Kotsch *et al.*, 1997, 1999). Some problems revolve around interpreting the sequences, and efficient software for HLA allele assignment has been developed (Sayer *et al.*, 2004; Sayer, Goodridge and Christian, 2004). Sequence-based typing identifies previously unknown HLA alleles.

Oligonucleotide ligation

For oligonucleotide ligation (OLA), two oligonucleotides adjacent to each other are ligated enzymatically when the bases next to the ligation position are fully complementary to the template strand by a DNA ligase (Barany, 1991; Samiotaki *et al.*, 1994; Baron *et al.*, 1996). Padlock is a variant of OLA, in which one oligonucleotide is circularized by ligation (Nilsson *et al.*, 1994, 1997; Landegren *et al.*, 1996). Allele-specific products can be visualized by separation on gels or in plate reader formats. These allele-discrimination formats have much unharvested potential and are being tapped for the new generation of SNP typing methods.

Cleavage

The restriction fragment length polymorphism (RFLP) is one of the most commonly used and the oldest format for SNP genotyping in a standard laboratory setting (Botstein, *et al.*, 1980). It far predates the coining of the term SNP. PCR products are digested with restriction endonucleases that are specifically chosen for the base change at the position of the SNP, resulting in a restriction cut for one allele but not the other. Fragment patterns are used for allele assignment after gel separation. Owing to the limited number of restriction enzymes, the complex patterns that may result and gel separation, it is a very labour-intensive method and does not lend itself to automation.

Alternatively, a flap endonuclease can be used for the discrimination between alleles of SNPs (Harrington and Lieber, 1994; DeFrancesco, 1998). This is known as the Invader assay. An invader oligonucleotide and a signal oligonucleotide with a 5′-overhang (flap) over the invader oligonucleotide are hybridized to a target sequence. Only in the case of a perfect match of the signal oligonucleotide with the target sequence is the flap cleaved off. Cleavage of the flap by this class of structure-sensitive enzymes can be linked to a change in fluorescence, for example with a fluorescence resonance energy transfer system.

1.4 Analysis Formats

Over recent years diverse analysis formats have been devised. They include gel electrophoresis, microtitre plate fluorescent readers with integrated thermocyclers, oligonucleotide microarrays (DNA chips), coded spheres with reading in a flow cytometer or on a bead array and mass spectrometers. Nearly all of the above-described methods for allele-distinction have been combined with all of these analysis formats (Figure 1.3).

Gel-based analysis

DNA fragments can be separated by size by electrophoretic migration through gels. The current state-of-the-art is separation in 'gel-filled' capillaries. Instrumentation with 96 or even 384 capillaries is commercially available. A great advantage of capillary systems over slab gels is the degree of automation these instruments come with that allows 24 h unsupervised operation. The cumbersome activity of gel pouring and loading, and lane tracking of slab gels is not required. This instrumentation is perfectly suited for sequence-based HLA typing as well as sequence-specific primer approaches.

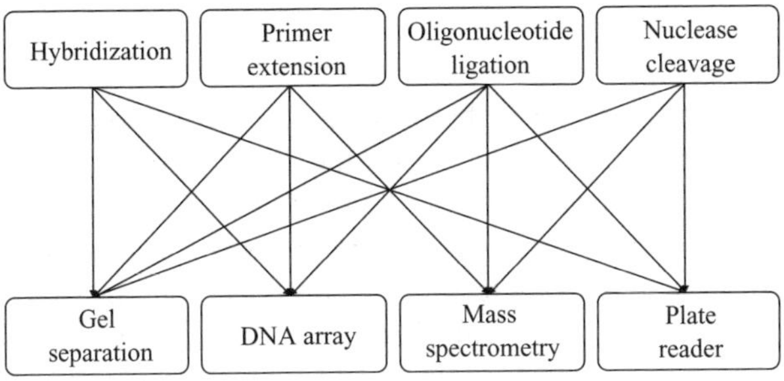

Figure 1.3 Methods used for allele distinction and detection.

Fluorescent-reader-based analysis

SNP typing methods with fluorescent endpoint readout, such as TaqMan™ or Amplifluor™, where fluorescence in two detection channels is quantified, are usually very easy to operate as they can often be combined with homogeneous preparation schemes.

Array-based analysis

For oligonucleotide microarrays (DNA-chips) a series of oligonucleotides (features) is chemically attached to a solid surface, usually a glass slide (Shalon *et al.*, 1996). The position of a specific oligonucleotide on the solid surface is used as the identifier. Far in excess of 10 000 different oligonucleotides can be arrayed on 1 cm^2 of the solid surface. The preparation of microarrays on a solid surface is done by one of two methods, either by spotting oligonucleotides or DNA (Cheung *et al.*, 1999), or by *in-situ* synthesis (Pease *et al.*, 1994; Lipshutz *et al.*, 1999). Hybridizing an unknown sequence to a known sequence on the DNA chip identifies a complementary sequence element in the unknown sequence. Microarrays primarily found their application in expression analysis (Brown and Botstein, 1999; Debouck and Goodfellow, 1999). In principle, there are three different formats that are used for SNP genotyping in an oligonucleotide microarray format:

1. Hybridization arrays – this format uses allele-specific hybridization. Oligonucleotides covering the complementary sequence of the two alleles of a SNP are on specific positions of the array. Fluorescently labelled PCR products containing the SNP sequences to be queried are hybridized to the array. This approach was applied by Patil *et al.* (2001) to resequence the entire chromosome 21 (www.perlegen.com). A practical difficulty is that the entire chromosome had to be amplified by PCR prior to hybridization, which is very labour-intensive. This is now available as a product from Affymetrix (www.affymetrix.com).

2. Arrays with enzymatic processing – arrayed primer extension (APEX) is a prominent example where single base primer extension is carried out using primers that are in an array format.

3. Tag arrays – an array approach that has become very popular recently is the tag array (Fan *et al.*, 2000). The features on the array represent tag oligonucleotides that are solely chosen based on their maximum sequence difference. Complementary tag sequences are included in the oligonucleotides for allele distinction and are used to capture allele-specific products on arrayed complementary features. All molecular biological procedures are carried out in solution using the complementary tag sequences. In the final step the samples are hybridized to the tag array and read. A completely integrated system of this kind is realized

in the Orchid SNPstream UHT system (www.orchid.com), which uses SNP-IT single-base primer extension for allele-discrimination. Genflex is an array marketed by Affymetrix. The feature density of this product is still increasing. This array is also used for genotyping by Parallele.

The direct use of genomic DNA on a microarray without amplification is still not possible, probably due to the high complexity and low concentration of target sequence. Thus all oligonucleotide array SNP genotyping approaches require the template DNA to be amplified, usually by PCR. To match the highly parallel analysis qualities of the microarrays, either highly multiplex PCRs have to be established or multiple PCR products have to be pooled. Multiplex PCR is difficult to control and the risk of cross-talk increases. PCR has a bias towards smaller products, which means that only very small PCR products can be generated in highly multiplex PCR. Multiplexes beyond 8 with small-fragment-size PCR products (<200 bases) require a significant optimization effort. An approach that was implemented recently is the restriction digest of genomic DNA, followed by size selection and the 5′ adaptation of universal primer sequences. Thereafter the entire genome can be amplified with universal PCR conditions, giving good representation of all regions of the genome (www.affymetrix.com).

Fluorescent emission from the features is recorded with an array scanner and used for allele calling. The majority of commercial array scanners are capable of distinguishing two different fluorescent emissions (Cy3 and Cy5). Recently, Xu *et al.* (2003) presented an SNP genotyping application with Q-dot beads. This effectively moves the array into solution. This bead system could potentially provide 20 000 different beads and could be used for 20 000-plex detection in conjunction with a flow cytometer for detection.

Mass spectrometry-based analysis

Mass spectrometry, specifically MALDI mass spectrometry (matrix-assisted laser desorption/ionization time-of-flight mass spectrometry) has been demonstrated as an analysis tool for SNP genotyping (for an exhaustive review see Tost and Gut, 2002). Allele-specific products are deposited with a matrix on the metal surface of a target plate. The matrix and the analyte are desorbed into the gas phase with a short laser pulse. Analytes are ionized and accelerated towards the detector. The time-of-flight of a product to the detector is directly related to its mass. The analysis is done serially (one sample after another), but at a very high rate. Mass spectrometers are capable of recording a single trace in much less than 1 s. Apart from the speed and accuracy, a major strength of mass spectrometric analysis is the number of available detection channels. Resolution of the current generation of mass spectrometers allows the distinction of base substitutions in the range 1000–6000 Da (this corresponds to product sizes of three to 20 bases; the smallest mass difference for a base change thymine to adenine is 9 Da). In principle this would allow the analysis of hundreds of products in a single trace. However, for reasons of difficulty in multiplexing the sample preparation, only SNP genotyping multiplexes on the order of up to 12 SNPs have been shown (Ross *et al.*, 1998). Sample consumption is very low, the data accumulation can be fully automatic and absolute mass information allows automatic allele calling with high confidence. In contrast to fluorescence-based methods, mass spectrometry allows the direct observation of the products. This makes the detection method far more reliable. Further, the interpretation of fluorescence-based methods relies on clustering of data points into classes (homozygous and heterozygous). A single independent result cannot be interpreted. SNPs where one allele is very rare and thus one of the homozygote classes might not be represented in the sample are difficult to score. In mass spectrometry this does not pose a problem. Each allele has a discrete mass. The presence of a peak surpassing a threshold at the discrete mass indicates the presence of an allele in a sample. An individual result can be scored independently of other results.

Many devised SNP genotyping methods tended to have a shelf-life and benefitted from different degrees of popularity. Some very effective methods did not achieve the popularity they deserved. The reasons for this can probably be found in some very subjective criteria, such as the ease of access. TaqMan, for example, benefitted from very effective support in getting started, easy operation and good quality data. Even though the cost per SNP is higher than many other methods, it is still very popular.

1.5 The Current Generation of Methods for SNP Genotyping

In the last few years SNP genotyping methods have started emerging that dramatically increase the number of SNPs that are analysed in a single reaction. They use the same principles as outlined above, but in some instances inverse the order of allele-distinction and amplification. Rather than amplifying genomic DNA (by PCR) to provide a sufficient amount of DNA template for allele query, the allele-distinction takes place at the level of genomic DNA by strategies with increased specificity. PCR is then used to provide a detectable amount of the allele-specific products using a universal amplification system. Multiplexing by these methods is 50–13 000-plex. Thus a 96-well plate can provide 1 000 000 genotypes. Manipulation of a single 96-well microtitre plate is easily achieved manually by a single technician without the need for extensive automation. On the other hand, with automation it is feasible to record millions of SNP genotypes per day.

SNPlex is a method for 50-plex genotyping using detection on a capillary sequencer. It relies on ligation for allele-distinction followed by universal amplification (www.appliedbiosystems.com). A higher degree of multiplexing is envisaged. Illumina (www.illumina.com) uses coded-sphere arrays (Healey *et al.*, 1997; Oliphant *et al.*, 2002). The code on the spheres consists of an oligonucleotide tag capture sequence. The coded-spheres are lodged in solid-phase wells in an array format. The Golden Gate Assay of Illumina combines allele-specific primer extension with ligation, a tag system and a universal amplification system. This system allows

1536-plex SNP genotyping. In a 96-well format this equates to about 150 000 genotypes. ParAllele is another company (www.parallelebio.com) that has focussed on increasing the multiplex factor for SNP genotyping. The molecular biological procedure that underlies the allele-distinction uses single-base primer extension in four individual reactions with each one of the four dNTPs, followed by ligation. This results in cyclization of probes. After amplification, the products are read on a Affymetrix Genflex array. According to ParAllele, 13 000-plex reactions are possible. In a 20 μl reaction volume this corresponds to a virtual reaction volume of less than 1 nl. With a read-time of a few minutes per array, millions of data points can be accumulated in a day.

Perlegen, a spin-off from Affymetrix, has adopted a brute force approach for re-sequencing entire genomes (www.perlegen.com). The entire sequence of a genome is synthesized on an array in short oligonucleotides. The DNA sequence of an individual is amplified by PCR and hybridized to the array. The absence of alleles of SNPs results in a decrease in the hybridization signal. Effectively the sequence of an individual is compared with a reference sequence. Data derived for chromosome 21 by this approach was presented by Patil *et al.* (2001) and was extended to the entire human genome. The technology developed by Perlegen is being distributed in products of Affymetrix. The HapMap project, a large systematic effort to define common haplotypes and determine discrete haplotype boundaries in the human genome, is well underway. It is based on the common disease–common variant hypothesis. It promises to provide a vehicle to facilitate whole-genome association studies based on choosing haplotype block-defining tag SNPs. This work is mainly being carried out on Illumina and ParAllele platforms (Couzin 2002).

Based on the HapMap project and based on the work of Perlegen and Affymetrix, Illumina and Affymetrix have developed products with up to 500 000 SNPs for genome-wide association studies. These products are currently being tested in many laboratories.

1.6 The Next Generation

Whole-genome association products do not take account of effects of rare variants. Chasing up rare variants will be difficult and cost-intensive. Essentially, sequencing can be considered a genotyping method with the highest resolution and does not require prior knowledge of the position and content of polymorphism. However, sequencing with the currently available methods for the identification of rare variants is economically not feasible. Thus there is an urgent need for cheaper, high-throughput sequencing methods. These will quite likely be based on other concepts than the ones that are currently used. Highly interesting methods for DNA sequencing are on the horizon. Most of these approaches are currently still very experimental, but promise to sequence an entire human genome in less than a day. The proponents are companies like Solexa (www.solexa.co.uk), 454 Corporation (www.454.com), US Genomics (www.usgenomics.com) and Visigen (www.visigenbio.com). In some instances the approaches differ dramatically from classical sequencing approaches and the emerging genotyping methods described above mainly in that they rely on sequencing single DNA molecules. They do not use standard laboratory tools and methods like microtitre plates or PCR. In some cases sequencing relies on directly monitoring a DNA polymerase incorporating dNTPs during DNA replication. Pulling DNA electrophoretically through a nanopore and measuring impedance changes is another option. In other cases a multitude of single oligonucleotides is used to capture fragments of genomic DNA and then sequence 25 bases from these primers. Thereafter the sequence tags are aligned and compared with a reference sequence. Alternatively, a pyrosequencing reaction can be carried out in a microwell format. Most of these innovative sequencing methods require extensive preparation of samples before the actual sequencing can start. However, once initiated they can produce vast amounts of DNA sequence data very rapidly. Promises go as far as accumulating the sequence of an entire human genome in a matter of a couple of hours at extremely low cost. Demands on informatics will be extensive for these methods. In any case these are exciting prospects for genetics.

1.7 Classical HLA Typing

Until the mid-1980s, typing of the human leukocyte antigens was carried out exclusively by serological methods. These methods are based on antibody screening approaches, such as complement-dependent microlymphocytoxicity. In operation they are comparatively inexpensive and today still appreciates widespread popularity. However, their main drawback is the high degree of cross-reactivity of antibodies, which leads to inaccurate and false positive results. They also only provide resolution at the two-digit level.

The introduction of PCR allowed the development of a variety of simple and rapid DNA sequence-based HLA typing methods (Erlich *et al.*, 1991; Erlich, Opelz and Hansen, 2001). These are now well established for four-digit typing. The most common methods in HLA typing laboratories are SSOP (Saiki *et al.*, 1986), reverse line blot (Saiki *et al.*, 1989), sequence-specific primers amplification (SSP; Welsh and Bunce, 1999), reference strand conformation analysis (RSCA; Arguello *et al.*, 1998) and sequence-based typing (SBT; Spurkland *et al.*, 1993).

SSOP uses a panel of labelled sequence-specific oligonucleotide probes. These probes are hybridized to a PCR product that has been immobilized on a nylon or nitrocellulose membrane. The hybridization and washing conditions are chosen so that only the complementary sequence binds to the immobilized PCR template. The probes are typically labelled with an enzyme (e.g. horseradish peroxidase) or biotin. The biotin labelled probes are interacted with streptavidin-conjugated enzyme or a fluorescent dye. The reverse line blot in principle is the same method, except that the oligonucleotide probes are immobilized on a membrane and the PCR template is labelled and hybridized.

For SSP a set of primers is used to amplify the template by PCR. Each primer has a specific sequence and just if the sequence is complementary to the sequence of the template amplification is possible. The success of amplification can be detected by gel electrophoresis. Effectively this is ARMS (Tonks *et al.*, 1999).

RSCA is a high-resolution HLA typing method which is used in quite a few HLA typing laboratories (Arguello *et al.*, 1998). Here the PCR product, which is to be analysed, is mixed with a reference product. Denaturation and renaturation result in a mixture of two homoduplexes and a heteroduplex. Sequence differences between the tester and the reference result in mismatches loop and bulges which affect the gel mobility of these products. The gel mobility is compared with a standard set of markers for each allele. It identifies previously unknown HLA types.

SBT is DNA sequencing of the different HLA genes (Kotsch *et al.*, 1997, 1999). It is the most labour-intensive method of the DNA-based methods for HLA typing, but at the same time a method that easily pinpoints so far unknown HLA alleles. However, heterozygous samples can pose many problems owing to ambiguous combinations of alleles and insertion/deletion polymorphisms resulting in a frame shift which cannot be called. A workaround is to generate homozygous samples by cloning. Key to SBT is the ability to call HLA alleles from the often very convoluted sequencing traces. Assign 2.0 an excellent software package for this was presented by Sayer *et al.* (2004) and Sayer, Goodridge and Christian (2004).

The DNA-based HLA typing methods have the use of polymorphisms to identify HLA types in common. Methods that rely on PCR amplification of the HLA genes all suffer from the difficulties associated with the efficiency of the PCR owing to the fact that there are very limited possibilities for positioning primers due to polymorphic positions. Sets of optimal primers are published by the International Histocompatibility Working Group (www.ihwg.org). These are subject to frequent change as better solutions are identified.

1.8 MHC Haplotypes

The MHC is the most gene dense regions of the human genome (Horton *et al.*, 2004). Owing to its relevance with respect to disease association and immunity, it has been very well studied and several common haplotypes of the entire MHC have been sequenced (Stewart *et al.*, 2004). The MHC is divided into three classes (I, II and III). Class III is between class I and class II. For disease the entire MHC is relevant. For tissue matching between a donor and recipient, genes in the class II region are primarily relevant followed by genes in class I. The number of alleles across the three classes of the MHC varies quite substantially. For the HLA-B, a class I gene, over 600 alleles are known, while HLA-DRB1, a class II gene, has nearly 400 known alleles. Interestingly the class III region has a comparatively limited number of haplotypes, maybe fewer than than 20 (Allcock *et al.*, 2004). This was determined by SNP genotyping. Gourraud *et al.* (2004) tried to establish a correlation between microsatellite markers and HLA types. Some correlation was established however, it was not a one-to-one correlation. Thus it is unlikely that microsatellite genotyping will be of great use for HLA typing.

1.9 Molecular Haplotyping

One of the major drawbacks of SNPs compared with microsatellites as genetic markers is the limited number of alleles they present – usually only two. A work-around is combining the information of several SNPs and determination of the phase haplotypes. A method to carry out molecular haplotyping at high throughput has been presented (Tost *et al.*, 2002). A multiplex primer extension SNP genotyping method with mass spectrometric detection was combined with allele-specific PCR. The allele-specific PCR amplifies each of the two parental alleles separately, while the multiplex SNP assay allows genotyping several polymorphisms on the amplified fragment. The phase of the SNP alleles is used to assemble the haplotype.

1.10 Microhaplotyping

Several years ago, under the auspices of an EU-funded programme (MADO; www.euromado.org) to develop an effective method to screen bone marrow donor registries, we decided to develop a set of SNPs that could be used for HLA typing. These were going to be the most indicative positions to distinguish 'frequent' and 'rare' HLA alleles in HLA-A, HLA-B and HLA-DRB1. In the course of this work we encountered difficulties with the high degree of polymorphism these genes presented, the number of known alleles and the amount of sequence variance. The use of a primer extension protocol coupled with mass spectrometric detection was literally impossible owing to the degree of polymorphism underneath the extension primers. As a consequence we started working with degenerate primers. We then discovered that, if there was polymorphism under the first eight nucleotides from the 3′ end of the primer, the primer would not be extended by DNA polymerase. As in the GOOD assay (Sauer *et al.*, 2000a, b), only at most the last four nucleotides of the primer plus the single extended base was carried to the analysis, the number of mass peaks being usually less than 8 with each mass representing a unique microhaplotype. Heterozygous samples resulted in two out of the possible masses, thus both parental microhaplotype alleles were resolved in a single reaction. By selecting 19 positions for HLA-A, 19 positions for HLA-B and 10 positions for HLA-DRB1 resolution of frequent and rare HLA alleles, and in some instances four-digit resolution, could be achieved (Kucharzak *et al.*, manuscript in preparation).

1.11 MHC and Disease Associations

Many diseases with a genetic component have been associated with the MHC. Once a disease gene association with the MHC has been established, for example by a whole genome scan, HLA typing is often used for further definition. Disease gene identification within the MHC is difficult due to gene density and the limited functional understanding of genes in this locus. A comprehensive list is of HLA alleles and disease association is given by Shiina, Inoko and Kulski: (2004).

1.12 Conclusions

The field of genotyping technologies still is in great flux. There is no telling which method will emerge as the best. Some methods that were hailed as unbeatable a few years ago are no longer available. On the other hand the field seemed content when methods with one allele-discrimination step and processing of a few SNPs in single reaction emerged and large-scale studies were initiated by automating this way of processing. Since then methods have emerged that combine multiple allele-discrimination procedures to increase the specificity, thus making it possible to improve multiplex preparation. Another advantage of multiplexing is that, apart from reducing DNA and reagent consumption, it takes the pressure off operational issues. Controls can be built into multiplexes that allow errors in manipulation to be pinpointed. A laboratory information management system (LIMS) for multiplex systems is far less complex. An important factor for any SNP genotyping method is its success rate – how many polymorphisms can be turned into working assays, and what percentage reliable genotype calls does an assay return. Considering that a few years ago people were struggling with multiplex reactions in the single digits, today multiplex factors of a few thousand are being achieved. This represents an increase of three orders of magnitude. The availability of the first whole-genome association genotyping systems with in excess of 100 000 SNPs is very exciting and a substantial advance. The prospect of obtaining whole genome sequences in a matter of a few hours is a great perspective for the future of genome variation analysis. However, complex regions of the genome such as the MHC will probably for some time to come require very sophisticated analysis tools using a less generic approach than the ones used in whole-genome association products.

Acknowledgements

We thank the European Community for funding under the programme QLG7-CT-2001-00065 and the French Ministry of Research for structural support.

References

Allcock RJ, Windsor L, Gut IG, Kucharzak R, Sobre L, Lechner D, Garnier JG, Baltic S, Christiansen FT, Price P. 2004. High-density SNP genotyping defines 17 distinct haplotypes of the TNF block in the Caucasian population: implications for haplotype tagging. *Hum Mutat* **24**: 517–525.

Arguello, JR, Little A, Pay AL, Gallardo D, Rojas I, Marsh SGE, Goldman JM, Madrigal JA. 1998. Mutation detection and typing of polymorphic loci through double-strand conformation analysis. *Nat Genet* **18**: 192–194.

Barany F. 1991. Genetic disease detection and DNA amplification using cloned thermostable ligase. *Proc Natl Acad Sci USA* **88**: 189–193.

Baron H, Fung S, Aydin A, Bahring S, Luft FC, Schuster H. 1996. Oligonucleotide ligation assay (OLA) for the diagnosis of familial hypercholesterolemia. *Nat Biotechnol* **14**: 1279–1282.

Botstein DR, White L, Skolnick M, Davies RW. 1980. Construction of a genetic linkage map in man using restriction fragment length polymorphisms. *Am J Hum Genet* **32**: 314–331.

Braun A, Little DP, Köster H. 1997. Detecting CFTR gene mutations by using primer oligo base extension and mass spectrometry. *Clin Chem* **43**: 1151–1158.

Brookes AJ. 1999. The essence of SNPs. *Gene* **234**: 177–186.

Brown PO, Botstein D. 1999. Exploring the new world of the genome with DNA microarrays. *Nat Genet Suppl* **21**: 33–37.

Cai H, White PS, Tosney D, Deshpande A, Wang Z, Marrone B, Nolan JP. 2002. Flow cytometry-base minisequencing: a new platform for high-throughput single-nucleotide polymorphism scoring. *Genomics* **66**: 135–143.

Cardon LR, Bell JI. 2001. Association study designs for complex diseases. *Nat Rev Genet* **2**: 91–99.

Chen X, Kwok P-Y. 1997. Template-directed dye-terminator incorporation (TDI) assay: a homogeneous DNA diagnostic method based on fluorescence resonance energy transfer. *Nucl Acids Res* **25**: 347–353.

Chen X, Zehnbauer B, Girnke A, Kwok P-Y. 1997. Fluorescence energy transfer detection as a homogeneous DNA diagnostic method. *Proc Natl Acad Sci USA* **94**: 10756–10761.

Chen X, Levine L, Kwok P-Y. 1999. Fluorescence polarization in homogeneous nucleic acid analysis. *Genome Res* **9**: 492–498.

Chen J, Iannone MA, Li M-S, Taylor D, Rivers P, Nelsen AJ, Slentz-Kesler KA, Roses A, Weiner MP. 2000. A microsphere-based assay for multiplexed single nucleotide polymorphism analysis using single base chain extension. *Genome Res* **10**: 549–557.

Cheung VG, Morley M, Aguilar F, Massimi A, Kucherlapati R, Childs G. 1999. Making and reading microarrays. *Nat Genet Suppl* **21**: 15–19.

Couzin J. 2002. Human Genome. HapMap launched with pledges of 100 M$. *Science* **298**: 941–942.

Day INM, Humphries SE. 1994. Electrophoresis for genotyping: microtitre array diagonal gel electrophoresis on horizontal polyacrylamide gels, hydrolink, or agarose. *Anal Biochem* **222**: 389–395.

Debouck C, Goodfellow PN. 1999. DNA microarrays in drug discovery and development. *Nat Genet Suppl* **21**: 48–50.

DeFrancesco L. 1998. The next new wave in genome analysis: Invader™ assays developed by Third Wave Technologies Inc. *Scientist* **12**: 16.

Edman CF, Raymond DE, Wu DJ, Tu E, Sosnowki RG, Butler WF, Nerenberg M, Heller MJ. 1997. Electric field directed nucleic acid hybridization on microchips. *Nucl Acids Res* **25**: 4907–4914.

Erlich, HA, Bugawan T, Begovich A, Scharf S, Griffith R, Higuchi R, Walsh PS. 1991. HLA-DR, DQ, and DP typing using PCR amplification and immobilized probes. *Eur J Immunogenet* **18**: 33–55.

Erlich, HA, Opelz G, Hansen J. 2001. HLA DNA typing and transplantation. *Immunity* **14**: 347–356.

Fan J-B, Chen X, Halushka MK, Berno A, Huang X, Ryder T, Lipshutz RJ, Lockhart DJ, Chakravarti A. 2000. Parallel genotyping of human SNPs using generic high-density oligonucleotide tag arrays. *Genome Res* **10**: 853–860.

Gelsthorpe AR, Wells RS, Lowe AP, Tonks S, Bodmer JG, Bodmer WF. 1999. High-throughput class I HLA genotyping using fluorescence resonance energy transfer (FRET) probes and sequence-specific primer-polymerase chain reaction (SSP-PCR). *Tissue Antigens* **54**: 603–614.

Germer S, Holland MJ, Higuchi R. 2000. High-throughput SNP allele-frequency determination in pooled DNA samples by kinetic PCR. *Genome Res* **10**: 258–266.

Gourraud PA, Mano S, Barnetche T, Carrington M, Inoko H, Cambon-Thomsen A. (2004). Integration of microsatellite characteristics in the MHC region: a literature and sequence based analysis. *Tissue Antigens* **64**: 543–555.

Griffin TJ, Tang W, Smith LM. 1997. Genetic analysis by peptide nucleic acid affinity MALDI-TOF mass spectrometry. *Nat Biotechnol* **15**: 1368–1372.

Griffin TJ, Hall JG, Prudent JR, Smith LM. 1999. Direct genetic analysis by matrix-assisted laser desorption/ionization mass spectrometry. *Proc Natl Acad Sci USA* **96**: 6301–6306.

Gut IG. 2001. Automation in genotyping of single nucleotide polymorphisms. *Hum Mutat* **17**: 475–492.

Haff L, Smirnov IP. 1997. Single-nucleotide polymorphism identification assays using a thermostable DNA polymerase and delayed extraction MALDI-TOF mass spectrometry. *Genome Res* **7**: 378–388.

Harrington JJ, Lieber MR. 1994. Functional domains within FEN-1 and RAD2 define a family of structure specific endonucleases: implications for nucleotide excision repair. *Genes Devl* **8**: 1344–1355.

Healey BG, Matson RS, Walt DR. 1997. Fiberoptic DNA sensor array capable of detecting point mutations. *Anal Chem* **251**: 270–279.

Holland PM, Abramson RD, Watson R, Gelfand DH. 1991. Detection of specific polymerase chain reaction product by utilizing the 5′–3′ exonuclease activity of Thermus aquaticus DNA polymerase. *Proc Natl Acad Sci USA* **88**: 7276–7280.

Horton R, Wilming L, Rand V, Lovering RC, Bruford EA, Khodiyar VK, Lush MJ, Povey S, Talbot CC Jr, Wright MW, Wain HM, Trowsdale J, Ziegler A, Beck S. 2004. Gene map of the extended human MHC. *Nat Rev Genet* **5**: 889–899.

Howell WM, Jobs M, Gyllensten U, Brookes AJ. 1999. Dynamic allele-specific hybridization. A new method for scoring single nucleotide polymorphisms. *Nat Biotechnol* **17**: 87–88.

Howell WM, Jobs M, Brookes AJ. 2002. iFRET: an improved fluorescence system for DNA-melting analysis. *Genome Res* **12**: 1401–1407.

Iannone MA, Taylor JD, Chen J, Li M-S, Rivers P, Slentz KA, Weiner MP. 2000. Multiplexed single nucleotide polymorphism genotyping by oligonucleotide ligation and flow cytometry. *Cytometry* **39**: 131–140.

Jobs M, Howell WM, Strömquist L, Brookes AJ. 2003. DASH-2: flexible, low-cost and high-throughput SNP genotyping by allele-specific hybridization on membrane arrays. *Genome Res* **13**: 916–924.

Kalinina O, Lebedeva I, Brown J, Silver J. 1997. Nanoliter scale PCR with TaqMan detection. *Nucl Acids Res* **25**: 1999–2004.

Khanna M, Park P, Zirvi M, Cao W, Picon A, Day J, Paty P, Barany F. 1999. Multiplex PCR/LDR for detection of K-ras mutations in primary colon tumors. *Oncogene* **18**: 27–38.

Kokoris M, Dix K, Moynihan K, Mathis J, Erwin B, Grass P, Hines B, Duesterhoeft A. 2000. High-throughput SNP genotyping with the masscode system. *Mol Diagnosis* **5**: 329–340.

Kotsch K, Wehling J, Blasczyk R. 1999. Sequencing of HLA class II genes based on the conserved diversity of the non-coding regions; sequencing based typing of HLA-DRB genes. *Tissue Antigens* **53**: 486–497.

Kotsch K, Wehling J, Kohler S, Blasczyk R. 1997. Sequencing of HLA class I genes based on the conserved diversity of non-coding regions: sequencing based typing of the HLA-A gene. *Tissue Antigens* **50**: 178–191.

Kruglyak L. 1997. The use of a genetic map of biallelic markers in linkage studies. *Nat Genet* **17**: 21–24.

Kruglyak L. 1999. Prospect for whole-genome linkage disequilibrium mapping of common disease genes. *Nat Genet* **22**: 139–144.

Landegren U, Samiotaki M, Nilsson M, Malmgren H, Kwiatkowski M. 1996. Detecting genes with ligases. *Meth Compan Meth Enzymol* **9**: 84–90.

Landegren U, Nilson M, Kwok P-Y. 1998. Reading bits of genetic information: methods for single-nucleotide polymorphism analysis. *Genome Res* **8**: 769–776.

Li J, Butler JM, Tan Y, Lin H, Royer S, Ohler L, Shaler TA, Hunter JM, Pollart DJ, Montforte JA, Becker CA. 1999. Single nucleotide polymorphism determination using primer extension and time-of-flight mass spectrometry. *Electrophoresis* **20**: 1258–1265.

Lipshutz RJ, Fodor SPA, Gingeras TR, Lockhart DJ. 1999. High density synthetic oligonucleotide arrays. *Nat Genet Suppl* **21**: 20–24.

Little DP, Braun A, Darnhofer-Demar B, Frilling A, Li Y, McIver RT, Köster H. 1997. Detection of RET proto-oncogene codon 634 mutations using mass spectrometry. *J Mol Med* **75**: 745–750.

Livak K, Marmaro J, Todd JA. 1995. Towards fully automated genome-wide polymorphism screening. *Nat Genet* **9**: 341–342.

Mein CA, Barrat BJ, Dunn MG, Siegmund T, Smith AN, Esposito L, Nutland S, Stevens HE, Wilson AJ, Phillips MS, Jarvis LS, deArruda M, Todd JA. 2000. Evaluation of single nucleotide polymorphism typing with invader on PCR amplicons and its automation. *Genome Res* **10**: 330–343.

'Myakishev MV, Khripin Y, Hu S, Hamer DH. 2001. High-throughput SNP genotyping by allele-specific PCR with universal energy-transfer-labeled primers. *Genome Res* **11**: 163–169.

Newton CR, Graham A, Heptinstall LE, Powell SJ, Summers C, Kalsheker N, Smith JC, Markham AF. 1989. Analysis of any point mutation in DNA. The amplification refractory mutation system (ARMS). *Nucl Acids Res* **17**: 2503–2516.

Nilsson M, Malmgren H, Samiotaki M, Kwiatkowski M, Chowdhary BP, Landegren U. 1994. Padlock probes: circularizing oligonucleotides for localized DNA detection. *Science* **265**: 2085–2088.

Nilsson M, Krejci K, Koch J, Kwiatkowski M, Gustavsson P, Landegren U. 1997. Padlock probes reveal single-nucleotide differences, parent of origin and in situ distribution of centromeric sequences in human chromosomes 13 and 21. *Nat Genet* **16**: 252–255.

Oliphant A, Barker DL, Stuelpnagel JR, Chee MS. 2002. BeadArray technology: enabling an accurate, cost-effective approach to high-throughput genotyping. *Biotechniques Suppl*: 56–61.

Ozaki K, Ohnishi Y, Iida A, Sekine A, Yamada R, Tsunoda T, Sato H, Hori M, Nakamura Y, Tanaka T. 2002. Functional SNPs in the lymphotoxin-alpha gene that are associated with susceptibility to myocardial infarction. *Nat Genet* **32**: 650–654.

Pastinen T, Partanen J, Syvänen A-C. 1996. Multiplex, fluorescent, solid-phase minisequencing for efficient screening of DNA sequence variation. *Clin Chem* **42**: 1391–1397.

Pastinen T, Kurg A, Metspalu A, Peltonen L, Syvänen A-C. 1997. Minisequencing: a specific tool for DNA analysis and diagnostics on oligonucleotide arrays. *Genome Res* **7**: 606–614.

Pastinen T, Raito M, Lindroos K, Tainola P, Peltonen L, Syvänen A-C. 2000. A system for specific, high-throughput genotyping by allele-specific primer extension on microarrays. *Genome Res* **10**: 1031–1042.

Patil N, Berno AJ, Hinds DA, Berrett WA, Doshi JM, Hacker CR, Kautzer CR, Lee DH, Marjoribanks C, McDonough DP, Nguyen BTN, Norris MC, Sheehan JB, Shen N, Stern D, Stokowski RP, Thomas DJ, Trulson MO, Vyas KR, Frazer KA, Fodor PA, Cox DR. 2001. Blocks of limited halotype diversity revealed by high-resolution scanning of human chromosome 21. *Science* **294**: 1719–1723.

Pease AC, Solas D, Sullivan EJ, Cronin MT, Holmes C, Fodor SPA. 1994. Light-generated oligonucleotide arrays for rapid DNA sequence analysis. *Proc Natl Acad Sci USA* **91**: 5022–5026.

Reich DE, Cargill M, Bolk S, Ireland J, Sebeti PC, Richter DJ, Lavery T, Kouyoumjian R, Farhadian SF, Ward R, Lander ES. 2001. Linkage disequilibrium in the human genome. *Nature* **411**: 199–204.

Ronaghi M. 2001. Pyrosequencing sheds light on DNA sequencing. *Genome Res* **11**: 3–11.

Ronaghi M, Karamohamed S, Pettersson B, Uhlén M, Nyren P. 1996. Real-time DNA sequencing using detection of pyrophosphate release. *Anal Biochem* **242**: 84–89.

Ronaghi M, Nygren M, Lundeberg J, Nyren P. 1999. Analyses of secondary structures in DNA by pyrosequencing. *Anal Biochem* **267**: 65–71.

Ross PL, Lee K, Belgrader P. 1997. Discrimination of single-nucleotide polymorphisms in human DNA using peptide nucleic acid probes detected by MALDI-TOF mass spectrometry. *Anal Chem* **69**: 4197–4202.

Ross P, Hall L, Smirnov I, and Haff L. 1998. High level multiplex genotyping by MALDI-TOF mass spectrometry. *Nat Biotechnol* **16**: 1347–1351.

Saiki R, Bugawan TL, Horn GT, Mullis KB, Erlich HA. 1986. Analysis of enzymatically amplified β-globin and HLA-DQ DNA with allele-specific oligonucleotide probes. *Nature* **324**: 163–166.

Saiki, RK, Walsh PS, Levenson CH, Erlich HA. 1989. Genetic analysis of amplified DNA with immobilized sequence-specific oligonucleotide probes. *Proc Natl Acad Sci USA* **86**: 6230–6234.

Samiotaki M, Kwiatkowski M, Parik J, Landegren U. 1994. Dual-color detection of DNA sequence variants by ligase-mediated analysis. *Genomics* **20**: 238–242.

Sauer S, Lechner D, Berlin K, Lehrach H, Escary J-L, Fox N, Gut IG. 2000a. A novel procedure for efficient genotyping of single nucleotide polymorphisms. *Nucl Acids Res* **28**: e13.

Sauer S, Lechner D, Berlin K, Plançon C, Heuermann A, Lehrach H, Gut IG. 2000b. Full flexibility genotyping of single nucleotide polymorphisms by the GOOD assay. *Nucl Acids Res* **28**: e100.

Sayer DC, Goodridge DM, Christiansen FT. 2004. Assign 2.0: software for the analysis of Phred quality values for quality control of HLA sequencing-based typing. *Tissue Antigens* **64**: 556–565.

Sayer DC, Whidborne R, De Santis D, Rozemuller EH, Christiansen FT, Tilanus MG. 2004. A multicenter international evaluation of single-tube amplification protocols for sequencing-based typing of HLA-DRB1 and HLA-DRB3,4,5. *Tissue Antigens* **63**: 412–423.

Shalon D, Smith SJ, Brown PO. 1996. A DNA microarray system for analyzing complex DNA samples using two-color fluorescent probe hybridization. *Genome Res* **6**: 639–645.

Shchepinov MS, Chalk R, Southern EM. 1999. Trityl mass-tags for encoding in combinatorial oligonucleotide synthesis. *Nucl Acids Symp Ser* **42**: 107–108.

Shiina T, Inoko H, Kulski JK. 2004. An update of the HLA genomic region, locus information and disease associations: 2004. *Tissue Antigens* **64**: 631–649.

Shumaker JM, Metspalu A, Caskey CT. 1996. Mutation detection by solid phase primer extension. *Hum Mutat* **7**: 346–354.

Sosnowski RG, Tu E, Butler WF, O'Connell JP, Heller MJ. 1997. Rapid determination of single base mismatch mutations in DNA hybrids by electric field control. *Proc Natl Acad Sci USA* **94**: 1119–1123.

Spurkland A, Knutsen I, Markussen G, Vartdal F, Egeland T, Thorsby E. 1993. HLA matching of unrelated bone marrow transplant pairs: direct sequencing of *in vitro* amplified HLA-DRB1 and -DQB1 genes using magnetic beads as solid support. *Tissue Antigens* **41**: 155–164.

Stewart CA, Horton R, Allcock RJ, Ashurst JL, Atrazhev AM, Coggill P, Dunham I, Forbes S, Halls K, Howson JM, Humphray SJ, Hunt S, Mungall AJ, Osoegawa K, Palmer S, Roberts AN, Rogers J, Sims S, Wang Y, Wilming LG, Elliott JF, de Jong PJ, Sawcer S, Todd JA, Trowsdale J, Beck S. 2004. Complete MHC haplotype sequencing for common disease gene mapping. *Genome Res* **14**: 1176–1187.

Syvänen A-C. 1999. From gels to chips: "Minisequencing" primer extension for analysis of point mutations and single nucleotide polymorphisms. *Hum Mutat* **13**: 1–10.

Syvänen A-C. 2001. Accessing genetic variation: genotyping single nucleotide polymorphisms. *Nat Rev Genet* **2**: 930–942.

Tonks S, Marsh SG, Bunce M, Bodmer JG. 1999. Molecular typing for HLA class I using ARMS-PCR: further developments following the 12th International Histocompatibility Workshop. *Tissue Antigens* **53**: 175–183.

Tost J, Gut IG. 2002. Genotyping single nucleotide polymorphisms by mass spectrometry. *Mass Spectrom Rev* **21**: 388–418.

Tost J, Brandt O, Boussicault F, Derbala D, Caloustian C, Lechner D, Gut IG. 2002. Molecular haplotyping at high throughput. *Nucl Acids Res* **30**: e96.

Tyagi S, Kramer FR. 1996. Molecular beacons: probes that fluoresce upon hybridization. *Nat Biotechnol* **14**: 303–308.

Tyagi S, Bratu DP, Kramer FR. 1998. Multicolor molecular beacons for allele discrimination. *Nat Biotechnol* **16**: 49–53.

Wang DG, Fan J-B, Siao C-J, Berno A, Young P, Sapolsky R, Ghandour G, Perkins N, Winchester E, Spencer J, Kruglyak L, Stein L, Hsie L, Topaloglou T, Hubbell E, Robinson E, Mittmann M, Morris MS, Shen N, Kilburn D, Rioux J, Nusbaum C, Rozen S, Hudson TJ, Lipshutz R, Chee M, Lander ES. 1998. Large-scale identification, mapping, and genotyping of single-nucleotide polymorphisms in the human genome. *Science* **280**: 1077–1082.

Weiner M, Hudson TJ. 2002. Introduction to SNPs: discovery of markers for disease. *Biotechniques Suppl* **32**: 4–13.

Weiss KM, Terwilliger JD. 2000. How many diseases does it take to map a gene with SNPs? *Nat Genet* **26**: 151–157.

Welsh K, Bunce M. 1999. Molecular typing for the MHC with PCR-SSP. *Rev Immunogenet* **1**: 157–176.

Whitcombe D, Theaker J, Guy SP, Brown T, Little S. 1999. Detection of PCR products using self-probing amplicons and fluorescence. *Nat Biotechnol* **17**: 804–807.

Xu H, Sha MY, Wong EY, Uphoff J, Xu Y, Treadway JA, Truong A, O'Brien E, Asquith S, Stubbins M, Spurr NK, Lai EH, Mahoney W. 2003. Multiplexed SNP genotyping using the Qbead system: a quantum dot-encoded microsphere-based assay. *Nucl Acids Res* **31**: e43.

2

Glycomics and the Sugar Code: Primer to their Structural Basis and Functionality

Hans-Joachim Gabius

Abstract

Cell surfaces are characterized by the presence of an array of glycoconjugates. This chapter highlights the remarkable talents of natural oligosaccharides as a high-density information coding system. The currently phenomenologically interpreted changes in the cell glycan profile (*glycome*) during, for example, differentiation or malignant transformation can thus underlie a change in the biological meaning of sugar-encoded signals, thereby attaining a functional dimension. The translation of this information into biological responses is carried out by lectins. Their binding to distinct epitopes with specificity to the carbohydrate structure, shape and spatial presentation is involved in innate immunity, regulation of cell adhesion, migration and apoptosis/proliferation as well as mediator release. Clinical perspectives focusing on tissue lectins such as galectins and their operative ligands are outlined.

2.1 Introduction

This chapter aims to raise the awareness for an emerging molecular aspect within the hardware panel assigned to functions in biological information storage and transfer. After all, the central dogma of molecular biology has limited our view of the downstream flow of genetic information to nucleic acids and proteins. Fittingly, these two substances gained a glamorous status which overshadowed other classes of biomolecules for decades (Sharon, 1998). Speaking of biological information automatically invoked the concept of the genetic code. This situation is becoming subject to a fundamental change because of two major lines of evidence. First, a systematic calculation of the upper limits of coding capacity for biomolecules by means of

Immunogenomics and Human Disease Edited by András Falus

oligomer formation has revealed a frontrunner: carbohydrates are second to no other compound class in this respect (Laine, 1997). While only 64 permutations (code words) are possible for a codon (nucleotide triplet) in protein biosynthesis with the four types of base, a total of 38 016 trisaccharides is already theoretically possible when starting with only three different monosaccharides as letters (Laine, 1997). Three different letters (amino acids, nucleotides) will only form 27 permutations. Even more intriguingly, 20 letters will lead to 6.4×10^7 different hexapeptides, a number which is orders of magnitude less than the 1.44×10^{15} theoretically possible. Ironically, the structural manifestation of this potential to enable high-density coding is at the heart of the problem of why recognition of the concept of the sugar code has apparently lagged so far behind the other fields, i.e. genomics and proteomics (Roseman, 2001). The plethora of oligosaccharides – valuable as it is for high-density coding – poses an analytical challenge of a new dimension compared with the structural analysis of nucleic acids and proteins. To clear this hurdle required development of novel strategies to integrate and refine different separation and analytical procedures such as high-resolution anion-exchange chromatography, mass spectrometry and NMR spectroscopy (Hounsell, 1997; Geyer and Geyer, 1998). Today, sophisticated analytical protocols have been successfully established and major guidelines, which govern glycan assembly and processing, have been delineated (Kobata, 1992; Brockhausen and Schachter, 1997; Sharon and Lis, 1997; Reuter and Gabius, 1999; Spiro, 2002). As an equivalent of the terms *genome* and *proteome*, the result of mapping the profile of glycan epitopes has been referred to as establishing the *glycome*. Naturally, the list of -omics terms has consequently been extended by the introduction of *glycomics*. With the demanding task to define glycan structures thus elegantly mastered, it is now possible to confidently introduce this topic into courses of basic biochemistry to let the message on this aspect of the sugar code gather momentum (Kobata, 1992; Sharon and Lis, 1997; Reuter and Gabius, 1999; Spiro, 2002).

The second line of reasoning reshaping our view on carbohydrates relates their diversity to a biological meaning. Interpreting glycan epitopes on the cell surface as biochemical signals, their high-density coding capacity is a real boon. Faced with stringent space limitation in the biological context, sugar coding facilitates presentation of a large number of signals in a minimum of space. That the inferred coding potential is actually utilized and is of medical relevance is attested by the clinical delineation of causal relationships between defects in glycosylation and disease. Proof-of-principle examples are lysosomal storage diseases or LAD II (leukocyte adhesion deficiency syndrome type II; see below), flanked by the insights from developmental biology that impaired glycosylation leads to embryonic defects and the list of examples for sugar compounds as pharmaceuticals (for a representative compilation, see Table 2.1; Reuter and Gabius, 1999; Gabius *et al.*, 2004; Haltiwanger and Lowe, 2004). It seems fair to say that further developments in drug design based on the sugar code can be anticipated by progress in basic research (Rüdiger *et al.*, 2000; Yamazaki *et al.*, 2000). At this stage, the reader may wonder why coding along the germline does not take advantage of the availability of the high-capacity hardware as utilized in the glycocalyx.

Table 2.1 Examples for sugar compounds as pharmaceuticals

Compound	Target	Disease
Acarbose	α-Glucosidases (amylases)	Diabetes mellitus
Heparin/heparinoids	Antithrombin III	Thrombosis
Heparin pentasaccharide (Fondaparinux)	Antithrombin III(factor Xa)	Thrombosis
Derivatives or mimetics of 2-deoxy-2,3-dehydro-N-acetylneuraminic acid	Neuraminidase	Viral infection
N-Butyldeoxynojirimycin	α-Glucosidases (N-glycan processing)	Viral infection
Derivatives or mimetics of milk oligosaccharides	Adhesins and toxins (lectins)	Bacterial infection
GlcN-(2-O-hexadecyl) phosphatidylinositol	GPI-mannosyltransferase I	Protozoan infection (e.g. African sleeping sickness)
Derivatives or mimetics of sialylated/sulfated $Le^{a/x}$-epitopes	Selectins	Inflammatory reaction
D-Man	Phosphomannose isomerase deficiency	Congenital disorder of glycosylation Ib
L-Fuc	GDP–fucose transport	Congenital disorder of glycosylation IIc (LAD II)
N-Butyldeoxygalactonojirimycin and properly glycosylated β-gluco(galacto) cerebrosidase	Glycosphingolipid synthesis and enzymatic degradation	Glycosphingolipid storage disorders

From Gabius *et al.* (2004).

Admittedly, information storage in a minimum of space would be facilitated by such a mode of coding. However, the genetic material does not serve as an immediate effector. Its information acts as a template, and it has to be copied as flawlessly as possible. The prime aim of optimizing copying fidelity in heredity accounts for limiting the structure to linearity and the alphabet size to four letters (Szathmáry, 2003). In contrast, the generation of directly read code words with carbohydrates as letters can take full advantage of all structural levels to achieve chemical diversity including anomeric variation and branching (note that starch and cellulose differ only in the anomeric position, an impressive illustration of what this single change means for the biochemical properties). In addition to anomeric variability, each hydroxyl group of a carbohydrate can in principle be engaged as an acceptor of a glycosidic bond, a situation totally different from nucleotides and amino acids and another reason for the unsurpassed structural variability. As a consequence, the number of enzymes responsible for generating the apparent glycan diversity must be considerably larger than in DNA/RNA and protein synthesis. Indeed, the elaborate system of glycosyltransferases ensures versatile glycan (code word) generation, although the synthesis of the complete panel of theoretically calculated linkage types is enzymatically not possible for each sugar moiety (Brockhausen and Schachter, 1997; Gabius *et al.*, 2002). As concrete examples for the attained levels of sophistication, 19 different α/β-galactosyltransferases, at least 18 sialyltransferases and 11 fucosyltransferases have already been detected in mammalian cells (Harduin-Lepers *et al.*, 2001; Hennet, 2002; Martinez-Duncker *et al.*, 2003). To start mucin-like O-glycan synthesis with conjugation of GalNAc in α-anomeric linkage to serine/threonine residues, the reader would therefore be surprised if only one enzyme were present. This seemingly simple reaction mechanism is indeed shared by 24 individual enzymes, as estimated by database mining, an indication for intricate regulation mechanisms (Ten Hagen, Fritz and Tabak, 2003).

This overall investment in genetic coding (i.e. genes for glycosyltransferases and supporting proteins such as transporters for the nucleotide sugars, for example GDP-fucose, a target for mutations causing LAD II; for details on this deficiency, see Wild *et al.*, 2002) pays off handsomely by rendering synthesis of a wide array of oligomers, especially at branch ends in glycans, feasible. However, their availability would be futile without intricate biochemical mechanisms for accurate translation of the high-density information coding of glycan determinants into a particular biological meaning. Towards this end, nature has devised carbohydrate recognition domains (CRDs) organized in the superfamily of lectins (Gabius, 1991, 1997; Lis and Sharon, 1998; Rüdiger and Gabius, 2001; Loris, 2002; Vasta *et al.*, 2004). This term denotes carbohydrate-binding (glyco)proteins. The following restrictions for its current use apply: proteins which enzymatically modify the bound carbohydrate ligand (e.g. glycosyltransferases or sulfotransferases; incidentally, these enzymes introduce substitutions into code words, increasing coding capacity), which are immunoglobulins or which physiologically interact with free mono- or disaccharides or their derivatives in transport or chemotaxis, are explicitly excluded (Rüdiger and Gabius, 2001). To describe noncatalytic recognition sites for plant cell wall glycans in bacterial glycoside hydrolases (currently listing 39 families) the term *carbohydrate-binding module* is in use (Boraston *et al.*, 2004). In a nutshell, lectins serve as an essential interface between glycan structure and function, and they also have been exploited in analytical methods (Spicer and Schulte, 1992; Gabius and Gabius, 1993, 1997; Cummings, 1997; Manning *et al.*, 2004). Notably, fucose-specific plant and animal (eel) lectins were instrumental in unraveling the structural basis of the AB0 blood group system, and the mitogenicity of plant lectins such as PHA for lymphocytes in a role model for illustrating cellular responses to lectin binding (Watkins and Morgan, 1952; Nowell, 1960; Kilpatrick and Green, 1992; Watkins, 1999; Rüdiger and Gabius, 2001; Gabius *et al.*, 2004). With the major players in information transfer by the sugar code thus identified, the next chapter will guide the reader to structural and functional aspects of lectins.

2.2 Lectins as Effectors in Functional Glycomics

The documented beginning of research on lectins dates back to 1860. Silas Weir Mitchell (1829–1913), later a leading expert in neurology and psychiatry in his time

```
           1                                                                                      90
hGal-1     MACGLVASNL ..NLKPGECL RVRGEVAPDA KSFVLNL... GKDSNNLCLH FNPRFNAHGD ANTIVCNSKD GGAWGTEQRE AVFPFQPGSV
hGal-2     MTGELEVKNM ..DMKPGSTL KITGSIADGT DGFVINL... GQGTDKLNLH FNPRFSE.... .STIVCNSLD GSNWGQEQRE DHLCFSPGSE
rGal-2     MSEKFEVTNL ..NMKSGMSL KIKGKIHNDV NSFTINL... GQGKETLNLH FNPRFNE.... .STIVCNTLD GSSWGQEQRE NHSCFSPGSE
hGal-7     MSNVPHKSSL PEGIRPGTVL RIRGLVPPNA SRFHVNLLCG EEQGSDAALH FNPRLDT... .SEVVFNSKE QGSWGREERG PGVPFQRGQP
Conse.     Ms.....sn$ ....kpG..L r!rG.!.p.a ..F..NL... g.....l.LH FNPRf..... .st!VcNsk# ggsWG.E#Re ...pFqpGs.

           91                                                  141
hGal-1     AEVCITFDQA NLTVKLPDGY EFKFPNRLNL EAINYMAADG DFKIKCVAFD .
hGal-2     VKFTVTFESD KFKVKLPDGH ELTFPNRLGH SHLSYLSVRG GFNMSSFKLK E
rGal-2     VKLTLTFQDK DFKVTLPDGH SLTFPNRLGH NHLRYLSM.D GLQISSFKLE .
hGal-7     FEVLIIASDD GFKAVVGDAQ YHHFRHRLPL ARVRLVEVGG DVQLDSVRIF .
Conse.     .ev.itf.dd .fkv.lpDg. ...FpnRL.l ...ry..v.g d.q..sv... .
```

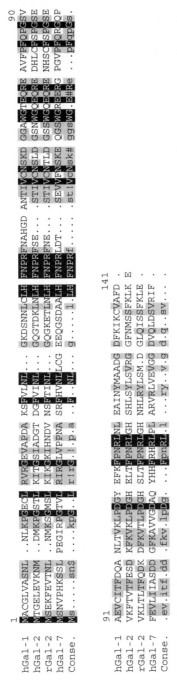

Figure 2.1 Inter-galectin and inter-species sequence comparison of subgroup of mammalian prototype galectins including galectins-1, -2 and -7. Amino acid sequences of human galectins-1, -2 and -7 as well as rat galectin-2 were aligned using the program Multalin (http://prodes.toulouse.inra.fr/multalin/multalin.html; version 5.4.1). Identical residues found in all four galectins are indicated as white letters on a black background, whereas residues that are identical or similar between at least two of the sequences are in black letters on a gray background. A consensus sequence calculated from the four galectin sequences is added to the alignment; consensus symbols represent: !, I or V; $, L or M; %, F or Y; and #, N, D, Q or E.

and also a successful novelist, observed that one drop of venom from the rattlesnake (*Crotalus durissus*) and a drop of blood from a pigeon's wounded wing 'coagulated firmly within three minutes' (Mitchell, 1860). More than 100 years after this pioneering observation, Marie Louise Ogilvie and T. Kent Gartner were able to conclusively prove Ca^{2+}-dependent lectins (later referred to as C-type lectins) in Crotalidae venoms to structurally embody this activity (Ogilvie and Gartner, 1984). Technically, sugar-dependent agglutination was the common assay to detect and monitor lectin activity, and lectins were then classified according to their monosaccharide specificity measured by systematic inhibition studies. The introduction of affinity chromatography to purify lectins in 1965 and the development of protein and gene sequencing to determine their primary structure as well as the application of biophysical techniques to move on to describe the folding have paved the way to refine the criteria to group lectins into families (for a detailed historical description, see Gabius, 2001a; Gabius *et al.*, 2004; Sharon and Lis, 2004). In this respect, sequence alignments have become a powerful tool to track down similarities (or even homologies) in proteins sharing a common sugar target. An example of the extent of identity scoring which is commonly encountered in these computer-supported calculations is shown for mammalian prototype galectins in Figure 2.1. We will come back to these lectins, which are potent growth regulators, in the next section. Such systematic computations in conjunction with biochemical properties, e.g. Ca^{2+}-dependence of sugar binding, led to the definition of five classes of animal lectins (Table 2.2). Crystal structures could thus be predicted to define characteristic folding patterns underlying this separation. The classical route from protein to crystal, its diffraction pattern and then to the resolved secondary/tertiary structures, shown in Figure 2.2 for the case of a galectin (Varela *et al.*, 1999), confirmed the validity of this classification system. Accounting for ongoing studies which are defining new folds for carbohydrate recognition in animals, we can conclude that the complexity of code word generation by glycosyltransferases is matched on the level of decoding devices. The crystallographic analysis of lectins not only aids establishment of a knowledge-based classification scheme. It has also led to an understanding of the positioning of invariant amino acid residues emerging from sequence alignments. An instructive example is delineated from looking at the sequences of galectins in Figure 2.1. Here the only Trp moiety can clearly be singled out as a constant feature. In principle, the binding site's topology should explain the epimer specificity of a galectin. As illustrated for the recognition of D-galactose in Figure 2.3, directional hydrogen bonds with distinct side chains and also the stacking and CH–π interactions with the Trp ring system govern the ability of galectins (or bacterial toxins such as the cholera toxin; see below) to distinguish β-galactosides from natural epimers such as mannose or glucose (note the involvement of the axial 4′-hydroxyl group in contacts shown in Figure 2.3; Lis and Sharon, 1998; Varela *et al.*, 1999; Sharon and Lis, 2001; Solís *et al.*, 2001). The involvement of hydrophobic patches in binding (here the C–H-rich bottom side of the galactose) documents the, at first sight, unexpected capacity of sugars to be engaged in different modes of contact in intermolecular interaction, a further reason for the versatility of the carbohydrates' role as ligand (Lemieux, 1996).

Table 2.2 Main families of animal lectins

Family	Structural motif	Carbohydrate ligand	Modular arrangements
C-type	Conserved CRD	Variable (mannose, galactose, fucose, heparin tetrasaccharide)	Yes
I-type	Immunoglobulin-like CRD	Variable ($Man_6GlcNAc_2$, HNK-1 epitope, hyaluronic acid, α2,3/α2,6-sialyllactose)	Yes
P-type	Conserved CRD	Mannose-6-phosphate-containing	Yes
Galectins (formerly S-type)	Conserved CRD	Galβ1,3(4)GlcNAc core structures with species- and galectin-type-dependent differences in affinity for extensions to, e.g., blood group A, B or H epitopes; internal stretches of poly(N-acetyllactosamine) chains	Variable
Pentraxins	Pentameric subunit arrangement	4,6-Cyclic acetal of β-galactose, galactose, sulfated and phosphorylated monosaccharides glycoproteins	Yes

CRD, carbohydrate recognition domain; from Gabius (2000) with modifications.

Beyond these insights structural lectin investigation has answered another pertinent question regarding the overall binding energy: how will the inherently low affinity of a single protein–carbohydrate contact be turned into strong binding when a lectin meets a binding partner?

Drawing an analogy, antibodies exploit the option of spatial clustering to attain this aim. As noted above, immunoglobulins are yet excluded from the lectin definition. Nonetheless, similar means to reach the same aim could be envisaged. Indeed, structural and cell biological studies have taught the amazing lesson that CRDs in a lectin often tend to be spatially associated, either noncovalently or by covalent bonds. The spatial topology is thus a factor which matters to a considerable extent, establishing a recurring theme in functional glycomics. Generation of multiple contacts by clustering is undoubtedly a general means of enhancing binding activity. To help the reader appreciate this strategy, Figure 2.4 presents an overview of CRD presentation in mammalian lectins. When the reader goes over this figure and its legend, the idea should be conveyed that the different modes of presentation of CRDs are tailored by evolution for distinct functions. Looking, for example, at lectins in innate immunity, the serum mannan-binding lectin (and also other collectins such as surfactant proteins-A and -D or the ficolins) and the tandem-repeat C-type

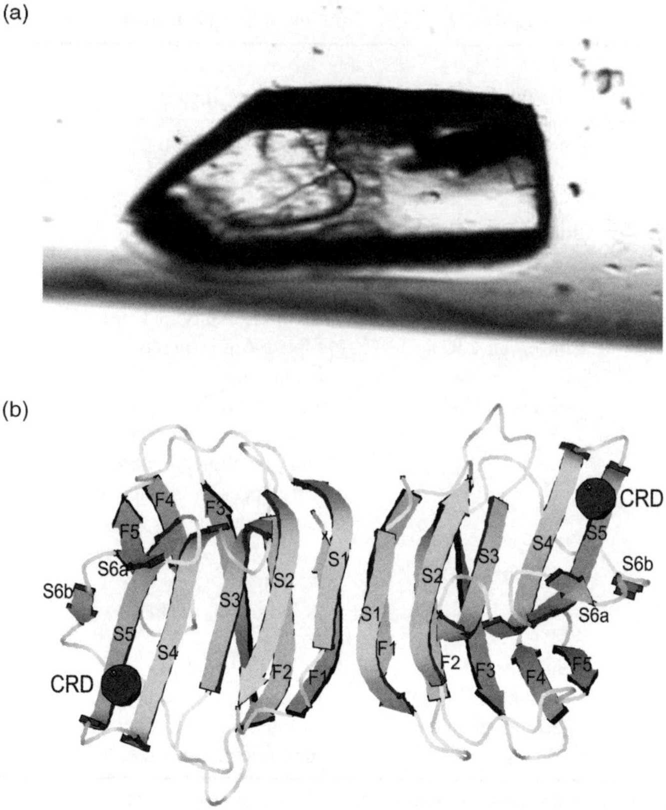

Figure 2.2 Orthorhombic crystal ($C222_1$) of the developmentally regulated homodimeric galectin from chicken liver (CG-16) grown in 2 M ammonium sulfate, 5% (v/v) isopropanol and 1% β-mercaptoethanol, pH 5.6, at an estimated final protein concentration of 10 mg/ml. The crystal size is $0.4 \times 0.6\ mm^3$ (a). Ribbon diagram of CG-16, prepared with MOLSCRIPT. The β-strands in the five-stranded (F1–F5) and six-stranded (S1–S6a/S6b) β-sheets are denoted by the letter–number code. The two carbohydrate-binding sites located at the opposite ends of the homodimer are indicated by spheres (b), adapted from Solís *et al.* (2001).

(macrophage) mannose receptor, their widely spaced CRDs are ideal sensors for the non-mammalian surface glycosignatures of infectious bacteria or yeasts, explaining the graphic term 'pattern recognition receptors' for these lectins (Lineham *et al.*, 2000; Kilpatrick, 2002; Lu *et al.*, 2002; McGreal *et al.*, 2004; Sano and Kuroki, 2005). Further informations on individual cases have been compiled in a recent collection of reviews on animal lectins (Gabius, 2002). As indicated, the presence of a CRD and the functional cooperation of CRDs by being positioned in spatial vicinity (or of a CRD in conjunction with other modules in mosaic-like proteins) is instrumental for lectin functionality. An overview on the current spectrum of documented lectin functions is given in Table 2.3. Its content underscores the validity of this chapter's heading, referring to lectins as effectors in functional glycomics.

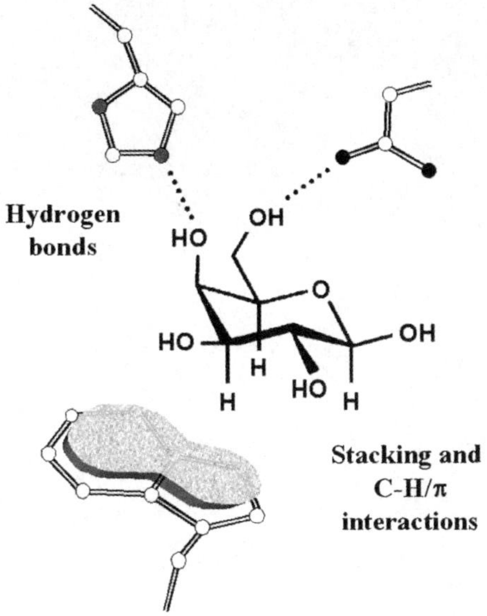

Figure 2.3 The main enthalpic features to facilitate binding of a sugar to a protein: hydrogen bonds using lone electron pairs of sugar oxygen atoms as acceptors and the hydrogen atoms of hydroxyl groups as donors as well as C–H/π-electron interactions between patches of a positively polarized character (in this case the B-face of D-Gal) and the delocalized π-electron cloud of a suitably positioned aromatic residue (here Trp).

Having accomplished the relation of lectin structure to function and defined guidelines to translate sequence information of newly discovered lectins into structural motifs by homology-based modeling, an emerging research topic is to address the following question: does glycan presentation in natural glycoconjugates regulate lectin affinity? Intuitively, it would mean missing opportunities not to let this parameter influence affinity. Thus, the aspect complementary to CRD presentation, i.e. the influence of distinct topological presentation of the lectin-reactive glycans by branching or clustering on the glycan's ligand properties, has become a promising research topic, and its assumed importance is in fact being unraveled. Systematic analysis of lectin binding to well-characterized glycoproteins with complex glycosylation profiles including multiantennary or branched glycan chains and to synthetic mimics of natural glycans proves indispensable in this respect (Wu *et al.*, 2001, 2002, 2004; Ahmad *et al.*, 2004). As will be discussed in the next but one chapter, we are now becoming aware of more than one level of specificity in lectin–carbohydrate recognition. Their definition, given after describing the properties of a selected lectin family, highlights the enormous potential to work with carbohydrate-based messages using the sugar code.

At this stage, we can reliably summarize that carbohydrates are 'ideal for generating compact units with explicit informational properties' (Winterburn and Phelps, 1972). Code word generation is achieved by a complex array of

Figure 2.4 Illustration of the strategies for how CRDs in animal lectins are positioned to reach optimal affinity, ligand selection (e.g. to separate self from nonself glycan profiles in innate immunity) and accessibility (modified from Gabius *et al.*, 2002). From left to right, the CRD display in the three subtypes within the galectin family (chimeric, prototype and tandem-repeat-type arrangements), the presentation of CRDs in serum and surfactant collectins or ficolins connected to their collagenous stalks and the noncovalent association of binding sites in transmembrane C-type lectins by α-helical coiled-coil stalks (e.g. asialoglycoprotein and Kupffer cell receptors, CD23, DC-SIGN or DC-SIGNR) are given. Next, the tandem-repeat display in mannose-specific macrophage receptor (also found on dendritic cells, hepatic endothelial cells, kidney mesangial cells, retinal pigment epithelial cells and tracheal smooth muscle cells) and related C-type subfamily lectins (e.g. DEC-205 or Endo-180) as well as in the cation-independent P-type lectin is presented. The occurrence of lectin activity for GalNAc-4-SO_4-bearing pituitary glycoprotein hormones in the cysteine-rich domain, a member of the β-trefoil familiy of protein modules, in the N-terminal section of the mannose receptor, which is linked via a fibronectin-type-II-repeat-containing module to the tandem-repeat section, is also included into the schematic drawing for these two classes of lectins/lectin-like proteins with more than one CRD per protein chain. Moving further to the right, the association of a distal CRD in selectins [attached to an epidermal-growth-factor-(EGF)-like domain and two to nine complement-binding consensus repeats] or in the siglec subfamily of I-type lectins using one to 16 C2-set immunoglobulin-like units as spacer equivalents to let the CRD reach out to contact ligands and to modulate the capacity to serve in *cis*- or *trans*-interactions on the cell surface is shown. In the matrix, the modular proteoglycans (hyalectans/lecticans: aggrecan, brevican, neurocan and versican) interact with hyaluronan (and also link protein) via the N-terminal loop assigned to the immunoglobulin superfamily, with receptors binding to the glycosaminoglycan chains in the central region and with carbohydrates or proteins (fibulins-1 and -2 and tenascin-R) via the C-type lectin-like domain flanked by EGF-like and complement-binding consensus repeat modules. For further information on individual lectin groups, see Gabius (2002).

Table 2.3 Functions of animal lectins

Activity	Example of lectin
Ligand-selective molecular chaperones in endoplasmic reticulum	Calnexin, calreticulin
Intracellular routing of glycoproteins and vesicles	ERGIC-53 and VIP-36 (probably also ERGL and VIPL), P-type lectins, comitin
Intracellular transport and extracellular assembly	Nonintegrin 67 kDa elastin/laminin-binding protein
Inducer of membrane superimposition and zippering (formation of Birbeck granules)	Langerin (CD207)
Cell-type-specific endocytosis	Hepatic and macrophage asialoglycoprotein receptors, dendritic cell and macrophage C-type lectins (mannose receptor family members (tandem-repeat type) and single CRD[a] lectins such as langerin/CD207), cysteine-rich domain of the dimeric form of mannose receptor for GalNAc-4-SO_4-bearing glycoprotein hormones in hepatic endothelial cells, P-type lectins
Recognition of foreign glycans(β1,3-glucans, LPS)	CR3 (CD11b/CD18), dectin-1, *Limulus* coagulation factors C and G, earthworm CCF
Recognition of foreign or aberrant glycosignatures on cells (including endocytosis or initiation of opsonization or complement activation)	Collectins, L-ficolin, C-type macrophage and dendritic cell receptors, α/θ-defensins, pentraxins (CRP, limulin), tachylectins
Targeting of enzymatic activity in multimodular proteins	Acrosin, laforin, *Limulus* coagulation factor C
Intra- and intermolecular modulation of enzyme activities *in vitro*	Porcine pancreatic α-amylase, galectin-1/α2-6-sialyltransferase
Bridging of molecules	Homodimeric and tandem-repeat-type galectins, cytokines (e.g. IL-2:IL-2R and CD3 of TCR), cerebellar soluble lectin
Induction or suppression of effector release (H_2O_2, cytokines etc.)	Galectins, selectins and other C-type lectins such as CD23, BDCA-2 and dectin-1
Cell growth control and induction of apoptosis/anoikis	Galectins, C-type lectins, amphoterin-like protein, hyaluronic acid-binding proteins, cerebellar soluble lectin
Cell migration and routing	Selectins and other C-type lectins, I-type lectins, galectins, hyaluronic acid-binding proteins (RHAMM, CD44, hyalectans/lecticans)
Cell–cell interactions	Selectins and other C-type lectins (e.g. DC-SIGN), galectins, I-type lectins (e.g. siglecs, N-CAM, P_0 or L1)
Cell–matrix interactions	Galectins, heparin- and hyaluronic acid-binding lectins including hyalectans/lecticans, calreticulin
Matrix network assembly	Proteoglycan core proteins (C-type CRD and G1 domain of hyalectans/lecticans), galectins (e.g. galectin-3/hensin), non-integrin 67 kDa elastin/laminin-binding protein

[a]carbohydrate recognition domain. From Gabius *et al.* (2004), extended and modified.

glycosyltransferases without the stringency of a template. The regulation of their expression and activity facilitates dynamic and reversible shifts in the glycome. Glycan diversity is matched on the level of receptor proteins (lectins). Strategic positioning of CRDs in lectins leads to high-affinity binding and a remarkably wide range of lectin functions. The classification of lectins is based on homology criteria in primary and secondary structures. Examples are given here in Figures 2.1 and 2.2 for galectins [lectins with Ca^{2+}-independent specificity to β-galactosides and the β-sandwich folding pattern shown in Figure 2.2(b)]. Because the case of the cell adhesion molecules, integrins, has epitomized the range of functional fine-tuning residing in intrafamily diversity, we next address this question on diversity and biological implications in the case of the lectin family listed in Table 2.2. With similar principles emerging from recent studies of different lectin families such as C-, I- and P-type lectins and even cytokines capable of bind to sugars (Kaltner and Stierstorfer, 1998; Angata and Brinkman-van der Linden, 2002; Cebo *et al.*, 2002; Dahms and Hancock, 2002; Crocker, 2004; Geijtenbeek *et al.*, 2004; Kanazawa *et al.*, 2004; Takahara *et al.*, 2004; McGreal *et al.*, 2005; van de Wetering *et al.*, 2005), we can select galectins as a representative example. They, like several C- and I-type lectins, home in on a class of glycans with distinguished properties, so we start the next part by explaining the structural significance of these determinants.

2.3 Galectins: Structural Principles and Intrafamily Diversity

A carbohydrate epitope should be spatially accessible to serve in information transfer. Its versatility will profit from the biosynthetic potential of introducing an array of substitutions, which modulates or even reversibly switches off its ligand activity. Which type of carbohydrate structure meets these two demands? Being positioned at ends of glycan branches, β-galactosides readily fulfill this topological requirement. Moreover, not a single but 13 different β-galactosyltransferases are responsible for synthesis of β1,3(4)-galactosides, an indication of intricate fine-tuning on the level of the acceptor structure (Hennet, 2002). Adding a variety of different substitutions introduced by α-fucosyl-, α-galactosyl- and *N*-acetylgalactosaminyltransferases and the switch-off signal α2,6-sialylation, β-galactosides in glycolipids and glycoproteins easily also satisfy the second requirement for structural diversity defined above (Gabius *et al.*, 2002). That being said, it becomes clear that β-galactosides are ideal to embody the prophetic statement that 'glycosyl residues' can 'impart a discrete recognitional role to the protein' (Winterburn and Phelps, 1972). By extending from a cell's or a protein's surface, these epitopes can act like sensors (or tentacles) in a sugar-based communication with galectins. Turning to Table 2.3, it thus comes as no surprise to find that galectins show up frequently in signaling and adhesion activities. The above-mentioned exquisite discrimination against other natural hexoses, explained in structural terms in Figure 2.3, excludes errors in decoding. On the molecular level, the axial 4′-hydroxyl group of D-galactose, as illustrated in Figure 2.3, is a major contact site for hydrogen bonding. This significance was verified independently by systematic chemical mapping with deoxy and fluoro derivatives of lactose as galectin ligands (Solís *et al.*, 1996; Solís and Díaz-Mauriño, 1997). Systematic work with the panel of synthetic lactose variants satisfactorily explains the above-mentioned selection against the naturally occurring epimers glucose and mannose and pinpoints major contact sites for hydrogen bonding.

As emphasized above when describing the first observation of activity for an animal lectin which later turned out to be a C-type lectin, galectins were also initially detected by the hemagglutination assay. The inhibition of lectin activity, i.e. agglutination of trypsin-treated rabbit erythrocytes, by lactose but not unrelated sugars led to the discovery of the activity of the first prototype member of this family. It was called electrolectin, because it was the lectin activity in extracts of electric organ tissue of *Electrophorus electricus* (Teichberg *et al.*, 1975). Acquisition of ability to cross-link cells or glycolipids/glycoproteins calls for bivalency with sites spatially separated from each other. The second panel of Figure 2.2 and the left section of Figure 2.4 illustrate how the two CRDs are positioned in a homodimeric prototype galectin to generate a cross-linking device (Gabius, 1997). Besides this noncovalent association of modules, two CRDs can be covalently linked by a connecting peptide establishing the tandem-repeat-type organization, and the third topological arrangement in the galectin family is found in galectin-3 with its collagenase-sensitive domain, which substantially contributes to account for oligomer formation (see Figure 2.4, left section). Purification of a galectin has become rather easy with custom-made resins.

Affinity chromatography with resin-immobilized sugar has become the method of choice for lectin purification, dating back to the landmark report that a cross-linked dextran gel is suited for one-step purification of concanavalin A (Agrawal and Goldstein, 1965). Oddly enough, this pioneering work was forthrightly rejected by the editors after submission to the first target journal (Sharon, 1998). Owing to the engineering of highly efficient prokaryotic expression vectors, bacterial extracts have replaced tissues as a convenient source for lectins. Because galectins are not glycosylated, it is not necessary in this case to work with mammalian cultures. Pure products are obtained after processing by affinity chromatography, e.g. using resin after divinyl sulfone activation (Gabius, 1990, 1999). The highly sensitive technique of nano-electrospray ionization mass spectrometry verifies the absence of any substitution by bacterial enzymology which might confound application of the lectins (Kopitz *et al.*, 2003). The latter method has even been adapted to determine the quaternary structure and is able to pick up dimers without signs of dissociation by sample processing. With the detailed information on structural aspects and the extensive sequence alignments, it was possible to assign each of the 14 currently known mammalian galectins to the three distinct subgroups (Kasai and Hirabayashi, 1996; Gabius, 2001a). Albeit practically useful, this classification system is not sufficiently detailed to mirror evolutionary traits. In order to depict such relationships between galectins, we constructed the genealogical tree, presented in Figure 2.5. An early separation between the prototype galectins-1 and -2 vs -7, the close similarity of prototype rat galectin-5 to the C-terminal domain of the tandem-repeat-type

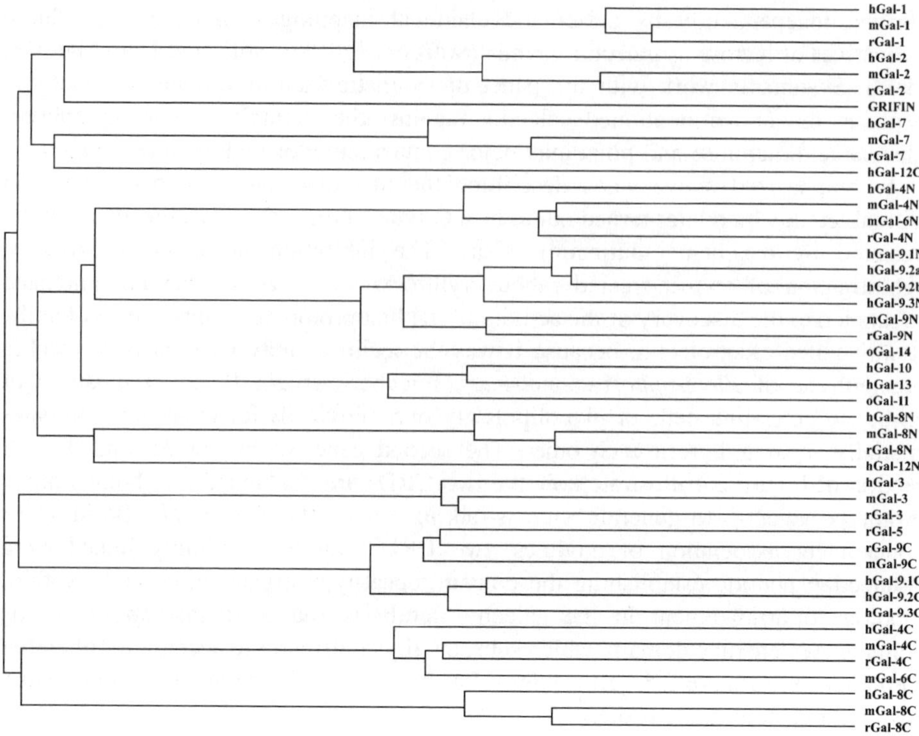

Figure 2.5 Genealogical tree of mammalian galectins.

galectin-9 and the difference between N- and C-terminal CRDs of any tandem-repeat-type galectin are noteworthy messages arising from this illustration. The latter point is taken into account in Figure 2.4 (left section). Here, the two CRDs of tandem-repeat-type galectins are sketched as different in contrast to those in homodimeric prototype family members. It is reasonable to assume that the origin of this subgroup can be attributed to merging two monomeric modules. This process facilitates the design of a protein for the cross-linking of two different ligand types, a special task for which homodimeric prototype galectins are not suited. Within this class, a close functional similarity was intuitively expected, which could reflect the sequence similarity (see also Figure 2.1). In other words, galectins-1 and -2 from the same branch of the evolutionary tree are surmised to be structural and functional homologs. It is pertinent to rebut this notion by referring to the observation that human galectins-1 and -2 activate disparate caspase profiles in T cell apoptosis (Sturm *et al.*, 2004). Galectin-3's position in the genealogical tree reflects its status as sole chimera-type family member.

At this stage, we can conclude that galectins establish a complex family of endogenous lectins with different degrees of sequence similarity and two principal types of structural organization: as cross-linking modules with identical or different

CRDs (after dissociation or proteolytic cleavage turned into monomers) or as chimeric protein with a collagenase-sensitive domain for reversible intermolecular self-aggregation (Gabius, 1997; Cooper, 2002; Ahmad *et al.*, 2004). A common feature is their secretion by a nonclassical pathway, by which intracellular association with glycoproteins originating from the Golgi apparatus is precluded. However, the intracellular presence of galectins can be detected. In this case, peptide motifs of proteins can serve as ligands such as oncogenic H-Ras for galectin-1 or β-catenin for galectin-3 (Elad-Sfadia *et al.*, 2002; Liu, 2004; Rotblat *et al.*, 2004; Shimura *et al.*, 2004; André *et al.*, 2005a). These telling examples require no further comment on the potential of the approach to turn endogenous lectins into tools in cyto- and histopathology (Gabius and Bardosi, 1991; Purkrábková *et al.*, 2003; Plzák *et al.*, 2004).

Of further general medical relevance in this context is the fact that a cell not only expresses a single galectin species but can harbor a characteristic, albeit often not yet fully explored profile of galectin expression. To determine a cell's lectin pattern, we have introduced fingerprinting by RT-PCR or by immunohistochemistry with non-crossreactive galectin-type-specific antibodies (Gabius *et al.*, 1986; Lahm *et al.*, 2001, 2003, 2004; Wollina *et al.*, 2002; Kayser *et al.*, 2003a, b; Nagy *et al.*, 2003). Intrafamily diversity even extends beyond the presence of several family members. Explicitly, variability of the length in linker peptides, detected for galectin-8 and also for galectin-9 with an in-frame insertion of 96 base pairs into the coding region of the linker (Lahm *et al.*, 2003), and alternative splicing [also known to be operative in C-type lectins such as dectin-1, CD23, DC-SIGN and DCL-1/CD205(DEC205)] add further mechanisms to increase intrafamily diversity (Gabius, 1997; Cooper, 2002).

The current status of immunohistochemical fingerprinting not only confirms the RT-PCR data but also provides new prognostic information in tumor diagnosis. In a case study on colon cancer, low indices of expression of galectins-1 and -4 (but not -7) were favorable prognostic markers for patients with Duke's A and B tumors (Nagy *et al.*, 2003). In parallel, studies in animal models together with histopathological monitoring have collected evidence for the functional basis of galectin-1 in tumor progression for another tumor type (for the crystal structure of human galectin-1, see López-Lucendo *et al.*, 2004). This galectin was found to be negatively correlated to prognosis in patients with glioblastoma, and it strongly stimulated cell motility and tissue invasion *in vitro* and in xenografts (Camby *et al.*, 2002). In order to preclude raising the impression of a generally negative effect of galectin-1 expression for the (tumor) patient, the next sets of discussed results teach the important lesson that the galectin-induced effect is specific for the cell type. On activated T cells and neuroblastoma (SK-N-MC) cells galectin-1 induces apoptosis or reduces cell growth, respectively, an emerging perspective for a new treatment modality in autoimmunity and this type of aberrant cell growth (Gabius, 2001b; Kopitz *et al.*, 2001, 2003; Rappl *et al.*, 2002). Notably, naive T cell survival is maintained by galectin-1, illustrating the dependence of the activity on the cell's physiological status (Endharti *et al.*, 2005). The availability of these experimental systems with a clear function of a certain galectin makes it possible to answer the question on functional overlap/divergence between galectin family members. As noted above, the presence of galectin-1 caused a drastic decrease in neuroblastoma cell proliferation *in vitro.* Galectin-3 blocks this effect, allowing tumor cells to maintain their growth pattern (Kopitz *et al.*, 2001). Competitive inhibition of cell surface binding of galectin-1 by galectin-3 is the reason for neutralizing the functionality of galectin-1 by the chimera-type family member, a case of functional divergence among galectins (Kopitz *et al.*, 1998, 2001, 2003). Likewise, a monomeric galectin also blocks galectin-1 activity on these cells, probably by making the essential cross-linking impossible (André *et al.*, 2005b). The galectins' major target in these cells is the pentasaccharide of ganglioside GM_1, a molecular rendezvous which will further be explained in structural terms in the next chapter. Evidently, galectin-3 is in this case not able to trigger signaling pathways as the cross-linking galectin-1 does (Brewer, 2001, 2002). This example highlights the critical importance of spatial factors in the realm of protein–carbohydrate recognition. So far, we have mainly focused on themes in CRD positioning summarized in Figure 2.4 and not looked at the glycans. Therefore, it is warranted to discuss this issue, already touched upon in the previous chapter, systematically and in more detail in the next part of our review.

2.4 Ligand-Dependent Levels of Affinity Regulation

The ongoing systematic scrutiny of all aspects of the molecular interaction between a lectin and its ligand is revealing an enormous intricacy in how spatial factors regulate lectin affinity. The way glycan structures are enzymatically substituted can thus have a biological meaning. As noted above, lectin activity was first detected by inhibition of hemagglutination by mono- or disaccharides. The first level of affinity regulation is embodied thus by mono- or disaccharide discrimination, detected for example in the hemagglutination assays. Commercially available substances generally suffice for this first-step characterization. Reflected in the term *galectin*, the β-galactoside lactose (or thiodigalactoside) is a commonly potent inhibitor for the members of this lectin family. However, this property by no means implies that the sugar specificity is an absolutely constant feature for all galectins. Among them, the selectivity for a distinct disaccharide, e.g. Galβ1,4GlcNAc (LacNAc, *N*-acetyllactosamine), may already differ markedly, endowing certain lectins with a distinctive capacity for ligand selection. This assumption was indeed validated in this case by measuring relative potency values of 5.6 and 700 compared with lactose in inhibition assays in three model studies on two different galectins (Wu *et al.*, 2001, 2002, 2004). While sharing the galactose unit as major contact point, selectivity already sets in on the level of the disaccharide. This result signifies that the elongation of the chain length of the sugar ligand guides us to the second level.

The current availability of a wide panel of oligosaccharides by chemical synthesis or biochemical purification from natural sources makes it possible to proceed to map the fine-specificity of any lectin at level 2. For example, it has become evident that human galectin-7 can bind internal LacNAc determinants (e.g. in α2,6-sialyl DiLacNAc), which are in contrast not accessible for galectin-1 (Ahmad *et al.*, 2002). Constituting level 2, the sequence of the carbohydrate structure, starting with linkage-point variations of the dimer, brings about affinity modulation. In detail, the selection of the two hydroxyl groups for the glycosidic linkage (1–2, 1–3, 1–4 or 1–6) and also the introduction of substitutions such as α1,2-fucosylation and then addition of α1,3-Gal(GalNAc) to obtain ABH-blood group substances can all account for affinity changes. Is the lectin a passive acceptor during binding? Intriguingly, lectins can respond to ligand contact. Binding of lactose triggered a significant change in galectin-1's shape in solution, revealing a structural impact of the ligand on the lectin (He *et al.*, 2003). Increasing contact complementarity beyond the basic structural unit of lactose offers the potential for enthalpic gains by new hydrogen bonds. Notably, the introduction of the term *complementarity* directly leads us to the next level. By moving from the initial depiction of a carbohydrate structure as sequence (in two dimensions) to the third dimension (the shape), we reach level 3 of affinity regulation. Here, a truly remarkable feature of glycans is disclosed.

The fact that the chair conformation of the hexopyranose rings is energetically preferred readily pinpoints the main source for flexibility of saccharides. It stems from the dihedral angles ϕ, Ψ of the glycosidic bond. The reader can visualize the basic principle of conformational mobility for a disaccharide by letting his thumbs touch and turn each hand (sugar unit) independently without losing thumb contact. Configurations where the pyranose rings come into spatial vicinity (clash) will resemble 'hills' in a topographical map, giving ϕ, Ψ, E-map contour lines for energy levels (von der Lieth *et al.*, 1998; Gabius *et al.*, 2004). Modeling studies and experimental monitoring by NMR spectroscopy have discovered that the accessible conformational space for a disaccharide is represented by a few 'valleys' so that carbohydrates, in contrast to peptides, harbor limited flexibility. Only distinct conformations are energetically favorable so that a small set of shapes constitutes a saccharide's three-dimensional repertoire (Gabius *et al.*, 2004). In this respect, carbohydrate epitopes resemble peptide motifs embedded in a protein. However, there is an important difference regarding flexibility, which we recently underscored by the study of a peptide free in solution and interacting with its sugar ligand (Siebert *et al.*, 2002). Whereas protein sections are conformationally restrained by the indispensable presence of neighbouring residues, such an expenditure to build special surroundings in order to let a bioactive section in a protein adopt its suitable shape is not necessary for carbohydrates: their preference to only distinct shapes simply comes by nature! This property is a valuable asset for a biomolecule to become a ligand in biorecognition. In the thermodynamic balance sheet, the entropic penalty incurred by binding such carbohydrates is obviously far less than what could be expected during the same process with a highly flexible ligand. Inevitably, it must lose its conformational freedom. As a result, we have pinpointed another feature rendering glycans predestined for a role in information transfer. With this background, we can proceed to look at the interaction process of a carbohydrate with a lectin in more detail.

In allegoric terms, drawing on E. Fischer's famous lock-and-key principle (Fischer, 1894), one carbohydrate conformation can be likened to a key. We are accustomed to think that one sequence will build only one key. The given results on bioactive carbohydrates have changed this notion completely. This discovery signifies that one sequence can form a 'bunch of keys' (Hardy, 1997). The conformational interconversion between low-energy conformers was described with its attractive consequence as follows: 'the carbohydrate moves in solution through a bunch of shapes, each of which may be selected by a receptor' (Hardy, 1997) – and, indeed, this is the case: depending on the nature of the receptor the same sugar sequence can either be bioactive or bioinert. In other words, a certain carbohydrate ligand is present in solution in a limited set of conformations. When we examined the bound-state conformation of a digalactoside in case studies with two lectins, each of them was found to exclusively host only a single shape ('key'), leaving the other aside (Gabius *et al.*, 2004). This principle is referred to as *differential conformer selection* (Gabius, 1998). A perspective for medical application exploiting this mechanism becomes apparent by looking at the growth-regulatory binding of the pentasaccharide of ganglioside GM_1 by galectin-1 (see above) and the cholera toxin (Kopitz *et al.*, 1998, 2001; Siebert *et al.*, 2003b). The pentasaccharide can form only a few distinct shapes in solution which are defined by the measured ϕ, Ψ-angle combinations at each glycosidic linkage. Besides crystallographic information, the strategic combination of NMR spectroscopy with computational calculations made it possible to define topological aspects of the lectin–ligand complex in solution (Siebert *et al.*, 2003a). The human lectin and the bacterial toxin both bind the pentasaccharide. In this context, the ensuing question is: will they select the same or two different conformations or alternatively permit the ligand to change shape while remaining bound? As explained with experimental detail recently (Siebert *et al.*, 2003b), galectin and toxin select two different conformers. The clinical perspective of this result is evident: by tailoring an inhibitor arrested in the toxin-binding shape, a potent pharmaceutical without side reactivity to the tissue lectin can become available.

Further structural scrutiny of the lectin–ligand complex in this case also taught the lesson that carbohydrate units other than the primary contact site (here GalNAc/sialic acid) can significantly increase the interaction energy in the overall thermodynamics of binding (Siebert *et al.*, 2003b). In this case, enthalpic gains assigned to level 2 underlie conformer selection at level 3. These additional contributions make it difficult to predict ligand affinity for oligosaccharides without experimental input, as exemplified by the discussed case of the ganglioside's pentasaccharide (Siebert *et al.*, 2003b). Caution is thus to be exercised when extrapolating binding data from simple structures (level 1) to linear or branched oligosaccharide shapes (levels 2 and 3) is attempted. However these insights could be highly relevant to explain the target specificity of a tissue lectin for only a few cell surface epitopes. In fact, despite the abundance of β-galactosides on the cell surface, a galectin appears to be very selective, associating with only few glycoconjugates. So far, we have only dealt with oligosaccharide chains, at best from a ganglioside. Besides extending a ligand's sequence and taking shape alterations into consideration, a look at the strategic presentation of CRDs in lectins (see Figure 2.4) implies that lectin-reactive epitopes could also be naturally displayed with impact on affinity. The choice for the branching/substitution mode of an N- or O-glycan would then mean more than a mere structural alteration picked from a panel of synthetic possibilities. Basically, we imply functional significance of this property arising from glycan synthesis. This assumption of nonrandom origin of glycan chain design leads us to further levels for the regulation of ligand affinity.

The basic structural motif of N-glycans is the biantennary chain. Further elaborating this element by enzymatic modifications yields (a) substituted versions, e.g. containing a core-fucose unit or a bisecting GlcNAc in the stem region, and (b) an increase in the number of branches by adding up to three antennae (Brockhausen and Schachter, 1997; Sharon and Lis, 1997; Reuter and Gabius, 1999). Initially viewed as a bothersome complication within the process of systematic structural analysis, the questions can now be addressed as to whether and to what extent these structural refinements serve to modulate the affinity for lectins. Progress in chemoenzymatic synthesis of the substituted N-glycans made it possible to start providing answers (André *et al.*, 1997, 2004; Unverzagt *et al.*, 2002). The reader may wonder whether the presence of such a substitution at a site distant from the lectin-reactive determinant in the branches can really affect the glycan's conformation. As proven by experimental and computational monitoring, presence of a substitution such as core fucosylation or bisecting GlcNAc is in fact able to alter the accessible conformational space of the N-glycan markedly, and recent reports substantiate that this effect on the conformational dynamics has an impact on lectin affinity, notably depending on the nature of the studied sugar receptor (Unverzagt *et al.*, 2002; André *et al.*, 2004). As a means of malleable glycoengineering, the substitutions also affect clearance of the N-glycans from circulation, shown by monitoring biodistribution in mice (Unverzagt *et al.*, 2002; André *et al.*, 2004). Thus, the consideration of the shape of the complete chain and the topological characteristics of ligand presentation influenced by substitutions leads us from level 3 to level 4.

This level is subdivided into two subcategories, because substitutions can go along with increases in the branching mode. The way this parameter acts upon lectin affinity is convincingly documented by proof-of-principle studies on glycan clearance from circulation *in vivo*. Citing the classical example of the hepatocyte C-type lectin (asialoglycoprotein receptor; see Table 2.3 and Figure 2.4), an increase in valency from mono- and bi- to trivalent glycans resulted in a geometrical increase in affinity, termed the *glycoside cluster effect* (Lee and Lee, 1997). Complementarity between the topology of presentation of the three CRDs of this lectin and sugar target presentation causes the relative affinity to increase from 1 to 1000 and finally reaching 1 000 000 with triantennary glycans (Lee and Lee, 1997). It goes without saying that this effect is of great potential for drug targeting to the liver (Yamazaki *et al.*, 2000; for a historical account of the prominent role of the hepatic asialoglycoprotein receptor in this area, see Gabius, 2004). Also, the availability of a triantennary glycan in a glycoprotein from human ovarian cyst fluid was instrumental in revealing the strong impact of branching on the avian galectin CG-16 and poor inhibitory capacity for a mammalian CRD, i.e. the N-terminal domain of galectin-4 (Wu *et al.*, 2001, 2002, 2004).

We stated above that glycans as information carriers offer the potential for dynamic and reversible structural alterations. Also, the activity profile of glycosyltransferases can be the object of modulation with implications for the glycome, as detected in various diseases, including inflammation (Brockhausen *et al.*, 1998; Rosen, 2004). So far, they have primarily been mapped and interpreted phenomenologically. The presented reasoning and evidence lead to the hypothesis that they can be functionally relevant. To give an example from clinical chemistry, the changes in glycosylation of α_1-acid glycoprotein with its five N-glycosylation sites in inflammation have been related to its anti-inflammatory effects on the level of interfering with leukocyte adhesion to endothelium by blocking the C-type lectin subfamily of the selectins (van Dijk *et al.*, 1998). Here, branching of O-glycans introduced by core 2 N-acetylglucosaminyltransferase and sulfation of GlcNAc moieties matters too (Hiraoka *et al.*, 2004; Ley and Kansas, 2004; Rosen, 2004). Needless to say, respective reports open a wide study area with clinical relevance.

As a message we can conclude that the spatial vicinity of branches from an individual glycan, a factor valid for N- and O-glycans, has a bearing on affinity for animal lectins. This consideration immediately guides us from level 4 to level 5. Of course, glycan clustering can be brought about by reaching spatial vicinity between different N- and/or O-glycans. Instructive examples are mucins with their high density of O-glycan chains, where branches of a single O-glycan and chains of different O-glycans are closely spaced together. Because mucin-like domains in glycoproteins are also not rare, it is straightforward to put the hypothesis to the test that this topological aspect of glycosylation has a functional meaning for protein–carbohydrate interaction. Equally important, different N-glycans or antennae from different N- and O-glycans of a glycoprotein can cooperate in such a setting. This property is lost when working with free glycans after digestion of the protein carrier. In order to examine the extent to which this level of affinity regulation is operative, a panel of well-characterized glycoproteins is indispensable, as are systematic studies in a lectin family, which are currently in progress for galectins (Wu *et al.*, 2001, 2002, 2004).

On the level of cell surfaces, we suggest encountering the sixth level of affinity regulation, i.e. the modulation of local density of lectin-reactive glycans. Microdomains, lipid rafts or glycosynapses are candidate structures to facilitate cell adhesion or signaling by lectin–carbohydrate recognition (Hakomori, 2002). Membrane fluidity, aggregate formation of glycoproteins/glycolipids and structural changes in glycan structures (sequence and shape) afford an array of possibilities for a wide range of regulatory events. A topological parameter is arresting or shifting conformational equilibria for glycan determinants by altering their density. To let this system work requires a conspicuous level of inherent specificity of a lectin toward certain cell surface glycans. In other words, despite binding a frequently present basic element on level 1, a lectin is expected to be very selective for ‘real’ cell surface ligands. Indeed, this is the case for galectins. A survey of the literature has attested that galectins-1 and -3 react only with distinct glycoproteins or glycolipids to trigger cell-type-specific activities (André *et al.*, 1999). Integrins such as the β_1-subunit or cell markers such as CD3, CD7 or CD45 are among these galectins ligands. Obviously, it is justified to define level 6, giving further research direction and aim. To avoid this work appearing purely academic, it is worth explicitly familiarizing the reader with perspectives for medical applications focusing on galectins.

2.5 Perspectives for Galectin-Dependent Medical Applications

The monitoring of the presence/absence of distinct lectins or lectin ligands is a current topic in tumor pathology, examples of new prognostic information having been given above. *In vitro* and animal models including cell clones with deliberate changes in galectin expression are and will be helpful in exploring the relevance of a certain lectin for tumor progression in distinct tumor classes. These experiments have a therapeutic dimension too. Merging the information on the importance of spatial factors for ligand presentation with the power of synthetic chemistry to prepare custom-made blocking substances, it has actually become feasible to synthesize target-selective inhibitors. The natural precedents of branched glycans inspire mimicry of their potency by devising suitable neoglycoconjugates (Gabius, 1988; Gabius *et al.*, 1988, 1991, 1993; Gabius and Bardosi, 1991; Bovin and Gabius, 1995; Lee and Lee, 1997; Roy, 2003). These tools can be employed not only to block CRDs but also to localize sugar-binding capacity on cells and in tissue sections cyto- and histochemically (Gabius, 2001a). Current experience in this area has shown that strong multivalency effects are attainable, especially for galectin-3 (André *et al.*, 2001, 2003). An enhancement of almost 4300-fold compared with the monomeric derivative was determined for the chimera-type galectin-3 with a tetravalent scaffold presenting lactose as ligand (Vrasidas *et al.*, 2003). The creativity of the chemist to supply suitable scaffolds is definitely not a limiting factor to turn this approach into further experiments (Lee and Lee, 1994; Bovin and Gabius, 1995; Roy, 2003). These tools can lead to advances in drug or enzyme targeting (see case study on asialoglycoprotein receptor above) or lectin-mediated vaccine delivery (Keler *et al.*, 2004). While such synthetic products are capable of localizing, targeting or blocking lectins, the development of vectors from molecular biological engineering is instrumental for another aim, i.e. rationally modulating lectin expression. Down-regulation of galectin-1 responsible for tissue invasion in glioblastoma is a perspective supported by current experimental evidence, as presented above. Similarly, the functional role of galectin-3 in cardiac dysfunction, an effector for increased mycocardial cyclin D_1 expression and collagen I:III ratio, can be counted as an incentive for taking the next steps to test therapeutic relevance (Sharma *et al.*, 2004). Establishing a local environment rich in pro-apoptotic activity on autoreactive T cells is a suggestion for the management of autoimmune disorders. Because any lectin effect is dependent on the highly specific selection of ligands, manipulating ligand presentation is an alternative route. A key role in this respect will be played by the α2,6-sialyltransferase, which introduces a switch-off signal to ligand properties for galectin-1. Consequently, regulating distinct glycosyltransferases is a long-range perspective as well, to deliberately channel signaling in the directions of clinical benefit. Glycomic profiling with plant/animal/human lectins or monoclonal antibodies in pathology, which maps disease-associated glycan changes, is thus on the way from phenomenology to defining functionally relevant shifts in the pattern of potential ligands for tissue lectins. With growing understanding of the functional relevance of disease-associated alterations in glycosylation, the envisaged interventions are likely to be more than symptomatic treatment modalities.

2.6 Conclusions

When asked to deliver input on how to construct an operative artificial cell, special attention will have to be paid to the design of cell surface determinants. They are required as sensors for communication with the environment. Spatial accessibility and a molecular vocabulary to encode a wide array of messages with high density (note that the space on the cell surface is limited) are salient parameters for signal functionality. When looking at natural role models, these prerequisites are completely fulfilled by the virtues of carbohydrates of the glycocalyx. In fact, the sugars of cellular glycoconjugates are second to no other class of biomolecules in terms of high-density information storage. The formation of only a few, energetically favored shapes is markedly suited for an intermolecular interaction. Fittingly, the emerging insights into the expression of binding partners for the information-storing sugar epitopes have given the concept of the sugar code its present shape: lectin–carbohydrate interactions transmit the information from oligosaccharides of glycoconjugates into diverse biological responses. The initially puzzling degree of diversity of glycan structures thereby gains a functional dimension. Topological aspects of the two sides in this recognition system, i.e. glycan and lectin, are presently being related to affinity modulation with perspectives for medical applications. As the current status of results presented herein attests, new diagnostic tools have already been established. Moreover, experimental approaches for examining therapeutic perspectives have been inspired. Merging the fields of synthetic and analytical chemistry, biochemistry and molecular cell biology with medical subjects such as immunology, oncology or pathology is sure to pave the way to a new level of understanding on the role of glycomics and functional lectinomics in disease (Gabius, 2000).

References

Agrawal BBL, Goldstein IJ. 1965. Specific binding of concanavalin A to cross-linked dextran gel. *Biochem J* **96**: 23c.

Ahmad N, Gabius H-J, Kaltner H, André S, Kuwabara I, Liu F-T, Oscarson S, Norberg T, Brewer CF. 2002. Thermodynamic binding studies of cell surface carbohydrate epitopes to galectin-1, -3 and -7. Evidence for differential binding specificities. *Can J Chem* **80**: 1096–1104.

Ahmad N, Gabius H-J, André S, Kaltner H, Sabesan S, Liu B, Macaluso F, Brewer CF. 2004. Galectin-3 precipitates as a pentamer with synthetic multivalent carbohydrates and forms heterogeneous cross-linked complexes. *J Biol Chem* **279**: 10841–10847.

André S, Unverzagt C, Kojima S, Dong X, Fink C, Kayser K, Gabius H-J. 1997. Neoglycoproteins with the synthetic complex biantennary nonasaccharide or its α2,3-/α2,6-sialylated derivatives: their preparation, the assessment of their ligand properties for purified lectins, for tumor cells *in vitro* and in tissue sections and their biodistribution in tumor-bearing mice. *Bioconjugate Chem* **8**: 845–855.

André S, Kojima S, Yamazaki N, Fink C, Kaltner H, Kayser K, Gabius H-J. 1999. Galectins-1 and -3 and their ligands in tumor biology. *J Cancer Res Clin Oncol* **125**: 461–474.

André S, Pieters RJ, Vrasidas I, Kaltner H, Kuwabara I, Liu F-T, Liskamp RMJ, Gabius H-J. 2001. Wedgelike glycodendrimers as inhibitors of binding of mammalian galectins to various glycoproteins, lactose maxiclusters and cell surface glycoconjugates. *ChemBioChem* **2**: 822–830.

André S, Liu B, Gabius H-J, Roy R. 2003. First demonstration of differential inhibition of lectin binding by synthetic tri- and tetravalent glycoclusters from cross-coupling of regidified 2-propynyl lactoside. *Org Biomol Chem* **1**: 3909–3916.

André S, Unverzagt C, Kojima S, Frank M, Seifert J, Fink C, Kayser K, von der Lieth C-W, Gabius H-J. 2004. Determination of modulation of ligand properties of synthetic complex-type biantennary N-glycans by introduction of bisecting GlcNAc *in silico, in vitro* and *in vivo*. *Eur J Biochem* **271**: 118–134.

André S, Arnusch CJ, Kuwabara I, Russwurm R, Kaltner H, Gabius H-J, Pieters RJ. 2005a. Identification of peptide ligands for malignancy- and growth-regulating galectins using random phage-display and designed peptide libraries. *Bioorg Med Chem* **13**: 563–573.

André S, Kaltner H, Lensch M, Russwurm R, Siebert H-C, Fallsehr C, Tajkhorshid E, Heck AJR, von Knebel-Döberitz M, Gabius H-J, Kopitz J. 2005b. Determination of structural and functional overlap/divergence of five proto-type galectins by analysis of the growth-regulatory interaction with ganglioside GM_1 *in silico* and *in vitro* on human neuroblastoma cells. *Int J Cancer* **114**: 46–57.

Angata T, Brinkman-van der Linden ECM. 2002. I-type lectins. *Biochim Biophys Acta* **1572**: 294–316.

Boraston AB, Bolam DN, Gilbert H-J, Davies GJ. 2004. Carbohydrate-binding modules: fine-tuning polysaccharide recognition. *Biochem J* **382**: 769–781.

Bovin NV, Gabius H-J. 1995. Polymer-immobilized carbohydrate ligands: versatile chemical tools for biochemistry and medical sciences. *Chem Soc Rev* **24**: 413–421.

Brewer CF. 2001. Lectin cross-linking interactions with multivalent carbohydrates. *Adv Exp Med Biol* **491**: 17–25.

Brewer CF. 2002. Binding and cross-linking properties of galectins. *Biochim Biophys Acta* **1572**: 255–262.

Brockhausen, I, Schachter H. 1997. Glycosyltransferases involved in N- and O-glycan biosynthesis. In Gabius H-J, Gabius S (eds), *Glycosciences: Status and Perspectives*. London: Chapman & Hall, pp 79–113.

Brockhausen I, Schutzbach J, Kuhns W. 1998. Glycoproteins and their relationship to human disease. *Acta Anat* **161**: 36–78.

Camby I, Belot N, Lefranc F, Sadeghi N, de Launoit Y, Kaltner H, Musette S, Darro F, Danguy A, Salmon I, Gabius H-J, Kiss R. 2002. Galectin-1 modulates human glioblastoma cell migration into the brain through modifications to the actin cytoskeleton and levels of expression of small GTPases. *J Neuropathol Exp Neurol* **61**: 585–596.

Cebo C, Vergoten G, Zanetta J-P. 2002. Lectin activities of cytokines: functions and putative carbohydrate recognition domains. *Biochim Biophys Acta* **1572**: 422–434.

Cooper DNW. 2002. Galectinomics: finding themes in complexity. *Biochim Biophys Acta* **1572**: 209–231.

Crocker PR. 2004. CD33-related siglecs in the immune system. *Trends Glycosci Glycotechnol* **16**: 357–370.

Cummings RD. 1997. Lectins as tools for glycoconjugate purification and characterization. In Gabius H-J, Gabius S (eds), *Glycosciences: Status and Perspectives*. London: Chapman & Hall, pp 191–199.

Dahms NM, Hancock MK. 2002. P-type lectins. *Biochim Biophys Acta* **1572**: 317–340.

Elad-Sfadia G, Haklai R, Ballan E, Gabius H-J, Kloog Y. 2002. Galectin-1 augments Ras activation and diverts Ras signals to Raf-1 at the expense of phosphoinositide 3-kinase. *J Biol Chem* **277**: 37169–37175.

Endharti AT, Zhou YW, Nakashima I, Suzuki H. 2005. Galectin-1 supports survival of naive T cells without promoting cell proliferation. *Eur J Immunol* **35**: 86–97.

Fischer E. 1894. Einfluss der Configuration auf die Wirkung der Enzyme. *Ber Dt Chem Ges* **27**: 2985–2993.

Gabius H-J. 1988. Tumor lectinology: at the intersection of carbohydrate chemistry, biochemistry, cell biology and oncology. *Angew Chem Int Edn* **27**: 1267–1276.

Gabius H-J. 1990. Influence of type of linkage and spacer on the interaction of β-galactoside-binding proteins with immobilized affinity ligands. *Anal Biochem* **189**: 91–94.

Gabius H-J. 1991. Detection and functions of mammalian lectins – with emphasis on membrane lectins. *Biochim Biophys Acta* **1071**: 1–18.

Gabius H-J. 1997. Animal lectins. *Eur J Biochem* **243**: 543–576.

Gabius, H-J. 1998. The how and why of protein–carbohydrate interaction: a primer to the theoretical concept and a guide to application in drug design. *Pharm Res* **15**: 23–30.

Gabius H-J. 1999. Lectins and glycoconjugates. In Kastner M (ed.), *Protein Liquid Chromatography*. Amsterdam: Elsevier, pp 619–638.

Gabius H-J. 2000. Biological information transfer beyond the genetic code: the sugar code. *Naturwissenschaften* **87**: 108–121.

Gabius H-J. 2001a. Glycohistochemistry: the why and how of detection and localization of endogenous lectins. *Anat Histol Embryol* **30**: 3–31.

Gabius H-J. 2001b. Probing the cons and pros of lectin-induced immunomodulation: case studies for the mistletoe lectin and galectin-1. *Biochimie* **83**: 659–666.

Gabius H-J. (ed.). 2002. Animal lectins. *Biochim Biophys Acta* **1572** (special issue): 163–434.

Gabius H-J. 2004. The sugar code in drug delivery. *Adv Drug Deliv Rev* **56**: 421–424.

Gabius H-J, Bardosi A. 1991. Neoglycoproteins as tools in glycohistochemistry. *Progr Histochem Cytochem* **22**(3): 1–66.

Gabius H-J, Gabius S. (eds). 1993. *Lectins and Glycobiology*. Heidelberg: Springer.

Gabius H-J, Gabius S. (eds). 1997. *Glycosciences: Status and Perspectives*. London: Chapman & Hall.

Gabius H-J, Brehler R, Schauer A, Cramer F. 1986. Localization of endogenous lectins in normal human breast, benign breast lesions and mammary carcinomas. *Virch Arch [Cell Pathol]* **52**: 107–115.

Gabius H-J, Bodanowitz S, Schauer A. 1988. Endogenous sugar-binding proteins in human breast tissue and benign and malignant breast lesions. *Cancer* **61**: 1125–1131.

Gabius H-J, Wosgien B, Hendrys M, Bardosi A. 1991. Lectin localization in human nerve by biochemically defined lectin-binding glycoproteins, neoglycoprotein and lectin-specific antibody. *Histochemistry* **95**: 269–277.

Gabius H-J, Gabius S, Zemlyanukhina TV, Bovin NV, Brinck U, Danguy A, Joshi SS, Kayser K, Schottelius J, Sinowatz F, Tietze LF, Vidal-Vanaclocha F, Zanetta J-P. 1993. Reverse lectin histochemistry: design and application of glycoligands for detection of cell and tissue lectins. *Histol Histopathol* **8**: 369–383.

Gabius H-J, André S, Kaltner H, Siebert H-C. 2002. The sugar code: functional lectinomics. *Biochim Biophys Acta* **1572**: 165-177.

Gabius H-J, Siebert H-C, André S, Jiménez-Barbero J, Rüdiger H. 2004. Chemical biology of the sugar code. *ChemBioChem* **5**: 740–764.

Geijtenbeek, TBH, van Vliet SJ, Engering A, 'tHart BA, van Kooyk Y. 2004. Self- and nonself-recognition by C-type lectins on dendritic cells. *A Rev Immunol* **22**: 33–54.

Geyer H, Geyer R. 1998. Strategies for glycoconjugate analysis. *Acta Anat* **161**: 18–35.

Hakomori S-i. 2002. The glycosynapse. *Proc Natl Acad Sci USA* **99**: 225–232.

Haltiwanger RS, Lowe JB. 2004. Role of glycosylation in development. *A Rev Biochem* **73**: 491–537.

Harduin-Lepers A, Vallejo-Ruiz V, Krzewinski-Recchi M-A, Samyn-Petit B, Julien S, Delannoy P. 2001. The human sialyltransferase family. *Biochimie* **83**: 727–737.

Hardy BJ. 1997. The glycosidic linkage flexibility and time-scale similarity hypotheses. *J Mol Struct* **395–396**: 187–200.

He L, André S, Siebert H-C, Helmholz H, Niemeyer B, Gabius H-J. 2003. Detection of ligand- and solvent-induced shape alterations of cell-growth-regulatory human lectin galectin-1 in solution by small angle neutron and X-ray scattering. *Biophys J* **85**: 511–524.

Hennet T. 2002. The galactosyltransferase family. *Cell Mol Life Sci* **59**: 1081–1095.

Hiraoka N, Kawashima H, Petryniak B, Nakayama J, Mitoma J, Marth JD, Lowe JB, Fukuda M. 2004. Core 2 branching β1,6-N-acetylglucosaminyltransferase and high endothelial venule-restricted sulfotransferase collaboratively control lymphocyte homing. *J Biol Chem* **279**: 3058–3067.

Hounsell EF. 1997. Methods of glycoconjugate analysis. In Gabius H-J, Gabius S (eds), *Glycosciences: Status and Perspectives*. London: Chapman & Hall, pp 15–29.

Kaltner H, Stierstorfer B. 1998. Animal lectins as cell adhesion molecules. *Acta Anat* **161**: 162–179.

Kanazawa N, Tashiro K, Miyachi Y. 2004. Signaling and immune regulatory role of the dendritic cell immunoreceptor (DCIR) family lectins: DCIR, DCAR, dectin-2 and BDCA-2. *Immunobiology* **209**: 179–190.

Kasai K, Hirabayashi J. 1996. Galectins: a family of animal lectins that decipher glycocodes. *J Biochem* **119**: 1–8.

Kayser K, Dünnwald D, Kazmierczak B, Bullerdiek J, Kaltner H, André S, Gabius H-J. (2003a). Chromosomal aberrations, profiles of expression of growth-related markers including galectins and environmental hazards in relation to the incidence of chondroid pulmonary hamartomas. *Pathol Res Pract* **199**: 589–598.

Kayser K, Höft D, Hufnagl P, Caselitz J, Zick Y, André S, Kaltner H, Gabius H-J. 2003b. Combined analysis of tumor growth pattern and expression of endogenous lectins as a prognostic tool in primary testicular cancer and its lung metastases. *Histol Histopathol* **18**: 771–779.

Keler T, Ramakrishna V, Fanger MW. 2004. Mannose receptor-targeted vaccines. *Expert Opin Biol Ther* **4**: 1953–1962.

Kilpatrick DC. 2002. Mannan-binding lectin: clinical significance and applications. *Biochim Biophys Acta* **1572**: 401–413.

Kilpatrick DC, Green C. 1992. Lectins as blood typing reagents. *Adv Lectin Res* **5**: 51–94.

Kobata A. 1992. Structures and functions of the sugar chains of glycoproteins. *Eur J Biochem* **209**: 483–501.

Kopitz J, von Reitzenstein C, Burchert M, Cantz M, Gabius H-J. 1998. Galectin-1 is a major receptor for ganglioside GM_1, a product of the growth-controlling activity of a cell surface ganglioside sialidase, on human neuroblastoma cells in culture. *J Biol Chem* **273**: 11205–11211.

Kopitz J, von Reitzenstein C, André S, Kaltner H, Uhl J, Ehemann V, Cantz M, Gabius H-J. 2001. Negative regulation of neuroblastoma cell growth by carbohydrate-dependent surface binding of galectin-1 and functional divergence from galectin-3. *J Biol Chem* **276**: 35917–35923.

Kopitz J, André S, von Reitzenstein C, Versluis K, Kaltner H, Pieters RJ, Wasano K, Kuwabara I, Liu F-T, Cantz M, Heck AJR, Gabius H-J. 2003. Homodimeric galectin-7 (p53-induced gene 1) is a negative growth regulator for human neuroblastoma cells. *Oncogene* **22**: 6277–6288.

Lahm H, André S, Höflich A, Fischer JR, Sordat B, Kaltner H, Wolf E, Gabius H-J. 2001. Comprehensive galectin fingerprinting in a panel of 61 human tumor cell lines by RT-PCR and its implications for diagnostic and therapeutic procedures. *J Cancer Res Clin Oncol* **127**: 375–386.

Lahm H, André S, Hoeflich A, Fischer JR, Sordat B, Kaltner H, Wolf E, Gabius H-J. 2003. Molecular biological fingerprinting of human lectin expression by RT-PCR. *Meth Enzymol* **362**: 287–297.

Lahm H, André S, Hoeflich A, Kaltner H, Siebert H, Sordat B, von der Lieth CW, Wolf E, Gabius H-J. 2004. Tumor galectinology: insights into the complex network of a family of endogenous lectins. *Glycoconjug J* **20**: 227–238.

Laine RA. 1997. The information-storing potential of the sugar code. In Gabius H-J, Gabius S. (eds), *Glycosciences: Status and Perspectives*. London: Chapman & Hall, pp 1–14.

Lee RT, Lee YC. 1997. Neoglycoconjugates. In Gabius H-J, Gabius S (eds). *Glycosciences: Status and Perspectives*. London: Chapman & Hall, pp 55–77.

Lee YC, Lee RT. (eds). 1994. *Neoglycoconjugates. Preparation and Applications*. San Diego, CA: Academic Press.

Lemieux RU. 1996. How water provides the impetus for molecular recognition in aqueous solution. *Acc Chem Res* **29**: 373–380.

Ley K, Kansas GS. 2004. Selectins in T-cell recruitment to non-lymphoid tissues and sites of inflammation. *Nat Rev Immunol* **4**: 1–11.

Lineham SA, Martínez-Pomares L, Gordon S. 2000. Macrophage lectins in host defence. *Microbes Infect* **2**: 279–288.

Lis H, Sharon N. 1998. Lectins: carbohydrate-specific proteins that mediate cellular recognition. *Chem Rev* **98**: 637–674.

Liu F-T. 2004. Double identity: galectins may not function as lectins inside the cell. *Trends Glycosci Glycotechnol* **16**: 255–264.

López-Lucendo MF, Solís D, André S, Hirabayashi J, Kasai K-i, Kaltner H, Gabius H-J, Romero A. 2004. Growth-regulatory human galectin-1: crystallographic characterization of structural changes induced by single-site mutations and their impact on thermodynamics of ligand binding increasing entropic penalty. *J Mol Biol* **343**: 957–970.

Loris R. 2002. Principles of structures of animal and plant lectins. *Biochim Biophys Acta* **1572**: 198–208.

Lu J, Teh C, Kishore U, Reid KB. 2002. Collectins and ficolins: sugar pattern recognition molecules of the mammalian innate immune system. *Biochim Biophys Acta* **1572**: 387–400.

Manning JC, Seyrek K, Kaltner H, André S, Sinowatz F, Gabius H-J. 2004. Glycomic profiling of developmental changes in bovine testis by lectin histochemistry and further analysis of the most prominent alteration on the level of the glycoproteome by lectin blotting and lectin affinity chromatography. *Histol Histopathol* **19**: 1043–1060.

Martinez-Duncker I, Mollicone R, Candelier J-J, Breton C, Oriol R. 2003. A new superfamily of protein-O-fucosyltransferases, α2-fucosyltransferases, and α6-fucosyltransferases: phylogeny and identification of conserved peptide motifs. *Glycobiology* **13**: 1C–5C.

McGreal EP, Martínez-Pomares L, Gordon S. 2004. Divergent roles for C-type lectins expressed by cells of the innate immune system. *Mol Immunol* **41**: 1109–1121.

McGreal EP, Miller JL, Gordon S. 2005. Ligand recognition by antigen-presenting cell C-type lectin receptors. *Curr Opin Immunol* **17**: 18–24.

Mitchell SW. 1860. Researches upon the venom of the rattlesnake. *Smithsonian Contrib Knowledge* **XII**: 89–90.

Nagy N, Legendre H, Engels O, André S, Kaltner H, Wasano K, Zick Y, Pector J-C, Decaestecker C, Gabius H-J, Salmon I, Kiss R. 2003. Refined prognostic evaluation in colon cancer using immunohistochemical galectin fingerprinting. *Cancer* **97**: 1849–1858.

Nowell PC. 1960. Phytohemagglutinin: an inhibitor of mitosis in cultures of normal human leukocytes. *Cancer Res* **20**: 462–466.

Ogilvie ML, Gartner TK. 1984. Identification of snake venoms. *J Herpetol* **18**: 285–290.

Plzák J, Betka J, Smetana Jr. K, Chovanec M, Kaltner H, André S, Kodet R, Gabius H-J. 2004. Galectin-3 – an emerging prognostic indicator in advanced head and neck carcinoma. *Eur J Cancer* **40**: 2324–2330.

Purkrábková T, Smetana Jr. K, Dvoránková B, Holíková Z, Böck C, Lensch M, André S, Pytlík R, Liu F-T, Klíma J, Smetana K, Motlik J, Gabius H-J. 2003. New aspects of galectin functionality in nuclei of cultured bone marrow stromal and epidermal cells: biotinylated galectins as tool to detect specific binding sites. *Biol Cell* **95**: 535–545.

Rappl G, Abken H, Muche JM, Sterry W, Tilgen W, André S, Kaltner H, Ugurel S, Gabius H-J, Reinhold U. 2002. $CD4^{+}CD7^{-}$ leukemic T cells from patients with Sézary syndrome are protected from galectin-1-triggered T cell death. *Leukemia* **16**: 840–845.

Reuter G, Gabius H-J. 1999. Eukaryotic glycosylation – whim of nature or multipurpose tool? *Cell Mol Life Sci* **55**: 368–422.

Roseman S. 2001. Reflections on glycobiology. *J Biol Chem* **276**: 41527–41542.

Rosen SD. 2004. Ligands for L-selectin: homing, inflammation, and beyond. *A Rev Immunol* **22**: 129–156.

Rotblat B, Niv H, André S, Kaltner H, Gabius H-J, Kloog Y. 2004. Galectin-1(L11A) predicted from a computed galectin-1 farnesyl-binding pocket selectively inhibits Ras-GTP. *Cancer Res* **64**: 3112–3118.

Roy R. 2003. A decade of glycodendrimer chemistry. *Trends Glycosci Glycotechnol* **15**: 291–310.

Rüdiger H, Gabius H-J. 2001. Plant lectins: occurrence, biochemistry, functions and applications. *Glycoconjug J* **18**: 589–613.

Rüdiger H, Siebert H-C, Solís D, Jiménez-Barbero J, Romero A, von der Lieth C-W, Díaz-Mauriño T, Gabius H-J. 2000. Medicinal chemistry based on the sugar code: fundamentals of lectinology and experimental strategies with lectins as targets. *Curr Med Chem* **7**: 389–416.

Sano H, Kuroki Y. 2005. The lung collectins, SP-A and SP-D, modulate pulmonary innate immunity. *Mol Immunol* **42**: 279–280.

Sharma UC, Pokharel S, van Brakel TJ, van Berlo JH, Cleutjens JPM, Schroen B, André S, Crijns HJGM, Gabius H-J, Maessen J, Pinto YM. 2004. Galectin-3 marks activated macrophages in failure-prone hypertrophied hearts and contributes to cardiac dysfunction. *Circulation* **110**: 3121–3128.

Sharon N. 1998. Lectins: from obscurity into the limelight. *Protein Sci* **7**: 2042–2048.

Sharon N, Lis H. 1997. Glycoproteins: structure and function. In Gabius H-J, Gabius S (eds), *Glycosciences: Status and Perspectives*. London: Chapman & Hall, pp 133–162.

Sharon N, Lis H. 2001. The structural basis for carbohydrate recognition by lectins. *Adv Exp Med Biol* **491**: 1–16.

Sharon N, Lis H. 2004. History of lectins: from hemagglutinins to biological recognition molecules. *Glycobiology* **14**: 53R–62R.

Shimura T, Takenaka Y, Tsutsumi S, Hogan V, Kikuchi A, Raz A. 2004. Galectin-3, a novel binding partner of β-catenin. *Cancer Res* **64**: 6363–6367.

Siebert H-C, Lü SY, Frank M, Kramer J, Wechselberger R, Joosten J, André S, Rittenhouse-Olson K, Roy R, von der Lieth C-W, Kaptein R, Vliegenthart JFG, Heck AJR, Gabius H-J. 2002. Analysis of protein–carbohydrate interaction at the lower size limit of the protein part (15-mer peptide) by NMR spectroscopy, electrospray ionization mass spectrometry, and molecular modeling. *Biochemistry* **41**: 9707–9717.

Siebert H-C, Frank M, von der Lieth C-W, Jiménez-Barbero J, Gabius H-J. 2003a. Detection of hydroxyl protons. In Jiménez-Barbero J, Peters T (eds), *NMR Spectroscopy of Glycoconjugates*. Weinheim: Wiley-VCH, pp 39–57.

Siebert H-C, André S, Lu S-Y, Frank M, Kaltner H, van Kuik JA, Korchagina EY, Bovin NV, Tajkhorshid E, Kaptein R, Vliegenthart JFG, von der Lieth C-W, Jiménez-Barbero J, Kopitz J, Gabius H-J. (2003b). Unique conformer selection of human growth-regulatory lectin galectin-1 for ganglioside GM1 versus bacterial toxins. *Biochemistry* **42**: 14762–14773.

Solís D, Díaz-Mauriño T. 1997. Analysis of protein–carbohydrate interaction by engineered ligands. In Gabius H-J, S Gabius (eds), *Glycosciences: Status and Perspectives*. London: Chapman & Hall, pp 345–354.

Solís D, Romero A, Kaltner H, Gabius H-J, Díaz-Mauriño T. 1996. Different architecture of the combining sites of two chicken galectins revealed by chemical-mapping studies with synthetic ligand derivatives. *J Biol Chem* **271**: 12744–12748.

Solís D, Jiménez-Barbero J, Kaltner H, Romero A, Siebert H-C, von der Lieth C-W, Gabius H-J. 2001. Towards defining the role of glycans as hardware in information storage and transfer: basic principles, experimental approaches and recent progress. *Cells Tissues Organs* **168**: 5–23.

Spicer SS, Schulte BA. 1992. Diversity of cell glycoconjugates shown histochemically: a perspective. *J Histochem Cytochem* **40**: 1–38.

Spiro RG. 2002. Protein glycosylation: nature, distribution, enzymatic formation, and disease implications of glycopeptide bonds. *Glycobiology* **12**: 43R–56R.

Sturm A, Lensch M, André S, Kaltner H, Wiedenmann B, Rosewicz S, Dignass AU, Gabius H-J. 2004. Human galectin-2: novel inducer of T cell apoptosis with distinct profile of caspase activation. *J Immunol* **173**: 3825–3837.

Szathmáry E. 2003. Why are there four letters in the genetic alphabet? *Nat Rev Genet* **4**: 995–1001.

Takahara K, Yashima Y, Omatsu Y, Yoshida H, Kimura Y, Kang Y-S, Steinman RM, Park CG, Inaba K. 2004. Functional comparison of the mouse DC-SIGN, SIGNR1, SIGNR3 and langerin, C-type lectins. *Int Immunol* **16**: 819–829.

Teichberg VI, Silman I, Beitsch DD, Resheff G. 1975. A β-D-galactoside binding protein from electric organ tissue of *Electrophorus electricus*. *Proc Natl Acad Sci USA* **72**: 1383–1387.

Ten Hagen KG, Fritz TA, Tabak LA. 2003. All in the family: the UDP-GalNAc:polypeptide N-acetylgalactosaminyltransferase. *Glycobiology* **13**: 1R–16R.

Unverzagt C, Andre S, Seifert J, Kojima S, Fink C, Srikrishna G, Freeze H, Kayser K, Gabius H-J. 2002. Structure-activity profiles of complex biantennary glycans with core fucosylation and with/without additional α2,3/α2,6 sialylation: synthesis of neoglycoproteins and their properties in lectin assays, cell binding, and organ uptake. *J Med Chem* **45**: 478–491.

van de Wetering JK, van Golde LMG, Batenburg JJ. 2005. Collectins: players of the innate immune system. *Eur J Biochem* **271**: 1229–1249.

van Dijk W, Brinkman-van der Linden ECM, Havenaar EC. 1998. Glycosylation of α_1-acid glycoprotein (orosomucoid) in health and disease: occurrence, regulation and possible functional implications. *Trends Glycosci Glycotechnol* **10**: 235–245.

Varela PF, Solís D, Díaz-Mauriño T, Kaltner H, Gabius H-J, Romero A. 1999. The 2.15 Å crystal structure of CG-16, the developmentally regulated homodimeric chicken galectin. *J Mol Biol* **294**: 537–549.

Vasta GR, Ahmad H, Odorn EW. 2004. Structural and functional diversity of lectin repertoires in invertebrates, protochordates and ectothermic vertebrates. *Curr Opin Struct Biol* **14**: 617–630.

von der Lieth C-W, Siebert H-C, Kozár T, Burchert M, Frank M, Gilleron M, Kaltner H, Kayser G, Tajkhorshid E, Bovin NV, Vliegenthart JFG, Gabius H-J. 1998. Lectin ligands: new insights into their conformations and their dynamic behavior and the discovery of conformer selection by lectins. *Acta Anat* **161**: 91–109.

Vrasidas I, André S, Valentini P, Böck C, Lensch M, Kaltner H, Liskamp RMJ, Gabius H-J, Pieters RJ. 2003. Rigidified multivalent lactose molecules and their interactions with mammalian galectins: a route to selective inhibitors. *Org Biomol Chem* **1**: 803–810.

Watkins WM. 1999. A half century of blood-group antigen research: some personal recollections. *Trends Glycosci Glycotechnol* **11**: 391–411.

Watkins WM, Morgan WTJ. 1952. Neutralisation of the anti-H agglutinin in eel serum by simple sugars. *Nature* **169**: 825–826.

Wild MK, Lühn K, Marquardt T, Vestweber D. 2002. Leukocyte adhesion deficiency II: therapy and genetic defect. *Cells Tissues Organs* **172**: 161–173.

Winterburn PJ, Phelps CF. 1972. The significance of glycosylated proteins. *Nature* **236**: 147–151.

Wollina U, Graefe T, Feldrappe S, André S, Wasano K, Kaltner H, Zick Y, Gabius H-J. 2002. Galectin fingerprinting by immuno- and lectin histochemistry in cutaneous lymphoma. *J Cancer Res Clin Oncol* **128**: 103–110.

Wu AM, Wu JH, Tsai MS, Kaltner H, Gabius H-J. 2001. Carbohydrate specificity of a galectin from chicken liver (CG-16). *Biochem J* **358**: 529–538.

Wu AM, Wu JH, Tsai M-S, Liu J-H, André S, Wasano K, Kaltner H, Gabius H-J. 2002. Fine specificity of domain-I of recombinant tandem-repeat-type galectin-4 from rat gastrointestinal tract (G4-N). *Biochem J* **367**: 653–664.

Wu AM, Wu JH, Liu J-H, Singh T, André S, Kaltner H, Gabius H-J. 2004. Effects of polyvalency of glycotopes and natural modifications of human blood group ABH/Lewis sugars at the Galβ1-terminated core saccharides on the binding of domain-I of recombinant tandem-repeat-type galectin-4 from rat gastrointestinal tract (G4-N). *Biochimie* **86**: 317–326.

Yamazaki N, Kojima S, Bovin NV, André S, Gabius S, Gabius H-J. 2000. Endogenous lectins as targets for drug delivery. *Adv Drug Deliv Rev* **43**: 225–244.

3

Proteomics in Clinical Research: Perspectives and Expectations

Ivan Lefkovits, Thomas Grussenmeyer, Peter Matt, Martin Grapow, Michael Lefkovits and **Hans-Reinhard Zerkowski**

Abstract

Proteomics is one of the most rapidly evolving fields of biomedical research. In this work we define the tools of proteomics and the perspectives of the clinical applications. For more than two decades, two-dimensional gel electrophoresis has been the primary tool for defining the constituents of the proteome and the alterations of its composition. More recently, powerful methods of protein identification have evolved based on measuring the mass of peptidic fragments. The synergism from the combined efforts of genomics, transcriptomics and proteomics is discussed, and projects that are under scrutiny in the cardiovascular research group. Elucidating the alterations of protein components in physiological and pathological processes is expected to provide both diagnostic biomarkers and predictors for adequate therapy and risk stratification.

3.1 Introduction

Patterns of protein expression are modulated and often profoundly modified in metabolic processes in healthy and diseased tissue of the human organism. Pathological changes are accompanied by alterations of protein expression that also leave traces in the bloodstream.

If these molecules are present at a high enough concentration, they can serve as biomarkers. Even minor lesions leave molecular marks that can be scrutinized. Thus both physiological and pathological events can be monitored provided we learn which entities are supposed to be monitored.

We have reasons to assume that events related to diseases that are studied in our group, e.g. acute coronary syndromes (conditions that are accompanied by platelet activation, inflammation, coagulation and myocyte necrosis) are traceable in the

Immunogenomics and Human Disease Edited by András Falus

blood circulation. Biomarkers of necrosis (cardiac troponin), inflammation (high-sensitivity C-reactive protein, CRP) or haemodynamic stress (brain natriuretic peptide, BNP, or the N-terminal portion of the precursor of BNP, NT-proBNP) identify patients who are at higher mortality risk. Elevated troponin in the setting of an acute coronary syndrome is a marker of elevated risk of death and mandates an early invasive therapeutic strategy (e.g. percutanous coronary intervention). Elevated CRP and BNP at presentation identify patients at higher risk, irrespective of troponin elevation. The therapeutic implications have yet to be established (James *et al.*, 2003; Lefkovits, 2004; Morrow and Braunwald, 2003).

A rapidly emerging set of biochemical methods is making it possible to identify large numbers of proteins in a complex mixture, to examine their identities, their interactions, their biological activation and especially processes in disease states, which in turn offers great promise for both diagnostics and therapy. The research endeavour that aims to elucidate the functional and structural aspects of the web of protein interactions in a systematic manner has been termed proteomics.

What is a proteome and what is proteomics about?

The term *proteome* (and the concept of proteome) in its strict sense is supposed to define the complete *set of proteins* expressed in a lifetime of a cell. Its everyday meaning is, however, the *set of proteins* involved in a certain complex cellular task and pathway or in a disease state.

The term *proteomics* (and the entire proteomic research) is defined as the systematic analysis and documentation of the overall distribution of proteins in biological samples, performed as a high-throughput analysis, aiming to elucidate the functional role of the proteins in physiological and pathological processes.

The emphasis on the analysis of *sets of proteins* indicates the expectation that proteomics will answer fundamental questions about biological mechanisms on a ‘systems level’ rather than on an ‘individual protein level’. To reiterate, proteomics includes not only the identification and quantitation of proteins, but also the modifications, interactions, tissue localization and ultimately determination of their function.

Proteomics in basic science, clinical research and drug discovery

Basic proteomics will allow a complex map of cell functions to be built up by analysing how changes in one signalling pathway affect other pathways. *Clinical Research* aims to discover the multiplicity of factors involved in disease. *The drug discovery effort* will attempt to define proteins with which drugs selectively interact to achieve a precise therapeutic response, and as a fine tuning the identified targets will allow identification of individuals who, owing to their proteome composition, would suffer side effects from drug therapy.

From genomics to transcriptomics to proteomics

Biological systems of higher organisms display regulation of functions on genomic, transcriptomic and proteomic levels. Although the blueprint of all functions is provided on the genomic level, fine-tuning of these functions occurs through transcriptomic regulation, and the final manifestation of biological functions occurs on the protein level.

The genome tells us what *could* 'theoretically' happen, the transcriptome tells us what *might* happen, and the proteome tells us what *does* happen. To reiterate, the entire proteomic endeavor is geared towards describing the phenotype instead of the genotype.

Thus, protein expression is intimately associated with the biological function, and any perturbation of a cellular function finds its counterpart in the translational or post-translational repertoire of proteins. Since it is the proteins that are ultimately responsible for all active life processes, the aim of proteomics, in clinical research, is to reveal changes that are correlated with state of health and disease. It is at the protein level where:

- most regulatory processes take place;
- disease processes manifest themselves; and
- most drug targets will eventually be found.

Synergism of the triad of '-omics'

Although each discipline is using its own tools, the resulting information feeds into one single system of life sciences, and the complementary information of the 'three -omics' synergizes the resulting 'knowledge base'.

Transcriptome provides a snapshot of the genome's plans for protein synthesis under the conditions of the study. Although the expression of mRNA is modulated during alterations of biological processes, and the up- or down-regulation of mRNA is of great importance for the resulting function, the conclusion of 'down-stream events', i.e. the protein synthesis, cannot be derived from the level of mRNA. The level of messages correlates poorly with the resulting abundance of protein species. Besides the above considerations, we have to be aware of the fact that transcriptomic results provide no clue as to the post-translational products.

3.2 Proteomics: Tools and Projects

Although the availability of a technology for protein separation and detection is a prerequisite for the identification of protein structure, there have been serious attempts to analyse entire mixtures of proteins without prior separation. For the time being, the two-dimensional gel electrophoresis and the mass spectrometry are the basic pillars of the proteomics.

Synopses of the relevant technologies

- for protein separation and detection; and
- for protein structure identification.

are given below.

Technology for protein separation and detection by two-dimensional gel electrophoresis

It is now accepted that the technique of two-dimensional gel electrophoresis was developed independently in the laboratories of O'Farrell (1975) and of Klose (1975). The first dimension of separation is based on differences in the charge of polypeptides; the second dimension is a molecular separation according to the mass of polypeptide molecules.

Two-dimensional gel electrophoresis: step one – charge separation

Isoelectric focusing (IEF) is an electrophoretic method that separates proteins according to their charge. Proteins are amphoteric molecules; they carry positive, negative or zero net charge, depending on their amino acid composition and the pH of their surroundings.

For many years the charge separation matrix was based exclusively on the use of carrier ampholytes (small, soluble, amphoteric molecules) with a high buffering capacity near their pI (Amersham Biosciences; www.electrophoresis.apbiotech.com), and usually it was a broad range of ampholytes covering pI values from 3 to 9 that was used. The era of ampholytes (the LKB brand was named Ampholines) was followed by the development of immobilized pH gradients (Righetti *et al.*, 1989; Görg *et al.*, 1995). There are several definite advantages of the use of Immobilines – three of them are mentioned here: first, owing to commercial availability of the Immobiline strips, a rigorous comparison of results among various laboratories is possible; second, the loading capacity of the Immobiline strips is considerably higher than that of Ampholine gels; third, narrow range separations are reasonably reproducible. A disadvantage is in the relatively high cost of the Immobilines strips, their limited shelf-life and also various technical inconveniences like the requirement for soaking the strips prior to use or the relatively long separation time (almost double the separation time required by the Ampholine system).

Two-dimensional gel electrophoresis: step two – size separation

The size 'separation is performed in a polyacrylamide matrix in SDS milieu. The most common procedure utilizes an acrylamide gradient of 10–20 per cent. Although several attempts were made to commercialize pre-cast gels, financial considerations (and the relatively short shelf-life of such gels) drive most of the researchers to cast their own gels.

Large-scale Isodalt separation system

Norman and Leigh Anderson have upgraded the method into a robust system capable of simultaneously performing analysis of 20 samples (Anderson and Anderson, 1978a,b, 1982; Anderson *et al.*, 1983). The Ampholine IEF gradient as well as the gels for the second dimension are cast in a single step for all 20 samples. For charge separation the tubes are immersed in a large container (2 l) of acidic buffer, and for size separation a tank with 30 l cooled electrolyte contents is used. The emphasis of the Isodalt system is not only to achieve reproducibility within the studied set of 20 samples, but also to obtain stable and robust experimental comparisons for consecutive experiments (Anderson, 1988; Lefkovits *et al.*, 1985).

Metabolic labeling and sensitivity of detection

The most common labeling procedures involve incorporation of ^{35}S methionine in to the *de novo* synthesized polypeptides. A spot containing as little as 0.3 dpm can be detected when a radiofluorographic readout is used, while about 3 dpm are required for autoradiographic detection. Autoradiographic readout yields spots with sharper boundaries, and closely localized spots of very different abundances can be well distinguished as distinct spots.

Sample preparation

It has been recognized that one of the prerequisites for successful analysis is an adequate preparation of samples that are to be applied on the separation matrix. A standard solubilizing buffer, containing NP-40 and urea (Anderson, 1988; Anderson *et al.*, 1983; Lefkovits *et al.*, 1985), is used in most instances, although a number of recipes have been worked out for the solubilization of various 'difficult' tissues (Santoni, Molloy and Rabilloud, 2000; Rabilloud, 1998; Chevallet *et al.*, 1998). It is our experience that the art of sample preparation remains the bottleneck of the whole separation system. Especially problematic are extracts of human aorta specimen that are used in our laboratory. Combinations of various homogenization procedures are under scrutiny at present (Anne von Orelli, dissertation in progress). In addition, samples in which there are predominant accompanying proteins, e.g. high actin, myosin light chain or serum albumin (Lucchiari *et al.*, 1992; Nezlin and Lefkovits, 1998) create serious obstacles to reproducible separation.

Staining of gels

Coomassie blue staining was developed some 40 years ago by Fazekas *et al.*, (1963), and it is used either in its original recipe or as a staining protocol using colloidal Coomassie staining (SLRI Proteomics Database; http://proteomics.mshri.on.ca/sample_preparation.php). A clear advantage of the colloidal staining is that excessive destaining is not necessary. Another suitable method for protein detection is based on staining with fluorescent dyes and ruthenium reagents (Lamanda *et al.*, 2004).

Silver staining is considerably more sensitive than Coomassie blue staining. In our hands, very good silver-stained gels (Switzer *et al.*, 1979; Kuhn, Kettman and Lefkovits, 1989) are obtained when about one-tenth of the amount of the polypeptide sample required for Coomassie staining is applied to the separation system.

Wet gels, dried gels

For drying, we use an apparatus constructed in the workshop of the Basel Institute for Immunology. The apparatus enables simultaneous drying of all 20 gels from the Isodalt system; it is constructed as a system of 10 drawers, each platform accommodating two gels, and during the (overnight) drying cycle it regulates heat and vacuum. Blueprints of the system are available from the author (I.L.) upon request (Anderson *et al.*, 1983). Wet gels are maintained in a sealed plastic folder, usually for further handling for mass spectrometry.

Autoradiography, radiofluorography

Metabolic labeling, followed by autoradiography or radiofluorography, is the most common readout in functional proteomics, while unlabeled material is used for structural proteomics. Usually several exposures are performed from each gel, and the most suitable ones are selected for image analysis. Radiofluorography is based on the impregnation of gels with diphenyloxazole (PPO) and on exposure of films at temperature of −70°C. The sensitivity of detection is considerably higher than with standard autoradiography, and the exposure time can be shortened by about 8-fold. The OD saturation curve for radiofluorography is different from autoradiography, and the two detection systems cannot be combined to evaluate gels within one experiment. We have recently started to use the phosphoimager readout. The high costs of the detection plates make this detection method prohibitive for many laboratories.

PDQuest image analysis

There are many sophisticated software systems for the evaluation of two-dimensional gel images. Until recently our laboratory has been experienced with the image analysis system, originally developed by John Taylor in Norman Anderson's team at the Argonne National Laboratory, under the name 'Tycho', (Taylor *et al.*, 1980) and the follow-up software 'Kepler' (Anderson, 1992). We currently use a PDQuest system that is similar to Kepler but with a more advanced user-friendly interface. The image files are processed for noise and streak removal and background correction, and then converted into spot files by spot modelling and fitting. In the final 'spot lists', each spot is defined by the x and y coordinates and by the spot volume. At the end of the matching process the *master* pattern contains all the spots occurring in each of the images.

Technology for protein structure elucidation by mass spectrometry and related methods

Each molecular species of a polypeptide is defined by its amino acid sequence. Methods for establishing such sequences were devised decades ago in which sequential cleavage of each consecutive N-terminal amino acid was performed. Once it became apparent that the mass of small peptides can be measured by mass spectrometry, and the amino acid composition can be deduced from the mass data, considerable effort was invested in adopting the methodology of working with 'true' polypeptides. Since the prerequisite for such analysis is conversion of the analyte into 'gas-phase ions', polypeptide molecules have to be cleaved into peptidic fragments, then converted into gas-phase ions and ejected to measure the 'time of flight'. The measured masses of the resolved analytes are then compared with those calculated for the predicted peptides from the genomic database, and the molecular identity of the analysed polypeptide is established. Compared with classical sequencing procedures, a million times less starting material is required, thus analysis of individual (excised) spots of proteomic pattern becomes feasible.

Mass spectrometers consist of three principal parts:

(1) an ionization source – converts molecules into gas-phase ions;

(2) a mass analyser – separates individual components according to their mass-to-charge ratio (usually measuring the 'time-of flight', TOF);

(3) an ion detector – measures the magnitude of the current (as a function of time).

Ionization by MALDI and ESI

Two ionization procedures are used in mass spectrometry:

(1) matrix-assisted laser desorption ionization (MALDI) – peptides are deposited on an energy-absorbing crystalline matrix and subsequently the laser energy strikes

the matrix and causes excitation of the matrix and ejection of the ions into the gas phase;

(2) electrospray ionization (ESI) – peptides in a solvent are induced by application of a potential to spray; the electrospray creates nanometer-size droplets, and subsequently the solvent is removed (by evaporation) and multiply charged ions are detected.

These and other approaches yield highly precise estimates of the molecular mass of the peptide sample. MALDI is used typically in conjunction with TOF mass analysers, while another type of information is obtained by tandem mass spectrometers (MS/MS), which produces a ladder of fragment ions that represent cleavage of amide bonds. Peptide molecular weight measurements are predictive of amino acid composition, and peptide fragmentation information relates to the amino acid sequence. Both types of information can be correlated to protein sequences of the database. Clearly, a single peptide mass is not unique to a specific protein, thus a collection of peptides (from the same tryptic digest) has to provide the required match. A peptide spectrum is obtained and this is a diagnostic identifier of the given protein sequence.

Peptide mass fingerprinting in single MS mode

This method is adequate to analyse material in single spots or in not overly complex mixtures. It does not provide optimal accuracy for protein identification.

Mass spectrometry – high throughput of excised spots

Mass spectrometry made it possible to establish molecular structures of a large number of two-dimensional gel spots. Although the methodology is a highly sophisticated one, and is an expensive undertaking, this approach has no competitive alternative. Because this 'one-by-one' analysis is based on robotic support, it provides a truly high-throughput system, and is capable of producing and integrating an enormous amount of data in a reasonably short time.

There is an ongoing discussion whether the methodology is also suitable for finding rare components, which are present at an abundance of only 10 polypeptide copies per cell. Gradually a consensus has started to take shape: standard proteomic procedures based on two-dimensional gels of entire cell extracts provide results for the most abundant set of proteins, while material from subcellular fractions furnishes data for low-abundance proteins. Special enrichment procedures will be needed for the above-mentioned example of the abundance of 10 polypeptide copies per cell. If indeed 10^{11} polypeptide copies extracted from a spot yield usable MS spectra, an extract of 10^{10} cells might be needed to enter the fractionation procedure.

Mass spectrometry – entire two-dimensional matrix

Another way to proceed, is to use an extension of the MS methodology to analyse the entire matrix of the two-dimensional gel. This approach, currently under scrutiny in Hochstrassers' laboratory, will probably give the final breakthrough, since in this approach the MS spectra are inspected in the context of all neighbouring MS spectra. This procedure, considered conceptually, will enable results to also be read from overlapping spot entities (that is minor spots masked by overlapping large spots).

It is probable that, owing to the enormous computing power of modern computers, it will become possible to deduce polypeptide structures from MS spectra obtained from polypeptide mixtures. Each month thousands of spectral prototypes are added to the databases of each MS laboratory, and each new sample has an increased chance of instantaneous structural identification.

Overview of projects under scrutiny

The major objective of proteomic studies in cardiovascular research is the identification of molecular changes causative or indicative for the disease under investigation. Biopsy samples from affected tissues represent precious material that can be subjected to proteomic scrutiny. The rather limited availability of biopsy samples from either normal or diseased persons, however, hampers adequate statistical coverage. Therefore, in addition to the investigation of human material (biopsies and body fluids), we include in our studies tissue samples from relevant animal models.

An important and well studied model for heart failure is the Dahl salt-sensitive rat. If these inbred animals are fed with a high-salt diet, they develop arterial hypertension and, as a consequence of this, left ventricular hypertrophy that in the time course further develops to left ventricular dilatation (Inoko *et al.*, 1994). Two-dimensional gel electrophoresis of ventricular tissues of diseased and control animals revealed 12 qualitative differences in protein expression (in preparation).

A valuable source of human heart tissue is the auricle of the right atrium, which has to been removed in bypass operations. From those auricles, complete and physiologically intact trabeculae can be obtained and cultured *in vitro*. We currently employ proteomic analysis to elucidate the effect of reactive oxygen stress on the examined trabeculae (Lampert, 2005). Proteomic projects in our group include studies of ischaemic heart disease (coronary artery disease), advanced heart valve disease, arterial hypertension, dilative cardiomyopathy and myocarditis. We attempt to establish proteomic patterns related to some of these diseases as well as progressive exhaustion of the myocardial energetic resources and alterations of the cardiomyocytes.

The proteomic approach might provide an important clue to the precise molecular pathways in the development and time course of these diseases. Our additional goal is to compare the proteome of cardiac tissue removed at implantation of LVAD (left ventricular assist devices) to the heart, combined with blood samples taken during the time course of circulatory support.

3.3 Discussion

We reported earlier on our attempts to establish proteomic patterns relevant to cardiovascular diseases (Zerkowski, 2004), while our interest in defining molecular components relevant to heart function (Zerkowski, 1991) paved the way for these studies.

The scientific community is certainly not discouraged by the finding that 'only' some 45 000 genes are present in the human genome (Lander and Weinberg, 2000; The Genome Issue, 2000). Owing to a profoundly large number of possible post-transcriptional and post-translational changes, we shall find that the 45 000 genes are only the basic blueprint for an excessive amplification of final products, where the resulting protein repertoire might be at least an order of magnitude higher (maybe 10^6 different proteins).

Annotation of genes, annotation of proteins

Those scientists who search literature for the key word 'proteomics' will realize that there are no papers on proteomics before 1994. Those thousands of pre-1994 papers have to be searched via different keywords: two-dimensional gel electrophoresis, protein catalogue, protein index, etc. Some 25 years ago Norman Anderson realized that the understanding of biological processes on a 'system level' would only be possible if a complete catalogue of human proteins was established. He called the catalogue the 'Human Protein Index' and, with hindsight, we recognize that this seminal paper pre-defined a new era of a holistic approach to molecular components of biological systems (Anderson and Anderson, 1981).

Biosignature of disease phenotypes

The combination of the two-dimensional gel components with mass spectrometry determination created a potent synergy of the two methods. Nevertheless, one should realize that, as the exploration of the complex samples from various disease conditions continues, there is (and there will be) a 'diminishing return' in the discovery of new proteins. Therefore one should not underestimate the unique (and not yet fully appreciated) aspect of the two-dimensional gel system, that the constellation of spots or alterations of entire clusters of spots might often be more significant than the knowledge of the structure of these proteins.

To obtain a meaningful biosignature it is of prime importance to achieve a proper quantitation, since the emphasis is not so much on 'all or none' findings, but rather on 'up and down' regulations, followed by a numerical taxonomy analysis (including 'principal component analysis' and 'cluster analysis').

Biomarkers as predictors to subsequent therapy and risk stratification

In present medical practice any (molecular) marker that enables a proper diagnosis is considered highly useful. There are only a few disease states in which several markers are available. Combining a biomarker of necrosis (cardiac troponin) with biomarkers of haemodynamic stress (brain natriuretic peptide BNP) or of inflammation (high sensitivity CRP) enhances the risk stratification of patients with acute coronary syndrome. The appropriate therapeutic response to an elevated troponin in the setting of acute coronary syndrome is well defined. However to date there is no consistent evidence to guide treatment on the basis of elevated BNP or CRP (James *et al.*, 2003; Lefkovits, 2004; Morrow and Braunwald, 2003). The combination of multiple molecular markers (for one disease) could lead to new definitions of differentiated therapeutic strategies.

Finally, we will attempt to generate a proteomic approach for patients after heart transplantation in peripheral blood samples with the aim of avoiding the need for biopsy and furthermore to find a proteomic pattern for the development of a new grading score for detection of chronic or acute rejection.

How much is 1 ng?

The experimenter has a good chance to detect a spot, if the spot contains 1 ng of proteins. Is 1 ng a lot or a little? [In order to be able to imagine such miniscule quantities, one should consider that the (average) mass of one single cell is 1 ng. The mass of all proteins in one cell is about 0.1 ng (100 pg)].

A total of 10 000 000 000 polypeptide molecules, each one 400 amino acid long, has a mass of 1 ng. If this number of molecules is 'focused' in a single spot of a two-dimensional gel, then it is readily detectable (by silver staining). Thus, when an extract of 1 million cells is used as a proteomic sample, then those molecular species which are present in the cell at an abundance of 10 000 molecules (or more) will be detectable ($10^6 \times 10^4$).

How much is 1 mg?

If we look at the number game from the other side, we can ask how many protein molecules have to be released into the blood stream (upon tissue damage such as infarction or injury) in order to be detected by the proteomic approach. From the calculation (above paragraph) it follows that 1 mg protein corresponds to

10^{16} polypeptide molecules, and if these molecules are dispersed in the circulation (5 l blood), then each 1 ml blood will contain some 2×10^{12} polypeptide molecules. This is not a bad premise for the search for biomarkers.

The above considerations might be useful when attempting to generate a proteomic approach in peripheral blood samples with the aim of avoiding invasive actions (biopsy) and furthermore to find a proteomic pattern for the development of a new grading score for the detection of chronic or acute heart transplant rejection.

Post-translational modifications

Proteins are much more 'sophisticated entities' than nucleic acids. Proteins are modified by many different means: they get glycosylated, phosphorylated, acetylated, ubiqitinated, farnesylated and sulfated, they get cleaved into fragments, they combine with other polypeptides or other biological entities, etc. The above modifications alter the half-life of the proteins, they switch *on* and *off* their function, they prepare the molecule to be removed from the 'stage' and often they determine their tissue localization.

Discovery of new proteins

It is becoming apparent that the 'easy proteins' (those present at high and intermediate abundances) have already been discovered and the high-throughput detection methods in proteomics will provide an ever-diminishing return of newly discovered proteins, the reason being that the concentrations of the yet to be discovered cytokines and lymphokines and regulatory proteins are considerably below the detection limit (of any proteomic method).

Detection limits

In the example below we consider a complex sample (cell lysate). Some proteins are present at 10^7 copies per cell while others are present at 10 copies or less per cell. The 100 most abundant proteins account for 90 per cent of the mass of cellular proteins. The 'remaining' species of proteins, maybe 3000–4000 of them, will account for the remaining 10 per cent of the mass of protein.

If we take into consideration a hypothetical regulatory protein present at an abundance of 100 copies (per cell), then the sample (extract of 10^6 cells) loaded on a gel will contain 10^8 polypeptides (of the mentioned molecular species), which will result in a two-dimensional-gel spot containing 10 pg polypeptides (1.5 fmol) – far below the detection limit.

3.4 Concluding Remarks

The above review would be incomplete if one would not add a note of caution about the use of variations of proteomic approaches. Immunoaffinity techniques for the removal of high-abundance proteins (e.g. albumin, immunoglobulins and transferrin) are useful in extending the dynamic range of the protein detection. There is a general consensus among practioners of two-dimensional gel methods that major proteins like albumin, transferrin, actin, tubulin and many others obscure the proteomic pattern, and should be removed prior to analysis. It turns out that some of the isoforms of these proteins might be characteristic for the studied disease! For example, there is an ischaemia-modified albumin that arises earlier than troponin and therefore is an ideal biomarker for early infarction (Apple *et al.*, 2005). The release of free radicals that activate enzymes that modify the N-terminal structure of albumin causes albumin to fail to bind metal. Removal of albumin would disguise the characteristics of the disease. This explicitly shows that any alteration in established proteomic protocols has to be well thought over and thoroughly justified.

Acknowledgements

We thank Emmanuel Traunecker for competent work in performing the two-dimensional gel electrophoresis.

References

Anderson, L.A. 1992. *Kepler Software Manual.* LSB: Rockville, MD.

Anderson NG, Anderson NL. 1978. Analytical techniques for cell fractions. XXI. Two-dimensional analysis of serum and tissue proteins, multiple isoelectric focusing. *Anal Biochem* **85**: 331–340.

Anderson NG, Anderson NL. 1981. The human protein index. *JAMA* **246**: 2621–2623.

Anderson NG, Anderson NL. 1982. The Human Protein Index. *Clin Chem* **28**: 739–748.

Anderson NL. 1988. Two-dimensional electrophoresis. In *Operation of the ISODALT System.* Large Scale Biology Press: Washington, DC.

Anderson NL, Anderson NG. 1978. Analytical techniques for cell fractions. XXII. Two-dimensional analysis of serum and tissue proteins: multiple gradient-slab gel electrophoresis. *Anal Biochem* **85**: 341–354.

Anderson NL, Hofmann J-P, Gemmell A, Taylor J. 1983. Global approaches to quantitative analysis of gene-expression patterns observed by use of two-dimensional gel electrophoresis. *Clin Chem* **30**: 2031–2036.

Apple FS, Wu AH, Mair J, Ravkilde J, Panteghini M, Tate J, Pagani F, Christenson RH, Mockel M, Danne O, Jaffe AS. 2005. Future biomarkers for detection of ischemia and risk stratification in acute coronary syndrome. *Clin Chem* **117**: 266–269.

Chevallet M, Santoni V, Poinas A, Rouquie D, Fuchs A, Kieffer S, Rossignol M, Lunardi J, Garin J, Rabilloud T. 1998. New zwitterionic detergents improve the analysis of membrane proteins by two-dimensional electrophoresis. *Electrophoresis* **19**(11): 1901–1909.

Fazekas de St, Groth S, Webster RG, Daytner A. 1963. On Coomassie-blue staining. *Biochem Biophys Acta* **71**: 377–391.

Görg A, Boguth G, Obermaier C, Posch A, Weiss W. 1995. Two-dimensional polyacrylamide gel electrophoresis with immobilized pH gradients in the first dimension (IPG-Dalt): the state of the art and the controversy of vertical versus horizontal systems. *Electrophoresis* **16**: 1079–1082.

Inoko M, Kihara Y, Morii I, Fujiwara H, Sasayama S. 1994. Transition from compensatory hypertrophy to dilated, failing left ventricles in Dahl salt-sensitive rats. *Am J Physiol* **267**: 2471–2482.

James SK, Lindahl B, Siegbahn A, Stridsberg M, Venge P, Armstrong P, Barnathan ES, Califf R, Topol EJ, Simoons ML, Wallentin L. 2003. N-terminal pro-brain natriuretic peptide and other risk markers for the separate prediction of mortality and subsequent myocardial infarction in patients with unstable coronary artery disease: a global utilization of strategies to open occluded arteries (GUSTO)-IV substudy. *Circulation* **108**: 275–281.

Klose J. 1975. Protein mapping by combined isoelectric focusing and electrophoresis of mouse tissues. A novel approach to testing for induced point mutations in mammals. *Humangenetik* **26**: 231–243.

Kuhn L, Kettman J, Lefkovits I. 1989. Consecutive radiofluorography and silver staining of two-dimensional gel electrophoretograms: application in determining the biosynthesis of serum and tissue proteins. *Electrophoresis* **10**: 708–713.

Lamanda A, Zahn A, Roder D, Langen H. 2004. Improved ruthenium II Tris (bathophenantroline disulfonate) staining and destaining protocol for a better signal-to-background ratio and improved baseline resolution. *Proteomics* **4**: 599–608.

Lampert F. 2005. *In vitro* cultivation of human atrial trabeculae. Dissertation, Albert-Ludwigs-Universität, Freiburg.

Lander ES, Weinberg RA. 2000. Genomics, journey to the center of biology. *Science* **287**: 1777–1782.

Lefkovits M. 2004. Ersetzen Biomarker die klinische Beurteilung des Herzens? *Schweiz Med Forum* **4**: 1051–1057.

Lefkovits I, Young P, Kuhn L, Kettman J, Gemmell A, Tollaksen S, Anderson L, Anderson N. 1985. Use of large scale two-dimensional ISODALT gel electrophoresis system in immunology. In *Immunological Methods III*. Academic Press: Orlando, FL; 163–185.

Lucchiari MA, Pereira CA, Kuhn L, Lefkovits I. 1992. The pattern of proteins synthesized in the liver is profoundly modified upon infection of susceptible mice with mouse hepatitis virus 3. *Res Virol* **143**: 231–240.

Morrow DA, Braunwald E. 2003. Future of biomarkers in acute coronary syndromes: moving toward a multimarker strategy. *Circulation* **108**: 250–252.

Nezlin R, Lefkovits I. 1998. Expressed immunoglobulin repertoire of LPS-stimulated splenocytes of unimmunized mice as studied by two-dimensional gel electrophoresis. *Mol Immunol* **35**: 1089–1096.

O'Farrell PH. 1975. High resolution two-dimensional electrophoresis of proteins. *J Biol Chem* **250**: 4007–4021.

Rabilloud T. 1998. Use of thiourea to increase the solubility of membrane proteins in two-dimensional electrophoresis. *Electrophoresis* **19**(5): 758–760.

Righetti PG, Gianazza E, Gelfi C, Chiari M, Sinha P. 1989. Isoelectric focusing in immobilized pH gradients: applications in clinical chemistry and forensic analysis. *Anal Chem* **61**: 1602–1612.

Santoni V, Molloy M, Rabilloud T. 2000. Membrane proteins and proteomics: un amour impossible? *Electrophoresis* **21**: 1054–1070.

Switzer RC, Merril CR, Shilfrin S. 1979. A highly sensitive silver stain for detecting proteins and peptides in polyacrylamide gels. *Anal Biochem* **98**: 231–237.

The Genome issue. 2000. *Science* **287**.

Taylor J, Anderson NL, Coulter BP, Scandora AE, Anderson NG. 1980. *Electrophoresis '79*, B Radola (ed.). de Gruyter: Berlin; 329–339.

Zerkowski HR. 1991. *Zur funktionellen Bedeutung myokardialer beta2-adrenorezeptoren des Menschen*. Thieme: V Stuttgart; 99.

Zerkowski HR, Grussenmeyer T, Matt P, Grapow M, Engelhardt S, Lefkovits I. 2004. Proteomics strategies in cardiovascular research. *J Proteome Res* **3**: 200–208.

4

Chemical Genomics: Bridging the Gap Between Novel Targets and Small Molecule Drug Candidates. Contribution to Immunology

György Dormán, Takenori Tomohiro, Yasumaru Hatanaka and **Ferenc Darvas**

4.1 Introduction of Chemical Genomics: Definitions

With the completion of the human gene map, understanding and healing a disease will require the integration of genomics, proteomics and metabolomics (system biology), with the early utilization of diverse compound libraries to create a more powerful 'total' drug discovery approach. The major task in the post-genomic discovery is to establish a synergy between the increasing number of targets and new chemical entities in many aspects, reducing the mismatch between the biological structure and chemical structure space. Chemical genomics/chemical biology could be the solution to the discovery and validation of novel targets, which could provide novel therapeutic interventions for traditional and newly emerging diseases, with more specific, efficient and safer, small molecule drug candidates interacting on them. Chemical genomics also contributes to the discovery of genetic variability accumulating knowledge about biological responses to chemical compounds. These approaches also contribute to the rapid development of diagnostics, which helps to find the most effective treatment.

The interaction of chemical entities with their biological partner can be studied in cell-based assays detecting expression or phenotype alterations. Alternatively, direct interaction with proteins can be investigated in affinity/activity-based screening methods.

Immunogenomics and Human Disease Edited by András Falus

These approaches are reflected in the major trends in chemical genomics (Darvas, Dormán and Guttman, 2004).

Chemical genomics

Chemical genomics is an extension of chemical genetics to a genome-wide scale:

1. (At cellular level.) Complex pathways are studied using a large number of small molecules in parallel experiments, identifying the effect of them on cell phenotypes and gene expression in cell-based assays.

2. (At proteome level.) Chemical genomics utilizes a large number of small molecules in a parallel manner early in the discovery pipeline to identify, validate and prioritize disease-associated proteins while simultaneously identifying compounds suitable for drug candidates. The determination and analysis of the function of proteins are carried out by analysing expression profiles in the proteome by chemical interaction with small molecules.

3. (In broad sense/post-genomic medicinal chemistry.) The new parallel approach to investigating and better understanding medicinal chemistry after we have the genome sequenced helps to reduce the mismatch between the chemical and biological structure space.

There are two major approaches, forward and reverse chemical genomics. *Forward chemical genomics* (Figure 4.1) starts with probing the genome using small molecule interaction at the cellular level (cell-based approach). Small molecules have been

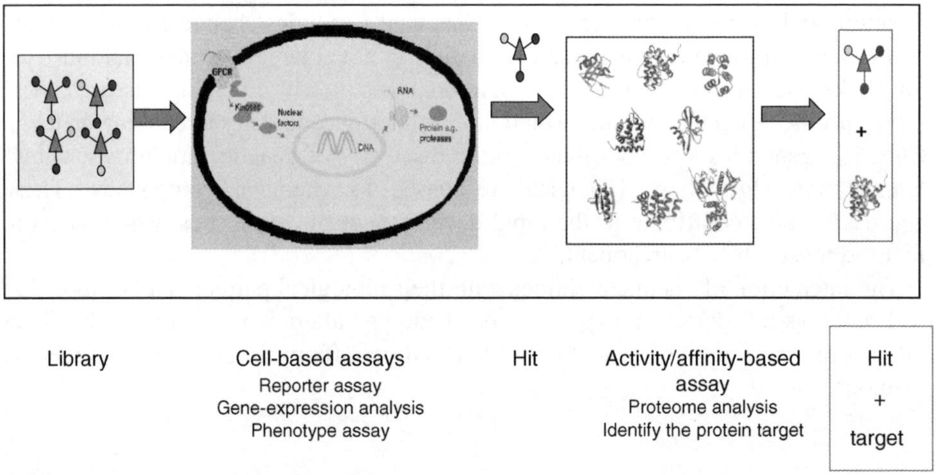

Figure 4.1 Forward chemical genetics/genomics.

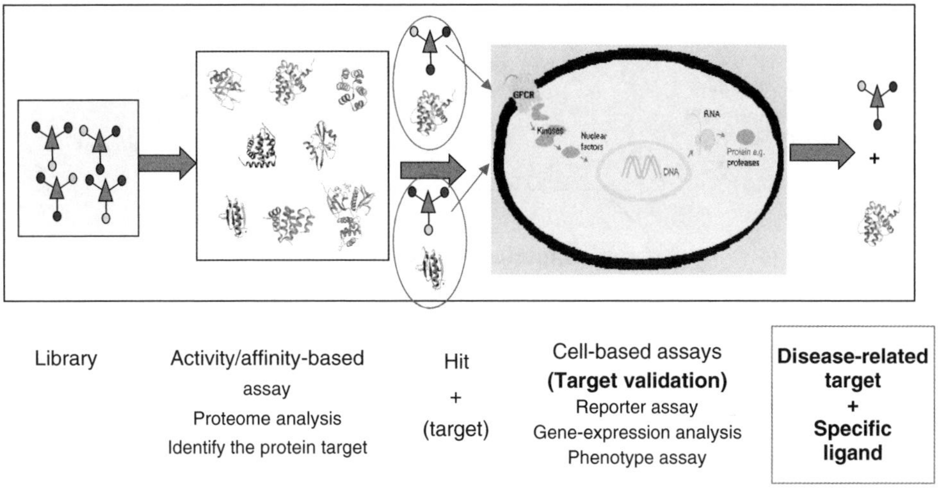

Figure 4.2 Reverse chemical genetics/genomics.

used to activate or inactivate several proteins within a particular pathway; thus, their overall cellular effects can be studied at the gene expression or protein expression level. Subsequently, or in parallel, the binding partners of the small molecules can be identified. *Reverse chemical genetics* (Figure 4.2) starts with an identified novel protein target and screens for small molecules that affect its activity and then tests whether the small molecule causes a phenotypic or gene-expression change. The initial target–ligand pairs can be linked to specific changes of the cell behaviour in disease-associated assays. The first step is identifying small molecule interaction in a genome-wide or proteome-wide manner using activity/affinity-based approaches, while the second step is a target validation in whole-cell assays. The first step is often referred to as chemical proteomics, where affinity-based chemical probes can reveal a binding affinity signature to proteins within a cell by investigating at the proteome level. Many elements of chemical genomics can be utilized in different areas of immunogenomics (Figure 4.3).

Immunogenomics is genome-wide analysis to define the genetic basis of susceptibility to complex (or polygenic) diseases (e.g. autoimmune and infectious diseases), in other words the identification of the genetic component of variable immune responses [the estimated effect of major histocompatibility complex (MHC) (Roopenian, Choi and Brown, 2002) and non-MHC chemokines and cytokine genes; Hill, 2001].

The determination of immunogenetic variants could be useful in identification of the physiological role of a large number of immunological mediators (chemokines, cytokines etc.) still seeking a therapeutic indication. The identification of their associations with receptors and signalling molecules in intracellular transduction could provide new target molecules for drug design in various diseases linked to defects of the immune system. Furthermore, immunogenetic profiling of patients

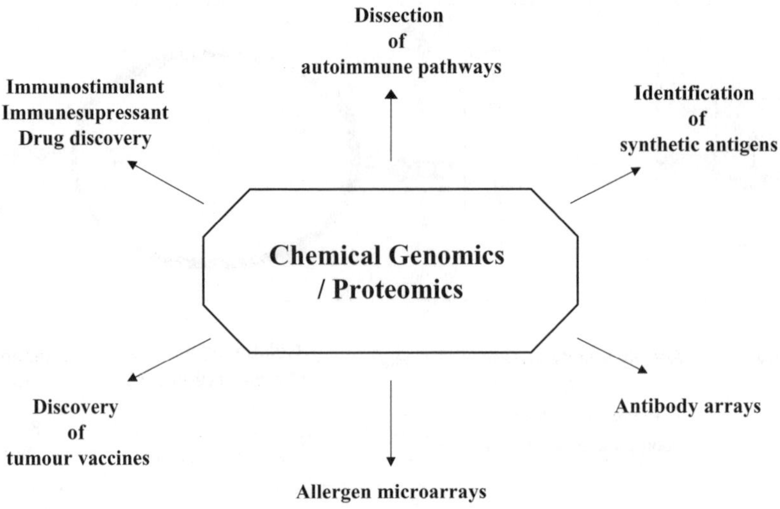

Figure 4.3 Contribution of chemical genomics to immunology.

leading to the recognition of their immunogenetic polymorphisms could be used to tailor immunotherapies. One of the rapidly developing fields of immunogenomics is immunoinformatics, which collects data on immune responses, antigen processing and generate mathematical models for host–pathogen interactions as well as the whole immune system (Petrovsky, Schönbach and Brusic, 2003).

Chemical genomics supports these efforts similarly as in the case of functional genomics, providing small or medium-sized chemical entities in this respect. The 'omics' approaches sooner or later complete the vision of their integration into 'systems biology' (Oltvai and Barabasi, 2002). Stuart Schreiber, the inventor of chemical genetics, uses the term 'chemical biology' by using the chemical entities 'to modulate the individual function of multifunctional proteins activating or inactivating individual functions' in complex pathways (Owens, 2004). In his pioneering work, which led to the designation of the approach, he elucidated the key signalling pathway of immunosuppression through the mechanism of action of cyclosporin A and FK506, and the identification of the interacting proteins (immunophilins, and the resulting complex protein phosphatase target, calcineurin; Brown and Schreiber, 1996).

Genes and mechanisms involved in autoimmune disorders affect approximately 5% of the population (Marrack, Kappler and Kotzin, 2001). The elucidation of the pathways and genetic and environmental factors is crucial to develop efficient therapy. Susceptibility to autoimmunity is described at three levels: the overall reactivity of the immune system, the specific antigen and its presentation and the target tissue (Ueda, 2003). Chemical agents that interact with the target proteins (e.g. several kinases or TGF receptor; Gorelik and Flavell, 2000) in the pathway could help the elucidation of the signalling path as well as providing a small molecule drug candidate for future therapy.

Typical chemical genomics approaches are employed in the understanding of anticancer immune responses through high-throughput gene expression profiling. Anticancer immune responses can be enhanced by immune manipulation. Molecules with immune modulatory properties are being produced by tumour and bystander cells within the tumour microenvironment (Wang *et al.*, 2004). Similarly, the effect of exogenous immunostimulant agents can be monitored.

Similarly, the innate immune response against microbial pathogens can be investigated by the gene expression analysis in human peripheral blood mononuclear cells responding to bacteria and bacterial products (Boldrick *et al.*, 2002). The addition of antibacterial agents can alter or recover the modified gene expression. In another paper, the kinetics of the immune response was identified by fuzzy clustering of gene expression profiles (Guthke *et al.*, 2005).

Gene expression cannot reveal post-translational modifications, like phosphorylation, glycosylation or changes in individual amino acids. For more precise elucidation of the immune response, proteomics is the right choice. Tumour immunology is a rapidly developing area aiming to discover cancer vaccines (Morse *et al.*, 2001; Mosca *et al.*, 2003). Malignant tumours express antigens that may stimulate and serve as targets for antitumour immunity. Tumours could also overexpress tissue differentiation antigens, which also have the potential to be recognized by the immune system (Shu *et al.*, 1997). Proteomics allows rapid serological screening of tumour antigens using two-dimensional (2-D) polyacrylamide gel electrophoresis, which allows simultaneously separation of several thousand individual proteins from tumour tissue or tumour cell lines (Le Naour, 2001). Autoantibodies to tumour antigens represent one type of markers that could be assayed in serum for the detection of cancer in individuals at risk. In lung cancer, as in other tumour types, the majority of tumour-derived antigens that elicit a humoural response are not the products of mutated genes. These antigens include differentiation antigens and other proteins that are overexpressed in tumours (Brichory *et al.*, 2001). Applied to different types of cancer, the proteome-based approach has allowed several tumour antigens to be defined. The common occurrence of autoantibodies to certain of these proteins in different cancers may be useful in cancer screening and diagnosis as well as for immunotherapy. As an extension of the concept, chemically synthesized proteins could serve as tumour-specific vaccines. Recently, Dutch scientists synthesized a synthetic E7 protein of the human papillomavirus type 16. This oncogene-derived protein is largely expressed in cervical cancer and thus can be considered as a potential target for tumour-specific immunity (Welters *et al.*, 2004). This approach has also been applied in other therapeutic areas like neurological diseases. Papini and co-workers identified synthetic glycopeptide antigens for the early diagnosis of multiple sclerosis (MS), Alzheimer's disease and prion diseases (Lolli *et al.*, 2003). Based on these synthetic peptides, a simple test was developed for MS (Papini, 2005). Recently several attempts were made to identify synthetic antigens for the immunodiagnosis of HIV-infected patients; the topic has been reviewed recently (Alcaro *et al.*, 2003). Advances in miniaturization and automation may also permit characterization of the immune response more rapidly and from smaller amounts of biological sample than is possible with existing assay systems (Mosca *et al.*, 2003).

Protein/antibody and small molecule microarrays have been developed as successors of DNA microarrays, allowing a proteome-wide screening of protein function in parallel (Glokler and Angenendt, 2003). A typical protein or small molecule microarray contains an ordered array of protein-specific ligands, typically antibodies, spotted onto a derivatized solid surface and studying protein–protein or protein–ligand interactions in a miniaturized, uniform, high-throughput manner. Derivatization chemistries of solid supports of chemical microarrays were mainly adapted from DNA-microarray technology.

Other companies have been exploring capture agents that include protein scaffolds, such as binding proteins, RNA or DNA aptamers – oligonucleotide sequences – and partial-molecule polymeric imprints (Gershon, 2003). MacBeath arrayed collections of proteins or protein domains onto a variety of chemically derivatized glass slides, typically using robotic arrayers using fluorescent dyes as labels. The slides were visualized using fluorescent scanners (MacBeath, 2002).

Microarrays of antibodies can measure the concentrations of many proteins quickly and simultaneously in health and disease (Wilson and Nock, 2003). The current generation of antibody microarrays has been developed as diagnostic tools as well as for protein profiling. Relatively small numbers of antibodies are currently available and sensitivity, specificity and signal-to-noise ratio are the critical issues (Taussig and Landegren, 2003). There are two major types of microarrays (Stoll *et al.*, 2004): direct protein capture on microarrays (e.g. glass slides) and the reversed screening approach. In the first case the detection of the interacting protein or other chemical species can be detected by direct fluorescent labelling, label-free methods or by the application of a second antibody if available (sandwich assays; Templin *et al.*, 2004). Another group is working on the development of antibody arrays for colon cancer screening (Potter, Lampe and Roth, 2005).

Ivanov and co-workers developed an antibody array system for detecting various post-translational modifications of proteins. In their approach, immunoprecipitated proteins were labelled with fluorescent dye followed by incubation over antibody arrays. The authors profiled protein tyrosine phosphorylation, ubiquitination and acetylation in mammalian cells under different conditions (Ivanov *et al.*, 2004). Antibody arrays or chips are on the way to be used in drug discovery and diagnostics. In a recent example, cytokine antibody array systems were used to identify various cytokines in different tissues (Huang, 2003). This approach was also reported as sandwich assays, where the antibody array system was combined with enzyme-linked immunosorbent assays (ELISA), sensitivity of enhanced chemiluminescence (ECL) and the high throughput of microspotting (Huang, 2004).

In reverse screening, cell lysates are immobilized by unspecific interactions and they interact with antibodies. Immobilized tissue samples or cells can be used to identify disease-specific antibodies that could serve as biomarkers. Reverse arrays were used to profile hundreds of antigens. Cell lysates, material from laser capture microdissection or serum samples were arrayed. The array was probed using a small number of antibodies in order to profile the signalling pathway of ovarian cancer (Sheehan *et al.*, 2005).

The experience with DNA and protein arrays was extended to small-molecules microarrays. Microarrays of small molecules have already been successfully applied in important areas ranging from protein profiling to the discovery of therapeutic leads (Uttamchandani *et al.*, 2005). One of the major applications in immunogenomics is the creation of allergen microarrays that allow determination of allergic patients, immunoglobulin E (IgE) reactivity profiles. The allergen components were microarrayed in triplicates onto glass slides in groups representing individual sources. Microarrayed allergen components were exposed to sera from allergic patients who were sensitized against a variety of allergen sources (Hiller *et al.*, 2002). Other investigators reported a rapid *in vitro* test system for allergy diagnosis, which is based on microscope glass slides activated with (3-glycidyloxypropyl)trimethoxysilane. Allergen extracts and solutions are immobilized as small droplets on the activated glass slides using an arrayer robot. With the disposable microarray slides, it was possible to distinguish between patients with and without elevated levels of allergen-specific IgE. Repeated measurements of serum samples demonstrated a sufficient reproducibility (Fall *et al.*, 2003).

4.2 Chemical Microarrays

One of the most promising tools in target identification and hit discovery is the use of chemical microarrays, which comprise thousands of small molecules attached to a solid surface in an ordered format. There are three ways to generate a collection of ligand molecules for immobilization on surfaces: combinatorial chemical library synthesis on micro or macro beads, solid-phase diversity-oriented combinatorial (preferably split-mix) synthesis of unique compounds separated by a bead-arrayer then cleaved and anchored again; solution-phase parallel synthesis, where the molecules are independently synthesized, separated, purified, analysed and finally uniformly conjugated with a tether, allowing efficient anchoring. The standard strategy to prepare microarrays is the attachment of compounds in solution to the reactive surface by microprinting or microspotting, in other words, creating covalent bonds with different groups (—OH, —SH or NH2) – positioned at the end of the linker molecules – by reacting with the active groups on the surface. It is evident that solid-phase combinatorial synthesis provides good opportunities for microprinting on glass slides, after removal from the support, making use of the linker-arm (tether) present. This approach can easily be used to prepare synthetic peptide antigens.

There are several reactive surfaces available for microspotting. MacBeath reported thiol-reactive surfaces (MacBeath, Koehler and Schreiber, 1999; Figure 4.4) while Hackler and co-workers developed a branched dendrimeric spacer system on glass slides for preparing chemical microarrays, which allows high loading capacity of the ligands with amino group-reactive anchoring groups (epoxy or acrylic; Figure 4.5).

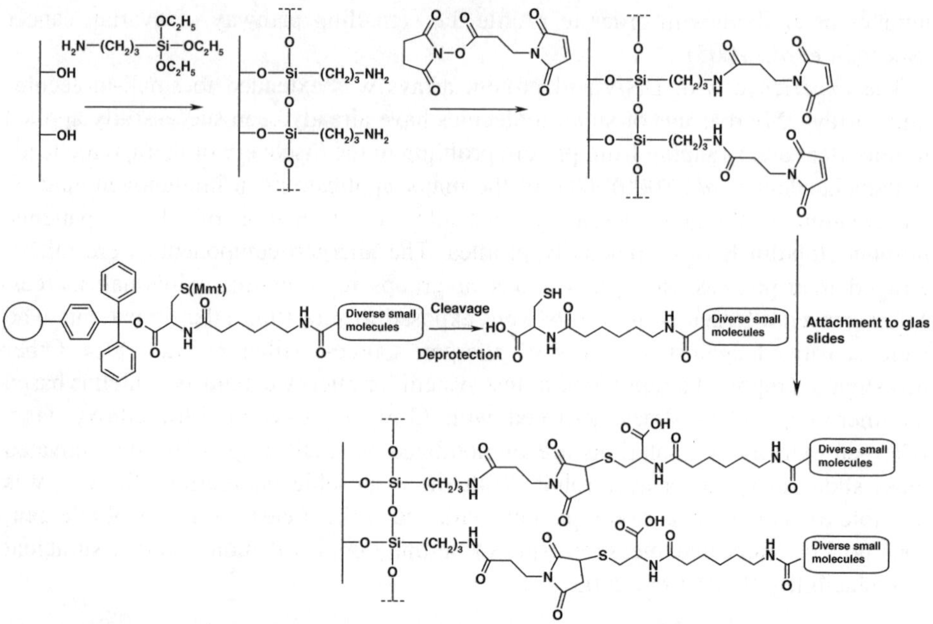

Figure 4.4 Chemistry of microprinting (from solid-phase synthesis to glass chips).

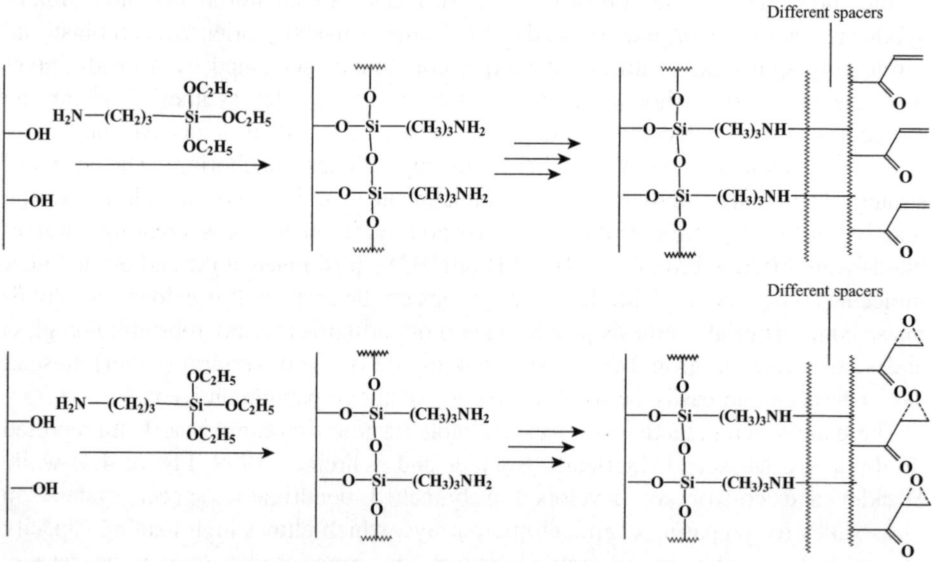

Figure 4.5 BRC/ComGenex approach: generation of high-density microarrays on glass slides.

Evaluation experiments were carried out in order to test the solid support for covalent attachment of ligands and the preparation of chemical arrays. A diverse subset of amino-alkyl ligands (600 compounds, >80% purity) were spotted onto chemically modified glass slides by means of a mechanical microspotter (Hackler *et al.*, 2003). The resulting chemical microarray is currently under validation in cancer proteomics studies using cell extracts in order to identify disease-specific (particularly overexpressed) molecular targets through the identification of their small molecule binders. For detection, two different dyes Cy-3 and Cy-5 are generally used with chemical microarrays, similarly to the detection of interactions with DNA chips.

In order to prepare proteins, and small molecules for microspotting, the compounds should be attached with a spacer or tether which is long enough to allow undisturbed binding interaction with the proteins.

For known drugs, allergens, peptides or chemical probes, the potential sites for tethering are frequently difficult to assess, thus, Darvas *et al.*, (2001) devised a combinatorial tethering approach, where the spacer arms were placed at various sites around the molecular framework with appropriate terminal functional groups.

Identification and selection of disease-specific targets is one of the most critical steps in post-genomic drug discovery. Overexpression of genes and their protein products is often the best starting point for dissecting the pathways that lead to the development of specific diseases. Such proteins often serve as molecular targets, particularly in cancer, where small molecules are designed to directly inhibit oncogenic proteins that are mutated and/or overexpressed. In a novel integrated approach, diverse small molecule microarrays are applied to identify such proteins directly from cell extracts/fractions using healthy counterparts as controls. The advantage of this approach lies in the fact that the 'positive' small molecules can serve as a tool for target isolation and identification through affinity-based methods as well as for target validation using phenotypic assays (Figure 4.6).

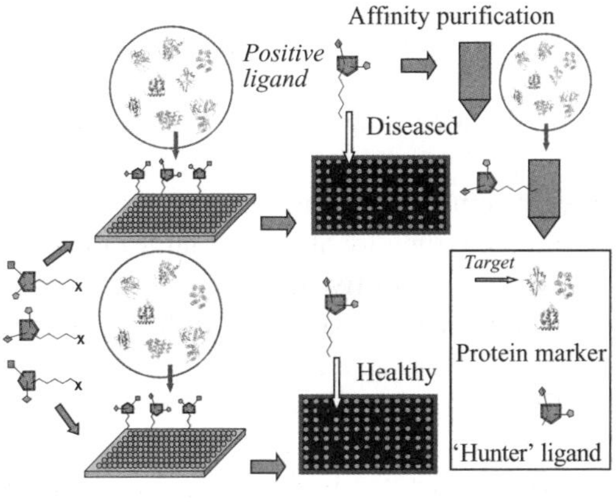

Figure 4.6 Identification of protein expression differences with chemical microarrays.

4.3 Small Molecule and Peptide Probes for Studying Binding Interactions Through Creating a Covalent Bond

Cystein-mutants as chemical sensors (SH-tagged libraries)

Goeldner introduced a novel method combining cysteine-scanning mutagenesis and SH-selective tagged site-directed reactive ligands to create covalent linkage between the receptor and ligands (Foucaud *et al.*, 2001). Covalent linkage is expected to form only if the binding site is in close proximity to one of the thiol-linked reactive sites, thus, around the bound ligand. This method has been applied successfully for the identification of several new compounds and binding sites at Sunesis Pharmaceuticals (www.sunesis.com). This method is particularly useful to study and inhibit protein–protein interactions in a specific biochemical pathway.

Irreversible inhibitors: activity-based probes for profiling catalytic activities

Cravatt and Bogyo reported this strategy independently, profiling the enzymatic activity of enzyme superfamilies in complex proteomes. These activity-based probes (ABPs) contain (1) a class-selective electrophyle that reacts primarily with a key amino acid in the active site of the proteins of a particular class and covalently attaches to it, (2) a reporter tag that allows detection and isolation and (3) a recognition element (binding group) that directs the probe more specifically to the active-site. ABPs have the potential to:

- identify the members of a given enzyme family or activity class;
- determine the activity levels of individual family members;
- localize active enzymes within cells;
- screen small molecule libraries in crude protein extracts for inhibitors.

These probes label the proteins based on their enzymatic activity rather than their abundance, thus monitoring the functional state of large enzyme families (Cravatt and Sorensen, 2000) and providing a sensitive detection of the alteration of the enzymatic activity in normal *vs* diseased state as well as in different tissues. This approach can be referred to as *activity-based proteomics*, which simplifies the proteome according to activity classes. ABPs react with a broad range of enzymes from a particular family, correlating with their catalytic activity and with minimal cross-reactivity with other protein classes. ABPs exist for numerous enzyme classes, such as cysteine proteases and serine hydrolases.

4.4 Photochemical Proteomics

In order to study enzymes where ABPs do no exist, a relatively classic approach provides opportunities, namely photocovalent cross-linking or photoafinity labelling. This is a powerful technique that has been used for identification and localization of binding proteins and mapping their contact region since the late 1960s. In photocovalent cross-linking, a covalent linkage is created between a light-sensitive, detectable ligand and a biopolymer upon irradiation in a reversibly bound state. As a result of the photoinitiated covalent coupling reaction, the functional biopolymer (receptor protein or enzyme) may undergo irreversible activation or inactivation, and at the same time, the occupied functional site is labelled with a detectable tag, which allows easy determination of its location.

4.5 General Aspects of Photoaffinity Labelling

The technique of photoaffinity labelling is currently used as a powerful method of chemical genomics for harnessing the target protein with its specific ligands by the introduction of photochemical cross-linking. The application of photoaffinity labelling includes two levels of analysis: the first is the identification of target proteins among crude protein mixtures using the photoreactive analogue of specific ligands; the second is the identification of peptides forming the ligand binding site after the HPLC separation of affinity-labelled peptides from the digest mixture of proteins (Figure 4.7). This strategy is based on the replacement of a reversible ligand–receptor interaction with a stable covalent bond, and indicates structural information on the transient complex. Therefore, it should be a powerful tool in structural analysis of the ligand–receptor interactions that are not suitable for crystallography or NMR analysis, and for elucidation of cellular responses by tracing protein networks involved in the binding of a specific ligand (Hatanaka and Sadakane, 2002; Kotzyba-Hibert, Kapfer and Goeldner, 1995; Brunner, 1993; Sadakae and Hatanaka, 2004; Fleming, 1995).

For the application of photoaffinity labelling, photoaffinity ligands are usually required to prepare by attaching a photophore, photoreactive group, to the specific ligands. Carbonyl, azide and diazirine groups are commonly used as the photophore for producing extremely reactive intermediates upon photolysis to yield an excited

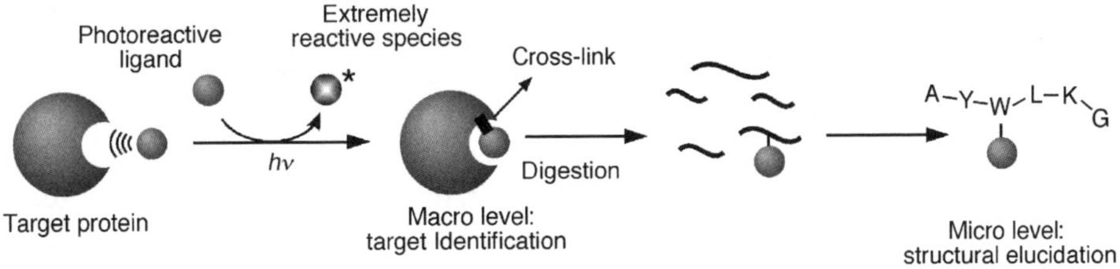

Figure 4.7 Two levels of application of photoaffinity labelling technique.

Major photophores and reactive species	Irradiation (nm)	Comments
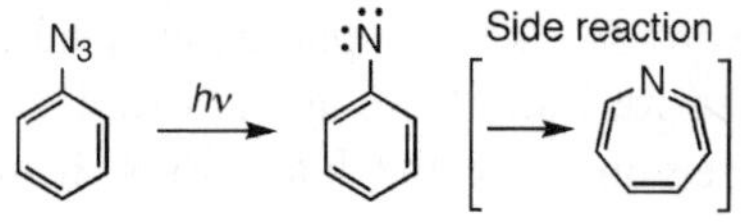 Nitrene; Side reaction; Phenylazide	250 – 290	Many derivatives are available Unstable to reductant (ex. thiol) Shorter wavelength for excitation Suspectable to nonspecific labelling Highly reactive to nucleophiles Relatively unstable cross-link
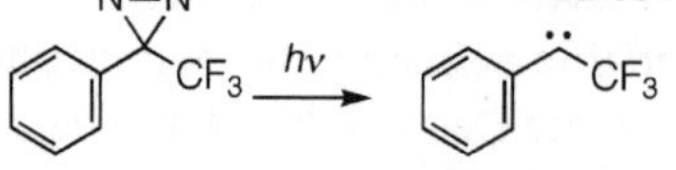 Carbene; (3-Trifluoromethyl)phenyldiazirine	350 – 380	Stable in synthetic conditions More reactive than nitrene Low cross-linking yield due to quenching with water Stable cross-link
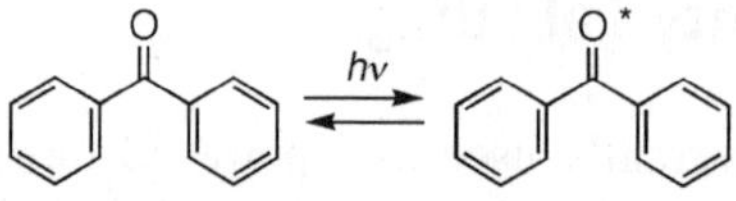 Excited carbonyl; Benzophenone	300 – 350	Very hydrophobic No quenching with water Longer irradiation time Radical-like reactivity Stable cross-link

Figure 4.8 General aspect of major photophores.

carbonyl, a nitrene and a carbene, respectively (Figure 4.2). Numerous derivatives have been synthesized, (Fleming, 1995) and these are chemically tethered into the molecules of interest such as peptide, nucleic acid, sugar, lipid and hormones to give photoreactive probes. Although some of these are already commercially available, the chemical properties of photoreactive probes should be considered to utilize them in affinity labelling; wavelength for activation, cross-linking yields, stability of labelled products, affinity of the photoreactive probes, cares for the preparation and handling of probes, etc. (Figure 4.8; Hatanaka and Sadanake, 2002; Kotzyba-Hibert, Kapfer and Goeldner, 1995; Brunner, 1993; Sadakae and Hatanaka, 2004; Fleming, 1995).

To define the specific interaction between interacting molecules, the cross-linking must occur entirely at the binding site, not at any other site or with any another protein and materials in the solution. Nonspecific labelling is, however, often a major problem in photoaffinity labelling, since ligands are able to dissociate from the binding site during photoaffinity labelling experiments especially in the case of low-affinity interaction. Thus, the photochemically generated species with longer lifetime or lower reactivity usually increase nonspecific labelling to the protein surface. Other side reactions also arise after irradiation due to the species generated from excess probes during heat denaturalization for SDS–PAGE analysis.

The most common photophore is the arylazide group, since a variety of compounds are known to be commercially available. However, the chemical processes of nitrenes generated from azide are considered to involve the side reactions for nonspecific labelling (Fleming, 1995). Furthermore, the reactivity of nitrene depends on the amino acid residues, where cysteine showed the highest reactivity and glycine was essentially nonreactive. (Schäfer and Schuhen, 1996). The shorter wavelength is usually required for the activation of azides, causing serious damage to biomolecules. Benzophenone photophore usually gives higher labelling efficiency, since the generation of water-stable radical-like species increases the capability of cross-linking. The longer irradiation also increases the nonspecifically formed cross-link at any site of the protein. Among the photophores, diazirine and diazo are less likely to produce undesired nonspecific labeling. Carbene species generated from the diazirine or diazo group are extremely reactive and immediately make an electrophilic insertion into the target proteins. Since they can react with water molecule, the unbound species are rapidly quenched, that results in giving the labelled protein cross-linked at the proper position with higher quality.

The following examples could be helpful in figuring out the proper design of photoaffinity probes. Modification of ligands by photoreactive moieties is an important matter affecting the affinity of ligand–receptor interaction, because the photoreactive groups are relatively hydrophobic. The photophore should be attached without significant loss of the ligand activity. A comparison of the binding affinity among tetrafluorophenyl-azide, (3-trifluoromethyl)phenyldiazirine and benzophenone derivatives has been reported for the interaction of ginkgolide derivatives and platelet-activating factor (PAF) receptor, for example (Figure 4.9; Strømgaard *et al.*, 2002). Ginkgolides are terpene trilactones and potent antagonists of PAFR. The binding ability of these photoreactive analogues was similar to each other and most of them were more potent antagonists than the original ginkgolides. However, different labelling results have been reported in several cases depending on the nature of the photophore. Photoaffinity labelling of glucose transporter proteins has been studied by three bis-mannose derivatives [Figure 4.10(a); Yang *et al.*, 1992]. The azide derivative gave nonspecific labelling and the benzophenone derivative required-longer irradiation time. The distribution of glucose transporter in cells was achieved using the diazirine derivative. Similar results have been reported for the interaction of α-cobratoxin (CTX) *Naja naja siamensis*, a long neurotoxin consisting of 70–73 amino acid residues [Figure 4.4(b)], and nicotinic acetylcholine receptor (AchR) (Utkin *et al.*, 1998). Photoreactive CTX analogues were prepared by introducing a

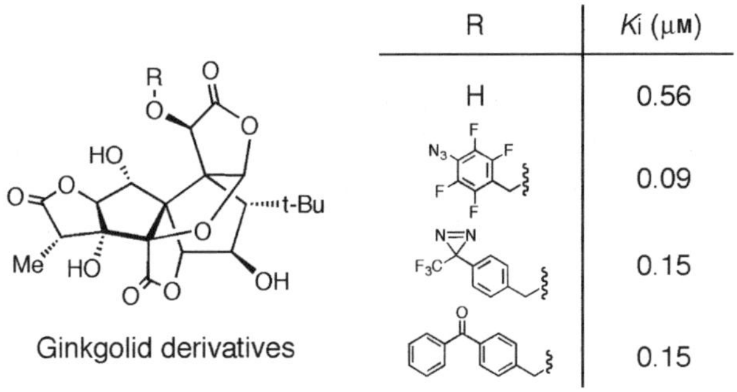

Figure 4.9 PAFR binding activity of Ginkgolide derivatives having a different photophore; inhibition of WEB 2086 as a potent and selective antagonist.

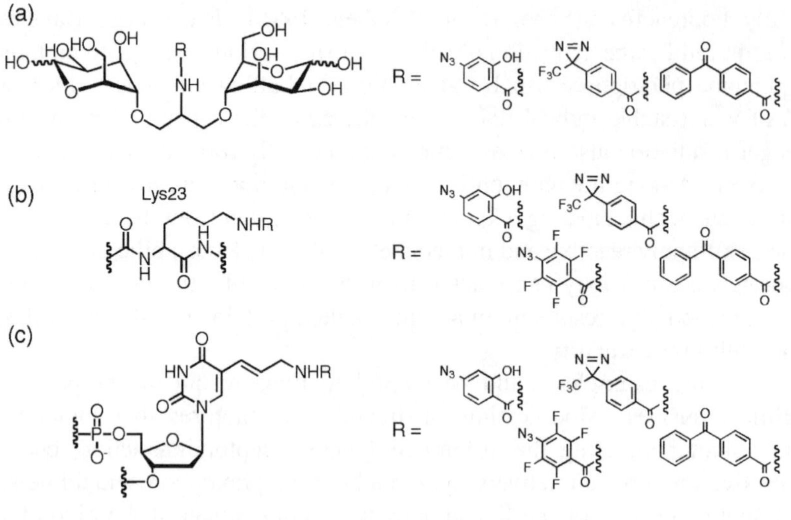

Figure 4.10 Photoaffinity probes of biomolecules having different photophores.

photophore at Tyr23. The results indicated that the different subunits were labelled depending on the photophore: the azido derivatives were labelled γ and δ subunits of AchR; the benzophenone derivative was labelled α and δ subunits; and the diazirine derivative was labelled α, γ and δ subunits. However, the azido derivatives proceeded as a side reaction and caused broadening of the HPLC peak compared with those derived from the diazirine derivative, and the benzophenone derivative needed a longer irradiation time that caused damage on the native toxin. In the case of DNA–protein interaction, nucleobases themselves possess reactivity toward nucleophiles upon photolysis at the shorter ultraviolet light region, which also damages biological molecules [Figure 4.4(c); Meisenheimer and Koch, 1997]. Thus, four different photoreactive nucleotides were incorporated into the *SUP4*tRNA^{Tyr} gene at −3/−2 or +11 bp and were tested for the photoaffinity labelling with Pol III initiation complexes (Tate, Persinger and Bartholomew, 1998). The azide derivative was incorporated at −3/−2 bp to cross-link only one subunit, whereas the result of labelling with the diazirine coded at +11 bp revealed that three subunits were contacted with DNA.

Although the arylazide group has been used as the most common photophore because of its synthetic ease and availablity, the 3-aryl-3-trifluoromethyldiazirine group has a very useful feature in the chemical and photochemical aspects and has given successful results in the characterization of ligand-binding sites, as described above.

4.6 Photoreactive Probes of Biomolecules

The photoreactive derivatives of major biomolecules such as nucleotides, peptides, oligosaccharides and lipids are the special interest of molecular probes for the

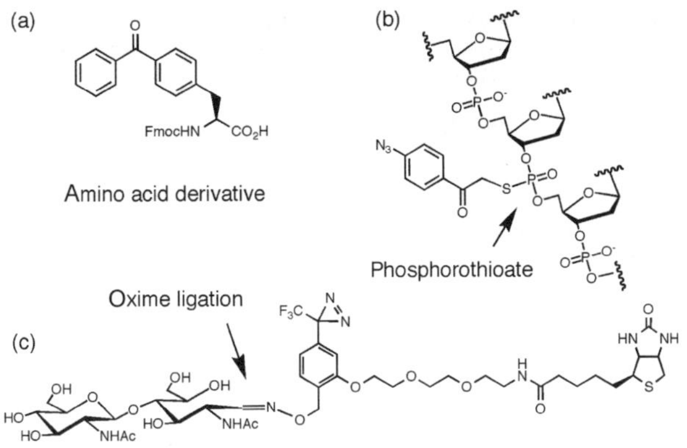

Figure 4.11 Simple and site-specific incorporation of photophores into biomolecules: (a) Fmoc phenylalanine derivative for automatic solid-phase peptide synthesis; (b) a photophore introduced in phosphate backbone by the one-step modification of phosphorothioate group; (c) a chitobiose photoprobe prepared by the one-step oxime ligation of a biotinyl diazirine at the reducing terminal of oligosaccharide.

analysis of the immunological network. Simple preparation of photoreactive biomolecules can be achieved using photoreactive groups bearing a proper functional group directional to the biomolecules of current interest. For example, the selective reaction of α-halomethylcarbonyl group to an SH group was applied for the one-step derivatization of DNA starting from a commercially available oligonucleotide carrying a phosphorothioate group at an appropriate position [Figure 4.11(b)]. Photoreactive moieties can be easily incorporated into the phosphorothioate group of DNA using 4-(α-iodoacetylamino)benzophenone (Musier-Forsyth and Schimmel, 1994) or 4-(α-bromoacetyl)phenylazide (Yang and Nash, 1994).

Immunologically important sugar ligands (Rudd *et al.*, 2001) are one of difficult biomolecules for the derivatization of the probes. Many synthetic steps are usually required for the conventional derivatization method, including protection and deprotection of a number of hydroxyl groups. In addition, natural oligosaccharides were mostly provided on a small scale, and the amount is not enough for the multiple-step synthesis. Recently, one-step derivatization of unprotected oligosaccharides with a biotinyl diazirine photophore was accomplished [Figure 4.5(c); Hatanaka, Kempin and Park, 2000]. The strategy utilized so-called oxime ligation by coupling aldehyde in the reducing terminal of sugar chains and aminooxy groups. Various photoreactive carbohydrates were synthesized and affinity labelling of the corresponding receptor proceeded successfully. Recently, an Na^+/K^+-ATPase β1 subunit was revealed to be a potassium-dependent lectin that bound β-GlcNAc-terminating glycans, which was confirmed using diazirine-conjugated chitobiose (Kitamura *et al.*, 2005).

The pioneering work of photoreactive polypeptides was achieved for site-specific incorporation of photoreactive moiety into synthetic peptides in the 1970s by the use of azidophenylalanine as a building block (Schäfer and Schuhen, 1996), which led to automated synthesis of the photoaffinity probes using solid-phase techniques.

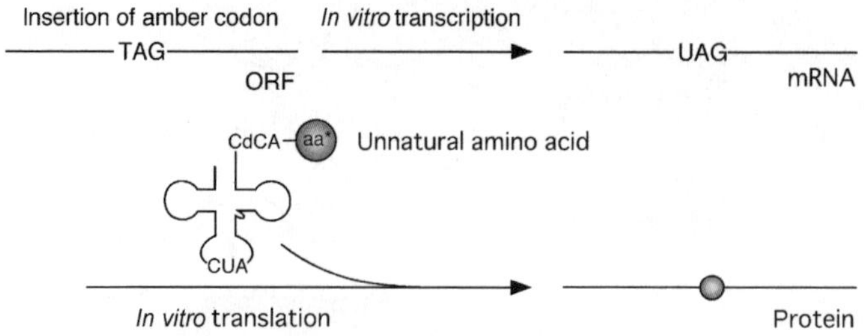

Figure 4.12 Site-specific incorporation of unnatural amino acid into protein using an *in vitro* translation system by the amber codon method.

4′-Benzoylphenylalanine [Figure 4.5(a)] (Kauer *et al.*, 1986) and 4′-(trifluoromethyl-diazirinyl)-phenylalanine (Falchetto *et al.*, 1991) had already been reported to improve analogue in the photophore. Furthermore, the site-directed, biosynthetic incorporation of photoreactive amino acid into proteins was reported (Figure 4.12; Johnson *et al.*, 1976; Hohsaka *et al.*, 2001). The strategy involved a cell-free translation supplied with an unnatural tRNA that charged with unnatural amino acid. A suppressor-tRNA-activated amino acid can be incorporated into the polypeptide chain in response to a stop codon at the desired position in the mRNA used. The *in vivo* system of unnatural amino acid mutagenesis was developed for surface-specific photocross-linking of interacting proteins inside a cell (Chin and Schultz, 2002). The approach of photoreactive proteins would be a powerful tool in chemical genomics for probing protein–protein interactions involved in the immunological network.

4.7 Application to the Immunobiology of Living Cells

To generate a pool of mature T cells that are self-MHC (major histocompatibility complex) restricted and self-MHC tolerant, thymocytes we subjected to positive and negative selection. T cell receptor (TCR)-mediated ligand recognition is essential for the selection that depends on the affinity, that is, low-affinity ligands induce positive selection and high-affinity ligands induce negative selection. Photoreactive (iodo, 4-azidosalicyloyl) peptide antigens were directly used in the molecular analysis of antigen presentation on MHC class I molecules in living cells and recognition by living cytotoxic T-lymphocytes (CTL; Romero, Maryanski and Luescher, 1993). The use of photoreactive peptides also gave information about functional CTL response correlated with the rate of TCR–ligand binding, and depended on the frequency of serial TCR engagement (Figure 4.13; Romero, Maryanski and Luescher, 1993). In fact, when TCR was cross-linked with the ligand, sustained intracellular calcium mobilization, which is required for T cell activation (Ag binding), was completely abolished. In addition, by peptide modification or blocking of CD8, the affinity of ligand and TCR largely decreased, resulting in inducing Fas-dependent cytotoxicity but not perforin-dependent cytotoxicity or cytokine or production (Kessler *et al.*, 1998).

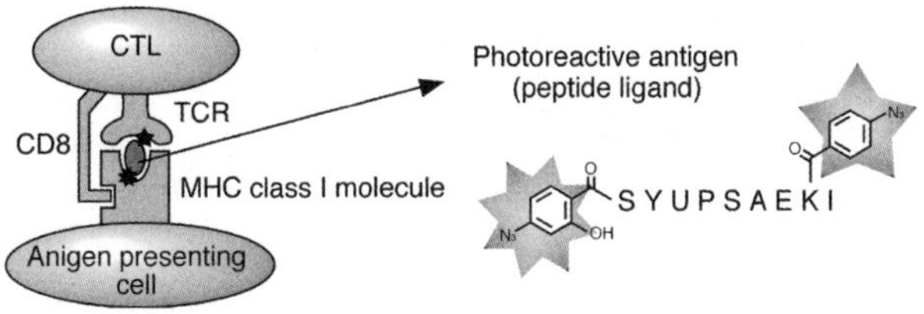

Figure 4.13 Usage of photoaffinity labelling technique for elucidation of TCR–antigen–MHC class I molecule interaction using living cells.

Chemokines are a large family of proinflammatory peptides and related to the recruitment and activation of leukocytes at the site of inflammation. Photoreactive (benzophenone photophore) derivatives of the 70-amino-acid chemokine macrophage inflammatory protein-1α (MIP-1α) were prepared and successfully cross-linked with the chemokine receptors CCR1 and CCR5 expressed on the CHO cells (Zoffmann *et al.*, 2001). Another photoreactive 23-amino acid cytokine was used for searching the target receptor using insect immune cells and found a single 190 kDa protein (Clark *et al.*, 2004).

4.8 Multifunctional Photoprobes for Rapid Analysis and Screening

Introduction of the radioisotopic (RI) tag at the binding site of the target molecule is a conventional application of affinity labelling to analyse and determine the interaction at a molecular level. Instead of using RI, photoaffinity biotinylation of receptors is attracting, since biotin binds to avidin tetramer, which provides non-RI chemiluminescent detection of the labelled products. However, the limitation of the method relies rather on the purification of the labelled peptides. As is distinct from PCR technology for the genome analysis, there are few methods for amplifying the photoaffinity identified binding site peptides. For the sequence analysis of binding sites, the labelled peptide fragments should be separated from a large number of unlabelled fragments before the mass spectrometric analysis. Immunoprecipitation is an approach for fishing out the corresponding peptide fragments after digestion, but sometimes requires great skills of the manipulation, and it takes time for the preparation of anti-peptide antibodies to purify the labelled peptide. Recent development of diazirine-based biotinylated probes provides an efficient solution to give labelled products with high quality, since biotin binds to tetramer avidin with extremely high affinity ($K_d = 10^{-15}$ M) (Bayer and Wilchek, 1990). The diazirine photophore contributes to the formation of irreversible cross-link; the biotinylated fragments can be isolated by single-step purification from avidin matrix. The application of an *N*-acetylglucosamine photoprobe carrying biotinylated diazirine provided the first information regarding acceptor site peptides of β1,4-galactosyltransferase (Figure 4.14; Hatanaka, Hashimoto and Kanaoka, 1998). In this case, a

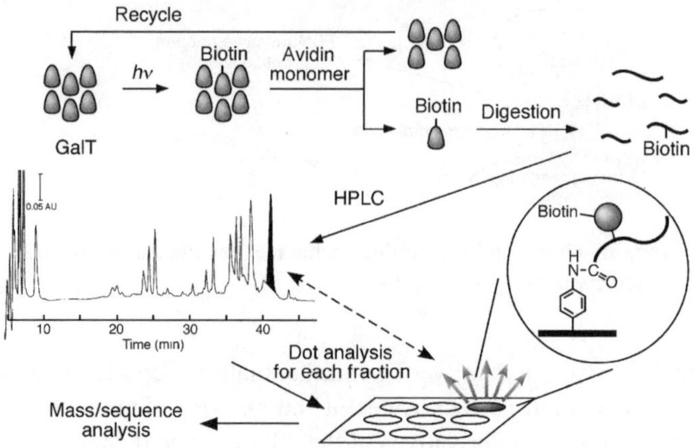

Figure 4.14 Multifunctional photoprobe for the rapid identification of a binding site.

simple but efficient detection system on solid surface has been developed for labelled peptide fragments coupled to the HPLC purification. Usually, it is quite difficult to know the peak of labelled fragments in HPLC after digestion, since the amounts are quite small compared with the amount of unlabelled fragments. Fractionated sample solutions by HPLC were subjected on a PVDF membrane and reacted with amino groups on the surface to immobilize them. When chemiluminescent detection is performed for each spot using avidin–HRP, only the spot including labelled (biotinylated) fragments should be emitted.

Although the biotin tag has shown great performance for reducing the time in the separation of labelled products, biotin–avidin binding is too stable to release trapped biotinylated components from an immobilized avidin matrix. The use of monomeric avidin contributes to the isolation of biotinylated products because of the lower affinity to biotin ($K_d = 10^{-8}$ M); however, it requires a high concentration of the biotinylated products for efficient trapping on the matrix (Schriemer and Li, 1996; Hashimoto and Hatanaka, 1999). Alternatively, biotinylated reagents with a scissile function have been considered for separation of the labelled proteins from a biotin–avidin complex. For displacing labile disulfide groups, (Shimkus, Levy and Herman, 1985), photoreactive (Olejnik *et al.*, 1995; Fang *et al.*, 1998), fluoride-sensitive (Fang and Bergstrom, 2003), alkali-sensitive, (Jahng *et al.*, 2003), enzyme-cleavable (Hashimoto *et al.*, 2004) and safe-catch (Park *et al.*, 2005) linkages have been developed. Cleavage of acylsufonamide groups known as safety-catch linkers needs an activation step by *N*-alkylation. Therefore, the linkage is very stable during probe synthesis, disulfide reducing conditions and photolysis, and then easily cleaved under mild activation conditions. Compared with a conventional heat denaturing method, the recovered samples did not contain a significant level of proteins nonspecifically adsorbed on the matrix (Figure 4.15).

Another application of biotin tag is rapid screening of molecules with high affinity (Morris *et al.*, 1998). SELEX (systematic evolution of ligands by exponential

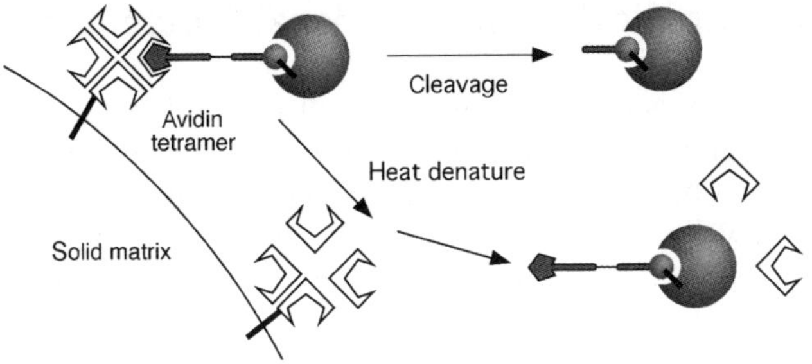

Figure 4.15 Efficient isolation of labelled molecules using cleavable linkage.

enrichment) methodology is applied to the selection of high-affinity oligonucleotides that bind specifically to any single protein of human red blood cell (RBC) membranes. The oligonucleotides with high affinity were isolated from a very large pool of random sequence molecules by reiterative rounds of selection and amplification. The selected oligonucleotides bearing biotin and cleavable azide derivative at the each end separately were incubated and irradiated with RBC ghost. The labelled membrane proteins were collected using streptavidin beads and isolated from the beads by dithiothraitol (DTT) treatment.

Also in response to the extensive study of proteins in proteomics, it was possible to use them for profiling interacting proteins of certain ligands and for screening of ligands/inhibitors for certain receptors. Photoaffinity labelling on a solid matrix can increase performance of throughput for rapid analysis as well as easy handling. An affinity-based screening system of inhibitors has been reported for glyceraldehyde-3-phosphate dehydrogenase (GAPDH) and Cibacron Blue 3GA as a typical ligand (Figure 4.16; Kaneda, Sadakane and Hatanaka, 2003). The GAPDH bearing a diazirine at the binding site was photochemically captured on the Cibacron Blue 3GA immobilized matrix whereas the capture was diminished when inhibitors were included in the solution. This method can be performed in a parallel fashion using a conventional 96 or 384 format for the efficient screening of inhibitors of target proteins. This will be a potential tool for the high-throughput affinity selection of active components from combinatorial ligand libraries.

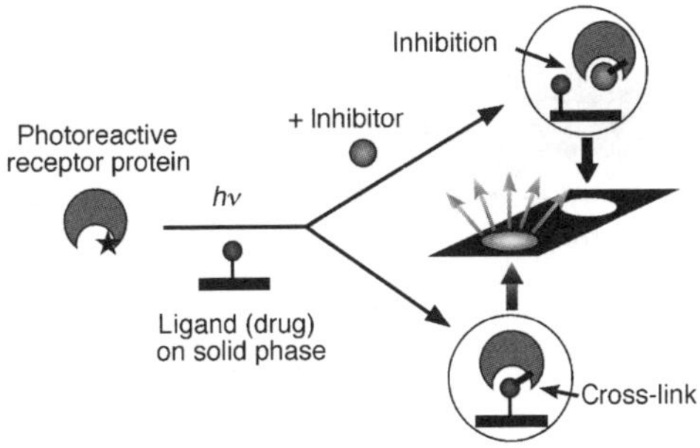

Figure 4.16 Affinity capture strategy for affinity-based drug screening on solid matrix.

4.9 Advanced Application to Functional Proteomics

Since proteomic analysis becomes a major target in bioinformatics, investigation of protein–protein interaction has been an inexhaustible challenge for chemical genomics. Chemical and biological approaches to studying multiprotein complexes and network have grown in popularity in a decade and given information on cellular responses via identification of binding proteins and visualization of the dynamic interaction, such as *in situ* imaging using fluorescence-labelled proteins, biopanning, quarts crystal microbalance (QCM) and atomic force microscope (AFM). Photocross-linking technique has been also applied to determine whish protein recognizes the target and how they fit together and change the affinity over the response. Photocross-linking label transfer technique has been employed for identifying the protein complexes. The reagents have heterobifunctional cross-linkers with a cleavable linkage connecting the two reactive functional groups (Figure 4.17). The method involves two steps of cross-linking at different conditions. A target protein chemically labelled with a heterobifunctional cross-linker at certain position was photochemically cross-linked to interacting molecules, and then a detection tag transferred onto the interacting molecules after the cleavage of the cross-link. It is basically valuable to identify proteins that interact weakly or transiently with the protein of interest. This strategy was applied to histone acetyltransferase (HAT, related to the regulation of gene expression) complexes (Brown *et al.*, 2001) and the pore-forming mechanism of intercellular membrane fusion using SASD-labelled calmodulin (CaM; Peters *et al.*, 2001). The latter indicated that the V0 sector of the vacuolar H^+-ATPase was the target of CaM, and the V0 *trans*-complex formed between docking and bilayer fusion made sealed channels, and expanded to form aqueous pores in a Ca^{2+}/CaM-dependent fashion. The cleavable linkage used in these cross-linkers is mainly the disulfide group; this means that the SH group remains at the interacting sites of these proteins after the cleavage reaction, and that may be useful for real-time analysis of protein interaction using post-labelling technique (Hamachi, Nagase and Shinkai, 2000).

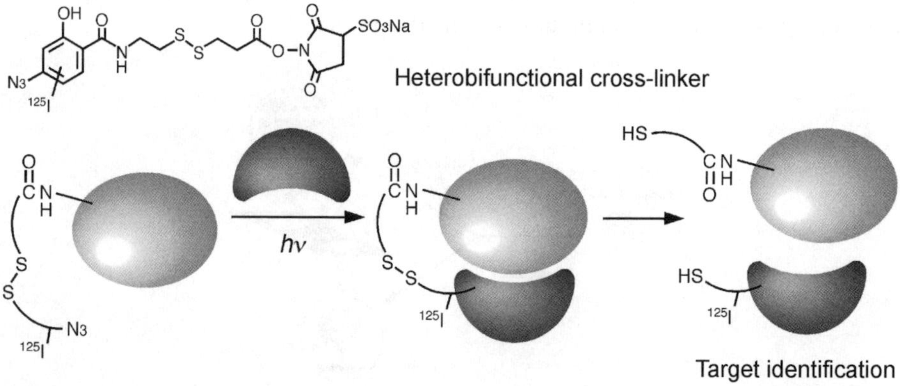

Figure 4.17 Label transfer method for the analysis of protein–protein interactions.

During the past two decades, the methodology of photoaffinity labelling has greatly advanced and has been applied extensively to the analysis of biosystems accompanying understanding of the photochemical and chemical properties of each photophore. The methods based on photoaffinity labelling will continue to be important tools for chemical genomics of the immune system. The method is a rapid and sensitive means to study structural aspects of biomacromolecules, dynamics of molecular interactions within the cell cycle, functional analysis of weak interactions, transcriptomics and proteomics at the cell surface as well as for the development of a rapid and reliable method of affinity-based ligand screening, leading to post-genomic drug discovery.

4.10 Summary

Chemical genomics/proteomics tools are widely used in the early stages of drug discovery. Although there is a large variety of techniques and their integration is still in progress, these approaches spread into other disciplines such as immunogenomics. The examples in the present account illustrate that there is a synergy between the methods, including microarray techniques, detection and miniaturization. In order to investigate rapidly the diversity and mechanism of the immune response and utilize the results in diagnosis and therapy, further improvements are needed in the areas of sensitivity, data-mining and throughput. In the first decade of the new millenium, rapid development is expected in immunology, and the chemical genomics/proteomics methods could certainly contribute to this trend.

References

Alcaro MC, Peroni E, Rovero P, Papini AM. 2003. Synthetic peptides in the diagnosis of HIV infection. *Curr Protein Pept Sci* **4**(4): 285–290.

Bayer EA, Wilchek M. 1990. Application of avidin-biotin technology to affinity-based separations. *J Chromatogr* **510**: 3–11.

Boldrick JC, Alizadeh AA, Diehn M, Dudoit S, Liu CL, Belcher CE, Botstein D, Staudt LM, Brown PO, Relman DA. 2002. Stereotyped and specific gene expression programs in human innate immune responses to bacteria. *Proc Natl Acad Sci USA* **99**(2): 972–977.

Brichory F, Beer D, Le Naour F, Giordano T, Hanash l S. 2001. Proteomics-based identification of protein gene product 9.5 as a tumor antigen that induces a humoral immune response in lung cancer. *Cancer Res* **61**: 7908–7912.

Brown EJ, Schreiber SL A. 1996. Signaling pathway to translational control. *Cell* **86**(4): 517–520.

Brown CE, Howe L, Sousa K, Alley SC, Carrozza MJ, Tan S, Workman JL. 2001. Recruitment of HAT complexes by direct activator interaction with the ATM-related Tra1 subunit. *Science* **292**: 2333–2337.

Brunner J. 1993. New photolabelling and crosslinking methods. *A Rev Biochem* **62**: 483–514.

Chin JW, Schultz PG. 2002. *In vivo* photocrosslinking with unnatural amino acid mutagenesis. *Chem Bio Chem* 2002; 1135–1137.

Clark KD, Garczynski SF, Arora A, Crim JW, Strand MR. 2004. Specific residues in plasmatocyte-spreading peptide are required for receptor binding and functional antagonism of insect immune cells. *J Biol Chem* **279**: 33246–33252.

Cravatt BF, Sorensen EJ. 2000. Chemical strategies for the global analysis of protein function. *Curr Opin Chem Biol* **4**: 663–668.

Darvas F, Dormán G, Guttman A (eds). 2004. *Chemical Genomics*. Marcel Dekker: New York.

Darvas F, Dormán G, Ürge L, Szabó I, Rónai Z, Sasváry-Székely M. 2001. Combinatorial chemistry. Facing the challenge of chemical genomics. *Pure Appl Chem* **73**: 1487–1498.

Falchetto R, Vorherr T, Brunner J, Carafoli E. 1991. The plasma membrane Ca^{2+} pump contains a site that interact with its calmodulin-binding domain. *J Biol Chem* **266**: 2930–2936.

Fall BI, Eberlein-Konig B, Behrendt H, Niessner R, Ring J, Weller MG. 2003. Microarrays for the screening of allergen-specific IgE in human serum. *Anal Chem* **75**(3): 556–562.

Fang S, Bergstrom DE. 2003. Fluoride-cleavable biotinylation phosphoramidite for 5′-end-labelling and affinity purification of synthetic oligonucleotides. *Nucl Acids Res* **31**: 708–715.

Fang K, Hashimoto M, Jockusch S, Turro NJ, Nakanishi K. 1998. A Bifunctional photoaffinity probe for ligand/receptor interaction studies. *J Am Chem Soc* **120**: 8543–8544.

Fleming SA. 1995. Chemical reagents in photoaffinity labelling. *Tetrahedron* **51**: 12479–12520.

Foucaud B, Perret P, Grutter T, Goeldner M. 2001. Cysteine mutants as chemical sensors for ligand–receptor interactions. *Trends Pharmac Sci* **22**: 170–173,

Gershon D. 2003. Probing the proteome, *Nature* **424**: 31 581–587.

Glokler J, Angenendt PJ. 2003. Protein and antibody microarray technology. *Chromatogr B Analyt Technol Biomed Life Sci* **797**(1–2): 229–240.

Gorelik L, Flavell RA. 2000. Abrogation of TGFbeta signaling in T cells leads to spontaneous T cell differentiation and autoimmune disease. *Immunity* **12**(2): 171–181.

Guthke R, Moller U, Hoffmann M, Thies F, Topfer S. 2005. Dynamic network reconstruction from gene expression data applied to immune response during bacterial infection. *Bioinformatics* (in press).

Hackler L, Dormán G, Puskás L, Darvas F, Ürge L. 2003. Development of chemically modified glass surfaces for nucleic acid, protein and small molecule microarrays. *Mol Div* **7**: 25–36.

Hamachi I, Nagase T, Shinkai S. 2000. A general semisynthetic method for fluorescent saccharide-biosensors based on a lectin. *J Am Chem Soc* **122**: 12065–12066.

Hashimoto M, Hatanaka Y. 1999. Identification of photolabelled peptides for the acceptor substrate binding domain of beta 1,4-galactosyltransferase. Chem Pharm Bull **47**: 667–671.

Hashimoto M, Okamoto S, Nabeta K, Hatanaka Y. 2004. Enzyme cleavable and biotinylated photoaffinity ligand with diazirine. *Bioorg Med Chem Lett* **14**: 2447–2450.

Hatanaka Y, Sadakane Y. 2002. Photoaffinity labelling in drug discovery and developments: chemical gateway for entering proteomic frontier. *Curr Top Med Chem* **2**: 271–288.

Hatanaka Y, Hashimoto M, Kanaoka Y. 1998. A rapid and efficient method for identifying photoaffinity biotinylated sites within proteins. *J Am Chem Soc* **120**: 453–454.

Hatanaka Y, Kempin U, Park J-J. 2000. One-step synthesis of biotinyl photoprobes from unprotected carbohydrates. *J Org Chem* **65**: 5639–5643.

Hill AV. 2001. Immunogenetics and genomics. *Lancet* **357**: 2037–2041.

Hiller R *et al.*, 2002. Microarrayed allergen molecules: diagnostic gatekeepers for allergy treatment. *FASEB J* **16**: 414–416.

Hohsaka T, Ashizuka Y, Taira H, Murakami H, Sisido M. 2001. Incorporation of nonnatural amino acids into proteins by using various four-base codons in an *Escherichia coli in vitro* translation system. Biochemistry **40**: 11060–11064.

Huang R-P. 2003. Cytokine antibody arrays: a promising tool to identify molecular targets for drug discovery. *Combin Chem High Throughput Screening* **6**(8): 769–775.

Huang R-P. 2004. Cytokine protein arrays. *Meth Mol Biol* **264**: 215–231.

Ivanov SS, Chung AS, Yuan ZL, Guan YJ, Sachs KV, Reichner JS, Chin YE. 2004. Antibodies immobilized as arrays to profile protein post-translational modifications in mammalian cells. *Mol Cell Proteomics* **3**(8): 788–795.

Jahng WJ, David C, Nesnas N, Nakanishi K, Rando RR. 2003. A cleavable affinity biotinylating agent reveals a retinoid binding role for RPE65. *Biochemistry* **42**: 6159–6168.

Johnson AE, Woodward WR, Herbert E, Menninger JR. 1976. *N*-Acetyllysine transfer ribonucleic acid: a biologically active analogue of aminoacyl transfer ribonucleic acids. *Biochemistry* **15**: 569–575.

Kaneda M, Sadakane Y, Hatanaka Y. 2003. A novel approach for affinity based screening of target specific ligands: application of photoreactive D-glyceraldehyde-3-phosphate dehydrogenase. *Bioconjugate Chem* **14**: 849–852.

Kauer JC, Erickson-Viitanen S, Wolfe HR Jr, DeGrado WF. 1986. *p*-Benzoyl-L-phenylalanine, a new photoreactive amino acid. *J Biol Chem* **261**: 10695–10700.

Kessler B, Hudrisier D, Schroeter M, Tschopp J, Cerottini J-C. and Luescher IF 1998. Peptide modification or blocking of CD8, resulting in weak TCR signaling, can activate CTL for Fas- but not Perforin-dependent cytotoxicity or cytokine production. *J Immunol* **161**: 6939–6946.

Kitamura N, Ikekita M, Sato T, Akimoto Y, Hatanaka Y, Kawakami H, Inomata M, Furukawa, K. 2005. Mouse Na^+/K^+-ATPase beta 1-subunit has a K^+-dependent cell adhesion activity for beta-GlcNAc-terminating glycans. *Proc Natl Acad Sci USA* **102**: 2796–2801.

Kotzyba-Hibert F, Kapfer I, Goeldner M. 1995. Recent trends in photoaffinity labelling. *Angew Chem Int Edn Engl* **34**: 1296–1312.

Le Naour F. 2001. Contribution of proteomics to tumour immunology. *Proteomics* **1**(10): 1295–1302.

Lolli F, Mazzanti B, Rovero P, Papini AM. 2003. Synthetic peptides in the diagnosis of neurological diseases. *Curr Protein Pept Sci* **4**(4): 277–284.

MacBeath G. 2002. Protein microarrays and proteomics. *Nat Genet* **32** (suppl): 526–532.

MacBeath G, Koehler AN, Schreiber SL. 1999. Printing small molecules as microarrays and detecting protein–ligand interactions en masse. *J Am Chem Soc* **121**: 7967–7968.

Marrack P, Kappler JJ, Kotzin BL. 2001. Autoimmune disease: why and where it occurs. *Nat Med* **7**(8): 899–905.

Meisenheimer KM, Koch TH. 1997. Photocross-linking of nucleic acids to associated proteins. Critical Review. *Biochem Mol Biol* **32**: 101–140.

Morris KN, Jensen KB, Julin CM, Weil M and Gold L. 1998. High affinity ligands from In Vitro selection: complex targets. *Proc Natl Acad Sci USA* **95**: 2902–2907.

Morse MA, Clay TM, Hobeika AC, Mosca PJ, Lyerly HK. 2001. Surrogate markers of response to cancer immunotherapy. *Expert Opin Biol Ther* **1**(2): 153–158.

Mosca PJ, Lyerly HK, Ching CD, Hobeika AC, Clay TM, Morse MA. 2003. Proteomics for monitoring immune responses to cancer vaccines. *Curr Opin Mol Ther* **5**(1): 39–43.

Musier-Forsyth K, Schimmel P. 1994. Acceptor helix interactions in a class II tRNA synthetase: photoaffinity cross-linking of an RNA miniduplex substrate. *Biochemistry* **33**: 773–779.

Olejnik J, Sonar S, Krzymanska-Olejnik E, Rothschild KJ. 1995. Photocleavable biotin derivatives: A versatile approach for the isolation of biomolecules. *Proc Natl Acad Sci* USA **92**: 7590–7594.

Oltvai ZN, Barabasi AL. 2002. Systems biology. Life's complexity pyramid. *Science* **25**(298): 763–764.

Owens J. 2004. Stuart Schreiber: biology from a chemist's perspective, *Drug Discov Today* **9**: 299–303.

Papini AM. 2005. Simple test for multiple sclerosis. *Nat Med* **11**(1): 13.

Park J-J, Sadakane Y, Masuda K, Tomohiro T, Nakano T, Hatanaka Y. 2005. Synthesis of diazirinyl photoprobe carrying a novel cleavable biotin. *Chem Bio Chem* **6** (in press).

Peters C, Bayer MJ, Bühler S, Andersen JS, Mann M, Mayer A. 2001. *Trans*-Complex formation by proteolipid channels in the terminal phase of membrane fusion. *Nature* **409**: 581–588.

Petrovsky N, Schönbach C, Brusic V. 2003. Bioinformatic strategies for better understanding of immune function. *In Silico Biol* **3**: 0034.

Potter DJ, Lampe DP, Roth BM. 2005. Antibody arrays for colon cancer screening (in press).

Romero P, Maryanski JL, Luescher IF. 1993. Photoaffinity labelling of the T cell receptor on living cytotoxic T lymphocytes. *J Immunol* **150**: 3825–3831.

Roopenian D, Choi EY, Brown A. 2002. The immunogenomics of minor histocompatibility antigens. *Immunol Rev* **190**(1): 86–95.

Rudd PM, Elliott T, Cresswell P, Wilson IA, Dwek RA. 2001. Glycosylation and the immune system. *Science* **291**: 2370–2376.

Sadakane Y, Hatanaka Y. 2004. Multifunctional photoprobes for rapid protein identification. In *Chemical Genomics: Advances in Drug Discovery and Functional Genomics Applications*, Darvas F, Guttman A, Dormán G (eds). Marcel Dekker: New York; 199–214.

Schäfer H-J, Schuhen A. 1996. Photoaffinity labelling and photoaffinity crosslinking of enzymes. *Biol Res* **29**: 31–46.

Schriemer DC, Li L. 1996. Combining avidin–biotin chemistry with matrix-assisted laser desorption/ionization mass spectrometry. *Anal Chem* **68**: 3382–3387.

Sheehan KM, Calvert VS, Kay EW, Liu Y, Fishman DS, Espina V, Aquino J, Speer R, Araujo R, Mills GB, Liotta LA, Petricoin EF, Wulfkuhle JD. 2005. Use of reverse phase protein microarrays and reference standard development for molecular network analysis of metastatic ovarian carcinoma. *Mol Cell Proteomics*. (in press)

Shimkus M, Levy J, Herman T. 1985. A chemically cleavable biotinylated nucleotide: Usefulness in the recovery of protein–DNA complexes from avidin affinity columns. *Proc Natl Acad Sci USA* **82**: 2593–2597.

Shu S, Plautz GE, Krauss JC, Chang AE. 1997. Tumour immunology. *JAMA* **278**: 1972–1981.

Stoll D, Bachmann J, Templin MF, Joos TO. 2004. Microarray technology: an increasing variety of screening tools for proteomic research. *Targets* **3**: 24–32.

Strømgaard K, Saito DR, Shindou H, Ishii S, Shimizu T, Nakanishi K. 2002. Ginkgolide derivatives for photolabelling studies: preparation and pharmacological evaluation. *J Med Chem* **45**: 4038–4046.

Tate JJ, Persinger J, Bartholomew B. 1998. Survey of four different photoreactive moieties for DNA photoaffinity labelling of yeast RNA polymerase III transcription complexes. *Nucl Acids Res* **26**: 1421–1426.

Taussig MJ, Landegren U. 2003. Progress in antibody arrays. *Targets* **2**(4): 169–176.

Templin MF, Stoll D, Bachmann J, Joos TO. 2004. Protein microarrays and multiplexed sandwich immunoassays: what beats the beads? *Combin Chem High Throughput Screen* **7**(3): 223–229.

Ueda H. 2003. Association of the T-cell regulatory gene CTLA4 with susceptibility to autoimmune disease. *Nature* **423**: 506–511.

Utkin YN, Krivoshein AV, Davydov VL, Kasheverov IE, Franke P, Maslennikov IV, Arseniev AS, Hucho F, Tsetlin VI. 1998. Labelling of *Torpedo californica* nicotinic acetylcholine receptor subunits by cobratoxin derivatives with photoactivatable groups of different chemical nature at Lys23. *Eur J Biochem* **253**: 229–235.

Uttamchandani M, Walsh DP, Yao SQ, Chang YT. 2005. Small molecule microarrays: recent advances and applications. *Curr Opin Chem Biol* **9**(1): 4–13.

Wang E, Panelli MC, Monsurro V, Marincola FM. 2004. Gene expression profiling of anticancer immune responses. *Curr Opin Mol Ther* **6**(3): 288–295.

Welters MJ, Filippov DV, van den Eeden SJ, Franken KL, Nouta J, Valentijn AR, van der Marel GA, Overkleeft HS, Lipford G, Offringa R, Melief CJ, van Boom JH, van der Burg SH, Drijfhout JW. 2004. Chemically synthesized protein as tumour-specific vaccine: immunogenicity and efficacy of synthetic HPV16 E7 in the TC-1 mouse tumour model. *Vaccine* 23(3): 305–311.

Wilson DS, Nock S. 2003. Recent developments in protein microarray technology. *Angew Chem Int Edn Engl* **42**(5): 494–500.

Yang J, Clark AE, Kozka IJ, Cushman SW, Holman GD 1992. Development of an intracellular pool of glucose transporters in 3T3-11 cells. *J Biol Chem* **267**: 10393–10399.

Yang S-W, Nash HA. 1994. Specific photocrosslinking of DNA-protein complexes: identification of contacts between integration host factor and its target DNA. *Proc Natl Acad Sci USA* **91**: 12183–12187.

Zoffmann S, Turcatti G, Galzi J-L., Dahl M, Chollet A. 2001. Synthesis and characterization of fluorescent and photoactivatable MIP-1 ligands and interactions with chemokine receptors CCR1 and CCR5. *J Med Chem* **44**: 215–222.

5

Genomic and Proteomic Analysis of Activated Human Monocytes

Ameesha Batheja, George Ho, Xiaoyao Xiao, Xiwei Wang and **David Uhlinger**

Abstract

We have used primary human monocytes as a model system for identification of novel or known genes differentially expressed in inflammation. Our approach combines classical cell biology with state of the art functional genomic, proteomic and analytical tools. Transcriptional profiling in cells and tissues is a valuable tool for following changes in gene expression in response to various biological and chemical effectors and models of pathophysiology. We have effectively used this approach to identify and elucidate the role of novel gene, Gene X, in inflammation. Owing to alternative splicing, post-translational modification and proteolytic processing, a single gene may yield multiple protein products with different functions/activities. We therefore performed proteomic profiling of activated primary human monocytes by capillary LC and quadropole/linear ion trap mass spectrometry. We demonstrate that genomics and proteomics coupled with bioinformatic capabilities can provide new dimensions to understanding complex biological systems such as inflammation.

This is an unprecedented time to study biological systems and signal transduction. Transcriptional profiling, functional genomic tools, sophisticated proteomic analysis and information from the human genome sequencing project have created a new paradigm for basic research. The availability of tools that combine informatics with genomics and proteomics allows us to look at a biological problem in many dimensions and from many different angles. With years of development and improvement, whole genome transcriptional profiling has evolved into a relatively mature tool for basic life science research in academic and drug discovery research in the pharmaceutical industry. Depending on the experimental system in question, one can either employ serial analysis of gene expression (SAGE) or cDNA- or oligo-based microarray chips to simultaneously observe changes in expression of thousands of genes. The advantage of using SAGE is ease of data comparison and ability to perform expression profiling

Immunogenomics and Human Disease Edited by András Falus

in organisms whose genomes have not been fully sequenced. Since SAGE requires the sequencing of thousands of concatemers in an experiment, its cost can be high and its high-throughput application limited. In comparison, microarrays are easier to use and more amenable to high-throughput applications (Polyak and Riggins, 2001). Bioinformatic tools have evolved to study and visualize extremely large and complex datasets, making microarrays an important tool for drug discovery research. There are many examples of successful application of this technology (Jain, 2004). In the near future, this technology will become an integral part of discovery research, making contributions to many steps along the drug discovery and development process.

5.1 Primary Human Monocytes, as a Model System

Bacterial infection followed by bacteraemia can result in septic shock, characterized by hypotension and multiorgan failure, resulting in death. Although staphylococci and *Candida* can also cause this condition, the majority of the septic shock cases are due to infection by Gram-negative bacilli (Bone, 1993; Paterson and Webster, 2000). Lipopolysaccharide (LPS), found in the cell wall of Gram-negative bacteria, potently stimulates inflammation, leading to septic shock. LPS associates with the circulating LPS binding protein (LBP) and, through the CD14–TLR4 complex in immune cells like monocytes and macrophages, signals the production of many pro-inflammatory cytokines such as IL-1β, tumour necrosis factor-α (TNF-α) and IL-6. LPS stimulates the production of chemokines such as IL-8 and monocyte chemotactic protein-1 (MCP-1), as well as induces endothelial cells to produce nitric oxide. The presence of all these inflammatory mediators is essential in orchestrating a balanced immune response (Dinarello, 1997; Gogos *et al.*, 2000; Rodriguez-Gaspar *et al.*, 2001; Boontham *et al.*, 2003).

In response to a chemotactic signal, such as MCP-1, and in a complex sequence of events, monocytes exit circulation in the blood vessels and are differentiated into macrophages. These primed macrophages reside in the tissues and, in response to secondary activation signals such as LPS through the CD14–TLR4 complex, become fully activated. The activated macrophages not only produce inflammatory cytokines and chemokines, but also produce anti-inflammatory cytokines such as IL-10, IL-18 and TGF-β. This is very important in generating a highly effective but tightly regulated immune response (Gordon, 1999). In a disease state, this regulation is often disrupted and discovering therapeutics that can help restore balance is the rationale for much current drug discovery research.

We have used primary human monocytes as a model system for identification of differentially expressed novel or known genes. The approach that we have taken combines classical cell biology with state-of-the-art analytical tools. Primary human monocytes were isolated from donated blood and treated with various pro-inflammatory agonists such as IL-1β, LPS, TNF-α, MCP-1 and PMA (phorbol myristate) + ionomycin for 4 h. As shown in Figure 5.1, we isolated cellular RNA and proteins, and used the culture supernatant to analyse secreted proteins.

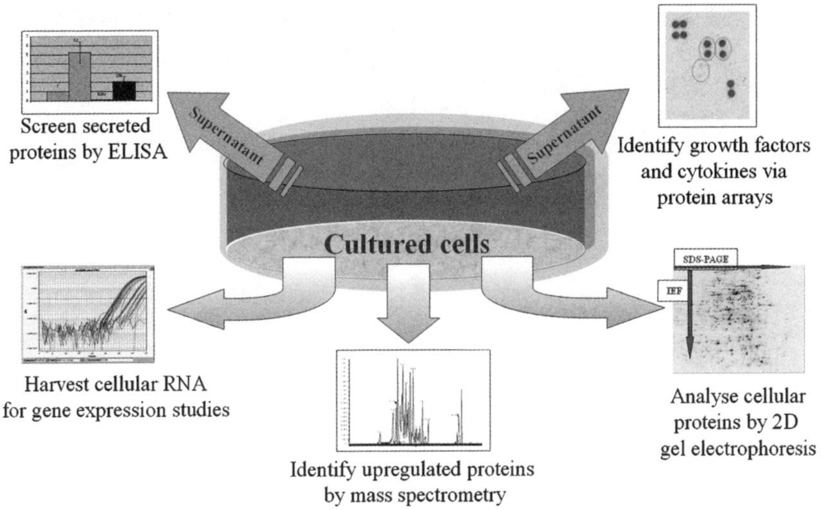

Figure 5.1 Analysis of activated human monocytes. Primary human monocytes were cultured and treated with various pro-inflammatory agonists. RNA was isolated from the cells for transcriptional profiling. Cellular and secreted proteins were investigated by 2D gel electrophoresis, mass spectrometry and ELISA as indicated in the figure.

5.2 Transcriptional Profiling of Activated Monocytes

Intrinsically, transcriptional profiling is a complex process involving many specialists, including platform engineers, therapeutic scientists, bioinformatics/informatics scientists and biostatisticians, contributing to experimental design, sample preparation, microarray chip processing, data handling, analysis and interpretation. Therefore it is not surprising to see a high degree of variation in different hands, primarily due to different chip platforms, different statistical rigour in design and data handling and interpretation. Moreover, there is no straightforward way to translate the high-throughput content of gene expression studies into reliable functional information without creating a downstream bottleneck. A multiplex approach is required to combine gene expression with information at other levels, such as proteomics and pathway information.

The RNA isolated from the activated, and control, monocytes was used for gene expression profiling using cDNA microarrays. The study was designed so that we could compare not only between the control and each agonist but also between different agonists. This comparison was performed using bioinformatic software such as OmniViz or Stanford tools. The microarray chip used in this study contains over 8000 clones representing over 5000 drug-able genes spanning many categories, including G-protein coupled receptors (GPCR), kinases, phosphatases, ion channels and membrane receptors. Hierarchical clustering was used to compare the expression changes of genes (rows) along the treatment conditions (columns), while a galaxy

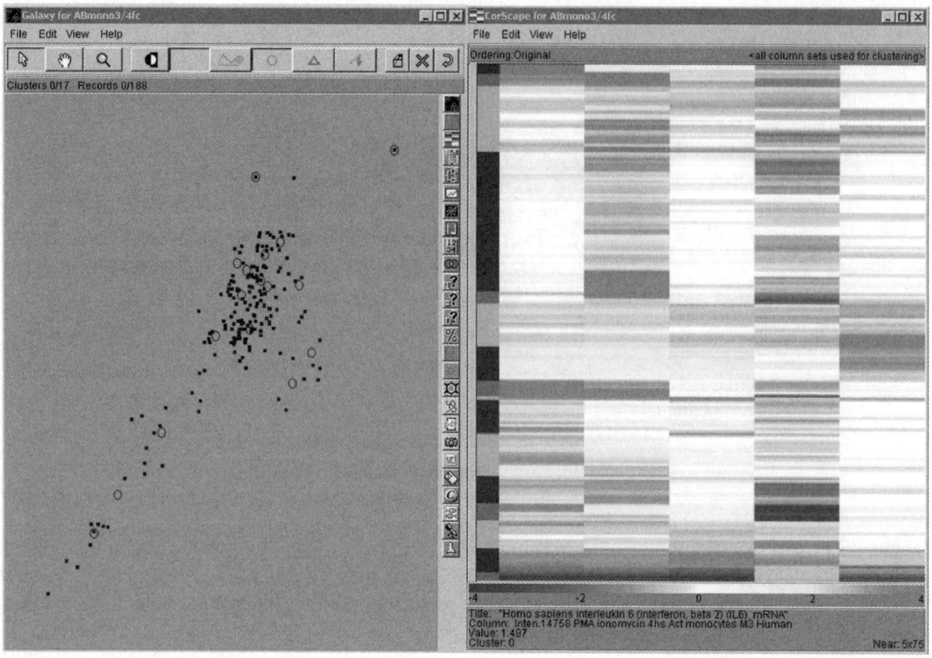

Figure 5.2 Monocyte transcriptional data analysis. A galaxy view of experimental samples and a heat map of monocyte genes. The left panel is the galaxy view mapping samples into the first two principle components. The right panel is the hierachical clustering of gene expression changes (rows) in different samples (columns). The colour scale is from 4-fold to –4-fold with red representing upregulation and blue representing downregulation. (A colour reproduction of this figure can be viewed in the colour plate section)

map based on the first two principle components was used to map experimental samples into two-dimensional space (Figure 5.2, left panel). We identified genes that were either upregulated or downregulated by at least 4-fold in comparison to the control. The results of this analysis are shown in the 'heat map' in the right panel of Figure 5.2. A novel gene was identified in the process and will be named 'Gene X' throughout this chapter. Gene X was found to be upregulated by about 8-fold in LPS-treated and PMA+ionomycin-treated monocytes. Its expression remained unaltered on treatment with TNF-α, MCP-1 and IL-1b. To ensure that the changes predicted by the microarray data were not artefacts, genes that were of interest, including Gene X, were analysed again in the context of the appropriate treatment using quantitative real-time polymerase chain reaction (PCR). PCR confirmation of differential gene expression predicted from microarray experiments is essential for many reasons, including limitations imposed by the specificity of the probes, accuracy of sequences being used on the chip and the calculation methods for fold changes (Kothapalli *et al.*, 2002).

Pathway tools help the integration of data from a large number of individual genes into biologically relevant information. This information can lead to understanding

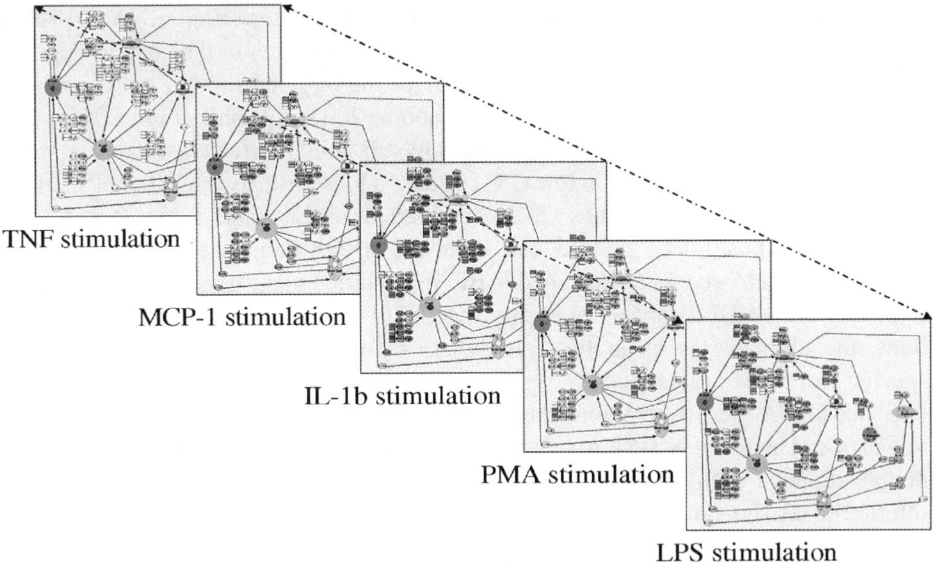

Figure 5.3 Cytokine signalling in activated monocytes. A cytokine network from BioCarta was used to create this series of snapshots of gene expression changes upon different experimental treatments.

and prediction of cellular responses. In our experiments, we were able to compare the effects of each treatment on different cellular pathways. Since some outcomes of agonists used in our study are already known, this tool was also important in determining whether the signature pathways were activated. For example, treatment of monocytes with LPS should upregulate the cytokine pathways, resulting in the expression of IL-6. When we linked the cytokine gene expression data from our experiments to the BioCarta cytokine pathway (Figure 5.3), we were able to demonstrate this effect. When data from each agonist was overlaid and only the cytokine pathway analysed, differences in signalling events leading to the induction of IL-6 could be identified.

Gene X overexpression has been correlated with LPS-induced IL-6 induction. To further understand the role of this gene in inflammation and gauge it as a target for drug discovery, we used the functional genomic tools discussed in the following section.

5.3 Functional Genomics

Following the identification of Gene X, we wanted to examine its role by either overexpressing or reducing gene expression. Classical tools require the generation of a cell line and finally an animal model where the gene of interest is deleted by homologous recombination. However, gene silencing can be accomplished by a phenomenon called RNA interference (RNAi). RNAi involves the targetting of specific mRNA using 19–21 nucleotide-long small interfering RNA (siRNA) sequences that engage naturally occurring enzyme complexes in the cells to degrade specific mRNA. As a consequence, the levels of the targeted protein are decreased, resulting in a functional knock-down (Dykxhoorn, 2003; Elbashir *et al.*, 2001). The rules for designing a good siRNA sequence are still being perfected. However, some rules recently suggested include low G/C content, a bias towards low internal stability at the sense strand 3′-terminus, lack of inverted repeats and sense strand base preferences at positions 3, 10, 13 and 19 in the siRNA sequence (Reynolds *et al.*, 2004). The siRNA sequence designed is BLAST queried against a sequence database (typically the NCBI database) to ensure specificity. However, despite all the precautions, the ultimate test is in the cells where off-target effects, toxicity and gene specific responses are gauged.

Using siRNA prediction algorithms, we designed a 19-nucleotide sequence specific to Gene X. The Gene X siRNA sequence was designed to contain a hairpin once the double-stranded RNA molecule was expressed in the cell. These sequences are referred to as short-hairpin RNA (shRNA). The Gene X shRNA was cloned into a plasmid vector called psilencer 1.0 (from Ambion) containing the human U6 promoter. Before such a knock-down construct can be used in a functional assay, it has to be validated for RNA as well as protein knock-down. Validation depends on many RNA and protein expression conditions such as abundance, stability and turnover rates for the molecule. Often, validation is done in relatively more artificial systems, where the gene is overexpressed either transiently or stably and the silencing ability of the shRNA construct/ sequence is determined.

Endogenous cellular levels of Gene X protein have been difficult to detect using western blotting, making validation of the Gene X–psil construct difficult in primary human monocytes or monocytic cell lines. Therefore, a construct encoding glutathione *S*-transferase (GST)-tagged Gene X was designed. The ability of Gene X shRNA construct (Gene X–psil) to knock down protein expression was tested by co-transfecting the GST-tagged Gene X expression vector along with Gene X–psil. Protein extracts from these cells were analysed with anti-Gene X antibodies to confirm protein knock-down. The Gene X–psil construct was then transfected into primary human monocytes and the transfected cells were challenged with LPS. Since the literature suggested that Gene X overexpression resulted in an increase in LPS-induced IL-6 expression, we used IL-6 secretion as the first functional assay for validating Gene X knock-down. Using ELISA we observed a 50–75 per cent decrease in IL-6 production in cultures transfected with the Gene X–psil construct.

Since IL-6 is one of many cytokines and chemokines produced by activated monocytes, the effect of Gene X knock-down on other inflammatory mediators was analysed (Figure 5.4). The availability of protein arrays allowed us to analyse multiple cytokines and chemokines in each sample simultaneously, saving us samples and time. Different arrays are categorized based on cell type (cytokines produced by a certain cell), pathway (NFkb pathway), function (apoptotic genes), etc. The cytokine array used in our experiments allowed us to analyse the differential expression of

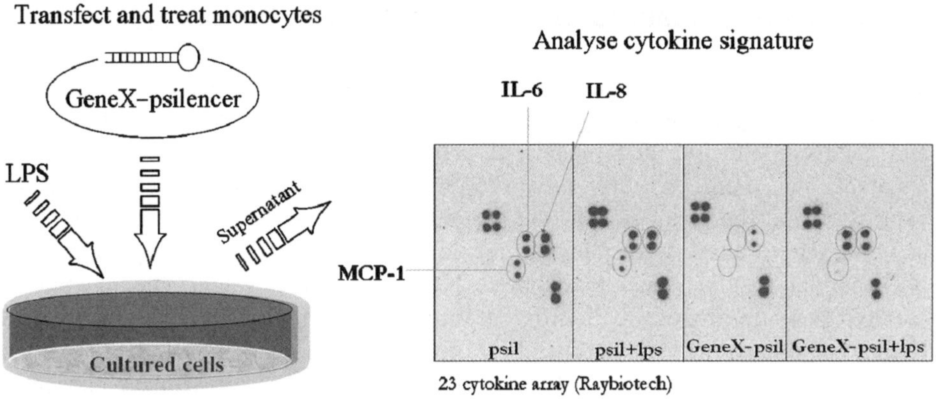

Figure 5.4 Cytokine profiling of human monocytes following Gene X knock-down. Following transfection of human monocytes with either the vector alone (psil) or Gene X–psil, in the presence or absence of LPS treatment, culture supernatants were harvested and profiled for differences in cytokine/chemokine secretion using 23 cytokine arrays (Copyright Ray Biotech Inc.).

23 cytokines/chemokines in each sample. We observed that, in cells transfected with Gene X-psil, the expression of IL-8 and MCP-1 was downregulated. On quantitation with ELISA, we found a 50–75 per cent decrease in the expression of both MCP-1 and IL-8.

Since primary human monocytes are extremely difficult to transfect, we have experimented with replication-deficient viral gene delivery systems such as adenoviruses and lentiviruses. The choice of a gene delivery system is crucial, especially when delivering siRNA or shRNA, because the knock-down sequence validated in one type of construct may not work as well in another vector. The reason for this discrepancy is not understood. Some advantages of lentiviruses over adenoviruses are the requirement of much lower virus-to-cell ratios, resulting in lower toxicity, and stable integration in the cell. These properties greatly facilitate the generation of cell lines stably overexpressing a gene or expressing shRNA. We observed that primary human monocytes and the pro-monocytic cell lines U937 and THP-1 could not be transduced with the tested adenoviral constructs, but were well transduced by lentiviruses. Thus, lentiviral vectors were the chosen gene delivery system and a lentiviral construct encoding Gene X was developed to generate a stable THP-1 clone overexpressing Gene X. This cell line will be a great tool in the identification and elucidation of Gene X functions.

Identifying proteins that interact with Gene X protein will be extremely valuable in understanding how Gene X functions. We have overexpressed Gene X with a 6 X His tag and purified it from *E. coli*. Through affinity coupling, this purified protein bound to magnetic beads can be used to probe for interacting proteins in cell lysates that have been induced by LPS. Furthermore, mass spectrometry allows us to identify these interacting proteins, both known and novel, with very high sensitivity.

5.4 Proteomic Analysis of Activated Human Monocytes

Transcriptional profiling of mRNA in cells and tissues has proven to be a useful tool for following changes in gene expression in response to a variety of biological and chemical effectors and models of pathophysiology. Recently it has become apparent that there is little direct correlation between differences in mRNA expression and the expression of the proteins they encode (Ideker *et al.*, 2001). In addition, due to alternative splicing, post-translational modification of proteins (e.g. phosphorylation, glycosylation) and proteolytic processing, a single gene may yield a variety of protein products with different functions/activities. Detection and analysis of these modifications may be critical for discovery of biomarkers and determination of their biological relevance. In light of this potential disparity we have chosen to examine a number of proteomic endpoints in addition to the transcriptional profiling presented above.

The most common proteomic platforms include protein arrays, two-dimensional gel electrophoresis (2D gels), ProteinChip arrays (Ciphergen) and liquid chromatography–tandem mass spectrometry or multidimensional liquid chromatography–tandem mass spectrometry (LC-MS/MS). Protein arrays consist of either 'native' protein bound to a support or capture antibodies directed against specific proteins/peptides. This approach is limited by the capture protein or antibody and relies upon prior knowledge of the biology of the system being studied. 2D gels are less biased but are relatively slow, labour-intensive and difficult to reproduce (Wagner *et al.*, 2002). The ProteinChip technology from Ciphergen is also a less biased approach and consists of performing chromatography on small surfaces then subjecting them to surface enhanced laser desorption ionization/time-of-flight mass spectrometry (SELDI-TOF). The resolution of this technology is best for proteins less than 20 kDa. The final platform, multidimensional liquid chromatography, in which separation mechanisms are orthogonal, along with MS/MS, have been widely used to identify proteins in complex protein digest mixtures (Wehr, 2003; Neverova and Eyk, 2005, in press). Two-dimensional (2D) peptide LC MS/MS combined with multiplexed iTRAQ labelling is an approach we are taking to provide higher-throughput proteomics allowing for both protein characterization and quantification.

In our study, proteomic profiling of activated primary human monocytes was performed by activating the cells for multiple time points, with different pro-inflammatory agonists, as described earlier in the chapter. At first, protein extracts from these samples were analysed using 2D gel analyses. Similar to microarrays, the dataset obtained was very large and complex. Comparisons were drawn between untreated samples and each agonist and also between the different agonists. In this study 26 spots were up/downregulated and 13 spots were further identified by matrix assisted laser desorption ionization (MALDI)/TOF or LC-MS/MS. Unlike transcription data, 2D gel electrophoresis data is difficult to discern due to the differences in mobility seen even between replicates. Also, the presence of some proteins in really high abundance can completely mask the detection of a potentially important but low-abundance protein.

We have developed a proteomic strategy utilizing iTRAQ, a new mass labelling reagent, and chromatographic supports. This paradigm entails relative quantitation and differential profiling of complex protein mixtures simultaneously. iTRAQ Reagent (Applied Biosystems) is a multiplexed set of four isobaric reagents which are amine specific and yield labelled peptides identical in mass, hence also identical in single MS mode, but produce strong, diagnostic, low-mass MS/MS signature ions allowing quantitation in up to four different samples simultaneously. In this example, human monocytes were activated with LPS for different lengths of time (0, 4, 10 and 20 h) and whole cell lysates were harvested for analysis. Each sample was independently reduced, alkylated, trypsin digested and labelled with iTRAQ Reagents 114, 115, 116 and 117, respectively. The labelled samples from different time points were then mixed and fractionated by strong cation exchange chromatography in the first dimension into eight fractions (0 to 400 mM KCl in 16 min), followed by C_{18} reverse-phase chromatography in the second dimension (5–40 per cent acetonitrile containing 0.1 per cent formic acid in 60 min). Both protein identification and relative quantification were performed using MS/MS (4000 QTRAP), Pro Quant software, Mascot search engine and NCBInr database. The experimental flow chart showing parallel workflow employed using 2D LC MS/MS and the iTRAQ Reagents is shown in Figure 5.5. Our analysis led us to a total of 120 proteins with scores greater than 55 ($p < 0.05$). Each protein has a specific accession number in the database. Twelve proteins identified on 2D gel analysis (described earlier) were part of the 2D LC-MS/MS dataset. Differential expression was indicated by the intensity ratio among *m/z*

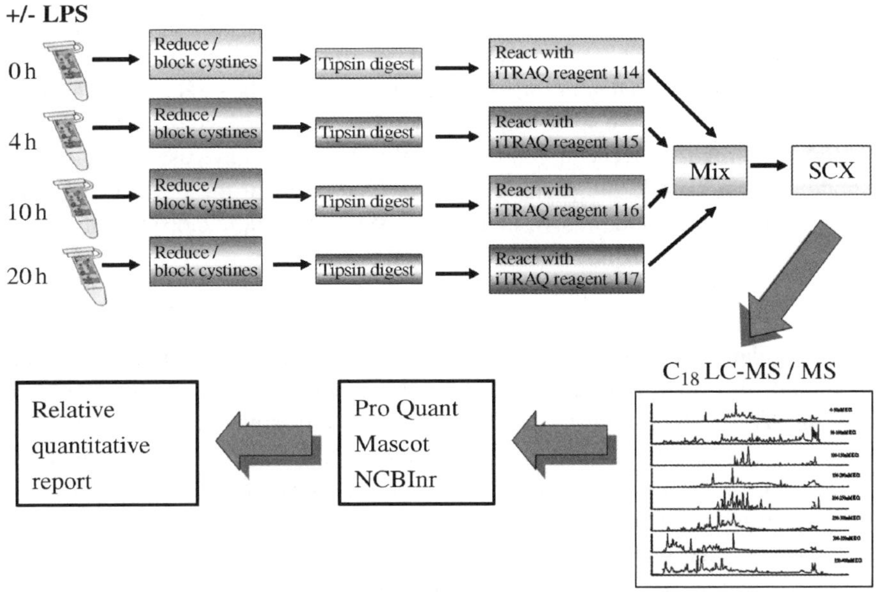

Figure 5.5 2D LC MS/MS experimental protocol. Flow chart showing the parallel workflow employed when using 2D LC MS/MS and the iTRAQ mass labelling reagents.

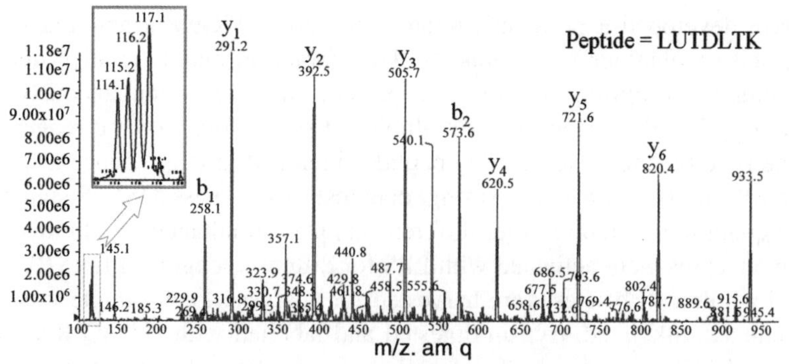

Figure 5.6 Analytic mass spectrometry of LPS-activated human monocytes. A typical MS/MS spectrum of a multiplexed iTRAQ reagent-labelled LPS-activated monocyte sample. The enlarged part shows diagnostic reporter ions at *m/z* 114, 115, 116 and 117.

114, 115, 116 and 117. This information included proteins that were upregulated or downregulated upon treatment or treatment time. A typical MS/MS spectrum of an LPS activated monocyte sample labelled with the multiplexed iTRAQ Reagent is shown in Figure 5.6. The inset panel shows diagnostic reporter ions at *m/z* 114, 115, 116 and 117, suggesting an increase in the expression of the protein with LPS treatment duration.

The availability of all the tools and technologies today has changed the speed and capability of drug discovery groups throughout the pharmaceutical industry. Functional genomics and proteomics platforms complement as well as strengthen each other. The presented study is a good example of this paradigm. On one hand, using microarrays and functional genomics in combination with classical molecular biology and cell biology tools, we have explored a relatively novel gene and analysed its potential as a drug discovery target. On the other hand, we have used state-of-the-art methodologies to profile the proteome of a primary cell in the presence of various agonists it might encounter *in vivo* in a disease state. A complete discussion of all the data we have collected is beyond the scope of this chapter.

In conclusion, we have employed a variety of methodologies to study inflammation in human monocytes. By transcriptional profiling we identified a number of novel genes upregulated in response to pro-inflammatory agonists, confirmed these findings by quantitative reverse transcriptase PCR and presented data for one of these genes. Using siRNA directed against GeneX or by overexpressing GeneX monocytic cell lines and primary human monocytes, we have analysed the role of Gene X in inflammation. To address the functional consequences of these perturbations we employed arrays to examine secreted cytokines from these cells in the presence or absence of agonists. Interestingly we observed not only a decrease in IL6 production but also IL8 and MCP1 in the knock-down experiments (see Figure 5.4). In parallel with the transcriptional profiling and cytokine arrays, we have examined the proteome in activated monocytes by 2D LC-MS/MS and iTRAQ labelling (see

Figures 5.5 and 5.6). Similar approaches used in the study of other disease systems and drug discovery can be easily envisioned.

References

Bone RC. 1993. Gram-negative sepsis: a dilemma of modern medicine. *Clin Microbiol Rev* 6–57.

Boontham P, Chandran P, Rowlands B, Eremin O. 2003. Surgical sepsis: dysregulation of immune function and therapeutic implications. *Surgeon* **1**: 187–206.

Dinarello CA. 1997. Proinflammatory and anti-inflammatory cytokines as mediators in the pathogenesis of septic shock. *Chest* **112**: 321S–329S.

Dykxhoorn DM. C. D. N. P. A. S. 2003. Killing the messenger: short RNAs that silence gene expression. *Nat Rev M Cell Biol* **4**: 457–467.

Elbashir SM, Harborth J, Lendeckel W, Yalcin A, Weber K, Tuschl T. 2001. Duplexes of 21-nucleotide RNAs mediate RNA interference in cultured mammalian cells. *Nature* **411**: 494–498.

Gogos CA, Drosou E, Bassaris HP, Skoutelis A. 2000. Pro- versus anti-inflammatory cytokine profile in patients with severe sepsis: a marker for prognosis and future therapeutic options. *J Infect Dis* **181**: 176–180.

Gordon S. 1999. Macrophages and the immune response. In *Fundamental Immunology*, Paul WE. (ed.) Lippincott-Raven: Philadelphia, PA; 533–545.

Ideker T, Thorsson V, Ranish JA, Christmas R, Buhler J, Eng JK, Bumgarner R, Goodlett DR, Aebersold R, Hood L. 2001. Integrated genomic and proteomic analyses of a systematically perturbed metabolic network. *Science* **292**: 929–934.

Jain KK. 2004. Applications of biochips: from diagnostics to personalized medicine. *Curr Opin Drug Discov Devl* **7**: 285–289.

Kothapalli R, Yoder SJ, Mane S. Loughran TP, Jr. 2002. Microarray results: how accurate are they? *BMC Bioinformatics* **3**: 22.

Neverova I, and Eyk, JEV. 2005. Role of chromatographic techniques in proteomic analysis. *J Chromatogr* (in press).

Paterson RL, Webster NR. 2000. Sepsis and the systemic inflammatory response syndrome. *J R Coll Surg Edinb* **45**: 178–182.

Polyak K, Riggins GJ. 2001. Gene discovery using the serial analysis of gene expresssion technique: implications for cancer research. *J Clin Oncol* **19**: 2948–2958.

Renu A, Heller MS, Chai A, Shalon D, Bedilion T, Gilmore J, Woolley DE, Davis RW. 1997. Discovery and analysis of inflammatory disease-related genes using cDNA microarrays. *Proc Natl Acad Sci USA* **94**: 2150.

Reynolds A, Leake D, Boese Q, Scaringe S, Marshall WS. Khvorova A. 2004. Rational siRNA design for RNA interference. *Nat Biotechnol* **22**: 326–330.

Rodriguez-Gaspar M, Santolaria F, Jarque-Lopez A, Gonzalez-Reimers E, Milena A, de la Vega MJ, Rodriguez-Rodriguez E, Gomez-Sirvent JL. 2001. Prognostic value of cytokines in SIRS general medical patients. *Cytokine* **15**: 232–236.

Wagner K, Miliotis T, Marko-Varga G, Bischoff R, Unger KK. 2002. An automated on-line multidimensional HPLC system for protein and peptide mapping with integrated sample preparation. *Anal Chem* **74**: 809–820.

Wehr T. 2003. Multidimensional liquid chromatography in proteomic studies. *LC.GC Europe* 2–8.

6

Bioinformatics as a Problem of Knowledge Representation: Applications to Some Aspects of Immunoregulation

Sándor Pongor and **András Falus**

Abstract

Bioinformatics uses a variety of models that fall into three broad categories such as linguistic, three-dimensional and interaction network models. Although the latter allow one to capture interactions among molecules and other cellular components, the underlying representations are predominantly static. The main molecular mechanisms of immunology, such as VJD recombination, cellular and molecular networks and somatic hypermutations, cannot be and are not adequately covered in current molecular databases. Other aspects, such as the maturation of single, monospecific immune response or that of immunological memory, apparently fall outside the scope of current molecular representations. The complexity of immunological regulation, such as the polarized T cell cytokine web, Treg subpopulation, idiotypic networks, etc., calls for a new generation of computational approach, leading to a new age of immunoinformatics ('immunomics').

6.1 Introduction

The growing network of biomedical databases and analysis programs constitutes one of the most sophisticated knowledge representation tools mankind has ever built. Bioinformatics differs from other informatics applications not so much in the amount of data but rather in the complexity and the depth of knowledge it communicates. As an example, bioinformatics deals with a wealth of molecular representations, such as

Immunogenomics and Human Disease Edited by András Falus

sequences, three-dimensional structures, symbolic diagrams (e.g. hydrophobicity plots, helical wheel diagrams), as well as a variety of group-wise representations such as multiple alignments, metabolic pathways, phylogenetic trees, etc., most of which could not have been conceived without computerized methods. It is customary to define bioinformatics as the informatics of biological data, but in fact it is not, or not exclusively, a specialized branch of science: it is rather a general approach to all life sciences that makes it possible to study problems previously inaccessible to systematic research. This aspect – the access to new domains of knowledge – is one of the common themes that link the current age of computerized resources to previous innovations in storing and representing information, and as representations of information are at the very heart of cultural evolution, it is appropriate to introduce our subject within a historic context.

Complexity leaps in evolution are known to be powered by improvements in the way genetic information is stored and transmitted (Szathmáry and Smith, 1995; Maynard Smith and Szathmáry, 1995). By analogy, one can point out that major improvements in scientific knowledge representation are correlated with innovations in the way information is shared within human societies, such as the appearance of writing, printing and the internet. In a traditional society, knowledge is exchanged mainly by repeated, face-to-face communication and is confirmed and stored by an entire community. Writing not only decoupled knowledge transfer from personal communication but it also created a powerful new medium for the storage and manipulation of complex symbols whose interpretation required, at the same time, an increased intellectual effort from the recipient. Few would doubt that the widespread use of written and especially printed information has been a prerequisite of the modern science that characterizes industrial societies. The paradigmatic knowledge source of this period is the *encyclopaedia*, an organized, searchable knowledge base that is, in many respects, the predecessor of current electronic databases.

The current age of bioinformatics is characterized by vast amounts of biological data collected by computerized methods and distributed via the internet and stored in electronic databases (Table 6.1). Knowledge in electronic databases is represented and transferred in ways that are radically different from those known before. While readers can directly interpret printed text, electronic databases can only be 'read' with the mediation of computer programs. Programs carry a large amount of implicit information in themselves that is not always transparent to the user. For instance, in order to draw a three-dimensional picture of a protein molecule from an input of atomic coordinates, a program needs to know how the atoms of various types of amino acids are connected to each other. This kind of implicit information represents an intermediate layer between the data and the program and is often organized into *ontologies*, i.e. formal sets of definitions and rules that are valid for the data domain (Table 6.2). Also, there are conspicuous changes in the way scientific information is confirmed. In the age of printed information, the quality of scientific discoveries was guaranteed by authoritative scientific societies, by the peer review of scientific

Table 6.1 Main types of bioinformatics databases[a]

Nucleotide sequence databases
International Nucleotide sequence database collaboration
Coding and noncoding DNA
Gene structure, introns and exons, splice sites
Transcriptional regulator sites and transcription factors
RNA sequence databases
Protein sequence databases
General sequence databases
Protein properties
Protein localization and targeting
Protein sequence motifs and active sites
Protein domain databases; protein classification
Databases of individual protein families
Structure databases
Small molecules
Carbohydrates
Nucleic acid structure
Protein structure
Genomics databases (nonvertebrate)
Genome annotation terms, ontologies and nomenclature
Taxonomy and identification
General genomics databases
Viral genome databases
Prokaryotic genome databases
Unicellular eukaryotes genome databases
Fungal genome databases
Invertebrate genome databases
Metabolic and signaling pathways
Enzymes and enzyme nomenclature
Intermolecular interactions and signalling pathways
Human and other vertebrate genomes
Model organisms, comparative genomics
Human genome databases, maps and viewers
Human open reading frames
Human genes and diseases
Model organisms, comparative genomics
Human genome databases, maps and viewers
Microarray data and other gene expression databases
Proteomics resources
Other molecular biology databases
Images of biological macromolecules
Bioremediation database
Drugs and drug design

(*Continued*)

Table 6.1 (*Continued*)

Molecular probes and primers
Organelle databases
Plant databases
General plant databases
Arabidopsis thaliana
Rice
Other plants
Immunological databases

[a]Based on the Database issue of *Nucleic Acids Research*, 2005; www3.oup.co.uk/nar/database/cat/12/

Table 6.2 Databases for annotation terms, onthologies and nomenclature used in bioinformatics

Genew the Human Gene Nomenclature Database, //www.gene.ucl.ac.uk/cgi-bin/nomenclature/searchgenes.pl
GO – Gene Ontology, www.geneontology.org/
GOA – Gene Ontology Annotation, www.ebi.ac.uk/GOA
IUBMB – nomenclature database for enzymes, www.chem.qmul.ac.uk/iubmb/
IUPAC – nomenclature database for organic and biochemistry, www.chem.qmul.ac.uk/iupac/
IUPHAR-RD – pharmacological nomenclature for receptors and drugs, www.iuphar-db.org/iuphar-rd/
PANTHER – gene products nomenclature, http://panther.celera.com/
STAR/mmCIF – an ontology for macromolecular structure, http://ndbserver.rutgers.edu/mmcif
UMLS – Unified Medical Language System (thesaurus, lexicon and semantic networks), http://umlsks.nlm.nih.gov

journals, and last but not least by the personal reputation of individual authors. In contrast, electronic databases are often produced by automated data collection, while textual annotations, such as the description of biological function, are added by anonymous teams of database annotators who often rely on computer-based prediction methods. In other words, the amount of data and the number of databases is growing while data quality is less transparent. However, there are signs of integration as well. Large efforts have been devoted to validating and interlinking biological data (sequences, structures), and, what is perhaps more important, highly complex scientific resources have been created wherein diverse data are controlled and accessed by uniform methods. In this setting, data integrity and quality will be increasingly controlled by autonomous agents, which will hopefully decrease the quality gap of current databases.

The first goal of this chapter is to provide an overview of how knowledge is represented in bioinformatics today and to show the cognitive roots of the underlying models. Bioinformatics is primarily concerned with the structure of protein and DNA molecules that fulfill functions in a series of interdependent systems such as pathways, cells, tissues, organs and organisms. This complex scenario can be best described with the concepts of systems theory.* Models in molecular biology are simplified systems that can be conveniently described in terms of entities and relationships (Pongor, 1988). We will deal with three main kinds of representational models, language-based models, three-dimensional models and networks, and briefly review the development of computational tools that operate on these models. The last sections of this chapter describe genomic and immunological resources. The second goal is to review the current knowledge on the information processing mechanism of the immune system. The immune system has specific algorithms for handling environmental information, and these mechanisms are among the most intensively researched and perhaps best understood phenomena in the life sciences that call for specific informatics approaches yet to appear.

6.2 Sequences and Languages

Language-based descriptions are broadly speaking those that use semantic definitions for entities and relationships. The term 'molecular biology' was independently coined by Warren Weaver and John Astbury in the 1930s, at a time when scientific methodology was dominated by linguistic theories (Wittgenstein, 1922; Carnap, 1939). Subsequent breakthroughs in information theory (Shannon, 1948a,b) and formal linguistics (Chomsky, 1957) all pointed towards a broad metaphoric context of language, communication and computation that provided the first framework within which genetics was discussed. Cryptography (Shannon, 1948c) and pattern recognition methods (Ripley, 1999), first developed within intelligence communities of the time, also contributed a great deal to the general view that biological sequences represent a code that carries information in a particular language, a metaphor reflected by such terms as the 'genetic code' or 'the book of life'.

The analysis of biological sequences first used the statistical tools developed to analyse character strings in linguistics (Konopka, 1994), and many of the first methods, such as those concerned with the string complexity, became standard bioinformatics tools in the later years. From the 1980s, as bioinformatics became part of laboratory routine, pattern recognition methodologies that used similarity measures and classification algorithms proved to be of immediate interest, and with the onset of the genomic era, heuristic methods of searching biological databases such as BLAST (Altschul *et al.*, 1990) became the most frequently used algorithms not only within bioinformatics, but allegedly in the entire field of scientific computing.

Margaret Dayhoff and her colleagues at the national Biomedical Research Foundation (NBRF), Washington, DC created the first sequence database in the 1960s, an Atlas of protein sequences organized into families and superfamilies, and their collection centre eventually became known as the PIR resource. Collections of

* According to systems theory, a system is a group of interacting elements functioning as a whole and distinguishable from its environment by recognizable boundaries (Csányi, 1989; Kampis, 1991). Molecules can be regarded as such systems. Generally speaking, structure is fixed state of a system, and the study of a system usually starts with its characteristic structures that are recurrent in space or time. Function on the other hand is not a property of the system, rather a role that the system plays in the context of a higher system.

DNA sequences (Table 6.1), started at the European Molecular Biology Laboratory (EMBL, Heidelberg) at Los Alamos National Laboratory (New Mexico) and DDBJ, Japan, gained importance with the spreading of productive DNA sequencing technologies. Initially, sequence records included only the sequence filename. These were eventually expanded to include annotation information such as references, function, regulatory sites, exons and introns, modified amino acids, protein domains, etc. The Swiss-Prot collection of protein sequences is an especially good example of a well-annotated sequence database wherein a uniform syntax was developed for annotations. Very soon, separate so-called secondary collections were created for annotated segments, such as the first protein domain sequence database (Pongor *et al.*, 1993), as well as for post-translational modifications, functional annotations, etc. (Table 6.3). Development of such secondary databases provided an important entry

Table 6.3 Examples of protein sequence databases

Primary protein sequence resources	
Uniprot/Swiss-Prot – annotated protein sequence db (University of Geneva, EBI)	www.expasy.org/sprot/
Uniprot/Trembl – computer annotated protein sequences (EBI)	www.ebi.ac.uk/trembl/
Uniprot/PIR – annotated protein sequences (Georgetown University)	http://pir.georgetown.edu/
Uniprot (Universal Protein database, Swiss-Prot + PIR + TREMBL)	www.expasy.uniprot.org/
Secondary protein sequence resources	
COG – clusters of orthologous groups of proteins	www.ncbi.nlm.nih.gov/COG
CDD – conserved domain database	www.ncbi.nlm.nih.gov/Structure/cdd/cdd.shtml
PMD – protein mutant database	http://pmd.ddbj.nig.ac.jp/
InterPro – integrated resources of proteins domains and functional sites	www.ebi.ac.uk/interpro/
PROSITE – PROSITE dictionary of protein sites and patterns	www.expasy.org/prosite/
BLOCKS – BLOCKS database	www.blocks.fhcrc.org/
Pfam – protein families database (HMM derived) [Mirror at St Louis (MO,USA)]	www.sanger.ac.uk/Pfam/ http://genome.wustl.edu/Pfam/
PRINTS – protein motif fingerprint database	http://bioinf.man.ac.uk/dbbrowser/PRINTS/
ProDom – protein domain database (automatically generated)	http://protein.toulouse.inra.fr/prodom.html
PROTOMAP – hierarchical classification of proteins	http://protomap.stanford.edu/
SBASE – SBASE domain database	www3.icgeb.trieste.it/~sbasesrv/
SMART – simple modular architecture research tool	http://smart.embl-heidelberg.de/
TIGRFAMs – TIGR protein families database	www.tigr.org/TIGRFAMs/
BIND – biomolecular interaction network database	www.bind.ca/
DIP – database of interacting proteins	http://dip.doe-mbi.ucla.edu/
MINT – molecular interactions	http://cbm.bio.uniroma2.it/mint/
ProNet – protein–protein interaction database	http://pronet.doubletwist.com/

for specialized information, and current databases such as PFAM are excellent examples of this tendency. An important step was the application of internet technology for cross-referencing the major databases with each other and subsequently with bibliographic database.

Information contained in current sequence databases (Tables 6.3 and 6.4) can be best pictured as an annotated sequence, a linear string of characters to which additional items of information are linked either as an added field within the same

Table 6.4 Examples of DNA sequence databases

Primary DNA sequence resources	
EMBL – EMBL nucleotide sequence database (EBI)	www.ebi.ac.uk/embl/
Genbank – GenBank nucleotide Sequence database (NCBI)	www.ncbi.nlm.nih.gov/Genbank/GenbankSearch.html
DDBJ – DNA Data Bank of Japan	www.ddbj.nig.ac.jp/
dbEST – dbEST (expressed sequence tags) database (NCBI)	www.ncbi.nlm.nih.gov/dbEST/
dbSTS – dbSTS (sequence tagged sites) database (NCBI)	www.ncbi.nlm.nih.gov/dbSTS/
Secondary DNA sequence resources	
NDB – nucleic acid databank (3D structures)	http://ndbserver.rutgers.edu/NDB/ndb.html
BNASDB – nucleic acid structure database from University of Pune	http://202.41.70.55/www/net/deva.html
AsDb – aberrant splicing database	www.hgc.ims.u–tokyo.ac.jp/~knakai/asdb.html
ACUTS – ancient conserved untranslated DNA sequences database	http://pbil.univ–lyon1.fr/acuts/ACUTS.html
Codon Usage database	www.kazusa.or.jp/codon/
EPD – eukaryotic promoter database	www.epd.isb–sib.ch/
HOVERGEN – homologous vertebrate genes database	http://pbil.univ–lyon1.fr/db/hovergen.html
ISIS – intron sequence and information system	www.introns.com/
RDP – ribosomal database project	http://rdp.cme.msu.edu/html/
gRNAs database – guide RNA database	http://biosun.bio.tu–darmstadt.de/goringer/gRNA/gRNA.html
PLACE – plant *cis*-acting regulatory DNA elements database	www.dna.affrc.go.jp/htdocs/PLACE/
PlantCARE – plant *cis*-acting regulatory DNA elements database	http://sphinx.rug.ac.be:8080/PlantCARE/
sRNA database – small RNA database	http://mbcr.bcm.tmc.edu/smallRNA/smallrna.html
ssu rRNA – small ribosomal subunit database	http://rrna.uia.ac.be/rrna/ssu/
lsu rRNA – large ribosomal subunit database	http://rrna.uia.ac.be/rrna/lsu/
5S rRNA – 5*S* ribosomal RNA database	http://rose.man.poznan.pl/5SData/

(*Continued*)

Table 6.4 (*Continued*)

tmRNA website	www.indiana.edu/~tmrna/
tmRDB – tmRNA dB	http://psyche.uthct.edu/db/tmRDB/tmRDB.html
tRNA – tRNA compilation from the University of Bayreuth	www.uni-bayreuth.de/departments/biochemie/sprinzl/trna/
uRNADB – uRNA database	http://psyche.uthct.edu/dbs/uRNADB/uRNADB.html
RNA editing – RNA editing site	www.lifesci.ucla.edu/RNA
RNAmod database – RNA modification database	http://medstat.med.utah.edu/RNAmods/
SOS-DGBD – Db of *Drosophila* DNA annotated with regulatory binding sites	http://gifts.univ–mrs.fr/SOS-DGDB/SOS-DGDB_home_page.html
TelDB – multimedia telomere resource	www.genlink.wustl.edu/teldb/index.html
TRADAT – transcription databases and analysis tools	www.itba.mi.cnr.it/tradat/
Subviral RNA database – small circular RNAs database (viroid and viroid-like)	http://nt.ars–grin.gov/subviral/
MPDB – molecular probe database	www.biotech.ist.unige.it/interlab/mpdb.html
OPD – oligonucleotide probe database	www.cme.msu.edu/OPD/

record or as a cross-reference to another database. Annotation item refer either to the entire sequence (global descriptors, such as protein name) or to a segment of it (local descriptors, such a protein domain or an exon). A database record itself can be an annotation item in a different database, for instance, a record in a protein domain database, or in a bibliographic database can be linked as an annotation item to a sequence database. In principle, annotated sequences should point to a unique protein or gene. In practice, a unique protein can be represented by many sequences, and databases differ in the way redundancy is handled. GenBank contains all published DNA sequences, so there is considerable redundancy. The same is true for the protein sequence collections prepared by automatic translation, or by automated experimental procedures such as expressed sequence tag (EST) sequencing. Maintenance of high-quality nonredundant databases requires a human overhead that is increasingly difficult to provide.

Finally, much of annotation information today is provided by automated procedures, such as similarity searches, HMM methods, etc. Even though these methods constantly improve, there is no absolute guarantee behind the information, so much of annotation information today is labelled as 'putative' or 'by homology'.

Interpretation of annotations and bibliographic records requires a uniform, computer-readable language. This need has fostered intensive research into the natural languages used in science. In early cultures, scientific language describing natural cultures was based on only four elements (earth, water, fire, wind), and a few

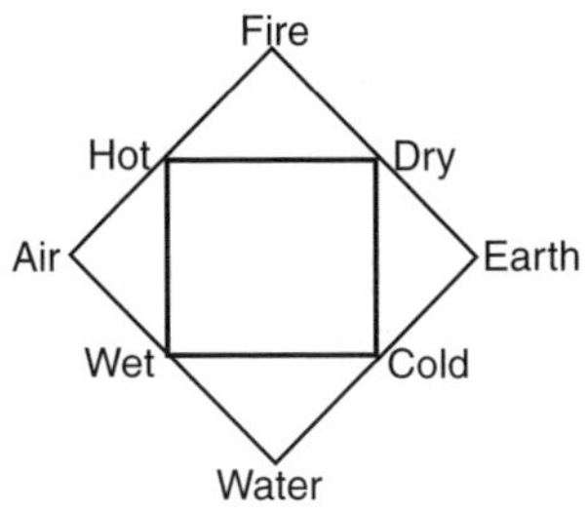

System	Entities	Relationships
Molecules	Atoms	Atomic interactions (chemical bonds)
Assemblies	Proteins, DNA	Molecular contacts
Pathways	Enzymes	Chemical reactions (substrates/products)
Genetic networks	Genes	Co-regulation
Protein structure	Atoms	Chemical bonds
Simplified rotein structure	Secondary structures	Sequential and topological vicinity
Folds	C_α atoms	3D vicinity
Protein sequence	Amino acid	Sequential vicinity

Figure 6.1 Examples of entities and relationships used in molecular models. Left: a simple 4-fold semantics (earth, water, air, fire) and binary relations (hot–cold, wet–dry) was used in most early cultures as a conceptual framework to describe nature. Right: entities and relationships of modern databases.

dichotomic relations (hot–cold, dry–wet, etc.) between them. Descriptions used in database annotations are based on a large number of models for atoms, molecules and reactions (Figure 6.1). This is a stripped-down language that can be kept uniform by the makers of the databases. On the other hand, scientific publications and abstracts use a 'free-style' scientific language that is hard to handle using computers. This problem has emphasized the need to develop common ontologies for molecular biology (Schulze-Kremer, 1997; Ashburner *et al.*, 2000). An ontology defines a common vocabulary and a shared understanding within a domain of knowledge such as molecular biology. An ontology is an explicit description of the knowledge domain using concepts, properties and attributes of the concepts, and constraints on properties and attributes. The Gene Ontology Annotation (GOA) database (www.ebi.ac.uk/GOA) is a system of concepts and relations that is designed to convert UniProt annotation into a recognized computational format. GOA provides annotated entries for nearly 60 000 species and is the largest and most comprehensive open-source contributor of annotations to the GO Consortium annotation effort.

6.3 Three-Dimensional Models

As sequence databases were born within molecular biology, three-dimensional databases were brought to life by chemistry and, later, structural biology. In the same way as language, they provide a conceptual framework for sequences; the metaphor for molecular models is common world objects whose handling and recognition is as at least as deeply rooted in human cognition as is language (Pinker, 2001). The first three-dimensional model of a molecule was constructed in 1874 by van t'Hoff, who recognized that optical isomerism can only be explained by a three-dimensional arrangement of the chemical bonds. As methods of X-ray crystallography became applicable to organic molecules, Olga Kennard and Desmond Bernal initiated the collection of three-dimensional structures what later became known as the Cambridge Crystal Structure Database, and access to three-dimensional information consolidated the use of molecular geometry in chemistry.

Three-dimensional descriptions of macromolecules are based on a series of concepts that resulted from several decades of scientific work. For instance, structural descriptions of proteins is based on the stereochemistry of the peptide bond, but elements of secondary structures, supersecondary structures and finally protein folds are the joint results of several scientific disciplines that form what is called structural biology today.

The common ancestor of structural databases is the Protein Data Bank (PDB; Berman *et al.*, 2000), which was established in 1971 and 6 years later contained only 77 atomic coordinate entries for 47 macromolecules (Bernstein *et al.*, 1977). Over the years, a conspicuous number of secondary databases has evolved from the PDB (see Table 6.5). Many of them concentrate on various classes of structural features, such as protein domains (Sander and Schneider, 1991; Orengo *et al.*, 1997; Murzin *et al.*, 1995; Siddiqui Dengler and Barten, 2001), loops (Donate *et al.*, 1996), contact surfaces (Jones and Thornton, 1996; Luscombe *et al.*, 2000), quaternary structure (Henrick and Thornton, 1998), small-molecule ligands (Kleywegt and Jones, 1998), metals (Castagnetto *et al.*, 2002) and disordered regions (Sim, Uchida and Miyano, 2001). Other databases concentrate on biological themes. The very concept of the 'protein fold' owes much of its existence to such protein domain databases as CATH, SCOP and FSSP.

The conceptual structure of current three-dimensional databases is similar to sequence databases, inasmuch as the records contain both a structural description and an annotation part. The definitions of the individual fields reflect the fact that PDB was originally created as a crystallographic database, and despite the fast-growing body of NMR data, this remains its legacy. Current databases are now linked to other molecular and bibliographic databases.

The development of ontologies has begun in structural biology. The STAR/mmCIF ontology (Westbrook and Bourne, 2000) of macromolecular structure is a description of structural elements and data items in the framework of X-ray crystallographic experiments, but it is extensible to other kinds of data collection techniques.

6.4 Genomes, Proteomes, Networks

Designing representations for genomes, proteomes and networks is a challenge as we deal with a wide variety of entities and relationships, partly predefined, partly discovered during the project. This class of representations can be called the 'general topological model', wherein the nature of entities and relationships is not limited either to semantic or to three-dimensional concepts, as in the previous chapters. The resulting general representation is a graph wherein the physical entities are the nodes and their relations are the edges. The common ancestor of this representation is the structural formula, and graph theory itself owes a great deal to the development of chemistry in the nineteenth century. The representations of genomes as a linear array

Table 6.5 Examples of three-dimensional databases

Primary three-dimensional resources	
Protein Data Bank	www.rcsb.org
Macromolecular Structure Database	www.ebi.ac.uk/msd/index.html
Nucleic Acid Database Project	http://ndbserver.rutgers.edu/NDB/index.html
BioMagResBank	www.bmrb.wisc.edu/Welcome.html
Protein domain/fold databases	
3Dee	http://jura.ebi.ac.uk:8080/3Dee/help/help_intro.html
CATH	www.biochem.ucl.ac.uk/bsm/cath
HSSP	www.sander.ebi.ac.uk/hssp
SCOP	http://scop.mrc-lmb.cam.uk.ac/scop
Examples of specialized resources	
BIND – binding database	www.bind.ca/index.phtml?page=databases
BindingDB – binding database	www.bindingdb.org/bind/index.jsp
Decoys 'R' Us	http://dd.stanford.edu
Disordered structures	http://bonsai.ims.u-tokyo.ac.jp/~klsim/database.html
DNA binding proteins	http://ndbserver.rutgers.edu/structure-finder/dnabind/
Intramolecular movements	http://molmovdb.mbb.yale.edu/MolMovDB/
Loops	www-cryst.bioc.cam.ac.uk/~sloop/Info.html
Membrane protein structures	http://blanco.biomol.uci.edu/Membrane_Proteins_xtal.html
Metal cations	http://metallo.scripps.edu/
P450 containing systems	www.icgeb.trieste.it/~p450srv/
Predicted protein models	http://guitar.rockefeller.edu/modbase
Protein–DNA contacts	www.biochem.ucl.ac.uk/bsm/DNA/server/
Protein–protein interfaces	www.biochem.ucl.ac.uk/bsm/PP/server/
ProTherm	www.rtc.riken.go.jp/jouhou/protherm/protherm.html
Quaternary structure	http://pqs.ebi.ac.uk
Small ligands	http://alpha2.bmc.uu.se/hicup/.
Small ligands	www.ebi.ac.uk/msd-srv/chempdb
The Protein Kinase Resource	http://pkr.sdsc.edu/html/index.shtml
Examples of search/retrieval facilities and database interfaces	
3DinSight – structure/function dbase	www.rtc.riken.go.jp/jouhou/3dinsight/3DinSight.html
BioMolQuest – structure/function dbase	http://bioinformatics.danforthcenter.org/yury/public/home.html
Entrez	www3.ncbi.nlm.nih.gov/entrez/query.fcgi
Image Library of Macromolecules	www.imb-jena.de/IMAGE.html
OCA	http://bioinfo.weizmann.ac.il:8500/oca-docs/
PDBSUM	www.biochem.ucl.ac.uk/bsm/pdbsum/
ProNIT – protein–nucleic acid interactions	www.rtc.riken.go.jp/jouhou/pronit/pronit.html
TargetDB	http://targetdb.pdb.org/
SRS	http://srs.ebi.ac.uk/

of genes and other DNA segments follow a similar tradition (Table 6.6). The entities, genes, are predicted with gene-prediction programs or are determined by experimental methods, and this adds a new layer of knowledge to the molecular data. The relationships are manifold but are predominantly binary in nature. Examples of

Table 6.6 Examples of genomic resources

Genomic databases for various organisms[a]	
Flybase – *Drosophila melanogaster*	http://flybase.bio.indiana.edu/
Subtilist – *Bacillus subtilis*	http://genolist.pasteur.fr/SubtiList/
Cyanobase – *Synechocystis* strain PCC6803	www.kazusa.or.jp/cyano/cyano.html
CYGD – *Sacharomyces cerevisiae*	http://mips.gsf.de/genre/proj/yeast/
ENSEMBLE – human and other invertebrate genomes	www.ensembl.org/
Comparative genomic visualization tools	
VISTA	www-gsd.lbl.gov/vista/
PipMaker	http://bio.cse.psu.edu/pipmaker/
Whole-genome annotation browsers	
NCBI Map Viewer	www.ncbi.nlm.nih.gov
UCSC genome browser	genome.ucsc.edu/
Ensembl	www.ensembl.org/
Whole-genome comparative genomic browsers	
UCSC genome browser	http://genome.ucsc.edu/
VISTA genome browser	http://pipeline.lbl.gov/
PipMaker	http://bio.cse.psu.edu/genome/hummus/
Custom comparisons to whole genomes	
GenomeVista (AVID)	http://pipeline.lbl.gov/cgi-bin/GenomeVista
UCSC genome browser (BLAT)	http://genome.ucsc.edu/
ENSEMBL (SSAHA)	www.ensembl.org/
NCBI (BLAST)	www.ncbi.nlm.nih.gov/blast/

[a]A more complete list is available at the websites of the EBI, NCBI and DDBJ as well as within the current database issue of *Nucleic Acids* research, cited in Table 1.

relations include physical vicinity, distance along the chromosome and regulatory links extracted from DNA chip data. The resulting picture is a graph of several tens of thousand nodes and relatively few edges per node denoting various relationships. The description of proteomes is somewhat different. The proteins are described in functional, biochemical and structural terms, and the relationships between proteins include metabolic relationships (sharing substrates in metabolic pathways) as well as structural relationships (sequence and structural similarities). Network models used in biology fall into two large categories. Dynamic models (such as metabolic network of a cell) are based on differential equations of the constituent enzymatic reactions. Topological models discussed here deal with the static properties of graphs, which can be undirected, directed and weighted.

From the computational point of view, genomes and proteomes are described as very large graphs in which the nodes (genes, proteins) and the edges (relations) are unknown or unsure. These large and fuzzy descriptions are in sharp contrast with the descriptions developed for well-defined graphs of molecular structures, but the methods are not dissimilar to those used in other applications of graph theory. Given the large and varied genome sizes as well as the uncertainties of the data, genomic networks are usually characterized and compared in terms of gross global descriptors such as the degree of distribution and composition (e.g. composition expressed in terms of gene or protein classes). Biological networks are also believed to contain recurrent local patterns (network motifs) that are analogous to sequence motifs found in biological sequences.

Even this sketchy introduction implies that we deal with new a kind of complexity that originates from the numerous and, to a large extent, unknown interactions between the entities. On the other hand, the study of network topology in various fields, such as internet, social, road and electrical networks, has provided interesting insights that have been successfully applied to genomes, proteomes and bibliographic networks. However these insights are limited by the fact that static network topology is only a very general description of the underlying biological phenomena; in fact it should rather be considered as a 'metamodel', i.e. a 'model of models'.

6.5 Computational Tools

Bioinformatics came to life at a time when computer technology reached the daily routine of scientific research. The development of bioinformatics tools (e.g. interfaces, database design, programming methods) is to some extent a mere reflection of the concomitant trends in informatics. On the other hand, bioinformatics software is peculiar because of its impact on how lay users access biological data today. In the 1970s and early 1980s, the first published programs were written in a basically sequential style for standalone computers such as campus mainframes. The second stage started with the recognition that the input and output of bioinformatics applications can be standardized, and modular packages based on the *software tools approach* (Knuth, 1998) were developed. The best known example of these, the GCG package of John Devereux (Devereaux and Haeberli, 1984), developed into a battery of several hundred programs over the years, covering virtually the entire scope of biological sequence analysis. However, such a large body of knowledge is difficult to maintain in a commercial context. EMBOSS, which is developed by a collaboration of academic researchers, was designed as an open source alternative to commercial programs (Rice, Longden and Bleasby, 2000). The development of Bioperl (Stajich *et al.*, 2002), a *Perl* library for bioinformatics applications, and Bioconductor (Gentleman *et al.*, 2004), a statistical programming package for microarray analysis based on the *R* programming language, are further examples of successful open source collaborations. Web servers developed by academic research groups represent a different trend since, in such cases, the source code is often not released and users can access the programs only on-line. Internet technology thus allows individual researchers to release their programs before the commercial or open source development stage, and the users interested in the most recent computational tools more and more accept the risk of using nontransparent programs.

The most visible tools of current bioinformatics are the complex knowledge resources composed of databases, analysis tools and internet interfaces that integrate various kinds of data into a navigable dataünetwork (Figure 6.2; Table 6.6).

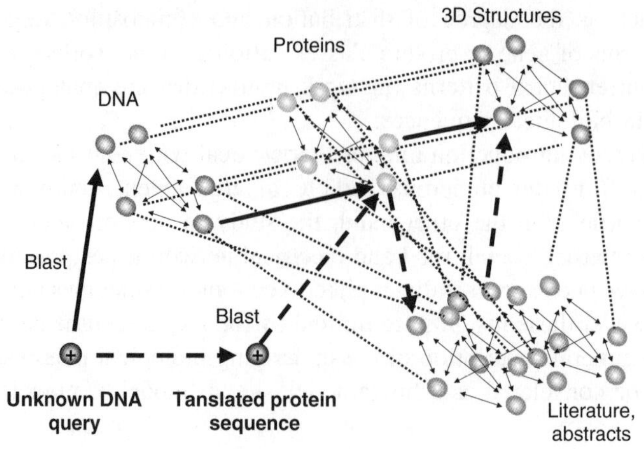

Figure 6.2 Search on an integrated database. Items in the individual databases (DNA sequence, protein sequence, three-dimensional structures, literature abstracts) are cross-referenced (dotted line) by internet links. Additional links (thin arrows) connect 'neighbourhoods', i.e. similar data items within each database. Consequently, if similarity search (thick arrows) points, for example, to an unannotated DNA entry, a member of its neighbourhood may help the user to find proteins or protein structures.

6.6 Information Processing in the Immune System

The immune response includes a number of regulating mechanisms that are organized into various networks affecting a wide range of phenomena ranging from uptake, processing and presentation of the antigens, to T and B cell activation and performing the effector functions. During the immune homeostasis, the spatial and temporal patterns of the cellular and soluble interaction networks develop the optimal qualitative and quantitative characteristics in starting, amplifying and finishing the immune response in an optimal way. The real understanding of immune response obviously requires a systems biology approach. There are four highly specific aspects of immune functions that warrant a specific immuno-informatic approach.

1. The immune systems itself functions as a highly regulated information processing system. The major informations are the sequence [αβ T cell receptor (TCR) recognition] and the conformation [B cell receptor (BCR)/Immunoglobulin (Ig) and γδTCR recognition] of the antigen/peptide molecule, the genetic (e.g. major histocompatibility complex, MHC) background of the antigen presenting cells, the activation of the innate immune systems [e.g. complement or natural killer (NK) pattern] and the actual environmental scenario [e.g. pathogen-associated molecular pattern (PAMP) or the local pattern of the inflammatory mediators).

2. The networking habits of the immune system, both at cellular level (enhancing and inhibitory effects, feedback regulations) and as Ig networks, such as the idiotypic-antiidiotypic webs.

3. The VJD recombinations and other mechanisms generating diversity of the antigen receptor repertoires are basically different from other, more simple nets. These molecular events result in rapidly evolving gene sets serving a more sophisticated recognition tool during the afferent input of the immune response.

4. Somatic hypermutations through *activation induced deamination (AID)* and repair machinery as highly effective ways to generate diversity belong to specific tools.

In the following issues some relevant representations of immune regulation are mentioned.

Information management by lymphocytes

What 'program' the given cell has, namely what the 'output' signal, the response, is, issues from its genetic characteristics and the features it acquired during its ontogenesis (Figure 6.3). In other words, after the appropriate co-stimulatory and cytokine effects, a cell-specific pattern of transcription factors develops. Upon their effect, the appropriate cell-response develops, thus, the cell divides or/and differentiates. Its fate will be the transformation into a memory cell or apoptosis; it releases antibody, cytokines or other secretion products into the outer world. In this way, it frequently participates in the activation or inhibition of another cell or functions as an effector. Sometimes it happens that the cell 'changes' its means of signal transduction during the regulation. For instance, the cAMP-signal transduction related to MHC class 2 on B cells 'switches' on to the tyrosine–kinase path.

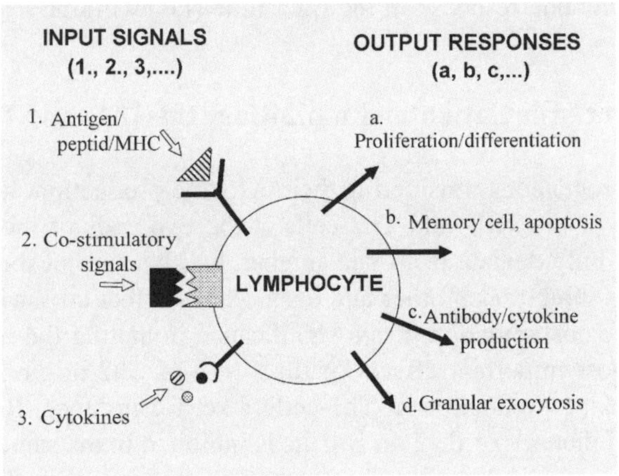

Figure 6.3 Information processing in lymphocytes.

The network of co-stimulatory effects

During the immune response, many cell–cell relations are formed and dissociated. Beyond the physical 'approach', these cell–cell relations may cause intracellularly created signs, which change the functioning of the cell.

Besides the antigen-specific interaction (TCR-MHC–peptide), the T and B cells must receive supplementary signals for carrying out a successful immune response. Without these, the antigen-specific interaction in most case produces exactly produces anergy, an incapability to respond. This anergy stands behind the phenomenon of immune tolerance. Next to the specific antigen-receptor, there are co-receptors (CD4, CD8), whose binding to the appropriate MHC-molecules orients T cells regarding the antigen-presenting cells (and exogenous or endogenous antigens) they should be connected to.

Our present knowledge lists numerous co-stimulatory (formed with adhesion molecules and other membrane-proteins) interactions of positive and negative effect between the T cell and the antigen-presenting cell. Some interactions increase the activation, e.g. the CD28–B7.1, CD2–LFA-3 and CD40–CD40L (CD154) interactions (Sadra, Cinek and Imboden, 2004). The CTLA-4–B7.2 linkage, for instance, may have a positive or negative effect depending on the circumstances (Zhang *et al.*, 2003).

The possibility in relation to the special characteristics of NK receptors raises a further regulation opportunity. The NKB1- and p58-receptors of NK-cells recognize MHC, but this recognition hinders the activity of NK-cells (KIR) (Wilson *et al.*, 2000). It has recently been stated that these NK-receptors are also found on a subgroup of $\alpha\beta$–and $\gamma\delta$-cells, where MHC recognition is exactly a (positive) condition for the stimulus realized through the $\alpha\beta$-TCR-receptors. The point is that, within one cell, two receptor structures (NK receptor and TCR) are oppositely regulated by the MHC; thus, the outcome of cell activation is influenced by the local proportion of the two kinds of receptors.

Presumably, the local, cell-level pattern of costimulating effects organized in time-order represents that fine regulation, which is optimal for starting, properly setting and concluding the immune response in the right time (Frauwirth and Thompson, 2002).

T-cell-dependent stimulation and inhibition, the Th1- and Th2-cytokines

Th-cells are heterogeneous considering their cytokine-production. It is proved that the cytokine-pattern of the Th1- and Th2-cells at the two ends of the polarized T-cell-lineage does not only deviate from one another, but, because of the cross-regulation, certain cytokines inhibit each other and each other's effect crosswise. This, together with the negative costimulation, has a significance in halting the response.

One of the most important effects of the IL-10 of Th2-origin is that it strongly inhibits the cytokine production of Th1-cells, like, for instance, IL-2-synthesis (and its effects). The influence of IL-2 on B-cells is inhibited in the same way by the IL-4, which also has mainly Th2-origin. The antagonism between the effects of IL-4 (Th2) and IFNγ (Th1) is extremely sharp and two-directional [e.g. on immunoglobulin E

(IgE) production, on delayed type hypersensitivity (DTH)]. Individual cytokines frequently affect the isotype of antibody production in a different way, stimulating one and inhibiting the other (e.g. IgG–IgE). The cytokine pattern determined by the tissue environment of B cells significantly influences the class-switch, that is, the development of the isotype (e.g. IgG or IgA antibodies are produced).

Usually it can be said (with exceptions) that T cells producing IFN-γ, IL-2 and TFN-β principally stimulate the cell-mediated immune response, while the T cells producing IL-4, IL-5, IL-9 and IL-6 mainly stimulate the humoral immune response. It is also interesting that chiefly B cells present the antigens to Th2-cells, while macrophages present the antigens to Th1-lymphocytes (Figure 6.4).

Elegant results have proved that the 'Th1–Th2' character, also related to chemokine patterns (Kim *et al.*, 2001), is not connected strictly to the CD4 marker, since this double nature has been detected in some CD8+ cells as well. These cells are called Tc1/Tc2 cells. We also know Tγδ-1, which mainly produces IFN-γ and Tγδ-2, primarily secreting IL-4.

Th3 cells represent a separate subset characterized by TGF-β production. They stimulate the functioning of Th1- and inhibit the functioning of Th2-subgroups. They also have an important role in the IgA production of the immune system attached to the gastro-intestinal system. Recently, markers characteristic of the Th1- (LAG-3) and Th2- (CD30) population have been found on the plasma membrane. LAG-3 is a molecule belonging to the immune-globulin supergene family, while CD30 is a protein belonging to the TNF-receptor family, already known on activated T and B cells as well as on the Sternberg–Reed cells typical of the Hodgkin lymphoma.

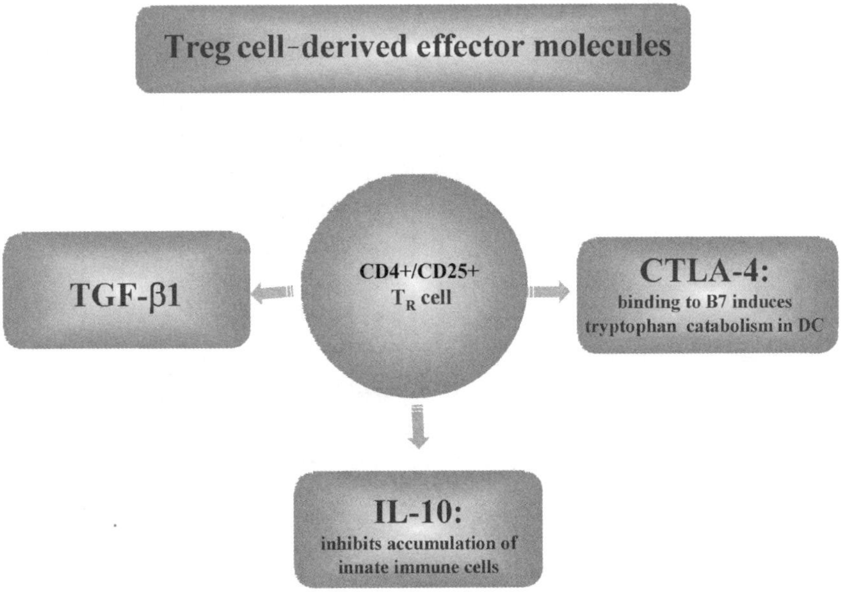

Figure 6.4 Cytokine patterns delivered by T cell polarization.

If it is indeed true that these molecules are markers of the two T cell populations, the cytofluorometry which detects the surface markers will be a supplementary method for measuring the Th1/Th2 rate besides the rather expensive cytokine-mRNA measurements. As a result of T cell polarization, an important 'division of labour' is formed while overcoming different infections. The role of the Th1-type Th cells is important against Gram-negative bacteria, while the role of the Th2-type Th cells is essential against parasite infections (Figure 6.4).

In these inhibiting and stimulating processes, the nonantigen-specific cells and products of the immune system also play an important role. For instance, if NK cells are activated in the local immune reaction because of the antigen's nature (tumour cell, virus) or other factors, this causes an increase in the local IL-12 and IFNγ-level, that is, a 'Th1'-like (+ and −) effect. According to present views, the IL-12 has a central role in the regulation of cellular (cell-mediated) immune response and promising results have come to light in the treatment of metastatic tumours and diseases caused by hepatitis B and C virus infections, with the help of IL-12.

If the number of basophilic granulocytes increases locally because of the antigen's nature (allergen, vermin) or other factors, the result will be the exact opposite: it asserts the increase of IL-4-level, that is, 'Th2'-like influences (+ and −); Figure 6.5). On the other hand, IL-4 plays the 'conductor' role regarding the humoral immune response. There is a therapeutic possibility in IL-10 (which is also supposed to have Th2-origin) in allergic diseases, because it inhibits the attractive effect of IL-5 on eosinophils.

Similar differences are caused by the locally effective prostaglandins produced by macrophages, fibroblasts and follicular dendritic cells. Prostaglandin E1 (PGE1) and PGE2 inhibit the cytokine-production of Th1 cells but do not influence Th2 lymphocytes. As a consequence, PGEs shift the balance locally in the direction of

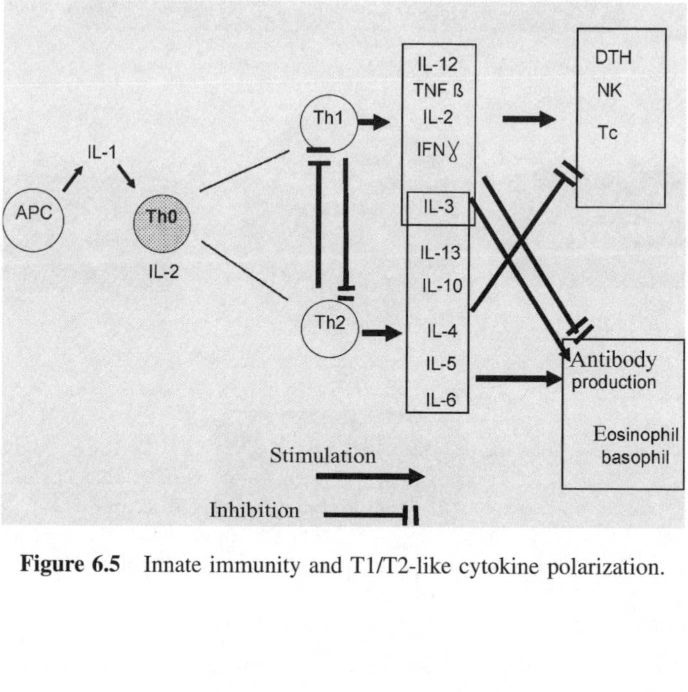

Figure 6.5 Innate immunity and T1/T2-like cytokine polarization.

humoral immune response. We have also come to know (e.g. from AIDS/HIV research) that corticosteroids principally inhibit Th1 cells (apoptosis induction), while certain androgen steroids (e.g. dehydro-epi-androsteron) inhibit the corticosteroids' Th1-blocking effect, that is, they antagonize it. β-Endorphin inhibits the rate of the Th1 and stimulates the rate of Th2 cells. β-Antagonists weaken the cellular immune response through IL-12 inhibition.

Generally we can say that, in the organism, the Th1 and Th2 found on the two ends of the polarized T-cell line participate in different processes as regulating cells. Nevertheless, we should never explain the effects of the Th1/Th2 cytokines dogmatically, since the same cytokine can often act oppositely, depending on the concentration, place and time.

The CD4+NK.1.1 and Treg subgroup

Recent results have reported on a T cell type called CD4+NK.1.1+. It can be considered the main source of IL-4 and plays central role in anti-microbial immunity. The NK.1.1+ subgroup is a population which is CD4 – CD8 – αβ in the thymus, CD4+CD8– αβ on the periphery, having numerous NK markers, cytotoxic ability and a relatively homogenous αβ repertoire. Its function is supposedly the regulation of haematopoesis, the development of T cell tolerance, the cytotoxic removal of virus-infected liver cells and the stimulation of the Th2 population's maturation by IL-4. It is presumed that these cells primarily recognize microbial antigens presented by the monomorphic CD1 (Jiang and Chess, 2004).

On the basis of all these, the CD4+NK1.1 cells could be considered cellular elements of the non-antigen-specific immunity, a new (regulating?) subclass of the T cells (Godfrey and Kronenberg, 2004).

Most recently a new concept of Treg cells (CD4+, CD25+) has been developed (Walsh, Taylor and Turka, 2004). These cells, representing over 10 per cent of CD4+ Th cells, are mostly silencing cells expressing foxp3 transcription factor acting in various regulatory networks of immune response, producing TGFb, IL-10 and negative co-stimulatory molecules such as CTLA4 (Figure 6.6). Treg cells are involved in transplantation tolerance, prevent pathological responses induced by the gut flora or microbial infections, play a role in maternal tolerance, can suppress antitumor immunity and protective immunity to pathogens and can lead to enhanced memory T cell response. Recently immunoregulatory disturbances of many autoimmune diseases (Frey *et al.*, 2005) have been coupled to dysfunction of Treg subsets.

Idiotype regulation, idiotype network (Poljak, 1994)

The potential of antibody and TCR diversity developing in the immune response is extremely high. About 10^{11} different antibody and 10^{15}–10^{18} different TCR specificities develop in a healthy adult immune system. Since a significant part of this huge repertoire is not expressed (or not in a significant degree) during maturation in the thymus, an autotolerance cannot be formed against them. As a consequence, the

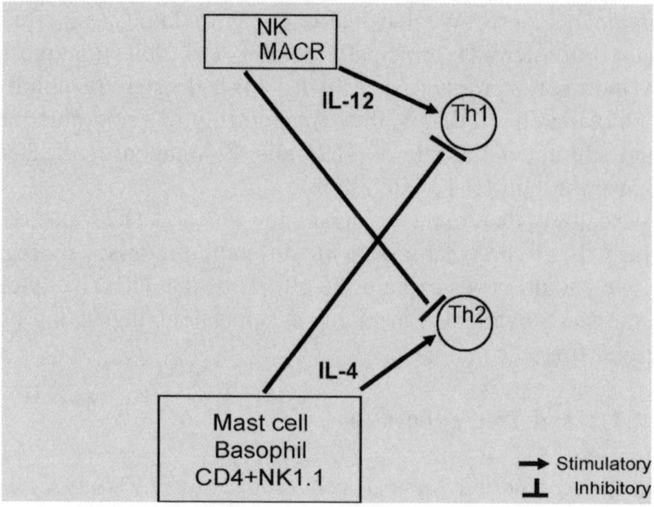

Figure 6.6 T regulatory cells.

segments of the variable region (paratope, complexity-determining region, CDR) appear as antigens (idiotypes) in the organism. In 1973, Niels Jerne announced his attractive theory according to which the organism produces anti-idiotype antibodies. Antibodies (anti-anti-idiotypes) are formed again against the antigens in the variable regions of these antibodies, and so on. In this way a network develops (Figure 6.7), where there is a possibility for every second element to contain similar epitopes (idiotopes) and have similar antibody specificity. Thus, the first antibody activates a (B and Th cell) self-regulating network and the antibody playing the role of 'antigen' activates the next element of the 'antibody' role. This again activates the next one,

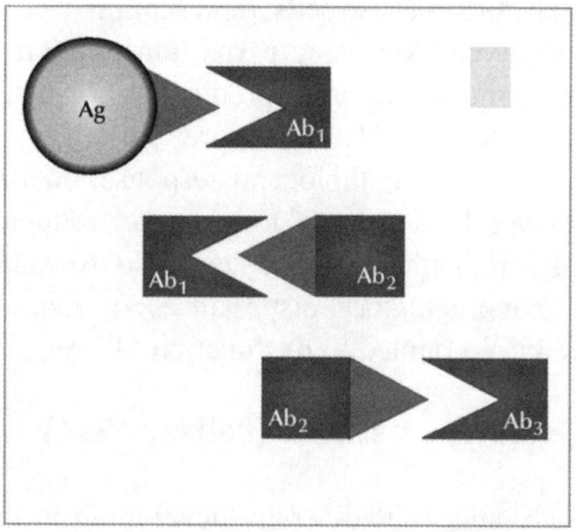

Figure 6.7 Idiotypic web. The antigen (Ag) is recognized by a specific antibody (Ab1) carrying idiotope stimulating asecond antibody (Ab2). Idiotypic determinant(s) of Ab2 may further activate Abs, etc. Parts of antigenic structure may be similar to that of Ab2. [Reproduced from Yehuda Shoenfeld, *Nature Medcine* **10**: 17–18 (2004).]

which may have 'first antibody'-like idiotopes among its idiotopes. The network starts to limit itself; smaller and smaller amounts of new antibodies are produced. Theoretically, the situation is similar in the variable regions of the TCR's αβ- and γδ-chains. The idiotype–anti-idiotype network represents a complex network of interacting T and B cells, which can equally stimulate or inhibit the immunological activation. One of the most important central principles of the network model is that the anti-idiotype antibody (second antibody), which reacts with an idiotope in the CDR of another antibody (first antibody), presents a similarity with the original epitope, being the inner image of the original antigen-determinant. This theory is supported by fact: 'second and third'-type anti-idiotype antibodies against monoclonal paraproteins have already been detected. By the improvement of detecting techniques, researchers have recently identified anti-idiotype antibodies during normal immune response as well. The idiotype network has a major significance in keeping the memory cells of the immune system in small but lasting excitement after the disappearance of the original antigen, in the presence of the second antigen, through an 'internal image'. This way, upon the repeated appearance of the 'real' antigen, they can react quickly to it.

The proof showing the presence of the idiotype network is the following: in a type of autoimmune thyroid disease, the anti-idiotype antibody acting against the autoantibody against the thyreoid-stimulating hormone (TSH) behaves like the TSH and binds to TSH receptors. In the future, this phenomenon can be possibly used in vaccination. Anti-idiotype antibodies will be produced in an experimental animal against the antibody (specific for a dangerous, living pathogen), also produced in an experimental animal. This second antibody may be similar to the original infectious antigen, therefore it can be used in immunization (securely, since it is an immunoglobulin), and thus, immunity develops in the organism.

Today conflicting opinions are published concerning the central role and size of a given idiotype network, but surely this system has a considerable role in the regulation of the immune response. There are proofs of the existence of idiotype-specific Th cells and idioype (antibody-, Th-) cross reactions are also assigned importance in certain autoimmune diseases.

6.7 Concluding Remarks

The general approach of bioinformatics emerged from a parallel growth in two major fields, life sciences and information technologies. This concomitant development provided access to several new fields, and it also resulted in new conceptual and technological tools for representing and manipulating scientific knowledge. Integrated databases, analysis programs and ontologies are typical results of this development.

Current bioinformatics deals with a large number of models that are basically static in nature. The molecular database represents information in terms of linguistic, three-dimensional and network models. The latter allow one to capture part of the interactions among the molecules and other cellular componenets; the descriptions are still predominantly static. A large part of the information is stored in cellular, tissue and systemic models that are not part of the databases but, if they are represented at all, are stored in ontologies or are part of the background knowledge of biologists that is not accessible to computers.

The current immunology databases are similar in structure and in philosophy to other molecular databases (Table 6.7). Nevertheless, immunology represents a

Table 6.7 Bioinformatics resources for immunology

ALPSbase – autoimmune lymphoproliferative syndrome database	http://research.nhgri.nih.gov/alps/
BCIpep – experimentally determined B-cell epitopes of antigenic proteins	http://bioinformatics.uams.edu/mirror/bcipep/
dbMHC – HLA sequences in human populations	www.ncbi.nlm.nih.gov/mhc/
FIMM – functional molecular immunology, T-cell response to disese-specific antigens.	http://research.i2r.a-star.edu.sg/fimm/
HaptenDB – hapten molecules	www.imtech.res.in/raghava/haptendb/
HLA ligand/motif database – a database and search tool for HLA sequences	http://hlaligand.ouhsc.edu/
IL2Rgbase – X-linked severe combined immunodeficiency mutations	http://research.nhgri.nih.gov/scid/
IMGT – integrated knowledge resource	http://imgt.cines.fr
IMGT-GENE-DB – genome database for immunoglobulin (IG) and T cell receptor (TR) genes from human and mouse	http://imgt.cines.fr/cgi-bin/GENElect.jv
IMGT/HLA – HLA sequence database	www.ebi.ac.uk/imgt/hla/
IMGT/LIGM-DB – immunoglobulin and T cell receptor nucleotide sequences, from human and other vertebrate species,	http://imgt.cines.fr/cgi-bin/IMGTlect.jv
Interferon stimulated gene database (by microarry)	www.lerner.ccf.org/labs/williams/xchip-html.cgi
IPD-ESTDAB – polymorphic genes in the immune system	www.ebi.ac.uk/ipd/estdab/
IPD-HPA – human platelet antigens	www.ebi.ac.uk/ipd/hpa/
IPD-KIR – killer-cell immunoglobulin-like receptors	www.ebi.ac.uk/ipd/kir/
IPD-MHC – polymorphic genes in the immune system	www.ebi.ac.uk/ipd/mhc
JenPep – peptide binding to biomacromolecules within immunobiology (epitopes)	www.jenner.ac.uk/Jenpep
Kabat – kabat database of sequences of proteins of immunological interest	http://immuno.bme.nwu.edu/
MHC – peptide interaction database	http://surya.bic.nus.edu.sg/mpid
MHCBN – peptides binding to MHC or TAP	www.imtech.res.in/raghava/mhcbn/
MHCPEP – MHC binding peptides	http://wehih.wehi.edu.au/mhcpep/
VBASE2 – germ-line V genes from the immunoglobulin loci of human and mouse	http://www.vbase2.org

specific case in many respects, and especially in the latter sense, since many of the models in immunology are different from those in other fields. This is on the one hand valid to the main molecular mechanisms of the immune system (such as VJD recombination, cellular and molecular networks, somatic hypermutations) which cannot be and are not adequately covered in current databases. On the other hand, major aspects of the immune system are not presently considered in molecular databases. As an example, the maturation of a single, monospecific immune response or that of immunological memory calls for a selectionist description of cell populations, or the similar structure–similar function paradigm that works magnificently at the level of most protein classes is blatantly invalid in discriminating IgG molecules specific for different epitopes.

The overview presented in this chapter suggests that the needs of immunological research will inevitably create a specific bioinformatics approach that will allow one to represent and access knowledge gained in these highly important fields. Finally we mention that many biological concepts have found their way back to informatics: artificial neural networks and genetic algorithms are success stories of the computer sciences. It is an intriguing possibility that the information processing methods of immunology, the *immunomics concept* will inspire novel computational approaches (De Groot, 2004; Wang and Falus, 2004; Brusic and Petrovsky, 2005).

References

Altschul SF, Gish W, Miller W, Myers EW, Lipman DJ. 1990. Basic local alignment search tool. *J Mol Biol* **215**: 403–410.

Ashburner M, Ball CA, Blake JA, Botstein D, Butler H, Cherry JM, Davis AP, Dolinski K, Dwight SS, Eppig JT, Harris MA, Hill DP, Issel-Tarver L, Kasarskis A, Lewis S, Matese JC, Richardson JE, Ringwald M, Rubin GM, Sherlock G. 2000. Gene ontology: tool for the unification of biology. The Gene Ontology Consortium. *Nat Genet* **25**: 25–29.

Berman HM, Westbrook J, Feng Z, Gilliland G, Bhat TN, Weissig H, Shindyalov IN. Bourne PE. 2000. The Protein Data Bank. *Nucl Acids Res* **28**: 235–242.

Bernstein FC, Koetzle TF, Williams GJ, Meyer EF JR, Brice MD, Rodgers JR, Kennard O, Shimanouchi T, Tasumi M. 1977. The Protein Data Bank: a computer-based archival file for macromolecular structures. *J Mol Biol* **112**: 535–542.

Brusic V, Petrovsky N. 2005. Immunoinformatics and its relevance to understanding human immune disease (in press).

Carnap R. 1939. *Foundations of Logics and Mathematics*. Chicago: University of Chicago Press.

Castagnetto JM, Hennessy SW, Roberts VA, Getzoff ED, Tainer JA, Pique ME. 2002. MDB: the metalloprotein database and browser at the Scripps Research Institute. *Nucl Acids Res* **30**: 379–382.

Chomsky N. 1957. *Syntactic Structures*. The Hague: Mouton.

Csányi V. 1989. *Evolutionary Systems and Society*. Durham, NC: Duke University Press.

De Groot AS. 2004. Immunome derived vaccines. *Expert Opin Biol Ther* **4**: 767–772.

Devereaux J, Haeberli P, O, S. 1984. A comprehensive set of sequence analysis programs for the VAX. *Nucl Acids Res* **12**: 387–395.

Donate LE, Rufino SD, Canard LH, Blundell TL. 1996. Conformational analysis and clustering of short and medium size loops connecting regular secondary structures: a database for modeling and prediction. *Protein Sci* **5**: 2600–2616.

Frauwirth KA, Thompson CB. 2002. Activation and inhibition of lymphocytes by costimulation. *J Clin Invest* **109**: 295–299.

Frey O, Petrow PK, Gajda M, Siegmund K, Huehn J, Scheffold A, Hamann A, Radbruch A, Brauer R. 2005. The role of regulatory T cells in antigen-induced arthritis: aggravation of arthritis after depletion and amelioration after transfer of CD4+CD25+ T cells. *Arthritis Res Ther* **7**: R291–301.

Gentleman RC, Carey VJ, Bates DM, Bolstad B, Dettling M, Dudoit S, Ellis B, Gautier L, Ge Y, Gentry J, Hornik K, Hothorn T, Huber W, Iacus S, Irizarry R, Leisch F, Li C, Maechler M, Rossini AJ, Sawitzki G, Smith C, Smyth G, Tierney L, Yang JY, Zhang J. 2004. Bioconductor: open software development for computational biology and bioinformatics. *Genome Biol* **5**: R80.

Godfrey DI, Kronenberg M. 2004. Going both ways: immune regulation via CD1d-dependent NKT cells. *J Clin Invest* **114**: 1379–1388.

Henrick K, Thornton JM. 1998. PQS: a protein quaternary structure resourse. *Trends Biochem Sci* **23**: 358–361.

Jiang H, Chess L. 2004. An integrated model of immunoregulation mediated by regulatory T cell subsets. *Adv Immunol* **83**: 253–88.

Jones S, Thornton JM. 1996. Principles of protein–protein interactions. *Proc Natl Acad Sci USA* **93**: 13–20.

Kampis G. 1991. *Self-modifying Systems in Biology and Cognitive Science*. Oxford: Pergamon Press.

Kim CH, Rott L, Kunkel EJ, Genovese MC, Andrew DP, Wu L, Butcher EC. 2001. Rules of chemokine receptor association with T cell polarization *in vivo*. *J Clin Invest* **108**: 1331–1339.

Kleywegt GJ, Jones DT. 1998. Databases in protein crystallography. *Acta Crystallogr D* **54**: 119–131.

Knuth DE. 1998. *The Art of Computer Programming*. Reading, MA: Addison-Wesley.

Konopka AK. 1994. Fundamentals of biomolecular cryptology. In Smith DW. (ed.), *Biocomputing: Informatics and Genome Projects*. San Diego, CA: Academic Press.

Luscombe NM, Austin SE, Berman HM, Thornton JM. 2000. An overview of the structures of protein–DNA complexes. *Genome Biol* 1.

Maynard Smith J, Szathmáry E. 1995. *The Major Transitions in Evolution*. Oxford: WH Freeman.

Murzin AG, Brenner SE, Hubbard T, Chothia C. 1995. SCOP: a structural classification of proteins database for the investigation of sequences and structures. *J Mol Biol* **247**: 536–540.

Orengo CA, Michie AD, Jones S, Jones DT, Swindells MB, Thornton JM. 1997. CATH – a hierarchic classification of protein domain structures. *Structure* **5**: 1093–1108.

Pinker S. 2001. *The Language Instinct: How the Mind Creates Language*. Perennial: New York.

Poljak RJ. 1994. An idiotope–anti-idiotope complex and the structural basis of molecular mimicking. *Proc Natl Acad Sci USA* **91**: 1599–1600.

Pongor S. 1988. Novel databases for molecular biology. *Nature* **332**: 24.

Pongor S, Skerl V, Cserzõ M, Hátsági Z, Simon G, Bevilacqua V. 1993. The SBASE domain library: a collection of annotated protein segments. *Protein Engrg* **6**: 391–395.

Rice P, Longden I, Bleasby A. 2000. EMBOSS: the European molecular biology open software suite. *Trends Genet* **16**: 276–277.

Ripley BD. 1999. *Pattern Recognition and Neural Networks*. Cambridge: Cambridge Univeristy Press.

Sadra A, Cinek T, Imboden JB. 2004. Translocation of CD28 to lipid rafts and costimulation of IL-2. *Proc Natl Acad Sci USA* **101**: 11422–11427.

Sander C, Schneider R. 1991. Database of homology-derived protein structures and the structural meaning of sequence alignment. *Proteins* **9**: 56–68.

Schulze-Kremer S. 1997. Adding semantics to genome databases: towards an ontology for molecular biology, *Proceedings of International Conference on Intelligent Systems and Molecular Biology*, Vol 5, pp 272–275.

Shannon CE. 1948a. Communication theory of secrecy systems. *Bell Syst Tech J* 29.

Shannon CE. 1948b. A mathematical theory of communication II. *Bell Syst Tech J* 27.

Shannon CE. 1948c. Prediction and entropy of printed English. *Bell Syst Tech J* 30.

Siddiqui AS, Dengler U, Barton GJ. 2001. 3Dee: A database of protein structural domains. *Bioinformatics* **17**: 200–201.

Sim KL, Uchida T, Miyano S. 2001. ProDDO: a database of disordered proteins from the Protein Data Bank (PDB). *Bioinformatics* **17**: 379–380.

Stajich JE, Block D, Boulez K, Brenner SE, Chervitz SA, Dagdigian C, Fuellen G, Gilbert JG, Korf I, Lapp H, Lehvaslaiho H, Matsalla C, Mungall CJ, Osborne BI, Pocock MR, Schattner P, Senger M, Stein LD, Stupka E, Wilkinson MD, Birney E. 2002. The Bioperl toolkit: Perl modules for the life sciences. *Genome Res* **12**: 1611–1618.

Szathmáry E, Smith JM. 1995. The major evolutionary transitions. *Nature* **374**: 227–232.

Walsh PT, Taylor DK, Turka LA. 2004. Tregs and transplantation tolerance. *J Clin Invest* **114**: 1398–1403.

Wang E, Falus A. 2004. Changing paradigm through a genome-based approach to clinical and basic immunology. *J Transl Med* **2**: 2.

Westbrook JD, Bourne PE. 2000. STAR/mmCIF: an ontology for macromolecular structure. *Bioinformatics* **16**: 159–168.

Wilson MJ, Torkar M, Haude A, Milne S, Jones T, Sheer D, Beck S, Trowsdale J. 2000. Plasticity in the organization and sequences of human KIR/ILT gene families. *Proc Natl Acad Sci USA* **97**: 4778–4783.

Wittgenstein L. 1922. *Tractatus Logico Philosophicus*. London: Routledge and Kegan Paul.

Zhang X, Schwartz JC, Almo SC, Nathenson SG. 2003. Crystal structure of the receptor-binding domain of human B7-2: insights into organization and signaling. *Proc Natl Acad Sci USA* **100**: 2586–2591.

7

Immune Responsiveness of Human Tumours

Ena Wang and **Francesco M. Marincola**

Abstract

Immune responsiveness of solid tumours may be dependent upon factors related to the genetic background of patients or to distinct characteristics of individual tumours that may alter their microenvironment to facilitate or inhibit immune responses occurring naturally in the tumour-bearing host. Whether either or both of these categories of factors play a prominent role in determining tumour rejection in natural conditions (spontaneous regression) or during immune therapy remains to be confirmed. Several lines of evidence derived from experimental animal or *in vitro* models suggest that immune responsiveness results from a combination of genetic and epigenetic factors that modulate the activation of effector immune responses at the tumour site. However, direct *ex vivo* observations in humans are scant, particularly when tumour–host interactions have been analysed in the target tissue: the tumour site. Available technology allows direct and kinetic analyses of such interactions in real time by serial sampling of lesions using minimally invasive techniques such as fine needle aspirates. Serial sampling permits the study of the biology of cancer while leaving the studied lesion in place. This permits prospective evaluation of the natural history of the lesion left in place to identify biomarkers as predictors of immune responsiveness to a given treatment. In addition, serial sampling during and after treatment may inform about the actual mechanisms of action of the treatment and its biological effects, including the induction of the tumour escape mechanism. In this chapter we will systematically review the tool and strategies available for this direct *ex vivo* analyses with particular attention on high-throughput hypothesis-generating methods. In addition, insights into the postulated mechanism of immune responsiveness in humans will be discussed based on pilot studies performed in this vein.

Immunogenomics and Human Disease Edited by András Falus

7.1 Introduction

The field of tumour immune biology stands at a delicate junction drawn by a positive current originating from astonishing progress in the understanding of tumour immune biology on one side and an upstream journey against the discontent of some who feel that disappointingly small improvements in patient care have resulted from scientific breakthroughs. A good example is the development of anticancer therapies specifically targeting tumour antigen recognized by T cells in the context of melanoma (Marincola, 2005; Marincola and Ferrone, 2003; Mocellin *et al.*, 2004; Rosenberg, Yang and Restifo, 2004; Slingluff and Speiser, 2005; Timmerman and Levy, 2004). The last two decades have seen enormous breakthroughs in the molecular understanding of autologous tumour cell recognition by the adaptive immune system. The identification and utilization of tumour antigen for active-specific vaccination has proven successful in expanding *in vivo* the number of antigen-specific T cells (Cormier *et al.*, 1997; Salgaller *et al.*, 1996). However, no correlation was noted between the extent of the tumour antigen specific immune response and its clinical effectiveness (Cormier *et al.*, 1997; Lee *et al.*, 1999; Salgaller *et al.*, 1996) and, more generally, very little clinical benefit in the form of rare cancer regressions was observed (Marincola and Ferrone, 2003; Rosenberg, Yang and Restifo, 2004). The ensuing paradoxical scepticism toward active-specific immunization (Rosenberg, Yang and Restifo, 2004), is, in our opinion, mostly the result of that hasty aspect of human nature which sometimes overtakes scientific thinking, allowing expectations to obfuscate cogent molecular realities. For this reason, we often naively expect an easy translation of novel therapeutic strategies crafted in animal or experimental models into improvements in patient care, disregarding that intrinsic property of human pathology that makes it elusive to over-simplification (Marincola, 2003). More specifically, a T cell-specific epitope can be logically expected to induce proliferation and activation of T cells when administered as an immunogen, but this cannot be expected to induce cancer regression simply because the downstream events related to T cell localization at the tumour site, proliferation and activation, are beyond the control of the effects of immunization on the afferent loop of the immune system.

In tune with the purpose of this book, we will discuss the reason why tumour deposits most often tolerate the presence of circulating tumour antigen-specific T cells. In addition, we will present some observations that might explain occasional tumour regression, framing the topic of this chapter with the enigmatic term ‘immune responsiveness’ (Wang and Marincola, 2001). In particular, we will discuss how hypothesis-generating tools may best apply nowadays to outlining logically the basis of human tumour immune responsiveness in contrast to academically more acceptable hypothesis-driven speculations. Finally, we will summarize the concept of immune responsiveness, taking into account global variables related to the polymorphic nature of human immune response (Jin, Marincola and Wang 2004; Jin and Wang 2003) and the extensive heterogeneity of human cancer (Wang *et al.*, 2004b).

Hypothesis-driven research with its comfortable minimalist approach to problem-solving has failed to produce results when the academic exercise reaches the clinical setting for the simple reason that most of the hypotheses have not been generated through direct clinical observations but through artificial systems (Marincola, 2003). Although this reflects a general problem confronted in translational medicine (Mankoff *et al.*, 2004), it particularly applies to tumour immune biology because this is a particularly complex system encompassing the complexity of human immune responses characterized by extensive species variability (polymorphism) and epigenetic adaptations on one side and the intrinsic heterogeneity of cancer and the rapid evolution of cancer phenotypes due to genetic instability on the other (Wang *et al.*, 2004b; Wang, Panelli and Marincola 2003). In this chapter, we will describe the application of high-throughput tools to address human reality with a hypothesis-generating attitude best fitted to address its heterogeneity. In particular, we will discuss tools that can allow the dynamic study of systemic and intratumoral immune phenomena through serial analysis of human samples before, during and after therapy.

7.2 Defining Tumour Immune Responsiveness

It should first be observed that tumour immune responsiveness is a general term that may include a variety of biological phenomena associated with the interactions between the host immune system and invading cancer cells in the context of distinct therapeutic approaches. For this reason, two logical approaches could be considered: the first is to categorize tumours of different histology and segregate individual treatments and analyse immune responsiveness in the context of each individual permutation. An approach that we have much preferred is to consider the phenomenon of immune responsiveness as part of a general acute immune reaction against self, similar to other autoimmune phenomena and look for commonalities among different cancers and their treatments. In particular, we have been interested in the identification of appropriate human cancer models to study immune responsiveness. For instance, testing in humans the effectiveness of TA-specific cancer vaccines has provided an unprecedented opportunity to dissect at the molecular level T cell–tumour interactions (Marincola and Ferrone, 2003). Most of this progress stemmed from the study of patients with advanced melanoma, while other immune responsive cancers remained unexplored. Among them, we have focused on basal cell carcinoma (BCC) and nasopharyngeal carcinoma (NPC). Yet we believe that there is a scientific need to study cancers other than metastatic melanoma. It could be argued that similar biological phenomena may result from convergence of different pathways into a final outcome represented by a least common denominator that, although masked by irrelevant biological variables, governs their occurrence. Thus, the immune responsiveness of melanoma could be compared with that of renal cell cancer or BCC, which are also quite responsive to immune manipulation in spite of distinct biological profiles. Our hypothesis is that, by identifying commonalities and diversities among various cancers, it will be possible to sieve those patterns that may be most relevant to immune responsiveness and/or lack of it.

7.3 Studying Immune Responsiveness in Human Tumours

Basal cell carcinoma and its responsiveness to toll receptor ligands

Imiquimod is a small molecule developed by 3-M (St Paul, MN, USA) as an anti-viral agent but soon recognized as possessing anticancer activity through stimulation of innate immune mechanism(s) (Richwald, 2003; Urosevic *et al.*, 2003). This drug is too toxic for systemic administration and it is predominantly used for the local treatment of skin cancers such as squamous cell carcinoma and BCC. In the case of BCC, Imiquimod induces an acute inflammatory process followed in the large majority of cases by complete regression of BCC. Like interleukin-2, Imiquimod has no direct anticancer activity and its mechanism(s) of action appears to be related to the induction of acute inflammation associated with massive infiltration of macrophages and dendritic cells (Urosevic *et al.*, 2003). It is reasonable to postulate that this treatment model could yield useful information about the postulated least common denominator responsible for tumour rejection. We therefore designed in collaboration with the National Naval Medical Center and the Pathology Department, NCI, a clinical protocol aimed at the study of the genetic and immunohistologic profile of BCC lesions undergoing treatment with Imiquimod to be compared with our previous findings in the context of melanoma (Panelli *et al.*, 2002; Wang *et al.*, 2002). The results of this study are currently been analysed and will, hopefully, shed light on the mechanism(s) of immune responsiveness.

Metastatic renal cell carcinoma and cutaneous melanoma

Another interesting phenomenon is the similarity in clinical behaviour of renal cell carcinoma and metastatic cutaneous melanoma. Both tumours are particularly sensitive to the administration of a systemic pro-inflammatory agent such as interleukin-2 or interferons. The example of the systemic administration of interleukin-2, a drug approved by the Food and Drug Administration specifically for the treatment of these diseases (Atkins *et al.*, 1998; Atkins, Regan and McDermott, 2004; Yang and Rosenberg, 1997), is outstanding. In spite of its commercial utilization, very little is known about the mechanism(s) inducing tumour regression and/or toxicity. Possibly, these two cancers (or the subgroup that responds to therapy) may share a number of intrinsic biologic characteristics conducive to immune attack. However, here is where similarities and yet interesting differences are also noted. For instance, melanoma is considered 'highly immunogenic' because of the relative ease with which tumour-reacting T cells can be expanded *in vitro* (Itoh, Tilden and Balch, 1986;

Kawakami *et al*., 1992; Wolfel *et al*., 1989) from metastatic lesions and the wealth of tumour-associated antigens that they recognize (Brinckerhoff, Thompson and Slingluff, 2000; Novellino, Castelli and Parmiani 2004; Renkvist *et al*., 2001). Immunogenicity or 'antigenicity' in itself does not seem, however, to explain the observed biological phenomenon of melanoma immune responsiveness to interleukin-2 administration since renal cell cancer arguably as sensitive to such treatment does not share this biological characteristic. Therefore, it is possible that the tendency of metastatic cutaneous melanoma to recruit tumour antigen-specific T cells is not the reason for its immune responsiveness but rather symptomatic of a biological *milieu* that favours tumour rejection in response to interlukin-2 administration, and at the same time favours the recruitment of T cells in melanoma lesions and their exposure to melanoma-restricted T cell epitopes. We have been interested in studying commonalities between these two cancers comparing their global transcriptional profile with other nonimmune responsive cancers. In a recent analysis of a large microarray data set, we identified a large number of genes whose expression is predominantly restricted to metastatic cutaneous melanoma. Among those genes, we identified a characteristic immunologic signature that robustly distinguished melanoma from any other cancer (Wang *et al*., 2004c). Analysis of the genes included in this melanoma-specific immunologic signature suggested a specific presence of natural killer cell function and in particular activation of immune effector function. Comparison of the expression profile of primary renal cell carcinoma specimens did not reveal a similar pattern (Wang *et al*., 2004a). Unfortunately, no metastatic renal cell samples were available for this study. Since primary renal cell cancer is not as sensitive to immunotherapy as metastatic lesions (Bex *et al*., 2002), we could not exclude the possibility that the immunologic signature specific of melanoma could explain its immune responsiveness as well as that of metastatic renal cells. Hopefully, future studies will elucidate this point. The same study, however, revealed a large proportion of genes coordinately expressed by both melanoma and renal cell cancer. Unfortunately, the function of most of these genes remains to be elucidated. Obviously, this line of study will need to be pursued more aggressively in the future by comparing these two diseases (and in particular material from metastatic renal cell carcinoma) to other nonimmunogenic cancers in natural conditions and in response to immune therapy. Another interesting observation is the association reported between the expression of Carbonic anhydrase IX and survival in advanced renal cell carcinoma (Bui *et al*., 2003). This is an interesting observation because others have reported an association between the expression of this enzyme and responsiveness to interleukin-2 therapy (Atkins, Regan and McDermott, 2004). Once again, comparison of the transcriptional profile of metastatic melanoma and renal cell carcinoma did not reveal similarities in the expression of this gene (Wang *et al*., 2004a). However, metastatic melanoma frequently expressed carbonic anhydrase III, opening up the possibility that enzymatic redundancy may explain their similar behaviour. Obviously, all these examples underline the hypothesis-generating attitude in which they were performed. More extensive analyses will need to be conducted to test whether any of these observations truly relate to immune responsiveness.

7.4 Immune Responsiveness in the Context of Therapy

The mechanism of action of interleukin-2

Interleukin-2 mediated tumour regression stands as a prototype phenomenon of immune responsiveness (Panelli *et al.*, 2003). When it occurs, tumour regression is a dramatic event characterized by tenderness and swelling of tumour lesions before they disappear. Although relatively rare, this phenomenon is quite remarkable and resembles in many ways an acute autoimmune reaction. Learning about this dramatic biological phenomenon may improve our chances of expanding its usefulness to other clinical settings and at the same time limiting its relatively high toxicity. We have recently re-framed the hypothesis of how intereleukin-2 acts when systemically administered, excluding a facilitator role in T cell migration to the tumour site as suggested by others (Rosenberg *et al.*, 1998). In addition, we could not document induction of T cell proliferation, activation at the tumour site or induction of tumour cell death through a cytotoxic cytokine cascade (Panelli *et al.*, 2002). This conclusion was based on direct analysis of transcriptional profile changes in melanoma metastases during systemic interleukin-2 therapy obtaining snapshots 3 h after the first and fourth dose of systemic cytokine administration through repeated fine needle aspirates (FNA). Real-time observation of human disease under therapy suggested that the predominant action of systemic interleukin-2 administration is to induce activation of mononuclear phagocytes with a bi-polar tendency toward a classic macrophage on one side and/or activated presenting cells on another. Subsequent *in vitro* studies suggested that this effect on tumour-associated monocytes probably results from a secondary release of an array of cytokines by circulating and/or resident cells in response to interleukin-2 rather than by a direct effect of this cytokine (Nagorsen *et al.*, 2004). A recent analysis of the protein profile of sera obtained 3 h after the first and fourth dose of systemic interleukin-2 administration outlined the complexity of the organism response to this treatment with a broad array of soluble factors increasing in concentration on a logarithmic scale, several of them with pleyotropic activities ranging from chemo attraction, pro- and anti-inflammatory properties, vascular adhesion and angioregulation and, finally, pro-apototic activity (Panelli *et al.*, 2004). Most importantly, several of the soluble factors released in response to interleukin-2 administration had the opposite biological activity on mononuclear phagocytes based on experimental models (Mantovani *et al.*, 2002; Nagorsen *et al.*, 2004) as well as possible clinical effects including toxicity or response to therapy. This type of study emphasizes the need for a broader, discovery-driven and/or hypothesis-generating approach to the study of human disease, taking advantage of modern, high-throughput methods. The identification of specific biological patterns whether in the systemic circulation or in the tumour microenvironment associated with toxicity pattern or therapeutic effectiveness of this cytokine will be important not only to make this therapy safer and more affective but also to shed light on the mechanism(s) mediating immune responsiveness in the context of these and possibly other cancers.

Although these studies were quite informative about the general mechanism of action of interelukin-2, they were not as informative with regard to the mechanism(s) occurring during tumour rejection. Interestingly, however, one of the six patients analysed in this pilot study convincingly responded to treatment (Panelli *et al.*, 2002). Analysis of the genetic profile of this single patient compared with that from samples obtained from nonimmune responsive lesions from other patients identified a small set of genes strongly upregulated during immune regression. Interestingly, several of the genes identified had been simultaneously reported by another group as associated with acute rejection of kidney allograft (another T cell-mediated phenomenon), suggesting that these mechanism(s) are similar if not identical. Interestingly, several of the genes identified suggested that tumour rejection in response to interleukin-2 therapy is associated with a switch from a chronic to an acute inflammatory process with expression of several genes associated with cytotoxic function such as NK4 and NK-G5 (Panelli *et al.*, 2002). In particular, NK4 may deserve special attention as this gene is already differentially expressed in melanoma metastases with a high degree of significance compared with other nonimmunogenic tumours (Wang *et al.*, 2004c).

7.5 The Spatial Dimension in the Quest for the Target

Localization of T cells at the tumour site

It been experimentally shown that localization at the tumour site is necessary for the fulfillment of the effector function of human tumour antigen-specific cytotoxic T cells. The administration of tumour-infiltrating lymphocytes labelled with ^{111}Indium demonstrated that tumour regression occurs only when the transferred cells localize within the lesion (Pockaj *et al.*, 1994). This important observation underlines the relevance of the presence of antigen-specific T cell at the tumour site as a mandatory contributor of tumour rejection. Although necessary, localization is not sufficient. Indium fact, in the same study several cases where observed of localization without tumour regression. Similar observation could be made in the context of active-specific immunization. A comparative expansion of tumour-infiltrating lymphocytes from biopsies obtained before and after immunization suggested that it was easier to expand immunization-specific T cells following immunization compared with before (Panelli *et al.*, 2000). An even more important observation came from the analysis of interferon-γ expression in melanoma metastases serially biopsied by FNA. Material obtained from FNA performed during immunization had a higher content of interferon-γ messenger RNA compared with those obtained before immunization (Kammula *et al.*, 1999). The increase in IFN-γ messenger RNA levels could be attributed to the localization of vaccination-specific T cells at the tumour site by identifying and enumerating them with tetrameric human leukocyte antigen (HLA) – epitope complexes. Furthermore, the levels of interferon-γ correlated tightly with the expression by tumour cells of the antigen targeted by the vaccine. This suggests that not only the T cells could localize at the tumour site but that they could also recognize tumour cells. Yet, this localization was not sufficient to induce tumour regression since most of the lesions analysed continued their unaltered growth.

A quiescent phenotype of immunization-induced T cells

The previous observations are very important because they confirm that tumour antigen-specific T cells induced by vaccination can reach the tumour site but they are not able to expand *in vivo* and to exploit their potential anticancer activity in the tumour microenvironment. If this observation is correct, various explanations could be considered as previously discussed. We will here focus on the possibility that circulating immunization-induced T cells rest in a physiologically quiescent state that allows them to leave the lymph nodes draining the immunization site where their loco-regional exposure to the immunogen occurred (Monsurro' *et al.*, 2004). This hypothesis suggests that, after the primary stimulus, T cells need to be re-activated in the target organ where they are re-exposed to the cognate antigen, and that such re-activation requires a co-stimulatory stimulus besides antigen recall for the induction of successful effector T cells. This is biologically intuitive in a teleological sense because effector T cells are most efficiently utilized if they are only activated in the presence of the relevant antigen and a danger signal that informs them of the necessity of action (Matzinger, 2001). In fact, in a recent analysis of vaccination-induced T cells, we observed in an *in vitro* model that full activation of effector function as well as massive proliferation could only be achieved when T cells were exposed to antigen recall plus interleukin-2 as a co-stimulating factor. In particular, we noted that both stimulations were required for the complete activation. If we are allowed to extrapolate the *in vitro* model to the *in vivo* situation, we could easily picture how, at the tumour site, tumour antigen-specific T cells may be exposed to antigen recall but seldom will sufficient co-stimulatory signals be present to induce their full activation (Fuchs and Matzinger, 1996). This hypothesis raises the obvious question: what are the requirements within the tumour microenvironment for full T cell activation that may influence the outcome of immunotherapy and may switch a dormant recognition of cancer cells into an acute rejection of cancer as observed in the context of acute rejection of solid transplant organs? This question points to the need for studying the tumour microenvironment and its complex biology.

7.6 Studying the Receiving end – Tumour as an Elusive Target for Immune Recognition

The role of escape, ignorance or immune editing

Cancer is a disease characterized by extreme genetic instability (Lengauer, Kinzler and Vogelstein, 1998). Individual differences in telomere size (Londono-Vallejo, 2004), different patterns of methylation (Szyf, Pakneshan and Rabbani, 2004) and other factors may influence the patterns of growth and adaptation to the surrounding environment of cancer cells. However, the role that environmental pressure mediated by immune selection or other biologically relevant conditions may have in influencing the rate of this natural process remains unclear. Animal models convincingly show that the immune system plays a major role in regulating and modifying tumour growth (Shankaran *et al.*, 2001). A term has been suggested recently for this concept: 'immune editing' (Dunn *et al.*, 2002; Dunn, Old and Schreiber, 2004), which differs from the original idea of immune surveillance (Burnet, 1970; Smyth, Godfrey and Trapani 2001) because it adds to the control of tumour growth by immune recognition a plasticity of interactions leading to the sculpting the immunogenic phenotype of tumour cells, which may in turn facilitate tumour progression (Dunn *et al.*, 2002). This process occurs in steps summarized by the acronym EEE, which includes at the beginning the elimination (of target cells by immune cells). Subsequently an equilibrium is stricken between the host immune response and the tumour cells, which eventually degenerates into a situation of escape when tumour cells with a phenotype capable of avoiding immune recognition take over the majority of the tumour mass (Dunn, Old and Schreiber, 2004).

Immune editing has never been definitively demonstrated in humans because the type of experimentation necessary to confirm that the de-differentiation naturally occurring during the neoplastic process is influenced by the surrounding environment is impractical and in some circumstances unethical. This is because clinical protocols would need to be designed where more effective therapies would be compared with less effective and tissues would be compared for evidence of escape with the hypotheses that more effective therapies would be more likely to induce escape phenomena (Khong and Restifo, 2002; Marincola *et al.*, 2003). Obviously, such studies would be unethical and could be only done retrospectively if appropriate material were available. Active-specific immunization may offer the opportunity to study immune editing by comparing antigen loss variance among groups of patients immunized against different antigens believed to have similar therapeutic potential. There are, however, three problems that limit this strategy: first, active-specific immunization in itself is not very effective in inducing tumour regression and changes in antigen-expression are only seen when this therapy is given in combination with cytokine stimulation (Jager *et al.*, 1997; Ohnmacht *et al.*, 2001), which in turn decreases the specificity of the immune response; second, epitope spreading *in vivo* may broaden the specificity of the immune response to antigens other than the other targeted by the vaccination (el-Shami *et al.*, 1999; Lally *et al.*, 2001; Vanderlugt and Miller, 2002); and finally, the expression of several types of tumour antigens might be cocoordinately regulated and, therefore, the growth of tumour cell phenotypes depleted of their expression may be favoured by an immune response directed against one of them (Wang *et al.*, 2004c). This last difficulty could be overcome by immunizing with antigens belonging to different categories that are not coordinately expressed, such as melanoma differentiation antigens on one side and tumour testing antigens on the other (Wang *et al.*, 2004c).

Nevertheless, some general observations support the concept that the evolution of the neoplastic phenotype is accelerated by immune selection: (1) the neoplastic process is by nature unstable (Lengauer, Kinzler and Vogelstein 1998) – gene expression profiling has shown that the transcriptional program of cancer cells evolves early in the determination of the disease and subsequent alteration in the behaviour of tumours occurs because the most de-differentiated and aggressive cellular phenotypes take over the other milder ones (Klein, 2003, 2004), offering a venue for immune selection; (2) immune suppression, owing to disease or to pharmaceutical manipulation to control transplant rejection, is associated with an increased rate of cancer (Abgrall *et al.*, 2002; Cottrill, Bottomley and Phillips 1997; Euvrard, Ulrich and Lefrancois, 2004; Grulich, 2000; Lutz and Heemann, 2003; Robertson and Scadden, 2003; Ulrich *et al.*, 2004); (3) anecdotal examples have been described by independent groups in which the phenotype of tumour cells changed to promote their escape from the immune therapies (Hicklin, Marincola and Ferrone 1999; Jager *et al.*, 1996, 1997; Lee *et al.*, 1998; Marincola *et al.*, 1994; Restifo *et al.*, 1996). Interestingly, most of these anecdotal observations were done on patients who responded to immune therapy.

The evolution of cancer cell phenotypes can be altered by treatment other than immune manipulation and some have argued that gene expression analysis should be applied to study the dynamic profile of cancers undergoing various forms of therapy (Ellis, 2003). Similarly, we have proposed that tools are now available to dissect in real time changes related and specific to a given therapy through serial fine needle aspirate (FNA; Wang and Marincola, 2000). With this strategy, we observed that the pre-treatment level of expression of the gene targeted by active-specific immunization (in this case the melanoma differentiation antigen, gp100/PMel17) did not predict the immune responsiveness of melanoma metastases (Ohnmacht *et al.*, 2001). However, following the short-term kinetics of tumour antigen expression during therapy was very informative and demonstrated a rapid loss of gp100/PMel17 in lesions undergoing clinical regression, while no changes were observed in progressing lesions. Tumour antigen loss was partially specific since other tumour-specific antigens belonging to the cancer/testis family and expressed selectively by tumour cells did not change their level of expression. Interestingly, the expression of the melanoma differentiation antigen MART-1/MelanA decreased in association with the expression of gp100/PMel17, although the former was not targeted by vaccination. Indeed, we had previously observed by immunohistochemical analysis that the expression of the two melanoma differentiation antigens is relatively although not absolutely coordinated in the majority of melanoma metastases (Cormier *et al.*, 1998a, b; Marincola *et al.*, 1996a). Recently, we compared the genetic profile of melanoma metastases and noted that the expression of melanoma differentiation antigens including gp100/PMle17, Mart-1/MelanA, tyrosinase, tyrosinase related protein-1 and preferentially expressed antigen in melanoma (PRAME) was tightly coordinated, suggesting that their downregulation or loss of expression during melanoma progression may be related to a central regulatory pathway not as yet identified (Wang *et al.*, 2004c). This finding may have important repercussions in the design of antigen-specific immunization protocols and at the same time may complicate the interpretation of tumour antigen loss variant analysis by broadening loss of expression to antigens other than those targetted by a given therapy.

Genetic profiling has been quite useful in analysing cancer progression in natural conditions and during therapy, demonstrating the rate in which cancer phenotypes can change and adapt to different environmental conditions. Several global transcript analysis surveys suggest that tumours of the same histology can be differentiated into molecularly defined subclasses beyond the discriminatory power of histopathological observation and such distinction may have diagnostic and prognostic value ('t Veer *et al.*, 2002; Alizadeh *et al.*, 2000; Bittner *et al.*, 2000; Boer *et al.*, 2001; Perou *et al.*, 2000; Takahashi *et al.*, 2003; Wang *et al.*, 2002; Wang, Panelli and Marincola, 2003; Young *et al.*, 2001). With minor exceptions (Perou *et al.*, 2000), these studies did not include a temporal dimension to the comparisons by evaluating the behaviour of the same cancer at different time points in its natural progression. We have recently argued that, while in some cases molecular subclasses may represent distinct disease taxonomies associated with specific biological behaviours, temporal changes associated with the dedifferentiation of cancer may contribute to the segregation of taxonomically identical entities into separate groups. For instance, although originally we noticed that melanoma metastases can be segregated into two molecular subclasses believed to represent distinct disease entities (Bittner *et al.*, 2000), serial sampling of identical metastases by FNA suggested that the two subclasses represented most likely temporal phenotypic changes since material from the same lesions could cluster within either subgroup with a unilateral shift of the later samples toward a less differentiated phenotype. This observation questions rigid classifications of morphologically identical diseases based on single time point observations (Marincola *et al.*, 2003; Wang *et al.*, 2002). Although this observation emphasizes that dynamic state in which tumour growth occurs, they do not provide conclusive evidence that the changes observed are due to the host's immune pressure. These studies only provide a potential molecular mechanism of evolution in the fast lane that could facilitate tumour adaptation to unfavourable growth conditions. Future studies should test whether a link exists between distinct global transcript patterns, specific responses and subsequent escape from therapy using serial sampling of identical lesions.

The basis of tumour phenotype change is not totally clear. To estimate the influence of ontogeny on the transcriptional profile of cancer we analysed a series of primary renal cell cancer accompanied by and paired with normal renal tissues subjected to identical surgical manipulation and experimental preparation. The transcriptional analysis of the paired specimens was compared with archival frozen samples of melanoma metastases representing a putative extreme of diversity with regard to ontogenesis and neoplastic progression as well as with various primary epithelial cancers to frame the origin of similarities and discrepancies in the transcriptional program. Primary renal cell cancers segregated into at least two molecular subgroups, one of which contained only cancer samples while the other contained a mixture of renal cell cancer samples and normal renal specimens. Gene signatures responsible for the segregation noted between the two subclasses of renal cell cancers demonstrated that the cancers segregating with the normal kidney tissue shared the expression of a large number of renal lineage-specific genes. If no information was available about the transcriptional profile of paired normal tissues, it might have been tempting to attribute such differences to separate taxonomies of disease as previously done in the context of metastatic melanoma (Bittner *et al.*, 2000; Wang *et al.*, 2002). This finding is very important because it suggests that the segregation of cancers of the same histology into distinct molecular entities may simply represent their level of de-differentiation at the particular time point evaluated rather than distinct biological entities originating from diverse oncogenic mechanisms. In addition, comparison of the genetic profile of renal cell cancers and other primary tumours suggested that only a minority of genes related to oncogenesis was renal-specific, while in the large majority the oncogenic process was shared not only among cancers of the same histology but also by most primary tumours studied independently of their tissue of origin. The implication of this observation is critical because it stresses the necessity of studying immune editing and cancer cell evolution in a dynamic fashion respectful of the time lines that might be most relevant to a particular therapy rather than adopting a static attitude when comparing biological findings with clinical data.

7.7 The Role of the Host in Determining Immune Responsiveness

Genetic background and immune responsiveness

Factor(s) independent of the individual tumour biology may also determine whether a patient is more likely to respond to one therapy than another. The influence that the genetic background of individual patients may have on the ability to respond to biological therapies remains largely unexplored. An oversimplification of the dynamics of the immune response may discriminate two distinct biological components that may influence immune responsiveness during biological therapy: (a) the genetic background of the patient; (b) the heterogeneous biology of individual tumours.

Although genetic background has often been proposed as a predictor of immune responsiveness, very little evidence exists that immune responsiveness is predetermined (Jin and Wang, 2003). Specific markers that may affect immune responsiveness, such as the HLA complex genes responsible for coding molecules involved in antigen presentation, have not been conclusively linked to prognosis, response to therapy and survival (Lee *et al.*, 1994; Marincola *et al.*, 1996b; Rubin *et al.*, 1991; Wang, Marincola and Stroncek, 2005). Associations between polymorphisms of cytokine genes and predisposition to develop melanoma or other cancers (Howell *et al.*, 2002; Howell, Calder and Grimble 2002) suggest a genetic influence in the immune modulation of cancer growth. However, the relevance of immune polymorphism as a modulator of the immune response to biological therapy has never been addressed in the context of tumour immunology (Howell, Calder and Grimble, 2002; Jin and Wang, 2003). This factor should be considered in future studies as techniques are becoming available that are suitable for the screening at a genome-wide level of polymorphisms in clinical settings (Jin and Wang, 2003; Wang *et al.*, 2003). Preliminary work in our laboratory suggests, for instance, that racial background may determine the transcriptional response of circulating immune cells to cytokine stimulation (unpublished data). Linking this information to genomic detection of polymorphic sites might explain the results. Obviously, such genetic predetermination needs to be explored in the future, not only comparing distinct ethnic backgrounds but various subsets of patients who demonstrate distinct responsiveness to immune therapy.

The mixed response phenomenon

The mixed response phenomenon is one of the most fascinating examples of the balanced influence that genetic predisposition and individual tumour biology may have on immune responsiveness. Very little has been documented in the literature about this biological phenomenon which is, however, relatively well known to clinicians treating with biological therapy metastatic cutaneous melanoma and renal cell carcinoma patients. Multiple lesions in patients with advanced forms of these cancers have a propensity to respond differently to immunotherapy with some regressing and others progressing in response to the same treatment. Considering the genetic background and immune status of the individual bearing such lesions as a constant at that point of time, the observation of the mixed responses suggests that different conditions in the microenvironment of distinct synchronous metastases present in the same individual may strongly affect the outcome. Obviously, the observation of the mixed response phenomenon does not exclude the possibility that genetic background could separately segregate individuals likely to respond from those who are unlikely to respond. It simply suggests that, among those likely to respond, some lesions may be more immune responsive than others. Because tumour immunology merges the intricacy of human immunology with the complex biology of cancer, the study of anticancer immune responses in humans should, therefore, include the analysis of the biology of individual tumours as, contrary to other immune phenomena, the immune systems confront in cancer a heterogeneous and potentially rapidly evolving target. This may also explain why experimental animal models created to bypass the complexity of humans through inbreeding and standardization of cancer cell clones rarely predict human response to therapy, although they have been extremely powerful tools for testing basic immunological concepts. Overall, this observation represents in our eyes the most powerful example of the complexity of tumour immune biology in humans and underlies the need to confront it with powerful tools that can address the genetic make-up of patients, their epigenetic adaptation to environmental stimulae and the evolving nature of tumour biology (Wang, Panelli and Marincola, 2003).

7.8 Concluding Remarks

To summarize, we discussed in this chapter various phenomena that exemplify the complexity of studying immune responsiveness in humans. In spite of such complexity, however, we believe that these examples are fundamental departure points for future investigations because they represent the reality of human disease. Experimental animal models have failed to shed light on human pathology because they are not designed based on true human observation. This decade and the following ones will hopefully mark a turn in the exploration of human tumour immune biology by shifting its attitude from an arbitrary hypothesis-driven approach to a hypothesis-generated approach based on direct human observation.

References

Abgrall S, Orbach D, Bonhomme-Faivre L, Orbach-Arbouys S. 2002. Tumors in organ transplant recipients may give clues to their control by immunity. *Anticancer Res* **22**(6B): 3597–3604.

Alizadeh AA, Eisen MB, Davis RE, Ma C, Lossos IS, Rosenwald A, Bedrick JC, Sabet H, Tran T, Xin Y, Powell JI, Yang L, Marti GE, Moore T, Hudson J Jr., Lisheng L, Lewis DB, Tibshirani R, Sherlock G, Chan WC, Greiner TC, Weisenburger DD, Armitage JO, Warnke R, Levy R, Wilson W, Grever MR, Byrd JC, Botstein D, Brown PO, Staudt LM. 2000. Distinct types of diffuse large B-cell lymphoma identified by gene expression profiling. *Nature* **403**: 467–578.

Atkins MB, Regan M, McDermott D. 2004. Update on the role of interleukin 2 and other cytokines in the treatment of patients with stage IV renal carcinoma. *Clin Cancer Res* **10**(18 Pt 2): 6342S–6346S.

Atkins MB, Lotze MT, Dutcher JP, Fisher RI, Weiss G, Margolin K, Abrams J, Sznol M, Parkinson DR, Hawkins M, Paradise C, Kunkel L, Rosenberg SA. 1998. High-dose recombinant interleukin-2 therapy for patients with metastatic melanoma: analysis of 270 patients treated between 1985 and 1993. *J Clin Oncol* **17**(7): 2105–2116.

Bex A, Horenblas S, Meinhardt W, Verra N, de Gast GC. 2002. The role of initial immunotherapy as selection for nephrectomy in patients with metastatic renal cell carcinoma and the primary tumour *in situ. Eur Urol* **42**(6): 570–574.

Bittner M, Meltzer P, Chen Y, Jiang E, Seftor E, Hendrix M, Radmacher M, Simon R, Yakhini Z, Ben-Dor A, Dougherty E, Wang E, Marincola FM, Gooden C, Lueders J, Glatfelter A, Pollock P, Gillanders E, Dietrich K, Alberts D, Sondak VK, Hayward N, Trent JM. 2000. Molecular classification of cutaneous malignant melanoma by gene expression: shifting from a countinuous spectrum to distinct biologic entities. *Nature* **406**: 536–840.

Boer JM, Huber WK, Sultmann H, Wilmer F, von Heydebreck A, Haas S, Korn B, Gunawan B, Vente A, Fuzesi L, Vingron M, Poustka A. 2001. Identification and classification of differentially expressed genes in renal cell carcinoma by expression profiling on a global human 31,500-element cDNA array. *Genome Res* **11**(11): 1861–1870.

Brinckerhoff LH, Thompson LW, Slingluff CL Jr. 2000. Melanoma vaccines. *Curr Opin Oncol* **12**(2): 163–173.

Bui MH, Seligson D, Han KR, Pantuck AJ, Dorey FJ, Huang Y, Horvath S, Leibovich BC, Chopra S, Liao SY, Stanbridge E, Lerman MI, Palotie A, Figlin RA, Belldegrun AS. 2003. Carbonic anhydrase IX is an independent predictor of survival in advanced renal clear cell carcinoma: implications for prognosis and therapy. *Clin Cancer Res* **9**(2): 802–811.

Burnet FM. 1970. The concept of immunological surveillance. *Prog Exp Tumor Res* **13**: 1–27.

Cormier JN, Salgaller ML, Prevette T, Barracchini KC, Rivoltini L, Restifo NP, Rosenberg SA, Marincola FM. 1997. Enhancement of cellular immunity in melanoma patients immunized with a peptide from MART-1/Melan A. [See comments.] *Cancer J Sci Am* **3**(1): 37–44.

Cormier JN, Abati A, Fetsch P, Hijazi YM, Rosenberg SA, Marincola FM, Topalian SL. 1998a. Comparative analysis of the *in vivo* expression of tyrosinase, MART-1/Melan-A, and gp100 in metastatic melanoma lesions: implications for immunotherapy. *J Immunother* **21**(1): 27–31.

Cormier JN, Hijazi YM, Abati A, Fetsch P, Bettinotti M, Steinberg SM, Rosenberg SA, Marincola FM. 1998b. Heterogeneous expression of melanoma-associated antigens (MAA) and HLA-A2 in metastatic melanoma *in vivo*. *Int J Cancer* **75**: 517–524.

Cottrill CP, Bottomley DM, Phillips RH. 1997. Cancer and HIV infection. *Clin Oncol (R Coll Radiol)* **9**(6): 365–380.

Dunn GP, Old LJ, Schreiber RD. 2004. The three Es of cancer immunoediting. *A Rev Immunol* **22**: 329–360.

Dunn GP, Bruce AT, Ikeda H, Old LJ, Schreiber RD. 2002. Cancer immunoediting: from immunosurveillance to tumour escape. *Nat Immunol* **3**(11): 991–998.

Ellis MJ. 2003. Breast cancer gene expression analysis – the case for dynamic profiling. *Adv Exp Med Biol* **532**: 223–234.

el-Shami K, Tirosh B, Bar-Haim E, Carmon L, Vadai E, Fridkin M, Feldman M, Eisenbach L. 1999. MHC class I-restricted epitope spreading in the context of tumour rejection following vaccination with a single immiunodominant epitope. *Eur J Immunol* **29**(10): 3295–3301.

Euvrard S, Ulrich C, Lefrancois N. 2004. Immunosuppressants and skin cancer in transplant patients: focus on rapamycin. *Dermatol Surg* **30**(4 Pt 2): 628–633.

Fuchs EJ, Matzinger P. 1996. Is cancer dangerous to the immune system? *Sem Immunol* **8**(5): 271–280.

Grulich AE. 2000. Update: cancer risk in persons with HIV/AIDS in the era of combination antiretroviral therapy. *AIDS Read* **10**(6): 341–346.

Hicklin DJ, Marincola FM, Ferrone S. 1999. HLA class I antigen downregulation in human cancers: T-cell immunotherapy revives an old story. *Mol Med Today* **5**(4): 178–186.

Howell WM, Calder PC, Grimble RF. 2002. Gene polymorphisms, inflammatory diseases and cancer. *Proc Nutr Soc* **61**(4): 447–456.

Howell WM, Bateman AC, Turner SJ, Collins A, Theaker JM. 2002. Influence of vascular endothelial growth factor single nucleotide polymorphisms on tumour development in cutaneous malignant melanoma. *Genes Immun* **3**: 229–232.

Itoh K, Tilden AB, Balch CM. 1986. Interleukin 2 activation of cytotoxic T-lymphocytes infiltrating into human metastatic melanomas. *Cancer Res*, **46**(6): 3011–3017.

Jager E, Ringhoffer M, Karbach J, Arand M, Oesch F, Knuth A. 1996. Inverse relationship of melanocyte differentiation antigen expression in melanoma tissues and CD8+ cytotoxic-T-cell responses: evidence for immunoselection of antigen-loss variants *in vivo*. *Int J Cancer* **66**(4): 470–476.

Jager E, Ringhoffer M, Altmannsberger M, Arand M, Karbach J, Jager D, Oesch F, Knuth A. 1997. Immunoselection *in vivo*: independent loss of MHC class I and melanocyte differentiation antigen expression in metastatic melanoma. *Int J Cancer* **71**(2): 142–147.

Jin P. Wang E. 2003. Polymorphism in clinical immunology. From HLA typing to immunogenetic profiling. *J Transl Med* **1**: 8.

Jin P, Marincola FM, Wang E. 2004. Cytokine polymorphism and its possible impact on cancer. *Immunol Res* **30**(2): 181–190.

Kammula US, Lee K-H, Riker A, Wang E, Ohnmacht GA, Rosenberg SA, Marincola FM. 1999. Functional analysis of antigen-specific T lymphocytes by serial measurement of gene expression in peripheral blood mononuclear cells and tumour specimens. *J Immunol* **163**: 6867–6879.

Kawakami Y, Zakut R, Topalian SL, Stotter H, Rosenberg SA. 1992. Shared human melanoma antigens. Recognition by tumor-infiltrating lymphocytes in HLA-A2.1-transfected melanomas. *J Immunol* **148**: 638–643.

Khong HT. Restifo NP. 2002. Natural selection of tumour variants in the generation of "tumor escape" phenotypes. *Nat Immunol* **3**(11): 999–1005.

Klein CA. 2003. The systemic progression of human cancer: a focus on the individual disseminated cancer cell – the unit of selection. *Adv Cancer Res* **89**: 35–67.

Klein CA. 2004. Gene expression sigantures, cancer cell evolution and metastatic progression. *Cell Cycle* **3**(1): 29–31.

Lally KM, Mocellin S, Ohnmacht GA, Nielsen M-B, Bettinotti M, Panelli MC, Monsurro' V, Marincola FM. 2001. Unmasking cryptic epitopes after loss of immunodominant tumour antigen expression through epitope spreading. *Int J Cancer* **93**: 841–847.

Lee JE, Reveille JD, Ross MI, Platsoucas CD. 1994. HLA-DQB1*0301 association with increased cutaneous melanoma risk. *Int J Cancer* **59**(4): 510–513.

Lee K-H, Panelli MC, Kim CJ, Riker A, Roden M, Fetsch PA, Abati A, Bettinotti MP, Rosenberg SA, Marincola FM. 1998. Functional dissociation between local and systemic immune response following peptide vaccination. *J Immunol* **161**: 4183–4194.

Lee K-H, Wang E, Nielsen M-B, Wunderlich J, Migueles S, Connors M, Steinberg SM, Rosenberg SA, Marincola FM. 1999. Increased vaccine-specific T cell frequency after peptide-based vaccination correlates with increased susceptibility to in vitro stimulation but does not lead to tumour regression. *J Immunol* **163**: 6292–6300.

Lengauer C, Kinzler KW, Vogelstein B. 1998. Genetic instabilities in human cancers. *Nature* **396**(6712): 643–649.

Londono-Vallejo, JA. 2004. Telomere length heterogeneity and chromosome instability. *Cancer Lett* **212**(2): 135–144.

Lutz J, Heemann U. 2003. Tumours after kidney transplantation. *Curr Opin Urol* **13**(2): 105–109.

Mankoff SP, Brander C, Ferrone S, Marincola FM. 2004. Lost in translation: obstacles to Translational Medicine. *J Transl Med* **2**(1): 14.

Mantovani A, Sozzani S, Locati M, Allavena P, Sica A. 2002. Macrophage polarization: tumor-associated macrophage as a paradigm for polarized M2 mononuclear phagocytes. *Trends Immunol* **23**(11): 549–555.

Marincola FM. 2003. Translational medicine: a two way road. *J Transl Med* **1**: 1.

Marincola FM. 2005. A balanced review of the status of T cell-based therapy against cancer. *J Transl Med* **3**: 16.

Marincola FM, Ferrone, S. 2003. Immunotherapy of melanoma: the good news, the bad news and what to do next. *Sem Cancer Biol* **13**(6): 387–389.

Marincola FM, Shamamian P, Alexander RB, Gnarra JR, Turetskaya RL, Nedospasov SA, Simonis TB, Taubenberger JK, Yannelli J, Mixon A, Restito NP, Herlyn M, Rosenberg SA. 1994. Loss of HLA haplotype and B locus down-regulation in melanoma cell lines. *J Immunol* **153**: 1225–1237.

Marincola FM, Hijazi YM, Fetsch P, Salgaller ML, Rivoltini L, Cormier JN, Simonis TB, Duray PH, Herlyn M, Kawakami Y, Rosenberg SA. 1996a. Analysis of expression of the melanoma associated antigens MART-1 and gp100 in metastatic melanoma cell lines and in in situ lesions. *J Immunother* **19**(3): 192–205.

Marincola FM, Shamamian P, Rivoltini L, Salgaller ML, Reid J, Restifo NP, Simonis TB, Venzon D, White DE, Parkinson DR. 1996b. HLA associations in the anti-tumor response against malignant melanoma. *J Immunother* **18**: 242–252.

Marincola FM, Wang E, Herlyn M, Seliger B, Ferrone S. 2003. Tumors as elusive targets of T cell-directed immunotherapy. *Trends Immunol* **24**(6): 334–341.

Matzinger P. 2001. Danger model of immunity. *Scand J Immunol* **54**(1-2): 2–3.

Mocellin S, Mandruzzato S, Bronte V, Marincola FM. 2004. Correspondence 1: Cancer vaccines: pessimism in check. *Nat Med* **10**(12): 1278–1279.

Monsurro' V, Wang E, Yamano Y, Migueles SA, Panelli MC, Smith K, Nagorsen D, Connors M, Jacobson S, Marincola FM. 2004. Quiescent phenotype of tumor-specific CD8+ T cells following immunization. *Blood* **104**(7): 1970–1978.

Nagorsen D, Wang E, Monsurro' V, Zanovello P, Marincola FM, and Panelli MC. 2004. Polarized monocyte response to cytokine stimulation. *Genome Biol* **6**(2): R15.

Novellino L, Castelli C, Parmiani G. 2004. A listing of human tumour antigens recognized by T cells: March 2004 update. *Cancer Immunol Immunother* **54**(3): 187–207.

Ohnmacht GA, Wang E, Mocellin S, Abati A, Filie A, Fetsch PA, Riker A, Kammula US, Rosenberg SA, Marincola FM. 2001. Short term kinetics of tumour antigen expression in response to vaccination. *J Immunol* **167**: 1809–1820.

Panelli MC, Riker A, Kammula US, Lee K-H, Wang E, Rosenberg SA, Marincola FM. 2000. Expansion of tumor/T cell pairs from fine needle aspirates (FNA) of melanoma metastases. *J Immunol* **164**(1): 495–504.

Panelli MC, Wang E, Phan G, Puhlman M, Miller L, Ohnmacht GA, Klein H, Marincola FM. 2002. Genetic profiling of peripharal mononuclear cells and melanoma metastases in response to systemic interleukin-2 administration. *Genome Biol* **3**(7): RESEARCH0035.

Panelli MC, Martin B, Nagorsen D, Wang E, Smith K, Monsurro' V, Marincola FM. 2003. A genomic and proteomic-based hypothesis on the eclectic effects of systemic interleukin-2 administration in the context of melanoma-specific immunization. *Cells Tissues Organs* **177**(3): 124–131.

Panelli MC, White RL Jr, Foster M, Martin B, Wang E, Smith K, Marincola FM. 2004. Forecasting the cytokine storm following systemic interleukin-2 administration. *J Transl Med* **2**: 17.

Perou CM, Sertle T, Eisen MB, van de Rijn M, Jeffrey SS, Rees CA, Pollack JR, Ross DT, Johnsen H, al-Katib A, Fluge O, Pergamenschikov A, Williams C, Zhu SX, Lenning PE, Berresen-Dale A-L, Brown PO, Botstein D. 2000. Molecular portraits of human breast tumorurs. *Nature* **406**: 747–752.

Pockaj BA, Sherry RM, Wei JP, Yannelli JR, Carter CS, Leitman SF, Carasquillo JA, Steinberg SM, Rosenberg SA, Yang JC. 1994. Localization of 111indium-labeled tumour infiltrating lymphocytes to tumour in patients receiving adoptive immunotherapy. Augmentation with cyclophosphamide and correlation with response. *Cancer* **73**: 1731–1737.

Renkvist N, Castelli C, Robbins PF, Parmiani G. 2001. A listing of human tumour antigens recognized by T cells. *Cancer Immunol Immunother* **50**(1): 3–15.

Restifo NP, Marincola FM, Kawakami Y, Taubenberger J, Yannelli JR, Rosenberg SA. 1996. Loss of functional beta 2-microglobulin in metastatic melanomas from five patients receiving immunotherapy. *J Natl Cancer Inst* **88**(2): 100–108.

Richwald GA. 2003. Imiquimod. *Drugs Today* **35**(7): 497–511.

Robertson P, Scadden DT. 2003. Immune reconstitution in HIV infection and its relationship to cancer. *Hematol Oncol Clin N Am* **17**(3): 703–716, vi.

Rosenberg SA, Yang JC, Restifo NP. 2004. Cancer immunotherapy: moving beyond current vaccines. *Nat Med* **10**(9): 909–915.

Rosenberg SA, Yang JC, Schwartzentruber D, Hwu P, Marincola FM, Topalian SL, Restifo NP, Dufour E, Schwartzberg L, Spiess P, Wunderlich J, Parkhurst MR, Kawakami Y, Seipp C, Einhorn JH, White D. 1998. Immunologic and therapeutic evaluation of a synthetic tumour associated peptide vaccine for the treatment of patients with metastatic melanoma. *Nat Med* **4**(3): 321–327.

Rubin JT, Adams SD, Simonis T, Lotze MT. 1991. HLA polymorphism and response to IL-2 bases therapy in patients with melanoma. *Society for Biological Therapy 1991 Annual Meeting*, Vol 1, p 18.

Salgaller ML, Marincola FM, Cormier JN, Rosenberg SA. 1996. Immunization against epitopes in the human melanoma antigen gp100 following patient immunization with synthetic peptides. *Cancer Res* **56**: 4749–4757.

Shankaran V, Ikeda H, Bruce AT, White JM, Swanson PE, Old LJ, Schreiber RD. 2001. IFN-g and lymphocytes prevent primary tumour development and shape tumour immunogenicity. *Nature* **410**: 1107–1111.

Slingluff CL, Jr. Speiser DE. 2005. Cancer vaccines targeting multiple antigens: elucidating cancer immunobiology while developing safe and accessible approaches for therapy and prevention of cancer. *J Transl Med* **3**: 18.

Smyth MJ, Godfrey DI, Trapani JA. 2001. A fresh look at tumour immunosurveillance and immunotherapy. *Nat Immunol* **2**(4): 293–299.

Szyf M, Pakneshan P, Rabbani SA. 2004. DNA methylation and breast cancer. *Biochem Pharmac* **68**(6): 1187–1197.

Takahashi M, Yang XJ, Sugimura J, Backdahl J, Tretiakova M, Qian CN, Gray SG, Knapp R, Anema J, Kahnoski R, Nicol D, Vogelzang NJ, Furge KA, Kanayama H, Kagawa S, Teh BT. 2003. Molecular subclassification of kidney tumours and the discovery of new diagnostic markers. *Oncogene* **22**(43): 6810–6818.

Timmerman JM, Levy R. 2004. Correspondence 2: cancer vaccines: pessimism in check. *Nat Med* **10**(12): 1279.

't Veer LJ, Dai H, van de Vijver MJ, He YD, Hart AA, Mao M, Peterse HL, van der KK, Marton MJ, Witteveen AT, Schreiber GJ, Kerkhoven RM, Roberts C, Linsley PS, Bernards R, Friend SH. 2002. Gene expression profiling predicts clinical outcome of breast cancer. *Nature* **415**(6871): 530–536.

Ulrich C, Schmook T, Sachse MM, Sterry W, Stockfleth E. 2004. Comparative epidemiology and pathogenic factors for nonmelanoma skin cancer in organ transplant patients, *Dermatol Surg* **30**(4 Pt 2): 622–627.

Urosevic M, Maier T, Benninghoff B, Slade H, Burg G, Dummer R. 2003. Mechanisms unerlying imiquimod-induced regression of basal cell carcinoma *in vivo*. *Arch Dermatol* **139**(10): 1325–1332.

Vanderlugt CL, Miller SD. 2002. Epitope spreading in immune-mediated diseases: implications for immunotherapy. *Nat Rev Immunol* **2**(2): 85–95.

Wang E, Marincola FM. 2000. A natural history of melanoma: serial gene expression analysis. *Immunol Today* **21**(12): 619–623.

Wang E, Marincola FM. 2001. cDNA microarrays and the enigma of melanoma immune responsiveness. *Cancer J* **7**(1): 16–23.

Wang E, Panelli MC, Marincola FM. 2003. Genomic analysis of cancer. *Princ Pract Oncol* **17**(9): 1–16.

Wang E, Marincola FM, Stroncek D. 2005. Human leukocyte antigen (HLA) and Human Neutrophil Antigen (HNA) systems. In Hoffman R, Benz EJ, Shattil SJ, Furie B, Cohen HJ, Silberstein LE, McGlave P (eds), *Hematology: Basic Principles and Practice*, 4th edn. Elsevier Science: Philadelphia, PA; 2401–2432.

Wang E, Miller LD, Ohnmacht GA, Mocellin S, Petersen D, Zhao Y, Simon R, Powell JI, Asaki E, Alexander HR, Duray PH, Herlyn M, Restifo NP, Liu ET, Rosenberg SA, Marincola FM. 2002. Prospective molecular profiling of subcutaneous melanoma metastases suggests classifiers of immune responsiveness. *Cancer Res* **62**: 3581–3586.

Wang E, Adams S, Zhao Y, Panelli MC, Simon R, Klein H, Marincola FM. 2003. A strategy for detection of known and unknown SNP using a minimum number of oligonucleotides. *J Transl Med* **1**: 4.

Wang E, Lichtenfels R, Bukur J, Ngalame Y, Panelli MC, Seliger B, Marincola FM. 2004a. Ontogeny and oncogenesis balance the transcriptional profile of renal cell cancer. *Cancer Res* **64**(20): 7279–7287.

Wang E, Panelli MC, Monsurro' V, Marincola FM. 2004b. Gene expression profiling of anti-cancer immune responses. *Curr Opin Mol Ther* **6**(3): 288–295.

Wang E, Panelli MC, Zavaglia K, Mandruzzato S, Hu N, Taylor PR, Seliger B, Zanovello P, Freedman RS, Marincola FM. 2004c. Melanoma-restricted genes. *J Transl Med* **2**: 34.

Wolfel T, Klehmann E, Muller C, Schutt KH, Meyer zum Buschenfelde KH, Knuth A. 1989. Lysis of human melanoma cells by autologous cytolytic T cell clones. Identification of human histocompatibility leukocyte antigen A2 as a restriction element for three different antigens. *J Exp Med* **170**: 797–810.

Yang JC, Rosenberg SA. 1997. An ongoing prospective randomized comparison of interleukin-2 regimens for the treatment of metastatic renal cell cancer. [See comments.] *Cancer J Sci Am* **3**(suppl. 1): S79–84.

Young AN, Amin MB, Moreno CS, Lim SD, Cohen C, Petros JA, Marshall FF, Neish AS. 2001. Expression profiling of renal epithelial neoplasms: a method for tumour classification and discovery of diagnostic molecular markers. *Am J Pathol* **158**(5): 1639–1651.

8

Chemokines Regulate Leukocyte Trafficking and Organ-specific Metastasis

Andor Pivarcsi, Anja Mueller and **Bernhard Homey**

Abstract

Recently, small chemotactic proteins called chemokines have emerged as critical regulators of organ-specific homing of both leukocytes and nonhaematopoietic cells during organogenesis and normal tissue homeostasis. Chemokines play important roles in orchestrating and integrating the afferent and efferent limbs of the immune system and they are implicated in the integration of innate and adaptive immune responses. Although it has been recognized for over a century that the organ preference for metastasis is a nonrandom process, only in the last decade have we started to understand that tumour dissemination is critically regulated by the chemokine system. Accumulating evidence suggests that the organ-specific metastasis of tumours is aided by interactions between functional chemokine receptors expressed on the metastatic tumour cells and the expression of their corresponding chemokine ligands in target organs. Hence, tumour cells seem to 'hijack' the chemokine system and utilize chemokine gradients for directing metastasis, growth and survival. This chapter aims to summarize the recent findings about the multiple roles of chemokines in cancer metastasis.

8.1 Chemokines and Chemokine Receptors

Chemokines are small (8–11 kDa) chemotactic proteins which play a central role in the organization of innate and adaptive immune responses (Homey and Zlotnik, 1999; Matsukawa *et al.*, 2000). The chemokine superfamily (45 members in the human) is thought to be among the first functional protein families completely characterized at the molecular level, offering a unique opportunity to systematically characterize their involvement in physiological and pathophysiological processes. Most chemokines

Immunogenomics and Human Disease Edited by András Falus

have four characteristic cysteines and, depending on the motif displayed by the first two cysteine residues, they have been classified into CXC (termed as CXC ligands, CXCL), CC (CCL), XC (XCL) and CX3C (CX_3CL) chemokine classes (Zlotnik and Yoshie, 2000). In addition, the CXC subfamily has been divided into two groups depending on the presence of the ELR motif preceding the first cysteine residue: the ELR-CXC and the non-ELR-CXC chemokines (Rossi and Zlotnik, 2000).

Chemokines bind to pertussis toxin-sensitive, 7-transmembrane-spanning G-protein coupled receptors (GPCRs; Zlotnik and Yoshie, 2000). To date 18 chemokine receptors have been identified in human: receptors for the CXC chemokines CXCR1–7; the CC chemokines CCR1–10; the receptor for CX3C chemokine CXCR1; and the C chemokine XCR1 (Murphy *et al.*, 2000). In addition, there are several 'orphan' GPCRs, which may also be chemokine receptors for as yet unknown ligands (Laitinen *et al.*, 2004). Chemokine receptor signalling activates different pathways that sustain cell survival, induce gene expression and most importantly enable directional cell migration.

Interestingly, there is a certain degree of promiscuity in the chemokine superfamily with many ligands binding different receptors or vice versa. So-called 'cluster' chemokines representing chemotactic proteins, which share a distinct chromosomal location, are likely to bind to the same receptors. In contrast, 'noncluster' chemokines are ligands, which demonstrate a unique chromosomal location and tend to present a restricted or even specific chemokine receptor interaction (Homey and Zlotnik, 1999; Zlotnik and Yoshie, 2000). Findings of recent studies show that, in contrast to cluster chemokines, noncluster chemokines mediate nonredundant biological functions.

Initially, chemokines have been functionally defined as soluble factors able to control the directional migration and organ-specific recruitment of leukocytes, in particular during infections and inflammation. It appears, however, that chemokines, expressed by virtually all cells including many tumour cell types, are also involved in a number of other physiological and pathological processes including inflammation, metastasis, angiogenesis, wound healing and lymphoid organ development (Rossi and Zlotnik, 2000).

During the multistep process of leukocyte trafficking comprising the floating, rolling, adhesion and extravasation of cells, chemokine ligand–receptor interactions mediate in an integrin-dependent manner the firm adhesion of rolling leukocytes to the endothelium and initiate transendothelial migration from the blood vessel into perivascular pockets (Tan and Thestrup-Pedersen, 1995; Moser and Willimann, 2004). From perivascular spaces, matrix-bound sustained chemokine gradients direct tissue-infiltrating leukocyte subsets to distinct anatomical locations within peripheral tissues (Homey and Bunemann, 2004). Chemokine receptors expressed on specific leukocyte subsets mediate directional migration of the cells toward tissues and organs expressing the corresponding chemokine (Table 8.1).

One of the best examples for the mechanism by which chemokine–chemokine receptor interactions mediate the organ-specific recruitment of leukocytes is CCR10, whose respective ligand CCL27/*CTACK* is produced by epidermal keratinocytes (Morales *et al.*, 1999; Homey *et al.*, 2000). The cutaneous lymphocyte-associated antigen (CLA)

Table 8.1 Chemokine receptors associated with organ-specific recruitment of leukocytes

Chemokine			
Receptor	Ligand	Biological function	Reference
CCR7	*CCL19* *CCL21*	Homing to secondary lymphoid organs, e.g. lymph nodes	(Forster *et al.*, 1999; Gunn *et al.*, 1999)
CCR9	*CCL25*	Immune cell trafficking to the gut	(Youn *et al.*, 1999; Zabel *et al.*, 1999)
CCR10	*CCL27*	Skin-selective recruitment of memory T cells	(Homey *et al.*, 2002)
CXCR4	*CXCL12*	Multiple; organogenesis, haematopoetic system, trafficking to lung, liver, brain, etc.	(Gunn *et al.*, 1999; Sallusto, Mackay and Lanzavecchia, 2000)
CXCR5	*CXCL13*	B cell homing to the lymph nodes	(Forster *et al.*, 1996)

identifies a subset of skin-homing memory T cells. Some 80–90 per cent of memory T cells in inflammatory skin lesions express CLA. In contrast, CLA^+ T lymphocytes represent only 10–15 per cent of the pool of circulating T cells and never exceed 5 per cent of lymphocytes within noncutaneous inflamed sites (Picker *et al.*, 1990; Berg *et al.*, 1991). These observations suggest that an active and specific recruiting process focused on CLA^+ memory T cells is present in inflammatory skin lesions. CLA interacts with its vascular ligand E-selectin and mediates the rolling of distinct leukocyte subsets along the vascular endothelium. However, E-selectin is not skin-specific but it is expressed on inflamed endothelium of various tissues. Therefore, other skin-specific cues must regulate the tissue-specific homing capacity of CLA^+ memory T cells.

CCL27 is abundantly expressed under homeostatic conditions and it is inducible by pro-inflammatory mediators produced during skin inflammation (Morales *et al.*, 1999; Homey *et al.*, 2002). The constitutive high level expression of CCL27 by human keratinocytes suggest that chemokine is an important mediator of the recruitment of CLA^+ memory T cells during the homeostatic trafficking of leukocytes as a part of the immune surveillance. Additionally, neutralization of CCL27, *in vivo*, significantly impairs inflammatory skin responses in mouse models, mimicking allergic contact dermatitis and atopic dermatitis, demonstrating a pivotal role for CCR10/CCL27 interactions in the recruitment and retention of CLA+ CCR10+ memory T cells in the skin (Morales *et al.*, 1999; Homey *et al.*, 2002). More recently, CCL27–CCR10 interactions have also been shown to be involved in the selective recruitment of cytotoxic lymphocytes to the skin in drug-induced cutaneous diseases, providing another example of the central role of chemokine–chemokine receptor interactions in organ-specific homing of leukocytes (Tapia *et al.*, 2004).

Furthermore, it has been demonstrated that CCR7–CCL21 interactions play a critical role in organizing the migration of antigen-loaded dendritic cells and naive T and B cells into the lymph nodes (Gunn *et al.*, 1999; Forster *et al.*, 1999). CCR7 is expressed on the cell surface of naive T and B cells and its expression is upregulated in dendritic cells during maturation, while the respective ligands, CCL21 and CCL19, are abundantly expressed by the secondary lymphoid organs including lymph nodes. Similarly, CXCR5/CXCL13 interactions play a pivotal role in the regulation of B cell migration and localization in secondary lymphoid organs (Forster *et al.*, 1996). CCL25 is abundantly expressed in the epithelial cells of the small intestine; it plays an important role in the selective extravasation of memory T lymphocytes and/or in the migration of $CCR9^+$ lymphocytes into the intestinal tissue (Zabel *et al.*, 1999; Papadakis *et al.*, 2000).

Interestingly, the ligand of CXCR4, CXCL12, is produced in multiple organs including liver, lung and bone marrow. In addition to mediating directional migration of leukocytes, CXCR4–CXCL12 interactions play a critical role in the embryonic development by regulating the migration of progenitor cells into the appropriate microenvironments, suggesting that CXCR4–CXCL12 interactions are vital for the migration of both haematopoietic and nonhaematopoietic cells, *in vivo* (Rossi and Zlotnik, 2000).

Taken together, circulating immune cells are known to use chemokine receptor-mediated mechanisms to home on specific organs (Table 8.1). Therefore, chemokines are critical regulators of organ-specific homing of both leukocytes and nonhaematopoietic cells during organogenesis and normal tissue homeostasis. Chemokines play important roles in orchestrating and integrating the afferent and efferent limbs of the immune system and they are implicated in the integration of innate and adaptive immune responses.

It has been recognized for over a century that the organ preference for metastatic colonization is influenced by interactions between the circulating tumour cell (the

Table 8.2 Chemokine receptors associated with organ-specific metastasis

Chemokine				
Receptor	Ligand	Tumour type(s)	Target organ(s)	Reference
CCR7	*CCL19* *CCL21*	Breast cancer, melanoma, gastric cancer, oesophageal SCC	Lymph nodes	(Muller *et al.*, 2001; Wiley *et al.*, 2001; Mashino *et al.*, 2002; Ding *et al.*, 2003).
CCR9	*CCL25*	Melanoma	Gut	(Youn *et al.*, 2001; Letsch *et al.*, 2004)
CCR10	*CCL27*	Melanoma, cutaneous T-cell lymphoma	Skin	(Muller *et al.*, 2001)
CXCR4	*CXCL12*	Breast cancer, glioblastoma, prostate cancer, melanoma	Multiple; lung, liver, brain, haematopoetic system, etc.	(Muller *et al.*, 2001; Murakami *et al.*, 2002; Balkwill, 2004).
CXCR5	*CXCL13*	Non-Hodgkin's lymphoma	Lymph nodes	(Durig, Schmucker and Duhrsen, 2001)

'seed') and the target host tissue (the 'soil'). Stephen Paget wrote 'An attempt is made in this paper to consider "metastasis" in malignant disease, and to show that the distribution of the secondary growth is not a matter of chance' in the *Lancet* in 1889 (Paget, 1889). Recent findings describing the expression of a specific, nonrandom pattern of functional chemokine receptors on tumour cells raise the possibility that cancer cells may use mechanisms of leukocyte trafficking for metastatic dissemination (Table 8.2).

8.2 Chemokine Receptors in the Organ-Specific Recruitment of Tumour Cells

Tumour metastasis represents the leading cause of mortality in most malignant tumours. Clinicians and pathologists have long known that metastasis formation is not a random but an organ-selective process (Paget, 1889). Most carcinomas preferentially metastasize to particular distant organs, suggesting a critical role of the organmicroenvironment for the localization of metastatic tumour cells. Similarly to tissue-specific homing of leukocytes, tumour metastasis consists of a series of sequential steps, all of which must be successfully completed. These include the release of cancer cells from the primary tumour, invasion into lymphatic or vascular vessels, adherence of a floating tumour cell to the endothelium at the target organs, extravasation of tumour cells from the vessels into the tissues, and ultimately the survival and proliferation of tumour cells in a new tissue environment (Chambers, Groom and MacDonald, 2002; Murakami, Cardones and Hwang, 2004). Accumulating evidence now indicates that organ-specific metastasis is aided by interactions between functional chemokine receptors expressed on the metastatic tumour cells and the expression of their corresponding chemokine ligands in target organs. Recent reports demonstrate that many tumour types express specific, nonrandom patterns of chemokine receptors, which play a central role in defining the metastatic characteristics of the given tumour entity.

CXCR4

Both CXCR4 and its ligand CXCL12 are widely expressed in normal tissues and play a critical role in foetal development, mobilization of haematopoetic cells and trafficking of lymphocytes (Rossi and Zlotnik, 2000). The importance of CXCR4–CXCL12 interactions is highlighted by the fact that targeted disruption of either CXCR4 or CXCL12 genes in knock-out mouse models results in a lethal phenotype at the embryonic stage due to profound defects in haematopoietic, cardiovascular and nervous systems (Rossi and Zlotnik, 2000). *In vivo* studies using neutralizing antibodies to CXCR4 demonstrated its function in the homing and repopulation of haematopoietic stem cells into the bone marrow. Collectively, these findings demonstrate a pivotal role for CXCR4–CXCL12 interactions in trafficking of progenitor cells and leukocytes.

Recently Muller *et al.* (2001) demonstrated that, among all chemokine receptors, human breast cancer cells specifically express CXCR4. The only known ligand for CXCR4 is CXCL12 (SDF-1), which is particularly highly expressed in target organs for breast cancer metastasis such as lung, liver and lymph nodes. Moreover, breast cancer cells migrated toward tissue extracts from these target organs and this migration could be partially inhibited by neutralizing antibodies against CXCR4. Furthermore, neutralizing the interactions of CXCL12–CXCR4 significantly impaired metastasis of breast cancer cells to regional lymph nodes and lung, demonstrating that chemokine–chemokine receptor interactions play a critical role in the organ-specific metastasis of tumors (Muller *et al.*, 2001). These findings suggest that cancer cells utilize chemokine networks involved in the organ-specific migration of leukocytes and normal cell types during organogenesis for metastasizing into distant organs. Behaving like leukocytes, CXCR4-expressing breast cancer cells follow chemotactic gradients formed by the specific ligand, CXCL12, that is released in high quantities only by certain organs such as lung, liver and bone marrow.

The involvement of CXCR4 in tumour metastasis to tissues highly expressing its ligand is not limited to breast cancer as CXCR4 is expressed in many cancer types including melanoma. Organ-specific patterns of melanoma metastasis also correlate with the expression of CXCR4 as high expression of CXCL12 is reported at major sites of metastasis, lung, liver and lymph nodes (Muller *et al.*, 2001). Confirming these observations, Murakami *et al.* (2002) demonstrated that overexpression of CXCR4 gene in melanoma cells dramatically enhanced pulmonary metastases of CXCR4-expressing cells following intravenous inoculation of tumour cells into syngeneic mice, demonstrating that CXCL4 plays an important role in promoting organ-specific metastasis.

Recent studies demonstrate that high levels of the chemokine receptor CXCR4 confer a more aggressive behaviour on prostate cancer cells and that CXCR4–CXCL12 interactions participate in localizing tumours to the bone marrow and the lymph nodes for prostate cancer (Darash-Yahana *et al.*, 2004; Sun *et al.*, 2005). Moreover, CXCR4 may act not only as a homing receptor for circulating tumour cells, but also as a positive regulator of tumour growth and angiogenesis, indicating the possibility of using CXCR4 as a future therapeutic interventional target, even in advanced cases of metastatic prostate cancer (Darash-Yahana *et al.*, 2004).

Expression analysis indicated that CXCR4 is highly expressed in 57 per cent of glioblastomas and in 88 per cent of the glioblastoma cell lines analysed. CXCR4 activation promotes survival, proliferation and migration of tumour cells through the activation of extracellular signal-regulated kinases 1 and 2 and Akt (Zhou *et al.*, 2002). Accordingly, overexpression of CXCR4 in glioblastoma cell lines enhances their soft agar colony-forming capability (Sehgal *et al.*, 1998). In contrast, systemic administration of CXCR4 antagonist AMD 3100 inhibits growth of intracranial glioblastoma and medulloblastoma xenografts by increasing apoptosis and decreasing the proliferation of tumour cells (Rubin *et al.*, 2003).

CXCL12, the specific ligand of CXCR4, upregulates the expression β1 integrin and increases the adhesion of melanoma cells to fibronectin (Robledo *et al.*, 2001). Thus, signalling via chemokine receptors can enhance adhesion of tumour cells to the extracellular matrix through regulation of the expression of adhesion molecules.

To date, the expression of functional CXCR4 has been described for at least 23 different types of cancer, including cancers of epithelial, mesenchymal and haematopoetic origin, showing that hijacking of leukocyte recruitment pathways is a common means of metastasis formation (Balkwill, 2004).

CCR7

CCR7–CCL21 interactions play a pivotal role in organization of migration of antigen presenting cells and naive T and B cells to lymph nodes, which are abundant sources of CCL21. This unique role of CCR7–CCL21 interactions is supported by the observation that, in *plt/plt* mice, which lack CCL21, maturing dendritic cells (DCs) are not able to migrate from skin to lymph nodes (Gunn *et al.*, 1999). Moreover, targeted disruption of the CCR7 gene results in the impaired homing of naive T and B cells into lymph nodes, demonstrating a central role for CCR7 during the recruitment of lymphocytes into secondary lymphoid organs (Forster *et al.*, 1999). Interestingly, certain viruses interfere with the antigen presentation by blocking the upregulation of CCR7 expression in DCs. Hence, virus may prevent DC migration to the draining lymph nodes and the initiation of antiviral immune responses (Sallusto, Mackay and Lanzavecchia, 2000).

Recently, elevated expression of CCR7 was detected in human malignant breast tumours in comparison to normal breast tissue (Muller *et al.*, 2001). The respective ligands for CCR7, CCL19 and CCL21 (SLC) are highly expressed in secondary lymphoid organs in which breast cancer metastases are frequently found. Studies analysing tissue biopsies demonstrated that the most important factor determining lymph node metastasis in gastric cancer was CCR7 expression in the primary tumour (Mashino *et al.*, 2002).

High CCR7 expression in oesophageal SCC correlates with lymphatic permeation, lymph node metastasis, tumour depth and tumour-node-metastasis stage and is associated with poor survival (Ding *et al.*, 2003). Moreover, *in vitro* studies demonstrated that CCL21 significantly increased the cell migration ability, pseudopodia formation and directional migration of oesophageal SCC cell lines, suggesting that the CCR7–CCL21 receptor ligand system plays an important role in the lymph node metastasis of esophageal SCC (Ding *et al.*, 2003).

Moreover, Wiley *et al.* demonstrated that melanoma cells expressing functional CCR7 metastasize 700-fold more efficiently to the draining lymph nodes following injection of tumour cells into syngeneic mice then do melanoma cells that do not express this receptor. Thus, cancer cells, including melanoma, appear to utilize the same receptor (CCR7) used by dendritic cells and T cells for homing to the lymph nodes (Wiley *et al.*, 2001; Murakami, Cardones and Hwang, 2004).

CXCR5

CXCR5 plays a pivotal role in the regulation of B cell migration and localization in secondary lymphoid organs. Disruption of CXCR5 in knock-out mice leads to a disturbed development of primary follicles (Forster *et al.*, 1996). As a consequence of impaired of B cell migration, these mice do not develop inguinal lymph nodes, have fewer Peyer's patches and show an altered architecture of the spleen (Forster *et al.*, 1996). In the mutant mice, activated B cells failed to migrate from the T cell-rich zone into B cell follicles of the spleen and, despite high numbers of germinal centre founder cells, no functional germinal centres develop in this organ (Forster *et al.*, 1996). The respective ligand of CXCR5, CXCL13, also known as B cell-attracting chemokine (BCA)-1, is expressed in lymphoid tissues and is a selective and highly efficacious chemoattractant for human blood B cells *in vitro* (Legler *et al.*, 1998).

Recent findings suggest that malignant B cells of different origin constitutively express CXCR5 (Trentin *et al.*, 2004). CXCR5 expression is particularly high in non-Hodgkin's lymphomas with lymph node involvement, supporting the concept that CXCR5 and its ligand may be involved in the homing of circulating malignant B cells into lymphoid tissues (Durig, Schmucker and Duhrsen, 2001). CXCL13 is able to trigger migration of malignant B lymphocytes isolated from other types of leukaemias, including hairy cell leukaemia (HCL), mantle cell lymphoma (MCL), chronic lymphocytic leukaemia (CLL) and marginal zone B cell lymphoma (MZL), indicating that chemokine receptor expression might play a role in malignant B cell circulation (Trentin *et al.*, 2004).

CCR10

Skin-infiltrating T cells play a pivotal role during the initiation and maintenance of inflammatory and autoimmune skin diseases such as atopic dermatitis, allergic contact dermatitis and psoriasis. Recent studies have identified the skin-specific chemokine CCL27 (CTACK), which is expressed by human keratinocytes, and its receptor CCR10 play a central role in the regulation of memory T cell recruitment to the skin (Homey *et al.*, 2002; Homey, 2004; Homey and Bunemann, 2004). In mouse models of allergic contact dermatitis and atopic dermatitis, neutralization of CCL27 significantly impairs inflammatory skin responses, demonstrating the central role of CCL27–CCR10 interaction in the organization of T cell recruitment to the skin (Homey *et al.*, 2002). Interestingly, human malignant melanoma cell lines and tumours frequently exhibit high levels of CCR10 expression in comparison to normal human melanocytes (Muller *et al.*, 2001; Murakami *et al.*, 2003). The association of CCR10 expression on malignant melanoma cells with the skin-specific expression of its ligand and the high incidence of skin metastases in this malignant disease support the involvement of CCR10 in melanoma metastasis (Muller *et al.*, 2001). Additionally, Murakami *et al.* demonstrated that overexpression of CCR10 in B16 melanoma cell line enhanced its tumour-forming capacity in mouse ear. Interestingly, CCR10-transfected melanoma cells had an increased resistance for Fas-mediated apoptosis and they were less susceptible to the host anti-tumour response, suggesting that CCR10–CCL27 interactions play an important role in tumour progression apart from their ability to affect skin-specific metastasis.

Cutaneous T cell lymphoma (CTCL) belongs to the large and heterogeneous group of mature (peripheral) T cell neoplasms characterized by clonal expansion of a mature CD4-positive clone with a T_{Helper} phenotype, putatively from a skin-homing subset of memory T cells. Moreover, CTCL is characterized by the recruitment of malignant T cell clones into the skin. Recent findings of Notohamiprodjo *et al.* (2005) strongly suggest that CCR10 and its ligand CCL27 may contribute to the skin infiltration of malignant T cells in this group of lymphoproliferative disorders.

CCR9

CCR9 is an excellent example for a chemokine receptor mediating the organ-specific recruitment of leukocytes, since its ligand, CCL25 (TECK), is selectively expressed by epithelial cells in the small intestine and thymus. CCR9–CCL25 interactions play a pivotal role in the regulation of T cell trafficking, during T cell development and mucosal immunity (Youn *et al.*, 1999; Zabel *et al.*, 1999). CCL25 is selectively expressed by thymic epithelial cortical cells and thymic epithelial medullary cells, therefore TECK/CCL25 may direct the centripetal intrathymic progression followed by developing T cells, and/or constitute a thymic retention factor. CCL25 is also abundantly expressed in the epithelial cells of the small intestine, therefore, in addition to its role in the intrathymic T cell development, CCL25 may play a role in the selective extravasation of memory intestinal T lymphocytes and/or in the migration of $CCR9^{+}$ lymphocytes into the intestinal tissue (Zabel *et al.*, 1999; Papadakis *et al.*, 2000).

CCR9 has been shown to be highly expressed on malignant melanoma cells isolated from small intestine metastases, suggesting that cell surface expression of organ-specific chemokine receptors is associated with targeted metastasis (Letsch *et al.*, 2004). In addition to directional migration of cells expressing the receptor, CCR9 activation also provides a cell survival signal (Youn *et al.*, 2001), suggesting that CCR9 and its ligands may promote survival or proliferation of cancer cells.

CXCR3

CXCR3 is expressed on the surface of a number of cell types, including activated T cells and NK cells, subsets of inflammatory dendritic cells, macrophages and B cells. CXCR3 is functionally active in a wide subset of T cells and is likely to be involved in the regulation of T cell-trafficking into the epidermis given that its ligands, CXCL9 (Mig), CXCL10 (IP-10) and CXCL11 (I-TAC), are known to be highly expressed in the epidermis under inflammatory conditions. Recent data suggest that CXCR3, in addition to other chemokine receptors, is also expressed on T and NK cell lymphomas and is likely to be involved in the epidermotropism of the malignant T cells (Ishida *et al.*, 2004; Notohamiprodjo *et al.*, 2005).

Recently, Kawada *et al.* (2004) demonstrated that melanoma metastasis to lymph nodes was largely suppressed when CXCR3 expression in B16F10 cells was reduced by antisense RNA or when mice were treated with specific antibodies against CXCL9 and CXCL10, suggesting that CXCR3 might be a novel therapeutic target to suppress lymph node metastasis.

8.3 Cancer Therapy Using Chemokine Receptor Inhibitors

The ability of chemokine–chemokine receptor interactions to direct tumour cells to target organs as well as their involvement in tumour growth and survival indicates that interference with these interactions may have therapeutic potential in human malignances. The results of studies using different cancer cell types show that inhibition of chemokine receptors reduces the frequency of metastasis. Therefore, academic and industrial research is now focusing on compounds blocking chemokine receptors using different approaches including neutralizing antibodies, small synthetic peptides and siRNAs.

Several years ago, it has been shown that treatment of SCID mice with anti-CXCR4 neutralizing mAbs after intravenous or orthotopic injection of human breast cancer tumour cells significantly inhibited lung and lymph node metastasis (Muller *et al.*, 2001). Recently, a small molecule antagonist of CXCR4, named TN14003, has been identified, which effectively blocked the binding of CXCL12 to CXCR4 and inhibited breast cancer cell invasion in *in vitro* assays (Liang *et al.*, 2004). *In vivo*, TN14003 significantly inhibited breast cancer metastasis to the lung, suggesting that CXCR4 antagonists can be useful therapeutic agents for the therapy of breast cancer metastasis (Liang *et al.*, 2004).

Systemic administration of another small molecule antagonist of CXCR4, AMD 3100, inhibited growth of intracranial glioblastoma and medulloblastoma xenografts by increasing apoptosis and decreasing the proliferation of tumour cells (Rubin *et al.*, 2003).

RNA interference technology, silencing targeted genes in mammalian cells, has become a powerful tool for studying gene function. Recently, independent groups successfully blocked metastasis of breast cancer cells by silencing CXCR4 gene expression with small interfering RNAs (siRNAs) specific to CXCR4 (Lapteva *et al.*, 2005; Liang *et al.*, 2005). CXCR4-specific siRNAs not only impaired the invasion of breast cancer cells in Matrigel invasion assay but also blocked breast cancer metastasis *in vivo*, owing to inhibited metastatic capability of tumour cells. These findings suggest that downregulation of CXCR4 gene expression by RNA interference can efficiently block cancer metastasis mediated by CXCR4–CXCL12 interaction (Lapteva *et al.*, 2005; Liang *et al.*, 2005).

Collectively, additional studies will lead to the identification of novel chemokine receptor antagonists, which may be used therapeutically in the treatment of tumours.

8.4 Conclusions

Breast cancer metastasis, as characterized by the original observation from Paget in the nineteenth century, has a distinct metastatic pattern that preferentially involves the regional lymph nodes, bone marrow, lung and liver, suggesting the existence of a specific pathway that governs the metastasis of tumour cells. Recent studies provide evidence that tumour cells use chemokine-mediated mechanisms during the process of metastasis. According to our current knowledge, the organ-specific metastasis of tumours is aided by interactions between functional chemokine receptors expressed on the metastatic tumour cells and the expression of their corresponding chemokine ligands in target organs. Investigation of metastatic pattern and chemokine expression profile of different human tumours suggests that CCR7 plays a central role in the formation of lymph node metastases, while CXCR4 and CCR10 are involved in pulmonary and skin metastases, respectively. However, expression of chemokine receptors on tumour cells provides not merely a cue for tumour cell migration. In addition to mediating directional migration of tumour cells toward chemokine gradients, chemokine–chemokine receptor interactions are likely to be involved in processes related to the invasion, adhesion, transendothelial migration, survival and growth of tumour cells in the target organ. Currently, small molecule antagonists against chemokine receptors are subject to intensive investigations since inhibition of these receptors may decrease the ability of tumour cells to disseminate, form metastases and survive. Taken together, chemokine receptors may represent promising targets for the development of novel strategies to interfere with tumour progression and metastasis.

References

Balkwill F. 2004. Cancer and the chemokine network. *Nat Rev Cancer* **4**(7): 540–550.

Berg EL, Yoshino T, Rott LS, Robinson MK, Warnock RA, Kishimoto L, Picker J, Butcher EC. 1991. The cutaneous lymphocyte antigen is a skin lymphocyte homing receptor for the vascular lectin endothelial cell-leukocyte adhesion molecule 1. *J Exp Med* **174**(6): 1461–1466.

Chambers AF, Groom AC, MacDonald IC. 2002. Dissemination and growth of cancer cells in metastatic sites. *Nat Rev Cancer* **2**(8): 563–572.

Darash-Yahana M, Pikarsky E, Abramovitch R, Zeira E, Pal B, Karplus R, Beider K, Avniel S, Kasem S, Galun E, Peled A. 2004. Role of high expression levels of CXCR4 in tumor growth, vascularization, and metastasis. *Faseb J* **18**(11): 1240–1242.

Ding Y, Shimada Y, Maeda M, Kawabe A, Kaganoi J, Komoto I, Hashimoto Y, Miyake M, Hashida H, Imamura M. 2003. Association of CC chemokine receptor 7 with lymph node metastasis of esophageal squamous cell carcinoma. *Clin Cancer Res* **9**(9): 3406–3412.

Durig J, Schmucker U, Duhrsen U. 2001. Differential expression of chemokine receptors in B cell malignancies. *Leukemia* **15**(5): 752–756.

Forster R, Mattis AE, Kremmer E, Wolf E, Brem G, Lipp M. 1996. A putative chemokine receptor, BLR1, directs B cell migration to defined lymphoid organs and specific anatomic compartments of the spleen. *Cell* **87**(6): 1037–1047.

Forster R, Schubel A, Breitfeld D, Kremmer E, Renner-Muller I, Wolf E, Lipp M. 1999. CCR7 coordinates the primary immune response by establishing functional microenvironments in secondary lymphoid organs. *Cell* **99**(1): 23–33.

Gunn MD, Kyuwa S, Tam C, Kakiuchi T, Matsuzawa A, Williams LT, Nakano H. 1999. Mice lacking expression of secondary lymphoid organ chemokine have defects in lymphocyte homing and dendritic cell localization. *J Exp Med* **189**(3): 451–460.

Homey B. 2004. Chemokines and chemokine receptors as targets in the therapy of psoriasis. *Curr Drug Targets Inflamm Allergy* **3**(2): 169–174.

Homey B, and Bunemann E. 2004. Chemokines and inflammatory skin diseases. *Ernst Schering Research Foundation Workshop* no. 45, pp 69–83.

Homey B, and Zlotnik A. 1999. Chemokines in allergy. *Curr Opin Immunol* **11**(6): 626–634.

Homey B, Wang W, Soto H, Buchanan ME, Wiesenborn A, *et al.* 2000. Cutting edge: the orphan chemokine receptor G protein-coupled receptor-2 (GPR-2, CCR10) binds the skin-associated chemokine CCL27 (CTACK/ALP/ILC). *J Immunol* **164**(7): 3465–3470.

Homey B, Alenius H, Muller A, Soto H, Bowman EP, Yuan W, McEvoy L, Laverma AI, Assmann T, Bunemann E, Lehto M, Wolff H, Yen D, Marxhausen H, To W, Sedgewick J, Ruzicka T, Lehman P, Zlotnik A. 2002. CCL27–CCR10 interactions regulate T cell-mediated skin inflammation. *Nat Med* **8**(2): 157–165.

Ishida T, Inagaki H, Utsunomiya A, Takatsuka Y, Komatsu H, Iida S, Takeuchi G, Eimoto T, Nakamura S, Ueda R. 2004. CXC chemokine receptor 3 and CC chemokine receptor 4 expression in T-cell and NK-cell lymphomas with special reference to clinicopathological significance for peripheral T-cell lymphoma, unspecified. *Clin Cancer Res* **10**(16): 5494–5500.

Kawada K, Sonoshita M, Sakashita H, Takabayashi A, Yamaoka Y, Marake T, Inaba K, Minato N, Oshima M, Taketo MM. 2004. Pivotal role of CXCR3 in melanoma cell metastasis to lymph nodes. *Cancer Res* **64**(11): 4010–4017.

Laitinen T, Polvi A, Rydman P, Vendelin J, Pulkkinen V, Salmikangas P, Makela S, Rehn M, Pirskanen A, Rautanen A, Zuchelli M, Gullsten H, Leino M, Alenius H, Petays T, Haahtela T, Laitinen A, Laprise L, Hudson TJ, Laitinen LA, Kere J. 2004. Characterization of a common susceptibility locus for asthma-related traits. *Science* **304**(5668): 300–304.

Lapteva N, Yang AG, Sanders DE, Strube RW, Chen SY. 2005. CXCR4 knockdown by small interfering RNA abrogates breast tumor growth *in vivo*. *Cancer Gene Ther* **12**(1): 84–89.

Legler DF, Loetscher M, Roos RS, Clark-Lewis I, Baggiolini M, Moser B. 1998. B cell-attracting chemokine 1, a human CXC chemokine expressed in lymphoid tissues, selectively attracts B lymphocytes via BLR1/CXCR5. *J Exp Med* **187**(4): 655–660.

Letsch A, Keilholz U, Schadendorf D, Assfalg G, Asemissen AM, Thiel E, Scheibenbogen C. 2004. Functional CCR9 expression is associated with small intestinal metastasis. *J Invest Dermatol* **122**(3): 685–690.

Liang Z, Wu T, Lou H, Yu X, Taichman RS, Lau SK, Nie S, Umbreit J, Shim H. 2004. Inhibition of breast cancer metastasis by selective synthetic polypeptide against CXCR4. *Cancer Res* **64**(12): 4302–4308.

Liang Z, Yoon Y, Votaw J, Goodman MM, Williams L, Shim H. 2005. Silencing of CXCR4 blocks breast cancer metastasis. *Cancer Res* **65**(3): 967–971.

Mashino K, Sadanaga N, Yamaguchi H, Tanaka F, Ohta M, Shibuta K, Inoue H, Mori M. 2002. Expression of chemokine receptor CCR7 is associated with lymph node metastasis of gastric carcinoma. *Cancer Res* **62**(10): 2937–2941.

Matsukawa A, Hogaboam CM, Lukacs NW, Kunkel SL. 2000. Chemokines and innate immunity. *Rev Immunogenet* **2**(3): 339–358.

Morales J, Homey B, Vicari AP, Hudak S, Oldham E, Hedrick J, Orozco R, Copeland NG, Jerkins NA, McEvoy LM, Zlotnik A. 1999. CTACK, a skin-associated chemokine that preferentially attracts skin-homing memory T cells. PG-14470–5. *Proc Natl Acad Sci USA* **96**(25): 14470–14475.

Moser B, Willimann K. 2004. Chemokines: role in inflammation and immune surveillance. *Ann Rheum Dis* **63**(2): ii84–ii89.

Muller A, Homey B, Soto H, Ge N, Catron D, Buchanan ME, McClanahan T, Murphy E, Yuan W, Wagner SN, Barrera JL, Mohar A, Verastegui E, Zlonik A. 2001. Involvement of chemokine receptors in breast cancer metastasis. *Nature* **410**(6824): 50–56.

Murakami T, Cardones AR, Hwang ST. 2004. Chemokine receptors and melanoma metastasis. *J Dermatol Sci* **36**(2): 71–78.

Murakami T, Maki W, Cardones AR, Fang H, Tun Kyi A, Nestle FO, Hwang ST. 2002. Expression of CXC chemokine receptor-4 enhances the pulmonary metastatic potential of murine B16 melanoma cells. *Cancer Res* **62**(24): 7328–7334.

Murakami T, Cardones AR, Finkelstein SE, Restifo NP, Klaunberg BA, Nestle FO, Cstillo SS, Dennis PA, Hwang ST. 2003. Immune evasion by murine melanoma mediated through CC chemokine receptor-10. *J Exp Med* **198**(9): 1337–1347.

Murphy PM, Baggiolini M, Charo IF, Hebert CA, Horuk R, Matsushima K, Miller LH, Oppenheim JJ, Power CA. 2000. International union of pharmacology. XXII. Nomenclature for chemokine receptors. *Pharmacol Rev* **52**(1): 145–176.

Notohamiprodjo M, Segerer S, Huss R, Hildebrandt B, Soler D, Difarzadeh R, Buck W, Nelson PJ, von Luettichau J. 2005. CCR10 is expressed in cutaneous T-cell lymphoma. *Int J Cancer* **7**: 7.

Paget S. 1889. The distribution of secondary growths in cancer of the breast. *Lancet* **1**: 99–101.

Papadakis KA, Prehn J, Nelson V, Cheng L, Binder SW, Ponath PD, Andreus DP, Targan SR. 2000. The role of thymus-expressed chemokine and its receptor CCR9 on lymphocytes in the regional specialization of the mucosal immune system. *J Immunol* **165**(9): 5069–5076.

Picker LJ, Terstappen LW, Rott LS, Streeter PR, Stein H, Butcher EC. 1990. Differential expression of homing-associated adhesion molecules by T cell subsets in man. *J Immunol* **145**(10): 3247–3255.

Robledo MM, Bartolome RA, Longo N, Rodriguez-Frade JM, Mellado M, Longo I, van Muijen GN, Sanchez-Mateos P, Teixido J. 2001. Expression of functional chemokine receptors CXCR3 and CXCR4 on human melanoma cells. *J Biol Chem* **276**(48): 45098–45105.

Rossi D, Zlotnik A. 2000. The biology of chemokines and their receptors. *A Rev Immunol* **18**: 217–242.

Rubin JB, Kung AL, Klein RS, Chan JA, Sun Y, Schmidt K, Kieran MW, Luster AD, Segal RA. 2003. A small-molecule antagonist of CXCR4 inhibits intracranial growth of primary brain tumors. *Proc Natl Acad Sci USA* **100**(23): 13513–13518.

Sallusto F, Mackay CR, Lanzavecchia A. 2000. The role of chemokine receptors in primary, effector, and memory immune responses. *A Rev Immunol* **18**: 593–620.

Sehgal A, Keener C, Boynton AL, Warrick J, Murphy GP. 1998. CXCR-4, a chemokine receptor, is overexpressed in and required for proliferation of glioblastoma tumor cells. *J Surg Oncol* **69**(2): 99–104.

Sun YX, Schneider A, Jung Y, Wang J, Dai J, Cook K, Osman NI, Koh-Paige AJ, Shim H, Pienta KJ, Keller ET, McCouley LK, Taichman RS. 2005. Skeletal localization and neutralization of the SDF-1(CXCL12)/CXCR4 axis blocks prostate cancer metastasis and growth in osseous sites *in vivo*. *J Bone Miner Res* **20**(2): 318–329.

Tan J, and Thestrup-Pedersen K. 1995. T lymphocyte chemotaxis and skin diseases. *Exp Dermatol* **4**(5): 281–290.

Tapia B, Padial A, Sanchez-Sabate E, Alvarez-Ferreira J, Morel E, Blanca M, Bellon T. 2004. Involvement of CCL27-CCR10 interactions in drug-induced cutaneous reactions. *J Allergy Clin Immunol* **114**(2): 335–340.

Trentin L, Cabrelle A, Facco M, Carollo D, Miorin M, Tosoni A, Pizzo P, Binotto G, Nicolasdi L, Zambello R, Adami F, Agostini C, Semenzato G. 2004. Homeostatic chemokines drive migration of malignant B cells in patients with non-Hodgkin lymphomas. *Blood* **104**(2): 502–508.

Wiley HE, Gonzalez EB, Maki W, Wu MT, Hwang ST. 2001. Expression of CC chemokine receptor-7 and regional lymph node metastasis of B16 murine melanoma. *J Natl Cancer Inst* **93**(21): 1638–1643.

Youn BS, Kim CH, Smith FO, Broxmeyer HE. 1999. TECK, an efficacious chemoattractant for human thymocytes, uses GPR-9-6/CCR9 as a specific receptor. *Blood* **94**(7): 2533–2536.

Youn BS, Kim YJ, Mantel C, Yu KY, Broxmeyer HE. 2001. Blocking of c-FLIP(L)-independent cycloheximide-induced apoptosis or Fas-mediated apoptosis by the CC chemokine receptor 9/TECK interaction. *Blood* **98**(4): 925–933.

Zabel BA, Agace WW, Campbell JJ, Heath HM, Parent D, Roberts AI, Ebert EC, Kassam N, Qin S, Zovko M, LaRosa GJ, Yang LL, Soler D, Butcher EC, Ponath PD, Parker CM, Andrew PD. 1999. Human G protein-coupled receptor GPR-9-6/CC chemokine receptor 9 is selectively expressed on intestinal homing T lymphocytes, mucosal lymphocytes, and thymocytes and is required for thymus-expressed chemokine-mediated chemotaxis. *J Exp Med* **190**(9): 1241–1256.

Zhou Y, Larsen PH, Hao C, Yong VW. 2002. CXCR4 is a major chemokine receptor on glioma cells and mediates their survival. *J Biol Chem* **277**(51): 49481–49487.

Zlotnik A, O Yoshie 2000. Chemokines: a new classification system and their role in immunity. *Immunity* **12**(2): 121–127.

9

Towards a Unified Approach to New Target Discovery in Breast Cancer: Combining the Power of Genomics, Proteomics and Immunology

Laszlo G. Radvanyi, Bryan Hennessy, Kurt Gish, Gordon Mills and **Neil Berinstein**

9.1 Introduction

The sequencing of the human genome capped off a series of milestones in human medicine and basic biology at the end of the last century (Venter *et al.*, 2001). For the first time we now have a roadmap of all the possible genes expressed in *Homo sapiens*, including their chromosomal location and promoter and mRNA sequences. In this new century our challenge is to meaningfully combine the information we have obtained from our human genome sequencing efforts with powerful new tools in bioinformatics, differential gene expression analysis and the emerging field of proteomics to create a powerful systems biology tool to identify new human disease markers as well as targets for disease therapy and disease prevention. This is especially critical for the complex problem of cancer, where many signalling pathways interact through the function of aberrant proteins modulating not only the transformed phenotype but also the extent of the antitumour immune response. A number of old as well as newly developed gene discovery technologies have capitalized on the availability of the full human genome sequence and have led the way in identifying new targets for both drug-based and immune-based therapy in multiple types of cancer. These genomic-based techniques are now increasingly being combined with other downstream technologies, such as proteomics and new

Immunogenomics and Human Disease Edited by András Falus

immunological assay tools, needed to validate both the specificity and extent of target gene expression at the protein level and for the identification of new targets for cancer immunotherapy based on active vaccination and monoclonal antibody treatment.

Our goal in this chapter is to describe some of the genomics-based approaches used to identify genes overexpressed in breast cancer and how these high-throughput genomics tools are being combined with new high-throughput proteomics methods in conjunction with immunological assays defining new target immunogenicity, together with mapping the 'immunome' of cancer cells, in a unified approach to new tumour antigen discovery and validation (Figure 9.1). We will illustrate how we have used both low-resolution chromosomal screens, such as comparative genomic hybridization (CGH), and high-resolution screens using differential gene expression analysis with DNA microarrays to find new targets in breast cancer. We will then describe how these genomic approaches can been combined with proteomic tools (e.g. immunohistochemistry and protein chips) as well as immunologic assay tools (T-cell reactivity and epitope identification) to further evaluate the suitability of new candidate genes as potential targets and some of the key issues involved. Using these approaches we have identified a number of different loci (e.g. 8q23.2) amplified in the genome of breast cancer cells associated with specific genes overexpressed in breast cancer that are not only tumour-specific but also potentially highly immunogenic. One of these genes, TRPS-1, mapped to chromosome 8q23.2, the region found most amplified in breast cancer through CGH analysis. Another similar highly overexpressed gene we identified was BFA5, which mapped to chromosome 10p11 and was identified as the NY-BR-1 gene found also at the Ludwig Institute for Cancer Research (New York) using a totally different proteomics-based screening method called SEREX (Jager *et al.*, 2001). This attests to how widely different methodologies can be used to discover common genes and antigens and how these different technologies can be combined to validate new target genes in cancer and other diseases.

In the final section we will look towards the future by addressing the emerging problem of information complexity or 'information overload' in new tumour target discovery and tumour immunology. We are clearly faced with a growing problem of combining large data sets obtained from a large variety of different systems bridging genomics, proteomics, immunology and initial clinical trials into a meaningful and manageable information resource. Data on target gene expression analysis in tumours under different conditions, proteomic assessments (e.g. IHC and tissue microarrays), epitope identification and immunogenicity studies, target gene interaction at various functional levels, how target expression and immunogenicity is altered under different condition and disease stages, and finally, how all this relates to clinical decision-making and the outcome of clinical trials, will eventually need to be captured in a useable format. A key question is how we can effectively accomplish this with manageable databases using some of the principles of systems biology. Moreover, how do we decide what data should be considered and what data be discarded (i.e. the quality or bar that a data set requires) when constructing such a

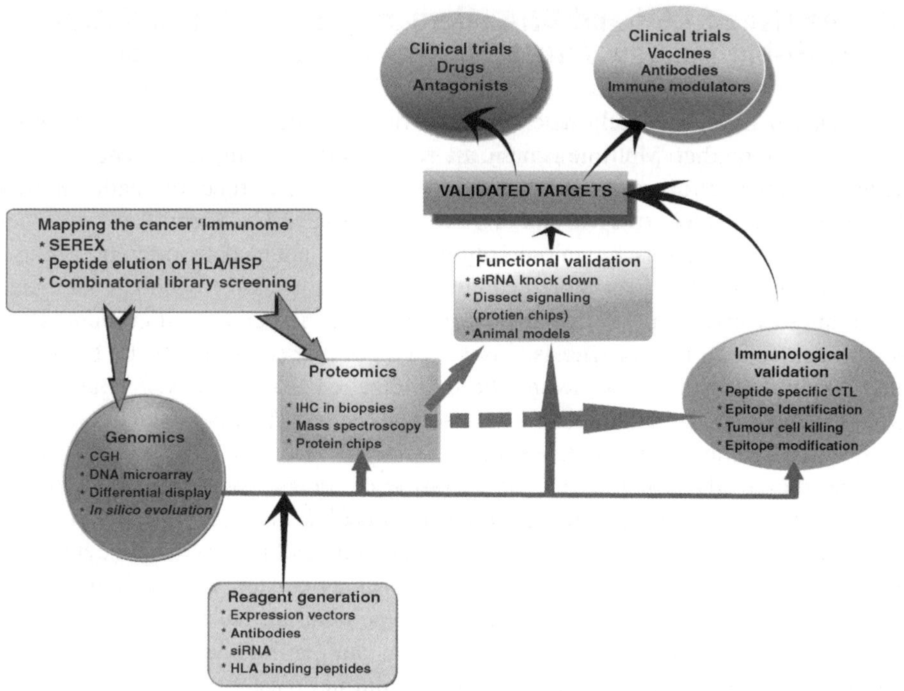

Figure 9.1 Approaches used for new target discovery in cancer for drug-based and antigen-specific immunotherapy. Genomic approaches have led the way together with more recent developments in proteomic technologies (indicated in the boxes). In addition to these technologies, the development of new immunological assays, such as SEREX and mapping peptides eluted from HLA molecules, have added an extra dimension to new target discovery for immunotherapy through direct identification immunogenic peptides and whole tumour antigens at the protein level. These technologies will serve to map the 'immunome' of cancer cells (all possible antigens in different types and stages of cancer) and are linked to genomics and proteomic tools to verify full-length sequence and target prevalence. A critical activity needed for further downstream work after initial antigen discovery is the generation of suitable reagents such as expression vectors, gene knockdown reagents and antibodies in order to validate target expression and tumour specificity, and for pre-clinical proof-of-concept studies. IHC is a critical for assessing target expression at the protein level. Target validation can also involve functional analysis of new target genes using siRNA knockdown when anticipating a drug-based approach for therapy. This validation needs to confirm that target gene function is needed for tumour cell survival and/or proliferation *in vitro* and in animal models. Immunological validation is needed when anticipating targetted antibody-based or active vaccination approaches against a new target. Identification of immunogenic epitopes capable of eliciting killing of tumour cells by CTL is a critical criterion that needs to be met. Only after all these parameters have been met can a target be considered 'validated' for further downstream testing in clinical trials. (A colour reproduction of this figure can be viewed in the colour plate section)

database. Inevitably, in order to reap the fruits of our labour in new target discovery in cancer we propose the development of a universal database using systems biology principles that can be updated regularly and easily queried for information at any level by any clinician or researcher in the world.

9.2 The Use of CGH and DNA Microarray-Based Transcriptional Profiling for New Target Discovery in Breast Cancer

Breast cancer is a genetically heterogeneous disease whose origin is still unknown and difficult to predict. Multiple genetic alterations are found in breast cancer tissues, including amplifications (gain of function), deletions (loss of function) and mutations in key cell cycle, cell survival and DNA repair genes. The identification of new chromosomal imbalances by microarray-based CGH coupled to microarray-based transcriptional profiling has identified tumour-specific genes, including oncogenes and suppressor genes whose gain or loss of function is associated with tumour stage, invasiveness and clinical prognosis. Classic cases of altered genes are the HER2/neu gene, found through fluorescence *in situ* hybridization (FISH) analysis (Band *et al.*, 1989; Coene *et al.*, 1997), amplified in 25–30 per cent of invasive ductal carcinomas and ductal carcinoma *in situ* (DCIS), and the $p16^{INK4a}$ gene loci, lost in more than 55 per cent of cases (Di Vinci *et al.*, 2005; Holst *et al.*, 2003; Yonghao *et al.*, 1999). In addition, many overexpressed genes are also targets of *de novo* cell-mediated immune responses detected in peripheral blood and tumour-infiltrating lymphocytes from breast cancer patients. The identification of genes associated with cell survival, or critical to maintaining the transformed phenotype, that are also immunogenic is of special interest because we would predict that antigen-loss variants that may be induced by immunotherapy or drug therapy would not exist.

Gene dosage variations occur in many diseases. Detection and mapping of copy number abnormalities provide an excellent approach for associating genetic aberrations with disease phenotype and for localizing critical disease-related genes. Deoxyribonucleic acid (DNA) copy number alterations are important in the pathogenesis of many solid tumours, including breast cancer, although the reasons for these changes are poorly understood (Devilee and Cornelisse, 1994). Oncogenes in amplified or gained sequences may become activated while tumour suppressor genes in deleted regions of the genome are lost. Thus, the identification of genes located in areas of genomic loss or gain in breast and other cancers will assist us in the elucidation of mechanisms of carcinogenesis and will allow us to determine new potentially important molecular markers and therapeutic targets. HER2/neu, PTEN, MYC and other important cancer-related genes were discovered through the identification of altered chromosomal regions in cancer (Li *et al.*, 1997; Little *et al.*, 1983). In addition to mapping copy number abnormalities, the detection of chromosomal breakpoint fusion genes due to chromosomal translocations in breast cancer and other solid tumours is of increasing interest. This approach is quite advanced in the study of leukaemia (e.g. PML–RAR and bcr–abl gene fusions), but so far has largely been ignored in the solid tumour field (Claxton *et al.*, 1994; Dong *et al.*, 1993). A number of different technologies have been employed to identify gene loss and gain of functions, including metaphase chromosome karyotyping and restriction-length polymorphism (RFLP) analysis. Loss of heterozygosity at key alleles such as those of tumour suppressor genes such as Rb and p53 were found using these methods (Chen *et al.*, 1991; Xing *et al.*, 1999). Spectral karyotyping (SKY) using fluorescence *in situ* hybridization (FISH) has improved the sensitivity and throughput of karyotype analysis and is applicable to interphase cells (Bayani and Squire, 2002). A newer method called comparative genomic hybridization (CGH) has recently been introduced (Pinkel *et al.*, 1998; Pollack *et al.*, 1999). CGH affords the ability to survey the genome with much higher resolution (<1 Mb) using overlapping genomic sequence libraries and microarray-based fluorescent technology on chips allowing for rapid high-throughput screening of many samples at a time.

Comparative genomic hybridization in breast cancer

CGH was developed to facilitate genome-wide DNA copy number determination (Pinkel *et al.*, 1998). Differentially labelled genomic DNA derived from a 'test' and a 'reference' cell population are co-hybridized to normal metaphase chromosomes with blocking DNA used to suppress signals from repetitive sequences. This results in a ratio of fluorescence intensities at locations on the 'cytogenetic map' of chromosomes that is approximately proportional to the ratio of copy numbers of corresponding DNA sequences in the test and reference genomes. However, there are a number of limitations to the use of metaphase chromosomes; changes involving small genomic regions (<20 Mb) and closely spaced abnormalities are not easy to detect, in part because of limited resolution, and while changes can be mapped with cytogenetic accuracy, it is difficult to link ratio changes to specific genomic markers.

Fluorescence *in situ* hybridization has higher resolution but is prohibitively labour-intensive on a genomic scale. Array-based CGH using readily available genomic BAC, P1, cosmid or cDNA sequences representing over 30 000 human genes provide a reproducible locus-by-locus measure of copy-number variation. This increases mapping resolution and is practically applicable to the entire genome (Figures 9.2 and 9.3; Pinkel *et al.*, 1998; Pollack *et al.*, 1999; Snijders *et al.*, 2001; Solinas-Toldo *et al.*, 1997). Arrays composed of large insert genomic clones (bacterial artificial chromosomes or BACs, P1s and cosmids) provide a resolution of approximately 1.4 Mb, although the use of overlapping clones allows mapping with a resolution of <50 kb (Albertson, 2003). The positions of the clones used on the array are accurately known relative to the genome sequence and genes mapping within altered regions can be identified with genomic databases. Using this assay, it is possible to identify genome-wide amplifications and deletions, even single copy gains and losses, with high resolution. More recently, oligonucleotide array-based single-nucleotide polymorphism (SNP) CGH has been developed and applied to the study of breast cancer, and can distinguish different genetic mechanisms that lead to loss of heterozygosity (Zhao *et al.*, 2004). Oligonucleotide-based CGH is able to detect amplifications with high accuracy and greater spatial resolution than currently used array CGH platforms (Carvalho *et al.*, 2004). Various statistical methods are used to analyse data derived from CGH. Copy number transition points in array-based CGH probably reflect breakage events leading to chromosome rearrangements.

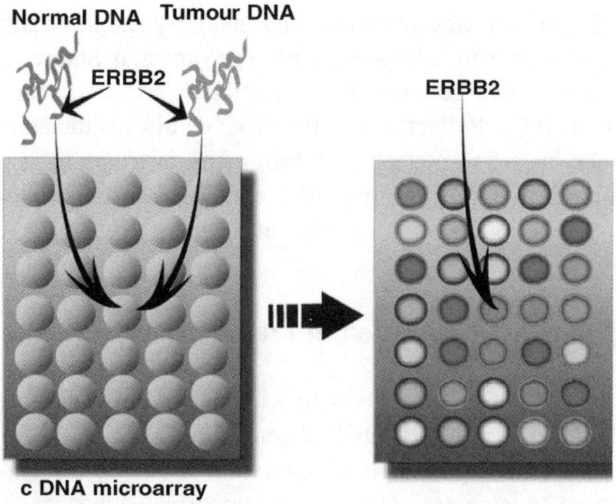

Figure 9.2 cDNA microarray DNA copy number analysis. Genomic DNA from tumour cells and normal blood leukocytes are labelled with Cy5 (red) and Cy3 (green), respectively, and hybridized together to arrayed cDNA elements and imaged by fluorescence confocal microscopy. The ratio of intensity of the fluorescence measured for the two fluorophores for each cDNA element/gene represents the relative DNA copy number in the two samples. Red represents gained (e.g. ERBB2 in the figure), green decreased and yellow unchanged DNA copy number in the tumour compared with normal cells. In fact, multiple independent elements represent the same gene on an array and an average fluorescence ratio from all of these is actually used. (A colour reproduction of this figure can be viewed in the colour plate section)

In addition to providing markers for localizing recurrent amplicons with high resolution, the identification of regions of DNA amplification and deletion in tumours has facilitated, and will continue to facilitate, the discovery and functional assessment of oncogenes and tumour suppressor genes important in cancer genesis and progression. For example, multiple recurrent regions of DNA amplification were detected in a study of three breast cancer cell lines (BT474, MCF7 and UACC-812) and one primary tumour (Pollack *et al.*, 1999). A recurrent amplicon at 17q12–q21 in BT474, UACC-812 and the primary tumour represents the HER2/neu amplicon and involves several genes in addition to HER2neu (GRB7, MLN64/CAB1 and various ESTs) which, though not as well studied as HER2neu, may also contribute to tumour progression (Stein *et al.*, 1994; Tomasetto *et al.*, 1995). An amplicon at 20q12 shared by BT474 and MCF7 includes the candidate oncogene AIB1 while BT474 and UACC-812 share an amplified sequence at 20q13 that contains the candidate oncogenes TFAP2C and STK15/BTAK/Aurora A (Anzick *et al.*, 1997; Sen, Zhou and White, 1997; Zhao *et al.*, 2003). Interestingly, it was noted in this study that the most highly amplified genes in both 20q sequences are in fact undetermined ESTs and not known candidate oncogenes at these regions. This suggests that the known genes may not be the most important at these amplicons in carcinogenesis.

Breast cancer cell lines are a practical starting point for discovery and functional analysis of important genes in breast cancer. Although the degree to which *in vitro*

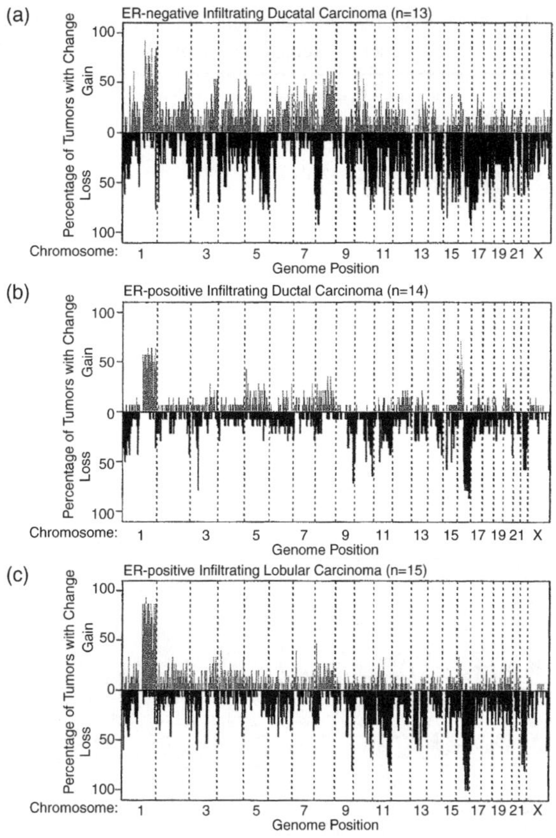

Figure 9.3 Examples of heterogeneous DNA instability in the primary breast cancer genome as determined by comparative genomic hybridization. The frequency of copy number gain and loss in subtypes of breast cancer are shown. The percentage of copy number gain (above 0, grey bars) and loss (below 0; black bars) were calculated for ER-1-negative infiltrating ductal carcinoma (a), ER-1-positive infiltrating ductal carcinoma (b), and ER-1-positive infiltrating lobular carcinoma (c) (Reproduced from Loo *et al.*, 2004 with permission).

studies reflect *in vivo* human breast tumours has long been debated, recent evidence from array-based CGH actually suggests that the extent and pattern of genomic instability in cell lines closely parallel human tumours (Lacroix and Leclercq, 2004). Cell line studies reveal many loci of chromosomal copy number abnormalities, including 35 recurrent high-level amplification sites (Table 9.1; detailed CGH profiles available at www.nhgri.nih.gov/DIR/CGB/++ +CR2000) and there is a substantial degree of overlap with human breast tumours (Forozan *et al.*, 2000). Using array-based CGH, candidate oncogenes and tumour suppressor genes at these sites of DNA copy number change can be identified. Other technologies being combined with CGH to facilitate the discovery of overexpressed genes and genetic alterations in breast cancer cells include expression profiling (Figure 9.4), suppression subtractive hybridization and multiplex FISH (Xie *et al.*, 2002). As discussed below, DNA

Table 9.1 The most common chromosomal changes in breast cancer cell lines and tumours and their frequency (reproduced from Forozan *et al.*, 2000)

Gains	Losses	High-level amplification
8q (75%)	8p (58%)	8q23 (37%)
1q (61%)	18q (58%)	20q13 (29%)
20q (55%)	1p (42%)	3q25–q26 (24%)
7p (44%)	Xp (42%)	17q22–q23 (16%)
3q (39%)	Xp (42%)	17q23–q24 (16%)
5p (39%)	4p (36%)	1p13 (11%)
7q (39%)	11q (36%)	1q32 (11%)
17q (33%)	18p (33%)	5p13 (11%)
1p (30%)	10q (30%)	5p14 (11%)
20p (30%)	19p (28%)	11q13 (11%)
		17q12–q21 (11%)
		7q21–q22 (11%)

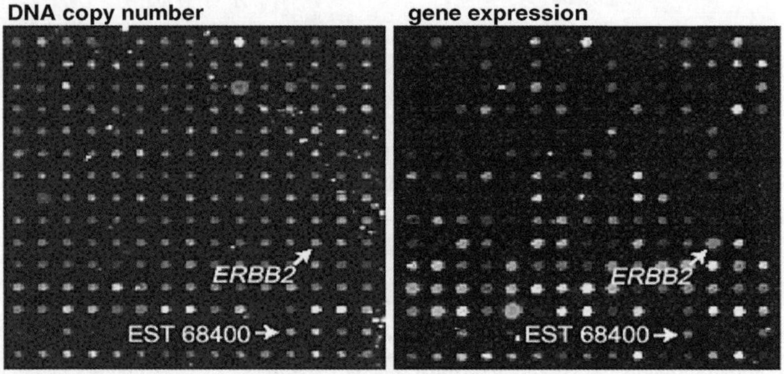

Figure 9.4 Concomitant assessment of DNA copy number changes and differential gene expression by microarray (transcriptional profiling). Identification of the ERBB2 gene (HER2/neu) is shown as an example of the use of DNA microarrays to validate CGH analysis. (Reproduced from Pollack *et al.*, 1998, with permission.) (A colour reproduction of this figure can be viewed in the colour plate section)

microarrays are an excellent way to delineate the specific genes found in amplified or deleted regions detected by CGH. There are 24 highly upregulated genes in the breast cancer cell line MCF7, 15 in ZR75-1, 80 in SUM-52, and 52 in SKBR3. The corresponding numbers of highly downregulated genes are 247, 257, 174 and 118. In these cell lines, several upregulated transcripts have chromosomal locations that match DNA amplification sites (for example, MYBL2 at 20q13, RCH1 at 17q23, TOPO II at 17q21–q22, NME1 at 17q21.3, and CAS at 20q13). FISH can be used to confirm that specific gene copy numbers are changed in tumours.

Using array-based CGH to analyse paraffin-embedded advanced breast cancers, it has been determined that the breast cancer cell genome contains on average 37 areas of genomic gain and 13 areas of loss, although there is considerable heterogeneity in genome copy number (Figure 9.3; Nessling *et al.*, 2005). Gains in more than 30 per cent of cases were found on 1p, 1q, 6p, 7p, 8q, 9q, 11q, 12q, 17p, 17q, 20q and 22q, and losses on 6q, 9p, 11q and 17p. Of 51 chromosomal regions found to be amplified, only 12 had been identified by metaphase CGH. Within these 51 amplicons, genome database information defined 112 candidate genes, 44 of which were validated by either polymerase chain reaction (PCR) amplification of sequence tag sites or DNA sequence analysis (Table 9.2). Thus, data from array-based CGH and other newer genomic technologies are beginning to provide us with a multitude of new potential markers and targets in breast cancer and other diseases. For example, one of the most amplified chromosomal regions found in breast is 8q23.3. Using microarray-based differential gene expression analysis we have found that one of the genes in this locus, TRPS-1 ('Trichorhinophalangeal Syndrome-1') is highly overexpressed in more than 90 per cent of breast cancers.

Differential gene expression methods to identify altered genes in breast cancer

As described above, transcriptional profiling is being used in collaboration with array-based CGH in an attempt to identify genes that are overexpressed or lost in areas of chromosomal amplification and deletion. This is based on the assumption that the expression of very important oncogenes and tumour suppressor genes in carcinogenesis at areas of DNA copy number change is likely to be more affected than other less important genes. In addition, the identification of deregulated genes at sites of recurrent tumourigenic viral integration may also alert us to potentially important candidate oncogenes as well as new tumour-associated antigens that can be targetted in therapeutic vaccines. Transcriptional profiling using cDNA-based and oligonucleotide-based microarrays (chips) has not only proven to be an excellent second-step method to corroborate and refine findings made using CGH in breast cancer, but has also been used as a first-line technology to directly screen RNA isolated from tumour cell lines and biopsy samples to identify new potential therapy targets (Jiang *et al.*, 2002; Mackay *et al.*, 2003; Seth *et al.*, 2003). Other differential gene expression profiling methods include differential gene display PCR and subtractive cDNA hybridization. The goal of all transcriptional profiling methodologies is to isolate and quantitate the expression of all RNA transcripts from tumour cells (whole tumour tissue or laser-guided microdissected cells) in comparison to RNA transcripts of non-tumour cells such as tissue samples from panels of normal adult organs, adjacent normal tissue in tumours and nontransformed cell lines.

In microarray analysis, samples of target cDNA and normal cDNA, labelled with different fluorochromes, are mixed in equal proportions and hybridized to the chip containing either full-length cDNAs or oligonucleotide sequences specific for the genes queried on the chip (Guo, 2003; Ochs and Godwin, 2003). The differences in signal intensities are used to detect and quantitate whether specific genes have

Table 9.2 Incidence of amplified candidate genes identified by array-based CGH in 31 cases of advanced breast cancer (reproduced from Xie *et al.*, 2002, with permission)

Location[a]	*n*	Genes verified by experiment[b]	Additional genes listed in GDB
1p36	6	*TP73, TNFRSF12, FGR, E2F2*	*CCNL2, ESPN, WDR8, ID3, UPLC1, TNFRSF25, FLJ20321, FLJ32825, FLJ37118, KIAA0495, KIAA0562, KIAA0720, KIAA1185*
1q25–q31	2	*LAMC2*	*LAMC1, RGS2, NMNAT2*
1q32	1	*MDM4*	
2p24	4	*MYCN*	*MYCNOS*
2p23	1		*RAB10*
3q26	3	*PIK3CA*	*MYNN, TERC, ARPM1, KCNMB3*
5p15	1	*BASP1*	
6q25	3	*ESR1*	
7p12	3	*EGFR*	
7q31	4	*MET*	
8p11	1	*FGFR1*	*FLJ25409, WHSC1L1*
8q24	6	*MYC*	
9q34	5		*LAMC3*
10p15	2	*PFKP, ADAR3, BS69, PRKCQ*	*ZMYND11*
11p15	3	*HRAS*	
11q13	13	*CCND1, FGF3, FGF4, RELA, BAD*	*RARRES3, ESRRA, FKBP2, PRDX5, PPP1R14B, DNAJC4, KCNK4, HSPC152, PLCB3, MGC13045, FGF19, GARP, ORAOV1, PAK1, EMS1, VEGFB*
12p12	1		*ETV6*
12q13–q15	9	*GLI, CDK4, SAS, TIP120A, MDM2*	*ERBB3, CPM*
15q25	2		*MAN2A2*
15q26	4	*IGF1R*	
17p11–p12	2		*GRAP*
17q12	10		
17q21–q22	12	*CAB2, ERBB2, RARA, TOP2A*	*STARD3, NEUROD2, PNMT, PPP1R1B, TCAP, PERLD1, BRCA1, CTEN, IGFBP4, ZNFN1A3, GRB7*
17q23	6		*ABC1, BCAS3, RPS6KB1, FLJ37451*
18p11	3	*YES1*	
19q13	2	*CCNE1, AKT2*	
20q12	7	*NCOA3 (AIB1)*	
20q13	7	*MYBL2, ZNF217, STK6*	*CSE1L, PTPN1, STAU, CSTF1*
21q22	2	*RUNX1*	*CLIC6*
22q11	3	*BCR*	
22q13	1	*SOX10*	

GDB, GenomeDataBase.

[a] Genomic amplicons with incidences of 20% and higher are in bold.

[b] STS-specific PCR or DNA sequence analysis.

increased, decreased or have not changed in expression levels. Gene chip profiling can either query sequences from all possible ORFs as well as numerous ESTs or can be more targetted using chips containing subsets of genes specific to certain signalling pathways or systems (e.g. immune response-related genes). The technology is more and more moving toward the use of oligonucleotide chips due to their ease of manufacture and consistency. One of the most versatile array technologies is the Affymetrix® chip. These chips are manufactured using a lithographic printing process to spot multiple 20–22-mer oligonucleotides from target gene sequences and ESTs on small silicon-based chips (Hardiman, 2004; Mei *et al.*, 2003).

Differential display PCR technology (DD) involves the amplification of cDNA from tumour and surrounding tissue using random primers (Liang, 2002; Martin and Pardee, 1999). Pools of PCR products are separated out on gels and differentially expressed genes in the tumour identified by sequencing. A powerful corollary method involves hybridizing amplified cDNA from the tumour target material against cDNA generated from surrounding normal cells or a pool of cDNA isolated from a panel of normal 'indispensable' tissues such as heart, kidney, lungs and brain. The nonhybridized fraction is then separated from the common hybridized sequences. A large number of tumour-associated antigens (TAA) and other tumour-specific genes have been found using these approaches in breast, lung, colon and other adenocarcinomas (Blok, Kumar and Tindall, 1995; Gardner-Thorpe *et al.*, 2002; Hubert *et al.*, 1999; Nagai *et al.*, 2003; Ree *et al.*, 2002). The use of subtractive cDNA hybridization has also become a useful first step in DNA microarray analysis in a method called a 'subtractive microarray' (Jiang *et al.*, 2002). Here cDNA from target tissues is first hybridized against cDNA libraries of a set of normal tissues. The nonhybridized portion of the cDNA is then assayed in the microarray step. This approach can remove much of the 'noise' due to low differential signal intensities found in nonsubtracted microarrays and a number of groups are beginning to use it to discover new less abundant tumour-specific genes and antigens (Beck, Holle and Chen, 2001; Jiang *et al.*, 2002).

Although DNA microarray analysis has gained the upper hand in the transcriptional profiling arena over DD and subractive hybridization methods, its one flaw is that it can only detect known gene sequences predicted from the 'unspliced' human genome, or previously identified splice variants, that are printed on to the chip. Thus, new gene sequences, not limited to the present-day probe sets, as well as new splice variants that may be critical in tumour biology, are missed. In this regard, DD and subtractive cDNA hybridization are still very powerful technologies that should not be dismissed. For example, a new antigen highly overexpressed in prostate cancer, STEAP-1, was recently discovered through DD methodology, while being missed in previous microarry-based screens (Hubert *et al.*, 1999). Inevitably, however, one must weigh the advantages and disadvantages of each technology and use the one most suited to the question or hypothesis posed. Although microarrays may miss some novel spliced sequences, the one critical advantage they have is their ease of use and the ability of the technology to rapidly screen large panels of diseased and normal tissue under different physiological conditions.

DNA microarrays have been applied to the study of breast cancer to identify tumour-specific genes as markers for diagnosis and prognosis (Pusztai *et al.*, 2003; Pusztai and Hess, 2004), monitor changes in gene expression profiles during chemotherapy and antihormone therapy as an attempt to establish predictors for outcome and survival (Chang *et al.*, 2003; Le *et al.*, 2005), and for the determination of specific gene expression differences between oestrogen receptor- and HER2/neu-positive and -negative tumours (Pusztai *et al.*, 2003). They have also been use to identify genetic signatures classifying breast tumours based on ductal epithelial subtypes (Perou *et al.*, 2000; Sorlie *et al.*, 2001). Another new and exciting application for microarrays is the characterization of earlier breast cancer lesions such as some forms of DCIS, preneoplastic syndromes such as atypical ductal hyperplasias, and areas of variant mammary epithelial cells characterized by $p16^{INK4a}$ gene hypermethylations and COX-2 gene induction (Crawford *et al.*, 2004; Gauthier *et al.*, 2005; Man *et al.*, 2005).

Application of DNA microarray-based gene profiling to identify new potential therapeutic vaccine targets in breast cancer

DNA microarrays are now beginning to be used in breast cancer to search new immunotherapy and small molecule drug targets. In this section, we will give an overview of a comprehensive screening that was performed at Sanofi-Pasteur, in which we used Affymetrix® oligonucleotide arrays to transcriptionally profile in a diverse panel of over 50 breast tumour surgical specimens. This included invasive ductal carcinoma, lobular and papillary carcinoma and DCIS. In parallel, we simultaneously arrayed a large ‘body atlas’ of 289 RNA samples collected from over 75 normal tissues and cells. This microarray screen was the first step in a comprehensive mutlistep target discovery programme (Figure 9.5) involving bioinformatics screening, the generation of reagents used to validate target expression at the protein level and immunological assays assessing target immunogenicity and potential suitability as a candidate vaccine antigen. Our final goal, as indicated in Figure 9.5, was the development of poxvirus-based active cancer vaccines encoding one or more of the validated genes overexpressed in breast cancer.

When attempting to perform such a large genetic screening programme, it is important to first establish a set of strict criteria to determine which genes are considered overexpressed and tumour-specific and which genes need to be discarded from further analysis. This includes strict statistical analysis of the microarray data, second tier bioinformatics screens and any other criteria decided by the user (e.g. patent databases). An improperly designed filtering process will easily yield a large and unmanageable list of potential genes from a microarray screen that will be impractical to deal with for any downstream validation work. An example of a filtering process that we have used to screen microarray data is illustrated Figure 9.6. One of the most important first steps in the analysis is to compile a list of ‘indispensable’ tissues, i.e. tissues such as heart, lung, brain, bone marrow, pancreas

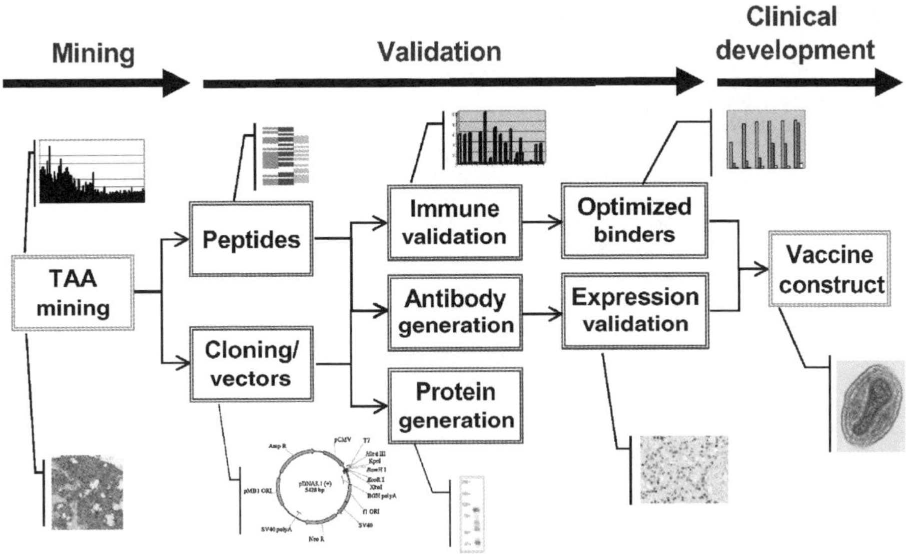

Figure 9.5 Overall pathway used in our breast cancer programme to discover and validate new breast cancer-specific target antigens for antitumour vaccination. The 'mining' stage was performed using differential gene expression on 54 breast tumour specimens in comparison to a large panel of normal tissue RNA using DNA microarrays. This approach identified 15 genes overexpressed in breast cancer, five of which we have taken forward to the 'validation' stage. This stage involved generating reagents to perform immunological validation (peptides and vaccine vectors) and antibodies for validation of expression at the protein level in primary tumour specimens using IHC. Specific sequence alterations are also made in immunogenic epitopes of the various candidate antigens in order to enhance HLA binding and T-cell activation. These sequences can be introduced into the sequence of the target genes in vaccine vectors such as poxviruses for the 'clinical development' stage.

and kidney in which no significant gene expression is tolerated. This is followed by a second list of so-called 'dispensable' tissues in which some gene expression (autoimmunity in a vaccine application) is tolerated (e.g. breast, skin, testis, prostate, etc.). In our case, we set the bar at only accepting genes that are overexpressed in at least 10 per cent of the breast tumour panel; these genes needed to have at least a 2.5-fold increase in expression over the average expression in all normal tissues and no significant expression in indispensable tissues. In this particular study, we also filtered out genes with known transmembrane and secretory domains because we concentrated our efforts on low-abundance intracellular targets (for $CD8^+$ T-cell targets) that would be less likely to be subjected to self-tolerance mechanisms in the immune system, a problem that plagues many other prospective active immunotherapy targets in breast cancer such as MUC-1 and HER2/neu. In addition, we were interested in genes having little or no evidence of extensive splice variation and which had no previous patent protection for application in a cancer vaccine.

An example of the type of microarray data generated in our screen after statistical analysis of all the chips is shown in Figure 9.7. RNA from each of the 54 breast

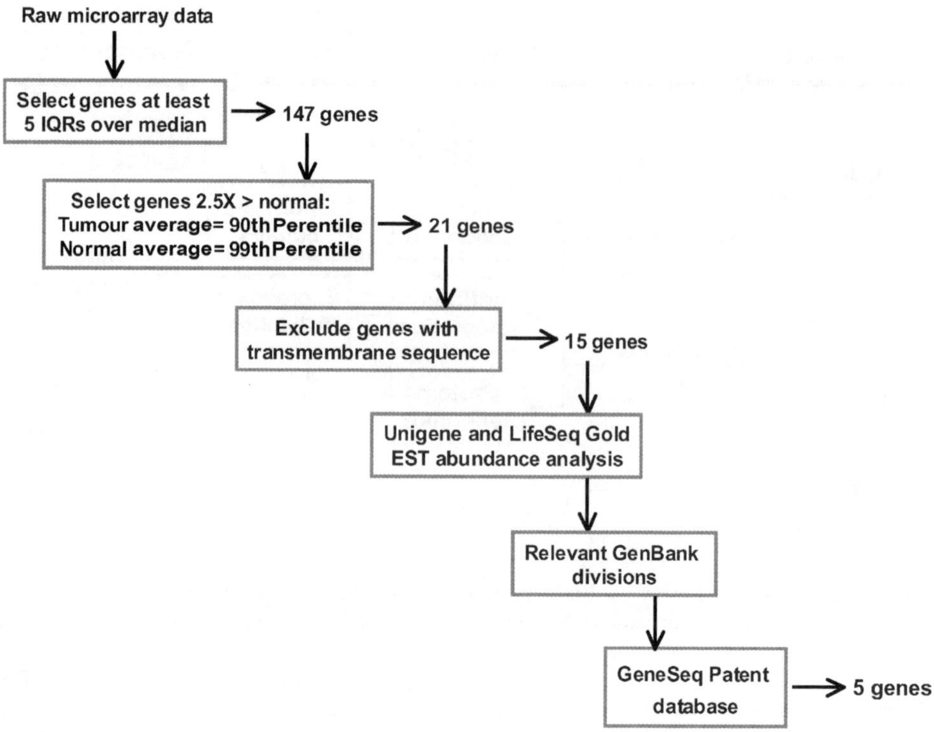

Figure 9.6 Filtering process used to analyse and select for new breast cancer-specific targets. The 54 breast tumour RNA samples and six normal breast RNA samples were run independently against 289 normal tissue RNA samples. Analysis of the microarray revealed that 147 genes had at least 5 IQR units above the mean relative gene expression in the 289 normal tissue RNA panel in at least five of the 54 breast cancers studied. This list was further filtered to include only genes encoding intracellular proteins (potential CTL-based vaccine targets) having a tumour-to-normal tissue expression ratio of at least 2.5 and at least 10% prevalence in the breast cancer panel and less than a 1% prevalence in the normal tissue panel. This resulted in a list of the 15 genes that were then further filtered by analysis of EST library sequence abundance (genes with high abundance in normal indispensable tissues were thrown out) and query of relevant GenBank divisions for splice variants and pseudogenes. Genes with excessive predicted splice variants were thrown out. A final screen was performed using relevant gene patent databases in order to rule out potential targets that had intellectual property protection barring the free development of the candidate as a cancer immunotherapy target. This search also includes any relevant published papers reporting the expression of the gene in breast cancer and its application for immunotherapy. This process ensured that only novel intracellular CTL targets, not previously proposed for use in breast cancer immunotherapy, were chosen for further downstream validation.

tumours and six normal breast specimens was co-hybridized with 289 different RNA from normal tissues. This was done numerous times to ensure that only reproducible gains or losses in gene expression were taken forward for further analysis. For each microarray assay run (over 1000 assays were run), a single custom oligonucleotide Affymetrix GeneChip® was used ('PDL-Hu03'), containing 400 000 perfect-match

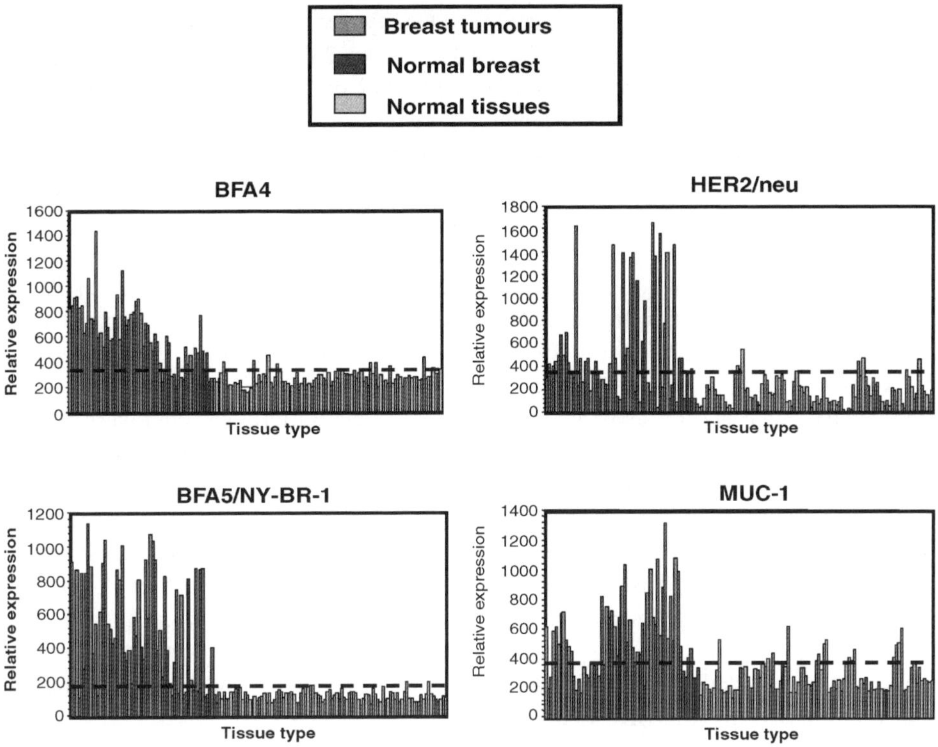

Figure 9.7 Results of microarray-based gene expression profiling on RNA isolated from 54 breast tumours specimens (red bars), six normal breast tissue specimens (blue bars) and over 200 normal tissue samples (grey bars). The arrow above the dashed line indicates the range of overexpression for the different genes shown. The gene array data was put through a strict filtering process (Figure 9.6) that selected for genes overexpressed at least 2.5-fold over the normal tissue panel where the average tumour expression was calculated as the 90th percentile value of the 54 breast tumours and the average normal tissue expression was the 99th percentile value of the 289-specimen body atlas. This method positively selected genes with at least 10% prevalence in the breast cancer cohort and negatively selected genes with greater than 1% prevalence in the normal body atlas. In all 15 genes passed this screen (see also Table 9.4), two of which are shown above – BFA4 and BFA5/NY-BR-1. Two other known breast cancer-specific genes, also identified in our microarray screen, are shown as a comparison. For example, HER2/neu was found to be over expressed in about 29.6% of the tumour panel, a figure that agrees well with the demonstrated prevalence of this gene in other studies. (A colour reproduction of this figure can be viewed in the colour plate section)

probe sets (about 59 000 probe sets), for interrogating essentially all known human genes in the public domain at the time of array design (September, 2001) (Henshall *et al.*, 2003).

To select for over expressed candidate breast cancer mRNAs, we derived a three-stage data-specific analysis process, using criteria of statistical significance, effect size and protein localization. To identify genes with significant overexpression in breast cancer, we converted the average intensity scores into inter-quartile range (IQR) units by subtracting the median value of the 289 nonpathogenic specimens from the average intensity, then dividing by the IQR of the same 289 normal tissues. This is similar to a conversion of standard deviations over the mean (z-scores), but is less susceptible to outliers. We initially selected genes that were at least 5 IQRs above the median in at least five of the 54 breast tumours. Only 147 unique genes passed this initial criterion, including many genes previously found to be overexpressed in primary breast cancer, such as HER-2/neu (Slamon *et al.*, 1987), MUC-1 (Ginestier *et al.*, 2002), mammaglobin (Watson and Fleming, 1996), NY-BR-1 (Jager *et al.*, 2001), oestrogen receptor (ER-1; Inoue *et al.*, 2002), pS2 (Rio *et al.*, 1987), progesterone receptor (Horwitz and McGuire, 1975) and SBEM (Miksicek *et al.*, 2002).

We then selected genes expressed at minimal levels in all normal tissues, and abundantly expressed in breast cancer. The 147 genes initially selected were further filtered such that the tumour-to-normal tissue expression ratio was greater than or equal to 2.5, where the average tumour expression was calculated as the 90th percentile value of the 54 breast tumours and the average normal tissue expression was the 99th percentile value of the 289-specimen body atlas. This method positively selected genes with at least 10 per cent prevalence in the breast cancer cohort and negatively selected genes with greater than 1 per cent prevalence in the normal body atlas. Of the 21 candidate TAA genes selected by this step, many 'tumour-specific' genes were excluded due to abundant expression in normal tissues (e.g. mammaglobin is expressed in skin; SBEM in skin and salivary glands).

Finally, we excluded overexpressed genes predicted to encode cell-surface accessible or extracellular proteins and concentrated on sequences predicted to encode intracellular proteins. A total of 15 sequences fulfilled these criteria. Of the 15 sequences that passed through this initial screening process, we further narrowed the list using a number of additional criteria to allow for a manageable number of genes to work with. This secondary filtering process is necessary for a number of reasons. Firstly, validating the expression of 15 new genes at the protein levels, developing separate reagents (cloning and antibodies), as well as immunological and pre-clinical proof-of-concept studies is impractical within a reasonable timeframe. This is especially pertinent when contemplating any downstream development as a new drug or therapeutic vaccine with all the production and regulatory issues involved. Secondly, we avoided genes with a significant number of predicted or known splice variants based on bioinformatics and mapping to EST databases. Thirdly, we used e-Northern using ORESTES and Unigene EST abundance analysis to confirm that the genes were enriched in tumour libraries and not highly represented in gene and EST libraries from normal (indispensable) tissues. Thirdly, an important issue for the vaccine industry, the entity that would eventually develop a new target through to Phase III trials, is intellectual property. Thus, we wanted to avoid target genes having a prior patent filing protecting their application as an active vaccine for cancer treatment. After taking into consideration these additional criteria, we ended up with a list of five new candidate breast cancer targets that were designated BFA4/TRPS-1, BFA5/NY-BR-1, BFY3/APA2, BCZ4/NAT-1 and BCY1.

Table 9.3 Many genes altered by DNA copy number changes in breast and other cancers influence functions involved in cancer progression (reproduced from Nessling et al., 2005, with permission)

Apoptosis	ELM02, EBAG9, TRIAD3, PDCD6, CSEIL, BAG3, FADD
Cytoskeleton	WDR1, CFL1, EMS1, RAE1, ELM02
Serine/threonine kinases	TRIO, MAPK1, STK3, STK24, CSNK1E, PCTK3, MAPKAPK2, IKKE, MAP3K4, TOPK, LOC57147, PTK2, KIAA0472, STK6, LIMK1, IKBKB
Intracellular signalling	ARHGEF11, SNX16, LOC59346, SCRIB, HSPC163, MY014, ENIGMA, PDZD2, PRKAR1A
Signal transduction	RALBP1, TXNL, PPP2R2A, MGST3, MDS026, MADH2, PTPN1, PIK4CB, LRDD, GNAS, PPFIA1, CNOT7, GNA13, NCOA3
Transcription factors	TFAP2C, HSF1, ADNP, FLJ11137, LASS2, ZNF42, FLT1, KIAA0863, ZNF211, SLC2A4RG, MBD1, ZNF217
Receptors	ERBB2, TRC8, FLJ11856, LOC51720, TOMM22, LGTN, TRIP13, SLC20A2, SRPR, C8FW, ATRN, MCP, FKBP1A

Figure 9.7 shows the expression profiles of BFA4/TRPS-1 and BFA5/NY-BR-1 in the 54 breast tumours relative to their expression in 289 normal tissue RNA samples (including normal breast parenchymal tissue) in comparison to a DNA microarray results for HER-2/neu, MUC-1. Table 9.4 lists these five genes along with GenBank Accession numbers, fold overexpression over the average gene expression in the 289 normal tissue RNA panel, and the proportion of tumours exhibiting at least the 2.5-fold increased expression we originally set as a cut-off in our analysis. Four of the candidates are known genes, of which three have been described to be associated with breast cancer (BFA5, BFY3 and BCZ4). As indicated, we found that BFA4 was the most prevalent gene expressed in breast cancer (83 per cent). Sequence analysis

Table 9.4 List of genes and their putative assignments found specifically overexpressed in a panel of 54 breast cancer specimens using microarray screening in relation to a panel of 289 normal tissue RNA samples

GenBank accession	Unigene ID	Cytoband	Gene title	Ratio (tumour/ normal)	Prevalence in 54 tumours
NM_052997	Hs.326736	10p11.21	Breast cancer antigen NY-BR-1 (NY-BR-1) (ANKRD30A)	52.4	0.76
AW248508	Hs.370809	1q22	Hypothetical gene XM_044166 (LOC92312); similar to PEM-3 (Ciona savignyi)/BCY1	4.7	0.46
NM_014112	Hs.253594	8q24	Trichorhinophalangeal syndrome I (TRPS1)	3.8	0.83
NM_003221	Hs.33102	6p12.3	AP-2 beta transcription factor	3.5	0.44
NM_000662	Hs.155956	8p22	*N*-acetyltransferase 1 (arylamine *N*-acetyltransferase/NAT-1)	3.2	0.41

found it to be identical to a previously known gene called TRPS-1, a new member of the family of transcriptional regulators with GATA domains (Momeni *et al.*, 2000). The gene had not been previously reported in breast cancer previous to our microarray findings at the time (2001), although a similar gene (GC79) was reported to be overexpressed in hormone-dependent prostate cancer (Chang *et al.*, 2002). TRPS-1 was originally described as an oncofoetal protein shown to be associated with three rare autosomal dominant genetic disorders called trichorhinophalangeal (TRP) syndromes caused by haploinsufficiency from inactivating mutations at the q23.3 locus on chromosome 8 (Hatamura *et al.*, 2001; Hilton *et al.*, 2002; Momeni *et al.*, 2000). Functional studies suggest that TRPS-1 functions as a transcriptional repressor inhibiting transcription of GATA-dependent genes during development (Malik *et al.*, 2001). In addition, it has been shown to interact with ring finger proteins such as RNF4 that bind to steroid receptors such as ER-1 (Kaiser *et al.*, 2003). Presently, we do not know the role of TRPS-1 in breast cancer; however, based on our target validation studies (see below), we suspect it has a critical function early in disease progression.

The fifth gene that made it to our microarray shortlist was BFA5. Interestingly, BFA5 is identical to a gene called NY-BR-1, found by SEREX at the Ludwig Institute for Cancer Research (Jager *et al.*, 2001, 2002). This overlapping finding attests to the power of a properly executed microarray screen and its associated filtering methodology in finding bonafide targets. NY-BR-1 was the second most highly represented gene in the panel of breast tumours tested (76 per cent of the tumours found to overexpress the gene at least 2.5-fold over the normal tissue panel). The function of NY-BR-1 is not known, but it is most likely a transcriptional regulator as suggested by the BLH and B-ZIP motifs in the mRNA sequence (Jager *et al.*, 2001). It is an exciting candidate from the immune therapy perspective because it induces a pre-existing humoural response in breast cancer patients, (hence, its dicovery through SEREX).

9.3 The Challenge of New Tumour Marker/Target Validation: Traditional Techniques Meet New Proteomics Tools

The known number of genetic abnormalities in cancer is large and continually increasing as new high-throughput technologies including array-based CGH and differential gene expression analysis through microarrays allow us to acquire an increasing understanding of the extreme molecular complexity of cancer (Albertson *et al.*, 2003; Futreal *et al.*, 2004; Suzuki *et al.*, 2000). More than 1500 genes may be altered in breast cancer by copy number aberration and mutations and many influence functions involved in cancer progression (Table 9.4). This has important implications for the development of targetted therapeutics that use either drugs to inhibit or knockout gene function, or immunotherapies that capitalize on the immunogenic properties of newly discovered tumour-specific gene products. In the first case, the development of drugs antagonizing the function of newly found targets necessitates the development of methods distinguishing between the role of a newly identified gene alterations in initiating and driving carcinogenesis vs genetic alterations that are only a consequence of underlying genomic instability. If a genetic aberration is to be an optimal target for molecular therapeutics, it should dominantly dysregulate a proliferative or metastatic pathway, bearing in mind that a role in tumour initiation does not necessarily imply a continued role in progression. In the case of immunotherapy, we need to get a thorough understanding of spectrum of cell-mediated immune responses against new targets that exist in cancer patients, or that can be induced through vaccination. This includes determining whether tumour cells process and present immunogenic peptides activating $CD8^+$ and $CD4^+$ T cells and whether vaccines (e.g. peptide/protein based or genetic vector-based) can adequately break tolerance. In either case target validation needs to start by first validating gene expression at the protein level. This can then be followed up by additional methodologies characterizing the signalling pathways involving target function and/or its immunogenicity.

Validation of gene expression at the protein level

The first step following any genomic-based target discovery approach must be to start the process of validation of gene expression at the protein level in primary tumour specimens and representative cell lines. IHC staining of formalin-fixed and paraffin-embedded tumour sections is still the ‘gold standard’ used to validate new target protein expression (Bodey, 2002; Hao *et al.*, 2004). It is important to begin this process as soon as possible owing to the sometimes difficult and lengthy process of generating suitable antibody reagents used for detection as well as the initial recombinant proteins and vectors needed for antibody generation. We have found that, for breast cancer, the use of traditional tumour blocks rather than tissue arrays of small bunch biopsies is the best starting point because it allows for staining of cancerous tissues together with surrounding normal tissues and also facilitates more accurate quantitation of the percentage of cells staining positive and the subcellular localization of the target protein. After this initial validation, work can proceed to tissue arrays containing many (up to hundreds of samples) of small tumour punches of 1 mm or less in diameter on each slide; these can be used for more high-throughput screens to establish prognostic correlations, for example (Zhang *et al.*, 2003). In addition, tumour tissue arrays coupled with IHC can been used for large-scale protein expression profiling on hundreds of specimens to detect molecular signatures for select kinases, growth factor receptors and apoptosis regulators in order subclassify breast tumours (Jacquemier *et al.*, 2005).

In our breast cancer studies, we performed an initial screen on 50 full-sized blocks covering a spectrum of different breast cancer histotypes, including lymph node positive and negative invasive ductal carcinoma, DCIS and lobular and papillary carcinomas. We also included a number of cases of early precancerous lesions, such as atypical ductal hyperplasia (ADH). These types of samples should also be screened whenever possible because they can yield insights into whether a new target gene plays a role in cancer development. After screening a library of hybridoma clones against a BFA4/TRPS-1 recombinant protein used as an immunogen in mice, two monoclonal antibodies were found to be superior at recognizing denatured protein in immunoblots and fixed protein in immunofluorescence microscopy studies with breast tumour cell lines expressing BFA4/TRPS-1. These antibodies were used to screen the tumour biopsy bank.

A properly performed IHC screen can lead to some pleasant surprises and, in this case, our staining data showed that BFA4/TRPS-1 is overexpressed in a significantly higher proportion of breast cancers than previously predicted by the original microarray findings. Our IHC analysis found BFA4/TRSP-1 protein expression in 90 per cent of the breast cancer blocks stained, with most specimens having expression in more than 75 per cent of the tumour cells. Another important finding was that both HER2/neu expressing and nonexpressing as well as ER-1 expressing and nonexpressing breast tumours were highly stained. We have also performed IHC staining for the other candidate genes found in our microarray screen, including BFY3 and BCZ4. Although these other targets were not expressed to the same extent as BFA4/TRPS-1 and in as many tumours, they did stain the small proportion of tumour specimens that were BFA4/TRPS-1 negative. This raises another critical issue in new target discovery for breast cancer and other malignancies given the high degree of heterogeneity found within and between tumours; that is the need for a multitarget approach to drug therapy or immunotherapy in order to eradicate as many tumour cells as possible and prevent the outgrowth of variants that lose expression of any one target. As seen in the Venn diagram in Figure 9.8, combining all the individual gene expression data from the five candidates that passed through all our

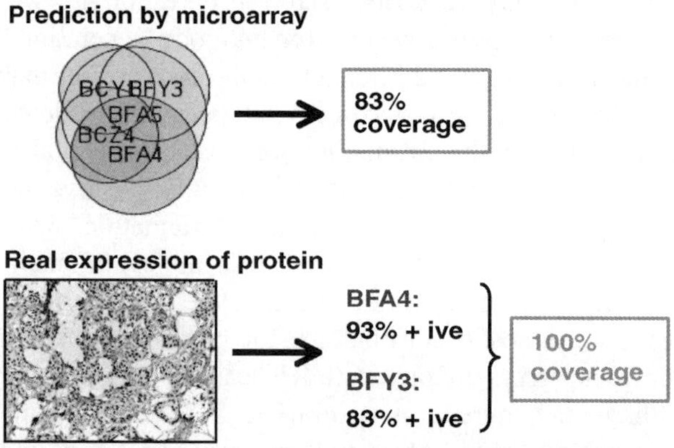

Figure 9.8 Comparison of the overall expression coverage in the 54 breast cancer samples of the five new candidate breast cancer targets as found in the original microarray analysis (BFA4/TRPS-1, BFA5/NY-BR-1, BFY3, BCZ4 and BCY1) to IHC results on two of the candidates (BFA4/TRPS-1 and BFY3), showing the importance of proper validation of gene expression at the protein level. Overall, a coverage of 83% was predicted when all five genes are combined based on mRNA expression data. However, after IHC screening for BFA4/TRPS-1 and BFY3 in a panel of 50 primary and metastatic breast cancer biopsies, coverage of 100% was found. This suggests that, in a therapeutic vaccine approach against breast cancer, combining these two genes in vaccine vectors should be sufficient for the vast majority of patients.

filtering process predicted an 83 per cent coverage of breast cancers if we were to therapeutically target all five of the genes. However, using IHC we found that only two of these targets, BFA4/TRPS-1 and BFY3 can cover 100 per cent of the breast tumours. This underscores the need to validate gene expression studies with properly designed protein expression screening since transcriptional and translational processes for any given gene may not always correlate.

Understanding the functional/signalling role of new targets in tumour cells

It is critical that we establish a clear understanding of the signalling pathways regulated by new target genes. It is necessary to understand signalling pathways and networks in sufficient depth so as to be able to identify ways to disrupt gene function and determine the consequences of this targetted inhibition. This is important not only for the development of new drugs but also for immunotherapy since many new target molecules can directly or indirectly alter the immunogenic properties of cancer cells through the regulation of antigen presentation, cell survival and the production of immunosuppressive factors. For example, the increased expression and activation of STAT3 in cancer cells has recently been found not only to increase cyclin D expression in tumours, but also induces the production of VEGF inhibiting dendritic cell differentiation and other factors such as NO, that activate suppressive macrophages and inhibit T-cells inside tumours (Gamero *et al.*, 2004).

It is critical that we identify and validate targets in clinical samples before investing time and money in the development of a drug; *in vitro* and animal models are useful but the information they provide may ultimately prove misleading in terms of efficacy and toxicity. A combination of genomic and proteomic tools is being tested for their ability to dissect gene function in cancer. At present, the available proteomic technologies lag somewhat behind genomic and transcriptional approaches but they continue to be developed and improved (Chan *et al.*, 2004a; Petricoin *et al.*, 2002a; Petricoin *et al.*, 2002b; Zangar *et al.*, 2004). One of the most powerful technologies currently being used to evaluate the function of new molecular targets is mRNA knockdown with small interference RNA (siRNA; Chiu and Rana, 2003). This has proven useful in identifying the function of a number of genes involved in cell survival, transformation, cell cycle control, tissue invasion and DNA repair such as Survivin, STAT3, raf-1, MMP-9, RAI3 and Fra-1 (Belguise *et al.*, 2005; Ling and Arlinghaus, 2005; Menendez and Lupu, 2005; Nagahata *et al.*, 2005; Pille *et al.*, 2005; Shen *et al.*, 2003; Thomas *et al.*, 2005). When targetting pathways in which key nodes are upregulated but not structurally altered, choosing one critical node (e.g. AKT/PKB) may be excessively toxic and several pathway targets may prove to be most effective and specific while minimizing toxicity.

In addition to siRNA methodologies, new proteomic technologies such as protein lysate microarrays are also now demonstrating great potential for new molecular marker and target validation in breast cancer and other tumours (Nishizuka *et al.*,

2003). Unlike immunohistochemistry, western blotting and tissue microarrays, protein microarrays allow for accurate protein quantitation in addition to identification. Protein arrays consist of multiple different antibodies spotted onto glass slides or membranes. After extensive blocking of the array, they are incubated with cell lysates and counter-stained with fluorescent secondary antibodies. At present, only one sample can be run on each particular microarray, unlike DNA microarrays, in which RNA from differentially labelled test and control samples are placed simultaneously on the chip surface. However, new methods of global protein labelling using small protein modifications that do not affect antigenic determinant structure should be introduced soon that will facilitate this approach in protein microarray analysis. Protein/antibody arrays are beginning to be utilized as a sensitive high-throughput platform for marker screening, pathophysiology studies, identification and validation of novel treatment targets, and therapeutic monitoring (Callagy *et al.*, 2005; Nishizuka *et al.*, 2003). These also allow the concurrent assessment of multiple signalling events and pathways in tumour samples, and thus enable the evaluation of the full complexity of signalling pathways involved in the transforming effects of important oncogenic events. One application of protein microarrays involves phosphoprotein profiling of signalling pathways using spotted antibodies specific for tyrosine and serine/threonine phosphorylated proteins in growth-related signalling pathways such as the ras-MAPK and SAPK pathways (Alessandro, Belluco and Kohn, 2005; Chan *et al.*, 2004b; Uttamchandani *et al.*, 2003). A number of antibodies have been validated for these applications. As more and more validated antibodies recognizing phosphorylated and native proteins become available in the ensuing years, it is easy to visualize how powerful this approach will become (after all we estimate that there are a maximum of about 30 000 proteins in the human proteome not counting splice variants).

9.4 Immunological Validation of New Target Genes in Breast Cancer: the Emerging Concept of the Cancer 'Immunome'

One of the key factors determining the suitability of a new tumour-specific gene target as a new cancer vaccine target is its immunogenicity in the human immune system and whether reactive epitopes presented on both HLA class I and class II molecules stimulating T-cell responses in cancer patients as well as normal donors can be identified. The determinants involved in regulating target immunogenicity include the frequency of $CD8^+$ and $CD4^+$ T cells capable of recognizing antigenic peptides (these should be elevated over normal control in cancer patients), the number and binding efficiency of target-derived peptides to HLA, the degree of tolerance to the potential antigen and the affinity of the TCR repertoire towards the potential target antigen. The latter parameter is of the biggest concern due to the 'self' nature of tumour-associated antigens and the different central and peripheral tolerance mechanisms that silence or suppress antiself antigen responses, such as T-cell deletion, anergy and natural or induced $CD4^+$ regulatory T cells. However, a significant number of tumour-associated antigens, especially of the CT family, such as NY-ESO-1, have proven to be remarkably immunogenic with $CD8^+$ and $CD4^+$ T-cell responses occurring *de novo* or capable of being triggered through vaccines (Jager *et al.*, 1998; Khong *et al.*, 2004). Other self-tumour-associated antigens such as CEA, MUC-1 and p53 are more weakly immunogenic and can also induce immune tolerance (Agrawal *et al.*, 1998; Agrawal and Longenecker, 2005; Pellegrini *et al.*, 1997).

The key interacting criteria we have used to evaluate new target immunogenicity and suitability for further downstream development as a vaccine are illustrated in Figure 9.9. The first criterion is to demonstrate that human T cells can recognize antigenic peptides from the target gene. This is accomplished using T-cell lines generated from either patient or normal donor peripheral blood mononuclear cells (PBMC) to determine the degree of $CD8^+$ T-cell reactivity against defined HLA class I binding peptides. The second area considers whether naturally processed and presented peptides can the presented to the immune system in a vaccine setting. Here, new types of transgenic mice expressing fusion proteins of HLA class I and class II, such as hybrids of HLA-A*0201 $\alpha1$ and $\alpha2$ with mouse $\alpha3$ K^b (e.g. HHD mice), have greatly facilitated the identification of processed and presented epitopes following genetic or protein-based immunization (Borenstein *et al.*, 2000; Carmon *et al.*, 2002; Firat *et al.*, 1999). The third criterion is the most critical when contemplating immunotherapy trials: tumour cells must inevitably present a set of immunogenic epitopes and exhibit sensitivity to lysis by antigen-specific CTL. Thus, a set of intersecting peptides meeting all three criteria must be found. Human HLA-A*0201 has been the classic HLA of choice for mapping HLA class I-restricted epitopes from tumour antigens due to the high proportion of this HLA type represented in the human population (25 to 40 per cent depending on geographic

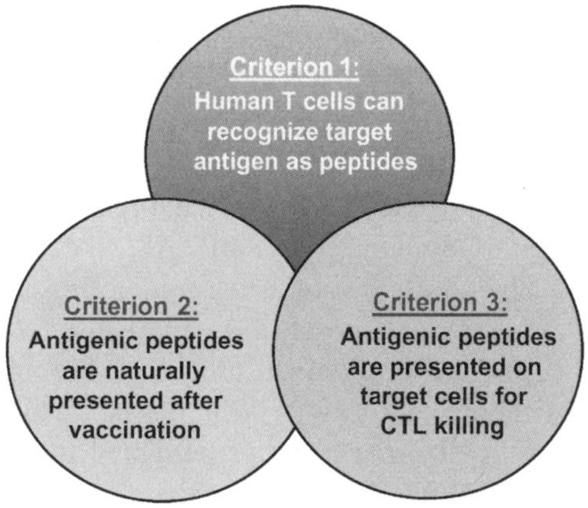

Figure 9.9 Three main criteria used to characterize the potential of a new candidate genes gene to be a target for a CTL-based vaccine against breast cancer. Three overlapping criteria are used that ultimately identify a small subset of peptide epitopes from a candidate gene that fulfills the three criteria as shown.

location). However, other HLA types (e.g. HLA-A1, -A3, -B1, -B27 and -B57) are also being increasingly studied as knowledge on the key anchor residues regulating peptide binding to these HLA subtypes is acquired.

Immunological validation is a laborious process and not for the researchers wanting a 'quick fix'. In many cases, it requires a careful screening of large peptide libraries (more than 100 peptides) using conventional cellular immunology approaches (e.g. cell line generation and screening using ELISPOT and CTL assays) and de-convolution of the libraries. It is only after peptides in the intersection of these three domains are found that we can consider a potential target validated from an immunological standpoint. In most cases, only one or a few peptides will meet these criteria due to self-tolerance and T-cell repertoire restrictions.

Another important area that has been gaining increasing attention is the critical role of $CD4^+$ T-helper responses in antitumour immunity. A flurry of recent activity mapping HLA class II-binding peptides using longer 15-mer or overlapping 20/22-mer or 30-mer peptides is now taking place (Bachinsky *et al.*, 2005; Hernandez *et al.*, 1998; Klyushnenkova *et al.*, 2005; Maecker *et al.*, 2001). Approaches similar to those used to identify HLA class I-restricted epitopes are being used for the study of HLA class II-restricted responses such as the screening of $CD4^+$ T-cell responses in patient blood and in antigen-specific T cell lines. A number of HLA-DRB1 and DP4 (represented in over 60 per cent of the human population) epitopes have been identified from a number of tumour antigens such as HER2/neu, PSMA, NY-ESO-1, CEA and MAGE 3 (Kobayashi *et al.*, 2000; Schroers *et al.*, 2003; Shen *et al.*, 2004; Zeng *et al.*, 2001).

Design and prescreening of peptide libraries used for cell-mediated immune response testing

A variety of approaches have been used to design peptide libraries for immunological testing, depending on the exact question being asked. For HLA class I peptides (9-mers or 10-mers), we have used a 'focused library' of predicted binders. A number of public databases of known HLA class I-binding peptides as well as binding algorithms are available in order to select predicted binders, such as 'SYFPEITHI' (University of Tübingen, Germany), the 'BIMAS' database (NIH, USA) and MHCPEP (Walter and Eliza Hall Institute, Australia; Brusic, Rudy and Harrison, 1998; Rammensee *et al.*, 1999; Schonbach, Kun and Brusic, 2002). These and other publicly accessible sites such as ProPred (Institute for Microbial Technology, India) can also be used to predict both short (9-mer) peptides predicted to bind to HLA-DRB alleles and longer 14-mer to 16-mer peptides predicted to bind to multiple HLA-DRB1 alleles and the HLA-DPB1 *0401 allele (Singh and Raghava, 2001; Sturniolo *et al.*, 1999). Identification of peptides for the latter is gaining importance in cancer vaccination due to the high representation of this allele (>65 per cent) in the human population. As experience with different peptide libraries and more data is gathered using both physical (e.g. plasmon resonance spectroscopy) and biological assays for

HLA binding, neural net programmes are being generated that further add to the predictive power to these databases (Hlavac *et al.*, 1996). It is envisioned that in the future these neural nets will be 'educated' with practically all proven biologically active peptides from any given antigen. One would then just simply HLA-type the patient and pick an off-the-shelf peptide cocktail for immunization.

Another approach to identifying reactive peptides for immunological validation is the use of overlapping peptide libraries. The most comprehensive method uses 9-mer peptides overlapping by eight amino acids from the NH_2 to the $COOH_2$ terminus of the protein that cover all possible epitopes for most HLA class I and some HLA-DRB alleles (Bachinsky *et al.*, 2005). These libraries, however, can be extensive and too expensive to synthesize and complicated to analyse. As a result, many have opted to use overlapping peptide libraries of 15-mers (overlapping by 11 amino acids) that have also proven adequate to narrow down the regions generating immunogenic epitopes (Martin *et al.*, 2004; Takahashi *et al.*, 2002). This is then followed up by synthesizing 9-mer peptides within the reactive 15-mer sequences.

Identification of immunogenic peptides from new breast cancer targets

We have performed extensive immunological characterization on the five new breast cancer targets identified in our initial microarray screen using the criteria shown in Figure 9.9. Many of the genes we were dealing with were quite large and not amenable to an overlapping peptide screening approach to find possible immunogenic epitopes owing to cost and technical issues. For this reason, we used focused libraries of predicted HLA-A*0201-binding peptides using the BIMAS (Schonbach, Kun and Brusic, 2002) and SYFPEITHI (Rammensee *et al.*, 1999), binding algorithms together with an in-house neural network programme that we have educated over the last 5 years. Using this approach, 100 predicted HLA-A*0201-binding nonamer peptides from each of our candidate antigens were synthesized. The peptides were chosen according to the following equation: (60 top-scoring peptides from BIMAS) + (30 top-scoring peptides from SYFPEITHI not found with BIMAS) + (10 top-scoring neural network peptides not found via BIMAS or SYFPEITHI).

The peptides were screened for T-cell reactivity by generating a series of HLA-A*0201-restricted T-cell lines against pools of peptides from each target (10 peptides/pool). The *in vitro* assay system involved pulsing mature dendritic cells generated from normal female donors with each different peptide pool and mixing these with autologous T cells. The T cells are allowed to expand for 10–12 days in IL-2 and IL-7 and then undergo a series of restimulations to increase the frequency of peptide-specific T cells from the original naive pool (usually starting from 1/100 000 to 1/1 000 000 peptide-specific T cells). We have found that autologous CD40-ligand-activated B cells can be used for these subsequent rounds of stimulation (Schultze *et al.*, 1997). Large numbers (over 50 million) of proliferating, activated B-cells expressing high levels of HLA class I and class II as well as the costimulatory

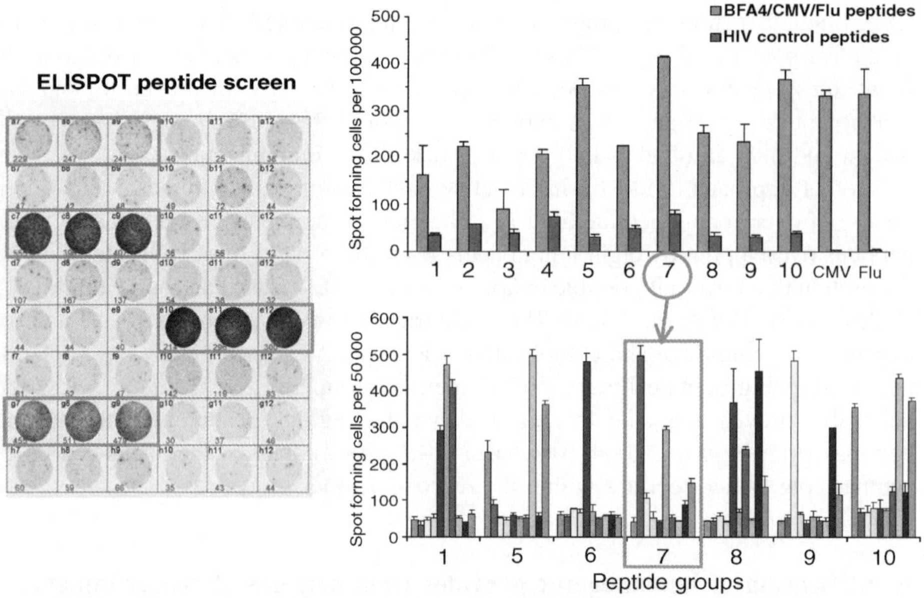

Figure 9.10 Enzyme-linked immunospot (ELISPOT) assay measuring IFN-γ secretion by $CD8^+$ T cells for screening large peptide libraries for immunoreactivity. ELISPOT is a highly sensitive technique using coated antibodies (in this case anti-IFN-γ) to detect the secretion of cytokines in response to antigen addition to T-cells placed in culture with relevant antigen-presenting cells. The example shown is a typical screening performed on a 100-peptide library from the BFA4/TRPS-1 breast cancer target. Human T-cell lines are generated using APC pulsed with pools of 10 peptides (10 lines generated in all). After a number of rounds of stimulation, these lines are screened for HLA class I-restricted $CD8^+$ T-cell reactivity using the respective peptide pools in ELISPOT assays. The most highly reactive pools are identified and de-convoluted in subsequent ELISPOT assays in which individual peptides are tested. The arrow and box on the right indicate the de-convolution of one such peptide pool (group 7). The panel on the left shows how a typical ELISPOT plate looks like after development. The wells indicated by the boxes show triplicate samples of single reactive peptides. Non-specific peptides (e.g. HIV peptides) binding to the same class I HLA that should not demonstrate any reactivity are used as background controls in all assays. (A colour reproduction of this figure can be viewed in the colour plate section)

molecules B7.1 and B7.2 can be generated from only a few million resting peripheral blood mature B-cells using multiple rounds of CD40 ligation and expansion with IL-4. These cells can also be cryopreserved and thawed whenever needed. These antigen-presenting cells (APC) are also a good alternative to Epstein–Barr virus (EBV)-transformed immortalized B-cell lines because they do not express EBV-related antigens and do not stimulate EBV-specific $CD8^+$ T-cell recall responses. To screen the peptide libraries and identify specific reactive pools, we have used IFN-γ ELISPOT. This is an excellent high-throughput assay capable of screening large numbers of peptides for immunoreactivity. In our case, we first screened the peptide pools for reactivity in our respective T-cell lines and then de-convoluted each reactive pool to identify the single reactive peptides in each pool (see

Figure 9.10). In this way, we have mapped a number of prospective epitopes from each of the new breast cancer target genes. Single reactive peptides were then further tested in CTL assays using peptide-loaded targets (e.g., HLA-A*0201^{+}, TAP$^{-/-}$ T2 cells) and cell lines endogenously expressing the candidate gene using viral-based or plasmid-based transformation.

After identifying potential epitopes, DNA vaccination in HLA transgenic mice (in this case the HLA-A2.1/K^{b} α_3 transgenic mice; Borenstein *et al.*, 2000) was used to identify epitopes processed and presented during vaccination (criterion 2, Figure 9.9). In most cases it has been demonstrated that murine proteasomes process and load similar peptides on MHC class I as the human antigen presentation machinery. Finally, the most important parameter from an immunotherapy point of view is to demonstrate the ability of peptide-specific CTL to kill HLA-matched breast tumour cells expressing the gene of interest. Using this approach we have found that a number of peptides initially identified in our ELISPOT screens for BFA4/TRPS-1 (Figure 9.10) may meet the three criterion we have established making BFA4/TRPS-1 a validated target from an immunological standpoint. Based on these results, combined with the genomic-based gene discovery results and the IHC analysis, we have designated BFA4/TRPS-1 as a validated target for inclusion in a multiantigen poxvirus-based vaccine approach for breast cancer.

Mapping the cancer immunome through 'reverse immunology' and new proteomics tools: linking genomics-based screening to antigen-mapping

New methods in proteomics driven by advances in mass spectrometry (MS) and high performance liquid chromatography (HPLC) together with high-throughput screening in immunology has created a powerful new approach to finding new genes involved in tumour initiation and progression that can be targets for cancer therapy. This approach has also been called 'reverse immunology' since it involves the detection of reactive antigens and peptide epitopes directly using immunological and proteomic analysis followed by back-tracking to the genome to pin-point the identity of the gene involved in the reaction. Together with genomics, the reverse immunology approach will prove invaluable to map what we call the cancer 'immunome', which would be an atlas of all antigenic targets processed and presented on cancer cells in context of different HLA molecules in the world.

Three main technologies are at the heart of the reverse immunology approach: (1) direct elution of peptides bound to cell surface or secreted HLA class I and class II molecules and sequence identification (Rotzschke *et al.*, 1990); (2) serological analysis of gene expression libraries (SEREX) (Chen *et al.*, 1997); and (3) screening of tumour-reactive CTL or TIL with position-scanning combinatorial peptide libraries and identification of the corresponding reactive protein antigen (Nino-Vasquez *et al.*, 2004; Pinilla *et al.*, 2001; Rubio-Godoy *et al.*, 2002). The latter two methodologies stem from the fact that most cancer patients generate *de novo* immune responses against a wide range of TAA; these are in the form of CD8^{+} T-cell and antibody responses, with increasing emphasis on the $CD4^+$ T-cell response. The $CD4^+$ T-cell 'immunome' is also critical from the point of view of immune suppression in cancer and the integral role of $CD4^+$, $CD25^+$ self-reactive T-regulatory cells. Recent evidence indicates that these suppressor cells are activated by different epitopes from the same or related antigens that stimulate productive CTL responses. For example, LAGE-1 a closely related protein to NY-ESO-1, has been shown to induce the activation of $CD4^+$ T-regulatory cells in melanoma in tumour infiltrates (Bolli *et al.*, 2005; Mandic *et al.*, 2003; Wang *et al.*, 2004). A class II-binding peptide from another gene found in melanoma, ARCT1, has also been recently shown to be associated with $CD4^+$ T-regulatory cells in melanoma (Wang *et al.*, 2005). These results predict that mapping the T-regulatory cell 'immunome' will become an important endeavour in dissecting T-cell tolerance in cancer and in identifying antigens and epitopes that may actually lead to the suppression of Th1 and CTL responses during antitumour vaccination by T-regulatory cell activation (Wang *et al.*, 2004, 2005). Thus, thus the choice of peptides used for antitumour vaccination needs to take this into account, especially in situations when the same region of an antigen or the same peptide can activate both suppressive and nonsuppressive $CD4^+$ T cells.

Direct detection of processed peptides

A number of academic groups and small biotech companies have developed acid wash-based techniques to elute bound peptides off HLA complexes on the cell surface of tumours or from isolated HLA molecules. HPLC-MS systems are then used to identify the peptide sequences. Hans Georg Rammensee's group at the University of Tübingen was one of the first to apply this technique to isolate peptides from common HLA class I and murine MHC class I molecules using cell lines (Rotzschke *et al.*, 1990). A whole library of peptides from housekeeping genes, cell proliferation-associated genes and TAA has been catalogued using this method. Public databases and a number of published compendia listing known peptides eluted from different HLA and major histocompatibility complex (MHC) subtypes are now available (Marsh, Parham and Barber, 2000; Papadopoulos *et al.*, 1997). Overall, this method has not received as much attention as other methods of target discovery due to its technical challenges and sensitivity issues. A large number of cells are required (millions to billions depending on the abundance of the target gene) to elute enough of a given peptide to make differential identification and sequencing possible. A newer method that may solve this problem uses cell lines transfected with a soluble HLA class I molecule of a given subtype (Barnea *et al.*, 2002; Buchli *et al.*, 2004). Tumour cell lines are transfected with these soluble HLAs and grown in small bioreactors, allowing for milligram quantities of HLA–peptide complexes to be isolated and peptide sequences deduced through MS. One potential drawback of this new method is that, owing to the shear scale at which HLA–peptide complexes can be isolated, many low-abundance peptides that may fall below the limit of detection by T cells may be isolated. This again points to the need to merge these approaches with proper immunological validation.

A new and interesting possible addition to the HLA elution-based technique is the use of tumour-derived heat shock proteins to map epitopes of tumour-associated genes. A number of heat shock proteins (HSP), some of which are overexpressed themselves in cancer, such as HSP70, gp96, calreticulin (p56), p90, p110 and p170, have been shown to intrinsically bind tumour-related peptides (Basu and Srivastava, 1999; Massa *et al.*, 2004; Nair *et al.*, 1999; Navaratnam *et al.*, 2001; Staib *et al.*, 2004). Isolated HSP are also being tested in clinical trials as cancer vaccines as a result. A number of research groups are translating the peptide elution-MS technology in the HLA field to the HSP arena in the hope of finding new tumour-specific T-cell epitopes (Belli *et al.*, 2002; Binder and Srivastava, 2005; Mazzaferro *et al.*, 2003; Srivastava, 2005; Srivastava and Amato, 2001). It will be interesting to follow the progress in this field and whether HSP vaccination will fulfill its promise as an effective cancer vaccine.

SEREX

SEREX (serological analysis of gene expression libraries) has received considerable attention as a reverse immunology tool following the discovery of one of the most immunogenic T-cell antigens in cancer, NY-ESO-1, using this technique at the Ludwig Institute for Cancer Research in New York (Chen *et al.*, 1997). The SEREX technology involves producing cDNA phage expression libraries using RNA isolated from tumour biopsy material or cell lines. Patient and normal sera-containing antibodies potentially reacting against different tumour-associated gene products are screened on membranes with thousands of expression phage clones using an immunoblotting technique. The positive spots are then de-convoluted to identify the reactive proteins by further subcloning of the reactive spots if needed (Li *et al.*, 2004; Tureci, Sahin and Pfreundschuh, 1997). Since its inception, SEREX has become a popular method with a large number of new potential targets identified in breast (Obata *et al.*, 1999), colorectal (Line *et al.*, 2002), renal (Scanlan *et al.*, 1999), gastric (Obata *et al.*, 2000), pancreatic (Nakatsura *et al.*, 2001) and liver cancers (Li *et al.*, 2003) as well as brain tumours (Behrends *et al.*, 2003), and leukaemia (Greiner *et al.*, 2003). For example, following SEREX identification, NY-BR-1 was found to be expressed in over 90 per cent of breast cancers at the mRNA level. Two other similar genes, NY-BR-62 and NY-BR-85, were also later discovered by SEREX (Jager *et al.*, 2002).

Although SEREX is a popular technology, it does have a number of disadvantages. A major assumption made by the technology is that antibody responses (mediated by $CD4^+$ T-cells) are always associated with $CD8^+$ T-cell responses. However, this is not always the case. Many intracellular proteins may be poor inducers of antibody-mediated responses yet are highly overexpressed and induce CTL responses in many patients. Interestingly, SEREX screens in breast cancer have not detected TRPS-1 as a potential breast cancer target despite our demonstration of its high degree of overexpression at the protein level and immunogenicity. Another issue with SEREX is that humans contain low to intermediate affinity IgG and IgM cross-reacting against self-proteins as well as tumour antigens (so-called 'immunological homunculus'; Cohen, 1993; Poletaev and Osipenko, 2003). This may result in large numbers of positive spots for normal proteins and false positive reactivity to tumour antigens due to antibody cross-reactivity. New methods to remove antigens bound to normal serum antibodies in columns or resins before screening with patient-derived sera will help to reduce the complexity and increase the reliability of the technique.

Combinatorial peptide libraries and mimotopes

Combinatorial peptide library screening has been successful in identifying potential new tumour antigen targets recently (Nino-Vasquez *et al.*, 2004; Pinilla *et al.*, 2001; Rubio-Godoy *et al.*, 2002). The technique involves the use of random libraries of peptides nonamers, decamers or longer peptides that are screened for reactivity in patient PBMC or TIL isolated from tumour biopsies and cultured in IL-2 (Matsushita *et al.*, 2001). The reactive pools are de-convoluted and single reactive peptides identified through ELISPOT screening and killing of tumour lines. The validity of the position-scanning combinatorial peptide library approach has recently been proven in a screening study performed with Melan A/MART-1-specific T cells in melanoma where both previously known and novel peptide epitopes were found (Pinilla *et al.*, 2001). A complicating factor in this approach, however, is that many positively reacting peptides are actually 'mimotopes' that activate T-cells against tumour antigens due to the promiscuousness of the TCR (Linnemann *et al.*, 2001; Tumenjargal *et al.*, 2003). Mimotopes differ in amino acid sequence from native tumour-associated antigen epitopes, but are nevertheless capable of activating T-cells through the same T-cell receptor (TCR) as the native epitope. This cross-reactivity sometimes enhances T-cell reactivity against a native epitope or altered peptide (Lawendowski *et al.*, 2002). Both mimotopes and altered peptides, found through combinatorial peptide library screening, are being investigated as potential cancer vaccines without the need for direct identification of the tumour antigen involved (Partidos, 2000). Peptides eliciting the most potent tumour cell killing by TIL or PBMC from cancer patients from these libraries are formulated in a vaccine. Proof-of-principle for this approach has been demonstrated in some tumour models (Blake *et al.*, 1996; Chung *et al.*, 2002; Tumenjargal *et al.*, 2003).

9.5 Future Prospects: Combining Target Discovery Approaches in Unified Publicly Accessible Databases

The genomics revolution, together with new technologies in proteomics, cell signalling pathway screening and immunological methods, has led to a huge wave of activity in new cancer target discovery over the last 10 years. However, there are serious deficiencies in the system that need be rectified in order for us to ultimately benefit from this ‘information explosion’. In addition, the system has emerged to be quite complex, with different techniques used to identify different potential targets. There is a serious lack of consistency or standardization in new target discovery methodologies, resulting in a great deal of information overload; one has to simply do a literature search to see the voluminous papers on new potential drug and immunotherapy targets. The question is: what is the real relevance of many of these ‘targets’ clinically and how can we establish a set of parameters that better predict the ultimate clinical relevance or efficacy of these new ‘targets’. One of the main problems at present is that most of new tumour-associated genes published in the literature have not been adequately validated using the methodologies we have described here such as IHC screening and verification of target function and immunogenicity. We need to bridge this gap by making sure new targets meet at least a minimal set of criteria to be accepted as new tumour-specific targets. This would help ensure that new therapeutic products launched into clinical use have the

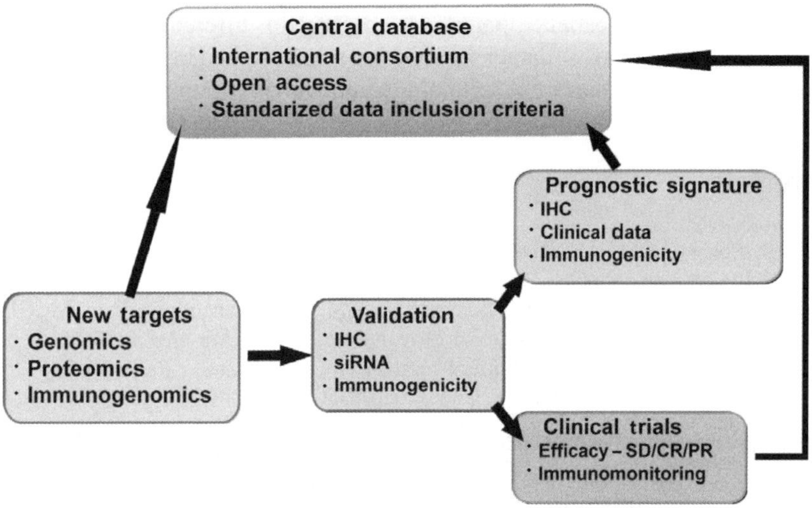

Figure 9.11 Proposed standardized international database to capture relevant data on new target genes in cancer. The database would capture data on the original genomics, proteomics and immunogenomics as well as data and methods relating to target validation for function and immunogenicity. Only data emerging from a strict filtering process ensuring data quality, consistency and statistical relevance would be included. A separate area of interacting information would capture data from clinical trials on the efficacy of a new tumour-specific gene as a therapy target for either drug-based inhibitors or vaccines. This would include data on disease stabilization (SD), complete responses (CR) and partial responses (PR) as well as the results of key immunomonitoring assays (in the case of vaccine trials) used as functional correlates. Such as system would require a high degree of international cooperation between chosen cancer institutes across the world and be openly accessible. The types of data to be included (e.g. types of assays and clinical response criteria) and rules regarding data quality would need to be fixed at the outset by an international consortium and any changes would have to be agreed upon at international conferences or meetings on database organization and logistics.

best chance of success. In addition, we also need to ensure that both basic scientists and clinicians have access to adequate and valid information on new targets and that information from clinical trials in different parts of the world testing validated targets are captured in a useful format.

Inevitably, we have to face the challenge of handling all this information and come up with some far-reaching solutions. One approach to this problem would be the creation of large internationally accessible databases that capture relevant data on new validated targets in cancer and other diseases together with key findings in clinical trials and the prognostic significance of target expression. These databases would need a strict set of standardized criteria establishing the conditions for what a validated target is and what type and quality of clinical and prognostic data can be included. Figure 9.11 illustrates how these interrelated groups of information can be linked into a central database freely accessible to researchers across the world. Ultimately, it will be through this type of information sharing that the relevance of new drug and immunotherapy targets can be tested in as wide a context as possible, especially given the genetic polymorphisms that exist in the human population that regulate how different patient populations respond to different targetted therapies. The generation of such a database is a tremendous undertaking, but it is not insurmountable and we have the technology and resources already at our disposal.

References

Agrawal B, Longenecker BM. 2005. MUC1 mucin-mediated regulation of human T cells. *Int Immunol* **17**: 391–399.

Agrawal B, Krantz MJ, Reddish MA, Longenecker BM. 1998. Cancer-associated MUC1 mucin inhibits human T-cell proliferation, which is reversible by IL-2. *Nat Med* **4**: 43–49.

Albertson DG. 2003. Profiling breast cancer by array CGH. *Breast Cancer Res Treat* **78**: 289–298.

Albertson DG, Collins C, McCormick F, Gray JW. 2003. Chromosome aberrations in solid tumors. *Nat Genet* **34**: 369–376.

Alessandro R, Belluco C, Kohn EC. 2005. Proteomic approaches in colon cancer: promising tools for new cancer markers and drug target discovery. *Clin Colorectal Cancer* **4**: 396–402.

Anzick SL, Kononen J, Walker RL, Azorsa DO, Tanner MM, Guan XY, Sauter G, Kallioniemi OP, Trent JM, Meltzer PS. 1997. AIB1, a steroid receptor coactivator amplified in breast and ovarian cancer. *Science* **277**: 965–968.

Bachinsky MM, Guillen DE, Patel SR, Singleton J, Chen C, Soltis DA, Tussey LG. 2005. Mapping and binding analysis of peptides derived from the tumor-associated antigen survivin for eight HLA alleles. *Cancer Immun* **5**: 6.

Band V, Zajchowski D, Stenman G, Morton CC, Kulesa V, Connolly J, Sager R. 1989. A newly established metastatic breast tumour cell line with integrated amplified copies of ERBB2 and double minute chromosomes. *Genes Chromosomes Cancer* **1**: 48–58.

Barnea E, Beer I, Patoka R, Ziv T, Kessler O, Tzehoval E, Eisenbach L, Zavazava N, Admon A. 2002. Analysis of endogenous peptides bound by soluble MHC class I molecules: a novel approach for identifying tumour-specific antigens. *Eur J Immunol* **32**: 213–222.

Basu S, Srivastava PK. 1999. Calreticulin, a peptide-binding chaperone of the endoplasmic reticulum, elicits tumor- and peptide-specific immunity. *J Exp Med* **189**: 797–802.

Bayani JM, Squire JA. 2002. Applications of SKY in cancer cytogenetics. *Cancer Invest* **20**: 373–386.

Beck MT, Holle L, Chen WY. 2001. Combination of PCR subtraction and cDNA microarray for differential gene expression profiling. *Biotechniques* **31**: 782–784, 786.

Behrends U, Schneider I, Rossler S, Frauenknecht H, Golbeck A, Lechner B, Eigenstetter G, Zobywalski C, Muller-Weihrich S, Graubner U. *et al.* 2003. Novel tumor antigens identified by autologous antibody screening of childhood medulloblastoma cDNA libraries. *Int J Cancer* **106**: 244–251.

Belguise K, Kersual N, Galtier F, Chalbos D. 2005. FRA-1 expression level regulates proliferation and invasiveness of breast cancer cells. *Oncogene* **24**: 1434–1444.

Belli F, Testori A, Rivoltini L, Maio M, Andreola G, Sertoli MR, Gallino G, Piris A, Cattelan A, Lazzari I. *et al.* 2002. Vaccination of metastatic melanoma patients with autologous tumor-derived heat shock protein gp96–peptide complexes: clinical and immunologic findings. *J Clin Oncol* **20**: 4169–4180.

Binder RJ, Srivastava PK. 2005. Peptides chaperoned by heat-shock proteins are a necessary and sufficient source of antigen in the cross-priming of CD8(+) T cells. *Nat Immunol.*

Blake J, Johnston JV, Hellstrom KE, Marquardt H, Chen L. 1996. Use of combinatorial peptide libraries to construct functional mimics of tumor epitopes recognized by MHC class I-restricted cytolytic T lymphocytes. *J Exp Med* **184**: 121–130.

Blok LJ, Kumar MV, Tindall DJ. 1995. Isolation of cDNAs that are differentially expressed between androgen-dependent and androgen-independent prostate carcinoma cells using differential display PCR. *Prostate* **26**: 213–224.

Bodey B. 2002. The significance of immunohistochemistry in the diagnosis and therapy of neoplasms. *Expert Opin Biol Ther* **2**: 371–393.

Bolli M, Schultz-Thater E, Zajac P, Guller U, Feder C, Sanguedolce F, Carafa V, Terracciano L, Hudolin T, Spagnoli GC, Tornillo L. 2005. NY-ESO-1/LAGE-1 coexpression with MAGE-A cancer/testis antigens: a tissue microarray study. *Int J Cancer.*

Borenstein SH, Graham J, Zhang XL, Chamberlain JW. 2000. CD8+ T cells are necessary for recognition of allelic, but not locus-mismatched or xeno-, HLA class I transplantation antigens. *J Immunol* **165**: 2341–2353.

Brusic V, Rudy G, Harrison LC. 1998. MHCPEP, a database of MHC-binding peptides: update 1997. *Nucl Acids Res* **26**: 368–371.

Buchli R, VanGundy RS, Hickman-Miller HD, Giberson CF, Bardet W, Hildebrand WH. 2004. Real-time measurement of *in vitro* peptide binding to soluble HLA-A*0201 by fluorescence polarization. *Biochemistry* **43**: 14852–14863.

Callagy G, Pharoah P, Chin SF, Sangan T, Daigo Y, Jackson L, Caldas C. 2005. Identification and validation of prognostic markers in breast cancer with the complementary use of array-CGH and tissue microarrays. *J Pathol* **205**: 388–396.

Carmon L, Bobilev-Priel I, Brenner B, Bobilev D, Paz A, Bar-Haim E, Tirosh B, Klein T, Fridkin M, Lemonnier F. *et al.* 2002. Characterization of novel breast carcinoma-associated BA46-derived peptides in HLA-A2.1/D(b)-beta2m transgenic mice. *J Clin Invest* **110**: 453–462.

Carvalho B, Ouwerkerk E, Meijer GA, Ylstra B. 2004. High resolution microarray comparative genomic hybridisation analysis using spotted oligonucleotides. *J Clin Pathol* **57**: 644–646.

Chan SM, Ermann J, Su L, Fathman CG, Utz PJ. 2004a. Protein microarrays for multiplex analysis of signal transduction pathways. *Nat Med* **10**: 1390–1396.

Chan SM, Ermann J, Su L, Fathman CG, Utz PJ. 2004b. Protein microarrays for multiplex analysis of signal transduction pathways. *Nat Med* **10**: 1390–1396.

Chang GT, van den Bemd GJ, Jhamai M, Brinkmann AO. 2002. Structure and function of GC79/TRPS1, a novel androgen-repressible apoptosis gene. *Apoptosis* **7**: 13–21.

Chang JC, Wooten EC, Tsimelzon A, Hilsenbeck SG, Gutierrez MC, Elledge R, Mohsin S, Osborne CK, Chamness GC, Allred DC, O'Connell P. 2003. Gene expression profiling for the prediction of therapeutic response to docetaxel in patients with breast cancer. *Lancet* **362**: 362–369.

Chen LC, Neubauer A, Kurisu W, Waldman FM, Ljung BM, Goodson W 3rd, Goldman ES, Moore D 2nd, Balazs M, Liu E. *et al.* 1991. Loss of heterozygosity on the short arm of chromosome 17 is associated with high proliferative capacity and DNA aneuploidy in primary human breast cancer. *Proc Natl Acad Sci USA* **88**: 3847–3851.

Chen YT, Scanlan MJ, Sahin U, Tureci O, Gure AO, Tsang S, Williamson B, Stockert E, Pfreundschuh M, Old LJ. 1997. A testicular antigen aberrantly expressed in human cancers detected by autologous antibody screening. *Proc Natl Acad Sci USA* **94**: 1914–1918.

Chiu YL, Rana TM. 2003. siRNA function in RNAi: a chemical modification analysis. *Rna* **9**: 1034–1048.

Chung J, Park S, Kim D, Rhim J, Kim I, Choi I, Yi K, Ryu S, Suh P, Chung D. *et al.* 2002. Identification of antigenic peptide recognized by the anti-JL1 leukemia-specific monoclonal antibody from combinatorial peptide phage display libraries. *J Cancer Res Clin Oncol* **128**: 641–649.

Claxton DF, Liu P, Hsu HB, Marlton P, Hester J, Collins F, Deisseroth AB, Rowley JD, Siciliano MJ. 1994. Detection of fusion transcripts generated by the inversion 16 chromosome in acute myelogenous leukemia. *Blood* **83**: 1750–1756.

Coene ED, Schelfhout V, Winkler RA, Schelfhout AM, Van Roy N, Grooteclaes M, Speleman F, De Potter CR. 1997. Amplification units and translocation at chromosome 17q and c-erbB-2 overexpression in the pathogenesis of breast cancer. *Virchow's Arch* **430**: 365–372.

Cohen IR. 1993. The meaning of the immunological homunculus. *Isr J Med Sci* **29**: 173–174.

Crawford YG, Gauthier ML, Joubel A, Mantei K, Kozakiewicz K, Afshari CA, Tlsty TD. 2004. Histologically normal human mammary epithelia with silenced p16(INK4a) overexpress COX-2, promoting a premalignant program. *Cancer Cell* **5**: 263–273.

Devilee P, Cornelisse CJ. 1994. Somatic genetic changes in human breast cancer. *Biochim Biophys Acta* **1198**: 113–130.

Di Vinci A, Perdelli L, Banelli B, Salvi S, Casciano I, Gelvi I, Allemanni G, Margallo E, Gatteschi B, Romani M. 2005. p16(INK4a) promoter methylation and protein expression in breast fibroadenoma and carcinoma. *Int J Cancer* **114**: 414–421.

Dong S, Geng JP, Tong JH, Wu Y, Cai JR, Sun GL, Chen SR, Wang ZY, Larsen CJ, Berger R. *et al.* 1993. Breakpoint clusters of the PML gene in acute promyelocytic leukemia: primary structure of the reciprocal products of the PML-RARA gene in a patient with t(15;17). *Genes Chromosomes Cancer* **6**: 133–139.

Firat H, Garcia-Pons F, Tourdot S, Pascolo S, Scardino A, Garcia Z, Michel ML, Jack RW, Jung G, Kosmatopoulos K. *et al.* 1999. H-2 class I knockout, HLA-A2.1-transgenic mice: a versatile animal model for preclinical evaluation of antitumor immunotherapeutic strategies. *Eur J Immunol* **29**: 3112–3121.

Forozan F, Mahlamaki EH, Monni O, Chen Y, Veldman R, Jiang Y, Gooden GC, Ethier SP, Kallioniemi A, Kallioniemi OP. 2000. Comparative genomic hybridization analysis of 38 breast cancer cell lines: a basis for interpreting complementary DNA microarray data. *Cancer Res* **60**: 4519–4525.

Futreal PA, Coin L, Marshall M, Down T, Hubbard T, Wooster R, Rahman N, Stratton MR. 2004. A census of human cancer genes. *Nat Rev Cancer* **4**: 177–183.

Gamero AM, Young HA, Wiltrout RH. 2004. Inactivation of Stat3 in tumor cells: releasing a brake on immune responses against cancer? *Cancer Cell* **5**: 111–112.

Gardner-Thorpe J, Ito H, Ashley SW, Whang EE. 2002. Differential display of expressed genes in pancreatic cancer cells. *Biochem Biophys Res Commun* **293**: 391–395.

Gauthier ML, Pickering CR, Miller CJ, Fordyce CA, Chew KL, Berman HK, Tlsty TD. 2005. p38 regulates cyclooxygenase-2 in human mammary epithelial cells and is activated in premalignant tissue. *Cancer Res* **65**: 1792–1799.

Ginestier C, Charafe-Jauffret E, Bertucci F, Eisinger F, Geneix J, Bechlian D, Conte N, Adelaide J, Toiron Y, Nguyen C. *et al.* 2002. Distinct and complementary information provided by use of tissue and DNA microarrays in the study of breast tumor markers. *Am J Pathol* **161**: 1223–1233.

Greiner J, Ringhoffer M, Taniguchi M, Hauser T, Schmitt A, Dohner H, Schmitt M. 2003. Characterization of several leukemia-associated antigens inducing humoral immune responses in acute and chronic myeloid leukemia. *Int J Cancer* **106**: 224–231.

Guo QM. 2003. DNA microarray and cancer. *Curr Opin Oncol* **15**: 36–43.

Hao X, Sun B, Hu L, Lahdesmaki H, Dunmire V, Feng Y, Zhang SW, Wang H, Wu C, Fuller GN. *et al.* 2004. Differential gene and protein expression in primary breast malignancies and their lymph node metastases as revealed by combined cDNA microarray and tissue microarray analysis. *Cancer* **100**: 1110–1122.

Hardiman G. 2004. Microarray platforms–comparisons and contrasts. *Pharmacogenomics* **5**: 487–502.

Hatamura I, Kanauchi Y, Takahara M, Fujiwara M, Muragaki Y, Ooshima A, Ogino T. 2001. A nonsense mutation in TRPS1 in a Japanese family with tricho-rhino-phalangeal syndrome type I. *Clin Genet* **59**: 366–367.

Henshall SM, Afar DE, Hiller J, Horvath LG, Quinn DI, Rasiah KK, Gish K, Willhite D, Kench JG, Gardiner-Garden M. *et al.* 2003. Survival analysis of genome-wide gene expression profiles of prostate cancers identifies new prognostic targets of disease relapse. *Cancer Res* **63**: 4196–4203.

Hernandez HJ, Edson CM, Harn DA, Ianelli CJ, Stadecker MJ. 1998. Schistosoma mansoni: genetic restriction and cytokine profile of the CD4 + T helper cell response to dominant epitope peptide of major egg antigen Sm-p40. *Exp Parasitol* **90**: 122–130.

Hilton MJ, Sawyer JM, Gutierrez L, Hogart A, Kung TC, Wells DE. 2002. Analysis of novel and recurrent mutations responsible for the tricho-rhino-phalangeal syndromes. *J Hum Genet* **47**: 103–106.

Hlavac F, Connan F, Hoebeke J, Guillet JG, Choppin J. 1996. Direct detection of peptide-dependent HLA variability by surface plasmon resonance. *Mol Immunol* **33**: 573–582.

Holst CR, Nuovo GJ, Esteller M, Chew K, Baylin SB, Herman JG, Tlsty TD. 2003. Methylation of p16(INK4a) promoters occurs *in vivo* in histologically normal human mammary epithelia. *Cancer Res* **63**: 1596–1601.

Horwitz KB, McGuire WL. 1975. Specific progesterone receptors in human breast cancer. *Steroids* **25**: 497–505.

Hubert RS, Vivanco I, Chen E, Rastegar S, Leong K, Mitchell SC, Madraswala R, Zhou Y, Kuo J, Raitano AB. *et al.* 1999. STEAP: a prostate-specific cell-surface antigen highly expressed in human prostate tumors. *Proc Natl Acad Sci USA* **96**: 14523–14528.

Inoue A, Yoshida N, Omoto Y, Oguchi S, Yamori T, Kiyama R, Hayashi S. 2002. Development of cDNA microarray for expression profiling of estrogen-responsive genes. *J Mol Endocrinol* **29**: 175–192.

Jacquemier J, Ginestier C, Rougemont J, Bardou VJ, Charafe-Jauffret E, Geneix J, Adelaide J, Koki A, Houvenaeghel G, Hassoun J. *et al.* 2005. Protein expression profiling identifies subclasses of breast cancer and predicts prognosis. *Cancer Res* **65**: 767–779.

Jager E, Chen YT, Drijfhout JW, Karbach J, Ringhoffer M, Jager D, Arand M, Wada H, Noguchi Y, Stockert E. *et al.* 1998. Simultaneous humoral and cellular immune response against cancer-testis antigen NY-ESO-1: definition of human histocompatibility leukocyte antigen (HLA)-A2-binding peptide epitopes. *J Exp Med* **187**: 265–270.

Jager D, Stockert E, Gure AO, Scanlan MJ, Karbach J, Jager E, Knuth A, Old LJ, Chen YT. 2001. Identification of a tissue-specific putative transcription factor in breast tissue by serological screening of a breast cancer library. *Cancer Res* **61**: 2055–2061.

Jager D, Unkelbach M, Frei C, Bert F, Scanlan MJ, Jager E, Old LJ, Chen YT, Knuth A. 2002. Identification of tumour-restricted antigens NY-BR-1, SCP-1, and a new cancer/testis-like antigen NW-BR-3 by serological screening of a testicular library with breast cancer serum. *Cancer Immun* **2**: 5.

Jiang Y, Harlocker SL, Molesh DA, Dillon DC, Stolk JA, Houghton RL, Repasky EA, Badaro R, Reed SG, Xu J. 2002. Discovery of differentially expressed genes in human breast cancer using subtracted cDNA libraries and cDNA microarrays. *Oncogene* **21**: 2270–2282.

Kaiser FJ, Moroy T, Chang GT, Horsthemke B, Ludecke HJ. 2003. The RING finger protein RNF4, a co-regulator of transcription, interacts with the TRPS1 transcription factor. *J Biol Chem* **278**: 38780–38785.

Khong HT, Yang JC, Topalian SL, Sherry RM, Mavroukakis SA, White DE, Rosenberg SA. 2004. Immunization of HLA-A*0201 and/or HLA-DPbeta1*04 patients with metastatic melanoma using epitopes from the NY-ESO-1 antigen. *J Immunother* **27**: 472–477.

Klyushnenkova EN, Link J, Oberle WT, Kodak J, Rich C, Vandenbark AA, Alexander RB. 2005. Identification of HLA-DRB1*1501-restricted T-cell epitopes from prostate-specific antigen. *Clin Cancer Res* **11**: 2853–2861.

Kobayashi H, Wood M, Song Y, Appella E, Celis E. 2000. Defining promiscuous MHC class II helper T-cell epitopes for the HER2/neu tumor antigen. *Cancer Res* **60**: 5228–5236.

Lacroix M, Leclercq G. 2004. Relevance of breast cancer cell lines as models for breast tumors: an update. *Breast Cancer Res Treat* **83**: 249–289.

Lawendowski CA, Giurleo GM, Huang YY, Franklin GJ, Kaplan JM, Roberts BL, Nicolette CA. 2002. Solid-phase epitope recovery: a high throughput method for antigen identification and epitope optimization. *J Immunol* **169**: 2414–2421.

Le XF, Lammayot A, Gold D, Lu Y, Mao W, Chang T, Patel A, Mills GB, Bast RC, Jr 2005. Genes affecting the cell cycle, growth, maintenance, and drug sensitivity are preferentially regulated by anti-HER2 antibody through phosphatidylinositol 3-kinase-AKT signalling. *J Biol Chem* **280**: 2092–2104.

Li B, Qian XP, Pang XW, Zou WZ, Wang YP, Wu HY, Chen WF. 2003. HCA587 antigen expression in normal tissues and cancers: correlation with tumour differentiation in hepatocellular carcinoma. *Lab Invest* **83**: 1185–1192.

Li G, Miles A, Line A, Rees RC. 2004. Identification of tumor antigens by serological analysis of cDNA expression cloning. *Cancer Immunol Immunother* **53**: 139–143.

Li J, Yen C, Liaw D, Podsypanina K, Bose S, Wang SI, Puc J, Miliaresis C, Rodgers L, McCombie R. *et al.* 1997. PTEN, a putative protein tyrosine phosphatase gene mutated in human brain, breast, and prostate cancer.[see Comment.] *Science* **275**: 1943–1947.

Liang P. 2002. A decade of differential display. *Biotechniques* **33**: 338–344, 346.

Line A, Slucka Z, Stengrevics A, Silina K, Li G, Rees RC. 2002. Characterisation of tumor-associated antigens in colon cancer. *Cancer Immunol Immunother* **51**: 574–582.

Ling X, Arlinghaus RB. 2005. Knockdown of STAT3 expression by RNA interference inhibits the induction of breast tumours in immunocompetent mice. *Cancer Res* **65**: 2532–2536.

Linnemann T, Tumenjargal S, Gellrich S, Wiesmuller K, Kaltoft K, Sterry W, Walden P. 2001. Mimotopes for tumor-specific T lymphocytes in human cancer determined with combinatorial peptide libraries. *Eur J Immunol* **31**: 156–165.

Little CD, Nau MM, Carney DN, Gazdar AF, Minna JD. 1983. Amplification and expression of the c-myc oncogene in human lung cancer cell lines. *Nature* **306**: 194–196.

Loo LWM, Grove DI, Williams EM, Neal CL, Cousens LA, Schubert EL, Holcomb IN, Massa HF, Glogovac J, Li CI. *et al.* 2004. Array comparative genomic hybridization analysis of genomic alterations in breast cancer subtypes. *Cancer Res* **64**: 8541–8549.

Mackay A, Jones C, Dexter T, Silva RL, Bulmer K, Jones A, Simpson P, Harris RA, Jat PS, Neville AM. *et al.* 2003. cDNA microarray analysis of genes associated with ERBB2 (HER2/neu) overexpression in human mammary luminal epithelial cells. *Oncogene* **22**: 2680–2688.

Maecker HT, Dunn HS, Suni MA, Khatamzas E, Pitcher CJ, Bunde T, Persaud N, Trigona W, Fu TM, Sinclair E. *et al.* 2001. Use of overlapping peptide mixtures as antigens for cytokine flow cytometry. *J Immunol Meth* **255**: 27–40.

Malik TH, Shoichet SA, Latham P, Kroll TG, Peters LL, Shivdasani RA. 2001. Transcriptional repression and developmental functions of the atypical vertebrate GATA protein TRPS1. *Embo J* **20**: 1715–1725.

Man YG, Zhang Y, Shen T, Zeng X, Tauler J, Mulshine JL, Strauss BL. 2005. cDNA expression profiling reveals elevated gene expression in cell clusters overlying focally disrupted myoepithelial cell layers: implications for breast tumor invasion. *Breast Cancer Res Treat* **89**: 199–208.

Mandic M, Almunia C, Vicel S, Gillet D, Janjic B, Coval K, Maillere B, Kirkwood JM, Zarour HM. 2003. The alternative open reading frame of LAGE-1 gives rise to multiple promiscuous HLA-DR-restricted epitopes recognized by T-helper 1-type tumor-reactive CD4+ T cells. *Cancer Res* **63**: 6506–6515.

Marsh SGE, Parham P, Barber LD. 2000. *The HLA Factsbook.* San Diego, CA: Academic Press.

Martin KJ, Pardee AB. 1999. Principles of differential display. *Meth Enzymol* **303**: 234–258.

Martin P, Parroche P, Chatel L, Barretto C, Beck A, Trepo C, Bain C, Lone YC, Inchauspe G, Fournillier A. 2004. Genetic immunization and comprehensive screening approaches in HLA-A2 transgenic mice lead to the identification of three novel epitopes in hepatitis C virus NS3 antigen. *J Med Virol* **74**: 397–405.

Massa C, Guiducci C, Arioli I, Parenza M, Colombo MP, Melani C. 2004. Enhanced efficacy of tumor cell vaccines transfected with secretable hsp70. *Cancer Res* **64**: 1502–1508.

Matsushita S, Tanaka Y, Matsuoka T, Nakashima T. 2001. Clonal expansion of freshly isolated CD4T cells by randomized peptides and identification of peptide ligands using combinatorial peptide libraries. *Eur J Immunol* **31**: 2395–2402.

Mazzaferro V, Coppa J, Carrabba MG, Rivoltini L, Schiavo M, Regalia E, Mariani L, Camerini T, Marchiano A, Andreola S. *et al.* 2003. Vaccination with autologous tumor-derived heat-shock protein gp96 after liver resection for metastatic colorectal cancer. *Clin Cancer Res* **9**: 3235–3245.

Mei R, Hubbell E, Bekiranov S, Mittmann M, Christians FC, Shen MM, Lu G, Fang J, Liu WM, Ryder T. *et al.* 2003. Probe selection for high-density oligonucleotide arrays. *Proc Natl Acad Sci USA* **100**: 11237–11242.

Menendez JA, Lupu R. 2005. RNA interference-mediated silencing of the p53 tumor-suppressor protein drastically increases apoptosis after inhibition of endogenous fatty acid metabolism in breast cancer cells. *Int J Mol Med* **15**: 33–40.

Miksicek RJ, Myal Y, Watson PH, Walker C, Murphy LC, Leygue E. 2002. Identification of a novel breast- and salivary gland-specific, mucin-like gene strongly expressed in normal and tumor human mammary epithelium. *Cancer Res* **62**: 2736–2740.

Momeni P, Glockner G, Schmidt O, von Holtum D, Albrecht B, Gillessen-Kaesbach G, Hennekam R, Meinecke P, Zabel B, Rosenthal A *et al.* 2000. Mutations in a new gene, encoding a zinc-finger protein, cause tricho-rhino-phalangeal syndrome type I. *Nat Genet* **24**: 71–74.

Nagahata T, Sato T, Tomura A, Onda M, Nishikawa K, Emi M. 2005. Identification of RAI3 as a therapeutic target for breast cancer. *Endocr Relat Cancer* **12**: 65–73.

Nagai MA, Ros N, Bessa SA, Mourao Neto M, Miracca EC, Brentani MM. 2003. Differentially expressed genes and estrogen receptor status in breast cancer. *Int J Oncol* **23**: 1425–1430.

Nair S, Wearsch PA, Mitchell DA, Wassenberg JJ, Gilboa E, Nicchitta CV. 1999. Calreticulin displays *in vivo* peptide-binding activity and can elicit CTL responses against bound peptides. *J Immunol* **162**: 6426–6432.

Nakatsura T, Senju S, Yamada K, Jotsuka T, Ogawa M, Nishimura Y. 2001. Gene cloning of immunogenic antigens overexpressed in pancreatic cancer. *Biochem Biophys Res Commun* **281**: 936–944.

Navaratnam M, Deshpande MS, Hariharan MJ, Zatechka DS, Jr, Srikumaran S. 2001. Heat shock protein–peptide complexes elicit cytotoxic T-lymphocyte and antibody responses specific for bovine herpesvirus 1. *Vaccine* **19**: 1425–1434.

Nessling M, Richter K, Schwaenen C, Roerig P, Wrobel G, Wessendorf S, Fritz B, Bentz M, Sinn HP, Radlwimmer B, Lichter P. 2005. Candidate genes in breast cancer revealed by microarray-based comparative genomic hybridization of archived tissue. *Cancer Res* **65**: 439–447.

Nino-Vasquez JJ, Allicotti G, Borras E, Wilson DB, Valmori D, Simon R, Martin R, Pinilla C. 2004. A powerful combination: the use of positional scanning libraries and biometrical analysis to identify cross-reactive T cell epitopes. *Mol Immunol* **40**: 1063–1074.

Nishizuka S, Charboneau L, Young L, Major S, Reinhold WC, Waltham M, Kouros-Mehr H, Bussey KJ, Lee JK, Espina V. *et al.* 2003. Proteomic profiling of the NCI-60 cancer cell lines using new high-density reverse-phase lysate microarrays. *Proc Nat Acad Sci USA* **100**: 14229–14234.

Obata Y, T TA, Tamaki H, Tominaga S, Murai H, Iwase T, Iwata H, Mizutani M, Chen YT, Old LJ, Miura S. 1999. Identification of cancer antigens in breast cancer by the SEREX expression cloning method. *Breast Cancer* **6**: 305–311.

Obata Y, Takahashi T, Sakamoto J, Tamaki H, Tominaga S, Hamajima N, Chen YT, Old LJ. 2000. SEREX analysis of gastric cancer antigens. *Cancer Chemother Pharmac* **46**(Suppl): S37–42.

Ochs MF, Godwin AK. 2003. Microarrays in cancer: research and applications. *Biotechniques* (Suppl): 4–15.

Papadopoulos KP, Suciu-Foca N, Hesdorffer CS, Tugulea S, Maffei A, Harris PE. 1997. Naturally processed tissue- and differentiation stage-specific autologous peptides bound by HLA class I and II molecules of chronic myeloid leukemia blasts. *Blood* **90**: 4938–4946.

Partidos CD. 2000. Peptide mimotopes as candidate vaccines. *Curr Opin Mol Ther* **2**: 74–79.

Pellegrini P, Berghella AM, Del Beato T, Maccarone D, Cencioni S, Adorno D, Casciani CU. 1997. The sCEA molecule suppressive role in NK and TH1 cell functions in colorectal cancer. *Cancer Biother Radiopharm* **12**: 257–264.

Perou CM, Sorlie T, Eisen MB, van de Rijn M, Jeffrey SS, Rees CA, Pollack JR, Ross DT, Johnsen H, Akslen LA. *et al.* 2000. Molecular portraits of human breast tumors. *Nature* **406**: 747–752.

Petricoin EF, Ardekani AM, Hitt BA, Levine PJ, Fusaro VA, Steinberg SM, Mills GB, Simone C, Fishman DA, Kohn EC, Liotta LA. 2002a. Use of proteomic patterns in serum to identify ovarian cancer. [See comment.] *Lancet* **359**: 572–577.

Petricoin EF, Zoon KC, Kohn EC, Barrett JC, Liotta LA. 2002b. Clinical proteomics: translating benchside promise into bedside reality. *Nat Rev Drug Discov* **1**: 683–695.

Pille JY, Denoyelle C, Varet J, Bertrand JR, Soria J, Opolon P, Lu H, Pritchard LL, Vannier JP, Malvy C. *et al.* 2005. Anti-RhoA and anti-RhoC siRNAs inhibit the proliferation and invasiveness of MDA-MB-231 breast cancer cells *in vitro* and *in vivo*. *Mol Ther* **11**: 267–274.

Pinilla C, Rubio-Godoy V, Dutoit V, Guillaume P, Simon R, Zhao Y, Houghten RA, Cerottini JC, Romero P, Valmori D. 2001. Combinatorial peptide libraries as an alternative approach to the identification of ligands for tumor-reactive cytolytic T lymphocytes. *Cancer Res* **61**: 5153–5160.

Pinkel D, Segraves R, Sudar D, Clark S, Poole I, Kowbel D, Collins C, Kuo WL, Chen C, Zhai Y. *et al.* 1998. High resolution analysis of DNA copy number variation using comparative genomic hybridization to microarrays. *Nat Genet* **20**: 207–211.

Poletaev A, Osipenko L. 2003. General network of natural autoantibodies as immunological homunculus (Immunculus). *Autoimmun Rev* **2**: 264–271.

Pollack JR, Perou CM, Alizadeh AA, Eisen MB, Pergamenschikov A, Williams CF, Jeffrey SS, Botstein D, Brown PO. 1999. Genome-wide analysis of DNA copy-number changes using cDNA microarrays. *Nat Genet* **23**: 41–46.

Pusztai L, Hess KR. 2004. Clinical trial design for microarray predictive marker discovery and assessment. *Ann Oncol* **15**: 1731–1737.

Pusztai L, Ayers M, Stec J, Clark E, Hess K, Stivers D, Damokosh A, Sneige N, Buchholz TA, Esteva FJ. *et al.* 2003. Gene expression profiles obtained from fine-needle aspirations of breast cancer reliably identify routine prognostic markers and reveal large-scale molecular differences between estrogen-negative and estrogen-positive tumors. *Clin Cancer Res* **9**: 2406–2415.

Rammensee H, Bachmann J, Emmerich NP, Bachor OA, Stevanovic S. 1999. SYFPEITHI: database for MHC ligands and peptide motifs. *Immunogenetics* **50**: 213–219.

Ree AH, Engebraaten O, Hovig E, Fodstad O. 2002. Differential display analysis of breast carcinoma cells enriched by immunomagnetic target cell selection: gene expression profiles in bone marrow target cells. *Int J Cancer* **97**: 28–33.

Rio MC, Bellocq JP, Gairard B, Rasmussen UB, Krust A, Koehl C, Calderoli H, Schiff V, Renaud R, Chambon P. 1987. Specific expression of the pS2 gene in subclasses of breast cancers in comparison with expression of the estrogen and progesterone receptors and the oncogene ERBB2. *Proc Natl Acad Sci USA* **84**: 9243–9247.

Rotzschke O, Falk K, Deres K, Schild H, Norda M, Metzger J, Jung G, Rammensee HG. 1990. Isolation and analysis of naturally processed viral peptides as recognized by cytotoxic T cells. *Nature* **348**: 252–254.

Rubio-Godoy V, Ayyoub M, Dutoit V, Servis C, Schink A, Rimoldi D, Romero P, Cerottini JC, Simon R, Zhao Y. *et al.* 2002. Combinatorial peptide library-based identification of peptide ligands for tumor-reactive cytolytic T lymphocytes of unknown specificity. *Eur J Immunol* **32**: 2292–2299.

Scanlan MJ, Gordan JD, Williamson B, Stockert E, Bander NH, Jongeneel V, Gure AO, Jager D, Jager E, Knuth A. *et al.* 1999. Antigens recognized by autologous antibody in patients with renal-cell carcinoma. *Int J Cancer* **83**: 456–464.

Schonbach C, Kun Y, Brusic V. 2002. Large-scale computational identification of HIV T-cell epitopes. *Immunol Cell Biol* **80**: 300–306.

Schroers R, Shen L, Rollins L, Xiao Z, Sonderstrup G, Slawin K, Huang XF, Chen SY. 2003. Identification of MHC class II-restricted T-cell epitopes in prostate-specific membrane antigen. *Clin Cancer Res* **9**: 3260–3271.

Schultze JL, Michalak S, Seamon MJ, Dranoff G, Jung K, Daley J, Delgado JC, Gribben JG, Nadler LM. 1997. CD40-activated human B cells: an alternative source of highly efficient antigen presenting cells to generate autologous antigen-specific T cells for adoptive immunotherapy. *J Clin Invest* **100**: 2757–2765.

Sen S, Zhou H, White RA. 1997. A putative serine/threonine kinase encoding gene BTAK on chromosome 20q13 is amplified and overexpressed in human breast cancer cell lines. *Oncogene* **14**: 2195–2200.

Seth A, Kitching R, Landberg G, Xu J, Zubovits J, Burger AM. 2003. Gene expression profiling of ductal carcinomas in situ and invasive breast tumors. *Anticancer Res* **23**: 2043–2051.

Shen C, Buck AK, Liu X, Winkler M, Reske SN. 2003. Gene silencing by adenovirus-delivered siRNA. *FEBS Lett* **539**: 111–114.

Shen L, Schroers R, Hammer J, Huang XF, Chen SY. 2004. Identification of a MHC class-II restricted epitope in carcinoembryonic antigen. *Cancer Immunol Immunother* **53**: 391–403.

Singh H, Raghava GP. 2001. ProPred: prediction of HLA-DR binding sites. *Bioinformatics* **17**: 1236–1237.

Slamon DJ, Clark GM, Wong SG, Levin WJ, Ullrich A, McGuire WL. 1987. Human breast cancer: correlation of relapse and survival with amplification of the HER-2/neu oncogene. *Science* **235**: 177–182.

Snijders AM, Nowak N, Segraves R, Blackwood S, Brown N, Conroy J, Hamilton G, Hindle AK, Huey B, Kimura K. *et al.* 2001. Assembly of microarrays for genome-wide measurement of DNA copy number. *Nat Genet* **29**: 263–264.

Solinas-Toldo S, Lampel S, Stilgenbauer S, Nickolenko J, Benner A, Dohner H, Cremer T, Lichter P. 1997. Matrix-based comparative genomic hybridization: biochips to screen for genomic imbalances. *Genes, Chromosomes Cancer* **20**: 399–407.

Sorlie T, Perou CM, Tibshirani R, Aas T, Geisler S, Johnsen H, Hastie T, Eisen MB, van de Rijn M, Jeffrey SS. *et al.* 2001. Gene expression patterns of breast carcinomas distinguish tumour subclasses with clinical implications. *Proc Natl Acad Sci USA* **98**: 10869–10874.

Srivastava PK. 2005. Immunotherapy for human cancer using heat shock protein–peptide complexes. *Curr Oncol Rep* **7**: 104–108.

Srivastava PK, Amato RJ. 2001. Heat shock proteins: the 'Swiss Army Knife' vaccines against cancers and infectious agents. *Vaccine* **19**: 2590–2597.

Staib F, Distler M, Bethke K, Schmitt U, Galle PR, Heike M. 2004. Cross-presentation of human melanoma peptide antigen MART-1 to CTLs from *in vitro* reconstituted gp96/MART-1 complexes. *Cancer Immun* **4**: 3.

Stein D, Wu J, Fuqua SA, Roonprapunt C, Yajnik V, D'Eustachio P, Moskow JJ, Buchberg AM, Osborne CK, Margolis B. 1994. The SH2 domain protein GRB-7 is co-amplified, overexpressed and in a tight complex with HER2 in breast cancer. *EMBO J* **13**: 1331–1340.

Sturniolo T, Bono E, Ding J, Raddrizzani L, Tuereci O, Sahin U, Braxenthaler M, Gallazzi F, Protti MP, Sinigaglia F, Hammer J. 1999. Generation of tissue-specific and promiscuous HLA ligand databases using DNA microarrays and virtual HLA class II matrices. *Nat Biotechnol* **17**: 555–561.

Suzuki S, Moore DH. 2nd, Ginzinger DG, Godfrey TE, Barclay J, Powell B, Pinkel D, Zaloudek C, Lu K, Mills G. *et al.* 2000. An approach to analysis of large-scale correlations between genome changes and clinical endpoints in ovarian cancer. *Cancer Res* **60**: 5382–5385.

Takahashi I, Sugiura S, Ohta H, Ozawa K, Kamiya T. 2002. Epitope analysis of antibodies in Japanese to human cytomegalovirus phosphoprotein 150 with synthetic peptides. *Biosci Biotechnol Biochem* **66**: 2402–2405.

Thomas P, Pang Y, Filardo EJ, Dong J. 2005. Identity of an estrogen membrane receptor coupled to a G protein in human breast cancer cells. *Endocrinology* **146**: 624–632.

Tomasetto C, Regnier C, Moog-Lutz C, Mattei MG, Chenard MP, Lidereau R, Basset P, Rio MC. 1995. Identification of four novel human genes amplified and overexpressed in breast carcinoma and localized to the q11-q21.3 region of chromosome 17. *Genomics* **28**: 367–376.

Tumenjargal S, Gellrich S, Linnemann T, Muche JM, Lukowsky A, Audring H, Wiesmuller KH, Sterry W, Walden P. 2003. Anti-tumour immune responses and tumour regression induced with mimotopes of a tumour-associated T cell epitope. *Eur J Immunol* **33**: 3175– 3185.

Tureci O, Sahin U, Pfreundschuh M. 1997. Serological analysis of human tumour antigens: molecular definition and implications. *Mol Med Today* **3**: 342–349.

Uttamchandani M, Chan EW, Chen GY, Yao SQ. 2003. Combinatorial peptide microarrays for the rapid determination of kinase specificity. *Bioorg Med Chem Lett* **13**: 2997–3000.

Venter JC, Adams MD, Myers EW, Li PW, Mural RJ, Sutton GG, Smith HO, Yandell M, Evans CA, Holt RA. *et al.* 2001. The sequence of the human genome. *Science* **291**: 1304–1351.

Wang HY, Lee DA, Peng G, Guo Z, Li Y, Kiniwa Y, Shevach EM, Wang RF. 2004. Tumour-specific human CD4+ regulatory T cells and their ligands: implications for immunotherapy. *Immunity* **20**: 107–118.

Wang HY, Peng G, Guo Z, Shevach EM, Wang RF. 2005. Recognition of a new ARTC1 peptide ligand uniquely expressed in tumour cells by antigen-specific CD4+ regulatory T cells. *J Immunol* **174**: 2661–2670.

Watson MA, Fleming TP. 1996. Mammaglobin, a mammary-specific member of the uteroglobin gene family, is overexpressed in human breast cancer. *Cancer Res* **56**: 860–865.

Xie D, Jauch A, Miller CW, Bartram CR, Koeffler HP. 2002. Discovery of over-expressed genes and genetic alterations in breast cancer cells using a combination of suppression subtractive hybridization, multiplex FISH and comparative genomic hybridization. *Int J Oncol* **21**: 499–507.

Xing EP, Yang GY, Wang LD, Shi ST, Yang CS. 1999. Loss of heterozygosity of the Rb gene correlates with pRb protein expression and associates with p53 alteration in human esophageal cancer. *Clin Cancer Res* **5**: 1231–1240.

Yonghao T, Qian H, Chuanyuan L, Yandell DW. 1999. Deletions and point mutations of p16, p15 gene in primary tumours and tumour cell lines. *Chin Med Sci J* **14**: 200–205.

Zangar RC, Varnum SM, Covington CY, Smith RD. 2004. A rational approach for discovering and validating cancer markers in very small samples using mass spectrometry and ELISA microarrays. *Dis Markers* **20**: 135–148.

Zeng G, Wang X, Robbins PF, Rosenberg SA, Wang RF. 2001. CD4(+) T cell recognition of MHC class II-restricted epitopes from NY-ESO-1 presented by a prevalent HLA DP4 allele: association with NY-ESO-1 antibody production. *Proc Natl Acad Sci USA* **98**: 3964–3969.

Zhang D, Salto-Tellez M, Putti TC, Do E, Koay ES. 2003. Reliability of tissue microarrays in detecting protein expression and gene amplification in breast cancer. *Mod Pathol* **16**: 79–84.

Zhao C, Yasui K, Lee CJ, Kurioka H, Hosokawa Y, Oka T, Inazawa J. 2003. Elevated expression levels of NCOA3, TOP1, and TFAP2C in breast tumours as predictors of poor prognosis. *Cancer* **98**: 18–23.

Zhao X, Li C, Paez JG, Chin K, Janne PA, Chen TH, Girard L, Minna J, Christiani D, Leo C. *et al.* 2004. An integrated view of copy number and allelic alterations in the cancer genome using single nucleotide polymorphism arrays. *Cancer Res* **64**: 3060–3071.

10

Genomics and Functional Differences of Dendritic Cell Subsets

Peter Gogolak and **Eva Rajnavölgyi**

Abstract

Dendritic cells (DC) represent a multifunctional population of cells with the capacity to prime and orchestrate antigen-specific immune responses. Human DC are classified into myeloid and plasmacytoid DC with distinct functional activities. Both subsets can be found as resting cells, acting as sensors of environmental changes. Uptake of exogenous material in combination with danger signals induces activation, migration and differentation of DC that transform them into potent antigen-presenting and secretory cells. DC at both activation states interact with other cells via direct cell-to-cell contacts and by released cytokines and chemokines. Depending on the combination of exogenous and endogenous stimuli, both DC subsets can initiate inflammatory or regulatory immune responses. The way in which the various DC subtypes become activated and collaborate with other cells determines the outcome of immune responses against pathogens or malignant cells. The enormous functional flexibility of DC offers new possibilities to manipulate antigen-specific immune responses.

Keywords

dendritic cells; phagocytosis; toll-like receptors; antigen processing and presentation; cross-presentation; NKT-cell; regulatory T-cells; cytokines and chemokines; dendritic cell-based tumour immunotherapy

10.1 Introduction

Transient activation, expansion and contraction of clonally selected antigen-specific T- and B-lymphocyte clones is the basic principle of adaptive immunity. Initiation of

Immunogenomics and Human Disease Edited by András Falus

this process, referred to as priming antigen-specific immune responses, requires the collaboration of cells and molecules of both innate and acquired immunity. Dendritic cells (DC) represent a rare, heterogeneous and multifunctional population of cells, which plays a pivotal role in initiating and orchestrating strictly controlled immune responses, restorating the resting state and maintaining self-tolerance (Moser, 2003; Morel *et al.*, 2003). DC are found throughout the body, but they are concentrated at all potential sites of pathogen entry.

Tissue-resident immature DC continuously take up, store and transport extracellular particles and soluble material. Constitutive engulfment and activation induce DC to leave peripheral tissues and transport the internalized material to draining lymph nodes. Activated DC home to peripheral lymphoid tissues, present their antigenic content to T-lymphocytes and act as highly potent professional antigen-presenting cells (APC). The physiological tissue environment is translated as tolerable and does not induce self-destructive inflammatory responses. Changes in the tissue environment, such as danger signals induced by traumatic or toxic shock, stress, inflammation or pathogenic invasion, alter the amount and composition of engulfed material and activate resident DC. During the priming process the degree, nature, combination and duration of stimulatory signals modulate the response of DC and consequently influence the outcome of T- and B-lymphocyte-mediated immune responses. Prevention and down-regulation of antigen-specific immune responses by regulatory T-lymphocytes is also mediated by their interaction with DC under the control of the actual tissue environment where these events take place (reviewed in Ardavin *et al.*, 2004; O'Neill, Adams and Bhardwaj, 2004; Gogolak *et al.*, 2003).

To our present view, the major function of DC is to protect self-tissues from damage by the induction and maintenance of self-tolerance and to alarm the immune system against foreign and dangerous interventions. Discovering the mechanisms by which these opposing tasks are accomplished may lead us to find *ex vivo* or *in vivo* means of manipulating DC function and thus regulating immune responses. A compelling strategy, which holds promise of manipulating anti-tumour immunity, is based on utilizing DC as adjuvants of antigen-specific preventive and/or therapeutic vaccines. The ultimate goal of active DC-based virus- or tumour-specific vaccination of patients with established disease is to elicit long-lasting protective immunity without causing adverse effects on healthy tissues (Mocellin *et al.*, 2004). This review focuses on the functional characteristics of human DC subsets with special emphasis on their collaboration with other cells of innate immunity and on their potential for clinical utility against tumours.

10.2 Origin, Differentiation and Function of Human Dendritic Cell Subsets

DC depelop from CD34+ haematopoietic stem cells (HSC) and can be distinguished by their cell surface molecules, master transcription factors and functional properties. On the basis of their origin and anatomical localization, both mouse and human DC

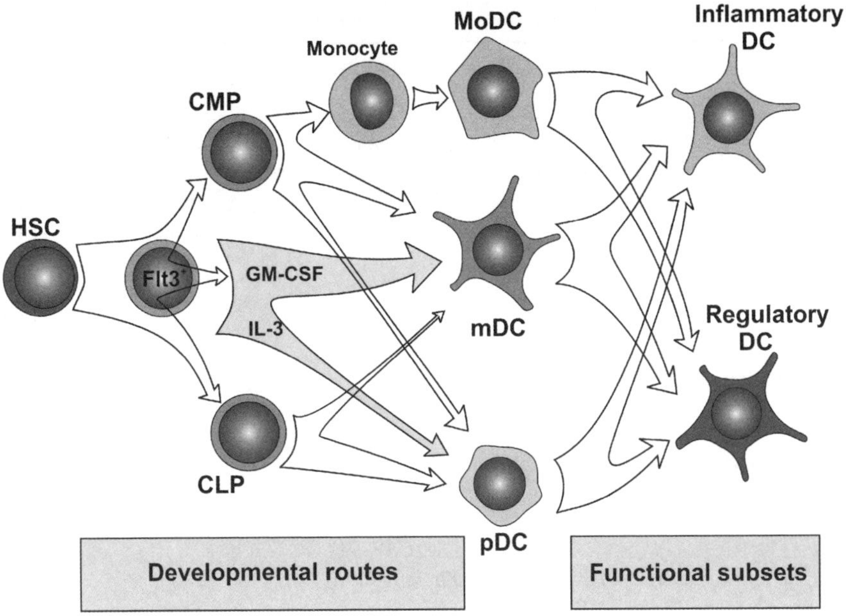

Figure 10.1 Differentiation of dendritic cells from bone marrow derived CD34+ haematopoietic stem cells. Human DC precursors differentiate in the bone marrow from haematopoietic stem cells and give rise to myeloid (mDC) and plasmacytoid (pDC) subsets. Differentiation of mDC is mediated by granulocyte–macrophage colony stimulatory factor (GM-CSF); differentiation of pDC is induced by IL-3. Both differentiation pathways are supported by Flt3L, which induces the expansion of the early Flt3+ progenitors of both subsets. mDC and monocyte-derived DC (MoDC) derived from the common myeloid precursor (CMP) and give rise to blood-circulating and tissue-resident mDC and pDC. The common lymphoid progenitor (pDC) differentiates to resting blood pDC. Both mDC and pDC are able to mature to inflammatory or regulatory DC.

are classified into two major subsets: myeloid and lymphoid DC (Shortman and Liu, 2002; Figure 10.1). This classification originates from early studies which identified two types of DC in the human blood: CD11c+CD123− myeloid DC (mDC) and CD11c−CD123+ plasma cell-like (plasmacytoid) DC (pDC). The majority of human DC derive from myeloid precursors and give rise to various subsets, such as Langerhans cells (LC) and interstitial/dermal/tissue DC. Blood CD14+ monocytes can also give rise to myeloid DC, which in their characteristics resemble interstitial/tissue DC (Figure 10.1, Table 10.1). Plasmacytoid DC comprise ~0.1 per cent of total peripheral blood mononuclear cells (PBMC) and to date represent a homogenous population of cells not segregated into subpopulations (Table 10.1). The origin of these two subsets is still controversial; they may derive from the common myeloid and common lymphoid progenitors or from FMS-like receptor tyrosine kinase 3 (Flt3)+ progenitors, suggesting that a DC developmental program can be induced in both lymphoid and myeloid precursors (Shortman and Liu, 2002; del Hoyo *et al.*, 2002; Shigematsu *et al.*, 2004; Karsunky *et al.*, 2003; Figure 10.1).

Table 10.1 Dendritic cell subsets in humans

Origin/subtype	Markers	Growth/differentiation factors	Localization	TLR expression
Myeloid progenitors *Langerhans cells*	CD11c+ DC-SIGN+ CD1a+ Langerin/(CD207)+ Birbeck granules	Flt-3 ligand GM-CSF + TNF-α GM-CSF + TNF-α+TGFβ	Epidermis Oral, respiratory, genital mucosa	TLR1 TLR2 TLR4 TLR5 TLR6 TLR7 TLR8
Myeloid progenitors *Tissue/interstitial*	CD11c+ DC-SIGN+ CD1a−	Flt-3 ligand GM-CSF + TNF-α	Dermis, submucosa	
Blood monocytes *Monocyte-derived*	CD11c+ DC-SIGN+ CD1a+/−	GM-CSF + IL-4 GM-CSF + IL-13 GM-CSF + IL-4+TGFβ	Peripheral tissue	
Myeloid and lymphoid progenitors *Plasmacytoid IFNα/β producing cell* (IPC)	CD123+/ IL-3Rα+ BDCA2+ BDCA4+ CD4+ Pre-Tα+ (thymus)	Flt-3 ligand IL-3	Blood Thymus Lymphoid organs	TLR1 TLR6 TLR7 TLR9 TLR10

Specificity of the toll-like receptors: TLR1, bacterial lipoprotein (with TLR2); TLR2, bacterial lipoprotein, peptidoglycane, lipoteicolic acid (heteromer with TLR1 and TLR6); TLR3, viral dsRNA; TLR4, bacterial LPS; TLR5, bacterial flagellin; TLR6, bacterial lipoprotein (with TLR2); TLR7, viral ssRNA; TLR8, GU-rich viral ssRNS, imidazoquinolin (anti-viral drug); TLR9, unmethylated CpG DNA.

Comparison of different DC subtypes by transcriptional profiling opened up new avenues for discovering regulatory circuits, which drive DC differentiation and subtype commitment. This approach revealed differential and/or subset-specific expression of selected genes, encoding numerous functionally different proteins. These included cell surface receptors, proteins involved in cellular adhesion and signalling, co-regulated genes related to antigen processing and presentation or organization of cell structure, chemokines, cytokines and their receptors, cytokine-induced genes and survival-related proteins (Tang and Saltzman, 2004; Lapteva *et al.*, 2001; Ahn *et al.*, 2002). These studies also identified transcription factors characteristic of different DC subsets. Comparison of human CD14−CD1a+ LC and CD14^{low}CD1a− tissue DC, differentiated form CD34+ HSC revealed the expression of common but also unique sets of genes, which were partially shared with monocyte-derived DC (MoDC; Ahn *et al.*, 2002; Tureci *et al.*, 2003; Angenieux *et al.*, 2001; Gatti *et al.*, 2000). TGFβ, which acts via the transcription factor Id2, is suggested as a key factor in the development of the CD1a+ and LC subsets (Ito *et al.*, 1999; Jaksits *et al.*, 1999; Hacker *et at.*, 2003). The TGFβ-induced gene-3 (betaig-h3)/TGFB1 was identified as a DC-associated gene, which was highly expressed in CD14+ tissue DC, suggesting the endogenous production of TGFβ by this DC subset (Ahn *et al.*, 2002). TGFB1 is a secreted protein with an Arg–Gly–Asp (RGD) motif, which is known to inhibit the adhesion of cells to plastic. IL-13, which can substitute for IL-4 to generate human MoDC *in vitro*, was also shown to induce TGFβ secretion in certain mouse myeloid cells (Terabe *et al.*, 2003). While supporting the expression of CD1a, TGFβ inhibits CD1d expression on human epidermal LC (Ronger-Savle *et al.*, 2005). Peroxisome proliferator-activated receptor-gamma (PPARγ), a ligand-induced nuclear hormone receptor, was demonstrated to promote MoDC differentiation to a unique cell type with low CD1a expression and with a tolerogenic potential (Szatmari *et al.*, 2004). RelB and PU1 were associated with both myeloid and plasmacytoid DC differentiation from CD34+CD1a−GATA-3- precursors. However, the expression pattern of both transcription factors was strongly dependent on the activation state of DC (Wu *et al.*, 1998; Fohrer *et al.*, 2004). The interferon consensus sequence-binding protein (ICSBP)/interferon regulatory factor-8 (IRF-8; Schiavoni *et al.*, 2002) has also been shown to affect DC development (Schiavoni *et al.*, 2002), and the ETS transcription factor Spi-B in cooperation with ICSBP/IRF-8 has been identified as a key regulator of human pDC development (Schotte *et al.*, 2004).

10.3 Tissue Localization of Dendritic Cell Subsets

DC are widely dispersed in all tissues; however, distinct migration programs have been described for the two major DC subsets. Owing to the expression of the CXCR4 chemokine receptor expressed on mDC and pDC, both cell types migrate to the lymph node homing chemokine CXCL12/SDF-1, which is produced by high endothelial venules (HEV), tonsillar epithelial, dermal endothelial and malignant cells (Zou *et al.*, 2001). Myeloid DC are actually detected in all tissues, while plasmacytoid DC are found in the thymus (Bendriss-Vermare *et al.*, 2001), in the peritoneal lavage fluid and reside around HEV in T-cell rich areas of lymph nodes (Curiel *et al.*, 2004). The constitutively generated mDC and pDC precursors and their specialized immature circulating counterparts are able to migrate to different anatomical sites and become educated by the actual tissue environment (Kelsall and Rescigno, 2004). Thus differentiation into various DC subsets may be due to the phenotypic and functional flexibility of these cell types and may reflect the heterogeneity of their actual tissue microenvironments (Hart, 1997). Both DC types tend to concentrate at the vicinity of the major antigenic portals, such as epithelial surfaces and the skin. At these sites environmental changes are monitored through epithelial, endothelial and stromal cells in collaboration with tissue-resident macrophages, DC and mast cells. DC stand out in particular for their unique capability to link and coordinate the function of natural and adaptive immune cells.

Two subsets of murine and human blood monocytes have been identified by the expression of FcγRIII/CD16 and the chemokine receptor CCR2 (Geissmann, Jung and Littman, 2003). Both types of monocytes differentiate to DC, but long-lived

CD16+CCR2− cells are capable of transendothelial migration, while short-lived CD16−CCR2+ cells have inflammatory properties (Randolph *et al.*, 1998, 2002). The MDC-8 monoclonal antibody (mAb) also defines a subpopulation of monocytes (~1% of PBMC), which differentiate into myeloid DC *in vitro* with the ability to produce high amounts of TNF-α in response to bacterial lipopolysaccharide (LPS) stimulation (Siedlar *et al.*, 2000). These cells can also be detected in the T-cell rich areas of inflammed tonsils and in the subepithelial dome of Peyer's patches. This cell population was also suggested to be responsible for the production of high amounts of pro-inflammatory cytokines in inflammatory bowel disease (de Baey *et al.*, 2003). In human tonsils five different DC subsets have been identified (Summers *et al.*, 2001), which try to cope with the high doses of antigenic stimuli, including antigens and commensal or pathogenic microorganisms of the nasopharyngeal and bronchial tracts.

The skin harbours at least two populations of DC: epidermal CD1a+ LC, which are characterized by langerin (CD207) and specialized intracellular Birbeck granules, and dermal DC, which resemble interstitial DC (Romani *et al.*, 2003; Lenz *et al.*, 1993). Low numbers of CD1a+ cells are also found in epithelial surfaces of the bronchoalveolar epithelium and in tonsils. Epidermal retention of LC is mediated by their interaction with keratinocytes through E-cadherin, which was shown to prevent LC maturation in the seady state, when LC constitutively migrate to skin-associated lymphoid organs (Tang *et al.*, 1993; Riedl *et al.*, 2000). The renewal of LC was recently shown to take place from resident precursors and not from the bone marrow. Circulating blood-borne LC precursors were recruited to the skin only as a consequence of dramatic LC loss induced by UV-irradiation. This process was mediated by CCR2 and the UV-induced inflmmatory chemokines CCL2/MCP-1 and CCL7/MCP-3 (Merad *et al.*, 2002).

DC, in close contact with epithelial cells, play a central role in orchestrating the response of the immune system to various stimuli. Based on this coordinated action immunological tolerance is induced to food, airborn antigens and commensal bacteria (Kelsall and Rescigno, 2004). DC, isolated from the gut or bronchial mucosal surfaces are considered tolerogenic, having a propensity to induce Th2 type responses *in vitro* and expressing anti-inflammatory cytokines such as IL-10 and TGFβ (Iwasaki and Kelsall, 1999). A well-defined DC subset with high expression of indolamine 2,3-dioxygenase (IDO) mRNA, a characteristic of tolerogenic DC, was also identified in the mouse gut (Kelsall and Rescigno, 2004). In contrast to commensal microbes, pathogens manage to pass through epithelial barriers and induce inflammatory responses (Macpherson and Uhr, 2004). This can be induced either by subepithelial DC, which are activated by TNF-α and/or type I IFN, produced by pathogen-activated epithelial cells or by direct activation of DC through microbial products. As an alternative, newly recruited bone marrow- or blood-derived inflammatory cells, among them DC not pre-conditioned by epithelial cells, migrate to the site of inflammation or pathogenic insult. This scenario proposes that resident DC/LC are adapted to the local environment, but mounting an inflammatory response at various body compartments relies on the recruitment of circulating, immature and flexible DC precursors (Sansonetti, 2004; Macpherson and Uhr, 2004).

10.4 Antigen Uptake by Dendritic Cells

Internalization of exogenous antigens by DC is the first step to initiate antigen-specific immune responses. DC capture antigens by pinocytosis, receptor-mediated endocytosis, phagocytosis or macro-pinocytosis (Brode and Macary, 2004). A wide array of internalizing receptors involved in the uptake of soluble or particulate antigens to endo/lysosomal compartments have been identified. Many of these receptors also function in DC signalling or cellular interactions (Table 10.2).

Uptake of soluble molecules

DC subsets express defined sets of pattern recognition receptors (PRR), which are also involved in antigen uptake. They are classified into various molecular families.

Calcium-dependent lectin-like receptors (CLL) represent the most prevalent PRR expressed on DC (Table 10.2). These type I membrane receptors are characterized by carbohydrate recognition domains (CRD), which bind carbohydrate-rich structures on microbes and self-antigens followed by internalization in an ATP-dependent manner. The usage of lectin-like receptors depends on the glycosylation pattern of the ligand, which involves pathogens and adhesion molecules expressed on various cell types. Internalization by C-type lectins is guided by leucine- or tyrosine-based internalization motifs in their cytoplasmic tail. This results in the transport of their captured antigen to selected subcellular compartments, where the corresponding presenting molecules recycle (Mizumoto and Takashima, 2004). Macrophage mannose receptor (MMR) is also expressed on myeloid DC but not on LC. It is internalized into coated pits, their ligand dissociates in early endosomes and the receptor recycles to the cell surface. DC-SIGN is expressed on myeloid DC but not on LC; it recycles to late endosomal compartments (Figdor, van Kooyk and Adema, 2002). DEC-205 is expressed on LC, dermal DC and MoDC and on thymic epithelial cells (Jiang *et al.*, 1995). Guided by an acidic EDE sequence it recycles into LAMP and MHC class II-containing compartments and facilitates antigen presentation (Mahnke *et al.*, 2000). However, this antigen-processing pathway induces tolerance instead of an antigen-specific immune response, mediated by the induction of regulatory T-cells and/or FasL-mediated deletion of T-cells. Langerin is restricted to LC and recognizes mannose residues through its single carbohydrate recognition domain. It acts as an internalizing receptor and targets mannose-containing ligands to Birbeck granules (Valladeau *et al.*, 2000). Langerin cooperates with CD1a molecules, which also have access to Birbeck granules, and thus facilitates CD1a-mediated glycolipid presentation (Hunger *et al.*, 2004). As a general rule, C-type lectins and CD1 molecules, both expressed in various combinations in DC, share targetting and sorting mechanisms (Moody and Porcelli, 2003) to ensure optimized presentation of peptide and nonpeptide antigens. The C-type lectin BDCA2 is expressed exclusively on pDC; Dectin-1 and LOX-1 belong to the newly discovered natural killer gene complex (NKC) (Yokota *et al.*, 2001). pDC express high levels of

Table 10.2

Name	Number of CRD[a] ligands	Expression	Function	Uptake	CP[b]
		C-type lectin like molecules			
		Type I			
DEC-205 (CD205)	10 Oligosaccharide glycoprotein	MDC,[c] Thymic epithelial	Recycling to late endo/lysosome, Clathrin, Tyr motif	•	•
Mannose receptor (CD206)	8/2 functional, single terminal mannose	IDC,[d] MDC, macrophage	Recycling to endosome, Clathrin, Tyr motif	•	
		Type II			
Langerin (CD207)	1 high mannose oligosaccharide glycoprotein	Skin epidermal LC[e]	Birbeck granules	•	
DC-SIGN (CD209)	1 high mannose glycoproteins, *M. tuberculosis*, HIV gp120	MDC, Skin dermis, Mucosal LP,[f] Tonsil T area	Adhesion, ICAM-3/2, T cell interaction, Internalization to late endo/lysosome, Clathrin, Tyr motif	•	
Dectin-1[t]	β-1,3 and β-1,6 linked glucans; MHC I-like structure?	pDC, Macrophage	ITAM, T cell interaction	•	
Dectin-2 [1] (CLECSF10)	EPN motif in the CRD: mannans?, endogenous ligand on T cells?	immature LC, mature MoDC, activated Th cells	UV induced tolerance, T cell interaction (Treg?)		
BDCA-2	EPN motif in the CRD, as Dectin-2	pDC	Suppresses IFNα/β production	•	
		Fc receptors			
FcγRI (CD64)	IgG, IgG-IC[g]	DC subpopulations	High-affinity IgG binding	•	
FcγRIIA/B (CD32)	IgG-IC	monocyte-derived DC	ITAM, ITIM	•	•
FcγRIII (CD16)	IgG-IC	DC subpopulation	ADCC[h]		

FcαR (CD89)	IgA-IC	interstitial DC	Induction of IL-10 production, activation	•	
FcεRI	IgE	Epidermal LC, Blood DC subpopulation	High-affinity IgE binding, activation		•
		Complement receptors			
CR2 (CD21-like)	C3d, C3dg	B lymphocyte, FDC,[i] IL-7 induced DC	B cell co-stimulation, EBV[j] receptor		
CR3 (CD11b-CD18)	iC3b, C3dg, C3d, LPS, fibrinogen, ICAM-1[k]	Monocytes, Macrophages, DC	Phagocytosis of opsonized Ag	•	
CR4 (CD11c-CD18)	iC3b, C3dg, C3d, fibrinogen, M. tubercolosis	Monocytes, Macrophages, DC	Phagocytosis of opsonized Ag		
		Scavenger receptors			
		Class A			
SR-A I and II	LPS[l] (lipid A), LTA[m]	Macrophages, DC	Phagocytosis of bacteria and apoptotic cells	•	•
MARCO	LPS, Acetylated LDL[n]	Macrophages	Phagocytosis of bacteria	•	
		Class B			
CD36	PS,[o] oxidized lipoproteins	Macrophages, monocytes, DC, EC[p]	Phagocytosis, lipid homeostasis	•	•
		Integrins			
αvβ5	RGD[q]	DC	Phagocytosis of apoptotic cells	•	•
		Other receptors			
PS-receptor[p]	PS	Macrophages, DC	Phagocytosis, lipid homeostasis	•	
CD14	LPS	Low expression on CD1a− MDC	Phagocytosis of bacteria and apoptotic cells	•	
CD47[r]	TSP[s]	Ubiquitous	Phagocytosis, DC inhibition	•	

[a]CRD, carbohydrate binding domain. [b]CP, cross presentation. [c]MDC, mature dendritic cell. [d]IDC, immature dendritic cell. [e]Langerhans cell. [f]Lamina propria. [g]IC, immune complex. [h]ADCC, antibody dependent cellular cytotoxicity. [i]FDC, follicular dendritic cell. [j]EBV, Epstein–Barr virus. [k]ICAM-1, intercellular adhesion molecule-1. [l]LPS, lipopolysacharide. [m]LTA, lipoteichoic acid. [n]LDL, low density lipoprotein. [o]PS, phosphatidylserine. [p]EC, endothelial cell. [q]RGD, Arg-Gly-Asp (RGD) motif. [r]CD47, Multispan transmembrane protein, thrombospondin receptor. [s]TSP, Thrombospondin.

Dectin-1, the main phagocytic receptor for yeast (Grunebach *et al.*, 2002; Herre *et al.*, 2004).

Scavanger receptors (SR) are involved in the uptake of lipid-like structures. The newly identified A type SR Marco was shown to be upregulated in mouse bone-marrow-derived DC after loading with tumor cell lysate of B6 melanoma tumor cells (Grolleau *et al.*, 2003). Expression of Marco was associated with changes in cell shape, rearrangement of the actin cytoskeleton and dependence on cell adhesion Pikkarainen, Brannstrom, and Tryggvason, 1999). SR-A receptors expressed by DC were also shown to be involved in 'nibbling', a process occurring when MoDC capture antigen from live cells through close cell-to-cell contacts (Harshyne *et al.*, 2003).

Heat shock protein (HSP) binding receptor CD91, a member of the lipoprotein receptor family, is expressed at low levels in DC. The most efficient HSP binding structure in DC is the scavenger receptor LOX-1 (Delneste *et al.*, 2002), which is the member of the NKC (Sobanov *et al.*, 2001). LOX-1 binds modified low-density lipoproteins (LDL), aged and apoptotic cells, platelets and bacteria.

To ensure the exclusive sampling of peripheral tissue environments, the expression of internalizing receptors is downregulated upon DC maturation. Some of these receptors, which are expressed by mDC and transport antigens to specialized intracellular compartments, are associated with 'cross-presentation', an alternative pathway for loading major histocompatibility complex (MHC) class I molecules.

Uptake of particles

Phagocytosis by DC is an efficient route to engulf extracellular particulate antigens. Internalization of microbes and apoptotic cells use the same intracellular machinery, but in macrophages they are targetted to distinct intracellular phagosomes. Signalling through TLR2 and TLR4 induces rapid maturation of the phagosome into late endosomes and lysosomes (Blander and Medzhitov, 2004). Internalizing receptors involved in antigen uptake influence subsequent differentiation of DC to stimulatory or tolerogenic cells (Mahnke, Knop and Enk, 2003). One of the earliest events in programmed cell death is the change in membrane structure, accompanied by expression of the apoptotic cell-associated membrane proteins (ACAMP), such as phosphatidylserine (PS) and modified carbohydrates (Platt, de Silva and Gordon, 1998; Larsson, Fonteneau and Bhardwaj, 2001). This altered membrane pattern is recognized directly by DC-associated receptors; in DC the uptake of apoptotic cells is mediated by the scavanger receptor CD36 and $\alpha v\beta 5$ integrin in collaboration with the bridging molecule thrombospondin-1 (TSP1), which interacts with CD47 and CD36 (Doyen *et al.*, 2003; Albert, Sauter and Bhardwaj 1998; Albert, Kim and Birge 2000; Rubartelli, Poggi and Zocchi, 1997). Internalization of apoptotic cells by DC is an efficient route of 'cross-presentation' for loading MHC class I molecules with exogenous protein-derived peptides antigens.

Uptake of opsonized cells and antigens

The altered lipid composition of the apoptotic cell membrane activates complement, which results in opsonization by iC3b and binding to complement receptors such as CR3. This pathway, however, does not activate DC but rather inhibits the production of inflammatory cytokines and induces tolerance (Verbovetski *et al.*, 2002; Morelli *et al.*, 2003). Antigens complexed with IgG type antibodies bind to and are taken up by various Fcγ receptors (FcγR). The expression of the high-affinity FcγRI/CD64 and FcγRIII/CD16 defines different DC precursors with different functional properties; the low-affinity FcγRII/CD32 is also expressed in mDC (Sanchez-Torres *et al.*, 2001; Grage-griebenow *et al.*, 2000, 2001; Siedlar *et al.*, 2000). Depending on the intracellular signalling sequence of the various Fcγ receptors (Table 10.2), internalization by this route may induce DC maturation (Regnault *et al.*, 1999). Internalization of immune complexes (IC) by FcγRII was shown to enhance peptide loading of MHC class II molecules, class I-restricted cross-presentation of a small amount of antigen and increase the efficacy of antigen presentation 100-fold as compared with the presentation of free antigen (Regnault *et al.*, 1999; Larsson *et al.*, 1997). As demonstrated by myeloma cells opsonized by anti-syndecan-1 antibodies, this mechanism could promote cross-presentation of intracellular tumour antigens (TA) and thus enhance the efficacy of therapeutical mAb-based therapies (Dhodapkar *et al.*, 2002).

FcαR (CD89) is expressed on CD14+ interstitial DC of CD34+ HSC origin, on immature MoDC but not on LC. CD89 expression is strongly decreased upon differentiation from monocyte to DC. DC efficiently internalize secretory but not serum IgA without any signs of DC maturation. This process, however, could not be inhibited by the anti-CD89 blocking antibody but was inhibited by specific sugars or by antibodies specific for MMR. These data indicate that IC, comprising secretory IgA, interact with DC via carbohydrate-specific receptors such as the MMR (Heystek *et al.*, 2002).

10.5 Antigen Processing and Presentation by Dendritic Cells

DC are up to 1000-fold more efficient in activating resting T-lymphocytes than other professional APC (Bhardwaj *et al.*, 1993). Beside highly efficient antigen internalization, the unique antigen-presenting capacity of mature DC is attributed to the special vesicular system and to the highly efficient antigen processing and presenting machinery operating in various DC subtypes (Guermonprez *et al.*, 2002). Membrane expression of MHC class II molecules is tightly linked to DC function and determines antigen presenting functions.

Loading MHC class II molecules

Exogenous antigens are processed within specialized MHC class II-rich compartments (MIIC) detected in DC as a result of activation and maturation (Brode and

Macary, 2004). During the maturation process DC translocate MHC class II–peptide complexes together with the CD86 costimulatory molecule to the cell surface, which results in the efficient activation of antigen-specific CD4+ helper T-lymphocytes (India *et al.*, 2002; Turley *et al.*, 2000). Several proteases with broad substrate specificity have been implicated in antigen processing. Cathepsin H and C act as amino-peptidases, whereas cathepsin B and Z function mainly as carboxy-exopeptidases. Antigen processing is tightly regulated in maturing DC by controlling protease inhibitors through inflammatory mediators (Watts, 2001). Immature mDC with high endocytic capacity are ineffective in loading MHC class II molecules with peptides unless they receive a maturation signal, thus the antigen-presenting function of mDC depends on microenvironmental factors influencing DC differentiation and maturation. This process is controlled by the regulation of cathepsin S expression responsible for the degradation of the invariant chain and subsequently for the availability of empty MHC class II peptide binding clefts (den Haan, Lehar and Bevan 2000).

The two major DC types, mDC and pDC, differ in the regulation of MHC class II molecules mediated by the cell type-specific promoters that control gene expression of the transcriptional co-activator CIITA, a master regulator of MHC class II synthesis (Leibund, Gut-Landmann *et al.*, 2004). Presentation of peptides, derived from the TA NY-ESO-1, by MHC class II molecules was two orders of magnitude more efficient by pDC than by B-cells but still one order of magnitude lower than by myeloid CD1c+ blood DC. It was also demonstrated that pDC cultured in IL-3 and triggered by CD40 ligand (CD40L) or TLR7/8 ligand (R-848) were equally effective in accomplishing this function (Schnurr *et al.*, 2005). These results show that various DC types differ substantially in their efficacy in antigen processing and presentation and their requirements for further activation.

Loading MHC class I molecules

Antigens expressed in the cytoplasm or the nucleus, such as TA, are processed by the proteasome and loaded onto MHC class I molecules. The majority of antigenic peptides are generated by the proteolytic cleavage of malfolded proteins, which are also referred to as defective ribosomal products (Princiotta *et al.*, 2003). As a result of DC activation, this fraction accumulates as aggregates and requires ubiquitination, enzymatic fragmentation by the proteasome and transportation to the endoplasmic reticulum by transporters for antigen presentation (TAP) (Lelouard *et al.*, 2004). Imported peptides require further trimming by ER-aminopeptidase I, and then are loaded onto MHC class I molecules with the assistance of the chaperons calnexin, calreticulin and tapasin. Peptide-loaded MHC class I molecules are transported to the cell surface through the secretory pathway. As a result of IFNγ-mediated signals, DC replace the standard 'housekeeping' proteasome with the immune proteasome, which contains three novel proteolytic subunits referred to as LMP2, LMP7 and MECL1 (Tanaka and Kasahara, 1998). Proteasomes have at least six active sites and three distinct cleavage specificities. The immune proteasome acquires altered cleavage specificity, so the spectrum of the generated antigenic peptides may change (Groettrup *et al.*, 1995). The appropriate cleavage of a given antigen by the IFNγ-induced immune proteasome can enhance the presentation of some epitopes or destroy others (Van den Eynde and Morel, 2001). Immature MoDC carry approximately equal amounts of the two types of proteasome, but as a result of maturation they upregulate the expression of the immune proteasome (Macagno *et al.*, 1999).

Cross-presentation as an alternative pathway for targetting antigens onto MHC class I molecules

DC are also unique in their capacity to direct exogeneous protein antigens to the MHC class I processing pathway, referred to as 'cross-presentation'. Antigens acquired from body fluids, or dead or live cells can be cross-presented and mediate tolerance or immunity *in vivo.* This pathway is restricted to mDC and to certain types of macrophages (Rock *et al.*, 1993). It allows APC to initiate CD8+ T cell responses against antigens that are not synthesized within the DC. It is dependent on the highly efficient uptake and processing of exogenous material and well-defined stimulatory signals (Larsson, Fonteneau and Bhardwaj, 2001, Albert, Sauter and Bhardwaj, 1998). The efficiency of exogenous antigen presentation by MHC class I and class II molecules is differentially regulated during DC maturation (Delamarre, Holcombe and Mellman, 2003). Soluble proteins internalized by immatue DC are stored intracellularly until activated by an appropriate cross-presentation signal, while increased expression of MHC class II molecules in the cell surface could be induced by various activation signals. Induction of cross-presentation requires CD40L or disruption of cell-to-cell contacts as a stimulatory signal. Comparative studies performed with various DC subtypes revealed that not all DC subsets are involved in cross-presentation. Myeloid DC generated *in vitro* from CD34+ progenitros in the presence of TGFβ, giving rise to LC-type DC, were inactive in cross-presentation (Nagata *et al.*, 2002). pDC are less potent to present MHC class II peptides for CD4+ T-lymphocytes than mDC and are able to present MHC class I epitopes only from intact (live or heat-inactivated) viruses (Fonteneau *et al.*, 2003). Certain internalizing receptors expressed in mDC were shown to be associated with cross-presentation and targetting antigens to intracellular compartments which have access to MHC class I molecules (Table 10.2).

Different routes of entry of exogenous material to the endogenous processing pathway have recently been described: (i) the cytoplasmic tail of MHC class I molecules with a highly conserved tyrosine sorting motif targets unfolded MHC class I molecules to the endosomal/lysosomal compartment, where peptide loading occurs in a TAP-independent manner (Heath and Carbone, 2001). The low pH of this environment ensures the exchange of bound endogenous peptides onto recycling MHC class I molecules (Kleijmeer *et al.*, 2001). (ii) Active phagocytosis of particles may result in the fusion of the phagolysosome with the endoplasmic reticulum (ER) membrane, by which the phagolysosome acquires the entire MHC class I loading complex, including newly synthesized MHC class I molecules, TAP and tapasin (Houde *et al.*, 2003; Guermonprez *et al.*, 2003; Ackerman and Cresswell, 2003). The proteasome complex was shown to accumulate at the outer surface of the phagolysosome, suggesting that proteasome-degraded peptides may be transported back to the phagolysosome. Degradation was shown to occur in the same compartment as utilized for internalization, suggesting that the phagolysosome can act as a specialized cross-presentation organelle and results in highly efficient peptide presentation by protecting peptides from cytosolic peptidases (Brode and Macary, 2004; York *et al.*, 2003).

The role of CD1 molecules in antigen presentation by dendritic cells

Nonpolymorphic CD1 molecules are structurally related to MHC class I molecules and are specialized for the presentation of modified self- and/or microbial lipids. The membrane-expressed human CD1a, CD1b, CD1c and CD1d molecules bind different types of lipid-like ligands and pass through distinct intracellular compartments controlled by targetting motifs in their cytoplasmic tail, which interact differently with the adaptor molecules AP2 and AP3. These and possibly other interactions result in continous recycling of CD1 molecules through the endocytic system (Sugita *et al.*, 2002). CD1b is internalized by clathrin-coated pits and is directed into MHC class II-rich compartments by a tyrosine-based motif in its cytoplasmic tail. CD1a lacks this targetting motif and thus recycles from the cell curface to early recycling endosomes. Owing to the diversity of their intracellular targeting motifs CD1c and CD1d molecules exhibit distinct trafficking patterns (Sugita *et al.*, 1999; Moody and Porcelli, 2003).

Various DC subtypes express different sets of CD1 molecules; therefore they are able to sample a wide range of lipid-like antigenic structures. LC express CD1a at an exceptionally high level, with very low co-expression of CD1b or CD1c molecules. As CD1b, CD1a is able to present mycobacterial glycolipids to T-lymphocytes. In contrast to LC, MoDC and dermal DC predominantly display CD1b with varying levels of CD1a and CD1c (Mizumoto and Takashima, 2004). Various CD1 molecules present their ligands to T-lymphocytes either with diverse or conserved $\alpha\beta$TCR. CD1d molecules bind various modified lipids, among them α-galactosylceramid (αGalCer) derived from a marine sponge and the tumour-associated disialoganglioside GD3 (Gumperz *et al.*, 2000; Wu *et al.*, 2003). These ligands are presented to a subpopulation of NKT cells characterized by an invariant TCR (iNKT). Opposing regulation of group I (CD1a, b, c, e) and group II (CD1d) molecules by the nuclear hormone receptor PPARγ has recently been shown. These results revealed that activation of PPARγ by specific ligand results in the increased expression of CD1d and decreased expression of group I CD1 molecules in MoDC (Szatmari *et al.*, 2004).

Similar to saponins, endosomal lipid transfer proteins (LTP) acts as important 'lipid chaperone'. They assist loading of glycolipid antigens onto CD1 molecules and are also able to pull glycosphyngolipids, such as gangliosides, out of the endosomal membrane and keep them accessible to hydrolases. Other saposins remove bound lipids from endosomal CD1 molecules and support the binding of new lipid antigens (De Libero, 2004; Zhou *et al.*, 2004).

10.6 Activation and Polarization of Dendritic Cells

DC can be activated by various stimuli, which transform immature DC, specialized in antigen capture, to mature DC which potentially activate DC, professional APC, which potently activate T-cells. These two functions are topographically separated and require the migration of DC from the site of inflammation or pathogenic insult to the draining lymph nodes guided by chemokine gradients. Gene expression analysis of various DC subsets, representing distinct activation and maturation stages, has been widely used to monitor the effect of various exogenous stimuli on DC function (Tang and Saltzman, 2004; Granucci *et al.*, 2001; Ricciardi-Castagnoli and Granucci, 2002; Hashimoto *et al.*, 2000; Huang *et al.*, 2001). This approach, combined with high-throughput proteomics and functional studies, holds promise to reveal novel regulatory circuits effecting DC differentiation, function and strategies for therapeutical utility of DC.

Maturation of DC can be induced by Toll-like receptor (TLR)-mediated microbial signals, host-derived pro-inflammatory mediators or cell bound receptors of the tumour necrosis (TNF) and TNF receptor (TNFR) family (Aderem and Ulevitch, 2000; Ulevitch, 2000). DC-activating signals can also be delivered by the interaction with natural T-lymphocytes such as NK- or NKT-cells (Mailliard *et al.*, 2004). Importantly, generation of fully active and stable mature DC requires the activation through multiple signalling pathways (Mailliard *et al.*, 2004). This suggests that signals through a single receptor may result in only partitial activation, which may be reverted by inhibitory signals which favour the differentiation of regulatory DC.

Toll-like receptors in dendritic cell activation

TLR are transmembrane signalling receptors, which are not directly involved in receptor-mediated internalization. They bind unique microbial compounds or self-ligands derived from stressed or demaged cells. Some TLR act on the cell membrane, while others (TLR3, TLR7, TLR8, TLR9) are intracellular receptors localized to subcellular compartments and are activated following internalization of their ligands by other receptors. All DC functions, which are crucial for T-cell activation, such as survival, proliferation, migration, cross-presentation, production of cytokines and chemokines, and expression of adhesion and co-stimulatory molecules, are modulated by TLR signalling (Kaisho and Akira, 2003; Iwasaki and Medzhitov, 2004). Despite their distinct specificities, TLR signal through the common adaptor protein MyD88 and induce NF-κB and mitogen-activated protein kinase activation. TLR-mediated signalling results in the activation of genes encoding for pro-inflammatory cytokines such as TNF-α, IL-1β and IL-6. Some TLR-mediated signalling is independent on MyD88 and may influence the T-cell polarizing activity of DC (Agrawal *et al.*, 2003). NFκB regulates the expression of numerous genes that are involved in DC survival and maturation, cell cycle progression or apoptosis (Rescigno *et al.*, 1998; Ardeshna *et al.*, 2000). Human mDC and pDC differ in their expression of TLR, which recognizes different sets of molecules (Table 10.1). Certain TLR-mediated signals were recently shown to induce cross-presentation in DC (Datta *et al.*, 2003; Table 10.2).

Cooperation of antigen internalization with activation signals

It is well established that antigen capture can modify DC maturation induced by other signals (Cambi and Figdor, 2003). For example, the binding of mycobacterial products to MMR or DC-SIGN induces IL-10 production, which inhibits TLR-mediated IL-12 production (Geijtenbeek *et al.*, 2003). Thus the simultaneous binding of mycobacterial components to MMR, DC-SIGN and TLR may divert the immune system towards a tolerogenic Th2 response, which facilitates immune escape of mycobacteria (Maeda *et al.*, 2003; Tailleux *et al.*, 2003). The natural ligand of BDCA2 has not yet been identified, but crosslinking of BDCA2 on pDC induces Ca^{2+} mobilization and tyrosine phosphorylation of cellular proteins. BDCA2 ligation, however, inhibits IFNαβ production of pDC induced by various stimuli such as the influenza virus or bacterial DNA (Dzionek *et al.*, 2001).

Antigen–antibody complexes internalized by FcγIIR of DC not only target MHC class I and class II molecules but also deliver activation signals, which sensitize DC for priming both CD4+ and CD8+ T lymphocytes (Regnault *et al.*, 1999). This pathway requires endosomal acidification but relies on cytosolic antigen processing. An inhibitory effect on TLR signalling has been attributed to hOSCAR, a human FcRγ-chain associated endocytic receptor. When cross-linked by specific antibodies, this receptor induced maturation of MoDC and increased IL-10 production. When hOSCAR signalling was triggered in the presence of LPS, production of IL12p70 by mDC decreased, while imidazoquinoline (R848) treatment induced IL-12 production (Merck *et al.*, 2004).

Ligation of FcεRI of primary human DC by polyvalent antigens results in NF-κB activation and the release of TNF-α and monocyte chemoattractant protein-1 (MCP-1) indicating the potency of FcεRI to induce inflammatory reactions (Kraft *et al.*, 2002). Cross-linking of the FcαR receptor induces internalization, IL-10 production, upregulation of MHC class II and CD86 costimulatory molecules and results in increased allostimulatory activity (Geissmann *et al.*, 2001).

The uptake of DNA-containing immune complexes in systemic lupus erythematosus by FcγRII of pDC resulted in the co-localization of unmethylated CpG as part of the IC and TLR9 in subcellular lysosomes, where binding and signal transduction could be initiated (Means *et al.*, 2005). These results revealed a novel functional interaction between FcγRII and TLR9 mediated by complexed TLR ligands, targetted towards DC via FcγR. This combined effect could induce intracellular signalling in pDC and result in robust IFNα production.

The role of cross priming and cross tolerance in anti-tumour immune responses

Cross-presentation has been considered as a major mechanism in the presentation of viral antigens and TA for CD8+ cytolytic T-lymphocytes (CTL). Animal studies have demonstrated that mDC are the most potent professional APC in mediating cross-priming of tumour-specific CTL (Albert, Sauter and Bhardwaj, 1998; den Haan, Lehar and Bevan, Kurts *et al.*, 2001; Heath and Carbone, 2001). Cross-presentation of exogenous antigens by mDC, however, does not necessarily result in efficient cross priming (Heath and Carbone, 2001), which requires proper maturation stimuli (Melief, 2003), a phenomenon called 'licencing' of DC for priming CTL. The activation signal can be provided through certain TLR ligands, type I interferons, or by the contact with other cells, such as NK-cells or activated CD4+ T-lymphocytes (Heath and Carbone, 2001). In the absence of these signals, DC induce tolerogenic signals by 'cross tolerance'. This is supported by autocrine negative regulators such as TSP secreted spontaneously by MoDC. TSP negatively regulates IL-12 and IL-10 secretion and is induced by PGE2 or TGFβ (Doyen *et al.*, 2003).

Signalling through TLR3 and TLR9 was recently shown to induce cross-presentation in DC (Datta *et al.*, 2003; Mohty *et al.*, 2003; Schulz *et al.*, 2005). The common features of these receptors are their intracellular localization and induction of type I IFN production upon ligand binding. It was postulated that activation and augmentation of cross-presentation of DC are either mediated by cytosolic TLR3-mediated signalling or by the indirect effect of type I interferons induced by this process. Enhanced cross-presentation is observed if DC maturation is triggered before antigen uptake, suggesting that cross-presentation is induced indirectly through IFNα/β. CpG acting through TLR9 was shown to enhance cross priming and its effects was largely dependent on type I interferons (Cho *et al.*, 2000). IFNα may induce cross-priming by upregulating TAP1 and MHC class I, but the key mechanism was proved to be alterations associated with DC maturation (Cho *et al.*, 2002; Luft *et al.*, 1998). Both stimuli were independent of CD4+ T-lymphocyte-mediated help or the interaction of CD40 with CD40L (Le Bon *et al.*, 2003).

Efficient and prolonged cross-presentation of the NY-ESO-1 TA by MoDC and CD1c+ blood DC could be achieved only through mAb-mediated targetting to FcγR, while pDC and B-cells were inactive in these settings (Schnurr *et al.*, 2005). The most efficient CD1c+ blood mDC rapidly and spontaneously acquired cross-presenting and migratory functions and were suggested as a promising candidate for DC-based immunotherapy (Schnurr *et al.*, 2005).

HSP, which act as chaperones for antigenic peptides and also for empty MHC class I molecules in the ER, may augment the efficiency of cross priming. HSP expressed by malignant cells induce protective cytotoxic T-lymphocyte responses against TA and act as carrier and/or adjuvant proteins in tumor vaccines (Suzue *et al.*, 1997). HSP carrying noncovalently bound TA-derived fragments are readily internalized by DC and co-localize with MHC class I molecules in early and late multivesicular endosomal vesicles (Blachere *et al.*, 1997; Arnold-Schild *et al.*, 1999). *In vivo* studies revealed that targetting model antigens to the LOX-1 receptor results in cross-presentation and triggers protective antitumour immune response (Delneste *et al.*, 2002).

Cytokine production and polarization of activated dendritic cell subsets

In addition to their endocytic and phagocytic activities, DC are potent secretory cells. Steady-state immature DC internalizing apoptotic cells and self-structures remain inactivated, secrete low amounts of cytokines and induce tolerance. As a result of activation they produce cytokines and chemokines, which induce and regulate local and systemic inflammatory immune responses. Induced by stimulatory signals DC also express different patterns of activation and co-stimulatory molecules, which not only determine the magnitude of the T-cell response but instruct DC to produce distinct combinations of cytokines and chemokines. Owing to their functional flexibility, DC may produce inflammatory, anti-inflammatory or regulatory cytokines and support the activation of Th1, Th2 or regulatory T-cells, respectively (Figure 10.1). All human DC subsets have been characterized by the capacity to mature to immunogenic, inflammatory or regulatory DC, suggesting that the outcome of the response depends on the finely tuned balance of various effector T-lymphocyte functions (Figure 10.1; Kalinski and Moser, 2005). The key mDC-derived cytokine required for the induction of inflammatory responses such as protective antitumour immune responses is IL-12, which promotes the differentiation of CD4+ Th1 cells. IL-12 production and the licence for priming CTL is provided by the interaction of CD40 expressed on activated mDC and CD40L on activated Th1 T-lymphocytes. Newly activated T cells are able to amplify but not initiate IL-12 secretion, therefore this process requires additional activation signals (Sporri and Reis e Sousa, 2003). Type I IFN, certain TLR-mediated signals or inflammatory cytokines may circumvent the requirement of Th1-mediated help in CTL priming, but the generation of long-lived functional memory cells still depends on this co-operation (Rajnavolgyi and Lanyi, 2003; Bevan, 2004). pDC are able to secrete large amounts of type I IFN upon virus infection or by activation through TLR7 or TLR9. Type I interferons are common mediators of innate immunity and have multiple effects on immune cells (Goodbourn, Didcock and Randall, 2000). IFNα is able to enhance the cytotoxic activity of NK-cells and macrophages, induce T-cell activation and maintain T-cell survival (Figure 10.2). IFNα skews monocyte differentiation to an inflammatory DC subtype with increased expression of MHC class I and multiple TLR, suggesting the

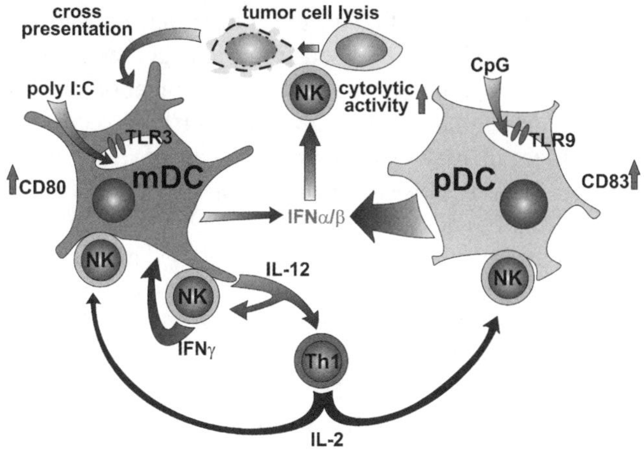

Figure 10.2 Collaboration of dendritic cells with natural killer cells. Activation of pDC by virus infection or through TLR9 results in the production of large amounts of IFNα/β, which triggers the cytolytic activity of NK cells. This mechanism results in the generation of apoptotic cells, which are taken up and utilized by myeloid DC for cross-presentation . IL-12 produced by mDC as a result of IL-2-induced NK cell triggers IFNγ secretion by NK cell, which activates mDC to express CD80. Thus a reciprocal interaction of NK cells with pDC and mDC leads to enhanced cytolytic activity induced by type I interferons and IFNγ production induced by the direct contact of NK cells activated by Th1-derived IL-2. NK cells activated by IL-2 have low cytolytic activity, but TLR3-mediated signalling of mDC results in IFN-α and TNF-α secretion (Gerosa *et al.* 2005).

cooperation of the two major DC subsets in inducing inflammatory responses (Mohty *et al.*, 2003; Figure 10.2). Both DC subsets can acquire tolerogenic potential; the characteristic cytokines of tolerogen DC are IL-10, TGFβ or IL-4 (Mills, 2004).

Migration and chemokine secretion of activated dendritic cell subsets

Activated DC migrate from non-lymphoid tissues to draining lymphoid organs via afferent lymphatic vessels (Viney, 2001). Migration is accompanied by downregulation of phagocytosis and receptor-mediated internalization, responsiveness to inflammatory chemokines, results in the re-distribution of MHC class II molecules from intracellular vesicles to the cell surface, re-organization of the cytoskeleton, formation of dendrites/pseudopodia, increased cell surface expression of co-stimulatory molecules and chemokine receptors. The recruitment to sites of inflammation in response to chemotactic stimuli and their migration to secondary lymphoid organs is essential for the initiation of optimal T-cell responses. Activation of both mDC and pDC results in the expression of CXCR4 and CCR7 chemokine receptors, which mediate DC attraction through CXCL-12/SDF-1α and CCDL19/MIP-3β, respectively. However, additional signals may also be involved in DC migration, such as prostaglandin E2, which has been shown as an important regulator of Mo-DC migration (Luft *et al.*, 2002; Scandella *et al.*, 2002). Nucleotides are also suggested as important soluble factors mediating the migration of various DC subsets. Nanomolar concentrations of ATP, released from damaged cells and from platelets, endothelial cells and T-cells, are involved in the recruitment of various immature DC subsets to the site of inflammation through P2X and P2Y receptors. Myeloid DC subtypes required orders of magnitude less ATP than pDC for mobilization. High concentration (μmol) of ATP, however, had an opposing effect and resulted in the arrest of immature MoDC and CD1a+ dermal DC movement, mediated by the P2Y11 receptor (Schnurr *et al.*, 2003; Idzko *et al.*, 2002). This effect may prolong DC residence in tissues to facilitate antigen uptake. High concentration of ATP also induces the maturation of MoDC (Wilkin *et al.*, 2001). Activation of mDC also results in the enhanced production of the homeostatic chemokines CCL17 and CCL22 and the attraction of Th2 and CD4+ Treg cells.

pDC do not migrate to common inflammatory cytokines such as CCL2/MCP-1, CCL5/RANTES and CXCL10/IP10, but express the chemokine-like receptor 1 (CMKLR1), expressed by both immature blood pDC and MoDC (Zabel, Silverio and Butcher, 2005). DC expressing CMKLR1 are attracted by chemerin, a factor detected in inflamed body fluids and triggered by blood coagulation, tissue damage or inflammation (Wittamer *et al.*, 2004). pDC are often detected in reactive tonsils, inflamed nasal mucosa or skin lesions associated with inflammation. pDC also express the adenosine receptor A1 and migrate to high concentrations of adenosine, released from damaged cells (Schnurr *et al.*, 2004). Activation of pDC by virus infection, signalling through TLR9, or by IL-3 and CD40L results in the maturation to fully potent APC expressing CCR7. pDC are also able to produce the pro-inflammatory chemokine CCL3 and CCL4, which attracts effector T-cells and Th1 cells through CCR1 and CCR5, respectively (Penna *et al.*, 2002). Ligands of CXCR4 and CXCR3, produced by Th1 cells, are able to enhance the responsiveness of pre-pDC to CCL12/SDF-1 and thus recruit pDC to lymphoid organs draining the site of inflammation (Vanbervliet *et al.*, 2003). Activation of pDC by CD40L alone results in the induction of CD8+ regulatory T cells (Gilliet and Liu, 2002).

10.7 Enhancement of Inflammatory Responses by NK Cells

Direct contact of DC with NK cells results in bi-directional signalling, which induces the proliferation and activation of NK cells as well as DC maturation (Degli-Esposti and Smyth, 2005). This mutual interaction is mediated by cell-to-contacts and induces the activation of NK cells for cytokine (IFNγ and TNF-α) secretion (Figure 10.2). Cell contact is mediated by the activating NKG2D receptor, which binds to MICA/B expressed on DC; additional contacts are provided by adhesion molecules. The communication between the two cell types is also mediated by cytokines. DC-derived IL-15 supports NK-cell survival, while IL-12 and IL-18 induce NK-cell proliferation and induction of cytolytic activity (Raulet, 2004) (Degli-Esposti and Smyth, 2005). The contact of NK cells with immature DC, depending on NK/DC ratios, may result in both DC activation and lysis. Sensitivity of immature DC to NK cell-mediated killing via TNF-related apoptosis-inducing ligand (TRAIL)-induced apoptosis is lost upon maturation, which may play a role in focusing antigen-specific immune responses to activated DC (Corazza *et al.*, 2004). An important, recently described function of DC is the recruitment of NK cells to lymph nodes, where they secrete IFNγ to direct Th1 polarization of primed antigen-specific T-lymphocytes (Martin-Fontecha *et al.*, 2004). NK cells are able to recognize and kill malignant cells in the absence of inflammatory stimuli (Karre *et al.*, 1986). NK cells, which contact tumour cells directly, 'licence' myeloid DC to present exogenous antigens derived from dead cells for CTL (Gerosa *et al.*, 2005). NK cell-derived perforin was shown to support the clonal expansion of CTL, which demonstrates the collaboration of NK cells and DC in eliciting inflammatory anti-tumour responses (Strbo *et al.*, 2003; Figure 10.2).

10.8 Suppression of Inflammatory Responses by Natural Regulatory T Cells

To control the potantially self-destructive action of effector mechanisms, T-lymphocyte activation is regulated by homeostatic mechanisms as well as by unique subsets of T-lymphocytes with regulatory function (Sakaguchi, 2004; Jiang and Chess, 2004; Mills, 2004). The homeostatic regulation of T-lymphocyte activation involves the elimination of antigen-specific T-lymphocyte clones with high-affinity TCR by activation-induced cell death (AICD; Lenardo *et al.*, 1999) and regulation of T cell costimulation by interaction of CTLA-4 with CD80/86 to induce functionally inactive T-lymphocytes (Phan *et al.*, 2003; Figure 10.3). The simultaneous differentiation of antigen-specific Th1 and Th2 CD4+ T-lymphocytes with opposing activities also exerts mutual regulatory functions. IFNγ produced by activated Th1 cells inhibits Th2 differentiation, while IL-4 secreted by Th2 cells has anti-inflammatory potential (O'Garra, 1998). In certain tissue environments other CD4+ T-cells, which secrete immunosuppressive IL-10 or IL-10 together with TGFβ, are also generated. Recent results revealed that the physiological role of these regulatory cells is to silence strong inflammatory responses and prevent tissue damage induced by pathogens or auto-reactive T-cell clones (O'Garra *et al.*, 2004).

Two major populations of regulatory T-lymphocytes have been identified in the thymic medulla. *Natural killer T cells* (NKT) express both NK- and T cell markers and exhibit functional properties of effector/memory T cells. NKT cells involve CD1d-restricted cells both with conserved (iNKT) and diverse αβTCR. Both types of NKT cells recognize nonpeptide ligands such as modified lipids. Depending on the nature and dose of the ligand, NKT cells respond with the rapid secretion of inflammatory (IFNγ) or anti-inflammatory (IL-4, IL-13, IL-10) cytokines (Sonoda *et al.*, 2001). Strong antigenic stimulation of NKT cells supports IL-12 secretion by mDC, while weak antigenic stimulation of NKT cells by self-antigens induces IL-10 production in mDC (Gumperz, 2004). *Natural regulatory T cells* (Treg), similarly to activated Th1 and Th2 cells, carry CD25, the α-chain of the IL-2 receptor. They are also characterized by the expression of CD38, CD62L, CD103, glucocorticoid-induced tumour necrosis factor (TNF) receptor (GITR), and by the forkhead transcription factor, FOXP3 (Fontenot, Gavin and Rudensky, 2003). Tolerogenic DC have been implicated in the induction of Treg in peripheral lymphoid tissues (Steinman and Nussenzweig, 2002). DC, which induce Treg cells, have an intermediate phenotype characterized by MHC class II and CD86 expression, but low level of CD40 and intercellular adhesion molecule 1 (ICAM-1) (Martin *et al.*, 2003). Treg cells exhibit diverse MHC class II-restricted αβTCR repertoire and induce tolerance by direct contact with activated T-lymphocytes to inhibit IL-2 gene transcription (Sakaguchi *et al.*, 2001; Figure 10.3). Both human mDC and pDC are able to induce CD4+CD25+ Treg cells; interestingly, their induction occurs together with DC activation (Moseman *et al.*, 2004).

Both NKT and Treg are continuously transported to the periphery and are referred to as natural regulatory T-lymphocytes. It has also been suggested that peripheral tolerogenic DC transport self- and nonself-antigens from the periphery back to the thymus, where they positively select and activate regulatory T-cells (Goldschneider and Cone, 2003). Indolamine 2,3-dioxygenase (IDO)-dependent suppression of T-cell

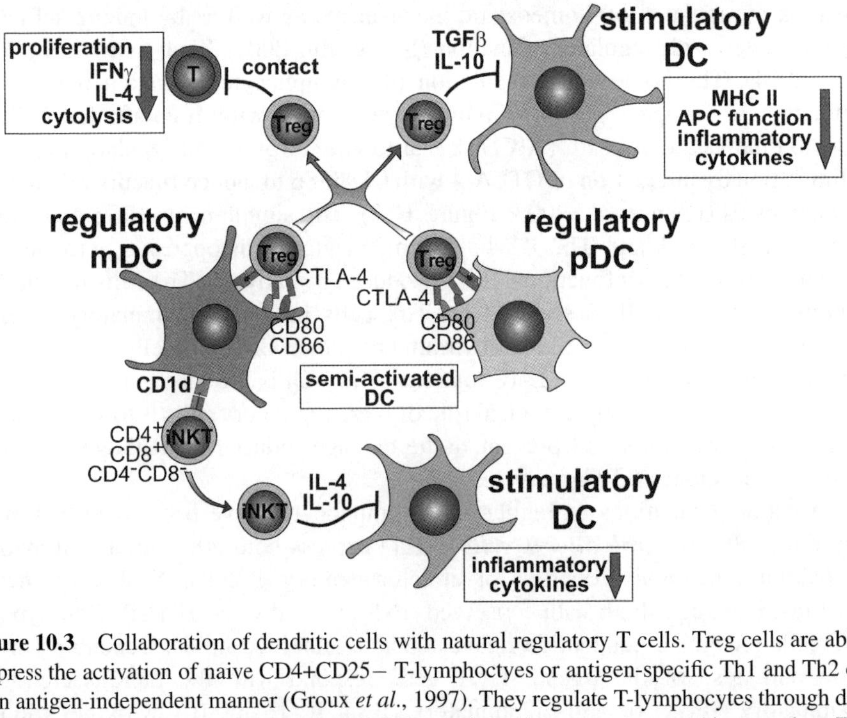

Figure 10.3 Collaboration of dendritic cells with natural regulatory T cells. Treg cells are able to suppress the activation of naive CD4+CD25− T-lymphoctyes or antigen-specific Th1 and Th2 cells in an antigen-independent manner (Groux *et al.*, 1997). They regulate T-lymphocytes through direct cell-to-cell contact with the postulated role of CTLA-4. Their suppressive cytokines, such as IL-10 and soluble or cell surface-bound TGFβ inhibit DC functions and induce tolerogenic DC. pDC are able to induce Treg but do not produce IL-10. Their effect on mDC could be indirect through the induction of Treg cells. iNKT cells interact with myeloid DC expressing DC1d, and through IL-4 and IL-10 they inhibit cytokine secretions of stimulatory DC.

responses was suggested as a natural immunoregulatory mechanism (Mellor and Munn, 2004). IDO, the rate-limiting tryptophan-catabolizing enzyme, is expressed in various cell types, including fibroblasts, macrophages, DC, trophoblasts and epithelial cells. DC expressing IDO contribute to the generation and maintenance of peripheral tolerance by depleting autoreactive T cells and by inducing Treg responses. Human DC can constitutively express immunoreactive IDO protein, yet it does not have functional enzymatic activity until these cells are activated by IFNγ and/or CD80/ CD86 ligation.

10.9 The Role of Dendritic Cells and T-Lymphocytes in Tumour-Specific Immune Responses

Eradication of tumours is mediated predominantly by highly potent CD8+ effector T-lymphocytes, the priming of which requires IFNγ produced by Th1 and NK cells (Yee *et al.*, 2002). Priming of naive CTL occurs exclusively in organized peripheral lymphoid organs and depends on the localization, concentration and persistence of the antigen (Karrer *et al.*, 1997). Co-stimulatory molecules expressed on professional APC may lower the threshold of antigen dose necessary for T cell activation (Ochsenbein *et al.*, 2001; Ludanyi *et al.*, 2004). A short encounter with the antigen is sufficient to induce cell division of antigen-specific CD8+ T-lymphocytes (Kaech and Ahmed, 2001), but continuous presence of antigen in lymphoid organs may result in T-cell exhaustion (Zinkernagel and Hengartner, 2001; Moskophidis *et al.*, 1993). This process is mediated by AICD of effector CTL highly sensitive to apoptosis.

Direct activation of tumor-specific CD8+ T-lymphocytes in lymphoid organs can be initiated by tumor cells migrated to the lymphoid organs, while indirect activation can be achieved by mobilized DC, which have taken up large amounts of TA in the tumor tissue. At low tumour burden TA do not reach the draining lymph nodes in sufficient quantity and will be ignored by T-lymphocytes. High tumour burden, however, may result in high antigen concentration being required for efficient cross-presentation (Spiotto *et al.*, 2002). Advanced necrotizing tumours may induce local inflammation and favour DC activation and migration, while metastatic tumour cells possess the capacity to infiltrate draining lymph nodes (Kurts *et al.*, 1998). The efficiency of CTL priming is highly dependent on the state of DC activation and the balance between stimulatory and tolerogenic signals (Rajnavolgyi and Lanyi, 2003).

NK and NKT cells may contribute to antitumour protection mediated by perforin and IFNγ release (Jameson, Witherden and Havran, 2003). The activity of NK cells is balanced by inhibitory and activating receptors, downregulation of MHC class I and upregulation of NKG2D favouring NK cell function. NK cells, activated by tumour cells with low MHC class I expression, are able to condition DC for priming antitumour CTL and enhance cross-presentation by mDC (Mocikat *et al.*, 2003; Figure 10.2). NKT cells were found to protect against chemically induced fibrosarcoma (Crowe, Smyth and Godfrey, 2002). They may support antitumour immune responses through the production of IFNγ, induced by αGalCer (van der Vliet *et al.*,

2004). A unique subset of MoDC with high CD1d expression could be induced by synthetic agonists of PPARγ. These immature DC, loaded with αGalCer, were able to expand and activate IFNγ secreting iNKT cells (Szatmari *et al.*, 2004). NKT cells were also shown to suppress antitumour immunity indirectly through the inhibition of CTL function. This regulatory circuit was mediated by the CD1d-restricted presentation of a tumour glycolipid to NKT cells (Terabe *et al.*, 2003; Terabe, Park and Berzofsky, 2004). In a tumour model NKT cells inhibited antitumour immune responses by modifying DC function through NKT cell-derived IL-13, which induced TGFβ and blocked IL-12 production of DC (Gumperz, 2004; Figure 10.3).

Tumour-specific antibodies can be detected in cancer patients, but their increased titre seems to correlate with tumour progression and poor prognosis mediated by competition of B cells and DC for TA (Houbiers *et al.*, 1995; Qin *et al.*, 1998). Natural IgM and high-affinity IgG-type tumour-specific antibodies concentrate TA to lymphoid organs and thus promote CTL priming (Ochsenbein and Zinkernagel, 2000). TA-specific antibodies also mediate antibody-mediated cellular cytotoxicity (ADCC) through FcγR expressed on NK cells and macrophages. The importance of this mechanism was demonstrated in patients treated with clinically relevant mAb (Carter, 2001), such as syndecan-specific mAb in myeloma (Dyall *et al.*, 1999).

Dendritic cells in tumour patients

Dendritic cell defects associated with the host immune system are currently considered escape mechanisms of tumour cells. Rapidly growing tumours were shown to contain only a small number of immature DC (Troy *et al.*, 1998), which in most tumours localized to the periphery of the tumour tissue and expressed low levels of co-stimulatory molecules (Vermi *et al.*, 2003). About 15–25 per cent of tumor-infiltrating CD4+ T-lymphocytes are CD25+ Treg cells in patients with epithelial ovarian cancer, and their number is correlated with reduced survival. Treg cells are recruited by tumour- and macrophage-derived CCL22 to the tumour through CCR4, and suppress local IFNγ- and IL-2-secreting tumour-specific effector cells (Curiel *et al.*, 2004). These results indicate that the number and activation state of DC at the tumour site are limiting factors of antitumour immunity. The presence of CD1a+ DC in various tumour tissues, however, was directly associated with prolonged survival and related to the capability of CD1a+ DC to present tumour-associated molecules (Coventry *et al.*, 2002; Coventry and Heinzel, 2004).

Blunted DC function of tumour patients is not restricted to tumour tissues but can be systemic. As compared with normal subjects, a 2-fold decrease in peripheral blood mDC numbers was shown in human neck squamous-cell carcinoma at the early stage of the disease, which reduced to 4-fold in subjects with metastatic or locally advanced tumours (Almand *et al.*, 2000; Della Bella *et al.*, 2003). The DC defect was attributed to arrested differentiation of myeloid progenitors resulting in a heterogenous population of immature myeloid cells and the inhibited maturation of immature DC generating functionally inactive mDC subsets (Gabrilovich, 2004).

Tumour-derived IL-6 switches DC differentiation to macrophages through upregulating M-CSF receptor (Menetrier-Caux *et al.*, 1998). It was also shown that surgical removal of tumours increased the number of myeloid DC in the peripheral blood (Hoffmann *et al.*, 2002). The number of pDC does not change in tumour patients, but pDC were shown to infiltrate primary (Salio *et al.*, 2003) and malignant melanoma (Vermi *et al.*, 2003), head and neck carcinoma (Hartmann *et al.*, 2003) and ovarian carcinoma-associated ascites (Zou *et al.*, 2001).

Development of cancer in most cases is not accompanied by strong inflammation but is associated with Treg cells, which prevent or dampen antitumour immune responses. Since TA recognized by autologous T-lymphocytes resemble self structures, the induction of TA-specific Treg cells can be considered as a normal reaction of the immune system (Figure 10.3).

Dendritic cell-based anti-tumour immunotherapy

Priming of cytotoxic T-lymphocytes does not occur if (i) the activation of DC is limited or inhibited, (ii) TA do not reach lymphoid organs or (iii) the tumor cells produce inhibitory molecules (Ochsenbein *et al.*, 1999, 2001). Ignorance of tumour cells can be circumvented by DC loaded with TA, but tumour-specific CTL tolerance can be reversed only after removal of Treg cells (Dalyot-Herman, Bathe and Malek, 2000; Yang *et al.*, 2004). The dual function of DC of inducing inflammatory or regulatory immune responses makes them potential targets of immunotherapy against cancer (Woltman and van Kooten, 2003). The major challenge of therapeutic vaccination against tumours is to overcome established immunological tolerance to tumours, release tumour-mediated immune suppression and provoke long-lasting tumour-specific inflammatory and cytotoxic responses (Yannelli and Wroblewski, 2004; Figdor *et al.*, 2004). Based on this concept, various strategies have been developed to utilize *in vitro* or *in vivo* manipulated human DC for antitumour therapy. In these studies blood DC or DC generated *in vitro* from CD34+ HSC or peripheral blood monocytes were loaded with tumour antigens and after appropriate activation were reintroduced to tumour-bearing patients. Immune responses to the loaded antigen and antitumour clinical responses could be detected (Mayordomo *et al.*, 1995; Davis *et al.*, 2003; Nestle *et al.*, 1998; Finn, 2003). However, the result of these preliminary studies showed that the efficacy of DC-based vaccines should be further improved. The origin and subset of DC, the right TA for different tumour types, the mode of TA introduction to the DC, the way of targetting to the endogenous antigen processing pathway and the optimal conditions of *in vitro* DC activation have still not been optimized (Figdor *et al.*, 2004). Better understanding of DC biology may help to design and develop new strategies to impove DC-based antitumour vaccines (Cerundolo, Hermans and Salio, 2004). In a recent study the superior activity of CD1c+ myeloid blood DC in cross-presentation and migratory functions has been reported (Schnurr *et al.*, 2005). Humanized mAb approved for clinical utility allow tumour cells to be targetted to FcγR to enhance the efficiency of cross-presentation and DC activation (Schuurhuis *et al.*, 2002). *In vitro* generated immature DC loaded with TA induce tumour-specific tolerance, while properly activated DC are able to elicit effective Th1-mediated antitumour immune responses (Fong and Engleman, 2000; Dhodapkar *et al.*, 2001). Targetting CCR6+ DC to the tumour site by intratumoural expression of CCL20/MIP-3α alone or in combination with CpG enhanced cross-presentation of TA and elicited strong antitumour immune responses (Furumoto *et al.*, 2004). The ability of NKT cells to support effective antitumour immune responses when activated by αGalCer offers a new strategy for imunotherapies. IFNα-based therapies have been approved for hepatitis C infection and for certain haematological tumours and may prove to be valuable in enhancing the efficacy of DC-based antitumour immune therapies. Immunodepletion of CD4+CD25+ natural Treg cells allowed an effective immune response against tumours to be provoked in nonresponder mice (Onizuka *et al.*, 1999).

These results altogether indicate that the biology of DC and their multifunctional role in immune responses is an intriguing field of immunology supported by novel technologies such as immunogenomics.

References

Ackerman AL, Cresswell P. 2003. Regulation of MHC class I transport in human dendritic cells and the dendritic-like cell line KG-1. *J Immunol* **170**: 4178–4188.

Aderem A, Ulevitch RJ. 2000. Toll-like receptors in the induction of the innate immune response. *Nature* **406**: 782–787.

Agrawal S, Agrawal A, Doughty B, Gerwitz A, Blenis J, Van Dyke T, Pulendran B. 2003. Cutting edge: different Toll-like receptor agonists instruct dendritic cells to induce distinct Th responses via differential modulation of extracellular signal-regulated kinase-mitogen-activated protein kinase and c-Fos. *J Immunol* **171**: 4984–4989.

Ahn JH, Lee Y, Jeon C, Lee SJ, Lee BH, Choi KD, Bae YS. 2002. Identification of the genes differentially expressed in human dendritic cell subsets by cDNA subtraction and microarray analysis. *Blood* **100**: 1742–1754.

Albert ML, Kim JI, Birge RB. 2000. Alphavbeta5 integrin recruits the CrkII–Dock180–rac1 complex for phagocytosis of apoptotic cells. *Nat Cell Biol* **2**: 899–905.

Albert ML, Sauter B, Bhardwaj N. 1998. Dendritic cells acquire antigen from apoptotic cells and induce class I-restricted CTLs. *Nature* **392**: 86–89.

Almand B, Resser JR, Lindman B, Nadaf S, Clark JI, Kwon ED, Carbone DP, Gabrilovich DI. 2000. Clinical significance of defective dendritic cell differentiation in cancer. *Clin Cancer Res* **6**: 1755–1766.

Angenieux C, Fricker D, Strub JM, Luche S, Bausinger H, Cazenave JP, Van Dorsselaer A, Hanau D, de La salle H, Rabilloud T. 2001. Gene induction during differentiation of human monocytes into dendritic cells: an integrated study at the RNA and protein levels. *Funct Integr Genomics* **1**: 323–329.

Ardavin C, Amigorena S, Reis e Sousa C. 2004. Dendritic cells: immunobiology and cancer immunotherapy. *Immunity* **20**: 17–23.

Ardeshna KM, Pizzey AR, Devereux S, Khwaja A. 2000. The PI3 kinase, p38 SAP kinase, and NF-kappaB signal transduction pathways are involved in the survival and maturation of lipopolysaccharide-stimulated human monocyte-derived dendritic cells. *Blood* **96**: 1039–1046.

Arnold-Schild D, Hanau D, Spehner D, Schmid C, Rammensee HG, de la Salle H, Schild H. 1999. Cutting edge: receptor-mediated endocytosis of heat shock proteins by professional antigen-presenting cells. *J Immunol* **162**: 3757–3760.

Bendriss-Vermare N, Barthelemy C, Durand I, Bruand C, Dezutter-Dambuyant C, Moulian N, Berrih-Aknin S, Caux C, Trinchieri G, Briere F. 2001. Human thymus contains IFN-alpha-producing CD11c(−), myeloid CD11c(+), and mature interdigitating dendritic cells. *J Clin Invest* **107**: 835–844.

Bevan MJ. 2004. Helping the CD8(+) T-cell response. *Nat Rev Immunol* **4**: 595–602.

Bhardwaj N, Young JW, Nisanian AJ, Baggers J, Steinman RM. 1993. Small amounts of superantigen, when presented on dendritic cells, are sufficient to initiate T cell responses. *J Exp Med* **178**: 633–642.

Blachere NE, Li Z, Chandawarkar RY, Suto R, Jaikaria NS, Basu S, Udono H, Srivastava PK. 1997. Heat shock protein-peptide complexes, reconstituted *in vitro*, elicit peptide-specific cytotoxic T lymphocyte response and tumor immunity. *J Exp Med* **186**: 1315–1322.

Blander JM, Medzhitov R. 2004. Regulation of phagosome maturation by signals from Toll-like receptors. *Science* **304**: 1014–1018.

Brode S, Macary PA. 2004. Cross-presentation: dendritic cells and macrophages bite off more than they can chew! *Immunology* **112**: 345–351.

Cambi A, Figdor CG. 2003. Dual function of C-type lectin-like receptors in the immune system. *Curr Opin Cell Biol* **15**: 539–546.

Carter P. 2001. Improving the efficacy of antibody-based cancer therapies. *Nat Rev Cancer* **1**: 118–129.

Cerundolo V, Hermans IF, Salio M. 2004. Dendritic cells: a journey from laboratory to clinic. *Nat Immunol* **5**: 7–10.

Cho HJ, Takabayashi K, Cheng PM, Nguyen MD, Corr M, Tuck S, Raz E. 2000. Immunostimulatory DNA-based vaccines induce cytotoxic lymphocyte activity by a T-helper cell-independent mechanism. *Nat Biotechnol* **18**: 509–514.

Cho HJ, Hayashi T, Datta SK, Takabayashi K, Van Uden JH, Horner A, Corr M, Raz E. 2002. IFN-alpha beta promote priming of antigen-specific CD8+ and CD4+ T lymphocytes by immunostimulatory DNA-based vaccines. *J Immunol* **168**: 4907–4913.

Corazza N, Brumatti G, Schaer C, Cima I, Wasem C, Brunner T. 2004. TRAIL and immunity: more than a license to kill tumor cells. *Cell Death Differ* **11**(suppl 2): S122–125.

Coventry B, Heinzel S. 2004. CD1a in human cancers: a new role for an old molecule. *Trends Immunol* **25**: 242–248.

Coventry BJ, Lee PL, Gibbs D, Hart DN. 2002. Dendritic cell density and activation status in human breast cancer – CD1a, CMRF-44, CMRF-56 and CD-83 expression. *Br J Cancer* **86**: 546–551.

Crowe NY, Smyth MJ, Godfrey DI. 2002. A critical role for natural killer T cells in immunosurveillance of methylcholanthrene-induced sarcomas. *J Exp Med* **196**: 119–127.

Curiel TJ, Coukos G, Zou L, Alvarez X, Cheng P, Mottram P, Evdemon-Hogan M, Conejo-Garcia JR, Zhang L, Burow M, Zhu, Y, Wei S, Kryczek I, Daniel B, Gordon A, Myers L, Lackner A, Disis ML, Knutson KL, Chen L, Zou W. 2004. Specific recruitment of regulatory T cells in ovarian carcinoma fosters immune privilege and predicts reduced survival. *Nat Med* **10**: 942–949.

Dalyot-Herman N, Bathe OF, Malek TR. 2000. Reversal of CD8+ T cell ignorance and induction of anti-tumor immunity by peptide-pulsed APC. *J Immunol* **165**: 6731–6737.

Datta SK, Redecke V, Prilliman KR, Takabayashi K, Corr M, Tallant T, Di Donato J, Dziarski R, Akira S, Schoenberger SP, Raz E. 2003. A subset of Toll-like receptor ligands induces cross-presentation by bone marrow-derived dendritic cells. *J Immunol* **170**: 4102–4110.

Davis ID, Jefford M, Parente P, Cebon J. 2003. Rational approaches to human cancer immunotherapy. *J Leukoc Biol* **73**: 3–29.

de Baey A, Mende I, Baretton G, Greiner A, Hartl WH, Baeuerle PA, Diepolder HM. 2003. A subset of human dendritic cells in the T cell area of mucosa-associated lymphoid tissue with a high potential to produce TNF-alpha. *J Immunol* **170**: 5089–5094.

De Libero G. 2004. Immunology. The Robin Hood of antigen presentation. *Science* **303**: 485–487.

Degli-Esposti MA, Smyth MJ. 2005. Close encounters of different kinds: dendritic cells and NK-cells take centre stage. *Nat Rev Immunol* **5**: 112–124.

del Hoyo GM, Martin P, Vargas HH, Ruiz S, Arias CF, Ardavin C. 2002. Characterization of a common precursor population for dendritic cells. *Nature* **415**: 1043–1047.

Delamarre L, Holcombe H, Mellman I. 2003. Presentation of exogenous antigens on major histocompatibility complex (MHC) class I and MHC class II molecules is differentially regulated during dendritic cell maturation. *J Exp Med* **198**: 111–122.

Della Bella S, Gennaro M, Vaccari M, Ferraris C, Nicola S, Riva A, Clerici M, Greco M, Villa ML. 2003. Altered maturation of peripheral blood dendritic cells in patients with breast cancer. *Br J Cancer* **89**: 1463–1472.

Delneste Y, Magistrelli G, Gauchat J, Haeuw J, Aubry J, Nakamura K, Kawakami-Honda N, Goetsch L, Sawamura T, Bonnefoy J, Jeannin P. 2002. Involvement of LOX-1 in dendritic cell-mediated antigen cross-presentation. *Immunity* **17**: 353–362.

den Haan JM, Lehar SM, Bevan MJ. 2000. CD8(+) but not CD8(−) dendritic cells cross-prime cytotoxic T cells *in vivo*. *J Exp Med* **192**: 1685–1696.

Dhodapkar KM, Krasovsky J, Williamson B, Dhodapkar MV. 2002. Antitumor monoclonal antibodies enhance cross-presentation ofcCellular antigens and the generation of myeloma-specific killer T cells by dendritic cells. *J Exp Med* **195**: 125–133.

Dhodapkar MV, Steinman RM, Krasovsky J, Munz C, Bhardwaj N. 2001. Antigen-specific inhibition of effector T cell function in humans after injection of immature dendritic cells. *J Exp Med* **193**: 233–238.

Doyen V, Rubio M, Braun D, Nakajima T, Abe J, Saito H, Delespesse G, Sarfati M. 2003. Thrombospondin 1 is an autocrine negative regulator of human dendritic cell activation. *J Exp Med* **198**: 1277–1283.

Dyall R, Vasovic LV, Clynes RA, Nikolic-Zugic J. 1999. Cellular requirements for the monoclonal antibody-mediated eradication of an established solid tumor. *Eur J Immunol* **29**: 30–37.

Dzionek A, Sohma Y, Nagafune J, Cella M, Colonna M, Facchett F, Gunther G, Johnston I, Lanzavecchia A, Nagasaka T, Okada T, Vermi W, Winkels G, Yamamoto T, Zysk M, Yamaguchi Y, Schmitz J. 2001. BDCA-2, a novel plasmacytoid dendritic cell-specific type II C-type lectin, mediates antigen capture and is a potent inhibitor of interferon alpha/beta induction. *J Exp Med* **194**: 1823–1834.

Figdor CG, de Vries IJ, Lesterhuis WJ, Melief CJ. 2004. Dendritic cell immunotherapy: mapping the way. *Nat Med* **10**: 475–480.

Figdor CG, van Kooyk Y, Adema GJ. 2002. C-type lectin receptors on dendritic cells and Langerhans cells. *Nat Rev Immunol* **2**: 77–84.

Finn OJ. 2003. Cancer vaccines: between the idea and the reality. *Nat Rev Immunol* **3**: 630–641.

Fohrer H, Audit IM, Sainz A, Schmitt C, Dezutter-Dambuyant C, Dalloul AH. 2004. Analysis of transcription factors in thymic and CD34+ progenitor-derived plasmacytoid and myeloid dendritic cells: evidence for distinct expression profiles. *Exp Hematol* **32**: 104–112.

Fong L, Engleman EG. 2000. Dendritic cells in cancer immunotherapy. *Rev Immunol* **18**: 245–273.

Fonteneau JF, Gilliet M, Larsson M, Dasilva I, Munz C, Liu YJ, Bhardwaj N. 2003. Activation of influenza virus-specific CD4+ and CD8+ T cells: a new role for plasmacytoid dendritic cells in adaptive immunity. *Blood* **101**: 3520–3526.

Fontenot JD, Gavin MA, Rudensky AY. 2003. Foxp3 programs the development and function of CD4+CD25+ regulatory T cells. *Nat Immunol* **4**: 330–336.

Furumoto K, Soares L, Engleman EG, Merad M. 2004. Induction of potent antitumor immunity by in situ targeting of intratumoral DCs. *J Clin Invest* **113**: 774–783.

Gabrilovich D. 2004. Mechanisms and functional significance of tumour-induced dendritic-cell defects. *Nat Rev Immunol* **4**: 941–952.

Gatti E, Velleca MA, Biedermann BC, Ma W, Unternaehrer J, Ebersold MW, Medzhitov R, Pober JS, Mellman I,. 2000. Large-scale culture and selective maturation of human Langerhans cells from granulocyte colony-stimulating factor-mobilized CD34+ progenitors. *J Immunol* **164**: 3600–3607.

Geijtenbeek TB, Van Vliet SJ, Koppel EA, Sanchez-Hernandez M, Vandenbroucke-Grauls CM, Appelmelk B, Van Kooyk Y. 2003. Mycobacteria target DC-SIGN to suppress dendritic cell function. *J Exp Med* **197**: 7–17.

Geissmann F, Jung S, Littman DR. 2003. Blood monocytes consist of two principal subsets with distinct migratory properties. *Immunity* **19**: 71–82.

Geissmann F, Launay P, Pasquier B, Lepelletier Y, Leborgne M, Lehuen A, Brousse N, Monteiro RC. 2001. A subset of human dendritic cells expresses IgA Fc receptor (CD89), which mediates internalization and activation upon cross-linking by IgA complexes. *J Immunol* **166**: 346–352.

Gerosa F, Gobbi A, Zorzi P, Burg S, Briere F, Carra G, Trinchieri G. 2005. The reciprocal interaction of NK cells with plasmacytoid or myeloid dendritic cells profoundly affects innate resistance functions. *J Immunol* **174**: 727–734.

Gilliet M, Liu YJ. 2002. Generation of human CD8 T regulatory cells by CD40 ligand-activated plasmacytoid dendritic cells. *J Exp Med* **195**: 695–704.

Gogolak P, Rethi B, Hajas G, Rajnavolgyi E. 2003. Targeting dendritic cells for priming cellular immune responses. *J Mol Recognit* **16**: 299–317.

Goldschneider I, Cone RE. 2003. A central role for peripheral dendritic cells in the induction of acquired thymic tolerance. *Trends Immunol* **24**: 77–81.

Goodbourn S, Didcock L, Randall RE. 2000. Interferons: cell signalling, immune modulation, antiviral response and virus countermeasures. *J Gen Virol* **81**: 2341–2364.

Grage-Griebenow E, Flad HD, Ernst M, Bzowska M, Skrzeczynska J, Pryjma J. 2000. Human MO subsets as defined by expression of CD64 and CD16 differ in phagocytic activity and generation of oxygen intermediates. *Immunobiology* **202**: 42–50.

Grage-Griebenow E, Zawatzky R, Kahlert H, Brade L, Flad H, Ernst M. 2001. Identification of a novel dendritic cell-like subset of CD64(+)/CD16(+) blood monocytes. *Eur J Immunol* **31**: 48–56.

Granucci F, Vizzardelli C, Virzi E, Rescigno M, Ricciardi-Castagnoli P. 2001. Transcriptional reprogramming of dendritic cells by differentiation stimuli. *Eur J Immunol* **31**: 2539–2546.

Groettrup M, Ruppert T, Kuehn L, Seeger M, Standera S, Koszinowski U, Kloetzel PM. 1995. The interferon-gamma-inducible 11 S regulator (PA28) and the LMP2/LMP7 subunits govern the peptide production by the 20 S proteasome *in vitro*. *J Biol Chem* **270**: 23808–23815.

Grolleau A, Misek DE, Kuick R, Hanash S, Mule JJ. 2003. Inducible expression of macrophage receptor Marco by dendritic cells following phagocytic uptake of dead cells uncovered by oligonucleotide arrays. *J Immunol* **171**: 2879–2888.

Groux H, O'Garra A, Bigler M, Rouleau M, Antonenko S, de Vries JE, Roncarolo MG. 1997. A CD4+ T-cell subset inhibits antigen-specific T-cell responses and prevents colitis *Nature* **389**: 737–742.

Grunebach F, Weck, MM, Reichert J, Brossart P. 2002. Molecular and functional characterization of human Dectin-1. *Exp Hematol* **30**: 1309–1315.

Guermonprez P, Valladeau J, Zitvogel L, Thery C, Amigorena S. 2002. Antigen presentation and T cell stimulation by dendritic cells. *A Rev Immunol* **20**: 621–667.

Guermonprez P, Saveanu L, Kleijmeer M, Davoust J, Van Endert P, Amigorena S. 2003. ER-phagosome fusion defines an MHC class I cross-presentation compartment in dendritic cells. *Nature* **425**: 397–402.

Gumperz JE. 2004. CD1d-restricted 'NKT' cells and myeloid IL-12 production: an immunological crossroads leading to promotion or suppression of effective anti-tumor immune responses? *J Leukoc Biol* **76**: 307–313.

Gumperz JE, Roy C, Makowska A, Lum D, Sugita M, Podrebarac T, Koezuka Y, Porcelli SA, Cardell S, Brenner MB, Behar SM. 2000. Murine CD1d-restricted T cell recognition of cellular lipids. *Immunity* **12**: 211–221.

Hacker C, Kirsch RD, Ju XS, Hieronymus T, Gust TC, Kuhl C, Jorgas T, Kurz SM, Rose-John S, Yokota Y, Zenake M. 2003. Transcriptional profiling identifies Id2 function in dendritic cell development. *Nat Immunol* **4**: 380–386.

Harshyne LA, Zimmer MI, Watkins SC, Barratt-Boyes SM. 2003. A role for class A scavenger receptor in dendritic cell nibbling from live cells. *J Immunol* **170**: 2302–2309.

Hart DN. 1997. Dendritic cells: unique leukocyte populations which control the primary immune response. *Blood* **90**: 3245–3287.

Hartmann E, Wollenberg B, Rothenfusser S, Wagner M, Wellisch D, Mack B, Giese T, Gires O, Endres S, Hartmann G. 2003. Identification and functional analysis of tumor-infiltrating plasmacytoid dendritic cells in head and neck cancer. *Cancer Res* **63**: 6478–6487.

Hashimoto SI, Suzuki T, Nagai S, Yamashita T, Toyoda N, Matsushima K. 2000. Identification of genes specifically expressed in human activated and mature dendritic cells through serial analysis of gene expression. *Blood* **96**: 2206–2214.

Heath WR, Carbone FR. 2001. Cross-presentation, dendritic cells, tolerance and immunity. *A Rev Immunol* **19**: 47–64.

Herre J, Marshall AS, Caron E, Edwards AD, Williams DL, Schweighoffer E, Tybulewicz V, Sousa CR, Gordon S, Brown GD. 2004. Dectin-1 uses novel mechanisms for yeast phagocytosis in macrophages. *Blood* **104**: 4038–4045.

Heystek HC, Moulon C, Woltman AM, Garonne P, van Kooten C. 2002. Human immature dendritic cells efficiently bind and take up secretory IgA without the induction of maturation. *J Immunol* **168**: 102–107.

Hoffmann TK, Muller-Berghaus J, Ferris RL, Johnson JT, Storkus WJ, Whiteside TL. 2002. Alterations in the frequency of dendritic cell subsets in the peripheral circulation of patients with squamous cell carcinomas of the head and neck. *Clin Cancer Res* **8**: 1787–1793.

Houbiers JG, van der Burg SH, van de Watering LM, Tollenaar RA, Brand A, van de Velde CJ, Melief CJ. 1995. Antibodies against p53 are associated with poor prognosis of colorectal cancer. *Br J Cancer* **72**: 637–641.

Houde M, Bertholet S, Gagnon E, Brunet S, Goyette G, Laplante A, Princiotta MF, Thibault P, Sacks D, Desjardins M. 2003. Phagosomes are competent organelles for antigen cross-presentation. *Nature* **425**: 402–406.

Huang Q, Liu D, Majewski P, Schulte LC, Korn JM, Young RA, Lander ES, Hacohen N. 2001. The plasticity of dendritic cell responses to pathogens and their components. *Science* **294**: 870–875.

Hunger RE, Sieling PA, Ochoa MT, Sugaya M, Burdick AE, Rea TH, Brennan PJ, Belisle JT, Blauvelt A, Porcelli SA, Modlin RL. 2004. Langerhans cells utilize CD1a and langerin to efficiently present nonpeptide antigens to T cells. *J Clin Invest* **113**: 701–708.

Idzko M, Dichmann S, Ferrari D, Di Virgilio F, la Sala A, Girolomoni G, Panther E, Norgauer J. 2002. Nucleotides induce chemotaxis and actin polymerization in immature but not mature human dendritic cells via activation of pertussis toxin-sensitive P2y receptors. *Blood* **100**: 925–932.

Inaba K, Turley S, Iyoda T, Yamaide F, Shimoyama S, Reis e Sousa C, Germain RN, Mellman I, Steinman RM. 2000. The formation of immunogenic major histocompatibility complex class II-peptide ligands in lysosomal compartments of dendritic cells is regulated by inflammatory stimuli. *J Exp Med* **191**: 927–936.

Ito T, Inaba M, Inaba K, Toki J, Sogo S, Iguchi T, Adachi Y, Yamaguchi K, Amakawa R, Valladeau J, Saeland S, Fukuhara S, Ikehara S. 1999. A CD1a+/CD11c+ subset of human blood dendritic cells is a direct precursor of Langerhans cells. *J Immunol* **163**: 1409–1419.

Iwasaki A, Kelsall BL. 1999. Freshly isolated Peyer's patch, but not spleen, dendritic cells produce interleukin 10 and induce the differentiation of T helper type 2 cells. *J Exp Med* **190**: 229–239.

Iwasaki A, Medzhitov R. 2004. Toll-like receptor control of the adaptive immune responses. *Nat Immunol* **5**: 987–995.

Jaksits S, Kriehuber E, Charbonnier AS, Rappersberger K, Stingl G, Maurer D. 1999. CD34+ cell-derived CD14+ precursor cells develop into Langerhans cells in a TGF-beta 1-dependent manner. *J Immunol* **163**: 4869–4877.

Jameson J, Witherden D, Havran WL. 2003. T-cell effector mechanisms: gammadelta and CD1d-restricted subsets. *Curr Opin Immunol* **15**: 349–353.

Jiang H, Chess L. 2004. An integrated view of suppressor T cell subsets in immunoregulation. *J Clin Invest* **114**: 1198–1208.

Jiang W, Swiggard WJ, Heufler C, Peng M, Mirza A, Steinman RM, Nussenzweig MC. 1995. The receptor DEC-205 expressed by dendritic cells and thymic epithelial cells is involved in antigen processing. *Nature* **375**: 151–155.

Kaech SM, Ahmed R. 2001. Memory CD8+ T cell differentiation: initial antigen encounter triggers a developmental program in naive cells. *Nat Immunol* **2**: 415–422.

Kaisho T, Akira S. 2003. Regulation of dendritic cell function through Toll-like receptors. *Curr Mol Med* **3**: 373–385.

Kalinski P, Moser M. 2005. Opinion: consensual immunity: success-driven development of T-helper-1 and T-helper-2 responses. *Nat Rev Immunol* **5**: 251–260.

Kanazawa N, Tashiro K, Miyachi Y. 2004. Signaling and immune regulatory role of the dendritic cell immunoreceptor (DCIR) family lectins: DCIR, DCAR, dectin-2 and BDCA-2. *Immunobiology* **209**(1–2): 179–190.

Karre K, Ljunggren HG, Piontek G, Kiessling R. 1986. Selective rejection of H-2-deficient lymphoma variants suggests alternative immune defence strategy. *Nature* **319**: 675–678.

Karrer U, Althage A, Odermatt B, Roberts CW, Korsmeyer SJ, Miyawaki S, Hengartner H, Zinkernagel RM. 1997. On the key role of secondary lymphoid organs in antiviral immune responses studied in alymphoplastic (aly/aly) and spleenless (Hox11(−)/−) mutant mice. *J Exp Med* **185**: 2157–2170.

Karsunky H, Merad M, Cozzio A, Wessman IL, Manz MG. 2003. Flt3 ligand regulates dendritic cell development from Flt3+ lymphoid and myeloid-committed progenitors to Flt3+ dendritic cells *in vivo*. *J Exp Med* **198**: 305–313.

Kelsall BL, Rescigno M. 2004. Mucosal dendritic cells in immunity and inflammation. *Nat Immunol* **5**: 1091–1095.

Kleijmeer MJ, Escola JM, UytdeHaag FG, Jakobson E, Griffith JM, Osterhaus AD, Stoorvogel W, Melief CJ, Rabouille C, Geuze HJ. 2001. Antigen loading of MHC class-I molecules in the endocytic tract. *Traffic* **2**: 124–137.

Kraft S, Novak N, Katoh N, Bieber T, Rupec RA. 2002. Aggregation of the high-affinity IgE receptor Fc(epsilon)RI on human monocytes and dendritic cells induces NF-kappaB activation. *J Invest Dermatol* **118**: 830–837.

Kurts C, Miller JF, Subramaniam RM, Carbone FR, Heath WR. 1998. Major histocompatibility complex class I-restricted cross-presentation is biased towards high dose antigens and those released during cellular destruction. *J Exp Med* **188**: 409–414.

Kurts C, Cannarile M, Klebba I, Brocker T. 2001. Dendritic cells are sufficient to cross-present self-antigens to CD8 T cells *in vivo*. *J Immunol* **166**: 1439–1442.

Lapteva N, Ando Y, Nieda M, Hohjoh H, Okai M, Kikuchi A, Dymshits G, Ishikawa Y, Juji T, Tokunaga K. 2001. Profiling of genes expressed in human monocytes and monocyte-derived dendritic cells using cDNA expression array. *Br J Haematol* **114**: 191–197.

Larsson M, Fonteneau JF, Bhardwaj N. 2001. Dendritic cells resurrect antigens from dead cells. *Trends Immunol* **22**: 141–148.

Larsson M, Berge J, Johansson AG, Forsum U. 1997. Human dendritic cells handling of binding, uptake and degradation of free and IgG-immune complexed dinitrophenylated human serum albumin *in vitro*. *Immunology* **90**: 138–146.

Le Bon A, Etchart N, Rossmann C, Ashton M, Hou S, Gewert D, Borrow P, Tough DF. 2003. Cross-priming of CD8+ T cells stimulated by virus-induced type I interferon. *Nat Immunol* **4**: 1009–1015.

LeibundGut-Landmann S, Waldburger JM, Reis e Sousa C, Acha-Orbea H, Reith W. 2004. MHC class II expression is differentially regulated in plasmacytoid and conventional dendritic cells. *Nat Immunol* **5**: 899–908.

Lelouard H, Ferrand V, Marguet D, Bania J, Camosseto V, David A, Gatti E, Pierre P. 2004. Dendritic cell aggresome-like induced structures are dedicated areas for ubiquitination and storage of newly synthesized defective proteins. *J Cell Biol* **164**: 667–675.

Lenardo M, Chan KM, Hornung F, McFarland H, Siegel R, Wang J, Zheng L. 1999. Mature T lymphocyte apoptosis – immune regulation in a dynamic and unpredictable antigenic environment. *A Rev Immunol* **17**: 221–253.

Lenz A, Heine M, Schuler G, Romani N. 1993. Human and murine dermis contain dendritic cells. Isolation by means of a novel method and phenotypical and functional characterization. *J Clin Invest* **92**: 2587–2596.

Ludanyi K, Gogolak P, Rethi B, Magocsi M, Detre C, Matko J, Rajnavolgyi E. 2004. Fine-tuning of helper T cell activation and apoptosis by antigen-presenting cells. *Cell Signal* **16**: 939–950.

Luft T, Pang KC, Thomas E, Hertzog P, Hart DN, Trapani J, Cebon J. 1998. Type I IFNs enhance the terminal differentiation of dendritic cells. *J Immunol* **161**: 1947–1953.

Luft T, Jefford M, Luetjens P, Toy T, Hochrein H, Masterman KA, Maliszewski C, Shortman K, Cebon J, Maraskovsky E. 2002. Functionally distinct dendritic cell (DC) populations induced by physiologic stimuli: prostaglandin E(2) regulates the migratory capacity of specific DC subsets. *Blood* **100**: 1362–1372.

Macagno A, Gilliet M, Sallusto F, Lanzavecchia A, Nestle FO, Groettrup M. 1999. Dendritic cells up-regulate immunoproteasomes and the proteasome regulator PA28 during maturation. *Eur J Immunol* **29**: 4037–4042.

Macpherson AJ, Uhr T. 2004. Induction of protective IgA by intestinal dendritic cells carrying commensal bacteria. *Science* **303**: 1662–1665.

Maeda N, Nigou J, Herrmann JL, Jackson M, Amara A, Lagrange PH, Puzo G, Gicquel B, Neyrolles O. 2003. The cell surface receptor DC-SIGN discriminates between Mycobacterium species through selective recognition of the mannose caps on lipoarabinomannan. *J Biol Chem* **278**: 5513–5516.

Mahnke K, Knop J, Enk AH. 2003. Induction of tolerogenic DCs: 'you are what you eat'. *Trends Immunol* **24**: 646–651.

Mahnke K, Guo M, Lee S, Sepulveda H, Swain SL, Nussenzweig M, Steinman RM. 2000. The dendritic cell receptor for endocytosis, DEC-205, can recycle and enhance antigen presentation via major histocompatibility complex class II-positive lysosomal compartments. *J Cell Biol* **151**: 673–684.

Mailliard RB, Wankowicz-Kalinska A, Cai Q, Wesa A, Hilkens CM, Kapsenberg, ML, Kirkwood JM, Storkus WJ, Kalinski P. 2004. Alpha-type-1 polarized dendritic cells: a novel immunization tool with optimized CTL-inducing activity. *Cancer Res* **64**: 5934–5937.

Martin E, O'Sullivan B, Low P, Thomas R. 2003. Antigen-specific suppression of a primed immune response by dendritic cells mediated by regulatory T cells secreting interleukin-10. *Immunity* **18**: 155–167.

Martin-Fontecha A, Thomsen LL, Brett S, Gerard C, Lipp M, Lanzavecchia A, Sallusto F. 2004. Induced recruitment of NK-cells to lymph nodes provides IFN-gamma for T(H)1 priming. *Nat Immunol* **5**: 1260–1265.

Mayordomo JI, Zorina T, Storkus WJ, Zitvogel L, Celluzzi C, Falo LD, Melief CJ, Ildstad ST, Kast WM, Deleo AB, et al. 1995. Bone marrow-derived dendritic cells pulsed with synthetic tumour peptides elicit protective and therapeutic antitumour immunity. *Nat Med* **1**: 1297–1302.

Means TK, Latz E, Hayashi F, Murali MR, Golenbock DT, Luster AD. 2005. Human lupus autoantibody-DNA complexes activate DCs through cooperation of CD32 and TLR9. *J Clin Invest* **115**: 407–417.

Melief CJ. 2003. Mini-review: Regulation of cytotoxic T lymphocyte responses by dendritic cells: peaceful coexistence of cross-priming and direct priming? *Eur J Immunol* **33**: 2645–2654.

Mellor AL, Munn DH. 2004. IDO expression by dendritic cells: tolerance and tryptophan catabolism. *Nat Rev Immunol* **4**: 762–774.

Menetrier-Caux C, Montmain G, Dieu MC, Bain C, Favrot MC, Caux C, Blay JY. 1998. Inhibition of the differentiation of dendritic cells from CD34(+) progenitors by tumor cells: role of interleukin-6 and macrophage colony-stimulating factor. *Blood* **92**: 4778–4791.

Merad M, Manz MG, Karsunky H, Wagers A, Peters W, Charo I, Weissman IL, Cyster JG, Engleman EG. 2002. Langerhans cells renew in the skin throughout life under steady-state conditions. *Nat Immunol* **3**: 1135–1141.

Merck E, Gaillard C, Gorman DM, Montero-Julian F, Durand I, Zurawski SM, Menetrier-Caux C, Carra G, Lebecque S, Trinchieri G, Bates EE. 2004. OSCAR is an FcRgamma-associated receptor that is expressed by myeloid cells and is involved in antigen presentation and activation of human dendritic cells. *Blood* **104**: 1386–1395.

Mills KH. 2004. Regulatory T cells: friend or foe in immunity to infection? *Nat Rev Immunol* **4**: 841–855.

Mizumoto N, Takashima A. 2004. CD1a and langerin: acting as more than Langerhans cell markers. *J Clin Invest* **113**: 658–660.

Mocellin S, Mandruzzato S, Bronte V, Lise M, Nitti D. 2004. Part I: Vaccines for solid tumours. *Lancet Oncol* **5**: 681–689.

Mocikat R, Braumuller H, Gumy A, Egeter O, Ziegler H, Reusch U, Bubeck A, Louis J, Mailhammer R, Riethmuller G, Koszinowski U, Rocken M. 2003. Natural killer cells activated by MHC class I(low) targets prime dendritic cells to induce protective CD8 T cell responses. *Immunity* **19**: 561–569.

Mohty M, Vialle-Castellano A, Nunes JA, Isnardon D, Olive D, Gaugler B. 2003. IFN-alpha skews monocyte differentiation into Toll-like receptor 7-expressing dendritic cells with potent functional activities. *J Immunol* **171**: 3385–3393.

Moody DB, Porcelli SA. 2003. Intracellular pathways of CD1 antigen presentation. *Nat Rev Immunol* **3**: 11–22.

Morel PA, Feili-Hariri M, Coates PT, Thomson AW. 2003. Dendritic cells, T cell tolerance and therapy of adverse immune reactions. *Clin Exp Immunol* **133**: 1–10.

Morelli AE, Larregina AT, Shufesky WJ, Zahorchak AF, Logar AJ, Papworth GD, Wang Z, Watkins SC, Falo LD Jr, Thomson AW. 2003. Internalization of circulating apoptotic cells by splenic marginal zone dendritic cells: dependence on complement receptors and effect on cytokine production. *Blood* **101**: 611–620.

Moseman EA, Liang X, Dawson AJ, Panoskaltsis-Mortari A, Krieg AM, Liu YJ, Blazar BR, Chen W. 2004. Human plasmacytoid dendritic cells activated by CpG oligodeoxynucleotides induce the generation of CD4+CD25+ regulatory T cells. *J Immunol* **173**: 4433–4442.

Moser M. 2003. Dendritic cells in immunity and tolerance – do they display opposite functions? *Immunity* **19**: 5–8.

Moskophidis D, Lechner F, Pircher H, Zinkernagel RM. 1993. Virus persistence in acutely infected immunocompetent mice by exhaustion of antiviral cytotoxic effector T cells. *Nature* **362**: 758–761.

Nagata Y, Ono S, Matsuo M, Gnjatic S, Valmori D, Ritter G, Garrett W, Old LJ, Mellman I. 2002. Differential presentation of a soluble exogenous tumor antigen, NY-ESO-1, by distinct human dendritic cell populations. *Proc Natl Acad Sci USA* **99**: 10629–10634.

Nestle FO, Alijagic S, Gilliet M, Sun Y, Grabbe S, Dummer R, Burg G, Schadendorf D. 1998. Vaccination of melanoma patients with peptide- or tumor lysate-pulsed dendritic cells. *Nat Med* **4**: 328–332.

Ochsenbein AF, Zinkernagel RM. 2000. Natural antibodies and complement link innate and acquired immunity. *Immunol Today* **21**: 624–630.

Ochsenbein AF, Klenerman P, Karrer U, Ludewig B, Pericin M, Hengartner H, Zinkernagel RM. 1999. Immune surveillance against a solid tumor fails because of immunological ignorance. *Proc Natl Acad Sci USA* **96**: 2233–2238.

Ochsenbein AF, Sierro S, Odermatt B, Pericin M, Karrer U, Hermans J, Hemmi S, Hengartner H, Zinkernagel RM. 2001. Roles of tumour localization, second signals and cross priming in cytotoxic T-cell induction. *Nature* **411**: 1058–1064.

O'Garra A. 1998. Cytokines induce the development of functionally heterogeneous T helper cell subsets. *Immunity* **8**: 275–283.

O'Garra A, Vieira PL, Vieira P, Goldfeld AE. 2004. IL-10-producing and naturally occurring CD4+ Tregs: limiting collateral damage. *J Clin Invest* **114**: 1372–1378.

O'Neill DW, Adams S, Bhardwaj N. 2004. Manipulating dendritic cell biology for the active immunotherapy of cancer. *Blood* **104**: 2235–2246.

Onizuka S, Tawara I, Shimizu J, Sakaguchi S, Fujita T, Nakayama E. 1999. Tumor rejection by *in vivo* administration of anti-CD25 (interleukin-2 receptor alpha) monoclonal antibody. *Cancer Res* **59**: 3128–3133.

Penna G, Vulcano M, Roncari A, Facchetti F, Sozzani S, Adorini L. 2002. Cutting edge: differential chemokine production by myeloid and plasmacytoid dendritic cells. *J Immunol* **169**: 6673–6676.

Phan GQ, Yang JC, Sherry RM, Hwu P, Topalian SL, Schwartzentruber DJ, Restifo NP, Haworth LR, Seipp CA, Freezer LJ, Morton KE, Mavroukakis SA, Duray PH, Steinberg SM, Allison JP, Davis TA, Rosenberg SA. 2003. Cancer regression and autoimmunity induced by cytotoxic T lymphocyte-associated antigen 4 blockade in patients with metastatic melanoma. *Proc Natl Acad Sci USA* **100**: 8372–8377.

Pikkarainen T, Brannstrom A, Tryggvason K. 1999. Expression of macrophage MARCO receptor induces formation of dendritic plasma membrane processes. *J Biol Chem* **274**: 10975–10982.

Platt N, da Silva RP, Gordon S. 1998. Recognizing death: the phagocytosis of apoptotic cells. *Trends Cell Biol* **8**: 365–372.

Princiotta MF, Finzi D, Qian SB, Gibbs J, Schuchmann S, Buttgereit F, Bennink JR, Yewdell JW. 2003. Quantitating protein synthesis, degradation, and endogenous antigen processing. *Immunity* **18**: 343–354.

Qin Z, Richter G, Schuler T, Ibe S, Cao X, Blankenstein T. 1998. B cells inhibit induction of T cell-dependent tumor immunity. *Nat Med* **4**: 627–630.

Rajnavolgyi E, Lanyi A. 2003. Role of CD4+ T lymphocytes in antitumor immunity. *Adv Cancer Res* **87**: 195–249.

Randolph GJ, Beaulieu S, Lebecque S, Steinman RM, Muller WA. 1998. Differentiation of monocytes into dendritic cells in a model of transendothelial trafficking. *Science* **282**: 480–483.

Randolph GJ, Sanchez-Schmitz G, Liebman RM, Schakel K. 2002. The CD16(+) (FcgammaRIII(+)) subset of human monocytes preferentially becomes migratory dendritic cells in a model tissue setting. *J Exp Med* **196**: 517–527.

Raulet DH. 2004. Interplay of natural killer cells and their receptors with the adaptive immune response. *Nat Immunol* **5**: 996–1002.

Regnault A, Lankar D, Lacabanne V, Rodriguez A, Thery C, Rescigno M, Saito T, Verbeek S, Bonnerot C, Ricciardi-Castagnoli P, Amigorena S. 1999. Fcgamma receptor-mediated induction of dendritic cell maturation and major histocompatibility complex class I-restricted antigen presentation after immune complex internalization. *J Exp Med* **189**: 371–380.

Rescigno M, Martino M, Sutherland CL, Gold MR, Ricciardi-Castagnoli P. 1998. Dendritic cell survival and maturation are regulated by different signaling pathways. *J Exp Med* **188**: 2175–2180.

Ricciardi-Castagnoli P, Granucci F. 2002. Opinion: Interpretation of the complexity of innate immune responses by functional genomics. *Nat Rev Immunol* **2**: 881–889.

Riedl E, Stockl J, Majdic O, Scheinecker C, Knapp W, Strobl H. 2000. Ligation of E-cadherin on *in vitro*-generated immature Langerhans-type dendritic cells inhibits their maturation. *Blood* **96**: 4276–4284.

Rock KL, Rothstein L, Gamble S, Fleischacker C. 1993. Characterization of antigen-presenting cells that present exogenous antigens in association with class I MHC molecules. *J Immunol* **150**: 438–446.

Romani N, Holzmann S, Tripp CH, Koch F, Stoitzner P. 2003. Langerhans cells – dendritic cells of the epidermis. *APMIS* **111**: 725–740.

Ronger-Savle S, Valladeau J, Claudy A, Schmitt D, Peguet-Navarro J, Dezutter-Dambuyant C, Thomas L, Jullien D. 2005. TGFbeta inhibits CD1d expression on dendritic cells. *J Invest Dermatol* **124**: 116–118.

Rubartelli A, Poggi A, Zocchi MR. 1997. The selective engulfment of apoptotic bodies by dendritic cells is mediated by the alpha(v)beta3 integrin and requires intracellular and extracellular calcium. *Eur J Immunol* **27**: 1893–1900.

Sakaguchi S. 2004. Naturally arising CD4+ regulatory T cells for immunologic self-tolerance and negative control of immune responses. *Rev Immunol* **22**: 531–562.

Sakaguchi S, Sakaguchi N, Shimizu J, Yamazaki S, Sakihama T, Itoh M, Kuniyasu Y, Nomura T, Toda M, Takahashi T. 2001. Immunologic tolerance maintained by CD25+ CD4+ regulatory T cells: their common role in controlling autoimmunity, tumor immunity, and transplantation tolerance. *Immunol Rev* **182**: 18–32.

Salio M, Cella M, Vermi W, Facchetti F, Palmowski MJ, Smith CL, Shepherd D, Colonna M, Cerundolo V. 2003. Plasmacytoid dendritic cells prime IFN-gamma-secreting melanoma-specific CD8 lymphocytes and are found in primary melanoma lesions. *Eur J Immunol* **33**: 1052–1062.

Sanchez-Torres C, Garcia-Romo GS, Cornejo-Cortes, MA, Rivas-Carvalho A, Sanchez-Schmitz G. 2001. CD16+ and CD16− human blood monocyte subsets differentiate *in vitro* to dendritic cells with different abilities to stimulate CD4+ T cells. *Int Immunol* **13**: 1571–1581.

Sansonetti PJ. 2004. War and peace at mucosal surfaces. *Nat Rev Immunol* **4**: 953–964.

Scandella E, Men Y, Gillessen S, Forster R, Groettrup M. 2002. Prostaglandin E2 is a key factor for CCR7 surface expression and migration of monocyte-derived dendritic cells. *Blood* **100**: 1354–1361.

Schiavoni G, Mattei F, Sestili P, Borghi P, Venditti M, Morse HC, 3rd Belardelli F, Gabriele L. 2002. ICSBP is essential for the development of mouse type I interferon-producing cells and for the generation and activation of CD8alpha(+) dendritic cells. *J Exp Med* **196**: 1415–1425.

Schnurr M, Toy, Stoitzner P, Cameron P, Shin A, Beecroft T, Davis ID, Cebon J, Maraskovsky E. 2003. ATP gradients inhibit the migratory capacity of specific human dendritic cell types: implications for P2Y11 receptor signaling. *Blood* **102**: 613–620.

Schnurr M, Toy T, Shin A, Hartmann G, Rothenfusser S, Soellner J, Davis ID, Cebon J, Maraskovsky E. 2004. Role of adenosine receptors in regulating chemotaxis and cytokine production of plasmacytoid dendritic cells. *Blood* **103**: 1391–1397.

Schnurr M, Chen Q, Shin W, Toy T, Jenderek C, Green S, Miloradovic L, Drane D, Davis ID, Villadangos J, Shortman K, Maraskovsky E, Cebon J. 2005. Tumor antigen processing and presentation depend critically on dendritic cell type and the mode of antigen delivery. *Blood* **105**: 2465–2472.

Schotte R, Nagasawa M, Weijer K, Spits H, Blom B. 2004. The ETS transcription factor Spi-B is required for human plasmacytoid dendritic cell development. *J Exp Med* **200**: 1503–1509.

Schulz O, Diebold SS, Chen M, Naslund TI, Nolte MA, Alexopoulou L, Azuma YT, Flavell RA, Liljestrom P, Reis e Sousa C. 2005. Toll-like receptor 3 promotes cross-priming to virus-infected cells. *Nature* **433**: 887–892.

Schuurhuis DH, Ioan-Facsinay A, Nagelkerken B, van Schip JJ, Sedlik C, Melief CJ, Verbeek JS, Ossendorp F. 2002. Antigen-antibody immune complexes empower dendritic cells to efficiently prime specific CD8+ CTL responses *in vivo*. *J Immunol* **168**: 2240–2246.

Shigematsu H, Reizis B, Iwasaki H, Mizuno S, Hu D, Traver D, Leder P, Sakaguchi N, Akashi K. 2004. Plasmacytoid dendritic cells activate lymphoid-specific genetic programs irrespective of their cellular origin. *Immunity* **21**: 43–53.

Shortman K, Liu YJ. 2002. Mouse and human dendritic cell subtypes. *Nat Rev Immunol* **2**: 151–161.

Siedlar M, Frankenberger M, Ziegler-Heitbrock LH, Belge KU. 2000. The M-DC8-positive leukocytes are a subpopulation of the CD14+ CD16+ monocytes. *Immunobiology* **202**: 11–17.

Sobanov Y, Bernreiter A, Derdak S, Mechtcheriakova D, Schweighofer B, Duchler M, Kalthoff F, Hofer E. 2001. A novel cluster of lectin-like receptor genes expressed in monocytic, dendritic and endothelial cells maps close to the NK receptor genes in the human NK gene complex. *Eur J Immunol* **31**: 3493–3503.

Sonoda KH, Faunce DE, Taniguchi M, Exley M, Balk S, Stein-Streilein J. 2001. NK T cell-derived IL-10 is essential for the differentiation of antigen-specific T regulatory cells in systemic tolerance. *J Immunol* **166**: 42–50.

Spiotto MT, Yu P, Rowley DA, Nishimura MI, Meredith SC, Gajewski TF, Fu YX, Schreiber H. 2002. Increasing tumor antigen expression overcomes 'ignorance' to solid tumors via crosspresentation by bone marrow-derived stromal cells. *Immunity* **17**: 737–747.

Sporri R, Reis e Sousa C. 2003. Newly activated T cells promote maturation of bystander dendritic cells but not IL-12 production. *J Immunol* **171**: 6406–6413.

Steinman RM, Nussenzweig MC. 2002. Avoiding horror autotoxicus: the importance of dendritic cells in peripheral T cell tolerance. *Proc Natl Acad Sci USA* **99**: 351–358.

Strbo N, Oizumi S, Sotosek-Tokmadzic V, Podack ER. 2003. Perforin is required for innate and adaptive immunity induced by heat shock protein gp96. *Immunity* **18**: 381–390.

Sugita M, Grant EP, van Donselaar E, Hsu VW, Rogers RA, Peters PJ, Brenner MB. 1999. Separate pathways for antigen presentation by CD1 molecules. *Immunity* **11**: 743–752.

Sugita M, Cao X, Watts GF, Rogers RA, Bonifacino JS, Brenner MB. 2002. Failure of trafficking and antigen presentation by CD1 in AP-3-deficient cells. *Immunity* **16**: 697–706.

Summers KL, Hock BD, McKenzie JL, Hart DN. 2001. Phenotypic characterization of five dendritic cell subsets in human tonsils. *Am J Pathol* **159**: 285–295.

Suzue K, Zhou X, Eisen HN, Young RA. 1997. Heat shock fusion proteins as vehicles for antigen delivery into the major histocompatibility complex class I presentation pathway. *Proc Natl Acad Sci USA* **94**: 13146–13151.

Szatmari I, Gogolak P, Im JS, Dezso B, Rajnavolgyi E, Nagy L. 2004. Activation of PPARgamma specifies a dendritic cell subtype capable of enhanced induction of iNKT cell expansion. *Immunity* **21**: 95–106.

Tailleux L, Schwartz O, Herrmann JL, Pivert E, Jackson M, Amara A, Legres L, Dreher D, Nicod LP, Gluckman JC, Lagrange PH, Gicquel B, Neyrolles O. 2003. DC-SIGN is the major *Mycobacterium tuberculosis* receptor on human dendritic cells. *J Exp Med* **197**: 121–127.

Tanaka K, Kasahara M. 1998. The MHC class I ligand-generating system: roles of immunoproteasomes and the interferon-gamma-inducible proteasome activator PA28. *Immunol Rev* **163**: 161–176.

Tang A, Amagai M, Granger LG, Stanley JR, Udey MC. 1993. Adhesion of epidermal Langerhans cells to keratinocytes mediated by E-cadherin. *Nature* **361**: 82–85.

Tang Z, Saltzman A. 2004. Understanding human dendritic cell biology through gene profiling. *Inflamm Res* **53**: 424–441.

Terabe M, Park JM, Berzofsky JA. 2004. Role of IL-13 in regulation of anti-tumor immunity and tumor growth. *Cancer Immunol Immunother* **53**: 79–85.

Terabe M, Matsui S, Park JM, Mamura M, Noben-Trauth N, Donaldson DD, Chen W, Wahl SM, Ledbetter S, Pratt B, Letterio JJ, Paul WE, Berzofsky JA. 2003. Transforming growth factor-beta production and myeloid cells are an effector mechanism through which CD1d-restricted T cells block cytotoxic T lymphocyte-mediated tumor immunosurveillance: abrogation prevents tumor recurrence. *J Exp Med* **198**: 1741–1752.

Troy AJ, Summers KL, Davidson PJ, Atkinson CH, Hart DN. 1998. Minimal recruitment and activation of dendritic cells within renal cell carcinoma. *Clin Cancer Res* **4**: 585–593.

Tureci O, Bian H, Nestle FO, Raddrizzani L, Rosinski JA, Tassis A, Hilton H, Walstead M, Sahin U, Hammer J. 2003. Cascades of transcriptional induction during dendritic cell maturation revealed by genome-wide expression analysis. *Faseb J* **17**: 836–847.

Turley SJ, Inaba K, Garrett WS, Ebersold M, Unternaehrer J, Steinman RM, Mellman I. 2000. Transport of peptide-MHC class II complexes in developing dendritic cells. *Science* **288**: 522–527.

Ulevitch RJ. 2000. Molecular mechanisms of innate immunity. *Immunol Res* **21**: 49–54.

Valladeau J, Ravel O, Dezutter-Dambuyant C, Moore K, Kleijmeer M, Liu Y, Duvert-Frances V, Vincent C, Schmitt D, Davoust J, Caux C, Lebecque S, Saeland S. 2000. Langerin, a novel C-type lectin specific to Langerhans cells, is an endocytic receptor that induces the formation of Birbeck granules. *Immunity* **12**: 71–81.

Vanbervliet B, Bendriss-Vermare N, Massacrier C, Homey B, de Bouteiller O, Briere F, Trinchieri G, Caux C. 2003. The inducible CXCR3 ligands control plasmacytoid dendritic cell responsiveness to the constitutive chemokine stromal cell-derived factor 1 (SDF-1)/CXCL12. *J Exp Med* **198**: 823–830.

Van den Eynde BJ, Morel S. 2001. Differential processing of class-I-restricted epitopes by the standard proteasome and the immunoproteasome. *Curr Opin Immunol* **13**: 147–153.

van der Vliet HJ, Molling JW, von Blomberg BM, Nishi N, Kolgen W, van den Eertwegh AJ, Pinedo HM, Giaccone G, Scheper RJ. 2004. The immunoregulatory role of CD1d-restricted natural killer T cells in disease. *Clin Immunol* **112**: 8–23.

Verbovetski I, Bychkov H, Trahtemberg U, Shapira I, Hareuveni M, Ben-Tal O, Kutikov I, Gill O, Mevorach D. 2002. Opsonization of apoptotic cells by autologous iC3b facilitates clearance by immature dendritic cells, down-regulates DR and CD86, and up-regulates CC chemokine receptor 7. *J Exp Med* **196**: 1553–1561.

Vermi W, Bonecchi R, Facchetti F, Bianchi D, Sozzani S, Festa S, Berenzi A, Cella M, Colonna M. 2003. Recruitment of immature plasmacytoid dendritic cells (plasmacytoid monocytes) and myeloid dendritic cells in primary cutaneous melanomas. *J Pathol* **200**: 255–268.

Viney JL. 2001. Immune fate decided by dendritic cell provocateurs. *Trends Immunol* **22**: 8–10.

Watts C. 2001 Antigen processing in the endocytic compartment. *Curr Opin Immunol* **13**: 26–31.

Wilkin F, Duhant X, Bruyns C, Suarez-Huerta N, Boeynaems JM, Robaye B. 2001. The P2Y11 receptor mediates the ATP-induced maturation of human monocyte-derived dendritic cells. *J Immunol* **166**: 7172–7177.

Wittamer V, Gregoire F, Robberecht P, Vassart G, Communi D, Parmentier M. 2004. The C-terminal nonapeptide of mature chemerin activates the chemerin receptor with low nanomolar potency. *J Biol Chem* **279**: 9956–9962.

Woltman AM, van Kooten C. 2003. Functional modulation of dendritic cells to suppress adaptive immune responses. *J Leukoc Biol* **73**: 428–441.

Wu DY, Segal NH, Sidobre S, Kronenberg M, Chapman PB. 2003. Cross-presentation of disialoganglioside GD3 to natural killer T cells. *J Exp Med* **198**: 173–181.

Wu L, D'Amico A, Winkel KD, Suter M, Lo D, Shortman K. 1998. RelB is essential for the development of myeloid-related CD8alpha- dendritic cells but not of lymphoid-related CD8alpha+ dendritic cells. *Immunity* **9**: 839–847.

Yang Y, Huang CT, Huang X, Pardoll DM. 2004. Persistent Toll-like receptor signals are required for reversal of regulatory T cell-mediated CD8 tolerance. *Nat Immunol* **5**: 508–515.

Yannelli JR, Wroblewski JM. 2004. On the road to a tumor cell vaccine: 20 years of cellular immunotherapy. *Vaccine* **23**: 97–113.

Yee C, Thompson JA, Byrd D, Riddell SR, Roche P, Celis E, Greenberg PD. 2002. Adoptive T cell therapy using antigen-specific CD8+ T cell clones for the treatment of patients with metastatic melanoma: *in vivo* persistence, migration, and antitumor effect of transferred T cells. *Proc Natl Acad Sci USA* **99**: 16168–16173.

Yokota K, Takashima A, Bergstresser PR, Ariizumi K. 2001. Identification of a human homologue of the dendritic cell-associated C-type lectin-1, dectin-1. *Gene* **272**: 51–60.

York IA, Mo AX, Lemerise K, Zeng W, Shen Y, Abraham CR, Saric T, Goldberg AL, Rock KL. 2003. The cytosolic endopeptidase, thimet oligopeptidase, destroys antigenic peptides and limits the extent of MHC class I antigen presentation. *Immunity* **18**: 429–440.

Zabel BA, Silverio AM, Butcher EC. 2005. Chemokine-like receptor 1 expression and chemerin-directed chemotaxis distinguish plasmacytoid from myeloid dendritic cells in human blood. *J Immunol* **174**: 244–251.

Zhou D, Cantu C, 3rd Sagiv Y, Schrantz N, Kulkarni AB, Qi X, Mahuran DJ, Morales CR, Grabowski GA, Benlagha K, Savage P, Bendelac A, Teyton L. 2004. Editing of CD1d-bound lipid antigens by endosomal lipid transfer proteins. *Science* **303**: 523–527.

Zinkernagel RM, Hengartner H. 2001. Regulation of the immune response by antigen. *Science* **293**: 251–253.

Zou W, Machelon V, Coulomb-L'Hermin A, Borvak J, Nome F, Isaeva T, Wei S, Krzysiek R, Durand-Gasselin I, Gordon A, Pustilnik T, Curiel DT, Galanaud P, Capron F, Emilie D, Curiel TJ. 2001. Stromal-derived factor-1 in human tumors recruits and alters the function of plasmacytoid precursor dendritic cells. *Nat Med* **7**: 1339–1346.

11

Systemic Lupus Erythematosus: New Ideas for Diagnosis and Treatment

Sandeep Krishnan and **George C. Tsokos**

Abstract

Systemic lupus erythematosus (SLE) is an autoimmune disease of unknown aetiology and complex pathogenesis. While genetic, hormonal and environmental factors have been implicated in the aetiopathogenesis of SLE, information about what exactly confers disease susceptibility to an individual remains obscure. In addition, the inadequacy of current diagnostic modalities has precluded early and definitive diagnosis of the disease, often prolonging initiation of treatment. The current therapeutic options for SLE are effective in controlling the inflammatory process. However, they often tamper with the normal functions of the immune system and thus could also possibly add to the morbidity and mortality that is common in SLE. Recent advances in biomedical research have provided new tools and technology to effectively tackle the genome and proteome, thereby raising new hopes of addressing these inadequacies and opening a new chapter in the diagnosis and treatment of SLE.

Keywords

systemic lupus erythematosus; autoimmunity; immune system; genetic markers; biomarkers; lipid rafts; T cells; gene therapy

11.1 Introduction

Systemic lupus erythematosus (SLE) is an autoimmune disease that affects multiple organs. The pathophysiology of SLE can be summed up as a culmination of complex interplay between multiple processes that include production of numerous types of autoantibodies, deposition of immune complexes in tissues and activation of

Immunogenomics and Human Disease Edited by András Falus

complement and autoreactive T cells. All these processes either singly or in combination induce varying degrees of organ inflammation and failure, most importantly of the kidneys, heart and nervous system (Kammer *et al.*, 2002).

The aetiology of SLE remains partially understood, with strong evidence of influence from genetic, environmental, hormonal and immunoregulatory factors in initiating and maintaining the pathogenic process (Kyttaris, Juang and Tsokos 2004). Over the past few decades, several genetic elements that may predispose to or modify the SLE disease process have been identified. However, despite this, our understanding of the genetic elements contributing to disease susceptibility has been complicated by several factors: (1) differences in the gene distribution pattern among different races; (2) genetic heterogeneity; (3) possible epistatic interactions between genetic elements; and (4) the fact that single genes may contribute to development of more than one autoimmune disease.

Currently, the diagnosis of SLE depends upon recognition of specific clinical signs and symptoms and detection of autoantibodies and markers in the serum that may appear much after the actual pathogenic process has set in. However, definitive biomarkers are lacking that may be used for accurately predicting an individual's risk of disease susceptibility before the breakage point of self-tolerance has reached. Furthermore, each individual displays a unique set of disease markers that could be easily overlooked or missed. As a result, patients often go undiagnosed for a long time and are denied the advantage of early therapeutic intervention. On the treatment front, the current options are mainly limited to corticosteroids and immunosuppressants that can achieve measurable control of the disease in a majority of patients. However, given the high relapse rate of the disease, these drugs often have to be continued for long periods of time (Tsokos and Nepom, 2000). The resultant morbidity and mortality stemming from disease complications and long-term adverse effects of drugs, combined with the cumulative economic burden imposed on patients, justify the dissatisfaction with current treatment options.

Thus, the need of the hour is a holistic approach aimed at dissecting and typifying the aetiopathology of SLE, establishing new diagnostic biomarkers and developing novel therapeutic modalities that tackle the disease process at the grass roots level. Recent advances in our understanding of the human genome and availability of newer sophisticated technology for the analysis of the genome and proteome, and development of newer methods of gene therapy have ushered in new avenues to identify diagnostic molecular markers of SLE and explore effective treatment options. This chapter will discuss the current direction and progress of research aimed at establishing new diagnostic markers and treatment strategies for SLE.

11.2 Strategies for Identifying Diagnostic Markers

The aim of identification of diagnostic markers of SLE can be stratified into two main components: (1) to identify genetic makers that determine the inheritance of disease susceptibility in order to predict the probability of disease manifestation in an individual before the onset of disease; (2) to identify protein markers that modify the disease process and dictate the prognosis of a patient once the disease has set in.

Genetic markers of SLE

There is uniform consensus that genetic determinants are the most important elements that confer disease susceptibility. The search for genes that predispose an individual to SLE has been largely made through association studies of candidate genes and genome-wide linkage analysis. These studies have yielded several disease susceptibility genes and genetic loci and are summarized in Tables 11.1 and 11.2. Let us consider a few important genes.

Table 11.1 Major genes associated with human SLE[a]

Gene		Loci	Functional effects
MHC class II	DR-3, DR-2	6p21	(?) Defective antigen presentation
Complement	C1q	1p36	Defective clearance of immune complexes and apoptotic debris
	C2, C4	6p21	
Fcγ receptor	Fcγ-RIIA, -RIIIA	1q23-24	Defective clearance of immune complexes
CTLA-4		2q33	Defective control of T cell activation
PDCD-1		2q37	(?) Defective control of activation-induced cell death
MBL		10q11.2–q21	(?) Defective clearance of immune complexes
ACE		17q23	(?) Role in renal pathology
Cytokines	IL-10	1q32	(?) Altered immune response
	TNF α	6q21	
	TNF α receptor	1p36	

[a]See text for references.

Table 11.2 SLE susceptibility loci in humans[a]

Locus	Associated genes
1q23-24	FcγRIIA, -IIB, -IIIA
1q41-42	PARP
2q35-37	PDCD-1
4p15-16	Unknown
5p15	Unknown
6p11-22	MHC class II, C2, C4
10q22	Unknown
11p13	Unknown
11q14	Unknown
12q24	Unknown
16q12-13	OAZ
17p12	Unknown

[a]References for this table include Kyttaris, Juang and Tsokos (2004); Nath, Kilpatrick and Harley (2004); Shen and Tsao (2004). PARP, poly(ADP-ribose) polymerase gene. OAZ, OLF1/EBF-associated zinc finger protein.

Major histocompatibility complex

Genes of major histocompatibility complex (MHC) have been associated with SLE for more than three decades. In the Caucasian population, HLA-DR2 and HLA-DR3 have been associated with a 2–3-fold relative risk conferred by each allele, while in the other population groups these findings are not well established (Graham *et al.*, 2002). Alterations in the nature of antigen presentation to helper T cells leading to abnormal T cell responses may be one mechanism by which these alleles contribute to the disease pathology.

Fcγ receptors

Allelic variants of Fcγ receptor genes can alter binding affinity to respective subclasses of IgG and affect the phagocytic immune response and thus predispose an individual to autoimmune diseases. Amongst these receptors, genetic polymorphisms of FcγRIIA, FcγRIIIA, FcγRIIIB and FcγRIIB, a cluster of four genes at 1q23 encoding for low-affinity IgG receptors, have been found to be associated with SLE (reviewed in Tsao, 2004). Of these genes, FcγRIIA and FcγRIIIA have the strongest association with SLE. (Magnusson *et al.*, 2004; Nath, Kilpatrick and Harley 2004).

Complement genes

Complement genes are components of the innate immunity and participate in clearing up the apoptotic fragments and cell debris that can otherwise form a source for autoantigens. Individuals with complete deficiency of complement components C1q, C4 and C2 have a high risk of developing SLE in an additive manner, with the risks being 90 per cent for C1q, 75 per cent for C4 and 10 per cent for C2 (Ghebrehiwet and Peerschke 2004; Manderson, Botto and Walport 2004; Slingsby *et al.*, 1996). Additionally, deficiencies of C1r/s, C5 and C8 also may induce SLE or SLE-like syndromes (reviewed in Nath, Kilpatrick and Harley, 2004).

Cytotoxic T lymphocyte antigen-4

Cytotoxic T lymphocyte antigen-4 (CTLA-4), a structural homolog of CD28, is a negative regulator of T cells and plays an important role in preventing autoimmune diseases. Several studies have found a strong association of CTLA-4 gene polymorphisms with susceptibility to SLE (Fernandez-Blanco *et al.*, 2004; Hudson *et al.*, 2002; Liu *et al.*, 2001). Among several polymorphisms associated with CTLA-4 gene, allelic variation characterized by T/C substitution at the -1722 site has been shown to specifically influence susceptibility to SLE (Hudson *et al.*, 2002).

Programmed cell death-1

Programmed cell death-1 (PDCD-1) is an immunorecepor of the CD28 family that bears a tyrosine-based inhibitory motif. It is normally expressed on the surface of activated T and B cells and regulates peripheral tolerance in both cell types. Intronic single-nucleotide polymorphisms (SNP) in PDCD-1 were found to be associated with SLE development in the European and Mexican population (Ferreiros-Vidal *et al.*, 2004; Prokunina *et al.*, 2002, 2004). They alter the binding site for the runt-related transcription factor 1 (RUNX1), causing aberrant regulation of PDCD-1 that induces the lymphocytic hyperactivity observed in SLE (Prokunina *et al.*, 2002).

Other genes

Gene polymorphisms that contribute to diminished expression of several other genes have been reported in SLE and association with disease has been found for these genes. These include C-reactive protein (CRP; Carroll, 2001; Russell *et al.*, 2004; Szalai *et al.*, 2002), mannose-binding lectin (MBL; Davies *et al.*, 1997; Huang *et al.*, 2003; Nath, Kilpatrick and Harley, 2004), and angiotensin-converting enzyme (ACE) inhibitor (Parsa *et al.*, 2002; Tsao, 2003).

Strategies for identifying new genetic markers

Despite advances in identification of several genes associated with SLE, extensive genetic heterogeneity and probable epistatic interactions between multiple genes necessary for disease development have complicated the elucidation of what exactly confers 'SLE susceptibility'. The current models of inheritance of disease susceptibility of multifactorial traits such as SLE and other autoimmune diseases largely follow the principle of 'threshold liability' (Tsao, 2003; Wandstrat and Wakeland, 2001). According to this model, the genome of each individual contains a certain number of SLE susceptibility genes that confer additive disease liability when their number exceeds a certain hypothetical threshold. This 'additive inheritance' may be modified further by 'multiplicative inheritance' wherein epistatic interactions among susceptibility alleles may further move the individual's disease liability toward disease threshold (Tsao, 2003; Wandstrat and Wakeland, 2001). These factors are further modified by environmental variables that the individual experiences in life.

In order to effectively develop diagnostic genetic markers, the first step would be to unequivocally establish the association of currently identified candidate markers with disease susceptibility through replicative studies from different sample population. It has begun to be recognized that an important goal of future efforts in the genetics of complex traits such as SLE will be a switch from linkage analysis and modelling to gene identification and pathway analysis (Wandstrat and Wakeland, 2001). Inclusion of larger patient population comprising of diverse races in each study and application of more powerful analytical procedures as they become available would be necessary. The extensive development of SNP technology may also facilitate gene identification.

Interestingly, genome-wide linkage analysis studies have revealed that disease susceptibility loci were colocalized in human and mouse (Becker *et al.*, 1998; Griffiths *et al.*, 1999; Jawaheer *et al.*, 2001; Wandstrat and Wakeland, 2001). The chance of success in identifying disease susceptibility genes is excellent in animal models owing to powerful tools such as congenic dissection, which permits characterization of phenotypes conferred by individual genetic components of a polygenic disease and alongside application of fine mapping studies that have the capability to track down the susceptibility interval to as little as 800–1000 kb (Wandstrat and Wakeland, 2001). It has been predicted that, with the ready availability of human genome data and with the mouse genome project poised to be completed soon, positional cloning efforts will receive a boost. Another advantage of utilizing animal models to identify targets in humans is that *in vivo* complementation studies using bacterial artificial chromosome (BAC) transgenic technologies can definitively establish the role of a gene in disease susceptibility (Wandstrat and Wakeland, 2001).

With the advent of gene expression microarrays, it may soon be possible to accelerate identification of SLE susceptibility alleles. Once such genes are identified, it would be possible to screen individuals to ascertain their disease susceptibility. Additionally, the relatives of SLE patients can also be screened to ascertain their disease susceptibility, thus providing them with a chance of preventive care. Once a database of SLE susceptibility genes is established, it is also expected to boost gene therapy-based therapeutic intervention.

Strategies for identification of protein markers

While the disease susceptibility genes may earmark an individual for development of SLE at birth, in most individuals the actual disease process does not set in until later in life. While it is known that there is contribution from multiple factors, including environmental and possibly hormonal, in triggering disease pathogenesis, information about when and how the breakage point of self tolerance is reached remains elusive. Ultimately, all the abnormal pathways are mediated by a complex interplay between several proteins that are turned on either as part of the disease process itself or as part of the body's attempt to contain the inflammatory process. These proteins may also be turned on and off differently in different stages of the disease. Thus, it would appear logical to identify and characterize such protein markers specific to SLE and establish a database that classifies their roles in various stages of SLE. Such an approach would be advantageous in several ways: (1) to predict an individual's immediate risk of disease development; (2) to detect the disease in an undiagnosed individual; (3) to ascertain the disease severity; (4) to predict relapses and remissions; (5) to design targeted gene- or drug-based therapy. Let us consider a few biomarkers that may serve as excellent diagnostic and prognostic tools.

Autoantibodies

Compared with other typical autoimmune diseases, SLE displays the highest number of autoantibodies produced against disparate antigens. At least 116 autoantibodies have been described in SLE patients that may target antigens present in the nucleus, cytoplasm and cell membrane as well as antigens associated with phospholipids, and cells of the endothelium, blood and nervous system (Sherer *et al.*, 2004). These antibodies display wide disparities in terms of patient distribution. While some antibodies such as antinuclear antibodies are found in nearly all SLE patients, most antibodies are found only in a small number of patients. In addition, some autoantibodies such as the rheumatoid factor might be found in other autoimmune diseases as well. The evidence that these antibodies correlate with disease activity is scanty and it unclear whether all of them contribute to the disease pathology. Thus, a systematic approach to characterizing their precise roles in the disease process will serve to delineate pathogenic from nonpathogenic antibodies. Their appearance in serum could then be utilized to denote various stages of the disease.

Cytokines

Cytokines can contribute to the pathology of SLE in several ways. For example, the absence of regulatory cytokines such as IL-2 may prevent effective activation and functioning of T cells, while the absence of IL-12 might affect differentiation of CD4 T cells into Th1 cells. High levels of cytokines such as IL-6 and IL-10 may promote antibody production by B cells whereas low levels of anti-inflammatory lymphokine transforming growth factor-β (TGF-β) might result in unregulated inflammation. An imbalance between type 1 (IFN-γ and IL-2) and type 2 (IL-4, IL-5, IL-6 and IL-10) cytokine production by peripheral blood mononuclear cells has also been reported in SLE. Similar to autoantibodies, there is disparity amongst patients in terms of cytokine expression pattern. Detecting fluctuations in the cytokine production by immune cells might be informative in predicting the immune status of the individual in various stages of the disease and planning appropriate treatment.

Cellular protein markers

Alterations in protein expression with in various cells of the immune system could be tapped to establish diagnostic markers. Such proteins could form either cell surface markers or intracellular markers that alter cellular physiology. For example,

Table 11.3 Altered expression of major proteins in SLE T and B cells[a]

Cell type	Molecule	Defect in expression
T cell (surface)	CD45	Decreased
	CD40 L	Increased
	CD70	Increased
T cell (intracellular)	TCR ζ chain	Decreased
	FcRγ chain	Increased
	LCK	Decreased
	Syk	Increased
	Protein kinase C	Decreased
	MAP kinase	Decreased
	NF-B, p65 subunit	Decreased
	Elf-1	Decreased
	CREM	Increased
B cells (surface)	CD40L	Increased
	CD80	Increased
	CD86	Increased
	CD21	Decreased

[a]Only selected proteins are included.

'rewiring' of T cell receptor (TCR) induced by reduced expression of TCR ζ chain and appearance and association of FcRγ chain with the TCR has been associated with increased excitability of SLE T cells (Krishnan, Farber and Tsokos 2003). Reduced expression of B7.1 on the surface of antigen-presenting cells (APC), such as B cells, might be responsible for the reduced recall responses demonstrated by SLE T cells (Tsokos *et al.*, 1996). Similarly, prolonged upregulation of CD40L expression on the surface of T cells might contribute to sustained stimulation of B cells to produce autoantibodies (Koshy, Berger and Crow, 1996). Table 11.3 highlights some of the important proteins of this class. Abnormal localization of proteins has also been reported in SLE T cells. In SLE, more than 50 per cent of T cells demonstrate preclustering of lipid rafts in their cell membranes, which also contain TCR and related signalling molecules (Figure 11.1). These observations were not found in other autoimmune diseases such as rheumatoid arthritis and Sjögren's syndrome, and identification of T cells bearing preclustered lipid raft could be a promising marker for diagnosis of SLE (Krishnan *et al.*, 2004).

In a complex disease such as SLE, there could be numerous abnormal proteins present. Identifying these proteins and their participation in the disease pathogenesis presents an enormous challenge. However, availability of new technology has eased this concern. For example, abundance-based protein microarrays could be used to measure the abundance of specific biomolecules using analyte-specific reagents (ASRs) such as antibodies. This technique can prove especially useful for detecting analytes such as cytokines. Function-based protein microarrays can be used to study protein interaction with other proteins, nucleic acids, lipids and small molecules.

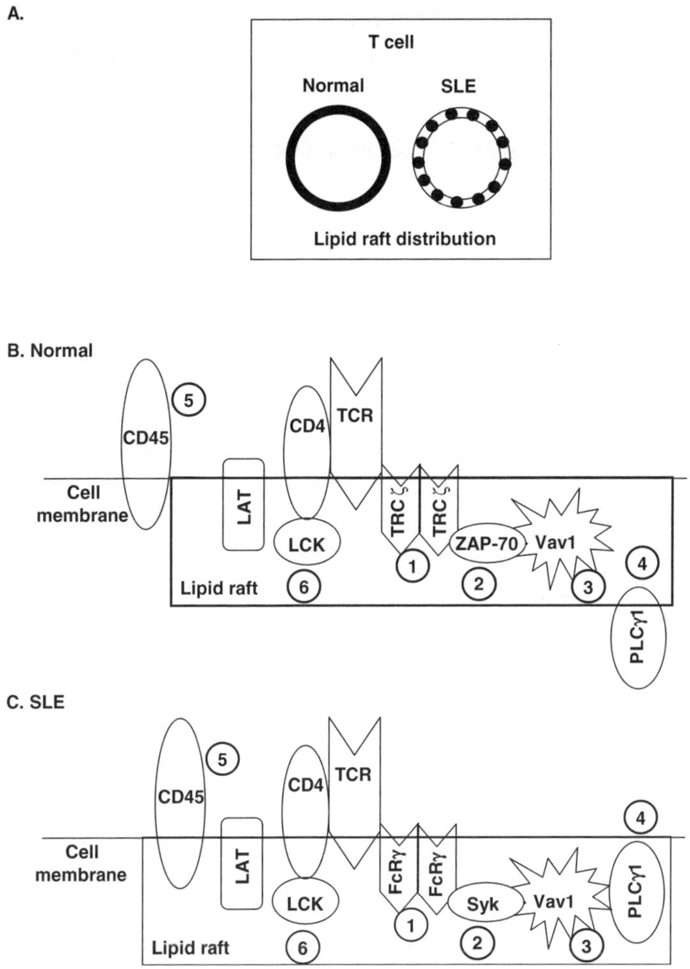

Figure 11.1 Defective lipid raft signalling in SLE. (A) In normal T cells, lipid rafts are homogenously distributed in the cell membrane, while in SLE T cells, lipid rafts are pre-clustered. (B, C) Differences in the protein distribution pattern between lipid rafts of normal and SLE T cells are shown. (1) In normal T cells, TCR ζ associates with TCR whereas in SLE, TCR ζ levels are reduced and instead FcRγ associates with TCR in the lipid rafts. (2) In SLE, Syk is the more dominant tyrosine kinase compared with ZAP-70 in normal T cells. (3) Vav1 is associated in greater amounts with in lipid rafts of SLE compared with normal T cells and undergoes greater tyrosine phosphorylation. (4) PLC-γ1 is localized to membrane rafts in freshly isolated SLE T cells whereas it is predominantly found in the cytoplasm in normal T cells. (5) CD45 is excluded from lipid rafts in normal T cells whereas it is found with in lipid rafts of SLE T cells. (6) Levels of LCK expression are reduced in lipid rafts of SLE T cells compared with normal T cells. Association of some proteins such as LAT with lipid rafts remains unaffected in SLE (Krishnan *et al.*, 2004; Jury *et al.*, 2004).

Using this technique, thousands of target proteins can be simultaneously screened for function. Thus, the goal of identifying biomarkers of SLE appears realistic.

11.3 Strategies for Gene Therapy for SLE

Given the complex nature of the disease process and given our limited understanding of it, development of successful therapeutic strategies faces a daunting task. Because the fundamental disease pathology involves the immune-regulatory system, it becomes necessary for any treatment plan designed to control the ongoing abnormal immune responses to avoid disruption of the normal immune surveillance functions (Tsokos and Nepom, 2000). Thus, therapies directed at common pathways of immune activation such as administration of cytokines and their antagonists, anti-T cell monoclonal antibodies, inhibitors of signal transduction and other pharmacologic agents used currently run the risk of inducing a state of global immune paralysis, sometimes more deleterious than the disease itself. In addition, they incur repeated expenses because of repeated administration owing to disease rebound upon stoppage of treatment. Therefore therapies that target only the abnormal immune pathways, are long-lasting with minimal side effects and at the same time cost-effective, are more desirable.

Gene therapy offers several advantages over conventional therapy. It can provide long-lasting yet safe and locally targetted responses, thereby avoiding widespread immune suppression, and could prove to be cost-effective in the long run. In addition, some of these methods provide the scope for designing 'molecular switches' that can be turned on and off during relapses and remission, respectively, and thus avoid long-term side effects. Avoidance of long-term effects of treatment is also possible by gene therapy with the option of designing molecules with shorter half-lives (Chernajovsky, Gould and Podhajcer, 2004; Neve *et al.*, 1996). However, today gene therapy for SLE remains largely experimental and restricted to animal models. Table 11.4 summarizes these approaches. Improvements in the existing modalities as well as development of newer ones is contingent upon our recognition of factors upon which their success is dependent, such as the right choice of the target molecule, efficient vector systems and improvements in vector delivery systems. Below we discuss current strategies being adopted to address these issues.

Strategies for improved gene delivery systems

Gene therapy involves expression of foreign DNA that has been inserted into a host cell. This DNA is usually delivered as an insert within a foreign vector. Thus, it is important that the chosen vector is nonimmunogenic as well as capable of accommodating various sizes of insert. In addition, the vector should be able to penetrate the target cell chosen for gene delivery. Two commonly used vectors are viruses and bacterial plasmid DNA vectors. Viral vectors that contain the gene sequence of

Table 11.4 Targeted gene therapy of SLE[a]

DNA construct	Vector	Animal model	**Mode of action**	Clinical response
MHC-1 binding epitopes of anti-DNA Ab	Plasmid	NZB/NZW F_1	Kills anti-DNA antibody producing B cells by CD8 cells	Retardation of nephritis, increase in survival
IFN-γ R/IgG1-Fc fusion protein	Plasmid	MRL lpr/lpr	Blocking IFN-γ	Improvement of nephritis, increase in survival
IL-2	S. typhimurium	MRL lpr/lpr	Increases IL-2 levels	Improvement of nephritis
	Vaccinia virus	MRL lpr/lpr	Increases IL-2 levels	Improvement of nephritis
IL-12	Plasmid	MRL lpr/lpr	Increases IL-12 levels	Inhibition of splenomegaly, improvement of nephritis, increase in survival
TGF-β	S. typhimurium	MRL lpr/lpr	Increases TGF-β levels	Minimal effects
	Plasmid	MRL lpr/lpr	Increases TGF-β levels	Reversal of nephritis, increase in survival
CTLA-4/IgG	Adenovirus	MRL lpr/lpr	Blocking CTLA-4	Suppression of nephritis

[a]See text for references.

interest but have been doctored to lose their capacity to spontaneously replicate can be safely transducted into host cells to deliver the genes. Some of the commonly used viral vectors include the adenoviral, lentiviral, retroviral, herpes virus and Epstein–Barr virus-based vectors. Bacterial plasmids can deliver either the naked DNA to the cell or DNA complexed with liposome which augments its uptake.

Two potential shortcomings of viral vectors are their potential immunogenicity leading to their rapid clearance, and the risk of induction of oncogenicity in the host (for example, the retrovirus-based vectors). Thus, strategies are necessary to minimize the effect of host on the vector and that of the vector on the host. Two promising advances in this regard are development of gutless adenoviral vectors that are nonimmunogenic, and development of nononcogenic retroviral vectors that lack the U3 region of the vector's 3′ long terminal repeats (LTR; Chernajovsky, Gould and Podhajcer, 2004; Yu *et al.*, 1986).

Some gene expression systems take advantage of the fact that some genes are overexpressed in the inflammatory process of autoimmune diseases. Systems have been designed in which anti-inflammatory transgene expression is autonomously regulated by an inflammation-inducible promoter. For example, in a rat model of streptococcal cell wall-induced arthritis, intra-articular injection of a recombinant adenoviral vector containing a two-component inflammation-inducible promoter [HIV Tat protein is placed under the control of complement 3 (C3) promoter and Tat then transactivates another cytokine gene inserted in the same vector] controlling the expression of anti-inflammatory cytokine, human IL-10 (hIL-10) cDNA, was used to achieve sufficient levels of hIL-10 expression within inflamed joints, which ameliorated the joint inflammation and swelling (Miagkov *et al.*, 2002).

An important aspect of SLE is its fluctuating course. Thus, a therapeutic modality that incorporates ‘molecular switches’ that can be ‘turned on and off at will’ during relapses and remission, respectively, offers the advantage of providing discontinuous need-based therapy as necessitated by the individual's clinical status rather than on a continuous basis. One such gene-therapy system incorporates a ‘molecular switch’ under the contol of the tetracycline analog doxycycline (Gossen *et al.*, 1995, reviewed in Chernajovsky, Gould and Podhajcer 2004). In this system, expression of the target gene is placed under the control of doxycycline. In the absence of doxycycline, a repressor (TetR) binds to the promoter and prevents transcription of the gene. When doxycycline is administered, TetR is removed and a transactivator activates transcription of the gene. Both plasmid vectors and adenoviral vectors have successfully incorporated the components of this system (Gould *et al.*, 2000; Lamartina *et al.*, 2003; Salucci *et al.*, 2002). Recently, manipulated dendritic cell (DC)-based immunotherapy was demonstrated in a mouse model of type II collagen-induced arthritis (Liu *et al.*, 2003). Injection of DC cells pulsed with type II collagen and transducted with an adenovirus-based vector capable of expressing TNF-related apoptosis-inducing ligand (TRAIL) under the control of doxycycline-inducible tetracycline response element (TRE) into the joints of mice with collagen-induced arthritis reduced the incidence of arthritis and infiltration of T cells in the joint. Given the function of DC as potent antigen-presenting cells, this approach offers tremendous potential in limiting the activity of autoreactive T cells in SLE. However, considering their high degree of rejection in nonhuman primates, first the nonimmunogenic potential of these vectors needs to be established in humans (Chernajovsky, Gould and Podhajcer 2004; Favre *et al.*, 2002; Latta-Mahieu *et al.*, 2002).

Strategies for gene delivery techniques

A gene can be either delivered directly *in vivo* to a patient or *ex vivo* by first extracting cells from the patient for gene delivery followed by their *in vitro* culture and transfer back to the patient. A naked plasmid vector containing the DNA sequence of interest can be transferred *in vivo* using a carrier such as cationic liposome that enhances its uptake. Intramuscular injection has been found to be reliable mode of delivery of plasmid vectors. In one study, vaccination of lupus-prone (NZB/NZW F_1) mice with plasmid DNA vectors encoding MHC-1-binding epitopes in the heavy chain variable region of anti-DNA antibodies induced $CD8^+$ T cells that killed anti-DNA antibody-producing B cells, retarded the development of nephritis and improved survival (Fan and Singh, 2002). It has been observed that electroporation at the injection site can enhance the intramuscular plasmid DNA transfection several fold. In MRL-Fas(lpr) mice, intramuscular delivery of an insert encoding IFN-γ-Receptor/Fc fusion protein was found to retard lupus development and progression (Lawson *et al.*, 2000). One disadvantage of *in vivo* gene delivery is that local injection at the site of inflammation/lesion is necessary. Thus, if a systemic rather than local access to the therapeutic gene is desired, treatment with an *in vivo* gene delivery system becomes more complicated.

Ex vivo gene therapy offers some advantages over *in vivo* techniques. By choosing the type of cell for gene delivery, one could achieve either local or systemic delivery. For example, by choosing cells such as DC or lymphocytes that widely traverse the body, one could administer genes to sites far from the injection site. By contrast, by choosing immobile cells such as pancreatic islet cells, one could achieve local gene delivery (Evans *et al.*, 2000). Moreover, because the cells injected into the patient are recognized as self, chances of rejection are minimized. Another major advantage is that patients are not exposed to the vectors directly thereby circumventing possible vector-induced harm. Advantages of choosing B and T lymphocytes for gene delivery include the capability of providing antigen-specific response in the scenario that the antigen is known. The lymphocytes containing the therapeutic gene will expand clonally only when they come into contact with the target antigen (Chernajovsky, Gould and Podhajcer 2004).

Strategies for identifying molecular targets for gene therapy

Despite the availability of efficient vector systems and sophisticated gene delivery techniques, gene therapy of SLE has lagged behind specifically for want of knowledge about which disease-specific pathological gene(s), protein(s) or pathway(s) to target. In SLE, abnormalities occur at the level of cell–cell interaction, signal transduction, functional response of cells to stimuli and cell turnover. The strategies employed to identify biomarkers that mediate these processes and could form targets for gene therapy have been discussed above. Let us examine some of the main targets currently being considered for gene therapy of SLE.

Gene therapy targetting cytokines

A major focus of gene therapy has been on targeting cytokines. Attempts have been made to block pro-inflammatory cytokines such as IFN-γ and enhance the levels of anti-inflammatory cytokines such as TGF-β. In some cases, pro-inflammatory cytokines have been found to induce anti-inflammatory responses, for example IL-12 (Hagiwara *et al.*, 2000). Levels of IFN-γ have been found to be elevated in experimental murine models of SLE. In an attempt to quell its effects, Lawson *et al.*, have used a plasmid that encoded IFN-γR/IgG1-Fc fusion protein that produces a disulfide-linked homodimer of IFN-γ receptor/IgG1Fc, a protein that can bind to circulating IFN-γ and neutralize its effects (Lawson *et al.*, 2000). This system has been tested in the lupus mouse MRL-Fas/lpr. The mice that received the plasmid either before or after the onset of disease survived longer and their renal manifestations of SLE were also reduced.

The importance of the right choice of target is evident from the studies reported by Raz *et al.* (1995). They demonstrated that repeated monthly intramuscular injections of cDNA expression vectors encoding for anti-inflammatory TGF-β between 6 and 26 weeks prolonged the mean survival of MRL/lpr/lpr mice in comparison to the control group and demonstrated improved renal function. By sharp contrast, IL-2 cDNA transferred in an identical manner appeared harmful with decreased mean survival and even enhanced autoantibody production. However, another group observed an almost opposite effect in mice replenished with TGF-β and IL-2 using a different system. In MRL/lpr/lpr mice orally treated with TGF-β gene-bearing nonpathogenic stains of *Salmonella typhimurium*, no improvement in pathology or clinical response was observed, whereas administration of bacterial vectors bearing IL-2 restored defective T cell functions such as mitogen-induced proliferative response and suppressed autoantibody production and renal abnormalities (Huggins *et al.,* 1997, 1999). A second group also observed beneficial effects of IL-2 replenishment in MRL/lpr/lpr mice in which live vaccinia virus-based vectors were used (Gutierrez-Ramos *et al.*, 1990). Thus, depending upon the mode of delivery of a gene, the immune response might vary and caution must be exercised in picking up the right method for gene delivery.

Another indirect approach to correct the levels of cytokines would be to correct factors that lead to their aberrant production. For example, let us consider IL-2, deficiency of which is induced by IL-2 gene transcription in SLE. NF-κB is a factor involved in positive regulation of this process. NF-κB consists of a heterodimer formed by p50 and p65. Around 80 per cent of SLE patients demonstrate reduced expression of p65 than healthy controls and T cells from these patients demonstrate reduced IL-2 production. Plasmid mediated overexpression of p65 in SLE T cells by electroporation demonstrated reconstitution of activation-induced IL-2 promoter activity (Herndon *et al.*, 2002). Another molecule c-AMP response element modulator (CREM) was found to be a negative regulator of IL-2 gene expression in SLE T cells (Solomou *et al.*, 2001). Using a modification of electroporation technique called nucleofection™, SLE T cells were transfected with a plasmid-encoding antisense CREM (Tenbrock *et al.*, 2002). This technique succeeded in eliminating CREM and normalized IL-2 production. As described below, replenishment of SLE T cells with TCR ζ chain, another protein that is expressed in low amounts, also rescued IL-2 production (Nambiar *et al.*, 2003b). Thus, gene therapy that replenishes missing molecules such as p65 and TCR ζ or eliminates gene transcription repressors such as CREM can restore IL-2 production and thus normalize the functions of T cells in SLE.

Gene therapy targetting cell–cell interaction

Another important treatment consideration has been modifying cell–cell interaction between T cells and APC. APC provide co-stimulatory signals to T lymphocytes via CD80 (B7-1) and CD86 (B7-2) that bind to CD28 on the surface of T cells. CTLA-4 is a molecule related to CD28 and can bind to CD80 and CD86 with higher affinity than CD28 but exerts the opposite action (Figure 11.2). Thus, the use of CTLA-4/IgG fusion protein leads to blockage of the positive effects of CD28 and induces T cell anergy. MRL/lpr/lpr mice intravenously injected with an adenovirus vector containing cDNA coding for CTLA-4/IgG fusion protein demonstrate complete suppression of nephritis (Takiguchi *et al.*, 2000). This method has also been tested in patients with rheumatoid arthritis.

Gene therapy targetting molecules involved in signal transduction

An indirect means of modifying interaction between T cells and APC and thereby normalizing cellular signal transduction would be to target gene products that regulate lipid raft signalling. The integrity of lipid raft dynamics is vital to sustaining the immunological synapse between T cells and APC and is thus critical to efficient downsteam signalling. As discussed above, in SLE T cells, lipid rafts are preclustered, and they contribute at least in part to aberrant T cell responses including heightened TCR-induced calcium responses (Krishnan *et al.*, 2004). SLE T cells demonstrate decreased levels of TCR ζ chain in the lipid rafts whereas a closely related molecule, FcRγ, is upregulated and associates with lipid rafts and contributes to the hyper-excitability of SLE T cells (Enyedy *et al.*, 2001; Krishnan *et al.*, 2004; Figure 11.1). Using nucleoporation™ technique, it has been shown that correction of TCR ζ levels in SLE T cells leads to reduction in the levels of FcRγ chain, reversal

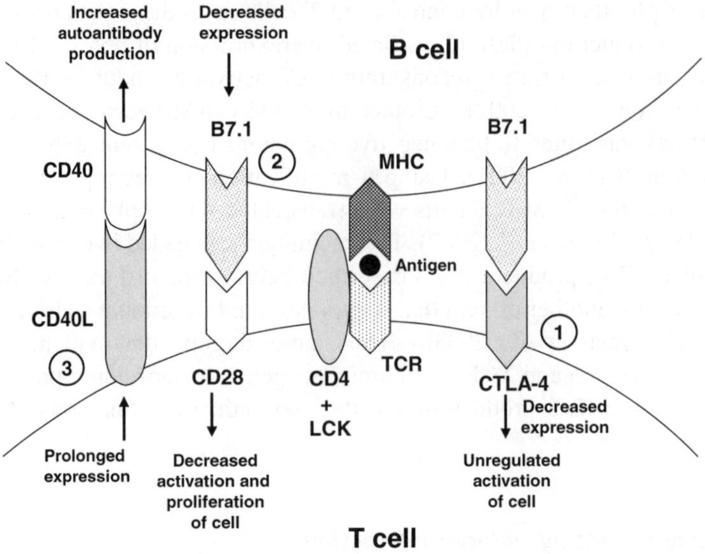

Figure 11.2 Defective B cell–T cell interaction in SLE. Normally, during antigen presentation by B cells to T cells in the context of MHC, binding of B7.1 and B7.2 on the surface of B cells to CD28 on the surface of T cells provides the costimulation necessary for optimum activation and proliferation of T cells. Subsequently, CTLA-4, a CD28-related molecule is upregulated on the surface of T cells and binds to B7.1 and B7.2 with higher affinity than CD28 and limits this activation. In SLE, CTLA-4 expression may be defective (1) causing uncontrolled T cell activation. Also, there is diminished expression of B7.1 (2) on the surface of B cells that leads to diminished signalling through CD28, resulting in diminished T cell activation. Another molecule on the surface of T cells, CD40L, is induced transiently upon activation of T cells and provides costimulation to B cells via CD40 on B cell surface to produce antibodies. In SLE, there is prolonged expression of CD40L (3) on the surface of T cells, thus aiding B cells in producing autoantibodies.

of preclustering of lipid rafts and lowering of the abnormal TCR-induced calcium response, thus reversing the hyperexcitable phenotype of T cells. In addition, these cells produced sufficient quantities of IL-2 and thus the functional response of SLE T cells were 'normalized' (Nambiar *et al.*, 2003b). Interestingly, introduction of FcRγ into normal T cells induced a hyperexcitable phenotype in the transfected cells akin to SLE T cells. These cells demonstrated higher levels of intracytoplasmic tyrosine phosphorylation events and greater TCR-induced calcium flux, and lowered IL-2 production (Nambiar *et al.*, 2003a). Thus, eliminating FcRγ by gene therapy would be a possible means of restoring normal T cell functions in SLE. Other signalling proteins, such as Syk, Vav1 and PLCγ1 which were also found to be increasingly associated with lipid rafts and to participate in TCR receptor cap formation and calcium signalling, may also be tried as candidates for gene modulation therapy aimed at disrupting interactions between T cells and APC (Krishnan *et al.*, 2004).

Gene therapy for restoring tolerance

Because loss of peripheral tolerance is an integral feature of autoimmune diseases such as SLE, attempts have also been made to restore tolerance to self antigens. In this context, administration of specific antigens has been tried because they would mediate activation-induced cell death of autoimmune cells (deletional tolerance), redirect cells away from the pathogenic pathways (immune deviation) or down-modulate the autoreactive immune response (immune regulation; Kyttaris, Juang and Tsokos, 2004; Tsokos and Nepom, 2000). *Ex vivo* gene delivery with autologous B cells transfected with constructs encoding fusion proteins of Fc fragment of IgG and a target antigen has been shown to suppress immune response to that antigen during subsequent challenges. This technique has been tried out with tangible success in animal models of several autoimmune diseases such as diabetes, encephalomyelitis and uveitis, where the dominant abnormal antigen is known (Agarwal *et al.*, 2000; Melo *et al.*, 2002). A practical limitation of this technique in SLE is that multiple autoantigens may be operative in SLE. However, as new technology becomes available, it may become possible to simultaneously deliver multiple target antigen genes.

Gene therapy for restoring effective clearance of antigens

Defective clearance of autoimmune cells and autoantigens derived from rapid cellular apoptosis remains a challenging problem in the treatment of SLE. Thus, components of the complement cascade and enzymes such as *DNaseI* involved in clearing proteins or protein–DNA complexes, respectively, could form excellent targets for gene therapy. Recently, SLE patients lacking *DNaseI* were reported (Yasutomo *et al.*, 2001). Mice lacking *DNaseI* also develop SLE-like syndromes (Napirei *et al.*, 2000), and administration of recombinant *DNaseI* was found to ameliorate disease symptoms in lupus murine models (Macanovic *et al.*, 1996; Pan *et al.*, 1998).

11.4 Conclusion and Future Direction

Today, the scorching pace of biomedical research has thrust us to the threshold of metamorphosing our approach toward treatment of autoimmune diseases in general and SLE in particular. The raw materials for understanding and manipulating the genome and proteome are all available in principle. From this crossroads, a multipronged strategy is envisioned that would encompass identification of precise genetic determinants that confer disease susceptibility, elucidation of specific biomarkers that shape inappropriate activation of the immune-mediated inflammation, and tapping of gene therapy for effective treatment. In the field of SLE genetics, with the availability of more sophisticated technology, a gradual switch from linkage analysis and modelling to disease-specific gene identification would be expected.

Gene and protein arrays would be predicted to expedite identification of biomarkers with diagnostic and prognostic potential. Acquisition of apt knowledge of the disease process combined with improved techniques of targetted gene delivery would be expected to upgrade the quality of gene therapy, which would become more focused, effective and with minimal side effects, and eventually shift from animal models to humans. While at present it would seem that we are far from this objective, the future of treatment of SLE does appear to be optimistic.

References

Agarwal RK, Kang Y, Zambidis E, Scott DW, Chan CC, Caspi RR. 2000. Retroviral gene therapy with an immunoglobulin–antigen fusion construct protects from experimental autoimmune uveitis. *J Clin Invest* **106**(2): 245–252.

Becker KG, Simon RM, Bailey-Wilson JE, Freidlin B, Biddison WE, McFarland HF, Trent JM. 1998. Clustering of non-major histocompatibility complex susceptibility candidate loci in human autoimmune diseases. *Proc Natl Acad Sci USA* **95**(17): 9979–9984.

Carroll M. 2001. Innate immunity in the etiopathology of autoimmunity. *Nat Immunol* **2**(12): 1089–1090.

Chernajovsky Y, Gould DJ, Podhajcer OL. 2004. Gene therapy for autoimmune diseases: *quo vadis*? *Nat Rev Immunol* **4**(10): 800–811.

Davies EJ, Teh LS, Ordi-Ros J, Snowden N, Hillarby MC, Hajeer A, Donn R, Perez-Pemen P, Vilardell-Tarres M, Ollier WE. 1997. A dysfunctional allele of the mannose binding protein gene associates with systemic lupus erythematosus in a Spanish population. *J Rheumatol* **24**(3): 485–488.

Enyedy EJ, Nambiar MP, Liossis SN, Dennis G, Kammer GM, Tsokos GC. 2001. Fc epsilon receptor type I gamma chain replaces the deficient T cell receptor zeta chain in T cells of patients with systemic lupus erythematosus. *Arthrit Rheum* **44**(5): 1114–1121.

Evans CH, Ghivizzani SC, Herndon JH, Wasko MC, Reinecke J, Wehling P, Robbins PD. 2000. Clinical trials in the gene therapy of arthritis. *Clin Orthop* **379**(suppl.): S300–S307.

Fan GC, Singh RR. 2002. Vaccination with minigenes encoding V(H)-derived major histocompatibility complex class I-binding epitopes activates cytotoxic T cells that ablate autoantibody-producing B cells and inhibit lupus. *J Exp Med* **196**(6): 731–741.

Favre D, Blouin V, Provost N, Spisek R, Porrot F, Bohl D, Marme F, Cherel Y, Salvetti A, Hurtrel B, Heard JM, Riviere Y, Moullier P. 2002. Lack of an immune response against the tetracycline-dependent transactivator correlates with long-term doxycycline-regulated transgene expression in nonhuman primates after intramuscular injection of recombinant adeno-associated virus. *J Virol* **76**(22): 11605–11611.

Fernandez-Blanco L, Perez-Pampin E, Gomez-Reino JJ, Gonzalez A. 2004. A CTLA-4 polymorphism associated with susceptibility to systemic lupus erythematosus. *Arthrit Rheum* **50**(1): 328–329.

Ferreiros-Vidal I, Gomez-Reino JJ, Barros F, Carracedo A, Carreira P, Gonzalez-Escribano F, Liz M, Martin J, Ordi J, Vicario JL, Gonzalez A. 2004. Association of PDCD1 with susceptibility to systemic lupus erythematosus: evidence of population-specific effects. *Arthrit Rheum* **50**(8): 2590–2597.

Ghebrehiwet B, Peerschke EI. 2004. Role of C1q and C1q receptors in the pathogenesis of systemic lupus erythematosus. *Curr Dir Autoimmun* **7**: 87–97.

Gossen M, Freundlieb S, Bender G, Muller G, Hillen W, Bujard H. 1995. Transcriptional activation by tetracyclines in mammalian cells. *Science* **268**(5218): 1766–1769.

Gould DJ, Berenstein M, Dreja H, Ledda F, Podhajcer OL, Chernajovsky Y. 2000. A novel doxycycline inducible autoregulatory plasmid which displays on/off regulation suited to gene therapy applications. *Gene Ther* **7**(24): 2061–2070.

Graham RR, Ortmann WA, Langefeld CD, Jawaheer D, Selby SA, Rodine PR, Baechler EC, Rohlf KE, Shark KB, Espe KJ, Green LE, Nair RP, Stuart PE, Elder JT, King RA, Moser KL, Gaffney PM, Bugawan TL, Erlich HA, Rich SS, Gregersen PK, Behrens TW. 2002. Visualizing human leukocyte antigen class II risk haplotypes in human systemic lupus erythematosus. *Am J Hum Genet* **71**(3): 543–553.

Griffiths MM, Encinas JA, Remmers EF, Kuchroo VK, Wilder RL. 1999. Mapping autoimmunity genes. *Curr Opin Immunol* **11**(6): 689–700.

Gutierrez-Ramos JC, Andreu JL, Revilla Y, Vinuela E, Martinez C. 1990. Recovery from autoimmunity of MRL/lpr mice after infection with an interleukin-2/vaccinia recombinant virus. *Nature* **346**(6281): 271–274.

Hagiwara E, Okubo T, Aoki I, Ohno S, Tsuji T, Ihata A, Ueda A, Shirai A, Okuda K, Miyazaki J, Ishigatsubo Y. 2000. IL-12-encoding plasmid has a beneficial effect on spontaneous autoimmune disease in MRL/MP-lpr/lpr mice. *Cytokine* **12**(7): 1035–1041.

Herndon TM, Juang YT, Solomou EE, Rothwell SW, Gourley MF, Tsokos GC. 2002. Direct transfer of p65 into T lymphocytes from systemic lupus erythematosus patients leads to increased levels of interleukin-2 promoter activity. *Clin Immunol* **103**(2): 145–153.

Huang YF, Wang W, Han JY, Wu XW, Zhang ST, Liu CJ, Hu QG, Xiong P, Hamvas RM, Wood N, Gong FL, Bittles AH. 2003. Increased frequency of the mannose-binding lectin LX haplotype in Chinese systemic lupus erythematosus patients. *Eur J Immunogenet* **30**(2): 121–124.

Hudson LL, Rocca K, Song YW, Pandey JP. 2002. CTLA-4 gene polymorphisms in systemic lupus erythematosus: a highly significant association with a determinant in the promoter region. *Hum Genet* **111**(4-5): 452–455.

Huggins ML, Huang FP, Xu D, Lindop G, Stott DI. 1997. Modulation of the autoimmune response in lupus mice by oral administration of attenuated *Salmonella typhimurium* expressing the IL-2 and TGF-beta genes. *Ann NY Acad Sci* **815**: 499–502.

Huggins ML, Huang FP, Xu D, Lindop G, Stott DI. 1999. Modulation of autoimmune disease in the MRL-lpr/lpr mouse by IL-2 and TGF-beta1 gene therapy using attenuated *Salmonella typhimurium* as gene carrier. *Lupus* **8**(1): 29–38.

Jawaheer D, Seldin MF, Amos CI, Chen WV, Shigeta R, Monteiro J, Kern M, Criswell LA, Albani S, Nelson JL, Clegg DO, Pope R, Schroeder HW, Jr, Bridges SL, Jr, Pisetsky DS, Ward R, Kastner DL, Wilder RL, Pincus T, Callahan LF, Flemming D, Wener MH, Gregersen PK. 2001. A genome-wide screen in multiplex rheumatoid arthritis families suggests genetic overlap with other autoimmune diseases. *Am J Hum Genet* **68**(4): 927–936.

Jury EC, Kabouridis PS, Flores-Borja F, Mageed RA, Isenberg DA. 2004. Altered lipid raft-associated signaling and ganglioside expression in T lymphocytes from patients with systemic lupus erythematosus. *J Clin Invest* **113**(8): 1176–1187.

Kammer GM, Perl A, Richardson BC, Tsokos GC. 2002. Abnormal T cell signal transduction in systemic lupus erythematosus. *Arthrit Rheum* **46**(5): 1139–1154.

Koshy M, Berger D, Crow MK. 1996. Increased expression of CD40 ligand on systemic lupus erythematosus lymphocytes. *J Clin Invest* **98**(3): 826–837.

Krishnan S, Farber DL, Tsokos GC. 2003. T cell rewiring in differentiation and disease. *J Immunol* **171**(7): 3325–3331.

Krishnan S, Nambiar MP, Warke VG, Fisher CU, Mitchell J, Delaney N, Tsokos GC. 2004. Alterations in lipid raft composition and dynamics contribute to abnormal T cell responses in systemic lupus erythematosus. *J Immunol* **172**(12): 7821–7831.

Kyttaris VC, Juang YT, Tsokos GC. 2004. Gene therapy in systemic lupus erythematosus. *Lupus* **13**(5): 353–358.

Lamartina S, Silvi L, Roscilli G, Casimiro D, Simon AJ, Davies ME, Shiver JW, Rinaudo CD, Zampaglione I, Fattori E, Colloca S, Gonzalez PO, Laufer R, Bujard H, Cortese R, Ciliberto G, Toniatti C. 2003. Construction of an rtTA2(s)–m2/tts(kid)-based transcription regulatory switch that displays no basal activity, good inducibility, and high responsiveness to doxycycline in mice and non-human primates. *Mol Ther* **7**(2): 271–280.

Latta-Mahieu M, Rolland M, Caillet C, Wang M, Kennel P, Mahfouz I, Loquet I, Dedieu JF, Mahfoudi A, Trannoy E, Thuillier V. 2002. Gene transfer of a chimeric *trans*-activator is immunogenic and results in short-lived transgene expression. *Hum Gene Ther* **13**(13): 1611–1620.

Lawson BR, Prud'homme GJ, Chang Y, Gardner HA, Kuan J, Kono DH, Theofilopoulos AN. 2000. Treatment of murine lupus with cDNA encoding IFN-gammaR/Fc. *J Clin Invest* **106**(2): 207–215.

Liu MF, Wang CR, Lin LC, Wu CR. 2001. CTLA-4 gene polymorphism in promoter and exon-1 regions in Chinese patients with systemic lupus erythematosus. *Lupus* **10**(9): 647–649.

Liu Z, Xu X, Hsu, HC, Tousson A, Yang PA, Wu Q, Liu C, Yu S, Zhang HG, Mountz JD. 2003. CII-DC-AdTRAIL cell gene therapy inhibits infiltration of CII-reactive T cells and CII-induced arthritis. *J Clin Invest* **112**(9): 1332–1341.

Macanovic M, Sinicropi D, Shak S, Baughman S, Thiru S, Lachmann PJ. 1996. The treatment of systemic lupus erythematosus (SLE) in NZB/W F_1 hybrid mice; studies with recombinant murine DNase and with dexamethasone. *Clin Exp Immunol* **106**(2): 243–252.

Magnusson V, Johanneson B, Lima G, Odeberg J, Alarcon-Segovia D, Alarcon-Riquelme, ME. 2004. Both risk alleles for FcgammaRIIA and FcgammaRIIIA are susceptibility factors for SLE: a unifying hypothesis. *Genes Immun* **5**(2): 130–137.

Manderson AP, Botto M, Walport MJ. 2004. The role of complement in the development of systemic lupus erythematosus. *Annu Rev Immunol* **22**: 431–456.

Melo ME, Qian J, El Amine M, Agarwal RK, Soukhareva N, Kang Y, Scott DW. 2002. Gene transfer of Ig-fusion proteins into B cells prevents and treats autoimmune diseases. *J Immunol* **168**(9): 4788–4795.

Miagkov AV, Varley AW, Munford RS, Makarov SS. 2002. Endogenous regulation of a therapeutic transgene restores homeostasis in arthritic joints. *J Clin Invest* **109**(9): 1223–1229.

Nambiar MP, Fisher CU, Kumar A, Tsokos CG, Warke VG, Tsokos GC. 2003a. Forced expression of the Fc receptor gamma-chain renders human T cells hyperresponsive to TCR/CD3 stimulation. *J Immunol* **170**(6): 2871–2876.

Nambiar MP, Fisher CU, Warke VG, Krishnan S, Mitchell JP, Delaney N, Tsokos GC. 2003b. Reconstitution of deficient T cell receptor zeta chain restores T cell signaling and augments T cell receptor/CD3-induced interleukin-2 production in patients with systemic lupus erythematosus. *Arthrit Rheum* **48**(7): 1948–1955.

Napirei M, Karsunky H, Zevnik B, Stephan H, Mannherz HG, Moroy T. 2000. Features of systemic lupus erythematosus in *DNase1*-deficient mice. *Nat Genet* **25**(2): 177–181.

Nath SK, Kilpatrick J, Harley JB. 2004. Genetics of human systemic lupus erythematosus: the emerging picture. *Curr Opin Immunol* **16**(6): 794–800.

Neve R, Kissonerghis M, Clark J, Feldmann M, Chernajovsky Y. 1996. Expression of an efficient small molecular weight tumour necrosis factor/lymphotoxin antagonist. *Cytokine* **8**(5): 365–370.

Pan CQ, Dodge TH, Baker DL, Prince WS, Sinicropi DV, Lazarus RA. 1998. Improved potency of hyperactive and actin-resistant human *DNaseI* variants for treatment of cystic fibrosis and systemic lupus erythematosus. *J Biol Chem* **273**(29): 18374–18381.

Parsa A, Peden E, Lum RF, Seligman VA, Olson JL, Li H, Seldin MF, Criswell LA. 2002. Association of angiotensin-converting enzyme polymorphisms with systemic lupus erythematosus and nephritis: analysis of 644 SLE families. *Genes Immun* **3**(suppl. 1): S42– S46.

Prokunina L, Castillejo-Lopez C, Oberg F, Gunnarsson I, Berg L, Magnusson V, Brookes AJ, Tentler D, Kristjansdottir H, Grondal G, Bolstad AI, Svenungsson E, Lundberg I, Sturfelt G, Jonssen A, Truedsson L, Lima G, Alcocer-Varela J, Jonsson R, Gyllensten UB, Harley JB, Alarcon-Segovia D, Steinsson K, Alarcon-Riquelme ME. 2002. A regulatory polymorphism in PDCD1 is associated with susceptibility to systemic lupus erythematosus in humans. *Nat Genet* **32**(4): 666–669.

Prokunina L, Gunnarsson I, Sturfelt G, Truedsson L, Seligman VA, Olson JL, Seldin MF, Criswell LA, Alarcon-Riquelme ME. 2004. The systemic lupus erythematosus-associated PDCD1 polymorphism PD1.3A in lupus nephritis. *Arthrit Rheum* **50**(1): 327–328.

Raz E, Dudler J, Lotz M, Baird SM, Berry CC, Eisenberg RA, Carson DA. 1995. Modulation of disease activity in murine systemic lupus erythematosus by cytokine gene delivery. *Lupus* **4**(4): 286–292.

Russell AI, Cunninghame Graham DS, Shepherd C, Roberton CA, Whittaker J, Meeks J, Powell RJ, Isenberg DA, Walport MJ, Vyse TJ. 2004. Polymorphism at the C-reactive protein locus influences gene expression and predisposes to systemic lupus erythematosus. *Hum Mol Genet* **13**(1): 137–147.

Salucci V, Scarito A, Aurisicchio L, Lamartina S, Nicolaus G, Giampaoli S, Gonzalez-Paz O, Toniatti C, Bujard H, Hillen W, Ciliberto G, Palombo F. 2002. Tight control of gene expression by a helper-dependent adenovirus vector carrying the rtTA2(s)-M2 tetracycline transactivator and repressor system. *Gene Ther* **9**(21): 1415–1421.

Shen N, Tsao BP. 2004. Current advances in the human lupus genetics. *Curr Rheumatol Rep* **6**(5): 391–398.

Sherer Y, Gorstein A, Fritzler MJ, Shoenfeld Y. 2004. Autoantibody explosion in systemic lupus erythematosus: more than 100 different antibodies found in SLE patients. *Sem Arthrit Rheum* **34**(2): 501–537.

Slingsby JH, Norsworthy P, Pearce G, Vaishnaw AK, Issler H, Morley BJ, Walport MJ. 1996. Homozygous hereditary C1q deficiency and systemic lupus erythematosus. A new family and the molecular basis of C1q deficiency in three families. *Arthrit Rheum* **39**(4): 663–670.

Solomou EE, Juang YT, Gourley MF, Kammer GM, Tsokos GC. 2001. Molecular basis of deficient IL-2 production in T cells from patients with systemic lupus erythematosus. *J Immunol* **166**(6): 4216–4222.

Szalai AJ, McCrory MA, Cooper GS, Wu J, Kimberly RP. 2002. Association between baseline levels of C-reactive protein (CRP) and a dinucleotide repeat polymorphism in the intron of the CRP gene. *Genes Immun* **3**(1): 14–19.

Takiguchi M, Murakami M, Nakagawa I, Saito I, Hashimoto A, Uede T. 2000. CTLA4IgG gene delivery prevents autoantibody production and lupus nephritis in MRL/lpr mice. *Life Sci* **66**(11): 991–1001.

Tenbrock K, Juang YT, Gourley MF, Nambiar MP, Tsokos GC. 2002. Antisense cyclic adenosine 5′-monophosphate response element modulator up-regulates IL-2 in T cells from patients with systemic lupus erythematosus. *J Immunol* **169**(8): 4147–4152.

Tsao BP. 2003. The genetics of human systemic lupus erythematosus. *Trends Immunol* **24**(11): 595–602.

Tsao BP. 2004. Update on human systemic lupus erythematosus genetics. *Curr Opin Rheumatol* **16**(5): 513–521.

Tsokos GC, Nepom GT. 2000. Gene therapy in the treatment of autoimmune diseases. *J Clin Invest* **106**(2): 181–183.

Tsokos GC, Kovacs B, Sfikakis PP, Theocharis S, Vogelgesang S, Via CS. 1996. Defective antigen-presenting cell function in patients with systemic lupus erythematosus. *Arthrit Rheum* **39**(4): 600–609.

Wandstrat A, Wakeland E. 2001. The genetics of complex autoimmune diseases: non-MHC susceptibility genes. *Nat Immunol* **2**(9): 802–809.

Yasutomo K, Horiuchi T, Kagami S, Tsukamoto H, Hashimura C, Urushihara M, Kuroda Y. 2001. Mutation of DNASE1 in people with systemic lupus erythematosus. *Nat Genet* **28**(4): 313–314.

Yu SF, von Ruden T, Kantoff PW, Garber C, Seiberg M, Ruther U, Anderson WF, Wagner EF, Gilboa E. 1986. Self-inactivating retroviral vectors designed for transfer of whole genes into mammalian cells. *Proc Natl Acad Sci USA* **83**(10): 3194–3198.

12

Immunogenetics of Experimentally Induced Arthritis

Tibor T. Glant and **Vyacheslav A. Adarichev**

Abstract

Astounding advances in biotechnology and swift accumulation of detailed information about genome sequences of different species, human and mouse in particular, lead us to believe that the time when we will know all genes controlling complex autoimmune diseases is very close. Indeed, our understanding of molecular mechanisms involved in immune regulation and immunogenetics has greatly improved during the last few years. However, the most recent progress rather underlines the distance between our skills showing the relevance of *known* or *artificial alteration* in a single gene for the formation of morbid phenotype, and our abilities to find out how *natural variations* in the genome might affect disease development. Thus, finding cryptic disease-susceptibility genes controlling *complex autoimmune diseases*, such as rheumatoid arthritis, is still an extremely challenging mission. Multiple-gene effects, phenocopies, heterogeneity of the human population, and the influence of poorly understood environmental factors make this task even more difficult. Experimentally induced autoimmune arthritis in murine strains, basically mimicking human disease, is based on a clean genetic background and a fully controllable environmental milieu. Through the dissection of the genetic background of the murine proteoglycan-induced arthritis we can bridge a gap to the understanding of human disease.

12.1 Rheumatoid Arthritis in Humans and Murine Proteoglycan-Induced Arthritis: Introduction

Rheumatoid arthritis (RA) is a complex autoimmune disease of unknown aetiology, affecting over 1 per cent of the human population (Hochberg, 1990; Hochberg and Spector, 1990; Lawrence *et al.*, 1989; Linos *et al.*, 1980). No other autoimmune disease appears in so many different clinical forms, characterized by such heterogeneous and diverse clinical symptoms and laboratory tests. Genome screening for

Immunogenomics and Human Disease Edited by András Falus

arthritis susceptibility genes in families originating from the UK, USA, Japan and Europe revealed a high degree of genetic heterogeneity of RA (Cornelis *et al.*, 1998; Jawaheer *et al.*, 2001, 2003; MacKay *et al.*, 2002; Shiozawa *et al.*, 1998). Each genome scan discovered about a dozen chromosome loci linked with disease susceptibility, albeit the overlap between any two pedigrees was only one or two loci. Even the major histocompatibility complex (MHC), which is a major genetic risk factor for RA in the general human population, was not linked to disease susceptibility in Japanese RA patients (Shiozawa *et al.*, 1998). Thus, human genetic studies support the idea that multiple, mostly unknown, aetiological factors are involved in RA pathogenesis.

There are many experimental animal models which attempt to mimic the multiple clinical symptoms of this disease. While none of the animal models of inflammatory arthritis (adjuvant-induced arthritis, antigen-induced arthritis, collagen-induced arthritis or proteoglycan aggrecan-induced arthritis) are identical to RA, these experimental models have provided important advances in understanding possible mechanisms for the human disease, and for the development of therapeutic agents. Based upon the clinical, immunological and genetic components, the most appropriate animal models for RA seem to be those applying cartilage matrix components such as type II collagen (Courtenay *et al.*, 1980; Remmers *et al.*, 1996), link protein (Zhang *et al.*, 1998), cartilage glycoprotein-39 (Verheijden *et al.*, 1997) or proteoglycan (PG) aggrecan (Glant *et al.*, 1987; Mikecz, Glant and Poole 1987) for systemic immunization of genetically susceptible mice or rats.

Proteoglycan-induced murine model for rheumatoid arthritis: the phenotype

Systemic immunization of genetically susceptible mouse strains (BALB/c and C3H) with human cartilage PG aggrecan, leads to the development of progressive polyarthritis (Glant *et al.*, 1987; Glant, Finnegan and Mikecz, 2003; Mikecz Glant and Poole, 1987). Murine PG-induced arthritis (PGIA) bears many similarities to RA, as indicated by clinical assessments, radiographic analyses, scintigraphic bone scans, biochemical tests and histopathology studies of diarthrodial joints. Joint inflammation starts as polyarticular synovitis in small peripheral joints after the second to fourth intraperitoneal injection of antigen emulsified in various adjuvants. During the early phase, mononuclear cells accumulate in the synovium, which induce massive proliferation of synovial lining cells and fibroblasts. Finally, repeated inflammatory episodes result in deterioration of articular cartilage, erosion of the subchondral bone and deformities of the peripheral joints and joint stiffness.

PGIA is a recessive polygenic disease: mode of inheritance

Polygenic autoimmune diseases, like RA or the corresponding animal models, are fundamentally influenced both by genetic components and environmental factors.

The proportion of genetic factors in the overall disease population variance varies greatly from 12 to 50 per cent depending on the family background. However, concordance of RA in identical twins is significantly higher than in dizygotic twins or nontwin siblings (Wanstrat and Wakeland, 2001), implying that RA is strongly influenced by genetic factors. These factors include both MHC and non-MHC genes, whereas the effect of the MHC is the most prominent. The MHC may affect predisposition to autoimmune diseases by several mechanisms, e.g. by shaping the T cell receptor (TCR) repertoire and thymic deletion, and peptide selection and presentation (Theofilopoulos, 1995). In context with non-MHC genes, a large number of factors (immunoglobulins and Fcγ receptors, chemokines/cytokines and their receptors, expression of adhesion molecules and their receptors, TCR and apoptosis proteins, hormonal components, etc.) may be involved in the control of disease development and severity. On the other hand, several genes inside arbitrary small chromosome regions can control disease susceptibility or other clinical or immunological traits, and this gene cluster might be accountable for a quantitative trait locus (QTL), which is considered as a single locus by conventional linkage analysis. Thus, composite QTL structure makes genetic analysis more difficult than originally expected. The familial segregation of different autoimmune diseases (Bergsteinsdottir *et al.*, 2000; Etzioni *et al.*, 1999; Theofilopoulos, 1995; Vyse and Todd, 1996; Wanstrat and Wakeland, 2001) and the prevalence of an autoimmune disease in a single family, however, do suggest that the same gene(s) and/or gene cluster(s) control susceptibility to different autoimmune diseases. Nevertheless, the identification of a controlling autoimmune disease gene cluster in genetically heterogeneous human population is a difficult task. Animal models may help us identify candidate genes linked to single or multiple autoimmune diseases, and then project these genes onto the syntenic region of the human genome. The completion of the human and mouse genome programs (Gregory *et al.*, 2002; Lander *et al.*, 2001; Venter *et al.*, 2001; Waterston *et al.*, 2002) and the high homology between the human and mouse genomes make this approach feasible.

Two strains, BALB/c and C3H/HeJCr, were found to be highly susceptible to PGIA among more than a dozen strains tested (Mikecz, Glant and Poole, 1987; Otto *et al.*, 2000; Figure 12.1). Interestingly, these two mouse strains are close relatives, since the Bagg mice were an early progenitor for both strains. BALB/c mice were bred from Bagg mice through the selection, and C3H was developed by Strong in 1920 from a cross of Bagg albino females with DBA male (Beck *et al.*, 2000). Despite century-long breeding, C3H/HeJCr and BALB/c strains are genetically still very similar and share high disease susceptibility (Figure 12.1). Logically, a high incidence of the disease has been observed also in F_1 and F_2 crosses between these two highly susceptible strains. Otherwise, when two genetically distant strains were crossed, PGIA was inherited as a recessive trait, since F_1 hybrids were all resistant to the disease. It is interesting, that all studied F_2 hybrid mice were always only partially susceptible to PGIA, and the incidence was lower, under 55 per cent in all crosses, even if they were based on PGIA-susceptible strains. Thus, crosses of almost 100 per cent susceptible BALB/c with C3H/HeJCr mice resulted in lower susceptibility

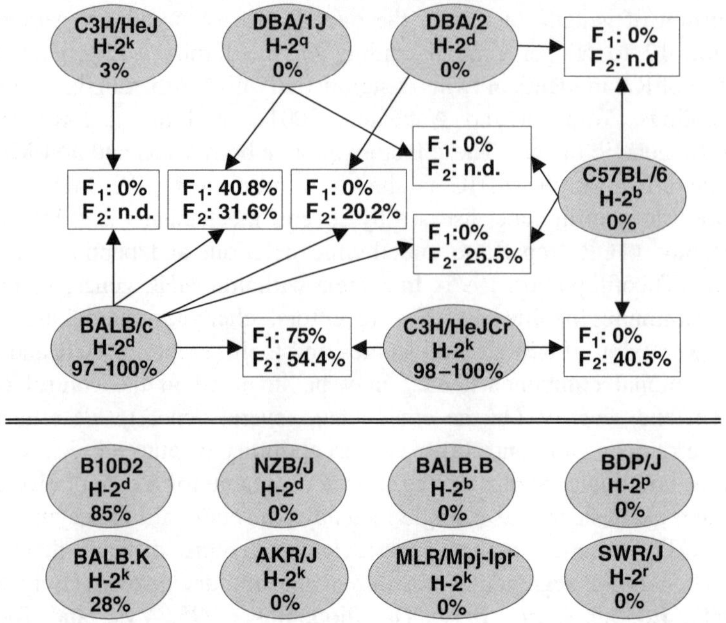

Figure 12.1 Incidence of the disease in association with the H-2 haplotypes of parental strains (ovals) and in their F_1 and F_2 hybrids (rectangular boxes).

in F_1 (75 per cent), then even lower disease incidence in F_2 hybrids (54.4 per cent; Figure 12.1). This pattern of inheritance is explained by the genetic model involving two major recessive arthritis genes, each originating from different parents. Disease incidence in other F_2 hybrids, which were based on the cross of highly susceptible mice with resistant counter-strains, varied greatly, from 20.2 to 40.8 per cent. Thus, a simple monogenic model should be further adjusted to the more complex genetic model involving many genes with variable penetrances, which are in play in parental strains, and in their F_1 and F_2 hybrids.

12.2 Genetic Linkage Analysis of PGIA

RA is a complex disease since (i) it clearly does not follow simple Mendelian pattern of inheritance for a single gene, but rather shows familial aggregation, (ii) RA affects approximately 1 per cent of the human population (Hochberg, 1990; Hochberg and Spector, 1990; Lawrence *et al.*, 1989; Linos *et al.*, 1980), which is far more frequent than any single-gene disorders and, accordingly, (iii) RA has a multifactorial genetic background without a clear major single-gene effect. As a consequence, the same morbid phenotype of a complex disease might be developed within different combinations of genes (*phenocopies*). Interestingly, the animal genetic model of RA, murine PGIA, fulfilled many criteria of a complex phenotype, since murine autoimmune arthritis is also polygenic and is a result of complex interactions between intermediate immunological phenotypes and multiple genes.

To dissect complex phenotypes of PGIA, we divided a major clinical trait (arthritis) into several independent subtraits that include both quantitative and qualitative traits. The main score of an autoimmune model is the qualitative (binary) trait, i.e. susceptibility to disease, which has only two values: '1' for positive arthritis-susceptible mice or '0' for nonarthritic resistant animals. All other components of the clinical score of arthritis are quantitative: onset time (length of the latent period prior to initial clinical symptoms), severity (i.e. magnitude of inflammation) and progression of the disease (increase or decrease of severity over time). Severity of the disease was defined and calculated in only arthritic mice, although disease onset and susceptibility scores were determined both for healthy and diseased mice of the F_2 population. Therefore, we separated the qualitative (susceptibility) and quantitative (other clinical) traits to characterize different features of the disease, and introduced additional scores based on the primary arthritis clinical score. The severity score of arthritis is the same as the arthritis score, but it is applied only to positive mice. Since the primary arthritis clinical score did not reflect the onset time, i.e. how early the arthritis develops following immunizations, we introduced the onset score, which reflected how quickly the animals developed arthritis. Thus, we built a set of independent clinical scores, each related to different sides of the disease physiology with minimum correlation between them, and we mapped genes responsible for these independent scores (Adarichev *et al.*, 2002a, 2003a,b).

In addition to the clinical scores, we determined a wide range of pathophysiological and immunological markers related to either B or T cell functions and serum levels of pro- and anti-inflammatory cytokines. The pathophysiological marker set included the levels of (i) antibodies (total IgG, and Th2-supported IgG1 and Th1-supported IgG2a isotypes) produced against the immunizing (human) and mouse (self) cartilage PGs, (ii) antigen-specific T-cell responses (IL-2, IFN-γ and IL-4 production and T-cell proliferation) measured *in vitro*, and (iii) a panel of cytokines (IL-1, IL-6, IL-10, IL-12, IFN-γ and TNFα) detected in sera of all arthritic and nonarthritic (resistant) mice. We determined whether any of these parameters could be used as phenotypic markers associated with disease susceptibility or severity, and then these immunological traits were mapped in the mouse genome. Eventually, we sought to determine whether QTLs controlling clinical symptoms, or any of the pathophysiological parameters, could be co-localized with QTLs previously identified (Jirholt *et al.*, 1998; McIndoe *et al.*, 1999; Otto *et al.*, 1999, 2000), and, if so, how these traits modify the clinical features of the original model.

Table 12.1 summarizes linkage analysis data based on five F_2 crosses accumulating a total population of more than 2200 mice. Up to 250 genomic markers were used for genome-wide linkage analysis in different murine crosses, thus creating the narrow net with average distance between markers as low as 9 cM. In these genome-wide screening studies, we identified a total of 29 *Pgia* loci in the mouse genome (Adarichev *et al.*, 2002a, 2003a,b; Otto *et al.*, 1999, 2000). Table 12.1 summarizes chromosome loci linked not only to clinical but also to immunological traits, which

Table 12.1 Summary of chromosome loci linked with clinical and immunological traits of PGIA in different murine crosses

			Clinical traits					Immunological traits	
Chromosome	QTL	cM	BALB/c × DBA/2 H-2^d/H-2^d	BALB/c × DBA/1 H-2^d/H-2^q	BALB/c × C57BL/6 H-2^d/H-2^b	BALB/c × C3H/HeJCr H-2^d/H-2^k	C3H/HeJCr × C57BL/6 H-2^k/H-2^b	BALB/c × DBA/2 H-2^d/H-2^d	BALB/c × DBA/1 H-2^d/H-2^q
1	*Pgia1*	33–106	Sev[3.1]					h/aG1[3.7]; Ab[4.3]	
2	*Pgia2*	85–112					Bin[2.5]	Ab[11.1]	
3	***Pgia26***	34–80	**Ons[4.9]**						
4	***Pgia13***	0–20	Sev[3.1]		Bin[3.1]	Bin[4.0]	**Bin[4.4]**	TNFα[5.1]	
	Pgia27	35–45	Sev[3.1]						
5	*Pgia18*	0–10	Sev[3.3]	Ons[3.6], Bin[2.7]				G1/G2a [2.8]	
	Pgia16	28–44					Bin[3.0]		CTLL[4.1], IL-1[3.1], IL-4[3.5] sCD44 [4.5]
6	*Pgia19*	19–37		Ons[3.1], Bin[3.4]					
7	***Pgia3***	23–44	**Bin[5.3]**			Bin[3.4]			
	Pgia21		**Bin[4.8],** Ons[2.5]						
8	*Pgia4*	1–22	Sev[3.9]			Bin[3.9]	Bin[2.1]	IL-2 [3.5], Prolif [3.5], Ab[18.9]	IL-1[4.6]
	Pgia22	47–59							CTLL[2.8], IL-1[3.1]
9	***Pgia5***	30–50	**Bin[10.1]**	**Ons[4.4], Bin[4.6]**			**Bin[4.4]**	G1/G2a [3.4]	
10	*Pgia6*	30–40	Ons[3.5]			Bin[3.4]	Bin[2.2]	Ab[7.1]	CTLL[3.2], IL-4[4.3]
11	*Pgia28*	0–28						G1/G2a[2.8]	
	Pgia7	42–70	Bin[3.1]		Bin[3.1]		Bin[2.6]	Ab[16.0]	

12	***Pgia14***	42–60			Bin[3.0]		**Bin[4.6]**	G1/G2a[9.1]	
13	*Pgia15*	0–47	Bin[4.1]				Bin[4.0]	IL-1[3.1]	
14	*Pgia29*	40–54	Sev[2.8]					Prolif [3.5]	
15	***Pgia8***	0–20	Sev[3.5], **Bin[8.4]**			Bin[4.0]			
	Pgia9	30–55	Sev[3.0], **Bin[6.0]**			Bin[4.2]			CTLL[4.2]
16	***Pgia10***	40–72	**Bin[7.1]**				Bin[3.9]	IL-1 [3.5], Ab[8.0]	hAb[2.8], IL-4[3.2]
17	***Pgia17***	15–25		**Ons[4.6], Bin[9.2]**	**Bin[4.8]**		**Bin[9.2]**	hG1[21.3]	CTLL[8.5]
	Pgia20	37–48		Ons[2.8], Bin[3.4]					
18	*Pgia11*	25–50		Sev[2.8], Bin[3.2]			Bin[2.4]	Ab[11.9]	IL-4[2.4]
19	***Pgia12***	11–33	**Sev[4.4], Bin[8.0]**				Bin[2.3]		IL-6[3.1]
	Pgia23								
20	*Pgia24*	19–30							hAb[5.6], IFN[3.2], G1/G2A[4.8]
	Pgia25	67–72	Ons[3.0]					h/aG1[3.4], IL-1[3.2], IL-6[3.5]	IL-1[7.2], sCD44[8.5]

Summary is based on previously published genome screenings (Adarichev *et al.*, 2001, 2002a,b, 2003b; Otto *et al.*, 1999, 2000). LOD score values are in brackets. The following arthritis clinical subphenotypes are listed: Bin, binary susceptibility trait; Sev, severity of the disease; Ons, disease onset index. Chromosomes and loci with LOD score greater than 4.3 threshold for genome-wide significant linkage are in bold and underlined. Immunological traits: h/aG1, ratio of antihuman PG to antimouse PG antibodies both of IgG1 isotype; G1/G2a, ratio of serum IgG1 to IgG2a isotype antibodies to human PG; Prolif, T cell proliferation in response to PG-stimulation; CTLL, production of IL-2 by PG-stimulated T cells *in vitro* measured by CTLL assay; sCD44, shed soluble CD44. IL-1, IL-2, IL-4, IL-6 and TNFα are serum cytokines.

can be considered as pre-clinical intermediate phenotypes of the disease. Several QTLs were consistently found in the same position in the mouse genome. For example, *Pgia13* on chromosome 4 was identified in four different crosses and controlled disease susceptibility, severity, and TNFα production (Table 12.1). *Pgia4* (chr 8), *Pgia5* (chr 9), *Pgia7* (chr11) and *Pgia17* (chr 17) were repeatedly found in three independent intercrosses and overlapped with a number of loci controlling different immune parameters. Other clinical QTLs were cross-specific.

Most of the clinical and immunological QTLs share the same chromosome position. Overlap of the chromosome loci indicates common entities of the disease mechanism, and underscores our hypothesis that certain chromosome regions linked to arthritis susceptibility, onset or severity should carry genes which control pathophysiological events, cellular functions or regulation of the immune homeostasis. In other words, altered cellular functions linked to certain QTLs may be strong indicators for selecting a QTL for more detailed studies and testing congenic lines. The difference between a clinical and a pathophysiological (immunological) trait is that, while a clinical trait is a 'forever' fact (animal did or did not develop arthritis), a serum marker (e.g. cytokine) or altered T cell function may be detectable only at certain time points of the disease, and thus the linkage with the disease may remain unexplored.

Effect of MHC on clinical and immunological traits of arthritis

The MHC is known as a major genetic risk factor of autoimmune diseases both in humans and corresponding animal models (Bali *et al.*, 1999; MacKay *et al.*, 2002; Marrack, Kappler and Kotzin, 2001; Merriman *et al.*, 2001; Shiozawa *et al.*, 1998; Wanstrat and Wakeland, 2001; Weiss, 1993), i.e. it controls a significant portion of the arthritis population variance. Whenever we wanted to exclude effects of the MHC (H-2 complex in mice) on linkage analysis and map non-MHC genes, we used murine crosses, wherein both parent strains carry similar H-2 haplotype. Thus, crosses of two types were generated to map MHC-related and non-MHC arthritis susceptibility genes: F_2 hybrids with H-2 complex that was matched between parental strains and H-2 complex-unmatched F_2 hybrid mice. In H-2 haplotype-unmatched F_2 populations, mouse MHC region on chromosome 17 was the strongest in genome scans with a logarithm of odds ratio (LOD) score of 4.8 in BALB/c × C57BL/6 cross, and even greater values of 9.2 in C3H/HeJ × C57BL/6 and BALB/c × DBA/1 crosses (Table 12.1).

Strong control of the MHC toward immunological traits was observed both in MHC-matched and -unmatched crosses. However, when strains with different haplotypes were intercrossed ($H\text{-}2^d/H\text{-}2^q$ combination in BALB/c × DBA/1 cross), the MHC controlled mostly T cell responses (LOD score 8.5 for CTLL and LOD score 3.2 for IL-4 production) and only suggestive linkage was observed for the control of antibodies (LOD score 2.8). In contrast, in MHC-matched BALB/c × DBA/2 cross ($H\text{-}2^d/H\text{-}2^d$), even if the linkage with the MHC in PGIA was undetectable, the

MHC had a very strong influence on serum antibodies (LOD score 21.3), but no control of T cell responses (Table 12.1). The MHC locus strongly controlled not only susceptibility to the disease, but also onset of PGIA in BALB/c × DBA/1 intercross (LOD score 4.6).

The major clinical parameter for arthritis (disease susceptibility) was calculated as the ratio of diseased mice in the F_2 population, i.e. the binary trait (healthy or arthritic) for each animal. However, PGIA can also be described by other important parameters, such as disease onset and severity of inflammation. Which clinical features of the arthritis are linked to the MHC? To address this question, we analysed linkage of H-2 loci with the disease in the total population of arthritic and healthy mice and in subpopulations of arthritic only mice. In BALB/c × DBA/1 crosses, the progenitor strains carried $H\text{-}2^d$ and $H\text{-}2^q$ haplotypes, thus making the H-2 complex the major QTL for both arthritis (LOD score 6.5) and antigen-induced T cell proliferation (LOD score 6.5; Figure 12.2). However, when analysis was performed for arthritic mice only, neither trait showed significant linkage with H-2 loci on chromosome 17; LOD scores were even lower than suggestive level 2.8. Thus, the murine MHC controls only susceptibility to PGIA, but not the magnitude of joint inflammation (Adarichev *et al.*, 2002a). In other words, while the murine MHC has a permissive role upon PGIA susceptibility and plays a critical role in the disease initiation, the MHC alone is insufficient for disease development and progression. As a result, severity of the disease is under the control of other loci/genes (Becker *et al.*, 1998; Bergsteinsdottir *et al.*, 2000; Gregersen, 1999; Lin *et al.*, 1998; Seldin *et al.*, 1999; Theofilopoulos 1995). The best illustration for the important role of non-MHC loci is the existence of murine strains with different PGIA susceptibilities but the same H-2 haplotype. PG-immunization of BALB/c, DBA/2 and NZB/J inbred strains and B10D2 mice that all carry $H\text{-}2^d$ haplotype varied from complete resistance in

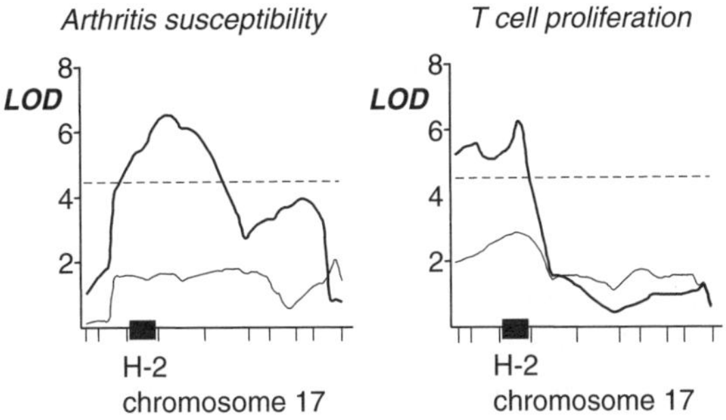

Figure 12.2 Arthritis susceptibility and PG antigen-specific T cell proliferation were mapped to chromosome 17 near the H-2 complex when linkage was calculated for the entire arthritic and nonarthritic F_2 population (solid thick line) of the BALB/c × DBA/1 cross ($H\text{-}2^d/H\text{-}2^q$). Linkage of the disease with H-2 loci was nonsignificant, when calculated for arthritic F_2 mice only (thin line). Small vertical tics indicate the positions of genetic markers on the *x*-axis, and the mouse H-2 locus is shown by a solid box. Dashed horizontal lines indicate the significant levels of LOD scores at a 4.3 cut-off value using interval mapping with free regression model.

DBA/2 and NZB/J strains to almost 100 per cent susceptibility in BALB/c and B10D2 mice. Another example is the high PGIA susceptibility in C3H/HeJCr strain and a virtual resistance to arthritis in C3H/HeJ mice (Glant *et al.*, 2001), although both strains carry H-2^{k} haplotype. Known *de novo* mutation in the toll-like receptor 4 (*Tlr4*) on chromosome 4 occurred only in the C3H/HeJ colony maintained by the Jackson Laboratory between 1960 and 1967 (Coutinho, Möller and Gronowicz, 1975; Watson and Riblet, 1974). However, this mutation is not associated with arthritis susceptibility, and we are compelled to believe that additional mutation(s) might occur in the C3H/HeJ mice, recalling susceptibility to arthritis.

Non-MHC loci

As stated above, PGIA has a complex genetic background, derived from a natural variation of multiple genes of unknown localization in the mouse genome. While linkage of RA to the MHC has been repeatedly confirmed, several possible linkages outside the MHC were also identified in human pedigrees (Bali *et al.*, 1999; Cornelis *et al.*, 1998; Jawaheer *et al.*, 2001; MacKay *et al.*, 2002; Ollier and Worthington, 1997; Weiss, 1993). The critical role of non-MHC-associated genes in arthritis has been further confirmed in numerous animal models such as collagen-, streptococcal cell wall-, pristane- and oil-induced arthritis in rats. While these studies have helped define the genetic relatedness and similarities of the available autoimmune models, none have successfully narrowed the genetic interval of any QTL to the point where positional cloning could be employed. Thus, the central problem of the identification of the disease-responsible genes remains. The use of different genetic crosses, increasingly dense genetic maps and congenic strains, as well as the completion of the human and mouse genome sequencing projects, will probably make this goal a reality.

Phenocopies are a common feature of PGIA: genetic heterogeneity of the disease

Genome scans of several crosses identified the different genetic backgrounds for qualitative and quantitative traits of PGIA. The major locus controlling the magnitude of joint inflammation (severity) was found on chromosome 19 (*Pgia12*), lesser QTL on chromosome 8 (*Pgia4*) and suggestive loci on chromosomes 1, 4, 5, 14, 15 and 18. The major binary QTL was mapped to chromosome 17 (H-2 complex, *Pgia17*); other significant QTLs were mapped on chromosomes 7, 9, 12, 13, 15, 16 and 19. The overlap between binary QTLs and severity QTLs were found on chromosomes 15, 18 and 19. Loci controlling onset of the disease showed no overlap with binary QTLs (Table 12.1). Probably, the most striking example of the variation for locus penetrance towards different arthritis sub-phenotypes is the H-2 complex. In the example mentioned above (Figure 12.2), we showed that the murine MHC (H-2^{d}/H-2^{q} combination) in BALB/c × DBA/1 cross has a crucial role in disease initiation and

H-2 loci-controlled PGIA onset and susceptibility. Nevertheless, these loci were not controlling inflammation itself, and other non-MHC genomic loci were linked to disease severity.

A relatively low disease incidence in F_1 hybrid mice derived from the cross of two highly susceptible murine strains, such as in BALB/c × C3H/HeJCr and BALB/c × DBA/1 crosses (Figure 12.1), clearly indicates the genetic heterogeneity of the disease. This pattern of disease inheritance can be described with two major genes using a simplified genetic model. Gene *A* is a cause of arthritis in the homozygous state (*AA*) of the susceptible strain A, and has no or little effect on arthritis in strain B (*aa*). In contrast, the genetic background of strain B makes another gene, *B*, mainly responsible for the disease development (*BB*), but gene *A* is silent in this strain (*aa*). Thus, when two arthritis genes are in a mixed heterozygous state in F_1 hybrids (*Aa*, *Bb*), the arthritis susceptibility is even lower than in parental mice.

Occurrence of phenocopies, when clinically identical phenotypes are induced by the variations in different genes, is a common feature of complex autoimmune diseases, including RA and PGIA in particular. Genome screening of various crosses supports this statement, since QTLs controlling PGIA were only partially overlapping when different crosses were genotyped (Table 12.1). For example, in C57BL/6 × C3H/HeJCr cross the major arthritis-susceptibility genes (LOD score > 4.3) were mapped on chromosomes 4, 9, 12 and 17; these loci were mapped on chromosomes 4 and 15 in BALB/c × C3H/HeJCr cross; only one major QTL was found on chromosome 17 in BALB/c × C57BL/6 cross; and two major binary loci were mapped on chromosomes 9 and 17 in BALB/c × DBA/1 hybrid mice. Susceptibility loci of PGIA in an MHC-matched BALB/c × DBA/2 intercross were localized on chromosomes 7, 9, 15, 16 and 19. Interestingly, many QTLs were found repeatedly in several crosses, although, the same chromosome segment might be associated with different clinical or immunological phenotypes of the disease depending on the genetic background. For example, *Pgia13* was found consistently in four crosses: BALB/c × DBA/2, BALB/c × C57BL/6, BALB/c × C3H/HeJCr and C3H/HeJCr × C57BL/6. This locus was the strongest in C3H/HeJCr × C57BL/6 and control disease susceptibility; however, in BALB/c × DBA/2 cross it was linked to PGIA severity and serum concentrations of TNFα (LOD score 5.1). The strongest QTLs, which were greater than LOD score significant threshold 4.3, are underlined in Table 12.1. The majority of the most reliable QTLs were found repeatedly in several crosses, further validating their significance.

Synteny mapping of autoimmune arthritis genes

Biochemical and cell signalling pathways are quite similar in primates and rodents, i.e. between humans and mice. This similarity is widely used to study mechanisms of action of newly designed compounds, drug toxicity and pharmacokinetics. It is believed that the pathogenesis of complex disease is also very similar in different species. Pursuing the ultimate goal of mapping disease susceptibility genes in the human genome, it is important to build a bridge between mouse and human genomes and map human genes using homologous disease loci in mice. The syntenic mapping for PGIA-susceptibility chromosome regions shows significant overlap with linkage analysis in humans (Figure 12.3). As sequencing projects both for human and mouse genomes are completed, the synteny maps can be constructed with great accuracy (Mapview at National Center for Biotechnology Information; www.ncbi.nlm.nih.gov/mapview/). Approximately half of the PGIA QTLs are in the regions of the human genome which have been identified as RA susceptibility genes or loci (Figure 12.3). The strongest murine locus controlling PGIA onset, *Pgia26*, overlaps with three human loci, which were found in pedigrees of various origin (Jawaheer *et al.*, 2001; John *et al.*, 2001; MacKay *et al.*, 2002). The homologous locus on human chromosome 1 (94–155 Mbp) carries lymphoid nonreceptor-type 22 protein–tyrosine phosphatase genes (*PTPN22*). A missense single-nucleotide polymorphism (SNP) inside *PTPN22* has been associated with susceptibility to rheumatoid arthritis (Begovich *et al.*, 2004). *Pgia13, Pgia3, Pgia14* and *Pgia12* were overlapped with at least one human RA-susceptibility QTL. The *Pgia14* locus on the telomeric part of chromosome 12 and corresponding human locus on chromosome 14 is mapped near the cluster of the immunoglobulin heavy chains. This locus controls PGIA in C3H/HeJCr × C57BL/6 cross probably through the regulation of the ratio of IgG1 to IgG2a antibody isotypes (LOD score 9.1). Thus, chromosome segments responsible for alteration in IgG expression or isotype switch are plausible candidates inside this locus in both mouse and human pedigrees. The *Pgia8* locus on mouse chromosome 15 is homologous to human chromosome 5, the region which carries the progressive ankylosis protein encoding gene (*ANKH*). The mutation inside this gene is a genetic basis for craniometaphyseal dysplasia and crystal deposition arthropathy (OMIM 118600, OMIM 605145), the monogenic disorder associated with tissue calcification and arthritis (Ho, Johnson and Kingsley, 2000; Reichenberger *et al.*, 2001). The *Pgia10* on the centromeric part of mouse chromosome 16 is homologous to human chromosomes 3 and 21. The corresponding human chromosome 3 region carries RA QTL (Cornelis *et al.*, 1998), and the human locus on chromosome 21 contains runt-related transcription factor 1 gene (*RUNX1*). The functional SNP is located in the binding site of *RUNX1* with *SLC22A4*, and strongly affects the expression level of the

▶

Figure 12.3 Murine autoimmune arthritis loci (*Pgia*) in the mouse genome and homologous chromosome segments in humans. Mouse chromosomes (c3–c19) are presented as G-banded ideograms with QTL-marked regions. Homologous chromosome segments in the human genome (h1–h21) are shown as open boxes. The size of human and mouse chromosome segments are shown using the same scale. Whether a homologous human QTL was identified is shown under the appropriate chromosome interval. *RA1–RA7* are QTLs from human genetic linkage studies (Cornelis *et al.*, 1998; Jawaheer *et al.*, 2001; John *et al.*, 2001; MacKay *et al.*, 2002; Shiozawa *et al.*, 1998; Wise *et al.*, 2000), and OMIM 8600 and OMIM 1605145. *PTPN22*, protein–tyrosine phosphatase nonreceptor-type 22; *RUNX1*, runt-related transcription factor 1; *HLA-DRB1*, major histocompatibility complex, class II, DR beta-1; *NFKBIL1*, nuclear factor of kappa light chain gene enhancer in B cells inhibitor-like 1.

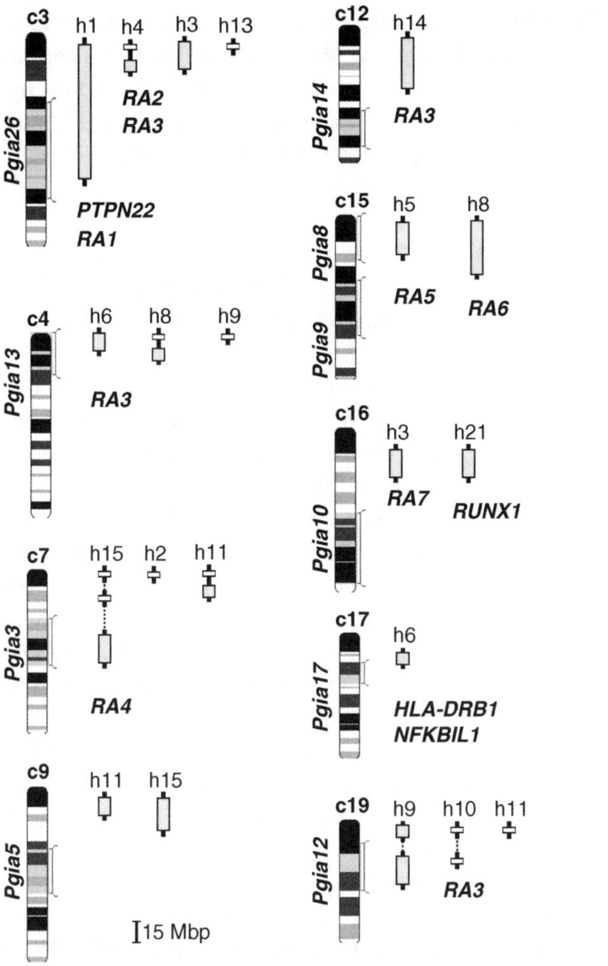

Mouse	Human	Mouse	Human
c3: *Pgia26:*	h1: 94-155	**c12: *Pgia14:***	h14; 58-97
	h3: 150-169		
	h4: 118-121, 152-161	**c15: *Pgia8:***	h5: 15-43
	h13: 36-39		
		c15: *Pgia9:*	h8: 107-145
c4: *Pgia13:*	h6: 88-100	**c16: *Pgia10*:**	h3: 86-107
	h8: 56-61, 88-97		h21: 14-36
	h9: 0.02-0.03		
		c17: *Pgia17:*	h6: 33-39
c7: *Pgia3*:	h2: 71.2-71.3		
	h11: 0.02-0.03, 84-89	**c19: *Pgia12:***	h9: 0-6, 68-78
	h15: 0.02-0.03, 29-30, 78-100		h10: 53.7-54.2
c9: *Pgia5*:	h11: 108-118		
	h15: 50-76		

latter gene. This SNP was strongly associated with RA in a case–control association test (Tokuhiro *et al.*, 2003). Finally, the strongest PGIA QTL on mouse chromosome 17 is homologous to the MHC locus on human chromosome 6p21 (Figure 12.3), which is strongly associated with the RA *HLA-DRB1* allele and newly discovered *NFKBIL1* gene (Okamoto *et al.*, 2003; Ota *et al.*, 2001).

12.3 Transcriptome Picture of the Disease: Gene Expression During the Initiation and Progression of Joint Inflammation

Genetic linkage studies in segregating human and mouse populations is an unique approach, which does not need any *a priori* knowledge about disease causative genes. Linkage analysis simply relies on the association between a morbid phenotype and DNA variations scattered throughout the genome. However, even the most extensive genome-wide screening and linkage analysis could never be able to identify the complete set of disease-responsible/inducing genes without physiological, biochemical and cell signalling studies. The completion of the human and mouse genome sequencing programs and subsequent annotation of previously unidentified genes have opened a new epoch in biology and biomedical sciences. The genetic information has greatly facilitated the discovery of novel disease-related genes and mapping of signature genes for early diagnosis. More specifically, poly- or oligonucleotide arrays have been applied in both human and experimentally induced disease conditions to determine characteristic expression patterns of signature genes, which then can be recognized as candidate genes of genetic loci (QTLs) of interest.

In an inflammatory disease such as RA, the gene expression profile is extremely complex due to the diversity of cell types involved in the pathology, and the polygenic character of the autoimmune disease (Barnes *et al.*, 2004; Grant *et al.*, 2002; Heller *et al.*, 1997; Watanabe *et al.*, 2002; Zanders *et al.*, 2000). The overall picture of molecular interactions in an inflamed joint, deduced from gene expression studies in both RA and its corresponding animal models, involves an enormous number of proteins participating in immunity, inflammation, apoptosis, proliferation, cellular transformation and cell differentiation, and other processes (Grant *et al.*, 2002; Maas *et al.*, 2002; Sweeney and Firestein, 2004; Thornton *et al.*, 2002; Watanabe *et al.*, 2002; Zanders *et al.*, 2000). The genetic heterogeneity of the human population, however, is a serious obstacle for correct data interpretation in gene expression studies. Again, RA animal models, such as PGIA, can facilitate the interpretation of genome-wide gene expression by providing genetic and clinical homogeneity, and an opportunity to monitor the onset and progression of the disease.

To monitor early inflammatory reactions in mouse paw and joint, we adoptively transferred the disease (PGIA) into syngeneic BALB/c^{SCID} mice that lack functional T and B cells. Severe combined immunodeficient (SCID) mice carry a natural mutation which prevents the V(D)J recombination in B and T lymphocytes, resulting in a failure to generate functional immunoglobulins and T cell receptors (Carlow *et al.*, 1995; Schaible *et al.*, 1989). Consequently, adoptively transferred arthritis in BALB/c^{SCID} mice is an ideal model where activated lymphocytes of arthritic donor BALB/c mice migrate and interact with an intact innate immunity environment in the joints of BALB/c^{SCID} mice (Bárdos *et al.*, 2002; Hanyecz *et al.*, 2003). Induction of arthritis in BALB/c^{SCID} mice was a multi-step process. First, donor BALB/c mice were immunized with cartilage PG to induce arthritis. Second, spleen cells from acutely arthritic donor mice were stimulated *in vitro* with cartilage PG, and live lymphocytes were isolated on a Lympholyte-M density gradient (Bárdos *et al.*, 2002; Hanyecz *et al.*, 2003). Third, these PG antigen-stimulated donor lymphocytes were injected into BALB/c^{SCID} mice. The gene expression profiles in normal (naive), pre-arthritic (not yet inflamed) and arthritic joints (either acutely or chronically inflamed) of the recipient BALB/c^{SCID} mice were determined using oligonucleotide DNA microarray technology (Affymetrix).

The mounting cascade of gene expression activities was observed during the initiation and progression phases of joint inflammation [Figure 12.4(B)]. At the initiation phase, when no clinical symptoms of inflammation were yet detected, only 37 genes were up- or downregulated (Adarichev *et al.*, 2005). However, differential expression of 277 genes was observed at the acute phase, and chronic inflammation was characterized by differential activity of 418 genes. Interestingly, most early arthritis signature genes (27 of 37) remained up- or downregulated in inflamed joints. A different set of genes was also involved in acute inflammation. At the chronic phase, less than half of acute phase-specific genes (127 of 277) were differentially expressed, and another half was chronic inflammation-specific. Eventually, a very limited number of transcripts ($n = 15$) remained up- or downregulated in all three phases of arthritis.

To identify genes whose expression is specific for each phase of arthritis, and combine transcripts by the pattern of their expression through all disease phases, we applied a hierarchical clustering technique (Eisen *et al.*, 1998). Genes, specific for pair-wise comparisons (pre-arthritic vs naive, acute vs naive, and chronic vs naive) were combined in one single file excluding redundant genes, and the merged set included 507 genes. Hierarchical clustering was performed for all experimental conditions studied, and four major gene clusters were identified, each with a distinct expression pattern [Figure 12.4(A), clusters I, II, III and IV]. Using further classification analysis with gene ontology terms, to examine the functions of genes inside each cluster, we identified genes encoding proteins whose biological functions were the most relevant to arthritis development and progression. Cluster I contained genes with major functions in collagen turnover and tissue repair, and the expression of these genes reached a peak in chronically inflamed joints. Cluster II was the largest cluster including approximately half of all phase-specific genes [Figure 12.4(A)]. Cluster II included genes with roles in immune, inflammatory and stress responses, extracellular matrix formation, cell growth and receptor activity. Expression of cluster II genes reached a peak at the acute phase of joint inflammation. Transcription of genes in cluster III and cluster IV gradually decreased during disease progression [Figure 12.4(A)]. These genes were mostly related to cytoskeleton remodelling,

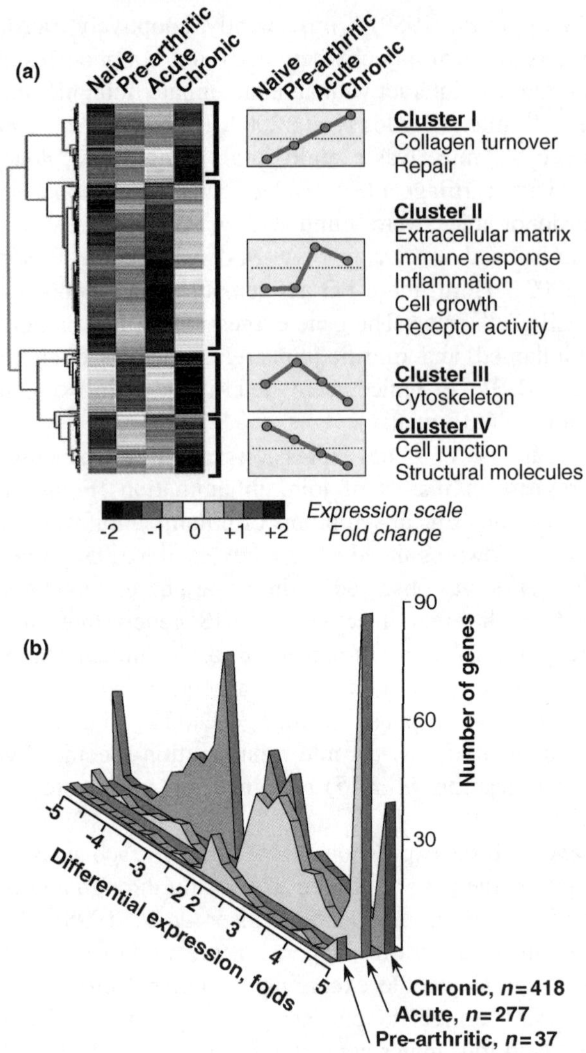

Figure 12.4 (a) Gene expression profiling in arthritis and hierarchical clustering of the expression profiles for signature genes at different phases of autoimmune arthritis. Hierarchical clustering was performed for genes whose expression significantly differed when joints (paws) of naive mice were compared with those in the pre-arthritic, acute or chronic phases of arthritis. The total number of genes ($n = 507$) is less than the sum of phase-specific genes because of the partial overlap. Rows represent individual genes; columns represent individual expression values for each gene at the indicated phase of arthritis. The major biological activities, specific for each cluster, were examined using functional clustering. This analysis yielded four different expression patterns (clusters I–IV). Upregulated genes (positive fold change) are shown in red, and downregulated genes (negative values of fold change) are shown in blue. (b) Histogram of the fold change values for genes differentially expressed at different phases of the disease. The size of the peaks indicates the number of genes within a given range of expression (fold increase and decrease). Spikes at ±5-fold-expression change combine all genes with differential expression level greater than ±5-fold. The number of genes differentially expressed in each phase is indicated. (A colour reproduction of this figure can be viewed in the colour plate section)

formation of cell junctions and production of structural molecules such as desmin, beta 3 laminin, envoplakin, dystonin and others. Genes associated with early arthritis (Adarichev *et al.*, 2005) were found in clusters III and IV, further underlining the importance of cell adhesion and cytoskeleton remodelling during the initiation phase of arthritis.

Arthritis signature genes in pre-inflamed joints

Paws of naive BALB/c^{SCID} mice and yet noninflamed pre-arthritic paws were clinically normal with no sign of inflammation, and comparison of these two experimental conditions identified a relatively small number of differentially expressed genes. Only 37 of the 36 000 screened genes were differentially expressed over a 2-fold threshold, of which 11 genes were over a ±3-fold threshold, and seven genes changed more than ±5-fold [Figure 12.4(B)]. The seven genes with the most significant change in expression levels encoded chemokine C-C motif receptor 5 (*Ccr5*), chemokine C-X-C motif ligand 1 (*Cxcl1*), IFN-γ-inducible protein (*Ifi47*), membrane-spanning 4-domains subfamily A member 6C (*Ms4a6c*), TNFα-induced protein 6 (*Tnfip6*), T cell receptor beta variable 13 (*Tcrbv13*) and Terf1-interacting nuclear factor 2 (*Tinf2*) (Adarichev *et al.*, 2005). While upregulation of Tcrbv13, *Tgtp* and IFN-induced genes may indicate the appearance of antigen-specific T cells in the synovium, significant upregulation of *Tnfip6* suggests the activation of an anti-inflammatory cascade (Glant *et al.*, 2002). Thus, gene expression related to pro- and anti-inflammatory events can be detected even before clinical evidence of the migration of inflammatory leukocytes into the joints.

Activated T cells must be present in the peripheral blood of recipient BALB/c^{SCID} mice after the transfer, but donor lymphocytes can be detected in joints as early as 3–5 days after the second transfer (Bárdos *et al.*, 2002). In earlier studies (Mikecz and Glant, 1994), and in control experiments (data not shown), using fluorescein- or isotope-labelled donor lymphocytes, only very few cells were found in joints during the first week of transfer, and a second cell transfer was needed to induce a significant influx of lymphocytes into the joints to challenge inflammation. In this study, we detected overexpression of a T cell specific GTPase (*Tgtp*) and T cell receptor beta (*Tcrbv13*) in yet noninflamed pre-arthritic paws of recipient BALB/c^{SCID} mice as early as 3–5 days after the second injection, indicating the presence of donor BALB/c lymphocytes. Thus, the initiation and development of arthritis in adoptively transferred PGIA must depend on the cooperation between adaptive immunity cells (represented by donor BALB/c lymphocytes) and cells of innate immunity (represented by nonlymphoid cells in the recipient BALB/c^{SCID} mice). Analysis of cellular and tissue specificity of gene expression, using public gene expression databases [the Bioinformatic Harvester (http://harvester.embl.de/), the Gene Expression Database (www.informatics.jax.org/mgihome/GXD) and the Gene Expression Omnibus (www.ncbi.nlm.nih.gov/geo)], indicated that genes encoding CD48 (*Cd48*), membrane-spanning 4A6B and 4A6C (*Ms4a6b, Ms4a6c*), EGF-like receptor-like protein 1 (*Emr1*) and interferon-induced 47 kDa protein (*Ifi47*) were most probably originating from donor lymphoid cells, while other early arthritis genes were related to the activation of the innate immune system represented by macrophages, dendritic cells and cells of myeloid lineage of recipient BALB/c^{SCID} mice.

Cooperation between transcriptomics and genomics to find autoimmune arthritis genes

We have analysed the genetic basis of the murine autoimmune arthritis and defined a list of the most important chromosome segments linked to PGIA in different crosses (Table 12.1). Alternatively, we know genes whose expression precedes or accompanies joint inflammation upon adoptive disease transfer [Figure 12.4(A)]; Adarichev *et al.*, 2005]. Since the primary (causative) disease gene(s) should contain genetic variation and this variation should carry biological function affecting either RNA concentration, protein-specific activity or protein–protein interaction, we analysed the overlap between differentially expressed genes and disease-linked chromosome loci (Figure 12.5). The small portion of transcriptome (37 out of 289 pre-arthritic- and acute phase-specific genes, 12.8 per cent) was localized inside arthritis-controlling chromosome loci. Most positional candidate genes were overexpressed during acute arthritis phase, and only CD53 leukocyte surface antigen was upregulated in both pre-inflamed and inflamed joints (Table 12.2).

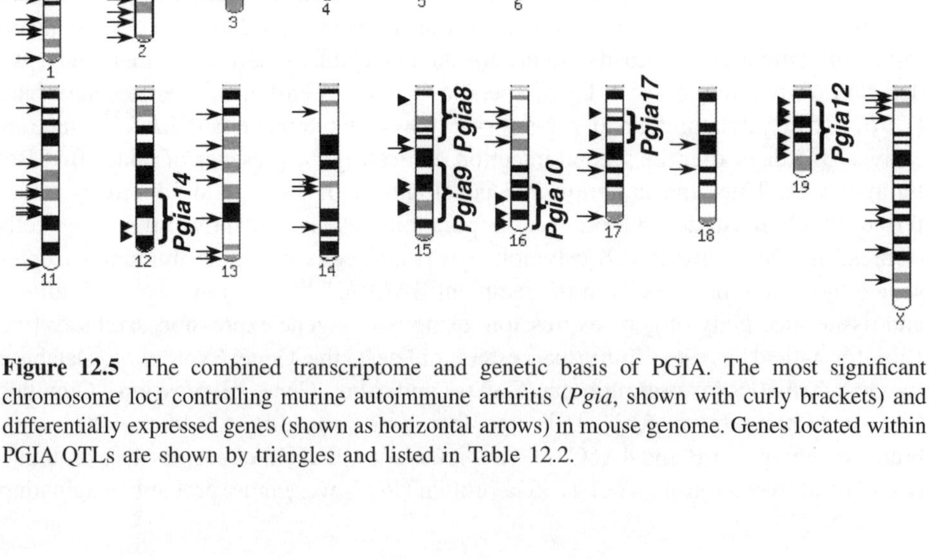

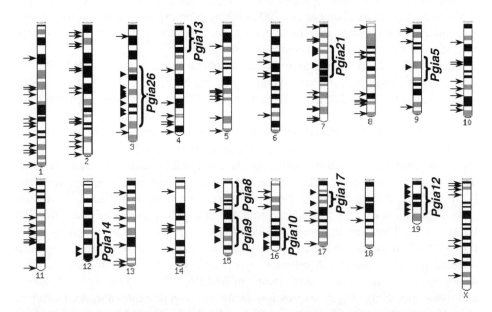

Figure 12.5 The combined transcriptome and genetic basis of PGIA. The most significant chromosome loci controlling murine autoimmune arthritis (*Pgia*, shown with curly brackets) and differentially expressed genes (shown as horizontal arrows) in mouse genome. Genes located within PGIA QTLs are shown by triangles and listed in Table 12.2.

Table 12.2 PGIA locus-specific genes differentially expressed prior to and during acute inflammation

Gene	Description	Phase	Fold change
	Chromosome 3		
Ptx3	Pentraxin-related gene	AA	15.3
Sfrp2	Secreted frizzled-related sequence protein 2	AA	3.1
S100a8	Migration inhibitory factor-related protein 8	AA	25.2
Sprr2a	Small proline-rich protein 2A	AA	3.2
Rptn	Repetin	AA	−13.3
Ctsk	Cathepsin K	AA	2.9
CD53	Leukocyte surface antigen CD53	PA, AA	2.2, 6.2
Vcam1	Vascular cell adhesion molecule 1	AA	5.4
Ddah1	NG, NG-dimethylarginine dimethylaminohydrolase 1	AA	4.2
	Chromosome 7		
Gmfg	Glia maturation factor gamma	AA	2.7
Chsy1	Carbohydrate (chondroitin) synthase 1	AA	2.4
Dkkl1	Soggy-1	AA	3.5
Saa3	Serum amyloid A-3	AA	34.1
Csrp3	Cysteine and glycine-rich protein 3	AA	−3.2
	Chromosome 9		
Pstpip1	Proline-serine-threonine phosphatase-interacting protein 1	AA	2.5
	Chromosome 12		
Gpr25	G-protein-coupled receptor 25	AA	5.9
Serpina3n	Clade A	AA	4.5
	Chromosome 15		
Dab2	Disabled homolog 2	AA	2.5
Fyb	FYN-T-binding protein	AA	2.8
Ncf4	Neutrophil cytosole factor 4	AA	4.8
Kdelr3	KDEL endoplasmic reticulum protein retention receptor 3	AA	3.5
	Chromosome 16		
Samsn1	SH3 domain and nuclear localization signals protein 1	AA	31.2
Adamts1	Disintegrin and metalloproteinase with thrombospondin motifs 1	AA	3.3
Ifnar2	Interferon-alpha/beta receptor 2	AA	3.0
Runx1	**Runt related transcription factor 1**	**AA**	**2.8**
	Chromosome 17		
Has1	Hyaluronan synthase 1	AA	3.1
Kifc1	Angiopoietin-like 4	AA	5.1
	Chromosome 19		
Ms4a4c	Membrane-spanning 4-domains subfamily A member 4C	AA	5.9
Mpeg1	Macrophage expressed gene 1	AA	2.9
Pcsk5	Proprotein convertase subtilisin/kexin type 5 precursor	AA	27.3
Ankrd1	Ankyrin-like repeat protein	AA	13.1
Entpd1	Ectonucleoside triphosphate diphosphohydrolase 1	AA	6.0

Column 'Gene' refers to short gene abbreviation. The list of overlapping genes also includes five RIKEN cDNA clones with uncharacterized functions; these transcripts were not included in this table. Chromosome assessment is shown for most significant and reproducible PGIA loci listed in Table 12.1. Listed locus-specific genes were differentially expressed either in acutely inflamed or in pre-arthritic joints of BALB/c^{SCID} mice with adoptively transferred arthritis. Positive values of fold change of differential expression stand for upregulated transcripts, negative values for downregulated genes. *Runx1* transcription factor is in bold because it was consistently associated with RA, overexpressed in acute inflammation in this murine model and located inside *Pgia10* QTL on chromosome 16 (Figure 12.5).

Human studies revealed several links to RA chromosome segments (Cornelis *et al.*, 1998; Jawaheer *et al.*, 2001, 2003; MacKay *et al.*, 2002; Shiozawa *et al.*, 1998). Syntenic mapping of the corresponding homologous regions in the mouse genome is a reliable and informative approach, owing recent progress in mouse and human genomics. Figure 12.3 combines data from the three gene-mapping approaches based on linkage studies of the human families, genetic linkage analysis of several crosses with PGIA and gene expression at early phases of autoimmune arthritis. Human chromosome loci (*RA1–RA7*) were not only overlapping with *Pgia* loci, but these homologous human–mouse loci in many cases contained genes which were over-expressed during joint inflammation (Table 12.2). Thus, murine *Pgia26* on chromosome 3 contains cluster of innate immunity-related genes, such as pentraxin-related gene (*Ptx3*), migration inhibitory factor 8 (*S100a8*), cathepsin K (*Ctsk*), CD53 and vascular cell adhesion molecule 1 (*Vcam1*).

Runt-related transcription factor 1 (*RUNX1*) gene was found to be consistently involved in autoimmune arthritis in human patients downregulating the expression of organic cationic transporter *SLC22A4* (Tokuhiro *et al.*, 2003). In the present PGIA murine model with adoptive disease transfer, we found similar pattern of expression for both genes: transcript concentration of *Runx1* was almost three times higher and *Slc22a4* gene was 2.5 times lower in inflamed joints of mice with adoptively transferred PGIA (Adarichev *et al.*, 2005). Remarkably, *Runx1* is localized in *Pgia10* QTL on chromosome 16, the locus which controls arthritis susceptibility and autoantibody production (Otto *et al.*, 1999). On the other hand, *Runx1* is located in a hot spot of recombination and often found as a fusion protein in acute myeloid leukaemias (OMIM 151385). Moreover, point mutations in this gene are associated with familial myeloid malignancies and platelet disorders (OMIM 601399), which is consistent with the major function of *Runx1* as an active repressor in CD4-negative/CD8-negative thymocytes (Taniuchi *et al.*, 2002).

12.4 Conclusions

Genetic and gene expression studies of PGIA showed a complex portrait of the disease, which is reliant on multiple and distinct sets of genes connected differently to the disease initiation or progression, and that gene activity depends on the genetic cross and phase of the joint inflammation. Despite the mapping of the disease susceptibility genes in human pedigrees having an obvious advantage of higher resolution due to accumulated multiple recombination events, at the same time human studies are greatly hampered by the high genetic heterogeneity of RA. Linkage analysis in animal models relies on a clean genetic background and less frequent DNA variations, but usually results in quite vague chromosome loci containing hundreds of genes, which require enormous and laborious efforts to complete detailed physical mapping and then confirm the role of the gene in autoimmune processes. The gene expression approach is an additional and powerful tool and offers a different kind of functional filter. Nevertheless, the density of genes, even in a very small (by genetic definition) human QTL, is still large enough to reveal several gene candidates in the locus, whose function might be relevant to the disease pathology. Thus, only complex and combined approaches may lead to the identification of primary genes involved in pathophysiology of autoimmune disorders.

References

Adarichev VA, Otto JM, Bárdos T, Mikecz K, Finnegan A, Glant TT. 2001. Collagen-induced and proteoglycan-induced arthritis: two murine models with diverse pathophysiologogy and genetic background. *Arthrit Rheum* **44**: S177.

Adarichev VA, Bárdos T, Christodoulou S, Phillips MT, Mikecz K, Glant TT. 2002a. Major histocompatibility complex controls susceptibility and dominant inheritance, but not the severity of the disease in mouse models of rheumatoid arthritis. *Immunogenetics* **54**: 184–192.

Adarichev VA, Bárdos T, Nesterovitch AB, Mikecz K, Finnegan A, Glant TT. 2002b. Gender effects on clinical and immunological traits of proteoglycan-induced arthritis. *Arthrit Rheum* **46**: S273.

Adarichev VA, Nesterovitch AB, Bárdos T, Biesczat D, Chandrasekaran R, Vermes C, Mikecz K, Finnegan A, Glant TT. 2003a. Sex effect on clinical and immunological quantitative trait loci in a murine model of rheumatoid arthritis. *Arthrit Rheum* **48**: 1708–1720.

Adarichev VA, Valdez JC, Bárdos T, Finnegan A, Mikecz K, Glant TT. 2003b. Combined autoimmune models of arthritis reveal shared and independent qualitative (binary) and quantitative trait loci. *J Immunol* **170**: 2283–2292.

Adarichev VA, Vermes C, Hanyecz A, Mikecz K, Bremer E, Glant TT. 2005. Gene expression profiling in murine autoimmune arthritis during the initiation and progression of joint inflammation. *Arthrit Res Ther* **7**: R196–R207.

Bali D, Gourley S, Kostyu DD, Goel N, Bruce I, Bell A, Walker DJ, Tran K, Zhu DK, Costello TJ, Amos CI, Seldin MF. 1999. Genetic analysis of multiplex rheumatoid arthritis families. *Genes Immun* **1**: 28–36.

Bárdos T, Mikecz K, Finnegan A, Zhang J, Glant TT. 2002. T and B cell recovery in arthritis adoptively transferred to SCID mice: antigen-specific activation is required for restoration of autopathogenic $CD4^+$ Th1 cells in a syngeneic system. *J Immunol* **168**: 6013–6021.

Barnes MG, Aronow BJ, Luyrink LK, Moroldo MB, Pavlidis P, Passo MH, Grom AA, Hirsch R, Giannini EH, Colbert RA, Glass DN, Thompson SD. 2004. Gene expression in juvenile arthritis and spondyloarthropathy: pro-angiogenic ELR^+ chemokine genes relate to course of arthritis. *Rheumatol (Oxford)* **43**: 973–979.

Beck JA, Lloyd S, Hafezparast M, Lennon-Pierce M, Eppig JT, Festing MFW, Fisher EMC. 2000. Geneologies of mouse inbred strains. *Nat Genet* **24**: 23–25.

Becker KG, Simon RM, Bailey-Wilson JE, Freidlin B, Biddison WE, McFarland HF, Trent JM. 1998. Clustering of non-major histocompatibility complex susceptibility candidate loci in human autoimmune diseases. *Proc Natl Acad Sci USA* **95**: 9979–9984.

Begovich AB, Carlton VE, Honigberg LA, Schrodi SJ, Chokkalingam AP, Alexander HC, Ardlie KG, Huang Q, Smith AM, Spoerke JM, Conn MT, Chang M, Chang SY, Saiki RK, Catanese JJ, Leong DU, Garcia VE, McAllister LB, Jeffery DA, Lee AT, Batliwalla F, Remmers E, Criswell LA, Seldin MF, Kastner DL, Amos CI, Sninsky JJ, Gregersen PK. 2004. A missense single-nucleotide polymorphism in a gene encoding a protein tyrosine phosphatase (PTPN22) is associated with rheumatoid arthritis. *Am J Hum Genet* **75**: 330–337.

Bergsteinsdottir K, Yang H-T, Pettersson U, Holmdahl R. 2000. Evidence for common autoimmune disease genes controlling onset, severity, and chronicity based on experimental models for multiple sclerosis and rheumatoid arthritis. *J Immunol* **164**: 1564–1568.

Carlow DA, Marth J, Clark-Lewis I, Teh HS. 1995. Isolation of a gene encoding a developmentally regulated T cell-specific protein with a guanine nucleotide triphosphate-binding motif. *J Immunol* **154**: 1724–1734.

Cornelis F, Faure S, Martinez M, Prud'homme JF, Fritz P, Dib C, Alves H, Barrera P, de Vries N, Balsa A, Pascual-Salcedo D, Maenaut K, Westhovens R, Migliorini P, Tran TH, Delaye A, Prince N, Lefevre C, Thomas G, Poirier M, Soubigou S, Alibert O, Lasbleiz S, Fouix S, Weissenbach J, for ECRAF. 1998. New susceptibility locus for rheumatoid arthritis suggested by a genome-wide linkage study. *Proc Natl Acad Sci USA* **95**: 10746–10750.

Courtenay JS, Dallman MJ, Dayan AD, Martin A, Mosedale B. 1980. Immunization against heterologous type II collagen induces arthritis in mice. *Nature* **282**: 666–668.

Coutinho A, Möller G, Gronowicz E. 1975. Genetical control of B-cell responses. IV. Inheritance of the unresponsiveness to lipopolysaccharides. *J Exp Med* **142**: 253–258.

Eisen MB, Spellman PT, Brown PO, Botstein D. 1998. Cluster analysis and display of genome-wide expression patterns. *Proc Natl Acad Sci USA* **95**: 14863–14868.

Etzioni A, Doerschuk CM, Harlan JM. 1999. Of man and mouse: leukocyte and endothelial adhesion molecule deficiencies. *Blood* **94**: 3281–3288.

Glant TT, Finnegan A, Mikecz K. 2003. Proteoglycan-induced arthritis: immune regulation, cellular mechanisms and genetics. *Cr Rev Immunol* **23**: 199–250.

Glant TT, Mikecz K, Arzoumanian A, Poole AR. 1987. Proteoglycan-induced arthritis in BALB/c mice. Clinical features and histopathology. *Arthrit Rheum* **30**: 201–212.

Glant TT, Bárdos T, Vermes C, Chandrasekaran R, Valdéz JC, Otto JM, Gerard D, Velins S, Lovász G, Zhang J, Mikecz K, Finnegan A. 2001. Variations in susceptibility to proteoglycan-induced arthritis and spondylitis among C3H substrains of mice. Evidence of genetically acquired resistance to autoimmune disease. *Arthrit Rheum* **44**: 682–692.

Glant TT, Kamath RV, Bárdos T, Gál I, Szanto S, Murad YM, Sandy JD, Mort JS, Roughley PJ, Mikecz K. 2002. Cartilage-specific constitutive expression of TSG-6 protein (product of tumor necrosis factor α-stimulated gene 6) provides a chondroprotective, but not anti-inflammatory, effect in antigen-induced arthritis. *Arthrit Rheum* **46**: 2207–2218.

Grant EP, Pickard MD, Briskin MJ, Gutierrez-Ramos JC. 2002. Gene expression profiles: creating new perspectives in arthritis research. *Arthrit Rheum* **46**: 874–884.

Gregersen PK. 1999. Genetics of rheumatoid arthritis: confronting complexity. *Arthrit Res* **1**: 37–44.

Gregory SG, Sekhon M, Schein J, Zhao S, Osoegawa K, Scott CE, Evans RS, Burridge PW, Cox TV, Fox CA, Hutton RD, Mullenger IR, Phillips KJ, Smith J, Stalker J, Threadgold GJ, Birney E, Wylie K, Chinwalla A, Wallis J, Hillier L, Carter J, Gaige T, Jaeger S, Kremitzki C, Layman D, Maas J, McGrane R, Mead K, Walker R, Jones S, Smith M, Asano J, Bosdet I, Chan S, Chittaranjan S, Chiu R, Fjell C, Fuhrmann D, Girn N, Gray C, Guin R, Hsiao L, Krzywinski M, Kutsche R, Lee SS, Mathewson C, McLeavy C, Messervier S, Ness S, Pandoh P, Prabhu AL, Saeedi P, Smailus D, Spence L, Stott J, Taylor S, Terpstra W, Tsai M, Vardy J, Wye N, Yang G, Shatsman S, Ayodeji B, Geer K, Tsegaye G, Shvartsbeyn A, Gebregeorgis E, Krol M, Russell D, Overton L, Malek JA, Holmes M, Heaney M, Shetty J, Feldblyum T, Nierman WC, Catanese JJ, Hubbard T, Waterston RH, Rogers J, De Jong PJ, Fraser CM, Marra M, McPherson JD, Bentley, DR. 2002. A physical map of the mouse genome. *Nature* **418**: 743–750.

Hanyecz A, Bárdos T, Berlo SE, Buzás E, Nesterovitch AB, Mikecz K, Glant TT. 2003. Induction of arthritis in SCID mice by T cells specific for the "shared epitope" sequence in the G3 domain of human cartilage proteoglycan. *Arthrit Rheum* **48**: 2959–2973.

Heller RA, Schena M, Chai A, Shalon D, Bedilion T, Gilmore J, Woolley DE, Davis RW. 1997. Discovery and analysis of inflammatory disease-related genes using cDNA microarrays. *Proc Natl Acad Sci USA* **94**: 2150–2155.

Ho AM, Johnson MD, Kingsley DM. 2000. Role of the mouse *ank* gene in control of tissue calcification and arthritis. *Science* **289**: 265–270.

Hochberg MC. 1990. Changes in the incidence and prevalence of rheumatoid arthritis in England and Wales, 1970–1982. *Sem Arthrit Rheum* **19**: 294–302.

Hochberg MC, Spector TD. 1990. Epidemiology of rheumatoid arthritis: update. *Epidemiol Rev* **12**: 247–252.

Jawaheer D, Seldin MF, Amos CI, Chen WV, Shigeta R, Monteiro J, Kern M, Criswell LA, Albani S, Nelson JL, Clegg DO, Pope R, Schroeder HW Jr, Bridges SL Jr, Pisetsky DS, Ward R, Kastner DL, Wilder RL, Pincus T, Callahan LF, Flemming D, Wener MH, Gregersen PK. 2001. A genome-wide screen in multiplex rheumatoid arthritis families suggests genetic overlap with other autoimmune diseases. *Am J Hum Genet* **68**: 927–936.

Jawaheer D, Seldin MF, Amos CI, Chen WV, Shigeta R, Etzel C, Damle A, Xiao X, Chen D, Lum RF, Monteiro J, Kern M, Criswell LA, Albani S, Nelson JL, Clegg DO, Pope R, Schroeder HW Jr, Bridges SL Jr, Pisetsky DS, Ward R, Kastner DL, Wilder RL, Pincus T, Callahan LF, Flemming D, Wener MH, Gregersen, PK. 2003. Screening the genome for rheumatoid arthritis susceptibility genes: A replication study and combined analysis of 512 multicase families. *Arthrit Rheum* **48**: 906–916.

Jirholt J, Cook A, Emahazion T, Sundvall M, Jansson L, Nordquist N, Pettersson U, Holmdahl R. 1998. Genetic linkage analysis of collagen-induced arthritis in the mouse. *Eur J Immunol* **28**: 3321–3328.

John S, Eyre S, Myerscough A, Barrett J, Silman A, Ollier W, Worthington J. 2001. Linkage and association analysis of candidate genes in rheumatoid arthritis. *J Rheumatol* **28**: 1752–1755.

Lander ES, Linton LM, Birren B, Nusbaum C, Zody MC, Baldwin J, Devon K, Dewar K, Doyle M, FitzHugh W, Funke R, Gage D, Harris K, Heaford A, Howland J, Kann L, Lehoczky J, LeVine R, McEwan P, McKernan K, Meldrim J, Mesirov JP, Miranda C, Morris W, Naylor J, Raymond C, Rosetti M, Santos R, Sheridan A, Sougnez C, Stange-Thomann N, Stojanovic N, Subramanian A, Wyman D, Rogers J, Sulston J, Ainscough R, Beck S, Bentley D, Burton J, Clee C, Carter N, Coulson A, Deadman R, Deloukas P, Dunham A, Dunham I, Durbin R, French L, Grafham D, Gregory S, Hubbard T, Humphray S, Hunt A, Jones M, Lloyd C, McMurray A, Matthews L, Mercer S, Milne S, Mullikin JC, Mungall A, Plumb R, Ross M, Shownkeen R, Sims S, Waterston RH, Wilson RK, Hillier LW, McPherson JD, Marra MA, Mardis ER, Fulton LA, Chinwalla AT, Pepin KH, Gish WR, Chissoe SL, Wendl MC, Delehaunty KD, Miner TL, Delehaunty A, Kramer JB, Cook LL, Fulton RS, Johnson DL, Minx PJ, Clifton SW, Hawkins T, Branscomb E, Predki P, Richardson P, Wenning S, Slezak T, Doggett N, Cheng JF, Olsen A, Lucas S, Elkin C, Uberbacher E, Frazier M, Gibbs RA, Muzny DM, Scherer SE, Bouck JB, Sodergren EJ, Worley KC, Rives CM, Gorrell JH, Metzker ML, Naylor SL, Kucherlapati RS, Nelson DL, Weinstock GM, Sakaki Y, Fujiyama A, Hattori M, Yada T, Toyoda A, Itoh T, Kawagoe C, Watanabe H, Totoki Y, Taylor T, Weissenbach J, Heilig R, Saurin W, Artiguenave F, Brottier P, Bruls T, Pelletier E, Robert C, Wincker P, Smith DR, Doucette-Stamm L, Rubenfield M, Weinstock K, Lee HM, Dubois J, Rosenthal A, Platzer M, Nyakatura G, Taudien S, Rump A, Yang H, Yu J, Wang J, Huang G, Gu J, Hood L, Rowen L, Madan A, Qin S, Davis RW, Federspiel NA, Abola AP, Proctor MJ, Myers RM, Schmutz J, Dickson M, Grimwood J, Cox DR, Olson MV, Kaul R, Raymond C, Shimizu N, Kawasaki K, Minoshima S, Evans GA, Athanasiou M, Schultz R, Roe BA, Chen F, Pan H, Ramser J, Lehrach H, Reinhardt R, McCombie WR, de la BM, Dedhia N, Blocker H, Hornischer K, Nordsiek G, Agarwala R, Aravind L, Bailey JA, Bateman A, Batzoglou S, Birney E, Bork P, Brown DG, Burge CB, Cerutti L, Chen HC, Church D, Clamp M, Copley RR, Doerks T, Eddy SR, Eichler

EE, Furey TS, Galagan J, Gilbert JG, Harmon C, Hayashizaki Y, Haussler D, Hermjakob H, Hokamp K, Jang W, Johnson LS, Jones TA, Kasif S, Kaspryzk A, Kennedy S, Kent WJ, Kitts P, Koonin EV, Korf I, Kulp D, Lancet D, Lowe TM, McLysaght A, Mikkelsen T, Moran JV, Mulder N, Pollara VJ, Ponting CP, Schuler G, Schultz J, Slater G, Smit AF, Stupka E, Szustakowski J, Thierry-Mieg D, Thierry-Mieg J, Wagner L, Wallis J, Wheeler R, Williams A, Wolf YI, Wolfe KH, Yang SP, Yeh RF, Collins F, Guyer MS, Peterson J, Felsenfeld A, Wetterstrand KA, Patrinos A, Morgan MJ, Szustakowki J. 2001. Initial sequencing and analysis of the human genome. *Nature* **409**: 860–921.

Lawrence RC, Hochberg MC, Kelsey JL, McDuffie FC, Medsger TA Jr, Felts WR, Shulman LE. 1989. Estimates of the prevalence of selected arthritic and musculoskeletal diseases in the United States. *J Rheumatol* **16**: 427–441.

Lin JP, Cash JM, Doyle SZ, Peden S, Kanik K, Amos CI, Bale SJ, Wilder RL. 1998. Familial clustering of rheumatoid arthritis with other autoimmune diseases. *Hum Genet* **103**(4): 475–482.

Linos A, Worthington JW, O'Fallon WM, Kurland LT. 1980. The epidemiology of rheumatoid arthritis in Rochester, Minnesota: a study of incidence, prevalence, and mortality. *Am J Epidemiol* **111**: 87–98.

Maas K, Chan S, Parker J, Slater A, Moore J, Olsen N, Aune TM. 2002. Cutting edge: molecular portrait of human autoimmune disease. *J Immunol* **169**: 5–9.

MacKay K, Eyre S, Myerscough A, Milicic A, Barton A, Laval S, Barrett J, Lee D, White S, John S, Brown MA, Bell J, Silman A, Ollier W, Wordsworth P, Worthington J. 2002. Whole-genome linkage analysis of rheumatoid arthritis susceptibility loci in 252 affected sibling pairs in the United Kingdom. *Arthrit Rheum* **46**: 632–639.

Marrack P, Kappler J, Kotzin BL. 2001. Autoimmune disease: why and where it occurs. *Nat Med* **7**: 899–905.

McIndoe RA, Bohlman B, Chi E, Schuster E, Lindhardt M, Hood L. 1999. Localization of non-*MHC* collagen-induced arthritis susceptibility loci in DBA/1j mice. *Proc Natl Acad Sci USA* **96**: 2210–2214.

Merriman TR, Cordell J, Eaves IA, Danoy PA, Coraddu F, Barber R, Cucca F, Broadley S, Sawcer S, Compston A, Wordsworth P, Shatford J, Laval S, Jirholt J, Holmdahl R, Theofilopoulos AN, Kono DH, Tuomilehto J, Tuomilehto-Wolf E, Buzzetti R, Marrosu MG, Undlien DE, Ronningen KS, Ionesco-Tirgoviste C, Shield JP, Pociot F, Nerup J, Jacob CO, Polychronakos C, Bain SC, Todd JA. 2001. Suggestive evidence for associatioin of human chromosome 18q12–q21 and its orthologue on rat and mouse chromosome 18 with several autoimmune diseases. *Diabetes* **50**: 184–194.

Mikecz K, Glant TT. 1994. Migration and homing of lymphocytes to lymphoid and synovial tissues in proteoglycan-induced murine arthritis. *Arthrit Rheum* **37**: 1395–1403.

Mikecz K, Glant TT, Poole AR. 1987. Immunity to cartilage proteoglycans in BALB/c mice with progressive polyarthritis and ankylosing spondylitis induced by injection of human cartilage proteoglycan. *Arthrit Rheum* **30**: 306–318.

Okamoto K, Makino S, Yoshikawa Y, Takaki A, Nagatsuka Y, Ota M, Tamiya G, Kimura A, Bahram S, Inoko H. 2003. Identification of I kappa BL as the second major histocompatibility complex-linked susceptibility locus for rheumatoid arthritis. *Am J Hum Genet* **72**: 303–312.

Ollier WER, Worthington J. 1997. New horizons in rheumatoid arthritis genetics. *J Rheumatol* **24**: 193.

Ota M, Katsuyama Y, Kimura A, Tsuchiya K, Kondo M, Naruse T, Mizuki N, Itoh K, Sasazuki T, Inoko H. 2001. A second susceptibility gene for developing rheumatoid arthritis in the human MHC is localized within a 70-kb interval telomeric of the TNF genes in the HLA class III region. *Genomics* **71**: 263–270.

Otto JM, Cs-Szabó G, Gallagher J, Velins S, Mikecz K, Buzás EI, Enders JT, Li Y, Olsen BR, Glant TT. 1999. Identification of multiple loci linked to inflammation and autoantibody production by a genome scan of a murine model of rheumatoid arthritis. *Arthrit Rheum* **42**: 2524–2531.

Otto JM, Chandrasekaran R, Vermes C, Mikecz K, Finnegan A, Rickert SE, Enders JT, Glant TT. 2000. A genome scan using a novel genetic cross identifies new susceptibility loci and traits in a mouse model of rheumatoid arthritis. *J Immunol* **165**: 5278–5286.

Reichenberger E, Tiziani V, Watanabe S, Park L, Ueki Y, Santanna C, Baur ST, Shiang R, Grange DK, Beighton P, Gardner J, Hamersma H, Sellars S, Ramesar R, Lideral AC, Sommer A, Raposo do Amaral CM, Gorlin RJ, Mulliken JB, Olsen BR. 2001. Autosomal dominant craniometaphyseal dysplasia is caused by mutations in the transmembrane protein ANK. *Am J Hum Genet* **68**: 1321–1326.

Remmers EF, Longman RE, Du Y, O'Hare A, Cannon GW, Griffiths MM, Wilder RL. 1996. A genome scan localizes five non-MHC loci controlling collagen- induced arthritis in rats. *Nat Genet* **14**: 82–85.

Schaible UE, Kramer MD, Museteanu C, Zimmer G, Mossmann H, Simon MM. 1989. The severe combined immunodeficiency (*scid*) mouse. A laboratory model for the analysis of lyme arthritis and carditis. *J Exp Med* **170**: 1427–1432.

Seldin MF, Amos CI, Ward R, Gregersen PK. 1999. The genetics revolution and the assault on rheumatoid arthritis. *Arthrit Rheum* **42**: 1071–1079.

Shiozawa S, Hayashi S, Tsukamoto Y, Goko H, Kawasaki H, Wada T, Shimizu K, Yasuda N, Kamatani N, Takasugi K, Tanaka Y, Shiozawa K, Imura S. 1998. Identification of the gene loci that predispose to rheumatoid arthritis. *Int Immunol* **10**: 1891–1895.

Sweeney SE, Firestein GS. 2004. Signal transduction in rheumatoid arthritis. *Curr Opin Rheumatol* **16**: 231–237.

Taniuchi I, Osato M, Egawa T, Sunshine MJ, Bae SC, Komori T, Ito Y, Littman DR. 2002. Differential requirements for Runx proteins in CD4 repression and epigenetic silencing during T lymphocyte development. *Cell* **111**: 621–633.

Theofilopoulos AN. 1995. The basis of autoimmunity: part II. Genetic predisposition. *Immunol Today* **16**: 150–159.

Thornton S, Sowders D, Aronow B, Witte DP, Brunner HI, Giannini EH, Hirsch R. 2002. DNA microarray analysis reveals novel gene expression profiles in collagen-induced arthritis. *Clin Immunol* **105**: 155–168.

Tokuhiro S, Yamada R, Chang X, Suzuki A, Kochi Y, Sawada T, Suzuki M, Nagasaki M, Ohtsuki M, Ono M, Furukawa H, Nagashima M, Yoshino S, Mabuchi A, Sekine A, Saito S, Takahashi A, Tsunoda T, Nakamura Y, Yamamoto K. 2003. An intronic SNP in a RUNX1 binding site of SLC22A4, encoding an organic cation transporter, is associated with rheumatoid arthritis. *Nat Genet* **35**: 341–348.

Venter JC, Adams MD, Myers EW, Li PW, Mural RJ, Sutton GG, Smith HO, Yandell M, Evans CA, Holt RA, Gocayne JD, Amanatides P, Ballew RM, Huson DH, Wortman JR, Zhang Q, Kodira CD, Zheng XH, Chen L, Skupski M, Subramanian G, Thomas D, Zhang J, Gabor Miklos GL, Nelson C, Broder S, Clark AG, Nadeau J, McKusick VA, Zinder N, Levine AJ, Roberts RJ, Simon M, Slayman C, Hunkapiller M, Bolanos R, Delcher A, Dew I, Fasulo D, Flanigan M, Florea L, Halpern A, Hannenhalli S, Kravitz S, Levy S, Mobarry C, Reinert K, Remington K, Abu-Threideh J, Beasley E, Biddick K, Bonazzi V, Brandon R, Cargill M, Chandramouliswaran I, Charlab R, Chaturvedi K, Deng Z, Di F, V, Dunn P, Eilbeck K, Evangelista C, Gabrielian AE, Gan W, Ge W, Gong F, Gu Z, Guan P, Heiman TJ, Higgins ME, Ji RR, Ke Z, Ketchum KA, Lai Z, Lei Y, Li Z, Li J, Liang Y, Lin X, Lu F, Merkulov GV, Milshina N, Moore HM, Naik AK, Narayan VA, Neelam B, Nusskern D, Rusch DB, Salzberg S, Shao W, Shue B, Sun J, Wang Z, Wang A, Wang X, Wang J, Wei M, Wides R, Xiao C, Yan C, Yao A, Ye J, Zhan M, Zhang W, Zhang H, Zhao Q, Zheng L,

Zhong F, Zhong W, Zhu S, Zhao S, Gilbert D, Baumhueter S, Spier G, Carter C, Cravchik A, Woodage T, Ali F, An H, Awe A, Baldwin D, Baden H, Barnstead M, Barrow I, Beeson K, Busam D, Carver A, Center A, Cheng ML, Curry L, Danaher S, Davenport L, Desilets R, Dietz S, Dodson K, Doup L, Ferriera S, Garg N, Gluecksmann A, Hart B, Haynes J, Haynes C, Heiner C, Hladun S, Hostin D, Houck J, Howland T, Ibegwam C, Johnson J, Kalush F, Kline L, Koduru S, Love A, Mann F, May D, McCawley S, McIntosh T, McMullen I, Moy M, Moy L, Murphy B, Nelson K, Pfannkoch C, Pratts E, Puri V, Qureshi H, Reardon M, Rodriguez R, Rogers YH, Romblad D, Ruhfel B, Scott R, Sitter C, Smallwood M, Stewart E, Strong R, Suh E, Thomas R, Tint NN, Tse S, Vech C, Wang G, Wetter J, Williams S, Williams M, Windsor S, Winn-Deen E, Wolfe K, Zaveri J, Zaveri K, Abril JF, Guigo R, Campbell MJ, Sjolander KV, Karlak B, Kejariwal A, Mi H, Lazareva B, Hatton T, Narechania A, Diemer K, Muruganujan A, Guo N, Sato S, Bafna V, Istrail S, Lippert R, Schwartz R, Walenz B, Yooseph S, Allen D, Basu A, Baxendale J, Blick L, Caminha M, Carnes-Stine J, Caulk P, Chiang YH, Coyne M, Dahlke C, Mays A, Dombroski M, Donnelly M, Ely D, Esparham S, Fosler C, Gire H, Glanowski S, Glasser K, Glodek A, Gorokhov M, Graham K, Gropman B, Harris M, Heil J, Henderson S, Hoover J, Jennings D, Jordan C, Jordan J, Kasha J, Kagan L, Kraft C, Levitsky A, Lewis M, Liu X, Lopez J, Ma D, Majoros W, McDaniel J, Murphy S, Newman M, Nguyen T, Nguyen N, Nodell M. 2001. The sequence of the human genome. *Science* **291**: 1304–1351.

Verheijden GFM, Rijnders AWM, Bos E, De Roo CJJC, van Staveren CJ, Miltenburg AMM, Meijerink JH, Elewaut D, de Keyser F, Veys E, Boots AMH. 1997. Human cartilage glycoprotein-39 as a candidate autoantigen in rheumatoid arthritis. *Arthrit Rheum* **40**: 1115–1125.

Vyse TJ, Todd JA. 1996. Genetic analysis of autoimmune disease. *Cell* **85**: 311–318.

Wanstrat A, Wakeland E. 2001. The genetics of complex autoimmune diseases: non-MHC susceptibility genes. *Nat Immunol* **2**: 802–809.

Watanabe N, Ando K, Yoshida S, Inuzuka S, Kobayashi M, Matsui N, Okamoto T. 2002. Gene expression profile analysis of rheumatoid synovial fibroblast cultures revealing the overexpression of genes responsible for tumor-like growth of rheumatoid synovium. *Biochem Biophys Res Commun* **294**: 1121–1129.

Waterston RH, Lindblad-Toh K, Birney E, Rogers J, Abril JF, Agarwal P, Agarwala R, Ainscough R, Alexandersson M, An P, Antonarakis SE, Attwood J, Baertsch R, Bailey J, Barlow K, Beck S, Berry E, Birren B, Bloom T, Bork P, Botcherby M, Bray N, Brent MR, Brown DG, Brown SD, Bult C, Burton J, Butler J, Campbell RD, Carninci P, Cawley S, Chiaromonte F, Chinwalla AT, Church DM, Clamp M, Clee C, Collins FS, Cook LL, Copley RR, Coulson A, Couronne O, Cuff J, Curwen V, Cutts T, Daly M, David R, Davies J, Delehaunty KD, Deri J, Dermitzakis ET, Dewey C, Dickens NJ, Diekhans M, Dodge S, Dubchak I, Dunn DM, Eddy SR, Elnitski L, Emes RD, Eswara P, Eyras E, Felsenfeld A, Fewell GA, Flicek P, Foley K, Frankel WN, Fulton LA, Fulton RS, Furey TS, Gage D, Gibbs RA, Glusman G, Gnerre S, Goldman N, Goodstadt L, Grafham D, Graves TA, Green ED, Gregory S, Guigo R, Guyer M, Hardison RC, Haussler D, Hayashizaki Y, Hillier LW, Hinrichs A, Hlavina W, Holzer T, Hsu F, Hua A, Hubbard T, Hunt A, Jackson I, Jaffe DB, Johnson LS, Jones M, Jones TA, Joy A, Kamal M, Karlsson EK, Karolchik D, Kasprzyk A, Kawai J, Keibler E, Kells C, Kent WJ, Kirby A, Kolbe DL, Korf I, Kucherlapati RS, Kulbokas EJ, Kulp D, Landers T, Leger JP, Leonard S, Letunic I, LeVine R, Li J, Li M, Lloyd C, Lucas S, Ma B, Maglott DR, Mardis ER, Matthews L, Mauceli E, Mayer JH, McCarthy M, McCombie WR, McLaren S, McLay K, McPherson JD, Meldrim J, Meredith B, Mesirov JP, Miller W, Miner TL, Mongin E, Montgomery KT, Morgan M, Mott R, Mullikin JC, Muzny DM, Nash WE, Nelson JO, Nhan MN, Nicol R, Ning Z, Nusbaum C, O'Connor MJ, Okazaki Y, Oliver K, Overton-Larty E, Pachter L, Parra G, Pepin KH, Peterson J, Pevzner P, Plumb R, Pohl CS, Poliakov A, Ponce TC, Ponting CP, Potter S, Quail M, Reymond A, Roe BA, Roskin KM, Rubin EM, Rust AG, Santos R, Sapojnikov V, Schultz B, Schultz J, Schwartz MS, Schwartz S, Scott C, Seaman S, Searle S, Sharpe

T, Sheridan A, Shownkeen R, Sims S, Singer JB, Slater G, Smit A, Smith DR, Spencer B, Stabenau A, Stange-Thomann N, Sugnet C, Suyama M, Tesler G, Thompson J, Torrents D, Trevaskis E, Tromp J, Ucla C, Ureta-Vidal A, Vinson JP, Von Niederhausern AC, Wade CM, Wall M, Weber RJ, Weiss RB, Wendl MC, West AP, Wetterstrand K, Wheeler R, Whelan S, Wierzbowski J, Willey D, Williams S, Wilson RK, Winter E, Worley KC, Wyman D, Yang S, Yang SP, Zdobnov EM, Zody MC, Lander ES. 2002. Initial sequencing and comparative analysis of the mouse genome. *Nature* **420**: 520–562.

Watson J, Riblet R. 1974. Genetic control of responses to bacterial lipopolysaccharides in mice. I. Evidence for a single gene that influences mitogenic and immunogenic responses to lipopolysaccharides. *J Exp Med* **140**: 1147–1161.

Weiss KM. 1993. *Genetic Variation and Human Disease: Principles and Evolutionary Approaches.* Cambridge: Cambridge University Press.

Wise CA, Bennett LB, Pascual V, Gillum JD, Bowcock AM. 2000. Localization of a gene for familial recurrent arthritis. *Arthrit Rheum* **43**: 2041–2045.

Zanders ED, Goulden MG, Kennedy TC, Kempsell KE. 2000. Analysis of immune system gene expression in small rheumatoid arthritis biopsies using a combination of subtractive hybridization and high-density cDNA arrays. *J Immunol Meth* **233**: 131–140.

Zhang Y, Guerassimov A, Leroux, J-Y, Cartman A, Webber C, Lalic R, de Miguel E, Rosenberg LC, Poole AR. 1998. Induction of arthritis in BALB/c mice by cartilage link protein. Involvement of distinct regions recognized by T- and B lymphocytes. *Am J Pathol* **153**: 1283–1291.

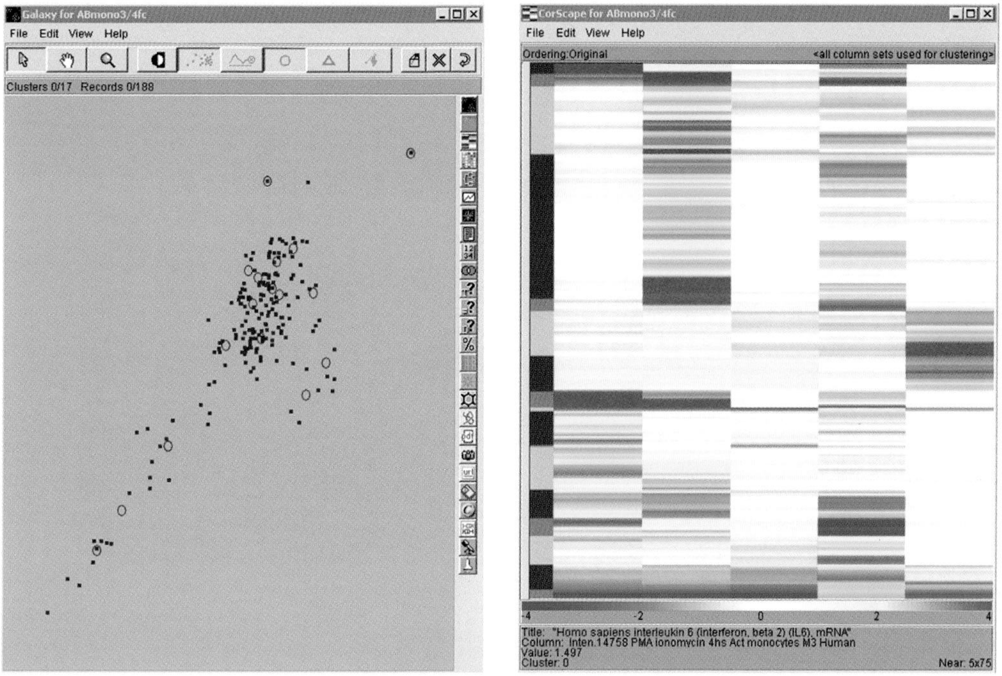

Figure 5.2

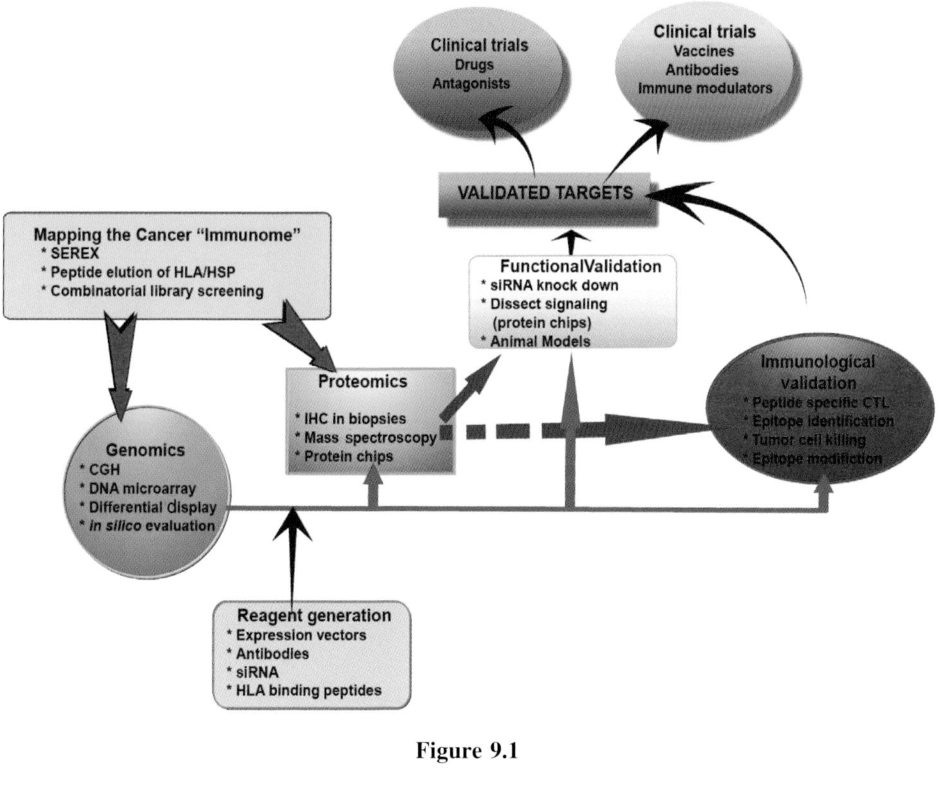

Figure 9.1

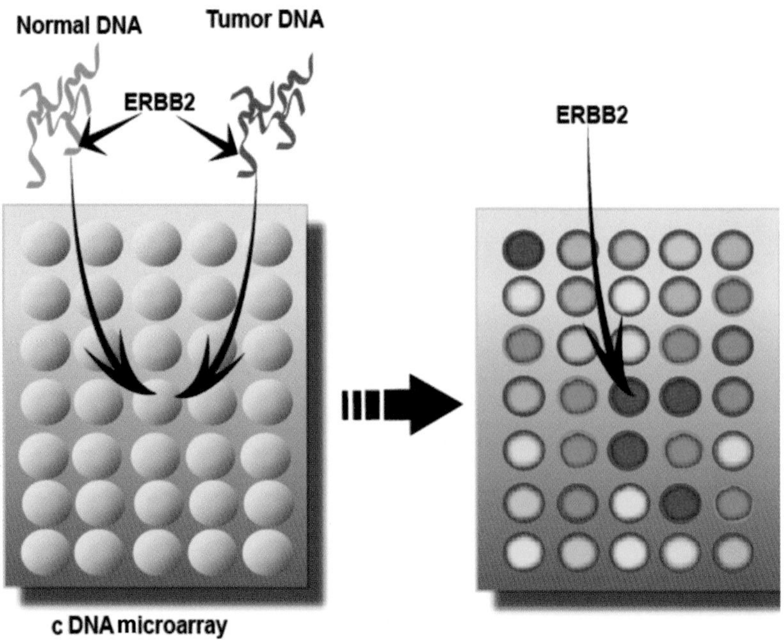

Figure 9.2

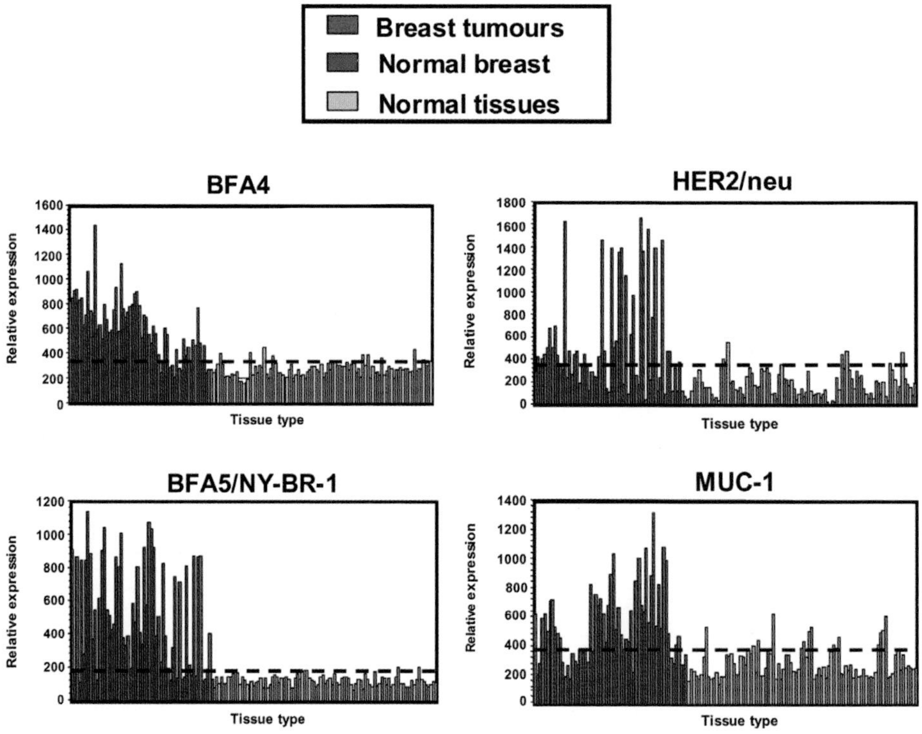

Figure 9.7

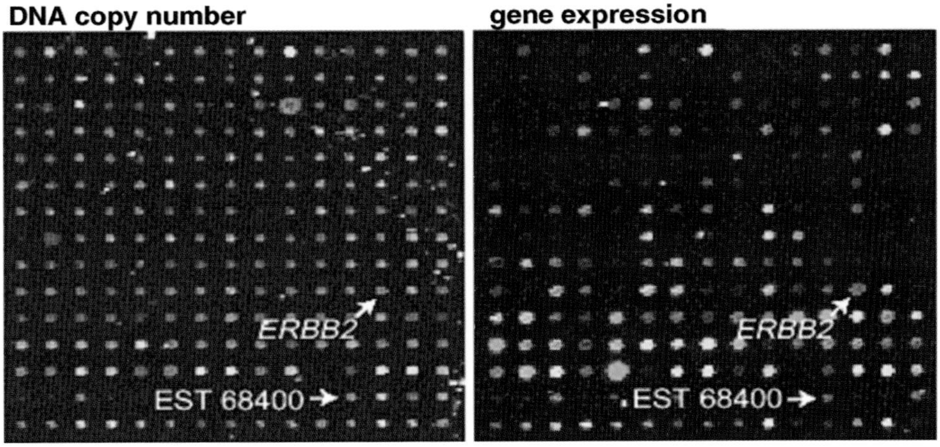

Figure 9.4

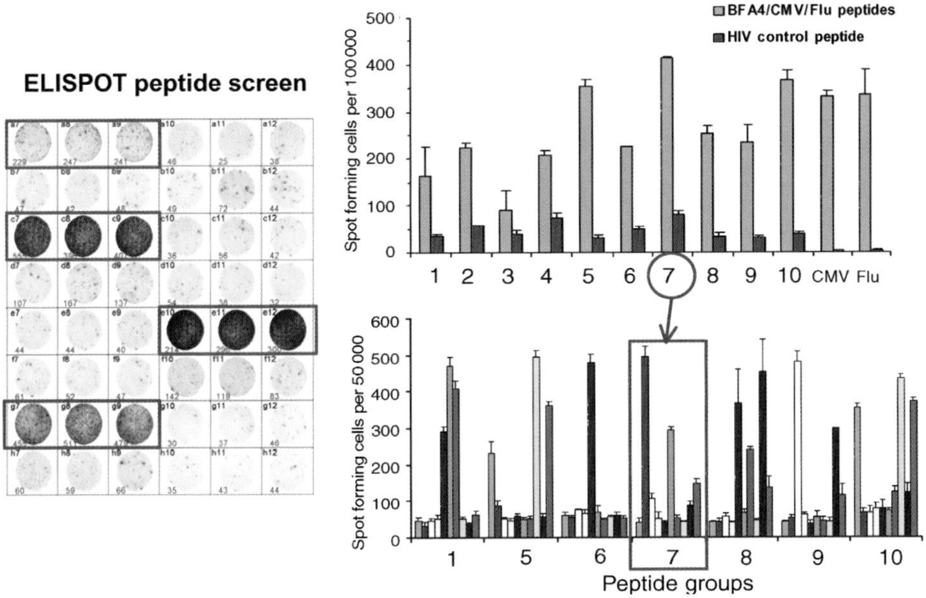

Figure 9.10

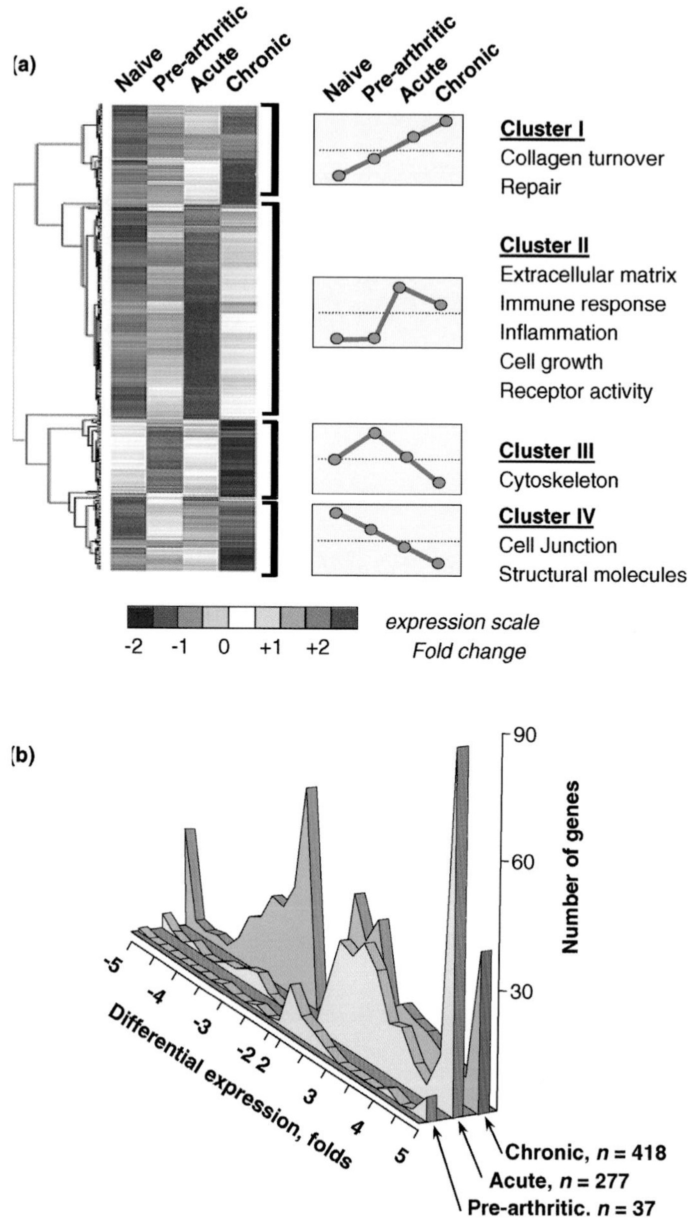
(a)
Naive
Pre-arthritic
Acute
Chronic
Naive
Pre-arthritic
Acute
Chronic
Cluster I
Collagen turnover
Repair
Cluster II
Extracellular matrix
Immune response
Inflammation
Cell growth
Receptor activity
Cluster III
Cytoskeleton
Cluster IV
Cell Junction
Structural molecules
-2 -1 0 +1 +2
expression scale
Fold change
(b)
90
60
30
Number of genes
-5 -4 -3 -2 2 3 4 5
Differential expression, folds
Chronic, n = 418
Acute, n = 277
Pre-arthritic. n = 37

Figure 12.4

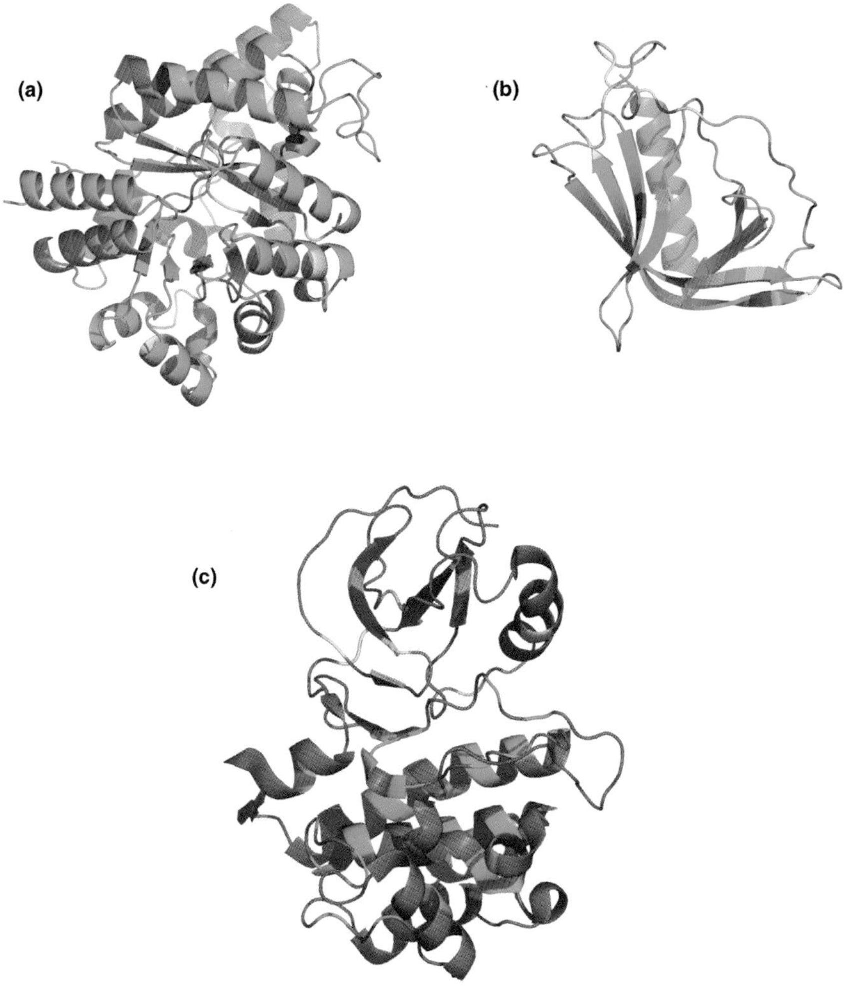

Figure 20.2

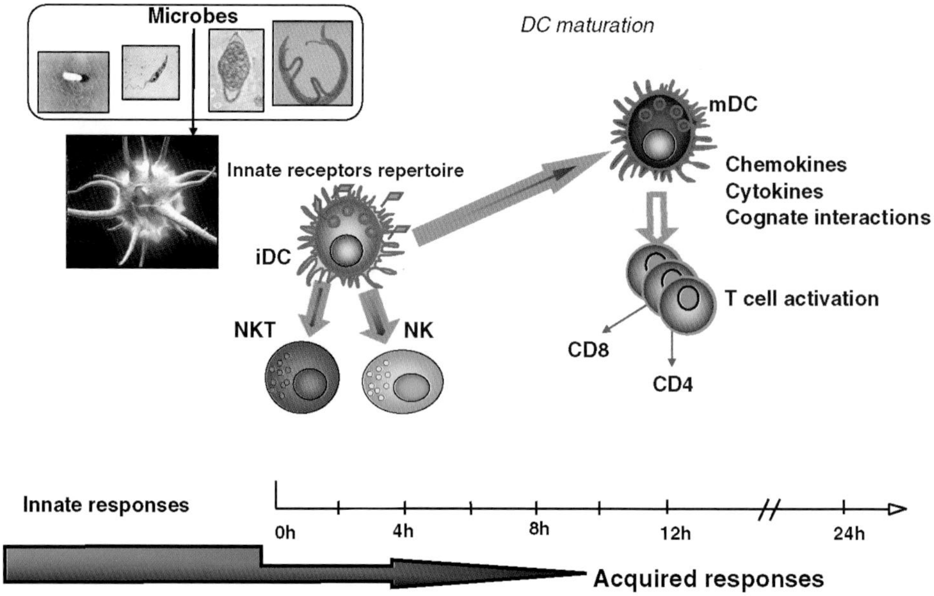

Figure 21.1

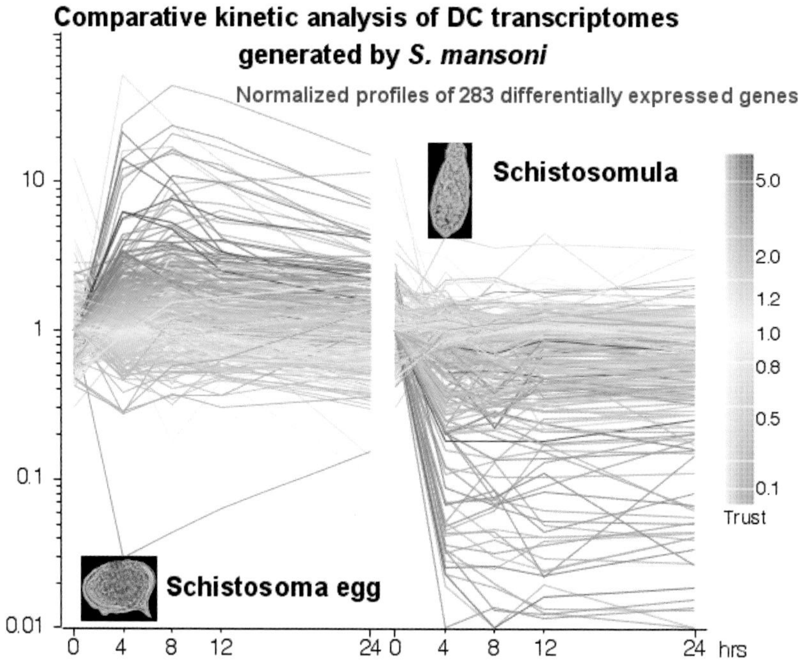

Figure 21.4

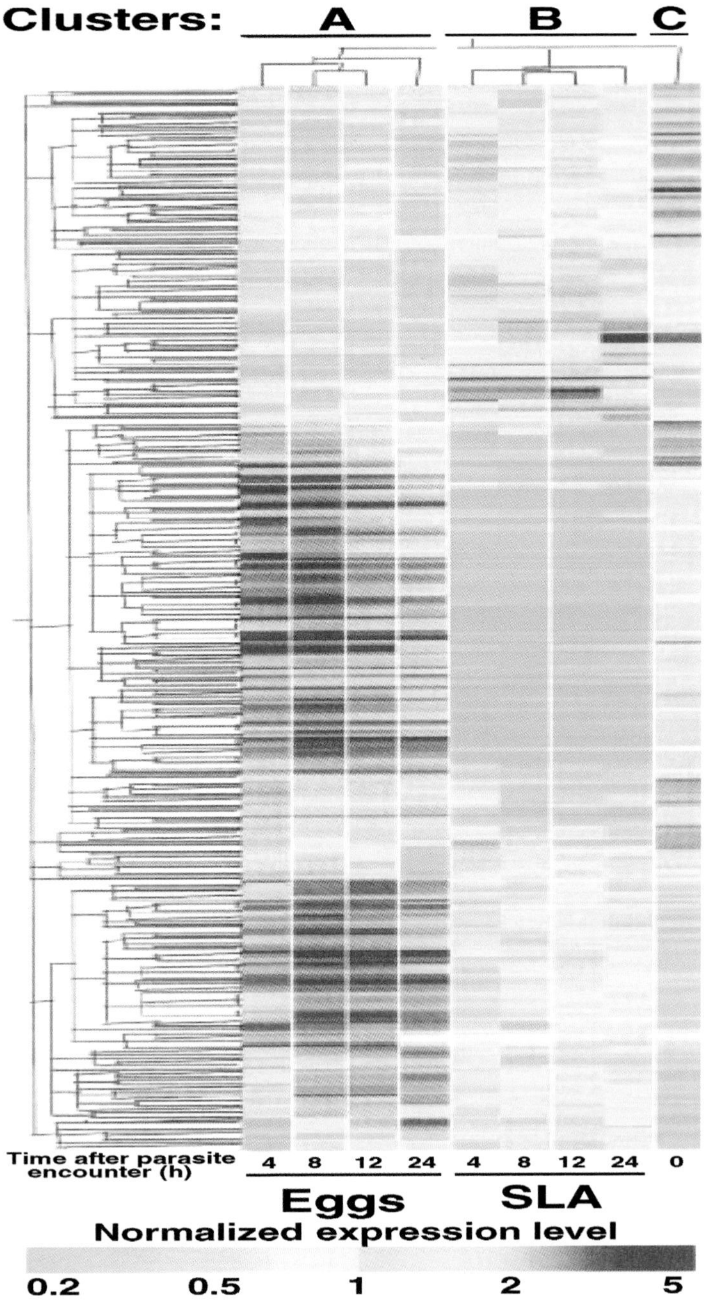

Clusters:
A
B
C
Time after parasite encounter (h)
4
8
12
24
4
8
12
24
0
Eggs
SLA
Normalized expression level
0.2
0.5
1
2
5

Figure 21.3

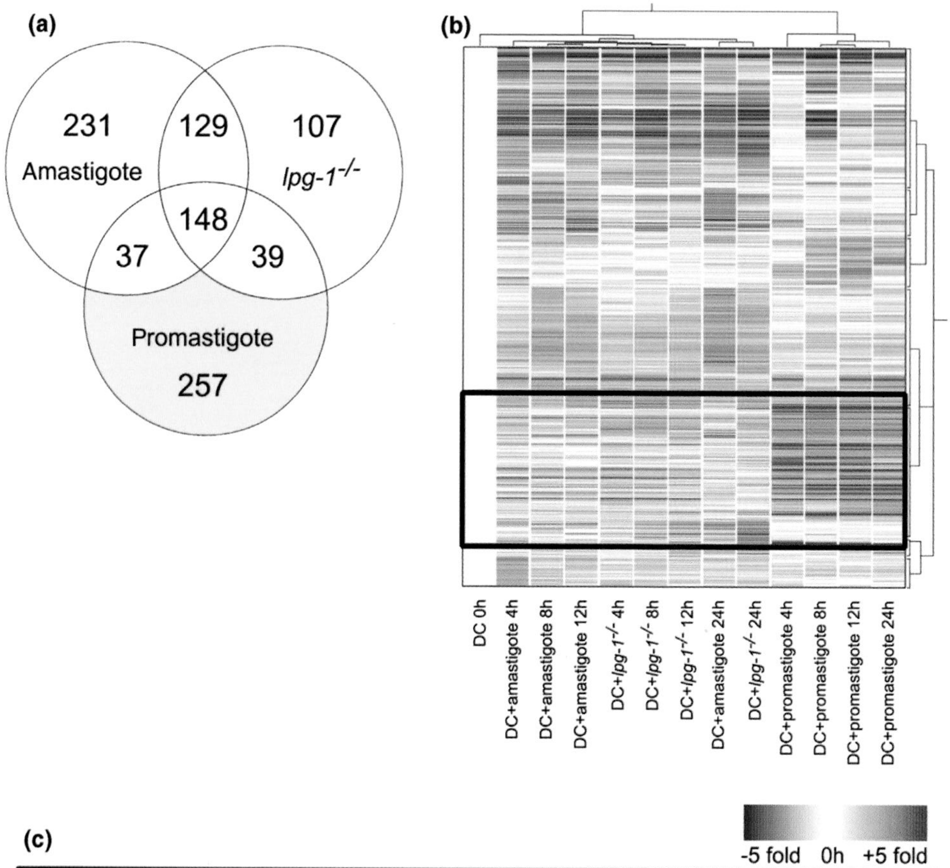

(c)

Interferon induced genes 17/260				Inflammatory cytokines 18/260			
name	**probeset**	**name**	**probeset**	**name**	**probeset**	**name**	**probeset**
Ifit2	103639_at	Irf1	102401_at	Il2	AFFX-MurIL2_at	Ccl6	92849_at
Ifi47	104750_at	Igtp	160933_at	Ccl4-Mip1b	94146_at	Ccl9-Mip1g	104388_at
Irf7	104669_at	Gtpi-p	98410_at	Ccl1-Tca3	94166_g_at	Il1b	103486_at
Isg15	98822_at	Ifrd1	160092_at	Cxcl10-IP10	93858_at	Ccl6	162198_f_at
Ifi204	98465_f_at	Isgf3	103634_at	Tnfa	102629_at	Ccl5-RANTES	98406_at
Ifit1	100981_at	Icsbp1	98002_at	Il12b-p40	100779_at	Ccl22	102310_at
Irf7	162202_f_at	Ifi30	97444_at	BC017607	96515_at	Cxcl2-Mip2a	101160_at
Ifi1	97409_at			Cxcl1	95349_g_at	Il1a	94755_at
Ifrg15	93997_at			Cxcl1	95348_at		
Ifi203	93321_at			Ccl3	102424_at		

Figure 21.5

13

Synovial Activation in Rheumatoid Arthritis

Lars C. Huber, Renate E. Gay and **Steffen Gay**

Abstract

Rheumatoid arthritis (RA) is a chronic polyarticular disorder leading to progressive destruction of the articular cartilage and the subchondral bone. The involvement of immune cells is a common hallmark of all systemic autoimmune diseases. What distinguishes, however, RA from other disorders is the finding of a unique cell type, the activated synovial fibroblast (RASF), which has been mainly associated with the pathogenic process. In recent years, remarkable progress has been made in elucidating the molecular features of these cells. This chapter highlights the complex interactions between the adaptive immune system and the RASFs as they occur during the pathogenesis of RA. Since the most relevant pathways of joint destruction are mainly mediated by RASFs, these cells represent an important target for novel therapeutic approaches, the application of 'functional genomics' should be considered as a potential tool for the design and investigation of novel pathways and disease-specific gene sequences.

13.1 Introduction

The population of industrialized countries is apparently confronted with an increase in the incidence of autoimmune disorders (Zinkernagel, 2001). Whether this is a true increase or attributable to improved diagnostic tools and more awareness of these diseases, remains unclear. It is a fact, however, that many autoimmune diseases, such as multiple sclerosis and type 1 diabetes mellitus, have been diagnosed substantially more frequently since World War II. In this context, rheumatoid arthritis (RA) represents one of the most common autoimmune-related inflammatory disorders.

Rheumatoid arthritis is a chronic polyarticular disorder that manifests primarily as a painful inflammation of the synovial tissues of joints, tendon sheaths and bursae. The progressive inflammation and the resulting erosion of the articular cartilage and

Immunogenomics and Human Disease Edited by András Falus

the adjacent subchondral bone causes severe pain, functional impairment and ultimately disability (Harris and Sledge, 1990). Considering that the prevalence of RA amounts to about 0.5–1% of the population and work disability affects as many as half of the patients at 10 years after onset of the disease, both the costs to public health systems and the social consequences for the individual patient are enormous (Gabriel, 2001; Lawrence *et al.*, 1998).

Like other autoimmune diseases, RA is a systemic disorder and its exact aetiology and its pathogenesis have not yet been determined. It is generally accepted, though, that autoimmune diseases emerge from a variable combination of individual genetic predisposition, environmental factors (e.g. infectious agents) and dysregulated immune responses (Ermann and Fathman, 2001; Smith and Haynes, 2002). A common hallmark of all systemic autoimmune disorders is the involvement of several types of immune cells, in particular T-cells and their pro-inflammatory mediators (Feldmann, Brennan, and Maini, 1996; Tarner and Fathman, 2001).

What distinguishes RA from other autoimmune disorders is the finding of a unique cell type, the transformed-appearing RA synovial fibroblast (RASF) that has been particularly associated with the pathogenic process (Muller-Ladner *et al.*, 1996; Pap *et al.*, 2000a). RASFs release large amounts of matrix-degrading enzymes and are thus considered as key cellular players in cartilage and bone destruction. Apart from this disease-driving role, RASFs represent an attractive target for novel therapeutic interventions.

Owing to our incomplete insight into the complex pathophysiology of joint destruction, the generation of specific therapies is still limited. Pharmacologic interaction with novel biologicals, in particular with tumour necrosis factor-α (TNF-α) blocking agents (such as the soluble receptor Etanercept and the monoclonal antibodies Infliximab and Adalimumab) has achieved impressive clinical success. However, there is a rather large number of patients (30–40 per cent) who are unresponsive to anti-TNF therapy (Maini *et al.*, 1999) and blocking TNF-α has been associated with an increased risk of certain uncontrollable granulomatous diseases (especially tuberculosis) and other opportunistic infections (Keane *et al.*, 2001).

Furthermore, initial failures and the high expectations put on scientific research have also caused disappointments, in particular among clinicians. Criticism has been raised that progress has been made at the bench rather than at the bedside. In addition, novel therapies are often challenged by public fears and confusing reports on potential risks. However, the research efforts of the last decade have led to the exponential development of new techniques in the field of molecular biology. Screening approaches such as gene arrays enable the fast and precise identification of genes that are differentially expressed in rheumatoid arthritis compared with the normal synovium and might be of functional importance for the progression and perpetuation of the disease. In this regard, the application of 'functional genomics' should be considered as a potential tool to design and investigate novel pathways and disease-specific gene sequences to support the identification of new therapeutic targets.

13.2 Synovial Activation in Rheumatoid Arthritis

The synovium is the key tissue for the initiation and the perpetuation of rheumatoid arthritis. However, it is yet unclear which of the different cell types that are involved in the pathogenesis of RA are active drivers of the disease and which act as mere responders (Firestein, 1996). Two scenarios are likely to address this question.

The first model focuses on T-cells, which home into the joint and are activated by a certain antigen. In this regard, a variety of possible antigens has been proposed, such as autoantigens, bacteria, viruses and components of the extracellular matrix, like collagen type II. Subsequently, the activated T-cells have been suggested to recruit macrophages and other immune cells, proliferate, differentiate and secrete pro-inflammatory cytokines and chemokines, which attract more inflammatory cells and, ultimately, activate synoviocytes and osteoclasts (Tarner *et al.*, 2003).

However, there is growing evidence for an alternative scenario: the first step in the development of arthritis could be the activation of synovial fibroblasts, e.g. by responses of the innate immune system. Once activated and in concert with the innate immune cells, RASFs start to produce cytokines, chemokines and matrix-degrading enzymes. Thus, various direct and indirect mechanisms would contribute to the progressive destruction of articular cartilage and adjacent bone (Geiler *et al.*, 1994; Muller-Ladner *et al.*, 1996).

In this review, we focus on the role of inflammatory cells and their cytokine profile, as well as on the activation and invasive properties of synovial fibroblasts.

Role of the adaptive immune system (T-cells and related cytokines)

Immune cells, in particular T-cells and T-cell-related cytokines, are important players in the pathogenesis of RA and represent promising targets for immunosuppressive and gene-transfer-based treatment.

$CD4^+$ cells, or T helper type cells (Th), differentiate from the unpolarized Th0-type cells into Th1-type and Th2-type cells. Figure 13.1 depicts the antigen-presenting cell (APC)-mediated differentiation from the Th0-type into polarized Th1- and Th2-type cells. Th1-type cells secrete primarily IL-2, IL-12, IFNγ and TNF-α, thus representing a strong pro-inflammatory cytokine profile, whereas the anti-inflammatory and pro-allergic Th2-type cells secrete mainly IL-4, IL-13, IL-10 and transforming growth factor β (TGFβ). More recent evidence by Schulze-Koops and Kalden (2001) suggests that in RA exists a defect in the generation of an approximate Th2-type response, resulting in an accelerated Th1-type disease. Experimental approaches with anti-CD4 antibodies prevented the development of arthritis in rodents and the use of an anti-α/β T-cell-receptor (TCR) reduced the severity of established disease in rats (Chiocchia, Boissier and Fournier, 1991; Goldschmidt, Jansson and Holmdahl, 1990; Ranges, Sriram and Cooper, 1985). Memory T-cells ($CD4^+$/$CD45R0^+$ phenotype) have been shown to accumulate in the RA synovial tissue (Thomas *et al.*, 1992). This has been associated with the release of

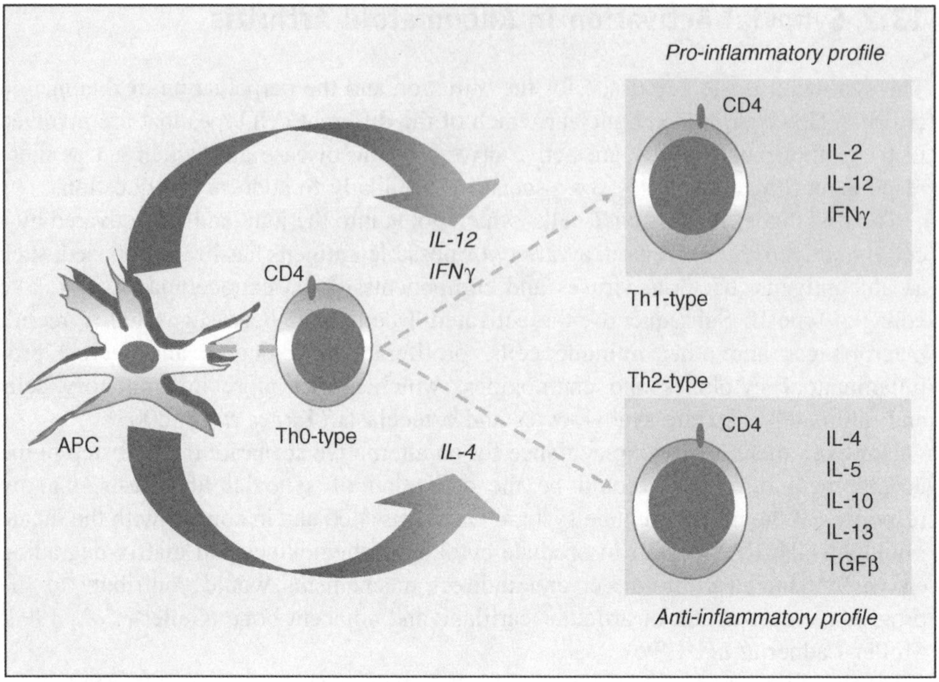

Figure 13.1 T-helper cell differentiation. Antigen-presenting cells (APC) and their respective cytokine-profile mediate the differentiation of unpolarized CD4$^+$-T cells into pro-inflammatory (Th1) and anti-inflammatory/pro-allergic (Th2) subtypes.

IL-17 and the differentiation of B-cells into (auto)antibody-producing plasma cells (Kehlen *et al.*, 2002; Thomas *et al.*, 1992). Autoantibodies (such as antibodies targetting cyclic citrullinated peptide, anti-CCP antibodies and rheumatoid factors, RF, which consist of various immunoglobulins directed against the constant region of IgGs) are also involved in the pathogenesis of RA and commonly used as clinical screening parameters.

Cytokines are key players in virtually every aspect of immunity. Mononuclear phagocytes produce mainly inflammation-promoting cytokines, including TNF and various Interleukins (IL). In this regard, the cytokine IL-12 mediates the maturation of Th1-type cells, which have been found to play a pivotal role in the development of autoimmune arthritides both in humans and various animal models. The production of IL-12 from stimulated macrophages (Hsieh *et al.*, 1993) and natural killer cells (Manetti *et al.*, 1993) is mainly a systemic process occurring in lymph nodes. However, it has also been observed as a local phenomenon in the synovium. IL-12 induces the release of pro-inflammatory cytokines (e.g. interferon gamma, IFNγ), several chemokines (MIP, MCP-1; Germann *et al.*, 1995; Parks *et al.*, 1998) and chemokine receptors (CCR-5), thereby enhancing the trafficking of inflammatory cells from the circulation into areas of inflammation (Kasama *et al.*, 1995; Koch *et al.*, 1992; Suzuki *et al.*, 1999) and the production of antibodies (Metzger *et al.*, 1997). IL-15 and IL-18 appear to act synergistically with IL-12 in the development of autoimmune arthritis (Gracie *et al.*, 1999). IL-15 has chemoattractive properties on T-cells and sustains interactions of T-cells and macrophages to promote activation and further cytokine release. Similarly, interactions between T-cells and RASFs with endogenous positive feedback loops have also been demonstrated (McInnes and Gracie, 2004; McInnes *et al.*, 2003).

In the collagen-induced murine arthritis (CIA) model, IL-18 accelerated the course of destructive arthritic disease, whilst IL-18 knockout animals showed a reduced incidence and severity of the disease. The application of either antibodies with blocking effects on IL-18 or a recombinant human IL-18 binding protein reduced established articular inflammation, cartilage destruction and secretion of the various cytokines from macrophages (Plater-Zyberk *et al.*, 2001). IFNγ, for example, released by IL-18 and/or IL-12, enhances the survival of CD68+ synovial macrophages (Gracie *et al.*, 1999) that, subsequently, secrete TNF-α, IL-6, various matrix degrading enzymes, and also IL-18 and IFNγ (Klimiuk *et al.*, 1999; Tanaka *et al.*, 2001).

The hallmark of RA that distinguishes it from other arthritic diseases is the progressive destruction of articular cartilage and adjacent bone structures. In this process, cytokines produced by macrophages (in particular TNF-α and, to a lesser extent, IL-1 and IL-6) play important roles.

TNF-α is a strong inductor of synovial proliferation and secretion of MMPs (Dayer, Beutler and Cerami, 1985), IL-1 (Brennan *et al.*, 1989; Dinarello *et al.*, 1986), IL-6, GM-CSF (Alvaro-Gracia *et al.*, 1991; Haworth *et al.*, 1991), and leukocyte-attractant chemokines (Chabaud, Page and Miossec, 2001; Taylor *et al.*, 2000). TNF-α also induces the transcription factor nuclear factor kappa b (NFκB) which was shown to be highly activated in RASFs (Han *et al.*, 1998; Makarov, 2001; Marok *et al.*, 1996; Miagkov *et al.*, 1998; Vincenti, Coon and Brinckerhoff, 1998). NFκB is an ubiquitously expressed transcriptional activator composed of DNA-binding heterodimers. Physiologically, NFκB is retained in the cytoplasm by its natural counterpart, IκB. In response to different stimuli such as cytokines or UV irradiation, IκB proteins are phosphorylated, polyubiquinated and finally undergo protein shredding by the 26 proteasome (Palombella *et al.*, 1998). This process unmasks the nuclear localization sequence of NFκB, enabling it to translocate into the nucleus and bind to the promoters of target genes such as IL-6, IL-8 and cyclooxygenase-2. However, NFκB not only regulates pro-inflammatory genes, but also the transcription of adhesion molecules and matrix degrading enzymes, and the synthesis of apoptosis-inhibiting proteins (Vincenti, Coon and Brinckerhoff, 1998). Upstream of NFκB, two IκB kinases (IKK1 and 2) regulate IκB activity (Aupperle *et al.*, 2001). IKK1- and 2 can be activated by the PTEN-dependent Akt serine–threonine kinase, which decreases the activity of pro-apoptotic proteins and increases the activity of anti-apoptotic proteins (Gustin *et al.*, 2004). In this regard, it was more recently reported that using a selective inhibitor of IKK-2 (TPCA-1) showed an attenuation of murine CIA by reducing the release of pro-inflammatory cytokines and antigen-induced

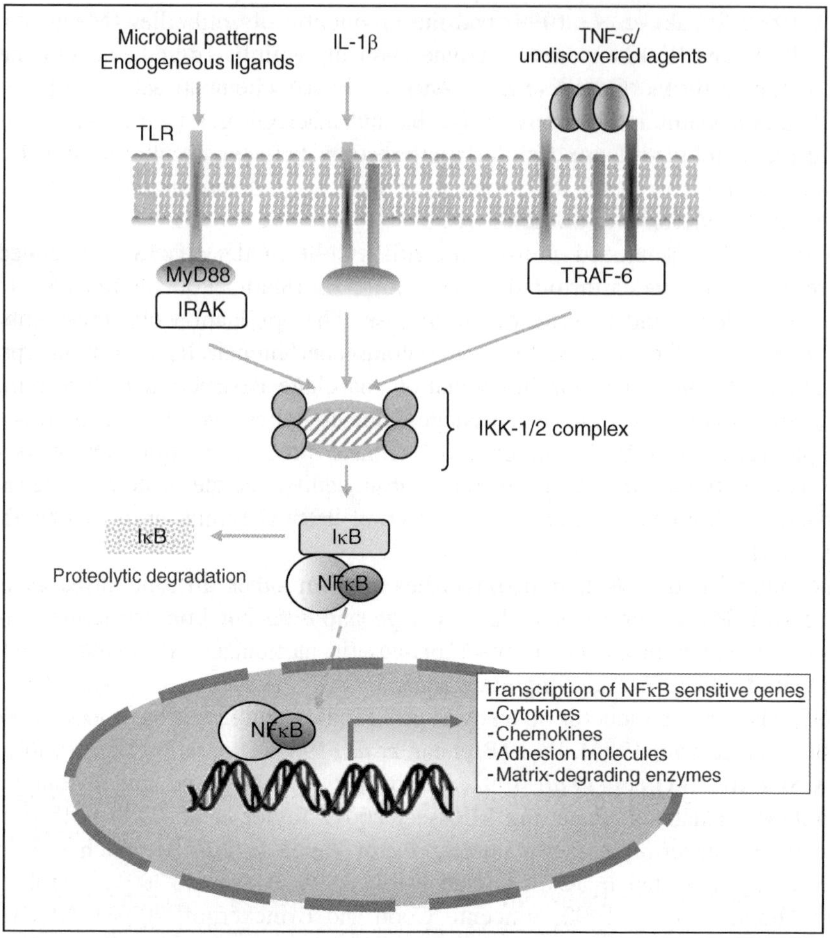

Figure 13.2 Schematic overview of the NFκB pathway. The downstream activation of NFκB involves the phosphorylation and proteosomal degradation of the inhibitory protein IκB by specific IκB kinases (IKK1/2). The unbound NFκB (usually a heterodimer of p50 and p65 subunits) then translocates into the nucleus, where it binds to the DNA in the promoter regions of pro-inflammatory genes.

T-cell proliferation (Podolin *et al.*, 2005). These pathways are summarized within Figure 13.2.

TNF-α, IL-1 and IL-6 have been demonstrated to activate osteoclasts and, together with IL-17 and RANKL (receptor activator of NFκB-ligand), to promote the differentiation of macrophages and other progenitor cells into bone-resorbing osteoclasts (Romas, Gillespie and Martin, 2002). Then, TNF-α, IL-1 and IL-17 can stimulate activated T-cells and osteoblasts to secrete RANKL, which, once released, interacts with its receptor (RANK) on myeloid precursor cells to promote osteoclastogenesis and bone destruction (Saidenberg Kermanac'h *et al.*, 2002). RANKL is a member of the TNF family and the RANKL–RANK interaction constitutes a signalling pathway that is important both in forming osteoclasts and enhancing their function (Burgess *et al.*, 1999; Tolar, Teitelbaum and Orchard, 2004). RANK signalling itself occurs through several intermediate molecules, including c-fos, NFκB and TNF-receptor-associated factor 6 (TRAF6; Lee and Kim, 2003). RANK–RANKL interaction can be inhibited by the decoy receptor osteoprotegerin (OPG), expressed on and secreted by osteoblasts. OPG binds RANKL with high affinity. Figure 13.3 shows the cellular interactions needed for osteoclastogenesis.

The Th2-type cytokines IL-4 and IL-10 have also been shown to prevent bone destruction. In addition, both cytokines reverse IL-17-induced IL-1β and TNF-α release (Saidenberg Kermanac'h *et al.*, 2002).

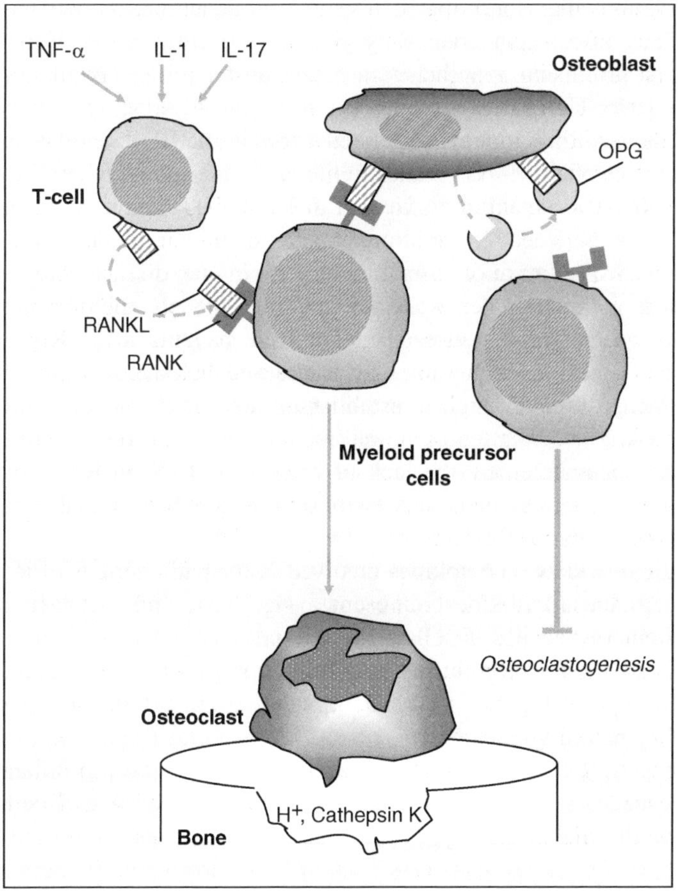

Figure 13.3 Cellular interactions involved in osteoclastogenesis. Receptor activator of nulcear factor-κB ligand (RANKL) is secreted by various cells, mainly T-cells and osteoblasts. Several cytokines promote the enhanced expression of RANKL, in particular TNF-α, IL-1, and IL-17. Osteoprotegerin is an endogenous inhibitor of osteoclastogenesis that is produced by osteoblasts and osteoclasts. It exerts its functions as a decoy molecule that prevents the interaction of RANK–RANKL. The downstream signalling events following the association of RANK and RANKL involve several pathways, including NFκB and mitogen-activated protein (MAP) kinases.

IL-17, released by $CD4^+$/$CD45R0^+$ T-cells in the synovium (Kehlen *et al.*, 2002; Kotake *et al.*, 1999), interacts with the IL-17R on RASFs, which, in turn, increases the production of the chemokines IL-8, GROα, GROβ and MIP-3α. The latter one mainly acts on T-cells and immature dendritic cells (Chabaud, Page and Miossec, 2001). Application of an IL-17R-IgG1 Fc fusion construct was shown to block the enhancer activity of IL-17 in rat adjuvant arthritis (Bush *et al.*, 2002).

Another cytokine that has been strongly associated with the pathogenesis of RA is IL-1 (Dayer, 2003). The IL-1 gene family has three members, namely IL-1α, IL-1β and the IL-1 receptor antagonist (IL-1Ra). While binding of IL-1 to its cell surface receptor activates downstream signal transduction, IL-1Ra (its naturally occurring inhibitor) binds competitively to IL-1R without inducing a cellular response. Based on the observations that inhibition of IL-1 in various animal models has resulted in major beneficial effects (van den Berg *et al.*, 1991), inhibition of IL-1 has been among the first promising approaches in gene transfer in RA (Bandara *et al.*, 1993; Evans *et al.*, 1996; Hung *et al.*, 1994).

Although these animal models suggested strongly that IL-1 regulates TNF-α (van den Berg and Bresnihan, 1999), studies in humans have shown otherwise and clinical studies using IL-1Ra (Anakinra, reviewed in Furst, 2004) have been disappointing. This discrepancy between observations made in animal models and rheumatoid arthritis shows the limitation of animal models for human disease. Although IL-1 has been thought to dominate other cytokines in animal models, the inhibition of TNF-α has been shown to be far more beneficial in patients with RA. Indeed, the development of anti-TNF-α therapies by Maini and Feldmann (Maini *et al.*, 1995) was a fundamental breakthrough in establishing a novel therapeutic strategy for the treatment of RA. The limitation of animal models for the design of novel therapeutic targets is also documented by the lack of benefit of IL-10 in RA (Neidhart *et al.*, 2004), although IL-10 has been successful administered in animal models (van de Loo and van den Berg, 2002).

A schematic overview on cytokines involved in the pathogenesis of RA is given in Table 13.1. All these cytokines represent a significant link between the adaptive immune system (especially T-cells), RASF and macrophages (innate immunity). While T-cells are the main producers of cytokines in the circulation and in lymphatic tissues, RASFs appear together with macrophages as the cellular source of cytokines within the rheumatoid synovium.

Taken together, T-cells and related cytokines show strong pro-inflammatory and pathogenic capacities in RA. However, it is still undefined *why* T-cells home and accumulate in the rheumatoid synovium.

Role of activated synovial fibroblasts in rheumatoid arthritis

Synovial fibroblasts (SFs) represent the centre of the local pathways operating in RA. Compared with SFs in healthy synovial tissue, RASFs comprise a distinguishable cell population within the synovial intimal lining (the interface between the synovium and

Table 13.1 Overview of cytokines involved in the pathogenesis of RA according to Borish and Steinke (2003)

Cytokine	Derivation	Effects
TNF-α	Mononuclear phagocytes Synovial macrophages (RASFs)	Release of IL-1, MMP-1, MMP-3, IL-6, GM-CSF Differentiation of osteoclasts Synovial proliferation Activation of NFκB and MAPK-pathways
IL-1β	Synovial macrophages Chondrocytes Endothelial cells RASFs	TNF-α dependent Activation of T-cells B-cell proliferation Production of antibodies Upregulation of endothelial adhesion molecules Activation of NFκB and MAPK-pathways
IL-2	T-cells	Pro-inflammatory Activation of T-cells Th1-type cell differentiation
IL-4	T-cells	Pro-allergic Anti-inflammatory Suppresses IL-1β release (IL-17 induced) Inhibits bone degradation
IL-6	Mononuclear phagocytes Synovial macrophages RASFs Endothelial cells	T-cell activation Maturation of B cells to plasma cells Synergistic to IL-1 Synthesis of acute-phase proteins Differentiation of osteoclasts (together with TNF-α) Inhibits synthesis of TNF-α and IL-1
IL-8	Mononuclear phagocytes (RASFs) Endothelial cells	Pro-angiogenic properties Most potent chemoattractant for neutrophils (CXCL8) Stimulates neutrophil degranulation, respiratory burst Mediates adherence of neutrophils to endothelial cells
IL-10	T-cells	Anti-inflammatory Suppresses IL-1β release (IL-17 induced) Inhibits bone degradation
IL-12	Mononuclear phagocytes B-cells Dendritic cells (APCs)	Differentiation of Th1-type cells Release of inflammatory cytokines, chemokines, chemokine receptors Trafficking of inflammatory cells Production of antibodies
IL-15	Mononuclear phagocytes RASFs Endothelial cells	Synergism with IL-18, IL-12
IL-16	T-cells RASFs	Chemotactic for $CD4^+$ cells
IL-17	T-cells ($CD4^+$/$CD45^+$) RASFs	Together with IL-1 and TNF: secretion of RANKL from activated T-cells Release of IL-8, GROα, β, and MIP-3α

(*Continued*)

Table 13.1 (*Continued*)

Cytokine	Derivation	Effects
IL-18		Survival of CD68+ cells Influences incidence and severity of disease
IL-21R	RASFs Synovial macrophages	Associated with the activated phenotype of RASFs independently of the major proinflammatory cytokines IL-1β and TNFα Correlates negatively with the destruction of articular cartilage and bone (Jungel *et al.*, 2004)
IFNγ	T-helper cells NK cells RASFs	Differentiation of Th1-type cells Survival of $CD68^+$ cells Production of reactive oxygen species Induction of endothelial adhesion molecules
TGFβ	Chondrocytes RASFs Osteocytes Th2-type cells	Stimulans of fibrosis Production of extracellular matrix Creation of an 'immunosuppressive' milieu
RANKL	T-cells Osteoblasts RASFs	Interaction with RANK Differentiation of myeloid progenitors into osteoclasts Promotes osteoclastogenesis
OPG	Osteoblasts	Inhibits osteoclastogenesis by high affinity interaction with RANKL

the intra-articular space) that shows both morphological and biological differences (Fassbender, 1983). At the morphological level, RASFs are characterized by a round-shaped and large pale nucleus with prominent nucleoli, indicating a high turnover of RNA metabolism. Of interest, RASFs can be expanded in cell culture over several passages. Moreover, they escape contact inhibition, which, among other specific changes in cellular activation, finally results in an aggressive and invasive behaviour of RASFs into adjacent cartilage and bone (Lafyatis *et al.*, 1989). Figure 13.4 shows the characteristic histological features of RASFs invading articular cartilage. One can note the pale nucleus with prominent nucleoli (with arrows). Of interest, nuclei appear to form the cellular frontline of invasion. White stars indicate chondrocytes.

There is growing evidence that the activation of SFs is the first and initiating step in autoimmune arthritis (Franz *et al.*, 1998b; Zvaifler *et al.*, 1997). In early disease, for example, hyperplastic and inflammatory changes in the synovial tissue (i.e. angiogenesis, overexpression of prostaglandins) have been detected before T-cells had accumulated (Zvaifler, Boyle and Firestein, 1994).

The majority of cytokines within the rheumatoid synovium are produced by synovial macrophages but also by RASFs, a finding that has also caused doubt regarding the T-cell-dependence of the disease (Fox, 1997). Moreover, the progressive destruction in late disease continues even in the absence of immune cells. Collagen-type II immunized mice that were deficient in mature T- and B-cells due to

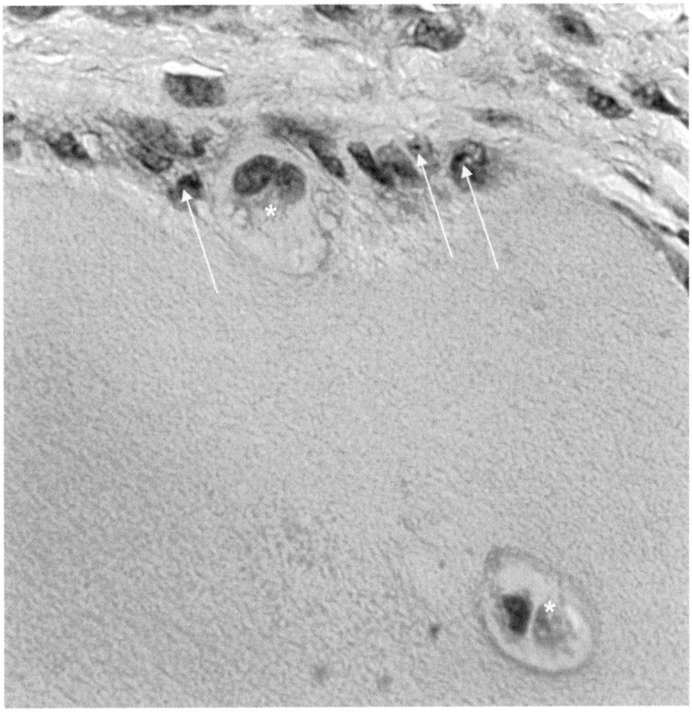

Figure 13.4 Histological characteristics of RASFs. In the severe combined immunodeficiency (SCID) mouse model, human RASFs and articular cartilage are co-implanted under the renal capsule of SCID mice. Since these mice are compromised in cellular and humoral immune responses, the behaviour of fibroblasts can be assesssed *in vivo* without the influence of pro-inflammatory cells and mediators. The figure shows the deep invasion of RASFs into the adjacent cartilage. RASFs are characterized by a round-shaped and large pale nucleus with prominent nucleoli. The nuclei appear to form the cellular frontline of invasion. White stars indicate chondrocytes. Reproduced by permission of Dr Astrid Jüngel.

a lacking RAG-1 (recombinant-activating gene 1) triggered cartilage and bone destruction without inflammation (Plows, Kontogeorgos and Kollias, 1999). These findings are also confirmed by clinical observations of RA patients infected with the human-immunodeficiency virus (HIV-1) who show progressive joint destruction even under low counts or complete systemic absence of $CD4^+$ cells (Muller-Ladner *et al.*, 1995a).

Blocking the activation of B-cells by Rituximab (a monoclonal anti-CD20 antibody that eliminates B-cells and thus disables their antigen-presenting function), has shown impressive scores in clinical responses. However, an ACR 70 response over 45 per cent has not been reported so far (Stahl *et al.*, 2003). This is another indication that there exist alternative pathways apart from the classical inflammation model of RA. T-cell-independent mechanisms thus play a significant role in the pathogenesis of rheumatoid arthritis and RASFs appear to be one of the pivotal mediators contributing to the progressive destruction in affected joints through both direct and indirect mechanisms.

Indirect mechanisms, for example, consist of governing the differentiation from macrophages into osteoclasts, for example through the upregulation of RANKL (Gravallese, 2002; Nakano *et al.*, 2004; Takayanagi *et al.*, 2000). Although the activation of RASFs can probably be maintained in the absence of inflammatory mediators, pro-inflammatory cytokines and growth factors still play an important role both in the continuous stimulation of RASFs towards an aggressive behavior and in the crosstalk between RASFs and other cell types in the synovium, namely chondrocytes. Interleukin-1 (IL-1) is thereby a key player in mediating the perichondrocytic cartilage damage (Joosten *et al.*, 1996; Neumann *et al.*, 2002). A likely scenario thus might be that RASFs invade human cartilage in RA at least in part through cytokine-independent pathways. In addition, certain cytokines (e.g. TNF-α, IL-1, -15, -17), produced by inflammatory cells in the synovium, stimulate RASFs further to release matrix-degrading enzymes, which in turn, amplify their invasive and degradative behavior.

Direct mechanisms on the other hand mainly comprise the attachment of fibroblasts to the articular cartilage by upregulation of cellular adhesion molecules (CAM) and the destruction of cartilage and bone by expression of matrix-degrading enzymes (Muller-Ladner *et al.*, 1996).

Subclinical arthritic disease is normally neglected both by the patient and his doctor. Animal models of RA, such as the aforementioned CIA, thus represent one possibility to investigate these early stages of disease. Despite the fact that animal models only reflect human disease in a limited manner, animal studies nevertheless gave important insights into the molecular patterns of the pathogenesis of RA. In the CIA model, for example, synovial tissue that was obtained before the onset of clinical symptoms (e.g. erythema and swelling) has raised concerns regarding the T-cell-dependent nature of RA. Histology from those samples revealed vasodilatation, fibrin deposition and a transformed-appearing phenotype of synoviocytes within a hyperplastic tissue that was lacking leukocytic infiltration (Caulfield *et al.*, 1982; Marinova-Mutafchieva *et al.*, 2002). These data raised the concept of an activation of the innate immune system as the initial event that progresses to stimulate T-cells and the production of cytokines. Similarly, incidental human biopsies have revealed that asymptomatic early synovitis precedes clinical disease, and macrophages and their cytokines predominate therein (Kraan *et al.*, 1998; Tak *et al.*, 1997).

Apart from cartilage destruction, angiogenesis is another hallmark of the inflamed synovial tissue in early RA. RASFs have been shown to release several pro-angiogenic factors such as IL-8, vascular endothelial growth factor (VEGF), basic fibroblast growth factor (bFGF) and TGFβ (Bodolay *et al.*, 2002). Since RASFs appear to be the major source of various chemokines including MCP-1, MIP-1α, MIP-3α and RANTES, and different cytokines such as IL-15, IL-16 and IL-17 (a potent chemoattractant for $CD4^+$ cells) in the RA synovium, it could be hypothesized that T-cells and macrophages expand in an antigen-independent way into the synovium. Both the upregulation of attachment molecules and the granzyme–perforine system of T-cells could pave the diapedesis through the vascular basement membranes into the synovium (Muller-Ladner *et al.*, 1995).

RASFs then are further stimulated by pro-inflammatory cytokines released from macrophages and T-cells. In particular, exposure to TNF-α and IL-1β has been associated with the release of matrix-degrading molecules. These mediators also induce high-level expression of IL-15 in synovial fibroblasts. IL-15-based interactions between T-cells and RASFs with endogenous positive feedback loops have also been demonstrated (Liew and McInnes, 2002; McInnes *et al.*, 2003) and, in this context, the application of an IL-15 monoclonal antibody (humax-IL-15) has shown promising results in early pharmacological trials (Baslund *et al.*, 2003).

These data mainly show that cytokines present in the inflamed synovium can stimulate RASFs and induce the release of matrix-degrading enzymes from these cells. On the other hand, it is interesting that RASFs produce IL-16, a cytokine shown to attract $CD4^+$ T cells (Franz *et al.*, 1998a) to the joint.

Evidence for a T-cell independent activation of RASFs was particularly derived from experiments with the severe combined immunodeficiency (SCID) mouse co-implantation model of RA. In this model, RASFs are implanted together with normal human cartilage in an inert gel sponge under the renal capsule of severe combined immunodeficient mice. Since these mice are not capable of rejecting xenografts due to their defects both in humoral and cellular immunity, the invasive potential of SFs into the co-implanted cartilage can be assessed histologically after a certain period of time, usually after 60 days. In this experimental setting (devoid of any component of the human immune system), it has been repeatedly observed that RASFs attach to and deeply invade normal articular cartilage, and, moreover, express matrix-degrading enzymes whereas normal SFs, osteoarthritis SFs or dermal fibroblasts showed no or only limited signs of such invasion or enzyme secretion (Muller-Ladner *et al.*, 1996). This *in vivo* model thus strongly suggests that the changes resulting in the activation of RASFs are not merely the responses to continuous stimulation by inflammatory mediators but are intrinsic features of these cells that do not require exogenous stimulation. Among others, crucial consequences of the activation of RASFs are two phenomena: the upregulation of adhesion molecules and tissue remodelling. The first step of the invasive process comprises the attachment of RASFs to the articular cartilage and is mediated by the upregulation of adhesion molecules on the surface of RASFs. Adhesion molecules mediate the anchoring of SFs to the extracellular matrix of the articular cartilage. Thereby, β1 integrins are most prominently involved (Ishikawa *et al.*, 1996; Kriegsmann *et al.*, 1995; Rinaldi *et al.*, 1997). However, integrins not only facilitate the adhesion of RASFs to cartilage. Upon matrix adhesion of RASFs, integrins and VCAM-1, also interact with various signalling cascades that regulate the early cell cycle and the expression of matrix metalloproteinases (MMPs; Schwartz, 1997). In this regard, c-fos [a component of the activator-protein (AP)-1 complex] and the proto-oncogene c-myc, which are both expressed within the RA synovium (Asahara *et al.*, 1997; Dooley *et al.*, 1996; Grimbacher *et al.*, 1997; Kontny *et al.*, 1995; Qu *et al.*, 1994; Trabandt *et al.*, 1992), were shown to be upregulated by integrin-mediated cell adhesion (Dike and Farmer,

1988). Together with other key molecules, these pathways play a pivotal role in tissue destruction of articular cartilage, the most crucial event involved upon activation of RASFs. Tissue degradation essentially contributes to the progressive loss of joint function. It comprises the following major pathophysiological phenomena: growth, spreading and invasion of inflamed synovial tissue, and destruction of cartilage and bone. All these processes have a common underlying mechanism, namely the degradation of extracellular matrix, which is mediated by MMPs, cathepsins (Pap, Gay and Schett, 2002), cathepsins and an activated plasmin system (van der Laan *et al.*, 2000).

MMPs are zinc-containing endopeptidases involved in the remodelling of extracellular matrix proteins. Their catalytic activity is regulated by the activity of endogenous inhibitors, the tissue inhibitors of matrix metalloproteinases (TIMPs). Apart from degradation and tissue remodelling, MMPs are important regulatory molecules acting on cytokines and adhesion molecules. Various extracellular signals, in particular pro-inflammatory cytokines, growth factors and matrix molecules, induce the expression of MMPs via transcriptional activation. The specific effect of such a signal on the expression of MMP though is highly variable and depends on the induced type of MMP, the cell type, and the signal transduction pathway. The aforementioned activator protein-1 (AP-1) appears to play a key role, since AP-1 binding sites have been found in the promoter region of all MMPs (Benbow and Brinckerhoff, 1997). Some of the MMP promoter sites additionally contain binding sites for NFκB (Barchowsky, Frleta and Vincenti, 2000; Mengshol *et al.*, 2000) and signal transduction and activation of transcription (STAT; Li, Dehnade and Zafarullah, 2001). Upstream of these transcription factors, all three mitogen-activated protein kinase p38 (MAPK), stress-activated protein kinase (SAPK) families, extracellular regulated kinase (ERK) and p38 kinase (which integrate extracellular signals upstream from the AP-1 forming molecules jun and fos) are involved in the regulation of MMP expression (Schett *et al.*, 2000).

Other enzymes involved in joint destruction are cysteine and aspartyl proteases. Concerning the destructive course of RA, cathepsin B and L are of special interest and cathepsin K has also been detected at the sites of invasion (Hummel *et al.*, 1998). Cathepsins B and L are cysteine proteases that participate in the lysosomal protein degradation system (Bohne *et al.*, 2004). While cathepsin B directly facilitates the degradation of ECM proteins, including fibronectin, collagen types I and IV, and laminin (Guinec, Dalet-Fumeron and Pagano, 1993), cathepsin L cleaves collagens type I, II, IX, XI and certain proteoglycans. Cathepsin B has additionally an indirect role that involves the activation of other enzymes, including MMPs and both the soluble and receptor-bound forms of the serine protease urokinase plasminogen activator (uPA) (Kobayashi *et al.*, 1992, 1993). MMPs and uPA have been shown to modulate the proteolytic cascade that mediates ECM degradation (Lakka *et al.*, 2004). The expression of both cathepsins B and L has been demonstrated in RASFs at sites of invasion into cartilage and bone. Stimulation of RASFs by proinflammatory cytokines such as TNF-α and IL-1 (Huet *et al.*, 1993; Lemaire *et al.*, 1997), and the expression of proto-oncogenes, lead to the release of cathepsins. In this regard, it was shown that the stable expression of constitutively active Ras resulted in increased levels of cathepsin L (Collette *et al.*, 2004). Other proteases, such as thrombin, have been asssociated with worsening of inflammation in RA joints through the secretion of IL-8 and the recruitment of leukocytes, which release cathepsin B into the synovial fluid (Mishiro *et al.*, 2004).

It was commonly thought that the expression of matrix-degrading enzymes by downstream activation of MAP-kinases is strongly dependent on pro-inflammatory cytokines. As shown more recently, however, certain members of the MAPK family can also be activated by cytokine-independent pathways. By exploring the role of retroviral sequences in the activation of RASFs, endogenous L1 elements have been detected in synovial fluids and on cultured RASFs (Neidhart *et al.*, 2000). Such mobile genetic elements have been shown to act as retrotransposons that are widely distributed within our genome. Subsequently, it was demonstrated that functional L1 elements induce the MAP kinase p38δ (also known as stress-activated kinase 4, SAPK4). Within the SAPK/p38 family (which is known to be a critical signalling pathway for pro-inflammatory cytokines), p38δ thus represents a cytokine-independent isoform. In turn, p38δ induces the production of MMP-1 (Kuchen *et al.*, 2001), MMP-3 (Ospelt *et al.*, 2004), IL-6 and IL-8 (Suzuki *et al.*, 2000).

From these data it can be concluded that RASFs are not only stimulated by proinflammatory cytokines but also by a cytokine-independent pathway through the activation of p38δ. This notion is also supported by reports on a substantial number of RA patients who show a progression of disease under TNF-blocking biologicals, even when combined with immunosuppressive drugs. Figure 13.5 depicts both cytokine-dependent and -independent pathways of MAP-kinases activation.

Another interesting approach to cytokine-independent activation of RASFs was the recent detection of members of the Toll-like receptor (TLR) family. TLRs comprise a variety of receptors that recognize conserved motifs on both microbial pathogens (lacking in higher eukaryotes) and endogenous ligands, such as heat-shock proteins and RNA molecules. Downstream, TLRs mediate phagocytosis and lead to the activation of pro-inflammatory pathways (Aderem and Ulevitch, 2000). For both TLR2 and TLR3, a strong expression in RA synovial tissue could be shown. In addition, expression was increased when RASFs were stimulated previously with TNF-α and IL-1 (Seibl *et al.*, 2003). More recently, it was also demonstrated that RASFs bear functional TLR4, -5 and -6, indicating a crucial role of TLRs in iniating and/or maintaining inflammation in RA joints (Brentano *et al.*, 2004). Of interest, ligands of TLR2 induced the release of chemokines (e.g. GCP-2, MCP-2), inflammatory cytokines (e.g. IL-6, IL-8), and various matrix-degrading enzymes (MMP-1, -3, -13) in RASFs (Pierer *et al.*, 2004; Seibl *et al.*, 2003).

As mentioned above, the ultimate cause of the activation of RASFs remains elusive. It is, though, well established that both the altered morphology and the aggressive behaviour of RASFs result from specific changes in the transcription of disease-relevant genes and in intracellular signalling pathways, including alterations in apoptotic cascades. In particular, upregulation of several proto-oncogenes as well as the downregulation or functional silencing of potentially protective tumour

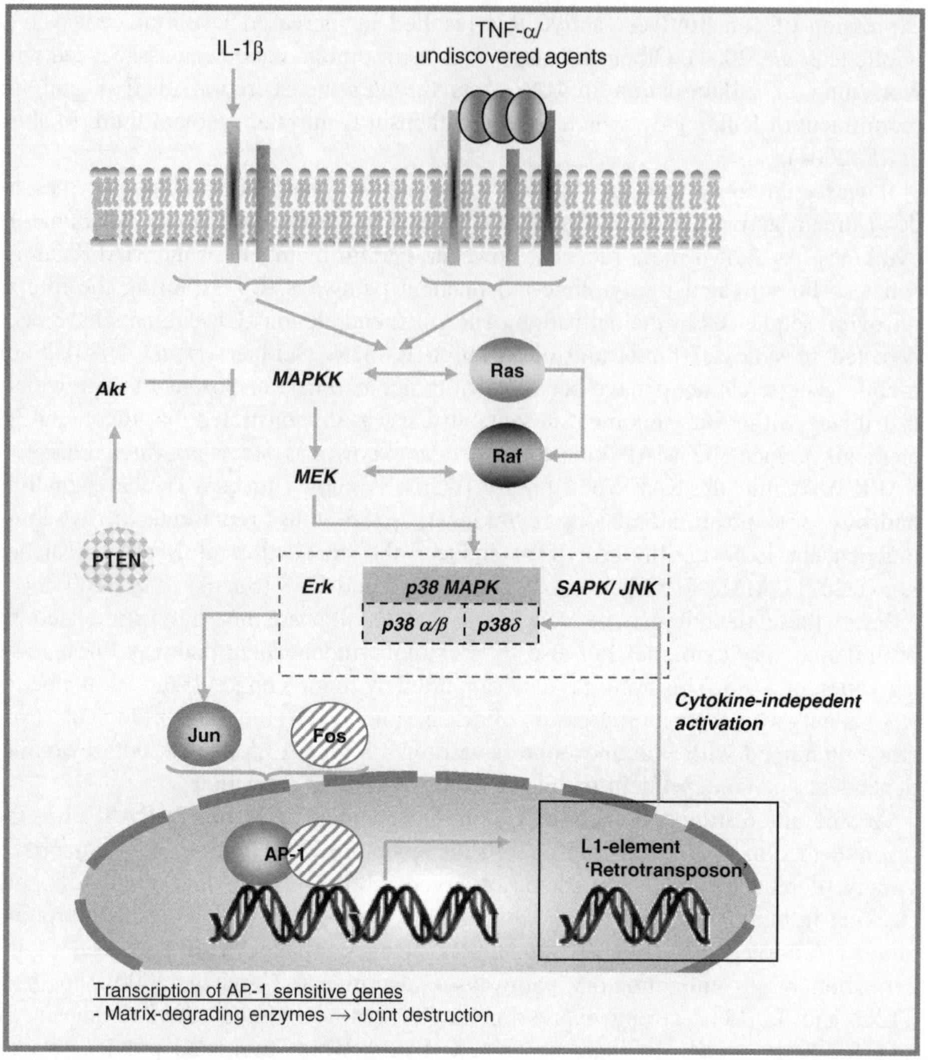

Figure 13.5 Interplay of cytokine-dependent and cytokine-independent pathways. The release of matrix-degrading enzymes upon activation of RASFs is triggered by several pathways. The cytokine-dependent cascade involves the downstream signalling by proto-oncogenes Ras–Raf and the activation of MAPK families, in particular ERK, JNK and p38α/β, leading to the association of Jun and Fos to form the transcription factor AP-1. This is also achieved in a cytokine-independent manner, namely by L1 elements activating the p38δ isoform.

suppressor genes – similarly to transformations seen in cancer cells – have been found. For example, various immediate early genes were shown to be upregulated in RASFs, including egr-1 (Grimbacher *et al.*, 1997; Trabandt *et al.*, 1992) and fos (Dooley *et al.*, 1996; Kontny *et al.*, 1995), as well as protooncogenes such as jun and myc (Pap *et al.*, 2004). Oncogenes such as ras, scr and raf appear to mediate the high expression of fos and jun, which are involved in the formation of the activator protein-1 (AP-1) transcription factor. On the other side, these oncogenic molecules are activation ligands for mitogen-activated protein kinase (MAPK) pathways, which are thought to regulate processes involved in apoptosis and proliferation.

Although still controversial, the impairment of apoptosis in RASFs and synovial macrophages appears to be responsible for the development of synovial hyperplasia. Deficient apoptosis and, thus, prolonged survival of RASFs, appears to be cause by the upregulation of anti-apoptotic molecules like bcl-2, sumo-1 (sentrin-1) and FLIP (Fas-associated death domain-like interleukin 1β converting enzyme inhibitory protein), protecting against FasL- and TNF-induced apoptosis (Catrina *et al.*, 2002; Franz *et al.*, 2000; Okura *et al.*, 1996; Perlman *et al.*, 2001; Schedel *et al.*, 2002). In addition to these data, alterations in the level of expression and function of the tumour suppressor PTEN (phosphatase and tensin homologue deleted from chromosome 10) have been investigated. PTEN exhibits a dual-specific tyrosine phosphatase activity and is functionally involved in cell cycle arrest and apoptosis. Compared with normal synovial tissue in which PTEN is homogeneously expressed, examination of cultured RASFs showed that only 40 per cent of cells expressed PTEN. In RASFs invading cartilage virtually no expression of PTEN was found, suggesting that the synovial hyperplasia in RA is due to defective apoptosis (Pap *et al.*, 2000b).

Finally, concerning the functional cross-talk between inflammatory T-cells and macrophages on the one side and RASFs on the other, an interesting approach has been made most recently. Microparticles are ultrastructural vesicles, which are released from activated inflammatory cells. Only few months ago, it was shown that immune cell-derived microparticles activate synovial fibroblasts in a dose-dependent manner to release matrix metalloproteinases and pro-inflammatory cytokines (e.g. TNF-α and IL-1β; Distler *et al.*, 2005). The observed accumulation of microparticles from various cellular origins in the synovial fluid (Berckmans *et al.*, 2002) thus may contribute to the individual course of the disease and the shedding of microparticles may represent an alternative stimulus in the complex cell–cell interaction of the pathogenesis of RA.

13.3 Conclusions/Perspectives

This review highlights the complex interactions between the adaptive immune system and activated synovial fibroblasts as they occur during the pathogenesis of RA. Since the most relevant pathways of joint destruction are mainly mediated by RASFs, these cells represent an important target for novel therapeutical approaches designed to inhibit the destruction of cartilage and bone in RA.

In the past decade, the exponential development in the field of molecular biology has shed some light on different pathomechanisms of the disease and, thus, led to novel therapeutic approaches in addition to corticosteroids and immunosuppressive drugs. The great success of TNF-α blocking agents was supplemented by various novel cytokine targets. In this regard, treatment with an anti-IL-6-receptor antibody

(MRA) has been encouraging by reducing both the mean activity score of the disease and inflammatory acute-phase proteins to normal values (Choy *et al.*, 2002). Promising results have also been achieved in a phase I/II trial with monoclonal antibodies against IL-15 (humax-IL-15; Baslund *et al.*, 2003). Application of an IL-17-Receptor–IgG1 fusion construct blocked the enhancer activity of IL-17 in rat adjuvant arthritis (Bush *et al.*, 2002). Furthermore, animal studies in the CIA model using a recombinant IL-18 binding protein showed a significant reduction of disease severity.

Cytokine-independent targets comprise the use of rituximab (monoclonal antibodies against CD20), which was originally used in non-Hodgkin's lymphomas (Cheson, 2002), and the application of the fusion protein CTLA4Ig, which blocks the binding of CD80/CD86 to the T-cell surface molecule CD28 and thereby prevents the activation of T-cells (Kremer *et al.*, 2003).

Gene transfer and gene targetting, finally, have led to remarkable advances in the development of novel therapies for arthritic diseases, which are based on our growing insights into the pathogenesis of RA, as well as progress in using gene transfer methods both to validate known molecular targets and to discover novel pathways. In this regard, the successful transfer of TIMP-3 into RASFs resulting in an inhibition of RASF-mediated destruction of cartilage in the SCID-mouse model of RA (van den Laan *et al.*, 2003) suggests that TIMP-3 might inhibit both the cytokine-dependent pathway of destruction mediated by TACE (TNFα-converting enzyme) and the cytokine-independent pathway of destruction mediated by the MMP-producing RASFs.

Since gene therapy is still far from clinical use, gene transfer techniques should be considered as potential tools to explore novel pathways involving newly described genes sequences to design novel therapeutic targets.

References

Aderem A, Ulevitch RJ. 2000. Toll-like receptors in the induction of the innate immune response. *Nature* **406**(6797): 782–787.

Alvaro-Gracia JM, Zvaifler NJ, Brown CB, Kaushansky K, Firestein GS. 1991. Cytokines in chronic inflammatory arthritis. VI. Analysis of the synovial cells involved in granulocyte-macrophage colony-stimulating factor production and gene expression in rheumatoid arthritis and its regulation by IL-1 and tumor necrosis factor-alpha. *J Immunol* **146**(10): 3365–3371.

Asahara H, Fujisawa K, Kobata T, Hasunuma T, Maeda T, Asanuma M, Ogawa N, Inoue H, Sumida T, Nishioka, K. 1997. Direct evidence of high DNA binding activity of transcription factor AP-1 in rheumatoid arthritis synovium. *Arthrit Rheum* **40**(5): 912–918.

Aupperle K, Bennett B, Han Z, Boyle D, Manning A, Firestein, G. 2001. NF-kappa B regulation by I kappa B kinase-2 in rheumatoid arthritis synoviocytes. *J Immunol* **166**(4): 2705–2711.

Bandara G, Mueller GM, Galea-Lauri J, Tindal MH, Georgescu HI, Suchanek MK, Hung GL, Glorioso JC, Robbins PD, Evans CH. 1993. Intraarticular expression of biologically active interleukin 1-receptor-antagonist protein by *ex vivo* gene transfer. *Proc Natl Acad Sci USA* **90**(22): 10764–10768.

Barchowsky A, Frleta D, Vincenti MP. 2000. Integration of the NF-kappaB and mitogen-activated protein kinase/AP-1 pathways at the collagenase-1 promoter: divergence of IL-1 and TNF-dependent signal transduction in rabbit primary synovial fibroblasts. *Cytokine* **12**(10): 1469–1479.

Baslund B, Tvede N, Danneskiold-Samsoe B, Peterson J, Peterson L, Schuurmann J. 2003. A novel human monoclonal antibody against IL-15 (humax-IL-15) in patients with active rheumatoid arthritis (RA): results of a double-blind, placebo-controlled phase I/II trial. *Arthrit Rheum* **48**(9): S653.

Benbow U, Brinckerhoff CE. 1997. The AP-1 site and MMP gene regulation: what is all the fuss about? *Matrix Biol* **15**(8–9): 519–526.

Berckmans RJ, Nieuwland R, Tak PP, Boing AN, Romijn FP, Kraan MC, Breedveld FC, Hack CE, Sturk A. 2002. Cell-derived microparticles in synovial fluid from inflamed arthritic joints support coagulation exclusively via a factor VII-dependent mechanism. *Arthrit Rheum* **46**(11): 2857–2866.

Bodolay E, Koch AE, Kim J, Szegedi G, Szekanecz Z. 2002. Angiogenesis and chemokines in rheumatoid arthritis and other systemic inflammatory rheumatic diseases. *J Cell Mol Med* **6**(3): 357–376.

Bohne S, Sletten K, Menard R, Buhling F, Vockler S, Wrenger E, Roessner A, Rocken C. 2004. Cleavage of AL amyloid proteins and AL amyloid deposits by cathepsins B, K, and L. *J Pathol* **203**(1): 528–537.

Borish LC, Steinke JW. 2003. 2. Cytokines and chemokines. *J Allergy Clin Immunol* **111** (2 suppl): S460–475.

Brennan FM, Chantry D, Jackson A, Maini R, Feldmann M. 1989. Inhibitory effect of TNF alpha antibodies on synovial cell interleukin-1 production in rheumatoid arthritis. *Lancet* **2**(8657): 244–247.

Brentano F, Schorr O, Simmen B, Gay R, Michel BA, Gay S, Kyburz D. 2004. TLR expression profiles of RA synovial fibroblasts compared to OA and normal skin fibroblasts. *Arthrit Rheum* (Abstract) **50**(9): S517.

Burgess TL, Qian Y, Kaufman S, Ring BD, Van G, Capparelli C, Kelley M, Hsu H, Boyle WJ, Dunstan CR, Hu S, Lacey DL. 1999. The ligand for osteoprotegerin (OPGL) directly activates mature osteoclasts. *J Cell Biol* **145**(3): 527–538.

Bush KA, Farmer KM, Walker JS, Kirkham BW. 2002. Reduction of joint inflammation and bone erosion in rat adjuvant arthritis by treatment with interleukin-17 receptor IgG1 Fc fusion protein. *Arthrit Rheum* **46**(3): 802–805.

Catrina AI, Ulfgren AK, Lindblad S, Grondal L, Klareskog L. 2002. Low levels of apoptosis and high FLIP expression in early rheumatoid arthritis synovium. *Ann Rheum Dis* **61**(10): 934–936.

Caulfield JP, Hein A, Dynesius-Trentham R, Trentham DE. 1982. Morphologic demonstration of two stages in the development of type II collagen-induced arthritis. *Lab Invest* **46**(3):321–343.

Chabaud M, Page G, Miossec P. 2001. Enhancing effect of IL-1, IL-17, and TNF-alpha on macrophage inflammatory protein-3alpha production in rheumatoid arthritis: regulation by soluble receptors and Th2 cytokines. *J Immunol* **167**(10): 6015–6020.

Cheson BD. 2002. CHOP plus rituximab–balancing facts and opinion. *New Engl J Med* **346**(4): 280–282.

Chiocchia G, Boissier MC, Fournier C. 1991. Therapy against murine collagen-induced arthritis with T cell receptor V beta-specific antibodies. *Eur J Immunol* **21**(12): 2899–2905.

Choy EH, Isenberg DA, Garrood T, Farrow S, Ioannou Y, Bird H, Cheung N, Williams B, Hazleman B, Price R, Yoshizaki K, Nishimoto N, Kishimoto T, Panayi GS. 2002. Therapeutic benefit of blocking interleukin-6 activity with an anti-interleukin-6 receptor monoclonal antibody in rheumatoid arthritis: a randomized, double-blind, placebo-controlled, dose-escalation trial. *Arthrit Rheum* **46**(12): 3143–3150.

Collette J, Ulku AS, Der CJ, Jones A, Erickson AH. 2004. Enhanced cathepsin L expression is mediated by different Ras effector pathways in fibroblasts and epithelial cells. *Int J Cancer* **112**(2): 190–199.

Dayer JM. 2003. The pivotal role of interleukin-1 in the clinical manifestations of rheumatoid arthritis. *Rheumatology (Oxford)* **42**(suppl 2): ii3–10.

Dayer JM, Beutler B, Cerami A. 1985. Cachectin/tumor necrosis factor stimulates collagenase and prostaglandin E2 production by human synovial cells and dermal fibroblasts. *J Exp Med* **162**(6): 2163–2168.

Dike LE, Farmer SR. 1988. Cell adhesion induces expression of growth-associated genes in suspension-arrested fibroblasts. *Proc Natl Acad Sci USA* **85**(18): 6792–6796.

Dinarello CA, Cannon JG, Wolff SM, Bernheim HA, Beutler B, Cerami A, Figari IS, Palladino MA, Jr, O'Connor JV. 1986. Tumor necrosis factor (cachectin) is an endogenous pyrogen and induces production of interleukin 1. *J Exp Med* **163**(6): 1433–1450.

Distler JH, Jungel A, Huber LC, Seemayer CA, Reich CF, 3rd, Gay RE, Michel BA, Fontana A, Gay S, Pisetsky DS, Distler, O. 2005. The induction of matrix metalloproteinase and cytokine expression in synovial fibroblasts stimulated with immune cell microparticles. *Proc Natl Acad Sci USA* **102**(8): 2892–2897.

Dooley S, Herlitzka I, Hanselmann R, Ermis A, Henn W, Remberger K, Hopf T, Welter C. 1996. Constitutive expression of c-fos and c-jun, overexpression of ets-2, and reduced expression of metastasis suppressor gene nm23-H1 in rheumatoid arthritis. *Ann Rheum Dis* **55**(5): 298–304.

Ermann J, Fathman CG. 2001. Autoimmune diseases: genes, bugs and failed regulation. *Nat Immunol* **2**(9): 759–761.

Evans CH, Robbins PD, Ghivizzani SC, Herndon JH, Kang R, Bahnson AB, Barranger JA, Elders EM, Gay S, Tomaino MM, Wasko MC, Watkins SC, Whiteside TL, Glorioso JC, Lotze MT, Wright TM. 1996. Clinical trial to assess the safety, feasibility, and efficacy of transferring a potentially anti-arthritic cytokine gene to human joints with rheumatoid arthritis. *Hum Gene Ther* **7**(10): 1261–1280.

Fassbender HG. 1983. Histomorphological basis of articular cartilage destruction in rheumatoid arthritis. *Coll Relat Res* **3**(2): 141–155.

Feldmann M, Brennan FM, Maini RN. 1996. Role of cytokines in rheumatoid arthritis. *Rev Immunol* **14**: 397–440.

Firestein GS. 1996. Invasive fibroblast-like synoviocytes in rheumatoid arthritis. Passive responders or transformed aggressors? *Arthrit Rheum* **39**(11): 1781–1790.

Fox DA. 1997. The role of T cells in the immunopathogenesis of rheumatoid arthritis: new perspectives. *Arthrit Rheum* **40**(4): 598–609.

Franz JK, Kolb SA, Hummel KM, Lahrtz F, Neidhart M, Aicher WK, Pap T, Gay RE, Fontana A, Gay S. 1998a. Interleukin-16, produced by synovial fibroblasts, mediates chemoattraction for CD4+ T lymphocytes in rheumatoid arthritis. *Eur J Immunol* **28**(9): 2661–2671.

Franz JK, Pap T, Muller-Ladner U, Gay R, Burmester G, Gay S. 1998b. T-cell independent joint destruction. In *T Cells in Arthritis* Miossec P, van-den BW, Firestein GS (eds). Basel: Birkenhäuser; 55–74.

Franz JK, Pap T, Hummel KM, Nawrath M, Aicher WK, Shigeyama Y, Muller-Ladner U, Gay RE, Gay S. 2000. Expression of sentrin, a novel antiapoptotic molecule, at sites of synovial invasion in rheumatoid arthritis. *Arthrit Rheum* **43**(3): 599–607.

Furst D. 2004. Anakinra: review of recombinant human interleukin-I receptor antagonist in the treatment of rheumatoid arthritis. *Clin Ther* **26**(12): 1960–1975.

Gabriel SE. 2001. The epidemiology of rheumatoid arthritis. *Rheum Dis Clin N Am* **27**(2): 269–281.

Geiler T, Kriegsmann J, Keyszer GM, Gay RE, Gay S. 1994. A new model for rheumatoid arthritis generated by engraftment of rheumatoid synovial tissue and normal human cartilage into SCID mice. *Arthrit Rheum* **37**(11): 1664–1671.

Germann T, Szeliga J, Hess H, Storkel S, Podlaski FJ, Gately MK, Schmitt E, Rude E. 1995. Administration of interleukin 12 in combination with type II collagen induces severe arthritis in DBA/1 mice. *Proc Natl Acad Sci USA* **92**(11): 4823–4827.

Goldschmidt TJ, Jansson L, Holmdahl R. 1990. *In vivo* elimination of T cells expressing specific T-cell receptor V beta chains in mice susceptible to collagen-induced arthritis. *Immunology* **69**(4): 508–514.

Gracie JA, Forsey RJ, Chan WL, Gilmour A, Leung BP, Greer MR, Kennedy K, Carter R, Wei XQ, Xu D, Field M, Foulis A, Liew FY, McInnes IB. 1999. A proinflammatory role for IL-18 in rheumatoid arthritis. *J Clin Invest* **104**(10): 1393–1401.

Gravallese EM. 2002. Bone destruction in arthritis. *Ann Rheum Dis* **61**(suppl 2): ii84–86.

Grimbacher B, Aicher WK, Peter HH, Eibel H. 1997. Measurement of transcription factor c-fos and EGR-1 mRNA transcription levels in synovial tissue by quantitative RT-PCR. *Rheumatol Int* **17**(3): 109–112.

Guinec N, Dalet-Fumeron V, Pagano M. 1993. '*In vitro*' study of basement membrane degradation by the cysteine proteinases, cathepsins B, B-like and L. Digestion of collagen IV, laminin, fibronectin, and release of gelatinase activities from basement membrane fibronectin. *Biol Chem Hoppe Seyler* **374**(12): 1135–1146.

Gustin JA, Ozes ON, Akca H, Pincheira R, Mayo LD, Li Q, Guzman JR, Korgaonkar CK, Donner DB. 2004. Cell type-specific expression of the IkappaB kinases determines the significance of phosphatidylinositol 3-kinase/Akt signaling to NF-kappa B activation. *J Biol Chem* **279**(3): 1615–1620.

Han Z, Boyle DL, Manning AM, Firestein GS. 1998. AP-1 and NF-kappaB regulation in rheumatoid arthritis and murine collagen-induced arthritis. *Autoimmunity* **28**(4): 197–208.

Harris WH, Sledge CB. 1990. Total hip and total knee replacement (1). *New Engl J Med* **323**(11): 725–731.

Haworth C, Brennan FM, Chantry D, Turner M, Maini RN, Feldmann M. 1991. Expression of granulocyte-macrophage colony-stimulating factor in rheumatoid arthritis: regulation by tumor necrosis factor-alpha. *Eur J Immunol* **21**(10): 2575–2579.

Hsieh CS, Macatonia SE, Tripp CS, Wolf SF, O'Garra A, Murphy KM. 1993. Development of TH1 CD4+ T cells through IL-12 produced by *Listeria*-induced macrophages. *Science* **260**(5107): 547–549.

Huet G, Flipo RM, Colin C, Janin A, Hemon B, Collyn-d'Hooghe M, Lafyatis R, Duquesnoy B, Degand P. 1993. Stimulation of the secretion of latent cysteine proteinase activity by tumor necrosis factor alpha and interleukin-1. *Arthrit Rheum* **36**(6): 772–780.

Hummel KM, Petrow PK, Franz JK, Muller-Ladner U, Aicher WK, Gay RE, Bromme D, Gay S. 1998. Cysteine proteinase cathepsin K mRNA is expressed in synovium of patients with rheumatoid arthritis and is detected at sites of synovial bone destruction. *J Rheumatol* **25**(10): 1887–1894.

Hung GL, Galea-Lauri J, Mueller GM, Georgescu HI, Larkin LA, Suchanek MK, Tindal MH, Robbins PD, Evans CH. 1994. Suppression of intra-articular responses to interleukin-1 by transfer of the interleukin-1 receptor antagonist gene to synovium. *Gene Ther* **1**(1): 64–69.

Ishikawa H, Hirata S, Andoh Y, Kubo H, Nakagawa N, Nishibayashi Y, Mizuno K. 1996. An immunohistochemical and immunoelectron microscopic study of adhesion molecules in synovial pannus formation in rheumatoid arthritis. *Rheumatol Int* **16**(2): 53–60.

Joosten LA, Helsen MM, van de Loo FA, van den Berg WB. 1996. Anticytokine treatment of established type II collagen-induced arthritis in DBA/1 mice. A comparative study using anti-TNF alpha, anti-IL-1 alpha/beta, and IL-1Ra. *Arthrit Rheum* **39**(5): 797–809.

Jungel A, Distler JH, Kurowska-Stolarska M, Seemayer CA, Seibl R, Forster A, Michel BA, Gay RE, Emmrich F, Gay S, Distler, O. 2004. Expression of interleukin-21 receptor, but not interleukin-21, in synovial fibroblasts and synovial macrophages of patients with rheumatoid arthritis. *Arthrit Rheum* **50**(5): 1468–1476.

Kasama T, Strieter RM, Lukacs NW, Lincoln PM, Burdick MD, Kunkel SL. 1995. Interleukin-10 expression and chemokine regulation during the evolution of murine type II collagen-induced arthritis. *J Clin Invest* **95**(6): 2868–2876.

Keane J, Gershon S, Wise RP, Mirabile-Levens E, Kasznica J, Schwieterman WD, Siegel JN, Braun MM. 2001. Tuberculosis associated with infliximab, a tumor necrosis factor alpha-neutralizing agent. *New Engl J Med* **345**(15): 1098–1104.

Kehlen A, Thiele K, Riemann D, Langner J. 2002. Expression, modulation and signalling of IL-17 receptor in fibroblast-like synoviocytes of patients with rheumatoid arthritis. *Clin Exp Immunol* **127**(3): 539–546.

Klimiuk PA, Yang H, Goronzy JJ, Weyand CM. 1999. Production of cytokines and metalloproteinases in rheumatoid synovitis is T cell dependent. *Clin Immunol* **90**(1): 65–78.

Kobayashi H, Ohi H, Sugimura M, Shinohara H, Fujii T, Terao T. 1992. Inhibition of in vitro ovarian cancer cell invasion by modulation of urokinase-type plasminogen activator and cathepsin B. *Cancer Res* **52**(13): 3610–3614.

Kobayashi H, Moniwa N, Sugimura M, Shinohara H, Ohi H, Terao T. 1993. Effects of membrane-associated cathepsin B on the activation of receptor-bound prourokinase and subsequent invasion of reconstituted basement membranes. *Biochim Biophys Acta* **1178**(1): 55–62.

Koch AE, Kunkel SL, Harlow LA, Johnson B, Evanoff HL, Haines GK, Burdick MD, Pope RM, Strieter RM. 1992. Enhanced production of monocyte chemoattractant protein-1 in rheumatoid arthritis. *J Clin Invest* **90**(3): 772–779.

Kontny E, Ziolkowska M, Dudzinka E, Filipowicz-Sosnowska A, Ryzewska A. 1995. Modified expression of c-Fos and c-Jun proteins and production of interleukin-1 beta in patients with rheumatoid arthritis. *Clin Exp Rheumatol* **13**(1): 51–57.

Kotake S, Udagawa N, Takahashi N, Matsuzaki K, Itoh K, Ishiyama S, Saito S, Inoue K, Kamatani N, Gillespie MT, Martin TJ, Suda T. 1999. IL-17 in synovial fluids from patients with rheumatoid arthritis is a potent stimulator of osteoclastogenesis. *J Clin Invest* **103**(9): 1345–1352.

Kraan MC, Versendaal H, Jonker M, Bresnihan B, Post WJ, Hart BA, Breedveld FC, Tak PP. 1998. Asymptomatic synovitis precedes clinically manifest arthritis. *Arthrit Rheum* **41**(8): 1481–1488.

Kremer JM, Westhovens R, Leon M, Di Giorgio E, Alten R, Steinfeld S, Russell A, Dougados M, Emery P, Nuamah IF, Williams GR, Becker JC, Hagerty DT, Moreland LW. 2003. Treatment of rheumatoid arthritis by selective inhibition of T-cell activation with fusion protein CTLA4Ig. *New Engl J Med* **349**(20): 1907–1915.

Kriegsmann J, Keyszer GM, Geiler T, Brauer R, Gay RE, Gay S. 1995. Expression of vascular cell adhesion molecule-1 mRNA and protein in rheumatoid synovium demonstrated by *in situ* hybridization and immunohistochemistry. *Lab Invest* **72**(2): 209–214.

Kuchen S, Seemayer CA, Kuenzler P, Pap T, Gay R, Michel BA, Neidhart M, Gay S. 2001. Cytokine-independent up-regulation of matrix metalloproteinase 1 mRNA expression by the stress activated protein kinase 4. *Arthrit Rheum* **44**(abstract): S183.

Lafyatis R, Remmers EF, Roberts AB, Yocum DE, Sporn MB, Wilder RL. 1989. Anchorage-independent growth of synoviocytes from arthritic and normal joints. Stimulation by exogenous platelet-derived growth factor and inhibition by transforming growth factor-beta and retinoids. *J Clin Invest* **83**(4): 1267–1276.

Lakka SS, Gondi CS, Yanamandra N, Olivero WC, Dinh DH, Gujrati M, Rao JS. 2004. Inhibition of cathepsin B and MMP-9 gene expression in glioblastoma cell line via RNA interference reduces tumor cell invasion, tumor growth and angiogenesis. *Oncogene* **23**(27): 4681–4689.

Lawrence RC, Helmick CG, Arnett FC, Deyo RA, Felson DT, Giannini EH, Heyse SP, Hirsch R, Hochberg MC, Hunder GG, Liang MH, Pillemer SR, Steen VD, Wolfe F. 1998. Estimates of the prevalence of arthritis and selected musculoskeletal disorders in the United States. *Arthrit Rheum* **41**(5): 778–799.

Lee ZH, Kim HH. 2003. Signal transduction by receptor activator of nuclear factor kappa B in osteoclasts. *Biochem Biophys Res Commun* **305**(2): 211–214.

Lemaire R, Huet G, Zerimech F, Grard G, Fontaine C, Duquesnoy B, Flipo RM. 1997. Selective induction of the secretion of cathepsins B and L by cytokines in synovial fibroblast-like cells. *Br J Rheumatol* **36**(7): 735–743.

Li WQ, Dehnade F, Zafarullah M. 2001. Oncostatin M-induced matrix metalloproteinase and tissue inhibitor of metalloproteinase-3 genes expression in chondrocytes requires Janus kinase/STAT signaling pathway. *J Immunol* **166**(5): 3491–3498.

Liew FY, McInnes IB. 2002. Role of interleukin 15 and interleukin 18 in inflammatory response. *Ann Rheum Dis* **61**(suppl 2): ii100–102.

Maini RN, Elliott MJ, Brennan FM, Feldmann M. 1995. Beneficial effects of tumour necrosis factor-alpha (TNF-alpha) blockade in rheumatoid arthritis (RA). *Clin Exp Immunol* **101**(2): 207–212.

Maini R, St Clair EW, Breedveld F, Furst D, Kalden J, Weisman M, Smolen J, Emery P, Harriman G, Feldmann M, Lipsky P. 1999. Infliximab (chimeric anti-tumour necrosis factor alpha monoclonal antibody) versus placebo in rheumatoid arthritis patients receiving concomitant methotrexate: a randomised phase III trial. ATTRACT Study Group. *Lancet* **354**(9194): 1932–1939.

Makarov SS. 2001. NF-kappa B in rheumatoid arthritis: a pivotal regulator of inflammation, hyperplasia, and tissue destruction. *Arthrit Res* **3**(4): 200–206.

Manetti R, Parronchi P, Giudizi MG, Piccinni MP, Maggi E, Trinchieri G, Romagnani S. 1993. Natural killer cell stimulatory factor (interleukin 12 [IL-12]) induces T helper type 1 (Th1)-specific immune responses and inhibits the development of IL-4-producing Th cells. *J Exp Med* **177**(4): 1199–1204.

Marinova-Mutafchieva L, Williams RO, Funa K, Maini RN, Zvaifler NJ. 2002. Inflammation is preceded by tumor necrosis factor-dependent infiltration of mesenchymal cells in experimental arthritis. *Arthrit Rheum* **46**(2): 507–513.

Marok R, Winyard PG, Coumbe A, Kus ML, Gaffney K, Blades S, Mapp PI, Morris CJ, Blake DR, Kaltschmidt C, Baeuerle PA. 1996. Activation of the transcription factor nuclear factor-kappaB in human inflamed synovial tissue. *Arthrit Rheum* **39**(4): 583–591.

McInnes IB, Gracie JA. 2004. Interleukin-15: a new cytokine target for the treatment of inflammatory diseases. *Curr Opin Pharmac* **4**(4): 392–397.

McInnes IB, Gracie JA, Harnett M, Harnett W, Liew FY. 2003. New strategies to control inflammatory synovitis: interleukin 15 and beyond. *Ann Rheum Dis* **62**(suppl 2): ii51–54.

Mengshol JA, Vincenti MP, Coon CI, Barchowsky A, Brinckerhoff CE. 2000. Interleukin-1 induction of collagenase 3 (matrix metalloproteinase 13) gene expression in chondrocytes requires p38, c-Jun N-terminal kinase, and nuclear factor kappaB: differential regulation of collagenase 1 and collagenase 3. *Arthrit Rheum* **43**(4): 801–811.

Metzger DW, McNutt RM, Collins JT, Buchanan JM, Van Cleave VH, Dunnick WA. 1997. Interleukin-12 acts as an adjuvant for humoral immunity through interferon-gamma-dependent and -independent mechanisms. *Eur J Immunol* **27**(8): 1958–1965.

Miagkov AV, Kovalenko DV, Brown CE, Didsbury JR, Cogswell JP, Stimpson SA, Baldwin AS, Makarov SS. 1998. NF-kappaB activation provides the potential link between inflammation and hyperplasia in the arthritic joint. *Proc Natl Acad Sci USA* **95**(23): 13859–13864.

Mishiro T, Nakano S, Takahara S, Miki M, Nakamura Y, Yasuoka S, Nikawa T, Yasui N. 2004. Relationship between cathepsin B and thrombin in rheumatoid arthritis. *J Rheumatol* **31**(7): 1265–1273.

Muller-Ladner U, Kriegsmann J, Gay RE, Koopman WJ, Gay S, Chatham WW. 1995a. Progressive joint destruction in a human immunodeficiency virus-infected patient with rheumatoid arthritis. *Arthrit Rheum* **38**(9): 1328–1332.

Muller-Ladner U, Kriegsmann J, Franklin BN, Matsumoto S, Geiler T, Gay RE, Gay S. 1996. Synovial fibroblasts of patients with rheumatoid arthritis attach to and invade normal human cartilage when engrafted into SCID mice. *Am J Pathol* **149**(5): 1607–1615.

Nakano K, Okada Y, Saito K, Tanaka Y. 2004. Induction of RANKL expression and osteoclast maturation by the binding of fibroblast growth factor 2 to heparan sulfate proteoglycan on rheumatoid synovial fibroblasts. *Arthrit Rheum* **50**(8): 2450–2458.

Neidhart M, Rethage J, Kuchen S, Kunzler P, Crowl RM, Billingham ME, Gay RE, Gay S. 2000. Retrotransposable L1 elements expressed in rheumatoid arthritis synovial tissue: association with genomic DNA hypomethylation and influence on gene expression. *Arthrit Rheum* **43**(12): 2634–2647.

Neidhart M, Jungel A, Comazzi M, von Knoch R, Ospelt C, Simmen B, Michel BA, Gay R, Gay S. 2004. Deficient expression of interleukin-10 receptor alpha – chain in RA synovium – limitations of inflammatory animal models. *Arthrit Rheum* (abstract) **50**(9): S112.

Neumann E, Judex M, Kullmann F, Grifka J, Robbins PD, Pap T, Gay RE, Evans CH, Gay S, Scholmerich J, Muller-Ladner U. 2002. Inhibition of cartilage destruction by double gene transfer of IL-IRa and IL-10 involves the activin pathway. *Gene Ther* **9**(22): 1508–1519.

Okura T, Gong L, Kamitani T, Wada T, Okura I, Wei CF, Chang HM, Yeh ET. 1996. Protection against Fas/APO-1- and tumor necrosis factor-mediated cell death by a novel protein, sentrin. *J Immunol* **157**(10): 4277–4281.

Ospelt C, Neidhart M, Michel BA, Simmen B, Gay RE, Gay S. 2004. p38delta induces MMP-3 in activated rhuematoid arthritis synovial fibroblasts. *Arthrit and Rheum* **10**(abstract) **50**(9): S155.

Palombella VJ, Conner EM, Fuseler JW, Destree A, Davis JM, Laroux FS, Wolf RE, Huang J, Brand S, Elliott PJ, Lazarus D, McCormack T, Parent L, Stein R, Adama J, Grisham MB. 1998. Role of the proteasome and NF-kappaB in streptococcal cell wall-induced polyarthritis. *Proc Natl Acad Sci USA* **95**(26): 15671–15676.

Pap T, Gay S, Schett G. 2003. Matrix Metalloproteinases. In: Smolen JS, Lipsky PE (eds). Martin Dunitz/Taylor & Francis: London, pp. 483–497

Pap T, Muller-Ladner U, Gay RE, Gay S. 2000a. Fibroblast biology. Role of synovial fibroblasts in the pathogenesis of rheumatoid arthritis. *Arthrit Res* **2**(5): 361–367.

Pap T, Franz JK, Hummel KM, Jeisy E, Gay R, Gay S. 2000b. Activation of synovial fibroblasts in rheumatoid arthritis: lack of Expression of the tumour suppressor PTEN at sites of invasive growth and destruction. *Arthrit Res* **2**(1): 59–64.

Pap T, Nawrath M, Heinrich J, Bosse M, Baier A, Hummel KM, Petrow P, Kuchen S, Michel BA, Gay RE, Muller-Ladner U, Moelling K, Gay S. 2004. Cooperation of Ras- and c-Myc-dependent pathways in regulating the growth and invasiveness of synovial fibroblasts in rheumatoid arthritis. *Arthrit Rheum* **50**(9): 2794–2802.

Parks E, Strieter RM, Lukacs NW, Gauldie J, Hitt M, Graham FL, Kunkel SL. 1998. Transient gene transfer of IL-12 regulates chemokine expression and disease severity in experimental arthritis. *J Immunol* **160**(9): 4615–4619.

Perlman H, Liu H, Georganas C, Koch AE, Shamiyeh E, Haines GK. 3rd, and Pope RM. 2001. Differential expression pattern of the antiapoptotic proteins, Bcl-2 and FLIP, in experimental arthritis. *Arthrit Rheum* **44**(12): 2899–2908.

Pierer M, Rethage J, Seibl R, Lauener R, Brentano F, Wagner U, Hantzschel H, Michel BA, Gay RE, Gay S, Kyburz D. 2004. Chemokine secretion of rheumatoid arthritis synovial fibroblasts stimulated by Toll-like receptor 2 ligands. *J Immunol* **172**(2): 1256–1265.

Plater-Zyberk C, Joosten LA, Helsen MM, Sattonnet-Roche P, Siegfried C, Alouani S, van De Loo FA, Graber P, Aloni S, Cirillo R, Lubberts E, Dinarello CA, van Den Berg WB, Chvatchko Y. 2001. Therapeutic effect of neutralizing endogenous IL-18 activity in the collagen-induced model of arthritis. *J Clin Invest* **108**(12): 1825–1832.

Plows D, Kontogeorgos G, Kollias G. 1999. Mice lacking mature T and B lymphocytes develop arthritic lesions after immunization with type II collagen. *J Immunol* **162**(2): 1018–1023.

Podolin PL, Callahan JF, Bolognese BJ, Li YH, Carlson K, Davis TG, Mellor GW, Evans C, Roshak AK. 2005. Attenuation of murine collagen-induced arthritis by a novel, potent, selective small molecule inhibitor of I{κ}B kinase 2, TPCA-1 (2-[(aminocarbonyl)amino]-5-(4-fluorophenyl)-3-thiophenecarboxamide), occurs via reduction of proinflammatory cytokines and antigen-induced T cell proliferation. *J Pharmac Exp Ther* **312**(1): 373–381.

Qu Z, Garcia CH, O'Rourke LM, Planck SR, Kohli M, Rosenbaum JT. 1994. Local proliferation of fibroblast-like synoviocytes contributes to synovial hyperplasia. Results of proliferating cell nuclear antigen/cyclin, c-myc, and nucleolar organizer region staining. *Arthrit Rheum* **37**(2): 212–220.

Ranges GE, Sriram S, Cooper SM. 1985. Prevention of type II collagen-induced arthritis by *in vivo* treatment with anti-L3T4. *J Exp Med* **162**(3): 1105–1110.

Rinaldi N, Schwarz-Eywill M, Weis D, Leppelmann-Jansen P, Lukoschek M, Keilholz U, Barth TF. 1997. Increased expression of integrins on fibroblast-like synoviocytes from rheumatoid arthritis *in vitro* correlates with enhanced binding to extracellular matrix proteins. *Ann Rheum Dis* **56**(1): 45–51.

Romas E, Gillespie MT, Martin TJ. 2002. Involvement of receptor activator of NFkappaB ligand and tumor necrosis factor-alpha in bone destruction in rheumatoid arthritis. *Bone* **30**(2): 340–346.

Saidenberg Kermanac'h N, Bessis N, Cohen-Solal M, De Vernejoul MC, Boissier MC. 2002. Osteoprotegerin and inflammation. *Eur Cytokine Netw* **13**(2): 144–153.

Schedel J, Gay RE, Kuenzler P, Seemayer C, Simmen B, Michel BA, Gay S. 2002. FLICE-inhibitory protein expression in synovial fibroblasts and at sites of cartilage and bone erosion in rheumatoid arthritis. *Arthrit Rheum* **46**(6): 1512–1518.

Schett G, Tohidast-Akrad M, Smolen JS, Schmid BJ, Steiner CW, Bitzan P, Zenz P, Redlich K, Xu Q, Steiner G. 2000. Activation, differential localization, and regulation of the stress-activated protein kinases, extracellular signal-regulated kinase, c-JUN N-terminal kinase, and p38 mitogen-activated protein kinase, in synovial tissue and cells in rheumatoid arthritis. *Arthrit Rheum* **43**(11): 2501–2512.

Schulze-Koops H, Kalden JR. 2001. The balance of Th1/Th2 cytokines in rheumatoid arthritis. *Best Pract Res Clin Rheumatol* **15**(5): 677–691.

Schwartz MA. 1997. Integrins, oncogenes, and anchorage independence. *J Cell Biol* **139**(3): 575–578.

Seibl R, Birchler T, Loeliger S, Hossle JP, Gay RE, Saurenmann T, Michel BA, Seger RA, Gay S, Lauener RP. 2003. Expression and regulation of Toll-like receptor 2 in rheumatoid arthritis synovium. *Am J Pathol* **162**(4): 1221–1227.

Smith JB, Haynes MK. 2002. Rheumatoid arthritis – a molecular understanding. *Ann Intern Med* **136**(12): 908–922.

Stahl H, L, S, J, S, Filipowicz-Sosnowska A, Edwards JCW. 2003. Rituximamb in RA: efficacy and safety from a randomised controlled trial. *Ann Rheum Dis* **62**(suppl 1): 65.

Suzuki N, Nakajima A, Yoshino S, Matsushima K, Yagita H, Okumura K. 1999. Selective accumulation of CCR5+ T lymphocytes into inflamed joints of rheumatoid arthritis. *Int Immunol* **11**(4): 553–559.

Suzuki M, Tetsuka T, Yoshida S, Watanabe N, Kobayashi M, Matsui N, Okamoto T. 2000. The role of p38 mitogen-activated protein kinase in IL-6 and IL-8 production from the TNF-alpha- or IL-1beta-stimulated rheumatoid synovial fibroblasts. *FEBS Lett* **465**(1): 23–27.

Tak PP, Smeets TJ, Daha MR, Kluin PM, Meijers KA, Brand R, Meinders AE, Breedveld FC. 1997. Analysis of the synovial cell infiltrate in early rheumatoid synovial tissue in relation to local disease activity. *Arthrit Rheum* **40**(2): 217–225.

Takayanagi H, Iizuka H, Juji T, Nakagawa T, Yamamoto A, Miyazaki T, Koshihara Y, Oda H, Nakamura K, Tanaka S. 2000. Involvement of receptor activator of nuclear factor kappaB ligand/osteoclast differentiation factor in osteoclastogenesis from synoviocytes in rheumatoid arthritis. *Arthrit Rheum* **43**(2): 259–269.

Tanaka M, Harigai M, Kawaguchi Y, Ohta S, Sugiura T, Takagi K, Ohsako-Higami S, Fukasawa C, Hara M, Kamatani N. 2001. Mature form of interleukin 18 is expressed in rheumatoid arthritis synovial tissue and contributes to interferon-gamma production by synovial T cells. *J Rheumatol* **28**(8): 1779–1787.

Tarner IH, Fathman CG. 2001. Gene therapy in autoimmune disease. *Curr Opin Immunol* **13**(6): 676–682.

Tarner IH, Muller-Ladner U, Gay RE, Fathman CG, Gay S. 2003. The pathogenesis of rheumatoid arthritis – gene transfer to detect novel targets for treatment. In: Zouli M (ed.). NATO Science Series: Life and Behavioural Sciences, IOS Press: Amsterdam, pp. 315–346

Taylor PC, Peters AM, Paleolog E, Chapman PT, Elliott MJ, McCloskey R, Feldmann M, Maini RN. 2000. Reduction of chemokine levels and leukocyte traffic to joints by tumor necrosis factor alpha blockade in patients with rheumatoid arthritis. *Arthrit Rheum* **43**(1): 38–47.

Thomas R, McIlraith M, Davis LS, Lipsky PE. 1992. Rheumatoid synovium is enriched in CD45RBdim mature memory T cells that are potent helpers for B cell differentiation. *Arthrit Rheum* **35**(12): 1455–1465.

Tolar J, Teitelbaum SL, Orchard PJ. 2004. Osteopetrosis. *New Engl J Med* **351**(27): 2839–2849.

Trabandt A, Aicher WK, Gay RE, Sukhatme VP, Fassbender HG, Gay S. 1992. Spontaneous expression of immediately-early response genes c-fos and egr-1 in collagenase-producing rheumatoid synovial fibroblasts. *Rheumatol Int* **12**(2): 53–59.

van de Loo FA, van den Berg WB. 2002. Gene therapy for rheumatoid arthritis. Lessons from animal models, including studies on interleukin-4, interleukin-10, and interleukin-1 receptor antagonist as potential disease modulators. *Rheum Dis Clin N Am* **28**(1): 127–149.

van den Berg WB, Bresnihan B. 1999. Pathogenesis of joint damage in rheumatoid arthritis: evidence of a dominant role for interleukin-I. *Baillières Best Pract Res Clin Rheumatol* **13**(4): 577–597.

van den Berg WB, van de Loo FA, Otterness I, Arntz O, Joosten LA. 1991. *In vivo* evidence for a key role of IL-1 in cartilage destruction in experimental arthritis. *Agents Actions Suppl* **32**: 159–163.

van der Laan WH, Pap T, Ronday HK, Grimbergen JM, Huisman LG, TeKoppele JM, Breedveld FC, Gay RE, Gay S, Huizinga TW, Verheijen JH, Quax PH. 2000. Cartilage degradation and invasion by rheumatoid synovial fibroblasts is inhibited by gene transfer of a cell surface-targeted plasmin inhibitor. *Arthrit Rheum* **43**(8): 1710–1718.

van der Laan WH, Quax PH, Seemayer CA, Huisman LG, Pieterman EJ, Grimbergen JM, Verheijen JH, Breedveld FC, Gay RE, Gay S, Huizinga TW, Pap T. 2003. Cartilage degradation and invasion by rheumatoid synovial fibroblasts is inhibited by gene transfer of TIMP-1 and TIMP-3. *Gene Ther* **10**(3): 234–242.

Vincenti MP, Coon CI, Brinckerhoff CE. 1998. Nuclear factor kappaB/p50 activates an element in the distal matrix metalloproteinase 1 promoter in interleukin-1beta-stimulated synovial fibroblasts. *Arthrit Rheum* **41**(11): 1987–1994.

Zinkernagel RM. 2001. Maternal antibodies, childhood infections, and autoimmune diseases. *New Engl J Med* **345**(18): 1331–1335.

Zvaifler NJ, Boyle D, Firestein GS. 1994. Early synovitis – synoviocytes and mononuclear cells. *Sem Arthritis Rheum* **23**(6 suppl 2): 11–16.

Zvaifler NJ, Tsai V, Alsalameh S, von Kempis J, Firestein GS, Lotz M. 1997. Pannocytes: distinctive cells found in rheumatoid arthritis articular cartilage erosions. *Am J Pathol* **150**(3): 1125–1138.

14

T Cell Epitope Hierarchy in Experimental Autoimmune Models

Edit Buzas

Abstract

During the course of an autoimmune inflammation the initial immune response focuses on dominant antigenic determinant(s). Endogenous self-priming subsequently leads to recognition of diverse cryptic epitopes of the same antigen (epitope spreading). Factors dictating epitope dominance/crypticity include thymic antigen expression, alternative splicing, thymic antigen processing and competitive interactions of antigenic peptides with self major histocompatibility complex, post-translational modifications, etc. Data from the experimental autoimmune encephalomyelitis model provided key insights into the rules governing establishment of epitope hierarchy. The chapter also summarizes data coming from the autoimmune aggrecan-induced murine arthritis model that suggest a significant role of glycosylation in establishment of epitope hierarchy. The initial immune response (focusing on the N-terminal G1 domain of aggrecan) later spreads to the C-terminal G3 domain. Intriguingly, both N- and C-terminal globular domains of aggrecan carry minimal glycosylation as compared with the long central carbohydrate attachment region, the epitopes of which are recognized only in acute arthritis.

14.1 Introduction

In spite of the fact that complex antigens may carry a variety of antigenic determinants, only a minor fraction of the potential epitopes are presented in an immunodominant manner, while the remaining peptides remain ‘silent’ (cryptic). This strong focus of the polyclonal response on one or a few particular epitopes of a given antigen is most evident at early phases of an immune response, while later, during intramolecular epitope spreading, recognition of previously hidden epitopes becomes increasingly evident. Understanding of the principles of this distinguished recognition of immunodominant antigenic determinants has an impact in vaccine

Immunogenomics and Human Disease Edited by András Falus

design against infectious agents as well as in development of novel therapeutic strategies to treat autoimmune diseases. This chapter focuses on data coming from experimental models of autoimmunity and discusses how the hierarchy of dominant and cryptic self-determinants is established and how immunogenomics and immunoinformatics can contribute to prediction of such hierarchic patterns.

14.2 Immunodominance and Crypticity

T cells recognize short linear peptide epitopes in context with appropriate major histocompatibility complex (MHC). Synthetic peptides have been widely used to map T cell epitopes in health and disease. However, the T cell repertoire activated by immunization with a synthetic peptide epitope is not identical with the repertoire activated if the corresponding intact antigen is used for immunization.

Peptides that are strongly immunogenic and also stimulate a recall response upon immunization with the intact antigen are considered *dominant* and result from natural intracellular processing of the corresponding antigen. In contrast, peptides are referred to as *cryptic* determinants if they themselves are strongly immunogenic, however fail to cross-react with cells primed with the intact antigen (as intracellular processing of the native antigen prevents such epitopes being displayed for T cell recognition; Vinear *et al.*, 1996). *Subdominant* epitopes are defined as determinants found in the epitope hierarchy between dominant and cryptic epitopes.

There are shifts in the hierarchy of display of dominant and cryptic determinants related to local inflammation, to changes in the state or type of the antigen-presenting cells (APCs), and to exogenous vs endogenous processing (reviewed by Sercarz *et al.*, 1993; Sercarz, 2002).

14.3 Epitope Spreading (Endogenous Self-Priming)

It is a phenomenon observed during chronic exposure to an antigen and autoimmune disease progression. It involves sequential activation of T cells that recognize epitopes distinct from and noncross-reactive with the inducing epitopes (Lehmann *et al.*, 1992, 1993). Thus, epitope spreading is a diversification of epitope specificity from the initial focused, dominant epitope-specific immune response, directed against a self or foreign protein, to subdominant and/or cryptic epitopes on the same protein (*intramolecular spreading*) or other proteins (*intermolecular spreading*).

Experimental autoimmune encephalomyelitis (EAE) is a widely used model of the human demyelinating disease, multiple sclerosis (MS) (Fressinaud, Sarlieve and Vincendon, 1990; Tsunoda and Fujinami, 1996). In EAE it has been shown that epitope spreading plays an important functional role in the pathomechanism of the ongoing disease. Determinant spreading within the same myelin protein as well as between myelin proteins may be responsible for different relapses (McRae *et al.*, 1995; Wang *et al.*, 2001).

There is an invariable sequence of epitope recognition during EAE progression in the SWXJ mouse when primed with proteolipid antigen (PLP) 139–151: PLP 139–151 →PLP 249–273 →myelin basic protein (MBP) 87–99→PLP 173–198 (Yu, Johnson and Tuohy, 1996). Induction of tolerance (activation of regulatory T cells) to spread myelin epitopes (Wildbaum, Netzer and Karin, 2002) or blocking costimulation of T cells (necessary for epitope spreading; Miller *et al.*, 1995) interferes with epitope spreading, and clinical relapses of EAE can be prevented.

In ongoing EAE of SWXJ mice, an immunoregulatory spreading repertoire was established by transferring T cells genetically modified in a way to secrete high levels of IL-10 in response to a dominant epitope spreading determinant. The Th2/Tr1-like spreading repertoire resulted in a significant and prolonged inhibition of disease progression and demyelination characterized by both bystander inhibition of the recall response to the priming antigen, and a Th1→Tr1 spreading response. Thus, targetting of the epitope spreading cascade with regulatory T cells induced an immune-deviated spreading response capable of blocking ongoing inflammatory autoreactivity and disease progression (Yin *et al.*, 2001).

It has been suggested that induction of TH2 responses via epitope spreading may be an important intrinsic immunoregulatory mechanism to limit tissue destruction (Vanderlugt and Miller, 2002).

Possibly naive T cells enter inflamed tissues and are activated by local APCs, presumably dentritic cells (DCs), to initiate epitope spreading. The endogenous self-priming possibly occurs when previously sequestered self-epitopes enter the inflammatory milieu as a result of tissue breakdown. Characteristically, under inflammatory conditions the cryptic determinants gain visibility and might play a role in determinant spreading and disease pathology.

14.4 Degenerate T Cell Epitope Recognition

Degeneracy has several synonymous terms, including promiscous recognition, cross-reactivity or polyreactivity and mimicry. These refer to the ability of an immune receptor to bind to many different ligands. This is an inherent property of the immune system and results in the ability of an individual to recognize almost any peptide–MHC complex within a limited time frame.

The concept of degenerate T cell recognition was strongly supported by data from many groups showing that T cells respond to epitope peptides as well as modified sequence variants (altered peptide ligands, APLs), and T cells have differential responses to these APLs. Some peptides inducing both proliferation and cytokine secretion of the T cells were called *full agonist* ligands as opposed to sequences that stimulated either cytokine response or proliferation (referred to as *partial agonists*). Peptides that triggered improved responses as compared with the natural peptide ligand were given the name *superagonist ligands.*

APLs that do not induce any responses are referred to as *null agonists*, while some APLs are able to specifically antagonize and inhibit T cell activation induced by the wild-type antigenic peptide; these are called *antagonist ligands* (Anderton and Wraith, 2002a).

It has been shown that APL can induce a qualitatively different pattern of signal transduction events from that induced by any concentration of the native ligand. Presumably several signalling modules are directly linked to the TCR–CD3 complex and they can be dissociated from each other as a result of the nature of the ligand (Sloan-Lancaster and Allen, 1996).

Surprisingly, it was found that cross-recognition of a microbial and a self epitope involved no similarity at all between the two molecules, suggesting that shape mimicry was sufficient for cross-recognition by a single T cell receptor (TCR; Bhardwaj *et al.*, 1993; Maverakis, van den Elzen and Sercarz, 2001).

A recent report showing CDR3 loop flexibility also provides structural data that might help explain TCR binding cross-reactivity (Reiser *et al.*, 2003). The unexpected estimate of the extent of degeneracy of a clonotypic TCR came from testing synthetic combinatorial libraries and appears to show that a single TCR might interact with at least 1 million different ligands (Sercarz and Maverakis, 2004).

Biometric analysis compares the information derived from libraries composed of trillions of decapeptides with the millions of decapeptides contained in a protein database to rank and predict the most stimulatory peptides for a given T cell clone. Such an approach might permit the quantitative analysis of specific and degenerate interactions between TCR and MHC peptide ligands (Zhao *et al.*, 2001; Sung *et al.*, 2002).

14.5 The Self-Reactive TCR Repertoire

Random generation of the TCR repertoire in the thymus is bias-free as far as epitope recognition is concerned. The shape and epitope-specificity of a future bulk peripheral immune response to a given antigen are outlined during the thymic selection steps.

In the avidity-based thymic selections antigen expression, processing and presentation by self MHC are the major factors predetermining the available repertoire. This repertoire is necessarily characterized by at least a low avidity self-reactiveness or else its cells did not receive survival signals during positive selection.

It is important to note that TCR cross-reactivity is ‘unfocused’, so the immune response to a foreign antigen activates a limited number of self-reactive cells within the large pool of foreign antigen-specific cells. Experiments testing T cell clones that respond to two closely related ligands (autoantigenic peptide variants of myelin proteolipid protein, PLP 139–151 with a single amino acid difference) showed that the patterns of responses of these clones to other structurally related ligands are random, unfocused and may limit nonspecific responses to autoantigens (Anderson *et al.*, 2000a).

14.6 Thymic Antigen Presentation

Thymic antigen presentation appears to be a major decision-making step in the establishment of epitope hierarchy. Efficient presentation of an autoepitope leads to negative selection and elimination of T cells with high TCR avidity to self peptide–MHC, whereas poor thymic presentation of an epitope renders it cryptic in the thymus, but, under inflammatory conditions, it could receive a significant recognition in the periphery.

Thymic expression of antigens

Recent advances in our understanding of promiscuous gene expression in the thymus revealed that there is a representative yet incomplete display of tissue-specific self antigens in the thymus (approximately 10 per cent of the known genes being expressed and presented for maturing T lymphocytes; (Kyewski and Derbinski, 2004).

Self molecules that are not expressed in or transported to the thymus for antigen presentation could be recognized in the periphery as dominant self molecules. T cells having TCR that recognizes such autoantigens with high avidity escape negative selection but might be rescued by receiving a survival signal during positive selection as a consequence of degenerate epitope recognition (that is by binding some other presented thymic peptide ligands with low avidity). As an example, immunodominance of the myelin-associated proteolipid antigen PLP 139–151 epitope in SJL mice appears to be due to the presence of expanded numbers of T cells [frequency of 1/20 000 CD4(+) cells] reactive to PLP 139–151 in the peripheral repertoire of naive mice. The lack of tolerance to this dominant peptide of the PLP is considered to be a consequence of the fact that the predominant form of PLP expressed in the thymus lacks residues 116–150 (Anderson *et al.*, 2000b).

Dual TCR expression

Dual TCR expression (with the reported two different productively rearranged TCR α alleles in $\sim$8 per cent of peripheral T cells (Corthay, Nandakumar and Holmdahl, 2001) could provide a mechanism by which some T cells bearing TCR with high avidity for self peptide–MHC can escape deletion in negative selection and get out to the periphery.

Alternative splicing

In the case of differential thymic and peripheral RNA splicing, autoantigens can be recognized with high avidity in the periphery. Alternative splicing, especially a noncanonical form, could result in the formation of nontolerated self protein isoforms. In a recent study the extent of alternative splicing within 45 randomly selected self-proteins associated with autoimmune diseases was compared with 9554 randomly selected proteins in the human genome. The authors have found alternative splicing in 100 per cent of the autoantigen transcripts. This was significantly higher than the approximately 42 per cent rate of alternative splicing observed in the randomly selected transcripts ($p < 0.001$). Furthermore, noncanonical alternative splicing was found to be significantly more frequent in autoantigen transcripts (Ng *et al.*, 2004).

Destructive processing

Enzymes of thymic antigen presenting cells are key factors dictating epitope display in context with MHC. The crypticity of an epitope can be a result of poor accessibility of a given determinant within a molecule or, where it carries an enzyme cleavage site, it could be destroyed within the APC. Manoury *et al.* (2002) recently provided convincing data on the role of such destructive thymic processing, showing an initial enzyme attack by asparagine endopeptidase interfering with the presentation of a T cell epitope (MBP peptide 85–99) that is immunodominant in the periphery.

It has been suggested that possible peripheral down-modulation of AEP expression or its activity could permit induction of MBP (85-99)-specific T cells. Alternatively, peripheral post-translational modification of the target peptide (deamidation) could prevent AEP cleavage. Finally, alternative initial endopeptidase cleavage (e.g. by cathepsin S) may alter epitope hierarchy in the periphery (Sercarz, 2002).

Therefore comparative enzymology within thymic and peripheral APCs and prediction of cleavage sites by thymic endosome–lysosomal proteases may shed light on the group of peptide epitopes that are possibly destroyed before they can be presented to T cells in the thymus. Thus, T cells reactive to determinants degraded in the thymus could escape central tolerance and may recognize the cognate epitope in the periphery as a dominant self epitope. Peripheral post-translational modifications of antigens (see later) might result in conformational changes or associations with other proteins that are known to alter antigen processing.

In a recent study, enhancement of antigen presentation in the presence of inhibitors of metallo-, aspartic and serine proteases was observed. This suggests that several enzyme families may play a role in destructive antigen processing. These enzymes are also considered to generate natural peptide ligands that block antigen presentation to CD4 T cells as competitive inhibitors (von Delwig *et al.*, 2003).

MHC binding

Thymic antigen presentation dictates positive and negatives selections of maturing T lymphocytes. Therefore, MHC binding as early as in the thymus might determine the composition of the T cell repertoire. The self peptides derived from the unfolding protein compete with each other for MHC binding. Also, the different MHC molecules can compete for determinants at an early stage of processing.

Such competitive interactions have a major impact on the determination of dominance hierarchy. The winner epitopes in the thymus delete specific T cells during negative selection; however, epitopes that lose the thymic competition (cryptic epitopes in the thymus) let specific clones go to the periphery, where, if the epitope is presented efficiently, it may become 'immunodominant' determinant. It has been shown clearly by Moudgil and Sercarz (1993): dominant determinants in foreign antigen, hen eggwhite lysozyme corresponded to the cryptic determinants within its self-homologue, mouse lysozyme.

14.7 Peripheral Antigen Presentation

Extracellular processing

There is increasing evidence for extracellular processing and MHC-loading of peptide epitopes. Several processes such as necrosis or frustrated phagocytosis are known to be associated with the release of enzymes. Also, numerous organs of the human body are characterized by high extracellular enzyme activities (e.g. the lumen of the gastrointestinal tract). Therefore local extracellular environment may significantly alter antigen determinants. The potential significance of extracellular processing is supported by several recent data.

Using a monoclonal antibody specific for the empty conformation of class II MHC molecules revealed the presence of abundant empty MHC molecules on the surface of spleen- and bone marrow-derived DCs. The empty class II MHC molecules were expressed predominantly on immature DC. They could capture peptide antigens directly from the extracellular space, consistent with the role of DCs as sentinels in the immune system (Santambrogio *et al.*, 1999a).

In accordance with these findings, in another report an unusual extracellular presentation pathway has been described for immature DC, in which antigen processing and peptide loading can occur entirely outside of the cell. Immature DCs express at the cell surface empty or peptide-receptive class II MHC molecules, as well as H-2M or HLA-DM. Secreted DC proteases act extracellularly to process intact proteins into antigenic peptides. Peptides produced by such activity are efficiently loaded onto cell surface class II MHC molecules (Santambrogio *et al.*, 1999b).

It has been shown that a 69-mer synthetic polypeptide carrying the optimal 9-mer K^d-restricted epitope from the *Plasmodium berghei* circumsporozoite protein did not require intracellular processing. Serum components, such as proteases and beta2 microglobulin, allowed the processing and loading of exogenous polypeptides onto empty cell surface class I molecules for presentation to CTL (Eberl *et al.*, 1999).

It has been demonstrated for an immunodominant HIV-1 gp160 epitope that a longer peptide P18IIIB (15 amino acids, P18IIIB: RIQRGPGRAFVTIGK) can bind to the class I molecules on the cell surface, and then be trimmed the minimal-sized epitope (I-10, RGPGRAFVTI) by angiotensin-1-converting enzyme (ACE), to which it is bound (Nakagawa *et al.*, 2000).

A peptide fragment of 33 residues [α(2)-gliadin 56–88] is produced by normal gastrointestinal proteolysis. This 33-mer is a potent T cell stimulator that does not require further processing within APC for T cell presentation and that binds to DQ2 with a pH profile that promotes extracellular binding (Qiao *et al.*, 2004).

Finally, it has been shown for immunostimulatory MHC class I-binding peptide sequences, that, if incorporated at the carboxy-terminal position, they did not require intracellular processing. Removal of the sterically hindering amino-terminal bulk of the protein via an extracellular Ag proteolysis by the T-cell- and/or APC-derived enzymes is required for effective T-cell stimulation (Diegel *et al.*, 2003).

Key initial steps in antigen processing

1. Recent identification of the AEP enzyme (see above) provided important information on candidate enzymes that could play a key role in initial unfolding of a compact globular antigen (Sercarz, 2002).

2. Similarly to the initial endopeptidase cleavage of a tightly folded protein, the activity of IFN-γ-inducible thiol reductase (GILT), which breaks disulfide bridges, can reveal further enzymatic target sites and can thus affect the processing of distant and previously inaccessible protein regions (Phan, Maric and Cresswell, 2002).

3. A member of the family of pro-protein convertases could be another candidate enzyme involved in the initial unfolding process (Seidah and Chretien, 1997). It has been demonstrated recently that such an unfolding antigen binds at an early stage to an MHC molecule and then is trimmed down to final size (Sercarz and Maverakis, 2003).

Prediction of MHC binding

An increasing number of prediction methods are becoming available for prediction of MHC I and II binding based on MHC-binding motifs (Rammensee, Friede and Stevanoviic, 1995). Also, some approaches use quantitative matrices (Parker, Bednarek and Coligan, 1994), artificial neuronal networks (Honeyman *et al.*, 1998), hidden Markov models (Mamitsuka, 1998), multivariate statistical approaches (Guan *et al.*, 2003), support vector machines (Zhao *et al.*, 2003), decision trees (Savoie *et al.*, 1999) etc.

Several methods predicting MHC–peptide binding and antigen presentation are currently available on the internet and are listed in recent reviews (Flower, 2003; De Groot and Rappuoli, 2004). The newly emergent complexity of T cell recognition (e.g. degenerate TCR binding, MHC register shifting, differential antigen processing, post-translational modifications of proteins), however, continually challenges predictions, and high-throughput functional tests might also be required to validate the results.

Peptide register shifting within the MHC groove

Single peptides can assume different positions (registers) in the groove of a single MHC molecule, as shown by several recent studies (Bhayani and Paterson, 1989; Li *et al.*, 2000; Anderton *et al.*, 2002). MHC register shifting multiplies the potential peptide–MHC surfaces that a TCR should recognize and could 'mask' autoantigenic epitopes during thymic selections. Furthermore, peptide–MHC binding prediction is highly complicated by the varying frameshifting within the MHC groove (Bankovich *et al.*, 2004).

Is there an advantage for repetitive sequences?

Proteins carrying tandem repeat sequences have been reported to elicit immune responses in which the majority of activated lymphocyte clones are reactive to immunodominant epitopes in the tandem repeat domains (Umezawa *et al.*, 1993; Singh *et al.*, 2001; Domenech, Henderson and Finn, 1995; Dailey and Alderete, 1991; Esen, 1990; Burchell *et al.*, 1989). An attractive explanation for this observed phenomenon is the potentially high copy number of tandem repeat region-derived peptide epitopes generated by antigen processing. Thus, tandem repeat peptides may have a selective advantage when they compete with other peptides for occupation of the peptide binding groove of the MHC molecule.

Post-translational modifications of antigens

During the past few years molecular genetics has revealed an unexpectedly low number of genes in the human genome (20 000–30 000 genes). This finding has shed light on the importance of post-translational modifications of proteins. It is estimated that 50–90 per cent of the proteins in the human body are post-translationally modified. These post-translational modifications can create neo self epitopes or mask antigens normally recognized by the immune system. They can interfere with intracellular processing by altering proteolytic cleavage sites. They can also cause hindrance by preventing a determinant binding to MHC or to a TCR (for reviews see Doyle and Mamula, 2001, 2002; Anderton, 2004).

Such post-translational modifications include well-known additions like glycosylation, phosphorylation, acetylation and methylation as well as amino acid conversions, e.g. deimination of arginine (citrullination). Post-translational modifications have been shown to have a strong impact on immunodominance/cripticity.

Glycosylation

Some of the clearest evidence for the role of glycosylation in autoimmunity has come from the murine model of collagen arthritis (CIA). In humanized mice expressing rheumatoid arthritis-associated DR4 and DR1 molecules, the immunodominant T cell epitope of type II collagen (CII) was identified as a peptide sequence 263–270 (Fugger, Rothbard and Sonderstrup-McDevitt, 1996; Rosloniec *et al.*, 1998; Andersson *et al.*, 1998).

Out of 29 T cell hybridomas derived from mice immunized with CII, 20 recognized CII(256–270) glycosylated with a monosaccharide (β-D-galactopyranose). Thus, this glycopeptide was considered immunodominant in CIA. Lys264 was shown to be a major TCR contact residue that can be hydroxylated and subsequently galactosylated (Corthay *et al.*, 1998). Type II collagen, in its glycosylated form, was much more efficient in arthritis induction than the carbohydrate-depleted one. Incidence, time of onset and severity of the disease were significantly affected by the elimination of carbohydrates (Michaelsson *et al.*, 1994). Neonatal tolerance to type II collagen was also efficiently induced by the galactosylated peptide and protected mice from CIA (Backlund *et al.*, 2002a).

Earlier human studies failed to identify DR4/DR1-restricted T cells specific for the immunodominant peptide epitope of type II collagen. Not only has glycosylation been shown to be important in collagen-induced arthritis, but there is a clear predominant selection of T cells specific for the glycosylated CII epitope (263–270) in humanized transgenic mice and in rheumatoid arthritis (Backlund *et al.*, 2002b).

Citrullination

Recent work has identified proteins containing citrulline (deiminated arginine) as specific immune targets in RA patients. There is currently a surge of publications in rheumatology demonstrating the specificity of immune reactivity to citrullinated proteins (e.g. filaggrin) in patients suffering from rheumatoid arthritis. In HLA-DRB1*0401 transgenic mice it has been demonstrated that the conversion of arginine to citrulline at the peptide side-chain position interacting with the 'shared epitope' significantly increases peptide–MHC affinity and leads to the activation of CD4(+) T cells (Hill *et al.*, 2003). Also, in multiple sclerosis it has been postulated that the generation of MBP peptides is directly related to the arginine–citrulline conversion of MBP (Cao *et al.*, 1999).

14.8 Epitope Hierarchy in Experimental Autoimmune Encephalomyelitis

Experimental EAE is induced in genetically susceptible mouse and rat strains by immunization with either MBP or PLP. Also, the model can be induced using synthetic peptides representing relevant T cell epitopes (Sakai *et al.*, 1988; Zamvil *et al.*, 1986; Chou *et al.*, 1989). Autoreactive CD4+ T cells infiltrate the central nervous system and initiate inflammation in which encephalopathogenic T cells recognize an immunodominant epitope of MBP (amino acids 89–101).

During the course of chronic EAE, a marked shift has been described in the epitope hierarchy from the initial focus on dominant towards subdominant and cryptic self determinants. In three commonly used EAE model strains of mice, i.e. B10.PL, SJL/J and (SJL × B10.PL)F_1, many cryptic proliferative T cell determinants have been detected (Bhardwaj *et al.*, 1994).

14.9 Epitope Hierarchy in Aggrecan-Induced Murine Arthritis

Large aggregating chondroitin sulfate-rich proteoglycan (aggrecan) has long been considered as a major structural component of hyaline cartilage that receives poor if any immune recognition. This cartilage proteoglycan (M_r 2–3 × 10^6 Da) consists of a central protein core (M_r 2.2 × 10^5 Da) to which glycosaminoglycan side chains of CS (M_r 2–3 × 10^4 Da) and KS (M_r 1–2 × 10^4 Da) are attached together with O-linked and N-linked oligosaccharides (Buzás, Mikecz and Glant, 1996; Glant *et al.*, 1998a).

Partial deglycosylation of human aggrecan by glycosidases such as chondroitinase ABC or testicular hyaluronidase, however, generated highly immunogenic carbohydrate stubs attached to the long core protein of human aggrecan. Glycosidase digestion not only rendered human foetal aggrecan strongly immunogenic in BALB/c mice but also generated arthritogenic epitopes that induced chronic, progressive autoimmune arthritis in the peripheral and axial joints (Glant *et al.*, 1987; Mikecz, Glant and Poole, 1987).

In spite of the slight methodological modifications that have been introduced to improve the efficacy of the induction protocol of aggrecan arthritis over the years (Glant *et al.*, 1998b, 2001; Hanyecz *et al.*, 2004), the principles remained the same: systemic hyperimmunization with partially deglycosylated heterologous aggrecan in adjuvant results in an increasing cross-reactive immune response with the autologous murine cartilage proteoglycan. On the basis of this cross-reactive humoural and cellular immune response, ultimately a robust polyarthritis is induced with close to 100 per cent incidence in susceptible mouse strains (Glant, Finnegan and Mikecz, 2003).

Aggrecan-induced arthritis proved to be an excellent and relevant model of human rheumatoid arthritis (and, in some aspects, ankylosing spondylitis as well) since systemic immunisation with a heterologous antigen leads to true autoimmunity to a joint autoantigen. Clinical manifestations including spontaneous remissions and exacerbations, and histopathological and radiographic findings show strikingly close resemblance to human rheumatoid arthritis and are reviewed elsewhere (Glant and Mikecz, 2004; Glant, Finnegan and Mikecz, 2003).

Here we overview data available on the T cell epitope hierarchy in this experimental model of arthritis.

We have shown earlier that that both Th1 and Th2 T cells play a role in the development of aggrecan arthritis (Hollo *et al.*, 2000). Generating T cell hybridomas from aggrecan-induced arthritic mice we isolated an I-A^d-restricted Th1 clone designated 5/4E8 that was reactive to both human and self (murine) aggrecans. Systemic injection of these cells into naive irradiated BALB/c mice induced clinical and histopathological signs of arthritis (Buzás *et al.*, 1995). Fulminant arthritis-like symptoms including swelling and redness of the paws dominated the clinical picture. Synovial cell proliferation, the accumulation of hybridoma and inflammatory cells in the joint space, the loss of glycosaminoglycans from the superficial layer of the articular cartilage, and the erosion of articular surface were the leading histopathological signs.

The aggrecan epitope recognized by this arthritogenic T cell hybridoma clone was mapped to the G1 domain of the core protein (GR/QVRVNSA/IY) of human/murine proteoglycan (Glant *et al.*, 1998a).

A sister clone of the arthritogenic 5/4E8 T cell hybridoma, TA20, recognized the same core epitope of cartilage aggrecan as 5/4E8, yet it failed to induce arthritis-like symptoms upon systemic injection to BALB/c mice. Testing a set of synthetic peptides we found that the two clones responded differentially to altered peptide ligands. Evidently, the arthritogenic T cell hybridoma line was characterized by higher degeneracy in epitope recognition (Buzás *et al.*, 2003).

The role of post-translational modification: glycosylation

The extremely heavily glycosylated aggrecan molecule provides a system that is particularly useful for studying the role of glycosylation: carbohydrate side chains constitute 90 per cent of the molecular mass.

We have shown that the lack of both chondroitin sulfate (CS) and keratan sulfate (KS) side chains is required for the induction of autoimmune responses and arthritis in mice. The presence of a KS side chain in adult proteoglycan inhibits the recognition of arthritogenic T cell epitopes, prevents the development of T cell response, while removal of CS side chains generates clusters of CS stubs and provokes a strong B cell response. These carbohydrate-specific B cells seem to be essential proteoglycan APCs (Glant *et al.*, 1998a).

KS consists of repeated disaccharide units of *N*-acetyl glucosamine and galactose; both components can be sulfated at either the C4 or C6 position. KS side chains can be O-linked to serine or threonine or N-linked to asparagine of the core protein of proteoglycan molecules. The presence of sulfate groups makes KS resistant to enzymatic degradation *in vivo* due to the lack of specific endo-β-galactosidases in eukaryotic cells. However, foetal and rheumatoid arthritic chondrocytes produce a KS-free *immature type* of aggrecan that may undergo proteolytic degradation in diseased cartilage. Thus, the potential exists that foetal-type core protein fragments, without KS chains, become accessible to the immune system in pathologic conditions.

It is not clear at present how KS interferes with T cell recognition. It has been shown that removal of KS results in increased cellular uptake by APCs *in vitro*. Moreover, after removal of KS by keratanase, the G1 domain of the core protein alone induced a severe erosive polyarthritis and spondylitis in BALB/c mice, identifying it as an immunodominant and arthritogenic domain of aggrecan (Leroux *et al.*, 1996; Zhang *et al.*, 1998).

The G1 domain of human aggrecan was also identified as an immunological hot spot in human rheumatoid arthritis after removal of KS cells from patients with RA showed enhanced recognition of G1 (Guerassimov *et al.*, 1998). These studies clearly identified the N-terminal G1 domain of aggrecan as a primary immunodominant region of the molecule.

Detailed T cell epitope mapping in aggrecan arthritis has been made possible by an array of 143 synthetic peptides spanning the core protein of human aggrecan (Buzás *et al.*, 1999, 2005). With this approach we have identified a total of 27 distinct T cell epitopes.

The N terminal G1 domain of aggrecan has indeed proved to contain a very high density of epitopes. While representing <15 per cent of the length of the core protein, it carries nine out of 27 T cell epitopes, and the earliest and strongest immune responses develop against these epitopes. Only a minor portion of the identified epitopes (five out of 27) proved to be full agonist peptides (that induced both cytokine secretion and proliferation). Interestingly, all full activator epitopes mapped to the G1 domain of aggrecan.

The secondary target region of the immune response was the C-terminal G3 domain of aggrecan. Intriguingly, both N- and C-terminal globular domains of aggrecan carry minimal glycosylation as compared with the KS and CS attachment regions (the latter two domains comprising approximately 60 per cent of the length of the core protein).

Strikingly, T cell epitope hierarchy underwent dynamic changes during the course of hyperimmunization with aggrecan to induce arthritis in BALB/c mice. As a result of epitope spreading, the initial strongly focused recognition (detected 9 days after aggrecan priming) changed to close to equal recognition of several further T cell epitopes in acute aggrecan arthritis.

Epitope hierarchy patterns clearly depended on the type of function assessed in the test: hierarchy, determined on the basis of induction of cytokine response, did not match the pattern determined on the basis of induction of cell proliferation. This is not surprising if we consider the relatively high number of partial activator determinants that could be identified in human aggrecan. The partial agonist peptides predominantly induced cytokine response but failed to elicit cell proliferation.

In our work we identified four epitopes of G1 and G3 domains of aggrecan, the T cell recognition of which was highly characteristic for the arthritic state ('arthritis associated epitopes'; Figure 14.1).

One of the most exciting findings of this work was the identification of an unusual set of epitopes (most of which were located in the most heavily glycosylated CS attachment region of aggrecan). For these epitopes we introduced a term 'conditionally

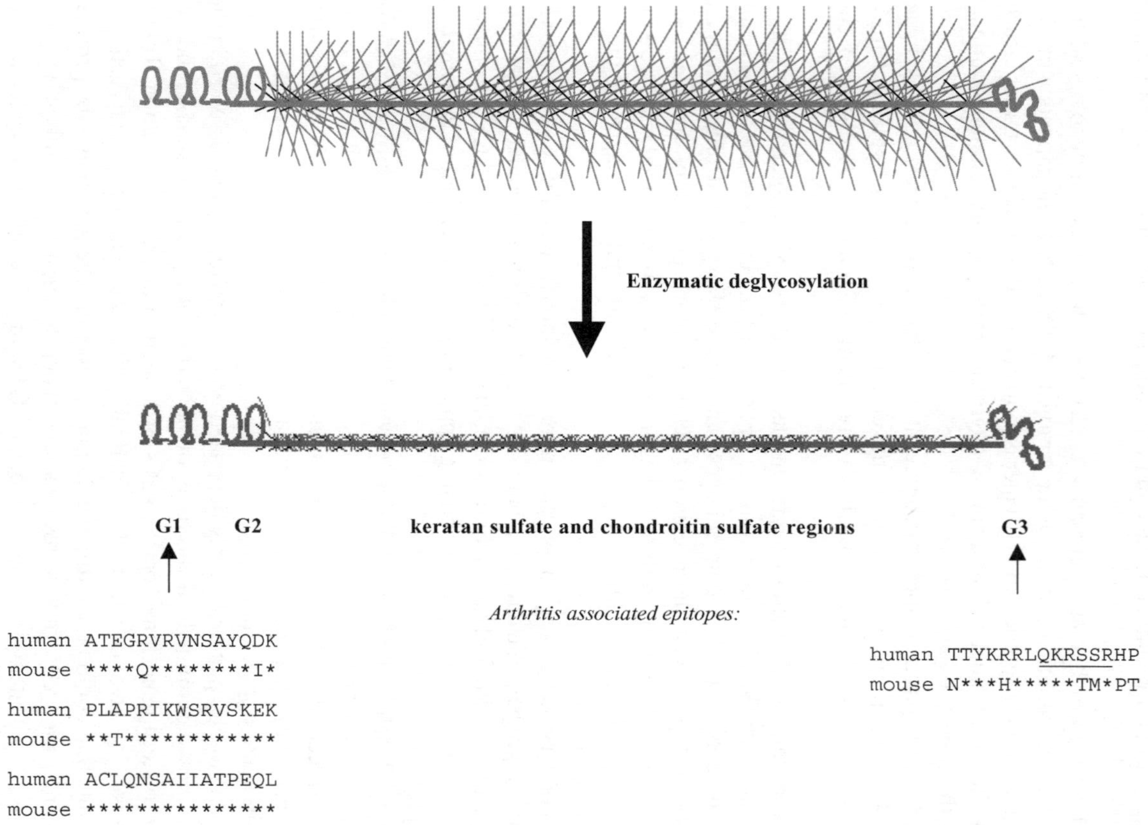

Figure 14.1 Cartilage aggrecan.

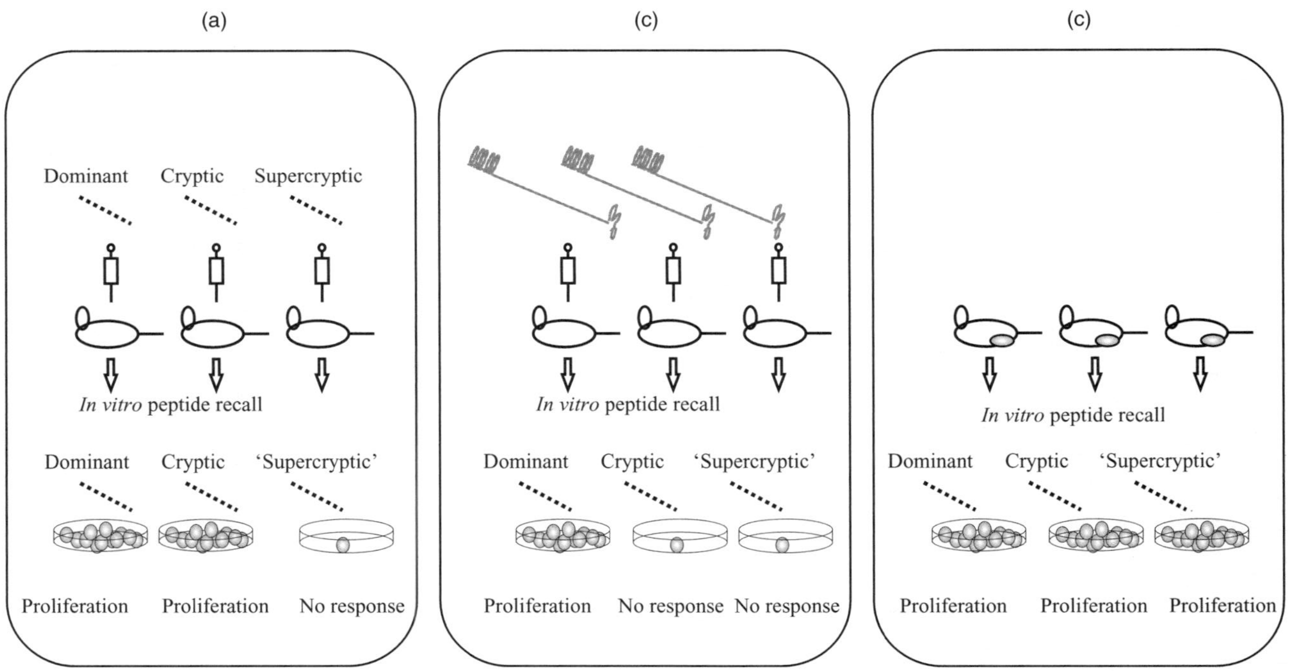

Figure 14.2 (a) Immunization with epitope peptide. (b) Immunization with aggrecan core protein. (c) Aggrecan-induced arthritis.

immunogenic' or 'supercryptic'. Such determinants failed to elicit a recall T cell response after immunization with the peptide in adjuvant (thus, these peptides did not even fulfil the criteria of cryptic epitopes, determinants eliciting response against themselves while not cross-reacting with the native antigen). These 'supercryptic' epitopes were readily recognized by spleen cells isolated from acutely arthritic mice and induced significant proliferation. Thus, in acute arthritis, unique antigen processing/presentation/costimulation conditions were established that could lead to the recognition of earlier hidden epitopes of the long carbohydrate attachment region (e.g. of tandem repeat sequences; Goodacre *et al.*, 1993). Such determinants induce proliferation comparable to that induced by the full agonist epitopes located in the immunodominant G1 domain of aggrecan and thus they could profoundly contribute to the autoimmune inflammation. (Figure 14.2).

We have recently shown very high exoglycosidase activities in the inflamed joints (Ortutay *et al.*, 2003). Extracellular processing by these exoglycosidases and possibly by other enzymes could make epitopes of the CS attachment region accessible to cells of the immune system in arthritic mice. (Figure 14.3).

Our work has also identified an epitope in the C terminus of aggrecan (P2373–2387 TTYKRRLQKRSSRHP), a sequence that differs in two conservatively substituted amino acids from the 'shared epitope' (QKRAA), the most common sequence motif in HLA–DR4 alleles, which predispose humans to the development of rheumatoid arthritis (RA). The 'shared epitope' is also overrepresented in bacterial heat-shock proteins and the envelope protein of human JC polyomavirus

Mice, hyperimmunized and presensitized with P2373–2387, required only a single dose of cartilage aggrecan in order to develop arthritis. (Hanyecz *et al.*, 2003). Synthetic peptides representing *Escherichia coli* heat-shock protein (DnaJ) or HLA–DR4 allele (both having the shared epitope sequence with different flanking regions) were also positive using the same protocol.

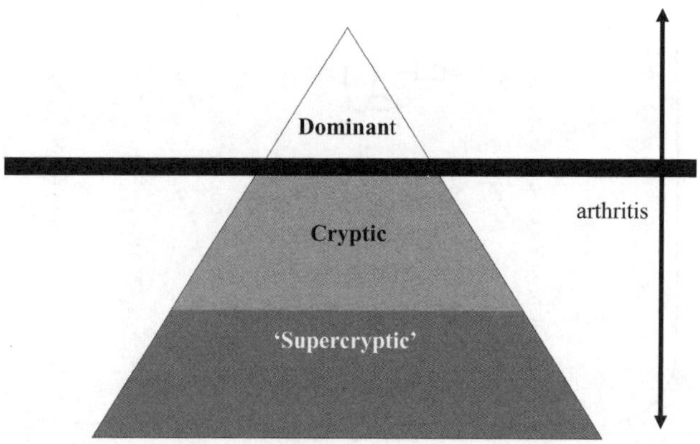

Figure 14.3 Epitopes of aggrecan recognized in primary immune response and in autoimmune inflammatory arthritis.

In search of human relevance of these findings we looked at the epitope hierarchy in mice transgenic for rheumatoid arthritis-predisposing MHC class II molecules ($DR4.Ab^0$, $DQ8.Ab^0$) and lacking their own (mouse) MHC II molecules. We found that five out of 27 $I\text{-}A^d$ associated aggrecan epitopes, identified in BALB/c mice, were also presented by at least one of the rheumatoid arthritis-associated HLA molecule-expressing mice. Interestingly, three of the four arthritis-associated peptide epitopes of BALB/c mice were also positive in DR4. Ab^0 and $DQ8.Ab^0$ mice.

The aggrecan G3 epitope with 'shared epitope'-like sequence (LQKRSSRHPR-RSRPST) proved to be an epitope eliciting strong immune response both in BALB/c and human RA-associated HLA-transgenic mice (Szanto *et al.*, 2004).

14.10 Summary

Epitopes that are not presented in the thymus fail to transmit survival signals to thymocytes having specific TCR. In contrast, peptides efficiently presented in the thymus induce profound nonresponsiveness in the periphery. Negative selection eliminates T cells that recognize self peptide–MHC complexes with high avidity (over a certain activation threshold). Efficient thymic presentation of a self epitope also leads to the selection of natural regulatory T cells (Tregs) in the case of peptides that activate thymocytes to a level just below the negative selection's threshold. It has been proposed recently that peripheral T cell activation processes mirror thymic events. Nonactivating epitopes (below a threshold) leave the specific T cells in a resting G_0 state, and partially activating epitopes might drive the T cells towards regulatory function, while full activator epitopes (over a threshold) induce effector cell differentiation (Graca *et al.*, 2005).

Innovative technologies such as peptide arrays and large peptides libraries are opening new vistas for epitope identification. A next generation of high-throughput systems will probably make peptides available also with all the potential post-translational modifications (e.g. arrays of glycopeptides).

Genomes and proteomes of further infectious microorganisms are becoming available these days for the scientific community for *in silico* analysis and epitope identification. Major integrative efforts are currently being made (like the Large Scale Antibody and T cell Epitope Discovery Program) to establish the Immune Epitope Database and Analysis Resource which will serve to make all newly identified epitopes freely available worldwide (Sette *et al.*, 2005).

With a deeper understanding of the human thymic antigen presentation and selection processes, with detailed information on the human proteome and its post-translational modifications, knowing an individual's HLA alleles, perhaps, in the future, personalized predictions could be made regarding the presentation of self epitopes and the consequential residual peripheral T cell reactivity. Cryptic (poorly presented) self epitopes would be of particular interest and potentially cross-reactive immunodominant foreign epitopes of infectious agents should be identified as possible triggers for autoimmunity.

Therapeutic interventions to autoimmune epitope hierarchies, blocking of epitope spreading or redirecting autopathogenic T cells towards regulatory pathways might be future immunotherapeutic strategies.

References

Anderson AC, Waldner H, Turchin V, Jabs C, Prabhu Das M, Kuchroo VK, Nicholson LB. 2000a. Autoantigen-responsive T cell clones demonstrate unfocused TCR cross-reactivity toward multiple related ligands: implications for autoimmunity. *Cell Immunol* **202**(2): 88–96.

Anderson AC, Nicholson LB, Legge KL, Turchin V, Zaghouani H, Kuchroo VK. 2000b. High frequency of autoreactive myelin proteolipid protein-specific T cells in the periphery of naive mice: mechanisms of selection of the self-reactive repertoire. *J Exp Med* **191**(5): 761–770.

Andersson EC, Hansen BE, Jacobsen H, Madsen LS, Andersen CB, Engberg J, Rothbard JB, McDevitt GS, Malmstrom V, Holmdahl R, Svejgaard A, Fugger L. 1998. Definition of MHC and T cell receptor contacts in the HLA-DR4restricted immunodominant epitope in type II collagen and characterization of collagen-induced arthritis in HLA-DR4 and human CD4 transgenic mice. *Proc Natl Acad Sci USA* **95**(13): 7574–7579.

Anderton SM. 2004. Post-translational modifications of self antigens: implications for autoimmunity. *Curr Opin Immunol* **16**(6): 753–758.

Anderton SM, Wraith DC. 2002a. Selection and fine-tuning of the autoimmune T-cell repertoire. *Nat Rev Immunol* **2**(7): 487–498.

Anderton SM, Viner NJ, Matharu P, Lowrey PA, Wraith DC. 2000. Influence of a dominant cryptic epitope on autoimmune T cell tolerance. *Nat Immunol* **3**(2): 175–181.

Backlund J, Treschow A, Bockermann R, Holm B, Holm L, Issazadeh-Navikas S, Kihlberg J, Holmdahl R. 2002a. Glycosylation of type II collagen is of major importance for T cell tolerance and pathology in collagen-induced arthritis. *Eur J Immunol* **32**(12): 3776–3784.

Backlund J, Carlsen S, Hoger T, Holm B, Fugger L, Kihlberg J, Burkhardt H, Holmdahl R. 2002b. Predominant selection of T cells specific for the glycosylated collagen type II epitope (263–270) in humanized transgenic mice and in rheumatoid arthritis. *Proc Natl Acad Sci USA* **99**(15): 9960–9965.

Bankovich AJ, Girvin AT, Moesta AK, Garcia KC. 2004. Peptide register shifting within the MHC groove: theory becomes reality. *Mol Immunol* **40**(14–15): 1033–1039.

Bhardwaj V, Kumar V, Geysen HM, Sercarz EE. 1993. Degenerate recognition of a dissimilar antigenic peptide by myelin basic protein-reactive T cells. Implications for thymic education and autoimmunity. *J Immunol* **151**(9): 5000–5010.

Bhardwaj V, Kumar V, Grewal IS, Dao T, Lehmann PV, Geysen HM, Sercarz EE. 1994. T cell determinant structure of myelin basic protein in B10.PL, SJL/J, and their F1S. *J Immunol* **152**(8): 3711–3719.

Bhayani H, Paterson Y. 1989. Analysis of peptide binding patterns in different major histocompatibility complex/T cell receptor complexes using pigeon cytochrome c-specific T cell hybridomas. Evidence that a single peptide binds major histocompatibility complex in different conformations. *J Exp Med* **170**(5): 1609–1625.

Burchell J, Taylor-Papadimitriou J, Boshell M, Gendler S, Duhig T. 1989. A short sequence, within the amino acid tandem repeat of a cancer-associated mucin, contains immunodominant epitopes. *Int J Cancer* **44**(4): 691–696.

Buzás EI, Mikecz K, Glant TT. 1996. Aggrecan: a target molecule of autoimmune reactions. *Pathol Oncol Res* **2**(4): 219–228.

Buzás EI, Brennan FR, Mikecz K, Garzo M, Negroiu G, Hollo K, Cs-Szabo G, Pintye E, Glant TT. 1995. A proteoglycan (aggrecan)-specific T cell hybridoma induces arthritis in BALB/c mice. *J Immunol* **155**(5): 2679–2687.

Buzás EI, Mikecz K, Finnegan A, Hudecz F, Glant TT. 1999. T-cell recognition of differentially tolerated epitopes of human cartilage proteoglycan (aggrecan) in arthritis. *Arthrit Rheum* **42**(suppl 9): S259.

Buzás EI, Hanyecz A, Murad Y, Hudecz F, Rajnavolgyi E, Mikecz K, Glant TT. 2003. Differential recognition of altered peptide ligands distinguishes two functionally discordant (arthritogenic and nonarthritogenic) autoreactive T cell hybridoma clones. *J Immunol* **171**(6): 3025–3033.

Buzás EI, Végvári A, Murad YM, Finnegan A, Mikecz K, Glant TT. 2005. T-cell recognition of differentially tolerated epitopes of cartilage proteoglycan aggrecan in arthritis. *Cell Immunol* (in press).

Cao L, Goodin R, Wood D, Moscarello MA, Whitaker JN. 1999. Rapid release and unusual stability of immunodominant peptide 45-89 from citrullinated myelin basic protein. *Biochemistry* **38**(19): 6157–6163.

Chou YK, Vandenbark AA, Jones RE, Hashim G, Offner H. 1989. Selection of encephalitogenic rat T-lymphocyte clones recognizing an immunodominant epitope on myelin basic protein. *J Neurosci Res* **22**(2): 181–187.

Corthay A, Nandakumar KS, Holmdahl R. 2001. Evaluation of the percentage of peripheral T cells with two different T cell receptor alpha-chains and of their potential role in autoimmunity. *J Autoimmun* **16**(4): 423–429.

Corthay A, Backlund J, Broddefalk J, Michaelsson E, Goldschmidt TJ, Kihlberg J, Holmdahl R. 1998. Epitope glycosylation plays a critical role for T cell recognition of type II collagen in collagen-induced arthritis. *Eur J Immunol* **28**(8): 2580–2590.

Dailey DC, Alderete JF. 1991. The phenotypically variable surface protein of Trichomonas vaginalis has a single, tandemly repeated immunodominant epitope. *Infect Immun* **59**(6): 2083–2088.

De Groot AS, Rappuoli R. 2004. Genome-derived vaccines. *Expert Rev Vaccines* **3**(1): 59–76.

Diegel ML, Chen F, Laus R, Graddis TJ, Vidovic D. 2003. Major histocompatibility complex class I-restricted presentation of protein antigens without prior intracellular processing. *Scand J Immunol* **58**(1): 1–8.

Domenech N, Henderson RA, Finn OJ. 1995. Identification of an HLA-A11-restricted epitope from the tandem repeat domain of the epithelial tumor antigen mucin. *J Immunol* **155**(10): 4766–4774.

Doyle HA, Mamula MJ. 2001. Post-translational protein modifications in antigen recognition and autoimmunity. *Trends Immunol* **22**(8): 443–449.

Doyle HA, Mamula MJ. 2002. PostTranslational protein modifications: new flavors in the menu of autoantigens. *Curr Opin Rheumatol* **14**(3): 244–249.

Eberl G, Renggli J, Men Y, Roggero MA, Lopez JA, Corradin G. 1999. Extracellular processing and presentation of a 69-mer synthetic polypeptide to MHC class I-restricted T cells. *Mol Immunol* **36**(2): 103–112.

Esen A. 1990. An immunodominant site of gamma-zein1 is in the region of tandem hexapeptide repeats. *J Protein Chem* **9**(4): 453–460.

Flower DR. 2003. Towards *in silico* prediction of immunogenic epitopes. *Trends Immunol* **24**(12): 667–674.

Fressinaud C, Sarlieve LL, Vincendon G. 1990. Multiple sclerosis: review of main experimental data and pathogenic hypotheses. *Rev Med Interne* **11**(3): 201–208.

Fugger L, Rothbard JB, Sonderstrup-McDevitt G. 1996. Specificity of an HLA-DRB1*0401-restricted T cell response to type II collagen. *Eur J Immunol* **26**(4): 928–933.

Glant TT, Mikecz K. 2004. Proteoglycan aggrecan-induced arthritis: a murine autoimmune model of rheumatoid arthritis. *Meth Mol Med* **102**: 313–338.

Glant TT, Mikecz K, Arzoumanian A, Poole AR. 1987. Proteoglycan-induced arthritis in BALB/c mice. Clinical features and histopathology. *Arthrit Rheum* **30**(2): 201–212.

Glant TT, Finnegan A, Mikecz K. 2003. Proteoglycan-induced arthritis: immune regulation, cellular mechanisms, and genetics. *Crit Rev Immunol* **23**(3): 199–250.

Glant TT, Buzas EI, Finnegan A, Negroiu G, Cs-Szabo G, Mikecz K. 1998a. Critical roles of glycosaminoglycan side chains of cartilage proteoglycan (aggrecan) in antigen recognition and presentation. *J Immunol* **160**(8): 3812–3819.

Glant TT, Cs-Szabo G, Nagase H, Jacobs JJ, Mikecz K. 1998b. Progressive polyarthritis induced in BALB/c mice by aggrecan from normal and osteoarthritic human cartilage. *Arthrit Rheum* **41**(6): 1007–1018.

Glant TT, Bardos T, Vermes C, Chandrasekaran R, Valdez JC, Otto JM, Gerard D, Velins S, Lovasz G, Zhang J, Mikecz K, Finnegan A. 2001. Variations in susceptibility to proteoglycan-induced arthritis and spondylitis among C3H substrains of mice: evidence of genetically acquired resistance to autoimmune disease. *Arthrit Rheum* **44**(3): 682–692.

Goodacre JA, Middleton S, Lynn S, Ross DA, Pearson J. 1993. Human cartilage aggrecan CS1 region contains cryptic T-cell recognition sites. *Immunology* **78**(4): 586–591.

Graca L, Chen TC, Le Moine A, Cobbold SP, Howie D, Waldmann H. 2005. Dominant tolerance: activation thresholds for peripheral generation of regulatory T cells. *Trends Immunol* **26**(3): 130–135.

Guan P, Doytchinova IA, Zygouri C, Flower DR. 2003. MHCPred: bringing a quantitative dimension to the online prediction of MHC binding. *Appl Bioinformat* **2**(1): 63–66.

Guerassimov A, Zhang Y, Banerjee S, Cartman A, Leroux JY, Rosenberg LC, Esdaile J, Fitzcharles MA, Poole AR. 1998. Cellular immunity to the G1 domain of cartilage proteoglycan aggrecan is enhanced in patients with rheumatoid arthritis but only after removal of keratan sulfate. *Arthrit Rheum* **41**(6): 1019–1025.

Hanyecz A, Bardos T, Berlo SE, Buzás E, Nesterovitch AB, Mikecz K, Glant TT. 2003. Induction of arthritis in SCID mice by T cells specific for the 'shared epitope' sequence in the G3 domain of human cartilage proteoglycan. *Arthrit Rheum* **48**(10): 2959–2973.

Hanyecz A, Berlo SE, Szanto S, Broeren CP, Mikecz K, Glant TT. 2004. Achievement of a synergistic adjuvant effect on arthritis induction by activation of innate immunity and forcing the immune response toward the Th1 phenotype. *Arthrit Rheum* **50**(5): 1665–1676.

Hill JA, Southwood S, Sette A, Jevnikar AM, Bell DA, Cairns E. 2003. Cutting edge: the conversion of arginine to citrulline allows for a high-affinity peptide interaction with the rheumatoid arthritis-associated HLA-DRB1*0401 MHC class II molecule. *J Immunol* **171**(2): 538–541.

Hollo K, Glant TT, Garzo M, Finnegan A, Mikecz K, Buzás E. 2000. Complex pattern of Th1 and Th2 activation with a preferential increase of autoreactive Th1 cells in BALB/c mice with proteoglycan (aggrecan)-induced arthritis. *Clin Exp Immunol* **120**(1): 167–173.

Honeyman MC, Brusic V, Stone NL, Harrison LC. 1998. Neural network-based prediction of candidate T-cell epitopes. *Nat Biotechnol* **16**(10): 966–969.

Kyewski B, Derbinski J. 2004. Self-representation in the thymus: an extended view. *Nat Rev Immunol* **4**(9): 688–698.

Lehmann PV, Forsthuber T, Miller A, Sercarz EE. 1992. Spreading of T-cell autoimmunity to cryptic determinants of an autoantigen. *Nature* **358**(6382): 155–157.

Lehmann PV, Sercarz EE, Forsthuber T, Dayan CM, Gammon G. 1993. Determinant spreading and the dynamics of the autoimmune T-cell repertoire. *Immunol Today* **14**(5): 203–208.

Leroux JY, Guerassimov A, Cartman A, Delaunay N, Webber C, Rosenberg LC, Banerjee S, Poole AR. 1996. Immunity to the G1 globular domain of the cartilage proteoglycan aggrecan can induce inflammatory erosive polyarthritis and spondylitis in BALB/c mice but immunity to G1 is inhibited by covalently bound keratan sulfate *in vitro* and *in vivo*. *J Clin Invest* **97**(3): 621–632.

Li Y, Li H, Martin R, Mariuzza RA. 2000. Structural basis for the binding of an immunodominant peptide from myelin basic protein in different registers by two HLA-DR2 proteins. *J Mol Biol* **304**(2): 177–188.

Mamitsuka H. 1998. Predicting peptides that bind to MHC molecules using supervised learning of hidden Markov models. Proteins **33**(4): 460–474.

Manoury B, Mazzeo D, Fugger L, Viner N, Ponsford M, Streeter H, Mazza G, Wraith DC, Watts C. 2002. Destructive processing by asparagine endopeptidase (AEP) limits presentation of a dominant T cell epitope in MBP. *Nat Immunol* **3**(2): 169–174.

Maverakis E, van den Elzen P, Sercarz EE. 2001. Self-reactive T cells and degeneracy of T cell recognition: evolving concepts-from sequence homology to shape mimicry and TCR flexibility. *J Autoimmun* **16**(3): 201–209.

McRae BL, Vanderlugt CL, Dal Canto MC, Miller SD. 1995. Functional evidence for epitope spreading in the relapsing pathology of experimental autoimmune encephalomyelitis. *J Exp Med* **182**(1): 75–85.

Michaelsson E, Malmstrom V, Reis S, Engstrom A, Burkhardt H, Holmdahl R. 1994. T cell recognition of carbohydrates on type II collagen. *J Exp Med* **180**(2): 745–749.

Mikecz K, Glant TT, Poole AR. 1987. Immunity to cartilage proteoglycans in BALB/c mice with progressive polyarthritis and ankylosing spondylitis induced by injection of human cartilage proteoglycan. *Arthrit Rheum* **30**(3): 306–318.

Miller SD, Vanderlugt CL, Lenschow DJ, Pope JG, Karandikar NJ, Dal Canto MC, Bluestone JA. 1995. Blockade of CD28/B7–1 interaction prevents epitope spreading and clinical relapses of murine EAE. *Immunity* **3**(6): 739–745.

Moudgil KD, Sercarz EE. 1993. Dominant determinants in hen eggwhite lysozyme correspond to the cryptic determinants within its self-homologue, mouse lysozyme: implications in shaping of the T cell repertoire and autoimmunity. *J Exp Med* **178**(6): 2131–2138.

Nakagawa Y, Takeshita T, Berzofsky JA, Takahashi H. 2000. Analysis of the mechanism for extracellular processing in the presentation of human immunodeficiency virus-1 envelope protein-derived peptide to epitope-specific cytotoxic T lymphocytes. *Immunology* **101**(1): 76–82.

Ng B, Yang F, Huston DP, Yan Y, Yang Y, Xiong Z, Peterson LE, Wang H, Yang XF. 2004. Increased noncanonical splicing of autoantigen transcripts provides the structural basis for expression of untolerized epitopes. *J Allergy Clin Immunol* **114**(6): 1463–1470.

Ortutay Z, Polgar A, Gomor B, Geher P, Lakatos T, Glant TT, Gay RE, Gay S, Pallinger E, Farkas C, Farkas E, Tothfalusi L, Kocsis K, Falus A, Buzás EI. 2003. Synovial fluid exoglycosidases are predictors of rheumatoid arthritis and are effective in cartilage glycosaminoglycan depletion. *Arthrit Rheum* **48**(8): 2163–2172.

Parker KC, Bednarek MA, Coligan JE. 1994. Scheme for ranking potential HLA-A2 binding peptides based on independent binding of individual peptide side-chains. *J Immunol* **152**(1): 163–175.

Phan UT, Maric M, Cresswell P. 2002. Disulfide reduction in major histocompatibility complex class II-restricted antigen processing by interferon-gamma-inducible lysosomal thiol reductase. *Meth Enzymol* **348**: 43–48.

Qiao SW, Bergseng E, Molberg O, Xia J, Fleckenstein B, Khosla C, Sollid LM. 2004. Antigen presentation to celiac lesion-derived T cells of a 33-mer gliadin peptide naturally formed by gastrointestinal digestion. *J Immunol* **173**(3): 1757–1762.

Rammensee HG, Friede T, Stevanoviic S. 1995. MHC ligands and peptide motifs: first listing. *Immunogenetics* **41**(4): 178–228.

Reiser JB, Darnault C, Gregoire C, Mosser T, Mazza G, Kearney A, van der Merwe PA, Fontecilla-Camps JC, Housset D, Malissen B. 2003. CDR3 loop flexibility contributes to the degeneracy of TCR recognition. *Nat Immunol* **4**(3): 241–247.

Rosloniec EF, Brand DD, Myers LK, Esaki Y, Whittington KB, Zaller DM, Woods A, Stuart JM, Kang AH. 1998. Induction of autoimmune arthritis in HLA-DR4 (DRB1*0401) transgenic mice by immunization with human and bovine type II collagen. *J Immunol* **160**(6): 2573–2578.

Sakai K, Zamvil SS, Mitchell DJ, Lim M, Rothbard JB, Steinman L. 1998. Characterization of a major encephalitogenic T cell epitope in SJL/J mice with synthetic oligopeptides of myelin basic protein. *J Neuroimmunol* **19**(1–2): 21–32.

Santambrogio L, Sato AK, Fischer FR, Dorf ME, Stern LJ. 1999a. Abundant empty class II MHC molecules on the surface of immature dendritic cells. *Proc Natl Acad Sci USA* **96**(26): 15050–15055.

Santambrogio L, Sato AK, Carven GJ, Belyanskaya SL, Strominger JL, Stern LJ. 1999b. Extracellular antigen processing and presentation by immature dendritic cells. *Proc Natl Acad Sci USA* **96**(26): 15056–15061.

Savoie CJ, Kamikawaji N, Sasazuki T, Kuhara S. 1999. Use of BONSAI decision trees for the identification of potential MHC class I peptide epitope motifs. *Pacific Symposium on Biocomputers* 182–189.

Seidah NG, Chretien M. 1997. Eukaryotic protein processing: endoproteolysis of precursor proteins. *Curr Opin Biotechnol* **8**(5): 602–607.

Sercarz EE. 2002. Processing creates the self. *Nat Immunol* **3**(2): 110–112.

Sercarz EE, Maverakis E. 2003. MHC-guided processing: binding of large antigen fragments. *Nat Rev Immunol* **3**(8): 621–629.

Sercarz EE, Maverakis E. 2004. Recognition and function in a degenerate immune system. *Mol Immunol* **40**(14-15): 1003–1008.

Sercarz EE, Lehmann PV, Ametani A, Benichou G, Miller A, Moudgil K. 1993. Dominance and crypticity of T cell antigenic determinants. *A Rev Immunol* **11**: 729–766.

Sette A, Fleri W, Peters B, Sathiamurthy M, Bui HH, Wilson S. 2005. A roadmap for the immunomics of category A–C pathogens. *Immunity* **22**(2): 155–161.

Singh KK, Zhang X, Patibandla AS, Chien P Jr, Laal S. 2001. Antigens of *Mycobacterium tuberculosis* expressed during preclinical tuberculosis: serological immunodominance of proteins with repetitive amino acid sequences. *Infect Immun* **69**(6): 4185–4191.

Sloan-Lancaster J, Allen PM. 1996. Altered peptide ligand-induced partial T cell activation: molecular mechanisms and role in T cell biology. *A Rev Immunol* **14**: 1–27.

Sung MH, Zhao Y, Martin R, Simon R. 2002. T-cell epitope prediction with combinatorial peptide libraries. *J Comput Biol* **9**(3): 527–539.

Szanto S, Bardos T, Szabo Z, David CS, Buzás EI, Mikecz K, Glant TT. 2004. Induction of arthritis in HLA-DR4-humanized and HLA-DQ8-humanized mice by human cartilage proteoglycan aggrecan but only in the presence of an appropriate (non-MHC) genetic background. *Arthrit Rheum* **50**(6): 1984–1995.

Tsunoda I, Fujinami RS. 1996. Two models for multiple sclerosis: experimental allergic encephalomyelitis and Theiler's murine encephalomyelitis virus. *J Neuropathol Exp Neurol* **55**(6): 673–686.

Umezawa ES, Shikanai-Yasuda MA, da Silveira JF, Cotrim PC, Paranhos G, Katzin AM. 1993. Trypanosoma cruzi: detection of a circulating antigen in urine of chagasic patients sharing common epitopes with an immunodominant repetitive antigen. *Exp Parasitol* **76**(4): 352–357.

Vanderlugt CL, Miller SD. 2002. Epitope spreading in immune-mediated diseases: implications for immunotherapy. *Nat Rev Immunol* **2**(2): 85–95.

Viner NJ, Nelson CA, Deck B, Unanue ER. 1996. Complexes generated by binding of free peptides to class II MHC molecules are antigenically diverse compared with those generated by intracellular processing. *J Immunol* **156**: 2365–2368.

von Delwig A, Musson JA, McKie N, Gray J, Robinson JH. 2003. Regulation of peptide presentation by major histocompatibility complex class II molecules at the surface of macrophages. *Eur J Immunol* **33**(12): 3359–3366.

Wang HB, Shi FD, Li H, Chambers BJ, Link H, Ljunggren HG. 2001. Anti-CTLA-4 antibody treatment triggers determinant spreading and enhances murine myasthenia gravis. *J Immunol* **166**(10): 6430–6436.

Wildbaum G, Netzer N, Karin N. 2002. Tr1 cell-dependent active tolerance blunts the pathogenic effects of determinant spreading. *J Clin Invest* **110**(5): 701–710.

Yin L, Yu M, Edling AE, Kawczak JA, Mathisen PM, Nanavati T, Johnson JM, Tuohy VK. 2001. Pre-emptive targeting of the epitope spreading cascade with genetically modified regulatory T cells during autoimmune demyelinating disease. *J Immunol* **167**(11): 6105–6112.

Yu M, Johnson JM, Tuohy VK. 1996. A predictable sequential determinant spreading cascade invariably accompanies progression of experimental autoimmune encephalomyelitis: a basis for peptide-specific therapy after onset of clinical disease. *J Exp Med* **183**(4): 1777–1788.

Zamvil SS, Mitchell DJ, Moore AC, Kitamura K, Steinman L, Rothbard JB. 1986. T-cell epitope of the autoantigen myelin basic protein that induces encephalomyelitis. *Nature* **324**(6094): 258–260.

Zhang Y, Guerassimov A, Leroux JY, Cartman A, Webber C, Lalic R, de Miguel E, Rosenberg LC, Poole AR. 1998. Arthritis induced by proteoglycan aggrecan G1 domain in BALB/c mice. Evidence for T cell involvement and the immunosuppressive influence of keratan sulfate on recognition of t and b cell epitopes. *J Clin Invest* **101**(8): 1678–1686.

Zhao Y, Gran B, Pinilla C, Markovic-Plese S, Hemmer B, Tzou A, Whitney LW, Biddison WE, Martin R, Simon R. 2001. Combinatorial peptide libraries and biometric score matrices permit the quantitative analysis of specific and degenerate interactions between clonotypic TCR and MHC peptide ligands. *J Immunol* **167**(4): 2130–2141.

Zhao Y, Pinilla C, Valmori D, Martin R, Simon R. 2003. Application of support vector machines for T-cell epitopes prediction. *Bioinformatics* **19**(15): 1978–1984.

15

Gene–Gene Interactions in Immunology as Exemplified by Studies on Autoantibodies Against 60 kDa Heat-shock Protein

Zoltán Prohászka

Abstract

The relationship between genotype and phenotype is expected to be nonlinear for most common, multifactorial human diseases, such as cancer, cardiovascular disease and systemic autoimmune diseases. The genomic research using high-throughput technology to generate genetic data on population level grows more rapid than the methods to analyse those data. The interpretation of vast quantities of genotype–phenotype information is even more complicated if one has the intention to explore effects of gene–gene or gene–environment interactios. The aim of this chapter is to summarize the basic features of gene–gene interactions (also called epistasis) together with appropriate strategies and methods to detect it. Epistasis is defined in its biological sense as interaction between two or more DNA variations either directly or indirectly, to alter disease risk separate from their independent effects. Epistasis is commonly found when properly investigated. However, taking multiple interactions into account, epistasis is difficult to detect and charaterize using traditional parametric methods such as logistic and linear regression because of the sparseness of the data in high dimensions. An example will be presented of how gene–gene interaction effects contribute to the determination of natural autoantibody levels, and thus possibly to risk of systemic autoimmune diseases. The success of our study to identify gene–gene interaction effects in association with autoantibody concentration was based on prior knowledge. However, owing to recent technological developments, researchers must face an incredible large amount of results and fast growing databases, which does not allow the testing of all prior hypotheses in a reasonable time with standard methods. Therefore, new methods are being developed and applied, such as the multifactor dimensionality reduction

Immunogenomics and Human Disease Edited by András Falus

method or logical analysis of data. These approaches are, besides finding the 'main effects', also suitable for analysis of interaction effects.

I wish I could think of one single colour which I have never seen

S. Weöres

15.1 Introduction

In additon to elucidating the basic mechanisms of biology, genetic studies offer insights into genotype–phenotype relationships that have the potential to improve our ability to diagnose, prevent and treat human diseases. One difficulty we will face in sifting through vast quantities of genetic data is that the relationship between genotype and phenotype is expected to be nonlinear for most common, multifactorial human diseases, such as cancer, cardiovascular disease and systemic autoimmune diseases. Part of this complexity can be attributed to epistasis or gene–gene interaction. Deciphering vast interconnected networks of genes and their relationships with disease susceptibility will be possible in the near future, given the recent technological advances enabling the study of hundreds and thousands of single-nucleotide polymorphisms (SNPs) at the population level (The International HapMap Consortium, 2003). Because strategies for analysing these data have not kept pace with the laboratory methods that generate data, however, it is unlikely that these advances will immediately lead to an improved understanding of the genetic contribution to common, multifactorial human diseases (Moore and Ritchie, 2004).

As the focus shifts away from rare Mendelian diseases towards common complex diseases, it is increasingly clear that gene–gene interactions and gene–environment interactions are important determinants of such diseases. There is, however, disagreement about how common such interactions are and what their importance is relative to independent main effects. This is largely due to the fact that gene–gene interactions are scarcely analysed in human genetic studies. If, for example, different SNPs do not have independent effects in a study, it is currently not obvious to search for interaction effects. This has to be done despite the fact (especially taking the very large number of SNPs into accout) that current computer technologies do not allow analysis of the astronomic number of possible combinations. However, in contrast to the above data-driven analyses, one might look for gene–gene interaction in a hypothesis-driven manner. In this chapter an example will be presented of how gene–gene interaction effects contribute to the determination of natural autoantibody levels, and thus possibly to systemic autoimmune diseases. However, before showing this example, let me highlight some basic features of gene–gene interaction (or epistasis) together with appropriate strategies and methods to detect it.

15.2 Basic Features of Gene–Gene Interactions

Very soon after the rediscovery of Mendel and his laws it was realized that the multilocus nature of inheritance could not be understood solely by examining the action of individual genes and then interpreting how these genes would behave in concert by simply combining the separate observations. Frequently genes interact with one another, leading to novel phenotypes. Some 100 years ago William Bateson coined the term 'epistasis' to describe this sort of deviation from simple Mendelian ratios (Bateson, 1909). Epistasis was meant to describe the situation in which the action of one locus masks the allelic effects at another locus. The locus being masked is said to be 'hypostatic' to the other locus. Not long after, Sir Ronald Fisher described epistasis as deviations from additivity in a linear statistical model (Fisher, 1918). More details on the development of the term epistasis are given elsewhere (Phillips, 1998).

The difference is that Bateson's definition is a biological–genetic one, whereas Fisher's is purely statistical. An important question is whether statistical evidence of epistasis at the population level (statistical epistasis) can be used to infer biological or genetical epistasis in a given individual. Conversely, does biological evidence of epistasis imply that statistical evidence will be found? Only a few studies have addressed this question directly; one was described by Cordell *et al.* (2001). The authors concluded that the degree to which statistical evidence of epistasis can elucidate underlying biological mechanisms may be limited and may require prior knowledge of the underlying aetiology. In agreement with this conclusion is our example on the genetic regulation of natural autoantibody levels, indicating that prior knowledge of basic physiological mechanisms of antibody production enhances the probability of finding statistical evidence of interaction at the population level. Today, epistasis is defined in its biological sense as *interaction* between two or more DNA variations either directly (DNA–DNA, DNA–RNA interactions), to change transcription or translation, or indirectly by way of their protein products, to alter disease risk separate from their independent effects (Figure 15.1; Cordell, 2002; Moore, 2004). Based on its biological features, epistasis is hypothesized to be a ubiquitous phenomenon (Moore, 2003), since biomolecular interactions are themselves a ubiquitous part of gene regulation, signal transduction, biochemical networks and homeostatic, developmental, immunological and physiological pathways. Epistasis in its statistical sense relates more to population genetics and genetic epidemiology. For a phenotype to be buffered against mutations, it must have an underlying genetic architecture that comprises networks of genes that are redundant and robust. As a result, substantial effects on the phenotype are observed only when there are multiple hits to the gene network. This sort of genetic buffering is realized as epistasis because it creates dependencies among the genes in the network. This stabilization or buffering of developmental pathways by multiple interacting genes has been referred to as 'canalization' (Waddington, 1942, 1957; Gibson and Wagner, 2000). As Rice (1998) suggests, nonlinear interactions among polymorphisms from multiple

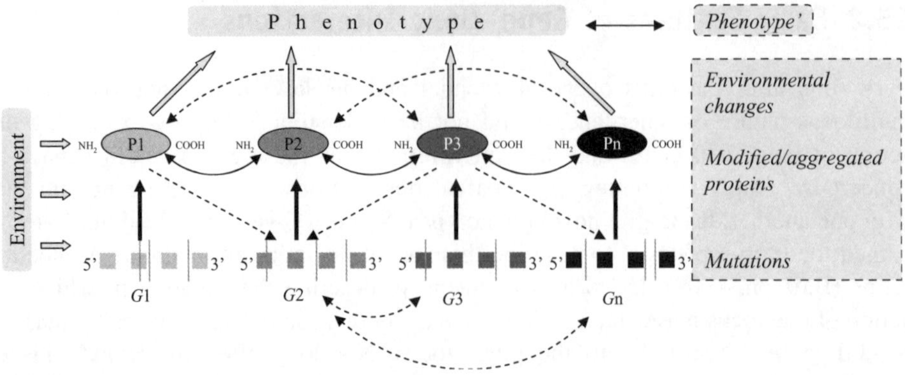

Figure 15.1 Genetical and biological epistatsis. The phenotype of a given individual is determined by a complex hierarchy of proteins (P). Genetic information affects phenotype through encoding proteins (thick arrows) by single genes (*G*). However, interindividual differences in the genes (also called polymorphisms, marked by vertical bars) give rise to particular differences in phenotypes by affecting the amount and function of proteins. Given the ubiquitous nature of biomolecular interactions (between DNA, RNA and proteins, thin arrows), combinations of individual variations result in the very complex regulation of phenotypes through interactions (grey arrows). In genetic studies where phenotypes are compared with genotype data, all kinds of macromolecular interactions affecting phenotype could be detected as a kind of gene–gene interaction, also called epistasis. Gene–gene interaction has two additional dimensions: time (developmental phase, cell-cycle, age, etc.) and localization (environmental factors). Disturbance in this model may arise at the level of genes (mutations) or proteins (aggregation, modification), or may come from the environment, resulting in altered phenotype (P′, i.e. disease). Based on Moore (2004), with modifications.

different pathways make it possible for canalization to evolve. One result of canalization is that the trait variability we observe in populations is partly the result of patterns of epistasis. The functional consequence of canalization is that extreme phenotypes due to single mutations are rare. Thus, mutations in genes from multiple pathways are necessary for extreme values of biological traits (disease endpoints). This suggests that multiple polymorphisms and their interactions will need to be considered if we are going to identify disease susceptibility loci or quantitative trait loci (QTL).

There is a growing awareness in genetic epidemiology that the results of genetic *linkage* and *association studies* of common diseases do not replicate across multiple samples. In fact, there are typically more negative than positive results for most candidate genes examined (reviewed by Hirschhorn *et al.*, 2002). Based on this ensemble of association and linkage studies, it appears as though there are few genes that have consistent large effects on multifactorial disease susceptibility in different populations. This can partly be explained by locus heterogeneity and partly by epistasis (Table 15.1). Many complicating factors causing heterogeneity can be addressed actively by a well-considered study design. The most important future task of researchers involved in the field is perhaps the collection of accurate, reliable

Table 15.1 Some factors complicating the genetic analysis of complex diseases

Name	Definition	Example
	Heterogeneity–related factors	
Allelic heterogeneity	Two or more alleles of a single locus are independently associated with the same trait	Hereditary angioneurotic oedema (HAE) type I (C1-esterase inhibitor deficiency, OMIM 106100) is an autosomal dominant disorder characterized by episodic local subcutaneous oedema and submucosal oedema involving the upper respiratory and gastrointestinal tracts. There are several published mutations affecting the gene of Cl-INH (SERPING1) causing C1-INH deficiency (Kalmar *et al.*, 2005)
Locus heterogeneity	Two or more DNA variations in distinct genetic loci are independently associated with the same trait	Severe combined immunodeficiency may be caused by different mutations. The T−, B+, NK− form is X-linked by mutation in the gene encoding the gamma subunit of the interleukin-2 receptor (OMIM 300400). The T−, B−, NK− form is caused by mutation in the adenosine deaminase gene (ADA; OMIM 102700)
Phenocopy	The presence of a disease phenotype that has a nongenetic (random or environmental) basis	Acquired HAE (OMIM 106100, type II) is characterized by normal or elevated levels of Cl-INH, but the protein is nonfunctional, and the SERPING1 gene is normal. The two types of HEA are clinically indistinguishable (Agostoni *et al.*, 2004)
Trait heterogeneity	Insufficient classification or definition of a trait or disease	There is complex classification for stroke (OMIM 601367) available (Bamford *et al.*, 1992), and it is increasingly clear that the different clinical forms of ischaemic strokes might have different genetic backgrounds (Harcos *et al.*, submitted)
Phenotypic variability	Variation in the degree, severity or age at onset of symptoms exhibited by persons who actually have the same trait or disease process	According to the recent classification of diabetes mellitus the latent autoimmune diabetes in adults (LADA) belongs to the group of type 1 autoimmune diabetes (OMIM 125853), as a slowly progressive form. A recent study (Vatay *et al.*, 2002), showing low prevalence of TNF2 allele in LADA, might explain some of the genetic differences between these two clinical forms of autoimmune diabetes

(*Continued*)

Table 15.1 (*Continued*)

Name	Definition	Example
	Interaction related factors	
Gene–gene interaction (epistasis)	Interaction between two or more DNA variations either directly (DNA–DNA, DNA–RNA interactions), to change transcription or translation, or indirectly by way of their protein products, to alter disease risk separately from their independent effects	A detailed example is given in this chapter
Gene–environment interaction	Interaction of DNA variation with an environmental factor, such that their combined effect is distinct from their independent effects	In coronary artery disease (OMIM 608320), the association between seropositivity to *Chlamydia pneumoniae* and CAD is restricted to patients carrying variant alleles of mannose-binding lectine (Rugonfalvi-Kiss *et al*., 2002)

C1-INH, C1-esterase inhibitor; LADA latent autoimmune diabetes in adults; CAD, coronary artery disease.

and abundant phenotypic (clinical) data. However, the above considerations make it equally essential to face the application of appropriate methods for the detection of epistasis in our genetic studies.

15.3 How to Detect Epistasis

Given the above considerations detection of epistasis in genetic epidemiology is not only a grain of comfort for researchers whose study failed to show independent main effects. Templeton (2000) has suggested that epistasis is commonly found when properly investigated. However, detecting interactions among variables is a well-known challenge in statistics and data-mining (Freitas, 2001). Methods for the detection of epistasis (reviewed recently by Hoh and Ott, 2003; and by Thornton-Wells, Moore and Haines, 2004) vary according to whether one is performing *association* or *linkage analysis*, and according to whether one is dealing with a quantitative or a qualitative (in particular a dichotomous) trait. For genetic association (such as case–control) studies, standard methods for epidemiological investigations may be employed, with genotypes at the various loci considered as risk factors for disease. For a given case–control data genotyped at two diallelic candidate loci we may fit *logistic regression models* using standard statistical software packages. This may provide an overall 4 degree-of-freedom (d.f.) test for interaction, but the interaction terms could each be tested individually on 1 d.f. by removal from the model, if required. Quantitative traits can be analysed in a similar way by the use of standard *multiple linear regression*. Note that these regression procedures are actually designed for testing epistasis between loci that have been genotyped. If it is believed that these loci are not themselves the aetiological variants but rather are in *linkage disequilibrium* (LD) with the true disease-causing variants, then epistasis between the surrogate genotyped loci is likely to be diluted compared with epistasis with true variants, although the extent to which this occurs will depend on the magnitude of LD.

Epistasis is relatively easily incorporated into standard nonparametric (model-free) methods of linkage analysis for quantitative traits. One popular method is the *variance components method*, in which the phenotypic covariance between relatives is modelled in terms of variance component parameters and underlying identity-by-descent sharing probabilities at one or more genetic loci. This method assumes underlying multivariate normality of the trait within pedigrees.

However, taking multiple interactions into account, epistasis is difficult to detect and characterize using traditional parametric methods such as logistic and linear regression models because of the sparseness of the data in high dimensions. When interactions among multiple polymorphisms are considered, there are many multi-locus genotype combinations that have very few or no data points. For example, if the number of marker loci is larger than the number of observations, these methods fail completely. This phenomenon has been referred to as the curse of dimensionality (Bellmann, 1961). Therefore, several alternatives to linear and logistic regression have been developed (Table 15.2). These methods can be distinguished whether they apply data-reduction techniques or not, or whether they are based on single-marker analysis or joint analysis of multiple loci or use pattern-recognition (data-mining).

For continuous outcome variables Cheverud and Routman (1995) developed a *parameterization method* for gene–gene interactions based on its effects on genetic-variance components (additive, dominance and interaction). However, it is limited to evaluating only two loci at a time and all possible genotypes must be present in the sample.

A more recently developed statistical method for evaluating gene–gene interactions is the *S-sum statistic*, a nested bootstrap (resampling) approach based on the selection of the 'best' SNP set (Zee *et al.*, 2002). This method is designed to overcome the curse of dimensionality and the multiple-testing problems by reducing any number of independent variable statistics to one sum statistic. After that it uses permutation testing to correct for an experiment-wise type I error rate. However, because the summed statistics are all single-marker statistics, set association analysis does not look at any specific (nonadditive) interactive effects among markers and would probably miss nonlinear or antagonistic types of gene–gene interactions.

Although the above methods are a clear improvement over traditional approaches because they combine information from multiple loci, a potential drawback is that they rely on single-locus effects, often also called main effects, in contrast to interaction effects. This shortcoming was avoided by the development of parameter-free approaches, also called exploratory data analysing methods. The *combinatorial partitioning method* (CPM) generates hypotheses about epistatic effects on quantitative trait variation (Nelson *et al.*, 2001). The CPM simultaneously considers

Table 15.2 Statistical methods suitable for the detection of gene–gene interactions

Name	Description	Allowed type of outcome variable	Does the method rely on single-locus effect?	Data reduction applied	Example/ reference
Multivariate analysis of variance (MANOVA)	Analysis of variance, designed for analysing data that contains more than one between-groups factor	Continuous	Yes	No	Several
Multiple linear regression	Mathematical modelling of the relationship between continuous variables and genetic factors and disease status	Continuous	Yes	No	Several
Parameterization	The method is based on the parameteriazation of physiological epistasis in a two-locus context	Continuous	Yes	No	Cheverud and Routman (1995) Leamy, Routman and Cheverud (2002)
Logistic regression	Mathematical modelling of the relationship of discrete variables to disease status	Dichotomous	Yes	No	Several
S-sum statistics (set-association analysis)	Selects the 'best' set of variables whose summary statistics are significant	Dichotomous	Yes	No	Zee *et al.*, (2002)
Multivariate adaptive regression splines	Generalization of stepwise linear regression particularly suited for high-dimensional problems with many independent variables	Continuous	No	Yes	Cook, Zee and Ridker (2004)
Combinatorial partitioning method	Utilizes data-reduction to investigate gene–gene interaction by identification of partitions of multilocus genotypes that predict interindividual variation in quantitative trait levels	Continuous	No	Yes	Nelson *et al.*, (2001)
Multifactor dimensionality reduction	Utilizes data-reduction to investigate gene–gene interaction by collapsing high-dimensional multilocus genetic data into a single dimension	Dichotomous	No	Yes	Ritchie, 2003a

Table 15.2 (*Continued*)

Name	Description	Allowed type of outcome variable	Does the method rely on single-locus effect?	Data reduction applied	Example/ reference
Logical analysis of data	Identification of patterns or 'syndromes' by mathematical optimization	Dichotomous	No	Yes	Alexe *et al.*, (2004)
Classification and regression trees	Iteratively subdivides data to build a hierarchical classification model	Dichotomous	No	Yes	Province, Shannon and Rao (2001)
Pattern recognition/data mining	Pattern detection by *a priori* algorithm with subsequent determination of pattern frequencies and relationships in the form of association rules	Dichotomous	No	Yes	Agrawal, Imielinski and Swami (1993) Agrawal and Srikant (1994)

multiple polymorphic loci to identify combinations of genotypes that are most strongly associated with variation in the quantitative trait. Genotypes from multiple loci are pooled into a smaller number of classes, thereby addressing increased dimensionality associated with modelling interactions. The application of CPM to study cardiovascular disease susceptibility genes and interindividual variability in plasma triglycerides (Nelson *et al.*, 2001) identified nonadditive, abundant epistatic interactions between multiple loci in the absence of independent main effects.

A modification or extension of the CPM is the *multifactor dimensionality reduction* (MDR) method (Ritchie, 2003a) designed specifically to improve the power to detect interactions in epidemiological studies over that provided by logistic regression. The MDR approach is nonparametric in that no parameters are estimated and is free of any assumed genetic model. In MDR, multilocus genotypes are pooled into high-risk and low-risk groups, effectively reducing the genotype predictors from *n* dimensions to one dimension. The new, one-dimensional multilocus-genotype variable is evaluated for its ability to classify and predict disease status through cross-validation and permutation testing. This data-reduction method has repeatedly been successful at finding gene–gene interactions in different diseases (Ritchie *et al.*, 2001; Tsai *et al.*, 2004; Cho *et al.*, 2004).

Methods aimed at the identification of specific *genotype patterns* partiy belong to the classification tree methods. *Multivariate adaptive regression splines* (MARS) is a generalization of stepwise linear regression that is particularly suited to high-dimensional problems in which many independent variables might be modelled (Cook, Zee and Ridker, 2004). MARS is also similar to *classification and regression trees* (CART), which iteratively subdivide data to build a hierarchical classification model (Province, Shannon and Rao 2001). MARS and CART suffer from the same problem of sequential conditioning that can plague many other regression-based methods, which makes it difficult to discover interactions (especially higher-order interactions) among predictor variables, depending on the strength of their individual effects. The binary nature of CART further limits its ability to model any additive interaction. Interestingly, in the artificial intelligence (machine learning) and in operation research, *pattern recognition* (also known as *data-mining*) methods that were described 10–15 years ago are now beginning to find their way into human genetics and might prove highly successful. An intriguing approach (Agrawal, Imielinski and Swami, 1993) formalizes pattern recognition (based on the *a priori algorithm*; Agrawal and Srikant, 1994) by defining pattern frequencies and relationships in the form of so-called 'association rules'. Several applications of this approach have been described to search for associations between potentially large numbers of SNPs and disease status (Toivonen *et al.*, 2000; Rodman *et al.*, 2001; Czika *et al.*, 2001). An even older method described in the late 1980s (Crama, Hammer and Ibaraki, 1988) is the *logical analysis of data* (LAD) approach using mathematical optimization on the basis of systematic (combinatorics supported) identification of patterns or 'syndromes'. The essence of LAD is to identify conditions which can collectively provide a classification system, to validate the classification models thoroughly and to extract from these models as much additional information as possible about the dataset. This approach has been applied earlier to problems in economics, seismology and oil exploration and also recently to medicine (Lauer *et al.*, 2002; Alexe *et al.*, 2004).

Neural networks (NN) have been also used for supervised pattern recognition in a variety of fields including genetic epidemiology. The success of the NN approach in genetics, however, varies a great deal from one study to the next, and has been applied with varied success. However, recent work has improved the reliability of artificial NN through their optimization by evolutionary computation algorithms (Ritchie, 2003b).

15.4 Autoimmunity to Heat-shock Proteins

Cellular response to stress is an evolutionarily ancient, ubiquitous and essential mechanism for survival. This mechanism protects cells from damage from environmental stress and associated misfolding (denaturation) of intracellular proteins. The molecular resources of this protective mechanism include a family of specialized proteins, *molecular chaperones*. These proteins are expressed in nonstressed cells at low levels and have essential functions in the cell cycle, as well as in cellular differentiation and growth. They are involved in metabolism, programmed cell death through protein assembly and transport, and influence the activation of enzymes and receptors. Molecular chaperones are often referred to as 'heat shock proteins' (Hsp) or Hsp stress proteins, as their expression can be induced by changes in environmental temperature (i.e. heat shock). Nonlethal heat shock (the most widely used experimental stimulus to model environmental stress) causes specific changes in cellular function and gene expression, that is, it elicits a *cellular stress response*. The changes comprise inhibition of DNA synthesis and transcription, as well as RNA processing and translation, arrest of the cell cycle, denaturation and misaggregation of proteins, enhanced degradation of proteins through both proteasomal and lysosomal pathways, derangement of cytosceletal structures, metabolic alterations that lead to a net reduction in intracellular ATP level and changes in membrane permeability that lead to the intracellular accumulation of Na^+, H^+ and Ca^{2+} (Sonna *et al.*, 2002). In mammalian cells, nonlethal heat shock alters gene expression and the activity of expressed proteins. Typically, this response enhances thermotolerance (i.e. the ability to survive subsequent, more severe heat stresses) and is temporally associated with the increased expression of Hsps. The same cellular stress response can be triggered by other stressors, including infections and exposure to toxins, and cellular reactions to a specific stressor often lead to cross-tolerance to others. Increasingly severe exposure to stress activates the apoptotic program and, under extreme conditions, cell death ensues.

Owing to their highly conserved and inducible nature, stress proteins are perfect mediators of cellular stress. Almost all pathogenic microorganisms studied hitherto possess heat-inducible genes of stress proteins and respond to thermal (and other) stresses of infection with enhanced Hsp expression. Higher organisms possess (by means of innate immunity) the inherent capability of responding to stress signals. On the other hand, essentially the same molecules can mediate stress of the host organism, making the altered self 'dangerous' in the case of cell necrosis, for example (Matzinger, 2002). However, the innate role of Hsps may become problematic in the 'new world' of adaptive immunity. With the appearance of specific receptors (i.e. antibodies and T cell receptors), overexpressed and conserved stress proteins become primary targets of autoimmunity, through infection-induced molecular mimicry. It is not surprising, therefore, that several autoantigens characterized in different autoimmune conditions (Jones, Coulson and Duff 1993) share some of their epitopes with Hsps. More surprising, however, is the fact that autoimmune diseases are rare and are characterized by well-defined autoimmune reaction to only a few autoantigens in a conserved nature. Furthermore, according to the idea that the source of self-tolerance is clonal deletion, one may speculate that, the closer a molecule is to self, the less immunogenic it should be. It is surprising, therefore, to find that among the major antigens recognized during a wide variety of infections many belong to conserved protein families sharing extensive sequence identity with the host's molecules, for example Hsps. One interesting hypothesis, dealing with these controversies was presented by Irun Cohen and co-workers, called *the immunological homunculus* (Cohen and Young, 1992). The idea is that some, perhaps all, major autoantigens are indeed dominant because each of them is encoded in the organizational structure of the immune system. Each dominant self-antigen is served by an interacting set of T and B cells that includes cells with receptors for the antigen (*antigen-specific*) and cells with receptors for the antigen-specific receptor (*anti-idiotypic*). Some of these lymphocytes suppress and others stimulate. Owing to the mutual connections between the various interacting lymphocytes in the network, some lymphocytes become activated even without being driven by contact with specific antigen in an immunogenic form. The state of autonomous activity defined by a pattern of interconnected lymphocytes constitutes a functional representation of the particular self-antigen around which the network is organized. In other words, the picture of the self-antigen is encoded within a cohort of lymphocytes.

Therefore, highly conserved Hsps (present in all mammalian cells) need special protection in the 'adaptive world'. Some recent results support the existence of regulating natural autoimmunity toward Hsps. Anti-Hsp90 reacting antibodies were recently shown to be part of the natural autoantibody repertoire and were characterized as broadly cross-reacting antibodies mainly belonging to the IgG2 subclass (Pashov *et al.*, 2002). Furthermore, in the study of Pashov *et al.* (2002), although the anti-Hsp90 antibodies bind to the same set of antigens, there were quantitative differences among the antibodies tested. In line with these observations, the concept of immunological homunculus may be extended to the B cell compartment as well. Thus, the natural autoantibody repertoire seems to be qualitatively invariable, directed against a highly conserved set of immunodominant antigens, including Hsps, but may vary quantitatively, in a throughout-life conserved manner (Pashov *et al.*, 2002; Mouthon *et al.*, 1995; Nobrega *et al.*, 1993). We recently reviewed evidence supporting our hypothesis that anti-Hsp60 antibodies are not simply remainders of reactive antibodies but belong to natural autoantibodies as well (Prohászka *et al.*, 2004). This evidence includes the presence of anti-Hsp60 IgG in all healthy individuals studied, even in children, the log–normal distribution of anti-Hsp60 antibodies in a given population basically independently from serological signs of chronic infections and the stability of anti-Hsp60 IgG levels in life.

Future research has to address the mechanisms by which the 'size of the homunculus' is regulated. It is tempting to speculate that gene–environment (colonization and infections) and gene–gene interactions are important factors in this regulation besides a few main effects of different loci. In the last part of this chapter I will summarize some of our recent results that lend credence to this speculation.

15.5 Epistatic Effect in the Regulation of Anti-Hsp60 Autoantibody Levels

Based on our hypothesis that anti-Hsp60 antibodies belong to the natural autoantibody repertoire and that a significant effect of gene–gene interaction in the regulation of their levels can be supposed, we started to search for genetic factors likely to be involved in this regulation. Since cytokines are major regulators of differentiation and activation of immune cells, including B lymphocytes, our first study was focused on the investigation of polymorphisms of certain cytokines in association with autoantibody titres. SNPs of proinflammatory cytokines [interleukin (IL)-6 gene at position-174, the biallelic exchange of the IL-1β gene at position -511 and the IL-1α gene at position -889] in parallel with antibody measurements were studied in a cohort of healthy men (Veres *et al.*, 2002). A strong association between IL-6-174 polymorphism and anti-Hsp60 IgG titre was seen; carriers of allele C at this position had a significantly lower level of anti-Hsp60 IgG as compared with carriers of GG. No other main effects with other polymorphisms studied were observed.

In a second attempt we aimed to determine whether immunoglobulin (Ig) GM and KM allotypes influence anti-Hsp60 autoantibody levels (Pandey *et al.*, 2004). The study included the same patient cohort as above and investigated whether allelic variation at the GM and KM loci is associated with autoantibody levels to Hsp60. A weak, but significant association of anti-Hsp60 with GM f,z genotypes was observed, subjects carrying allele z had approximately 1.5 lower antibody titres than subjects who were noncarriers of allele z. No other associations with GM b,g or KM 1,3 genotypes were observed. However, most importantly, significant interactive effects of GM and IL-6 genotypes were noted for anti-Hsp60 levels (Figure 15.2). Subjects carrying IL-6 -174 C and GM z alleles appeared to have the lowest levels of autoantibodies. Compared with these subjects, those who lacked the IL-6 -174 C allele, with or without the GM z allele, had significantly higher levels of anti-Hsp60. However, subjects with genotype GM f,f together with allele C at IL-6 -174, did not have elevated concentrations of autoantibody. Thus, based on these results we conclude that the IL-6 -174 locus has epistatic effect on GM f,z locus. In other words, allele C at IL-6 -174 masks the association of GM f,f genotype with high anti-Hsp60 IgG levels. It should be noted that all of the investigated genotypes fulfilled the HW-equilibrium and there was no significant population association between IL-6 and GM alleles.

15.6 Conclusions

The success of our study to identify gene–gene interaction effect in association with autoantibody concentration was based on prior knowledge. Among only six loci investigated, we could recognize interaction between two, IL-6 -174 and Ig GM f,z. This observation of interaction, however, was based on significant main effects.

Interleukin-6 was originally identified as a B-cell differentiation factor (Hirano *et al.*, 1986), but it is now known to be a multifunctional cytokine that regulates immune response, haematopoiesis, the acute phase response, and inflammation (Hirano, 1998). The gene encoding IL-6 in humans (*IL-6*) is organized in five exons and four introns, and it maps to the short arm of chromosome 7 (Yasukawa *et al.*, 1987). The regulatory and promoter region of *IL-6* reveals a complex organization which may account for a considerable part of the multifunctionality

(a)

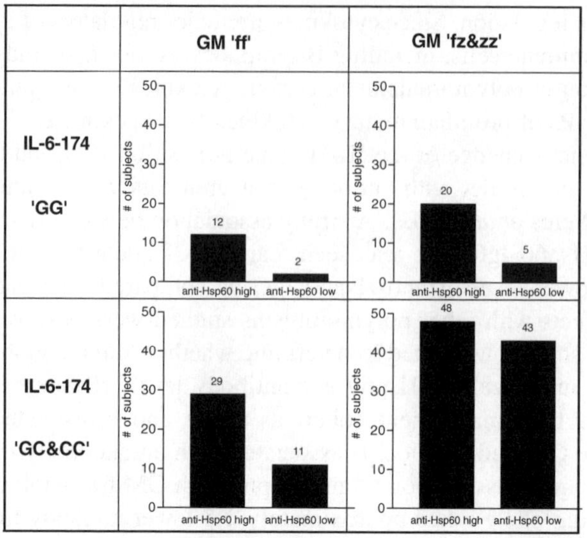

(b)

Genotype	Anti-Hsp60 high (*n*)	Anti-Hsp60 low (*n*)	Anti-Hsp60 [AU/ml, median (interquartile range)]
IL-6-174-C and GM-z	48	43	21.3 (12.339.6)
IL-6-174-C and GM-non z	29	11	29.8 (15.7–72.4)
IL-6-174-GG and GM-z	20	5	36.6 (20.8–79.0)*
IL-6-174-GG and GM-non z	12	2	55.6 (25.4–75.4)*
	Chi-square test for trend: p=0.0009		ANOVA p=0.002

Figure 15.2 Anti-heat-shock protein 60 (Hsp60) autoantibody levels stratified according to interleukin 6 -174 promoter C/G and immunoglobulin GM f/z allele carrier status (a). Anti-Hsp60 antibody was measured as described earlier (Prohászka *et al.*, 1999) and was expressed in arbitrary unit/mi values (b). Low anti-Hsp60 was defined as equal to or less than 17.66 AU/ml, the 33th percentile of the sample (total $n = 170$). $^*p < 0.05$ in the post-test as compared with IL-6-174C/GM z.

of IL-6. Several transcription factors mediate the activation of the *IL-6* promoter, whereas steroids and retinoblastoma control element result in suppression of it. The involvement of IL-6 in many biological functions is paralleled by genetic association of allelic variants of *IL-6* with several physiological and pathophysiological conditions. Two of several polymorphic sites were used most frequently for genetic association studies: a multiallelic variable number of tandem repeats (VNTR) polymorphism in the 3′ flanking region of *IL-6*, and a biallelic G to C polymorphism of the promoter at position -174. This latter polymorphism has been associated with the prevalence, incidence and prognosis of a variety of disease states, such as Alzheimer disease, cardiovascular disease, cancer, osteoporosis, type-2 diabetes mellitus, sepsis and rheumatoid arthritis (reviewed by Papassotiropoulos, Hock and

Nitsch, 2001). The G allele of the -174 SNP has been associated with an increased transcriptional response to endotoxin or IL-1β *in vitro* (Fishman *et al.*, 1998). Studies, however, investigating the role of the -174 G>C promoter polymorphism for the plasma levels of IL-6 concentrations *in vivo* have produced conflicting results. However, in a recent, well-designed study, Benneromo *et al.* (2004) report that individuals homozygous for the -174 G allele had significantly higher plasma IL-6 levels after vaccination than individuals with the CC genotype. However, no relationship between basal IL-6 concentrations and genotype was observed, which may partly explain the previously reported inconsistency.

Taken together, IL-6 -174 GG genotype may contribute to higher antibody production after an antigenic stimulus by influencing peak IL-6 levels and thus increasing activation of B lymphocytes. Supporting this conclusion is the observation that, in patients with systemic sclerosis, continuous B cell activation is paralleled by high IL-6 concentrations and abundant autoantibody formation (Sato *et al.*, 2004). In concert with this study we have indeed observed elevated amounts of anti-Hsp60 autoantibodies in patients with systemic sclerosis and undifferentiated connective tissue disease (Horváth *et al.*, 2001).

Immunoglobulin GM and KM allotypes are antigenic determinants of γ chains and κ-type light chains, respectively (Grubb, 1995). The marked differences in the frequencies of Ig allotypes among races, strong linkage disequilibrium within a race and racially restricted occurrence of GM haplotypes all suggest that differential selection over many generations may have played an important role in the maintenance of polymorphism at these loci. One likely selective force may be the association between GM and KM allotypes and specific antibody responses to pathogens, resulting in differential immunity to infectious diseases. Indeed, several data have been reported on the associations between certain GM and KM allotypes and immune responsiveness to polysaccharide vaccines and to particular infectious agents (Pandey, 2001). Importantly, evidence has been published on the association of GM z carrier status and low antibody responses to meningococcal polysaccharide A (Pandey *et al.*, 1982). It is important to note that Ig allotype-associated immune response genes primarily influence antibody responses to T-independent antigens and may require interaction with other genes – perhaps HLA or IL-6 – to confer immunity to T-dependent antigens.

Possible ways in which constant-region allotypes could contribute to antibody reactivity and specificity include the following. First, certain alleles coding for allotypes may be in linkage disequilibrium with particular variable-region determinants associated with immune responsiveness. Furthermore, they could directly contribute to the formation of specific idiotypes. In addition, allotype-associated structural variability in the constant region could modulate the kinetic competence of the antigen binding sites.

There are apparently at least two ways of identifying gene–gene interactions in genetic studies: hypothesis (knowledge)-driven and data-driven. The first may be used in 'traditional' experiments dealing with small or moderate-sized databases. The SNPs of choice and type of analyses (mainly standard methods such as ANOVA and regression) in such studies are based on prior knowledge. However, based on recent technological developments, researchers must face an incredibly large quantity of results and fast-growing databases, which does not allow all prior hypotheses to be tested in a reasonable time with standard methods. Therefore, new methods have been developed and applied for these tasks, including different exploratory and data-mining approaches. These approaches are, besides finding the 'main effects', also suitable for analyses that were not possible earlier, including the testing of gene–gene effects.

Nevertheless, the two ways did not converge until now, one reason being the lack of appropriate methods to study large data bases (and interactions) in a hypothesis-driven way. This situation can also be translated in the following way: studies yielding significant main effects should not stop at that level but have to step forward to determine interaction effects. If an advance in method development could be reached allowing hypothesis-driven analysis of interactions, the redundancy and the unavoidable false-positivity could be controlled and we would not be failing to see the forest for the trees.

Appendix

Interaction	In statistics, an *interaction* effect occurs when the relation between (at least) two variables is modified by (at least one) other variable. In other words, the strength or the sign (direction) of a relation between (at least) two variables is different depending on the value (level) of some other variable(s). Note that the term 'modified' in this context does not imply causality but represents a simple fact that, depending on what subset of observations [regarding the 'modifier' variable(s)] you are looking at, the relation between the other variables will be different.
Association analysis	The analysis of population-based (case–control) or family-based genotypic data to detect the association of a disease with a specific allele *across cases* (*or families*).
Linkage analysis	The analysis of family-based data to detect linkage of a disease locus with one ore more loci *within a family*.
Linkage disequilibrium	If a genetically encoded trait is only partially explained by the presence of allele considered in the analysis (A), one can expect that individuals affected with the trait tend to share a different set of alleles at loci around the mutated locus (A′) than do unaffected individuals. Allele A is said to be in linkage disequilibrium with alleles A′.

References

Agostoni A, Aygoren-Pursun A, Binkley KE, Blanch A, Bork K, Bouillet L, Bucher C, Castaldo AJ, Cicardi M, Davis AE, De Carolis C, Drouet C, Duponchel C, Farkas H, Fay K, Fekete B, Fischer B, Fontana L, Fust G, Giacomelli R, Groner A, Hack CE, Harmat G, Jakenfelds J, Juers M, Kalmar L, Kaposi PN, Karadi I, Kitzinger A, Kollar T, Kreuz W, Lakatos P, Longhurst HJ, Lopez-Trascasa M, Martinez-Saguer I, Monnier N, Nagy I, Nemeth E, Nielsen EW, Nuijens JH, O'grady C, Pappalardo B, Penna V, Perricone C, Perricone R, Rauch U, Roche O, Rusicke E,

Spath PJ, Szendei G, Takacs E, Tordai A, Truedsson L, Varga L, Visy B, Williams K, Zanichelli A, Zingale L. 2004. Hereditary and acquired angioedema: problems and progress. In Proceedings of the Third C1 Esetrase Inhibitor Deficiency Workshop and Beyond. *J Allergy Clin Immunol* **114**: S51–S131.

Agrawal R, Srikant R. 1994. Fast algorithms for mining association rules. *Proceedings of the 20th International Conference on Very Large Databases*; www.almaden.ibm.com/people/ragrawal/papers/vldb94_rj.ps

Agrawal R, Imielinski T, Swami A. 1993. *Proceedings of ACM SIGMOND Conference on Management of Data*, (Buneman P, Jajodia S) (eds). Washington, DC: Association for Computing Machinery; 207–216.

Alexe G, Alexe S, Liotta LA, Petricoin E, Reisss M, Hammer PL. 2004. Ovarian cancer detection by logical analysis of proteomic data. *Proteomics* **4**: 766–783.

Bamford JM, Sandercock P, Dennis M, Burn M, Warlow C. 1991. Classification and natural history of clinically identifiable subtypes of acute cerebral infarction. *Lancet* **337**: 1521–1526

Bateson W. 1909. *Mendel's Principles of Heredity*. Cambridge: Cambridge University Press.

Bellmann R. 1961. *Adaptive Control Processes*. Princeton: Princeton University Press.

Benneromo M, Held C, Stemme S, Ericsson CG, Silveria A, Green F, Tornvall P. 2004. Genetic predisposition of the interleukin-6 response to inflammation: Implications for a variety of major diseases? *Clin Chem* **50**: 2136–2140.

Cheverud J Routman EJ. 1995. Epistasis and its contribution to genetic variance components. *Genetics* **139**: 1455–1461.

Cho YM, Ritchie MD, Moore JH, Park JY, Lee KU, Shin HD, Lee HK, Park KS. 2004. Multifactor-dimensionality reduction shows a two-locus interaction associated with type 2 diabetes mellitus. *Diabetologia* **47**: 549–554.

Cohen IR, Young DB. 1992. Autoimmunity, microbial immunity and the immunological homunculus. *Immunol Today* **12**: 105–110.

Cook NR, Zee RY, Ridker PM. 2004. Tree and spline based association analysis of gene–gene interaction models for ischaemic stroke. *Stat Med* **23**: 1439–1453.

Cordell HJ. 2002. Epistasis: what it means, what it doesn't mean, and statistical methods to detect it in humans. *Hum Mol Genet* **11**: 2463–2468.

Cordell HJ, Todd JA, Hill NJ, Lord CJ, Lyons PA, Peterson LB, Wicker LS, Clayton DG. 2001. Statistical modeling of interlocus interactions in a complex disease: rejection of the multiplicative model of epistasis in type 1 diabetes. *Genetics* **158**: 357–367.

Crama Y, Hammer PL, Ibaraki T. 1988. Cause–effect relationships and partially defined Boolean functions. *Ann Opns Res* **16**: 299–326.

Czika WA, Weir BS, Edwards SR, Thompson RW, Nielsen DM, Brocklebank JC, Zinkus C, Martin ER, Hobler KE. 2001. Applying data mining techniques to the mapping of complex disease genes. *Genet Epidemiol* **21**: S435–S440.

Fisher RA. 1918. The correlation between relatives on the supposition of Mendelian inheritance. *Trans R Soc Edinburgh* **52**: 399–433.

Fishman D, Faulds G, Jeffery R, Mohamed-Ali V, Yudkin JS, Humphries S, Woo P. 1998. The effect of novel polymorphisms in the interleukin-6 (IL-6) gene on IL-6 transcription and plasma IL-6 levels, and an association with systemic-onset juvenile chronic arthritis. *J Clin Invest* **102**: 1369–1376.

Freitas AA. 2001. Understanding the crucial role of attribute interaction in data mining. *Artif Intell Rev* **16**: 177–199.

Gibson G, Wagner G. 2000. Canalization in evolutionary genetics: a stabilizing theory? *BioEsseys* **22**: 372–380.

Grubb R. 1995. Advances in human immunoglobulin allotypes. *Exp Clin Immungenet* **12**: 191–197.

Hirano T. 1998. Interleukin 6 and its receptor: ten years later. *Int Rev Immunol* **16**: 249–284.

Hirano T, Yasukawa K, Harada H, Taga T, Watanabe Y, Matsuda T. 1986. Complementary DNA for a novel human interleukin (BSF-2) that induces B lymphocytes to produce immunoglobulin. *Nature* **324**: 73–76.

Hirschhorn JN, Lohmuller K, Byrne E, Hirschhorn K. 2002. A comprehensive review of genetic association studies. *Genet Med* **4**: 45–61.

Hoh J, and Ott J. 2003. Mathematical multi-locus approaches to localizing complex human trait genes. *Nat Rev Genet* **4**: 701–709.

Horváth L, Czirjak L, Fekete B, Jakab L, Prohaszka Z, Cervenak L, Romics L, Singh M, Daha MR, Fust G. 2001. Levels of antibodies against C1 q and 60 kDa family of heat shock proteins in the sera of patients with various autoimmune diseases. *Immunol Lett* **75**: 103–109.

Jones DB, Coulson AF, Duff GW. 1993. Sequence homologies between Hsp60 and autoantigens. *Immunol Today* **14**: 115–118.

Kalmár L, Hegedüs T, Farkas H, Nagy M, Tordai A. 2005. HAEdb: a novel, interactive, locus-specific mutation database for the C1 inhibitor gene. *Hum Mutat* **25**: 1-5.

Lauer MS, Alex S, Pothier Sander CE, Blackstone EH, Ishwaran H, Hammer PL. 2002. Use of the logical analysis of data method for assessing long-term mortality risk after exercise electrocardiography. *Circulation* **106**: 685–690.

Leamy LJ, Routman EJ, Cheverud JM. 2002. An epistatic genetic basis for fluctuating asymmetry of mandible size in mice. *Evol Int Org Evol* **56**: 642–653.

Matzinger P. 2002. The danger model: a renewed sense of self. *Science* **296**: 301–305.

Moore JH. 2003. The Ubiquitous nature of epistasis in determining susceptibility to common human diseases. *Hum Hered* **56**: 73–82.

Moore JH. 2004. A global view of epistasis. *Nat Genet* **37**: 13–14.

Moore JH, Ritchie MD. 2004. The challenges of whole-genome approaches to common diseases. *JAMA* **291**: 1642–1643.

Mouthon L, Nobrega A, Nicolas N, Kaveri SV, Barreau C, Coutinho A, Kazatchkine MD. 1995. Invariance and restriction towards a limited set of self antigens characterize neonatal IgM antibody repertoires and prevail in autoreactive repertoires of healthy adults. *Proc Natl Acad Sci USA* **92**: 3839–3843.

Nelson MR, Karida SLR, Ferrell RE, Sing CF. 2001. A combinatorial partitioning method to identify multilocus genotypic partitions that predict quantitative trait variation. *Genome Res* **11**: 458–470.

Nobrega A, Haury M, Grandien A, Malanchere E, Sundblad A, Coutinho A. 1993. Global analysis of antibody repertoires. II. Evidence for specificity, self-selection and the immunological 'homunculus' of antibodies in normal serum. *Eur J Immunol* **23**: 2851–2859.

Pandey JP. 2001. Immunoglobulin GM and KM allotypes and vaccine immunity. *Vaccine* **19**: 613–617.

Pandey JP, Ambrosch F, Fudenberg HH, Stanek G, Wiedermann G. 1982. Immunoglobulin allotypes and immune response to meningococcal polysaccharides A and C. *J Immungenet* **9**: 25–29.

Pandey JP, Prohászka Z, Veres A, Füst G, Hurme M. 2004. Epistatic effects of genes encoding immunoglobulin GM allotypes and interleukin-6 on the production of autoantibodies to 60- and 65-kDa heat-shock proteins. *Genes Immun* **5**: 68–71.

Papassotiropoulos A, Hock C, Nitsch R. 2001. Genetics of interleukin 6: implications for Alzheimer's disease. *Neurobiol Aging* **22**: 863–871.

Pashov A, Kenderov A, Kyurkchiev S, Kehayov I, Hristova S, Lacroix-Desmazes S, Giltiay N, Varamballi S, Kazatchkine MD, Kaveri SV. 2002. Autoantibodies to heat shock protein 90 in the human natural antibody repertoire. *Int Immunol* **14**: 453–461.

Phillips PC. 1998. The language of gene interaction. *Genetics* **149**: 1167–1171.

Prohászka Z, Duba J, Lakos G, Kiss E, Varga L, Janoskuti L, Csaszar A, Karadi I, Nagy K, Singh M, Romics L, Fust G. 1999. Antibodies against human Hsp60 and mycobacterial Hsp65 differ in their antigen specificity and complement activating ability. *Int Immunol* **11**: 1363–1370.

Province MA, Shannon WD, Rao DC. 2001. Classification methods for confronting heterogeneity. *Adv Genet* **42**: 273–286.

Rice SH. 1998. The evolution of canalization and the breaking of von Buer's laws: modeling the evolution of development with epistasis. *Evolution* **52**: 647–656.

Ritchie MD, Hahn LW, Moore JH. 2003. Power of multifactor-dimensionality reduction for detecting gene–gene interactions in the presence of genotyping error, missing data, phenocopy, and genetic heterogeneity. *Genet Epidemiol* **24**: 150–157.

Ritchie MD, Hahn LW, Roodi N, Bailey LR, Dupont WD, Parl FF, Moore JH. 2001. Multifactor-dimensionality reduction reveals high-order interactions among estrogen- metabolism genes in sporadic breast cancer. *Am J Hum Genet* **69**: 138–147.

Ritchie MD, White BC, Parker JS, Hahn LW, Moore JH. 2003. Optimization of neural network architecture using genetic programming improves detection and modeling of gene–gene interactions in studies of human diseases. *BMC Bioinformatics* **4**: 28; www.biomedcentral.com/1471-2105/4/28

Rodman P, Macula AJ, Spance MA, Tomey DC. 2001. Preliminary implementation of new data mining techniques for the analysis of simulation data from Genetic Analysis Workshop 12: Problem 2. *Genet Epidemiol* **21**: S390–395.

Rugonfalvi-Kiss S, Endresz V, Madsen HO, Burian K, Duba J, Prohaszka Z, Karadi I, Romics L, Gonczol E, Fust G, Garred P. 2002. Association of *Chlamydia pneumoniae* with coronary artery disease and its progression is dependent on the modifying effect of mannose-binding lectin. *Circulation* **106**: 1071–1076.

Sato S, Fujimoto M, Hasegawa M, Takehara K, Tedder T. 2004. Altered B lymphocyte function induces systemic autoimmunity in systemic sclerosis. *Mol Immunol* **41**: 1123–1133.

Sonna LA, Fujita J, Gaffin SL, Lilly CM. 2002. Molecular biology of thermoregulation. Invited review: effects of heat and cold stress on mammalian gene expression. *J Appl Physiol* **92**: 1725–1742.

Templeton AR. 2000. Epistasis and complex traits. In *Epistasis and the Evolutionary Process*, Wolf J, Brodie III B, Wade M (eds). New York: Oxford University Press.

The International HapMap Consortium. 2003. The International HapMap Project. *Nature* **426**: 789–796.

Thornton-Wells TA, Moore JH, Haines JL. 2004. Genetics, statistics and human disease: analytical retooling for complexity. *Trends Genet* **20**: 640–647.

Toivonen HT, Onkamo P, Vasko K, Ollikainen V, Sevon P. Mannila H. 2000. Data mining applied to linkage disequilibrium mapping. *Am J Hum Genet* **67**: 133–145.

Tsai CT, Lai LP, Lin JL, Chiang FT, Hwang JJ, Ritchie MD. 2004. Renin–angiotensin system gene polymorphisms and atrial fibrillation. *Circulation* **109**: 1640–1646.

Vatay Á, Rajczy K, Pozsonyi E, Hosszúfalusi N, Prohászka Z, Füst G, Karadi I, Szalai C, Grosz A, Bartfai Z, Panczel P. 2002. Differences in the genetic background of latent autoimmune diabetes in adults (LADA) and type 1 diabetes mellitus. *Immunol Lett.* **84**: 109–115.

Veres A, Prohaszka Z, Kilpinen S, Singh M, Fust G, Hurme M. 2002. The promoter polymorphism of the IL-6 gene is associated with levels of antibodies to 60 kDa heat shock proteins. *Immunogenetics* **53**: 851–856.

Waddington CH. 1957. *The Strategy of the Genes*. New York: Macmillan.

Waddington CH. 1942. Canalization of development and inheritance of acquired characters. *Nature* **150**: 563–565.

Yasukawa K, Hirano T, Watanabe Y, Muratani K, Matsuda T, Nakai S, Kishimoto T. 1987. Structure and expression of human B cell stimulatory factor-2 (BSF-2/IL-6) gene. *EMBO J* **6**: 2939–2945.

Zee RYL, Hoh J, Cheng S, Reynolds R, Grow MA, Silbergleit A, Walker K, Steiner L, Zangenberg G, Fernandez-Ortiz A, Macaya C, Pintor E, Fernandez-Cruz A, Ott J, Lindpainter K. 2002. Multilocus interactions predict risk for post-PTCA restenosis: an approach to the genetic analysis of common complex disease. *Pharmacogenomics J* **2**: 197–201.

16

Histamine Genomics and Metabolomics

András Falus, Hargita Hegyesi, Susan Darvas, Zoltan Pos and **Peter Igaz**

Abstract

Histamine, one of the most abundantly studied general mediators, is a 112 Da biogenic amine generated by histidine decarboxylase (HDC) from L-histidine. Histamine binds to at least four different G-protein-associated plasma membrane receptors (H1–H4) with dissimilar tissue expression patterns and various biological functions. It plays an important role in the regulation of immune and inflammatory processes and, based on recent studies, is involved in malignant and normal cell proliferation. Both synthetizing and degradating enzymes, as well as histamine receptors, reveal significant, functionally relevant genetic polymorphism. The phenotype of the gene-targeted HDC-'knock-out' mice suggests that histamine, part of metabolome, is a major player in the regulation of mammalian physiology and pathophysiology.

16.1 Introduction

Histamine is one of the smallest biomolecules (Figure 16.1). It consists of only 17 atoms, of which as few as five carbons and three nitrogens constitute the molecular skeleton. Histamine is, however, one of the most extensively studied chemical entities due to its enormous significance in biology and medicine, having over 30 identified physiological functions (Hill *et al.*, 1997). Remarkably enough, all the histamine-driven biological reactions leave the molecular backbone of histamine totally unchanged. How can a compound of 112 Da molecular weight influence that large a number of biochemical processes, without any rupture in its covalent bonds? What properties qualify histamine to participate in so many biological functions, from contracting smooth muscles through stimulating NO formation up to radical shifting of T cell polarization toward Th2?

Immunogenomics and Human Disease Edited by András Falus

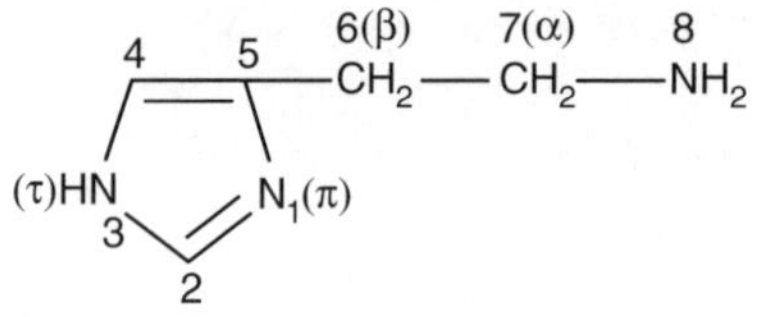

Figure 16.1 Constitutional formula of histamine with numbering and indices.

16.2 Chemistry

Taking into account states of protonation, ring tautomerism and side-chain conformation, the number of nonidentical histamine species is 14. They all correspond to intramolecular energy minima, with small differences. Such species have very short ($\sim 10^{-9}$ s) individual lifetimes in solution.

16.3 Biosynthesis and Biotransformation

This paragraph covers the chemical 'birth' and 'death' of histamine, which constitute its covalent biochemistry, as depicted in Figure 16.2. The formation of histamine in the body occurs in one step, in which L-histidine, one of the 20 'classical' amino acids, undergoes decarboxylation. This reaction is catalysed by histidine decarboxylase (HDC), a pyridoxal phosphate-containing enzyme that differs from aromatic amino acid decarboxylase. Histidine is therefore considered a specific substrate of HDC, presumably due to the imidazole nitrogens. Concerning the bioenergetics of decarboxylation, it breaks a strong, covalent carbon–carbon bond, nevertheless it is an energy-releasing (exergonic) reaction (Roskoski, 1996). The metabolic pathways of histamine are different in the brain and the periphery, with an extra 3-methylation reaction in the brain. This methyl moiety is provided by *S*-adenosyl-methionine. Both types of transformation result in derivatives of imidazole–acetic acid. These compounds contain an anionic carboxylate site, which enhances water solubility, promoting in this way the excretion of histamine in the urine.

Biosynthesis and biotransformation are processes with ruptures and formations of covalent bonds in the 60–120 kcal/mol energy range. Thus, the formation and decomposition of histamine are accompanied by high-energy changes. Strictly speaking, however, the biosynthesis and biotransformation are previous and subsequent stages of histamine biology. The actual biological chemistry of histamine is noncovalent. This conclusion can be drawn, for example, from observations on the receptorial behaviour: histamine is able to activate at least four types of receptor, while its agonists and antagonists can typically bind to one type of receptor only. Consequently, the different biological functions of histamine belong to distinct molecular shapes and ionization forms within the covalent frame. In fact, the

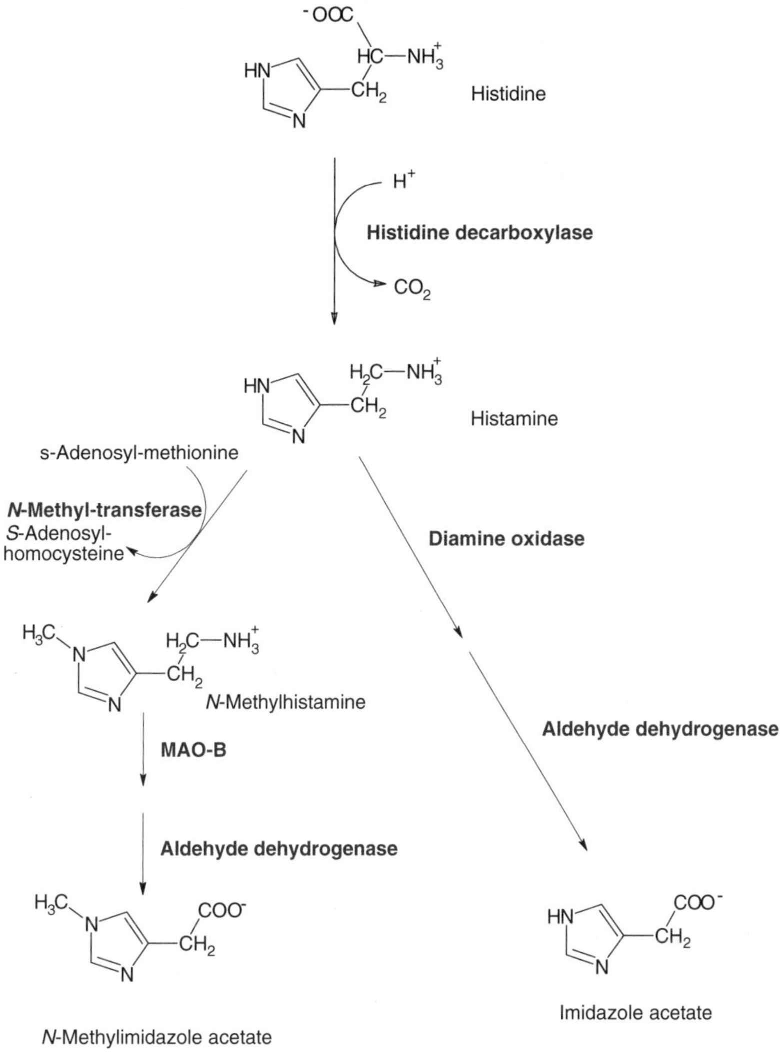

Figure 16.2 Biosynthesis and biotransformation of histamine.

biological chemistry of histamine is a wide variety of delicately governed weak interactions in the 0.1–20 kcal/mol energy range, under subtle stereochemical control.

So far no method exists that could provide direct information on the *in vivo*, moiety-specific weak interactions of histamine or other biomolecules (Noszál, Kraszni and Rácz, 2004).

The most promising technique, which will probably be able to offer direct insight into molecular events in the body, is magnetic resonance spectroscopy (MRS), the *in vivo* version of nuclear magnetic resonance (NMR), the most powerful method for elucidation of molecular structure and dynamics *in vitro*. The delicately modulated weak interactions of histamine are based on its acid–base and complexation behaviour and charge distribution.

16.4 Histidine Decarboxylase – Gene and Protein

The rate limiting enzyme in histamine production is L-HDC (EC 4.1.1.22), which catalyses the decarboxylation of L-histidine to histamine. HDCs can be devided into two groups: those that contain pyridoxal-phosphate (PLP) coenzyme and those that contain a pyruvoyl residue at the active site. Gram-positive bacteria and mammals contain PLP-dependent HDC enzymes. Human HDC was mapped to chromosome 15, mouse HDC gene to chromosome 2 and rat HDC gene to chromosome 3. HDC gene contains 12 exons and 11 introns, spanning 24 kb, and the eighth exon encodes the PLP-binding domain. Mammalian HDC gene promoters have similar and different motifs which are responsible for tissue-specific gene expression. HDC proteins contain different domains which are involved in protein targetting, digestion and regulation of activity. The kinetical parameters and pH sensitivity of mammalian and Gram-positive bacteria HDCs are very similar.

Mammalian HDC genes, promoters and tanscriptional regulation

Mammalian HDC gene sequences have been cloned and characterized from human and rodents (rat, mouse). Full-length cDNA of human HDC was identified, characterized and mapped to chromosome 15 in the 15q15–q21 region (Yamauchi *et al.*, 1990; Zahnow *et al.*, 1991). Genomic DNA blot analysis has revealed that the HDC gene is present as a single copy in the human genome. Genetic analysis of the isolated human HDC gene has revealed that this relatively large gene contains 12 exons and 11 introns, spanning approximately 24 kb. Almost half of the total mRNA sequence is encoded by the last extremely large exon (exon 12; base pairs from 1349 to 2404). The consensus PLP-binding domain can be found in exon 8 (Yatsunami *et al.*, 1994).

Mouse HDC gene was mapped to the E5-G region of murine chromosome 2 by *in situ* hybridization (Malzac *et al.*, 1996). Genetic analysis of the HDC gene from the genomic libraries has revealed that mouse gene contains 12 exons and 11 introns spanning approximately 24 kb, like the human HDC gene. The nucleotide sequence of exons shows high overall sequence identity to the human exons (84–92%), although lower homology (64–75%) can be detected between human and mouse exons 1 and 12 (Suzuki-Ishigaki *et al.*, 2000). Rat HDC gene is mapped to chromosome 3 and is present as a single copy in the rat genome (Joseph *et al.*, 1990).

Among the mammalian HDCs the rat cDNA was the first to be isolated and the deduced amino acid sequence of HDC protein was given (Joseph *et al.*, 1990). Heterogeneity of mRNA transcripts was found because two mRNA species (a minor 3.5 kb and a major 2.7 kb) were detected following an alternative splicing (Joseph *et al.*, 1990). Nucleotide sequence analysis of cDNAs from non-inbred Sprague–Dawley rats has revealed that some of these cDNAs contained three residues that were different from the foetal rat liver cDNA library clone. This finding represents genetic polymorphism of the gene (Joseph *et al.*, 1990). HDC mRNA in most human cell and tissue types was found to be approximately 2.4 kb in length. The observed heterogeneity of the HDC mRNA (2.4 and 3.4 kb) in some human cell lines is caused by alternative splicing, because the longer transcript contains a mis-spliced insert and an alternatively spliced exon of the HDC gene (Yatsunami *et al.*, 1994). In mouse the only HDC transcription product is a 2.7kb mRNA, so alternative splicing is not detected (Suzuki-Ishigaki *et al.*, 2000; Yamamoto *et al.*, 1990).

HDC gene expression and its regulation at transcriptional level by the different transcription factors seem to be tissue- and cell type-specific. Most mammalian tissues that have been studied have shown some HDC activity, albeit most of them at low level. Analysis of the HDC gene promoter sequence upstream or downstream of the transcription start site reveals different gene regulatory motifs in the three mammalian HDCs.

In the human promoter a TATA-like sequence, a CG box and CACC boxes were found upstream from the transcription start site. In addition, sequences match the GATA consensus sequences and leader-binding protein-1 motifs were detected (Yatsunami *et al.*, 1994). Deletion analysis has revealed *cis*-acting elements (23 nucleotides in length) at different places downsteam of the transcriptional start site in the human HDC promoter. This promoter motif is called gastrin responsive element (GAS-RE; Zhang *et al.*, 1996). In mouse HDC promoter region the same promoter motifs and three other consensus sequences for transcription factors were detected (Joseph *et al.*, 1990). These are the interferon-γ responsive element, the glucocorticoid responsive element and an interleukin-6 nuclear factor binding site, which have an important role in the tissue specific gene regulation.

Sequence analysis of rat HDC promoter has revealed GRE-like consensus sites, CCAAT-boxes, and some other putative regulatory elements (Zahnow *et al.*, 1998). Comparative analysis of HDC promoter deleted mutants in human mast cell line HMC1, basophil cell line KU 812-F, erythroleukaemia cell line K562 and cervical cancer cell line Hela has revealed that there are differences in their HDC regulation. Deletion (at −4687 bp) decreases, while deletion (at −153 bp) increases the gene activity in the mast cell line. This finding shows the presence of cell-specific positive and negative regulatory elements in this region. Removal of CG box containing sequences abolished the gene activity, emphasizing its importance in HDC regulation.

Expression of the HDC gene is influenced by the methylation level of the gene. The importance of the methylation state in gene activity was proved with HDC-expressing cell lines where the CpG sites around the transcription initiation site were unmethylated (Kuramasu *et al.*, 1998). Role of the methylation pattern in gene activity was further supported by the results of the mouse HDC promoter analysis where the gene activity was repressed by patch methylation and its activity was restored following treatment with the demethylating agent, 5′-azacytidine (Suzuki-Ishigaki *et al.*, 2000).

Further analysis of transfected human basophil and Hela cell lines has revealed the presence of two positive and one negative gene regulatory elements. One of these positive elements contains a c-Myb transcription factor binding motif. Myb proteins participate in the differentiation of basophil and mast cell lineages. Nuclear extract of the basophil cell line contains Myb factor, suggesting that this motif is involved in the tissue-specific expression of HDC gene (Kuramasu *et al.*, 1998).

The regulation of HDC gene expression is well characterized in gastric mucosa and in the gastric cancer cell line expressing the recombinant human cholecystokinin B/gastrin receptor (Zhang *et al.*, 1996; Höckr *et al.*, 1998; Raychowdhary *et al.*, 2002). GATA-4 and GATA-6 transcription factors acting on GATA binding sites of HDC promoter could negatively regulate the HDC gene expression in this tumour cell line (Watson, Kiernan and Dimaline, 2002).

Gastrin and phorpbol-12-myristate-13-acetate (PMA) could stimulate transcriptional activity of human HDC via the GAS-RE found in the promoter. Their effects are mediated by different signal transduction pathways. Deletion analysis of mouse HDC promoter has revealed that the sequence from –297 to –43 is essential for regulation and increased transription of HDC gene to mediate the effects of dexamethasone and PMA treament, but GAS-RE is not found in this region. In rat lung, down-regulation of HDC transcription was found following dexamethasone (73%) or corticosterone (57%) treatment (Ohgon *et al.*, 1993).

HDC protein was found to influence its own transcription rate through down-regulation of ERK activity in the same gastric tumour cell line. An enzymatically inactive region at the amino terminal part of the HDC protein (residues 1–271) was responsible for this inhibition (Collucci *et al.*, 2001).

Examination of androgen-binding protein (ABP) expression in foetal rat liver gave a surprising result. Analysis of cDNA libraries derived from foetal rat liver revealed a fused transcript of the ABP gene and the HDC gene encoding a 98 kDa precursor protein. The two protein domains were joined at splice-junctions, indicating a *trans*-splicing mechanism. Until now this has been the only description of a trans-splicing event found between HDC and other gene, and it remains to be determined if this phenomenon is unique in foetal rat liver, if it could be any enzyme activity of the fusion protein, and whether this phenomenon has any biological significance (Sullivan *et al.*, 1991). Characteristic features of the mammalian HDC genes, promoter regions and mRNAs and the genetic polymorphisms of the coding region of HDC the are summarized in Tables 16.1 and 16.2, respectively.

HDC protein and enzyme (Figure 16.3)

The deduced 662 amino acid residue-containing sequence of human HDC shows high overall homology with the mouse (662 amino acid residues) and rat (665 amino acid residues) proteins (Joseph *et al.*, 1990; Mamune-Sato *et al.*, 1992; Ohmori *et al.*, 1990). The amino-terminal parts of the three proteins show a higher 92% homology, while the carboxyl-terminal parts have lower homology (70%). Moreover, the high

Table 16.1 Parameters of mammalian HDC genes and mRNAs

HDC	Human	Mouse	Rat
Gene	15q15–q21 1 copy	2 E5-G 1 copy	3 chromosome 1 copy
mRNA	24 kb DNA, 12 exon 2.4 kb mRNA Alternative splicing 3.4 kb mRNA	24 kb DNA, 12 exon 2.7 kb mRNA No alternative splicing	2.6 kb mRNA Alternative splicing 3.5 kb mRNA *trans*-Splicing
Promoter regulatory sequences	TATA-like box GC box CACC box GATA consensus sequence LBP-1 binding motif c-Myb binding motif GAS-RE	TATA-like box GC box GATA consensus sequence γ-IFN-RE GRE-like NF-IL-6 binding site PEA-3 binding site	CCAAT- boxes AP-1 and AP-2 binding sites Okt-1 binding site Sp1 binding site HIF-1 factor binding sites GRE-like

Table 16.2 Genetic polymorphisms of coding sequences found in the histidine decarboxylase gene

Sequence	Accession no.	Nucleotide change	Amino acid change
HDC mRNA	X54297.1	AGC493CAG	Ser148Gln
H. Sapiens cDNA FLJ23568 fis, clone LNG11092	AK027221.1	T1250A A1900T	Ile400Asn Arg617Trp

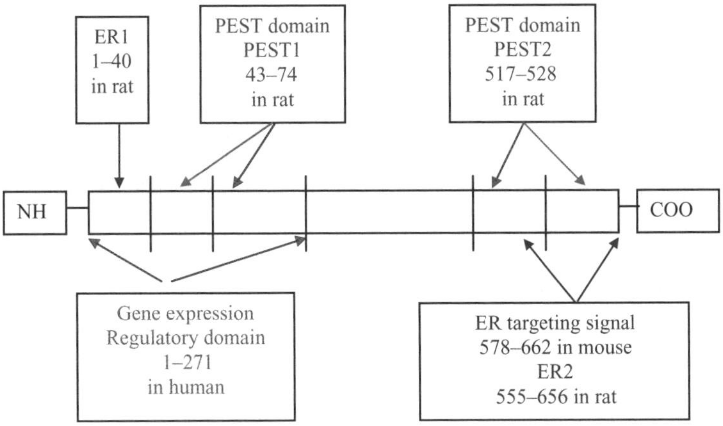

Figure 16.3 Different domains of mammalian histidine decarboxylase protein.

overall homology among the three mammalian HDCs and the amino acid sequences of these proteins shows a high degree of sequence identity to the dopa-decarboxylases of rat and *Drosophila*. The calculated molecular weight of the enzymes is very similar (~74 kDa), but they are post-translationally processed to yield the mature smaller (53–54 kDa) subunits which form homodimers in most of the cells (Joseph *et al.*, 1990; Zahnow *et al.*, 1998; Mamune-Sato *et al.*, 1992; Ohmori *et al.*, 1990; Yatsunami *et al.*, 1995). The pyridoxal phosphate-binding domain of seven amino acids includes a lysine residue at position 305 or 307 and this binding site is a conserved region of the three mammalian HDC proteins (Ohmori *et al.*, 1990).

Concerning post-translational modifications, four potential *N*-glycosylation sites and two putative cAMP-dependent protein kinase (protein kinase A) phosphorylation sites were found in mouse HDC sequence, while sequence analysis of rat protein reveals similar potential sites for *N*-glycosylation, as well as two sequences fitting consensus phosphorylation sites of protein kinase A. One of these protein kinase A consensus phosphorilation sites is adjacent to the putative PLP-binding site lysine (Lys-307), and would inactivate the enzyme if glycosylated (Joseph *et al.*, 1990; Mamune-Sato *et al.*, 1992).

To compare the activity, formation and localization of the 74 and 53–55 kDa proteins in human, mouse and rat, many experiments have been done (Yatsumani *et al.*, 1995; Yamamoto *et al.*, 1993; Tanaka *et al.*, 1998; Hirasawa, Murakami and Ohuchi, 2001). Studing the human recombinant HDCs, both sizes were expressed using an *in vitro* expression system, which also revealed that 87% of the bigger form was in an insoluble form while the 54 kDa form was found in the soluble fraction and both forms had similar enzyme activity to the native purified enzymes (Yatsunami *et al.*, 1995). Similar distribution of the two proteins was observed using mouse recombinant HDC in the same expression system, but here the 74 kDa form had a lower enzyme activity compared with the 53 kDa form (Yamamoto *et al.*, 1993). In contrast, in mouse macrophage (cell line RAW 264.7) the only active form was the bigger protein, which had a basal and inducible enzyme activity (Hirasawa, Murakami and Ohuchi, 2001). Both forms of HDC show enzymed activity in rat where the bigger HDC was found in the cytosol, while the processed smaller form was detected in the lumen of endoplasmic reticulum (ER). It was assumed that the 74 kDa form which is synthesized in the cytosol is translocated to the lumen of the ER where it is converted to a 53 kDa form. Regarding these findings, not only does the regulation of HDC gene expression seem to be cell- and tissue-specific, but also the processing of HDC proteins and the enzyme activity of the different forms.

Studies on mammalian HDC proteins suggest the presence of a number of functional domains with different functions in the enzyme targetting, degradation or regulation of gene expression. An example for domain function in gene regulation was mentioned earlier in this chapter, where it was reported that HDC protein could influence its own transcription rate through a region at the amino terminal part of the molecule (Collucci *et al.*, 2001).

Processing of HDC protein is an adenosine triphosphate (ATP)- and proteasome-dependent pathway, because proteasome inhibitors could inhibit it (Tanaka *et al.*,

1997; Suzuki *et al.*, 1998). This step occurs in the lumen of the ER and for this reason the protein has to be targetted to the ER. Membrane targetting and binding of mouse 74 kDa protein was examined to clarify which domain of HDC is important in this process (Suzuki *et al.*, 1998). Deletion analysis has revealed that the carboxyl-terminal part of the protein is involved, especially a 20 amino acid sequence within this domain, because its removal abrogated the targetting to the ER. *In vitro* translation assays showed that the binding of labelled 74 kDa HDC to microsome fraction is a post-translation process (Suzuki *et al.*, 1998).

Most of the HDC enzymes are very unstable proteins as well as the other short-lived amino acid–decarboxylases, like mammalian ornithine decarboxylase (ODC). PEST regions (domains rich in proline, glutamine, serine and threonine) of ODC were described as signals for ATP-dependent degradation by 26S proteasome. Sequence analysis of mammalian HDCs exhibits the presence of one PEST region in the amino terminal and at least one or more PEST region close to the carboxyl terminal. To study the role of PEST regions in the degradation, three different lengths of HDC were used by *in vitro* proteolytic degradation assays. Results indicate that only the amino terminal PEST region of rat HDC is involved in an ATP-dependent degradation process (Olmo *et al.*, 2000). Moreover, it was found that a significant ATP-independent pathway also existed (Rodriguez-Agudo *et al.*, 2000). Using similar *in vitro* degradation assays, this other pathway was found to be an *m*-calpain-dependent process. *m*-Calpain is an intracellular calcium-dependent cystein protease which is involved in limited or total proteolysis of many cytoskeletal or regulatory proteins. Degradation by *m*-calpain was quick, efficient and strongly inhibited by calpeptin, a highly specific calpain inhibitor (Rodriguez-Agudo *et al.*, 2000).

It was shown that gastrin could increase the steady-state level of at least six HDC isoforms in rat. Gastrin could regulate this stabilization of HDCs through two transferable and sequentially unrelated PEST domains (namely PEST1 and PEST2) involved in HDC protein degradation. Another region of protein at the amino terminal part, called ER2, has a role in targetting HDC into the ER. This region is influenced by gastrin as well (Fleming and Wang, 2000). These results revealed a novel regulatory mechanism by which the peptide hormone gastrin can disrupt the degradation function of a PEST-domain-containing proteins and can influence the tissue- or cell-specific activity of these proteins (Fleming and Wang, 2000). The characterized domains of mammalian HDC proteins are visible in Figure 16.3.

PLP-dependent HDCs are the only and rate-limiting enzymes in mammals involved in the formation of histamine from its precursor L-histidine under physiological conditions. The role of PLP has been considerably clarified through studies of nonenzymatic model reactions and various PLP-dependent enzymes reactions. PLP is present in the enzyme and an internal aldimide group is formed with the amino group of a specific lysine residue in the conserved PLP domain. This internal aldimide group reacts with the amino-acid substrate by *trans*-aldimination to form an external aldimide, which, due to the strongly electrophilic character of its pyridoxylidene moiety, weakens the bond of the carboxyl group, resulting in loss of CO_2. Finally, a proton adds to the resulting carbanion and the product is released (Hayashi and Watanabe, 1993).

Specific inhibitors of histidine decarboxylase are developed because inhibition of HDC activity that leads to depletion of histamine *in vivo* is useful in studying biological processes. The most widely used inhibitor of HDC is α-fluoromethylhistamine (α-FMH), which belongs to the group of so-called k_{cat} inhibitors, suicide substrates or enzyme-activated irreversible inhibitors (Kubota *et al.*, 1984).

HDC enzyme was characterized in many different cell types and tissues (Yamada *et al.*, 1984; Savany and Cronenberger, 1982; Mamune-Sato *et al.*, 1990; Yamauchi *et al.*, 1993; Ohmori *et al.*, 1990; Watanabe *et al.*, 1979). Purification of HDC is difficult because the protein is unstable and the enzyme activity is very low in most mammalian tissues. Purified HDC enzymes from adult rat stomach, brain and from foetal rat showed similar K_m values at pH 6.8, while isoelectric focusing revealed that pI values of the enzymes were different (Yamada *et al.*, 1984; Savany and Cronenberger, 1982). Analysis of the nucleotide sequence of HDC suggested some explanation for this heterogeneity, because different levels of *N*-glycosylation or phosphorilation can be responsible for these differences. HDC activity of human basophil cell line was similar to HDCs from other sources (Mamune-Sato *et al.*, 1990; Yamauchi *et al.*, 1993). Purified HDC enzyme from mouse mastocytoma has an isoelectric point, optimal pH, K_m value and V_{max} value which fit to the values of other mammalian HDC enzymes (Ohmori *et al.*, 1990). All these results support mammalian HDC enzymes from different species and tissues having many common or similar features. The differences between the enzymes can arise from different levels of purification or different degrees of *N*-glycosylation or phosphorylation. Moreover other parameters can change these values as well, for example K_m values and V_{max} values of rat HDC are influenced by the ionic strength and composition of the incubation medium (Watanabe *et al.*, 1979).

16.5 Catabolic Pathways of Histamine

The action of histamine, released by histamine-producing cells or taken up exogenously, can be terminated by two alternative pathways. Histamine inactivation can occur either by methylation of the imidazole ring or by oxidative deamination of the primary amino group (Maslinski and Fogel, 1991). The enzyme catalysing histamine methylation, histamine *N*-methyltransferase (HNMT), will be discussed in the next chapter of this book, and this section will focus on diamine oxidase, the enzyme responsible for the direct oxidation of histamine. Diamine oxidase (DAO, EC 1.4.3.6) was first characterized as an enzymatic activity-inactivating histamine and was therefore originally named histaminase. Subsequent biochemical characterization revealed that histaminase and DAO are identical proteins that catalyse the oxidative deamination of various diamines besides histamine (Buffoni, 1966) and the generally accepted name is now DAO.

An alternative route of histamine inactivation besides direct oxidation is methylation of the imidazole ring, catalysed by histamine *N*-methyltransferase (HNMT). HNMT is a soluble, cytosolic protein of M_R 33 000 that is encoded by a single gene of six or seven exons in mammals. HNMT primary structures are highly conserved among different species. In contrast to diamine oxidase, HNMT is widely expressed in mammalian tissues with particularly high expression levels in kidney, liver, spleen, colon, prostate, ovary and spinal cord cells. Determination of the crystal structure of recombinant human HNMT revealed a two-domain structure with the large domain representing a classic methyltransferase fold.

16.6 Histamine Receptors

The first three histamine receptor subtypes (H1–H3) are well characterized and have been established to be responsible for the majority of histamine actions. The H4 receptor is the most recent member of this group, but its functional significance is not yet entirely clear. Histamine receptor numbering refers to the chronology of their discovery. The H1, H2 and H3 receptors are well characterized and various histamine actions have been shown to be exerted by activation of these receptors.

Among the diverse actions of histamine, the most relevant responses realized via H1 receptors – from a clinical point of view – include acute inflammatory and allergic processes, whereas H2 receptors participate in the regulation of gastric acid secretion. H1 receptor antagonists are widely used in the treatment of allergic diseases, whereas H2 antagonists are efficient drugs for the treatment of peptic ulcers. Both H1 and H2 receptors transmit important signals to immune cells. H3 receptors are predominantly involved in central nervous system (CNS) functioning, and various biological responses (e.g. arousal and circadian rhythm) are supposed to be associated with histamine acting in the CNS. Besides the H3 receptor, the H1 and H2 receptors are also widely expressed in the mammalian brain. Whereas the H1, H2 and H3 receptors were initially characterized by pharmacological means and the cloning of their respective genes was only performed later, the H4 receptor has been found recently using cloning approaches, i.e. *in silico* screening of DNA libraries and databases for H3 receptor-like fragments. (The term *in silico* refers to the application of computational techniques, databases and bioinformatic tools.) Its functions are not yet characterized in detail, but its principal expression in the bone marrow, on leukocytes (particularly mast cells, eosinophils, neutrophils and dendritic cells), may allude to the possibility of its involvement in the regulation of hematopoiesis and immune response. A summary of the most important genomic, proteomic and signalling characteristics of histamine receptors is presented in Table 16.3.

Table 16.3 Histamine receptors and some elements of their signal pathways

Receptor type	Chromosomal localization	Amino acids	G protein type	Main elements of signalling pathway
H1	3	487	Gq/11	PKC; PLC; PLA NOS; MAP kinases, e.g. ERK 1/2; p38
H2	5	359	Gsα and Gq	PKA; PLC; c-fos
H3	20	445	$G_{i/0}$	MAP kinase, e.g. p44/42 kinase, high constitutive activity
H4	18	390	$G_{i/0}$	PLC MAP kinase, e.g. p44/42 kinase

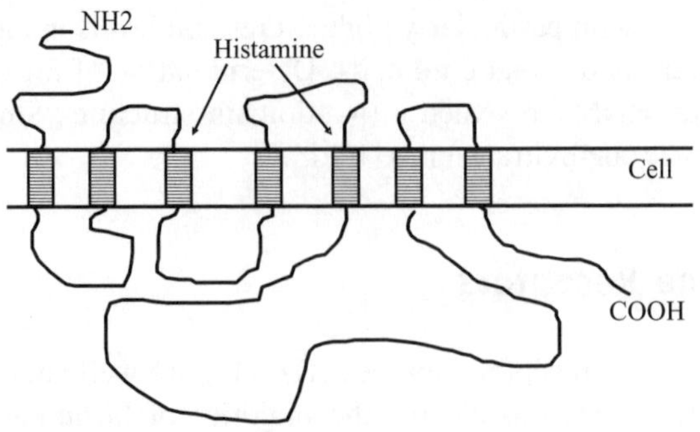

Figure 16.4 General structure of histamine receptors.

Histamine receptors belong to the group of membrane receptors composed of seven transmembrane domains that are coupled to G-protein-mediated signal transduction pathways (Figure 16.4). The transmembrane domains contain stretches of 20–25 hydrophobic amino acids that are predicted to form α-helices that span the cell membrane. The third and fifth transmembrane domains of heptahelical receptors appear to be responsible for ligand binding. In the following, we present a brief summary of the histamine receptors, with an emphasis on human ones, including the coding genes, protein products and genomics issues (Smit *et al.*, 1999).

H1 receptor

The H1 receptor was the first membrane protein to be discovered as a transmitter of histamine action. Its main activities include smooth muscle contraction as well as interaction with the endothelium, thus leading to vasodilatation and increased vascular permeability. It is the chief histamine receptor subtype involved in acute inflammatory and allergic disorders. H1 receptors have been identified in a wide variety of tissues, including mammalian brain, airway smooth muscle, gastrointestinal tract, genitourinary system, cardiovascular system, adrenal medulla, endothelia and various immune cells (Hill *et al.*, 1997). Enhanced expression of H1 receptors is observed, for example, in the nasal mucosa of patients with allergic rhinitis, in cultured aortic intimal smooth muscle cells of patients suffering from atherosclerosis and in the inflamed joints of rheumatoid arthritis patients. H1 receptor knockout mice have been developed recently. These mice are viable and fertile, but show various alterations in immune responses and impaired locomotor activity and exploratory behaviour (Inonue *et al.*, 1996). These findings hint at the involvement of the H1 receptor in a multitude of histamine actions.

The H1 receptor was the first histamine receptor to be characterized as a heptahelical G-protein coupled receptor. Using site-directed mutagenesis techniques, agonist binding and activation were shown to require two amino acid residues, Asp^{107} in the third and Asn^{198} in the fifth transmembrane domain (Smit *et al.*, 1999; Hill *et al.*, 1997). In one of the prototypes of G-protein-coupled receptors, the β2-adrenergic receptor, the third intracellular loop was identified as the region involved in coupling to G-proteins. A remarkable conservation of this region is observed in G-protein-coupled receptors, including H1 and H2 receptors.

The H1 receptor gene is located on chromosome 3. An intron was found in the 5′ untranslated region, immediately upstream of the start codon. The coding sequence of the H1 receptor gene is intronless. The promoter sequence appears to share similarities with other G-protein-coupled receptor genes, i.e. it was shown to lack TATA and CAAT sequences at the appropriate locations (Smit *et al.*, 1999; De Backer *et al.*, 1998).

Some nucleotide polymorphisms have already been described in the H1 receptor gene, e.g. a cytosine to thymine substitution at nucleotide position -17 bp from the transcription initiation site and an Asp349Asn change (Sasaki *et al.*, 2000). These polymorphisms do not seem to be associated with atopic asthma (Sasaki *et al.*, 2000). Using an *in silico* approach, we found a nucleotide sequence variant resulting in a Gly252Val amino acid change located in the third intracellular loop of the receptor protein. Since this region is implicated in the signal-transduction machinery, this variant – if verified by experimental studies – may also be functionally relevant (Igaz *et al.*, 2002) (Table 16.4).

H2 receptor

The sole existence of the H1 receptor could not account for the effects of histamine on cardiac muscle and gastric secretion. Therefore a second type, termed the H2 receptor, was anticipated and subsequently confirmed. In addition, it is involved in a wide array of physiological histamine actions, including the regulation of right atrial and ventricular muscle of the heart, inhibition of basophil chemotactic responsiveness, various actions on immune cells and inhibition of prostaglandin E2-stimulated duodenal epithelial bicarbonate secretion (Del Valle and Gantz, 1997). Its principal action from a clinical point of view is related to its role in stimulating gastric acid secretion. H2 receptor knockout mice have been developed recently that appear to be viable and fertile, but show considerable alterations in the morphology and structure of the gastric mucosa (Kobayashi *et al.*, 2000). Several selective H2 receptor antagonists have been developed that have turned out to be powerful agents in treating peptic ulcers.

Table 16.4 Polymorphisms (sequence variants) of the coding sequence for histamine receptors: experimental and *in silico* data

Receptor type	Variants of the amino acid sequence
H1	Asp349Asn, Gly252Val[a]
H2	Val133Ala, Lys175Asn, Lys207Arg, Asn217Asp, Lys231Arg, Val268Met
H3	Asp19Glu[a]
H4	Val138Ala[a], Arg206His[a], Arg253Gln[a]

[a]*In silico* data.

The ligand binding site of the H2 receptor appears to be similar to the corresponding region of the H1 receptor, with Asp^{98} of the third and Asp^{186} and Thr^{190} of the fifth transmembrane domain supposed to constitute the histamine binding site (Figure 16.4) and the third intracellular loop constituting the main region involved in signal transduction. The most notable difference between the structures of H1 and H2 receptors is the much shorter third intracellular loop and longer C-terminus of the H2 receptor (Hill *et al.*, 1997; Del Valle and Gantz, 1997).

The H2 receptor gene was cloned at the beginning of the 1990s. The 5′ untranslated region of the H2 receptor gene appears to harbour several regulatory elements, e.g. cAMP response elements, GATA motifs and AP2 sites, but apparently without any TATA box-like sequences. Furthermore, multiple transcription initiation sites were revealed in the H2 receptor gene. Similar to the H1 receptor gene, the coding part lacks intronic sequences (Del Valle and Gantz, 1997).

Several nucleotide sequence variants of the H2 receptor gene have been described (Table 16.4). Among these, the variants Lys207Arg, Asn217Asp and Lys231Arg appear to be particularly interesting, since they are located in the third intracellular loop of the receptor. The Val133Ala may also be relevant, since the second intracellular loop, where it is situated, also appears to have some role in the signal-transduction pathways. In individuals suffering from schizophrenia, the Asn217Asp allele of the H2 receptor was found to be 1.8 times more frequent than in the normal population (Figure 16.5). This may be a relevant observation considering the possible role of the H2 receptor in schizophrenia, as indicated by the efficacy of H2 receptor

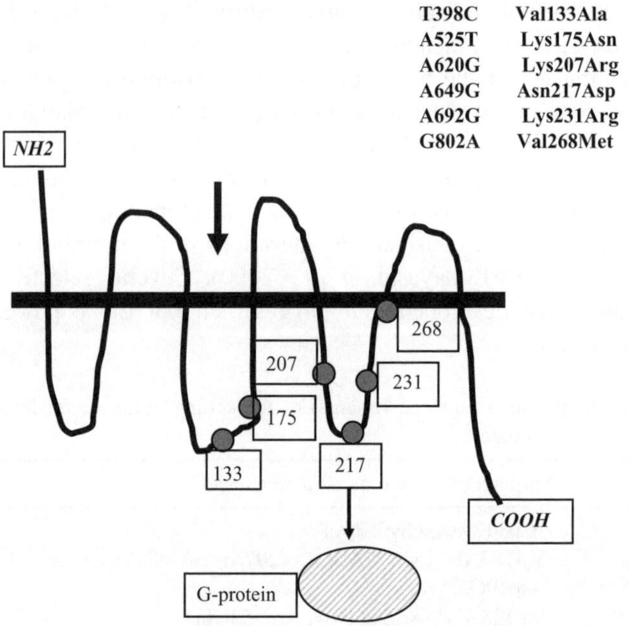

Figure 16.5 Polymorphic hot spots on histamine H2 receptor.

antagonists in the treatment of this disorder. In a later study, however, no definitive correlations were found (Igaz *et al.*, 2002; Orange *et al.*, 1996).

H3 receptor

The H3 receptor was identified at the beginning of the 1980s as a receptor subtype involved in the regulation of nervous system functioning as a presynaptic autoreceptor influencing neuronal histamine release. Later studies revealed that, in addition to histamine, the release of other neurotransmitters is also influenced by this receptor subtype (Leurs *et al.*, 1998). The H3 receptor is widely distributed in the CNS, e.g. in hypothalamus, frontal cortex, hippocampus and the caudate nucleus, where it is supposed to modify a multitude of processes. Among these, the regulation of the circadian rhythm of sleep and wakefulness, cognitive and memory processes and food intake are affected. Although drugs with a direct effect on H3 receptor mediated histamine actions have not been introduced in clinical practice yet, it is highly conceivable that H3 receptor agonists and antagonists may be effective in treating obesity and sleep, age-related memory and attention disorders (Leurs *et al.*, 1998). Most recently, H3 receptor-deficient mice have been generated that show a decrease in overall locomotion, reduced body temperature during the dark phase, but maintained normal circadian rhythmicity (Toyota *et al.*, 2002).

In addition to its primary CNS activities, the H3 receptor appears to be involved in the regulation of peripheral histamine actions as well. Histamine via H3 receptors also affects the function of the cardiovascular system, e.g. it inhibits sympathetic neurotransmission in the right atrium of the heart (Leurs *et al.*, 1998). In cats, dogs and rabbits, histamine seems to inhibit gastric acid secretion via the H3 receptor.

The human H3 receptor gene, located on chromosome 20, was cloned in 1999 (Lovenberg *et al.*, 1999). In contrast to the H1 and H2 receptor genes, whose coding sequences are intronless, the H3 receptor gene contains four exons and at least three introns. In addition, several isoforms (splice variants) of the single copy H3 receptor gene have also been characterized (e.g. one harbouring a deletion in the second putative intracellular domain and another with a variable deletion in the third intracellular loop), that appear to show different agonist binding and signal-transduction properties (Cogé *et al.*, 2001a).

H4 receptor

The histamine H4 receptor has been cloned recently by *in silico* approaches. Several groups independently identified a DNA sequence with homology to the H3 histamine receptor (37–43%), which was cloned and later termed the histamine H4 receptor. The H4 receptor was found to be expressed predominantly on haematopoietic and immunocompetent cells. It appears to be highly expressed in mast cells, eosinophils, neutrophils, dendritic cells and to a smaller extent $CD4^+$ T-cells, but only weakly in brain, liver and lungs. The predominant expression of H4 receptors on haematopoietic cells raises the possibility that this receptor mediates some actions of histamine affecting hematopoiesis, e.g. its capacity to promote terminal myeloblast and promyelocyte differentiation (Schneider *et al.*, 2002).

The H4 receptor gene comprises three introns, like the H3 receptor, and is located on chromosome 18 in humans. No TATA or CAAT promoter elements were found adjacent to the putative transcription initiation site, similar to the human H1 and H2 receptor genes (Cogé *et al.*, 2001b). By performing *in silico* homology searches, three nucleotide variants of the H4 receptor gene leading to amino acid changes Val138Ala, Arg206His and Arg253Gln were found (Table 16.4). Since no structure–function data are available, their possible biological significance remain to be elucidated (Igaz *et al.*, 2002).

Analysis of the promoter region of H4 receptor gene has identified consensus sequences for several transcription regulatory factor binding motifs, including NF-IL6, ISRE, NF-kB, uteroglobin, IRF and several γ-IRE. This suggests that H4 receptor gene expression might be stimulated by factors such as interferon, TNF-α or IL-6 involved in viral host defence or inflammatory responses, or might also be modulated by anti-inflammatory uteroglobin-like proteins, particularly in the lung, where both H4 receptors and uteroglobin-like proteins are expressed (Cogé *et al.*, 2001b).

The expression of H4 receptors in mononuclear cells is regulated in connection with cell activation, in a manner depending on the particular cell type, e.g. activated monocytes only express H4 receptors when neutralizing antibodies to IL-10 are present, whereas the H4 receptor expression is down-regulated in activated Th2 cells, which express IL-10 and IL-13. In addition, resting bone marrow-derived mast cells express H4 receptors which decrease dramatically upon activation with PMA and ionomycin, along with an increased IL-13 expression in these cells (Morse *et al.*, 2001). Furthermore, H4 receptor expression is found to be higher in resting $CD4^+$ and $CD8^+$ T cells than in the activated cells, and very low levels are detected in $CD19^+$ B cells and resting $CD14^+$ monocytes.

The putative 'Hic' (intracellular) receptor

In 1985 Brandes *et al.* proposed that histamine may bind intracellularly to sites in microsomes, nuclei and chromatin. Later they found that a major proportion of these microsomal histamine binding sites represents P450 enzymes, which are the principal participants of steroid synthesis, drug metabolism, etc. [Brandes, Queen and LaBella, 1998]. This intracellular histamine binding site has been termed the 'Hic' receptor.

The discovery of intracellular histamine binding sites followed the synthesis of *N,N*-diethyl-2-[4-(phenylmethyl) phenoxy] ethanamine HCl (DPPE), a tamoxifen derivative structurally similar to various arylalkylamines. DPPE was proposed to be an antagonist of intracellular histamine binding (Brandes, Queen and LaBella, 1998).

'Hic' cloning efforts, however, have been unsuccessful to date. The existence of intracellular histamine recptors is not unanimously accepted, and other mechanisms, e.g. interactions with the effects of histamine on polyamine sites and on hetero-trimeric G-proteins, might explain some of the phenomena attributed to the activity of the 'Hic' (Hill *et al.*, 1997).

16.7 Histamine and Cytokines, Relation to the T Cell Polarization of the Immune Response

One of the most fundamental and best characterized cytokine networks involves the regulation of T-helper cell polarization, which is an essential phenomenon in the adaptive immune system. Two main types of T-helper cells are known, the Th1 and Th2 cells. Whereas Th1 cells produce mainly IL-2, IL-12 and interferon (IFN)-γ and promote cellular/cytotoxic T-cell responses, the Th2 cells, producing IL-4, IL-6, IL-10 and IL-13, act in a way to enhance humoral, antibody responses, particularly the production of immunoglobulin E (IgE). The Th1 and Th2 subclasses mutually inhibit the actions of each other via the production of cytokines. IFN-γ, IL-4 and IL-10 are the most significant in this respect.

Histamine inhibits IL-12 mRNA expression in human monocytes and the production of IL-12 in whole blood cultures, while it increases the production of IL-10 in the same experimental system. These effects appear to be mediated via H2 receptors. Histamine also inhibits the lipopolysaccharide (LPS)-induced IFN-γ production of peripheral blood mononuclear (PBM) cells (Igaz *et al.*, 2001; Azuma *et al.*, 2001).

Dendritic cells are very important in the processes of T-cell maturation. It is therefore interesting that histamine via H2 receptor appears to alter the T-cell polarizing capacity of immature dendritic cells, by increasing IL-10 and decreasing IL-12 secretion. Histamine-matured dendritic cells polarized naive $CD4^+$ T-cells toward the Th2 phenotype, as compared with dendritic cells matured in the absence of histamine (Mazzoni *et al.*, 2001).

The *in vitro* data show a complicated array of histamine actions regulating the Th1–Th2 network. Although a few Th1-promoting and Th2-inhibiting effects of histamine have been reported, the majority of *in vitro* data appear to support the concept that histamine mainly promotes Th2 responses, thereby shifting the Th1–Th2 balance in the Th2 direction.

Recent data, however, make the situation even more complicated (Schneider *et al.*, 2002). These data show that histamine acting via the H1 receptor can enhance Th1 responses, whereas both Th1 and Th2 responses were found to be inhibited via the H2 receptor (Jutel *et al.*, 2001a). The H1 receptor was shown to be predominantly expressed by Th1 cells, whereas Th2 cells appear to express principally H2 receptors. Histamine appears to enhance anti-CD3-induced Th1 proliferation, whereas it inhibits Th2 cell proliferation. Proliferation of specific allergen-driven cells from house dust mite- and bee venom-allergic individuals was found to be suppressed by histamine. It has been hypothesized that the mechanism of allergen-specific immunotherapy, when high doses of allergen result in peripheral Th2 tolerance, may be explained by the Th1-promoting effect of histamine, which could redirect the immune response from a dominant Th2-type towards a Th1 profile (Jutel *et al.*, 2001a). These findings revealed a new aspect of histamine actions, as, besides the Th2-promoting actions of histamine, Th1 promotion may also occur, and emphasize the importance of different signals via the H1 and H2 receptors.

The newly developed H1 and H2 knockout mice add important data for the clarification of the *in vivo* importance of histamine in the regulation of the Th1–Th2 balance. H1-deleted mice show lower, H2-deleted mice higher percentages of IFN-γ-producing T-cells compared with wild-type mice. The frequency of IL-4 producing cells was higher both in H1 and H2 receptor knockout mice. These results suggest the importance of H1 receptor-mediated signalling in promoting Th1 functions, whereas H2 receptor appears to provide negative signals for T-cells (Jutel *et al.*, 2001b). These findings underline the immune modulatory actions of histamine. Based on the currently available experimental data, it is difficult to deduce a clear picture. It can, however, be stated that histamine appears to enhance and to suppress both Th1 and Th2 responses depending on the receptor type involved and the surrounding microenvironment.

Bilateral regulation of histamine and cytokine network is shown in Tables 16.5 and 16.6. A multitude of experimental data seem to support the idea that a complex interactive network is present between histamine and cytokines, both parties influencing the synthesis and release of the other. Enhancing as well as inhibitory actions of histamine on numerous cytokines are known, whereas histamine release in turn seems to be modulated by various cytokines.

Table 16.5 Bilateral regulation between histamine and cytokines: effect of cytokines on histamine release

Change in histamine release	Cytokine acting
Enhancement	IL-1, IL-3, IL-5. GM-CSF, TNF-α,[a] RANTES,[a] MCP-1[a]
Inhibition	TNF-α,[a] RANTES,[a] MCP-1[a]

[a] TNF-α, RANTES and MCP-1 were found to have both enhancing and inhibitory effects on histamine release

Table 16.6 Bilateral regulation between histamine and cytokines: effects of histamine on cytokine secretion

Change in cytokine secretion	Cytokines affected
Enhancement	IL-1, IL-2,[a] IL-5, IL-6, IL-11, IL-8, IL-16, GM-CSF, RANTES
Inhibition	IL-2,[a] IL-4, IL-12, IFN-γ, TNF-α

[a]The secretion of IL-2 can either be enhanced or inhibited depending on the experimental setting.

16.8 Histamine and Tumour Growth

Overexpression of HDC has been detected in a wide range of tumours, including melanoma, colon, breast, stomach, lung cancer and leukaemia. On the other hand the neoangiogenesis and the antitumoural immune responses are also clearly influenced by histamine modifying the local tumour growth. In experimental carcinomas, the endogenous histamine may act as an autocrine growth factor and its major effect seems to be that it stimulates proliferation via H2 receptors. Newly formed histamine is probably not stored in the cells but rapidly released or metabolized. Moreover, the fact that histamine behaves as a strong inhibitor of Th1-type cytokines represents an indirect means of histamine playing a modulatory role in weakening the regulation of anti-tumour immunity (Pos, Hegyesi and Rivera, 2004). Recent *in vivo* findings on a murine melanoma cell line transfected by HDC sense constructs provide direct evidence for the enhancing effect of histamine on tumour growth through H2 histamine receptor (Pos *et al.*, 2005).

16.9 Histamine Research: an Insight into Metabolomics, Lessons from HDC-Deficient Mice

HDC knockout mice (Hegyi *et al.*, 2001; Ohtsu *et al.*, 2001) are histamine-free and exhibit an exciting palette of unexpected phenotypes, including low mast cell count with poor granulation (Wiener *et al.*, 2002), resistance to ovariectomy-induced osteoporosis (Fitzpatrick *et al.*, 2003), experimental asthma with low eosinophil count (Kozma *et al.*, 2003), very high leptin level (Fulo *et al.*, 2003) and high autoantibody titer (Quintana *et al.*, 2004) with characteristic Th1 shift, characterized by very low IL-10 and IL-6, elevated IFNg, high abortion rate and elevated natural killer cell (NK) count (Igaz *et al.*, 2001; Pár *et al.*, 2003; Bene *et al.*, 2004; Garaczi *et al.*, 2004).

Using expression microarray we found more than 400 genes (mostly nonannotated as yet), showing significantly different expression between histamine-deficient and normal mice with identical genetic background (unpublished data).

Owing to the absence of histamine, a low molecular weight (112 Da) substance, upon feeding with histamine, many phenotypes of the endogenously H deficient mice

Table 16.7 Database parameters of histamine metabolism and receptors of coding sequences

Gene	Search sequence	Accession number
HDC	Human HDC mRNA	NM_002112.1
H1R	Human H1R mRNA	NM_000861.1
H2R	Human H2R complete cds	M64799.1
H3R	Human H3R mRNA	NM_007232.1
H4R	Human H4R mRNA	AB044934.1
DAO	Human HP-DAO2 mRNA, complete cds	U11863.1
HNMT	Human HNMT exons 1–6	U44106.1–U44111.1

cds, coding sequence.

can be reversed just by feeding the mice with histamine-rich food. This situation results in a diet-inducible 'conditioned' phenotype.

Uncovering the relationship between decarboxylase (including HDC) mRNAs, enzyme expression levels with the primer amine metabolome patterns at a genome-wide context seem to be one of the most exciting 'metabolomic adventures' in the very close future of a post-genomic era.

16.10 Histamine Genomics on Databases

Database parameters of histamine metabolism and receptors of coding sequences are provided in Table 16.7.

References

Azuma Y, Shinohara M, Wang PL, Hidaka A, Ohura K. 2001. Histamine inhibits chemotaxis, phagocytosis, superoxide anion production and the production of TNFα and IL-12 by macrophages via H2-receptors. *Int Immunopharmac* **1**: 1867–1875.

Bene L, Sapi Z, Bajtai A, Buzas E, Szentmihalyi A, Arato A, Tulassay Z, Falus A. 2004. Partial protection against dextran sodium sulphate induced colitis in histamine-deficient, histidine decarboxylase knockout mice. *J Pediatr Gastroenterol Nutr* **39**: 171–176.

Brandes LJ, Queen GM, LaBella FS. 1998. Potent interaction of histamine and polyamines at microsomal cytochrome P450, nuclei, and chromatin from rat hepatocytes. *J Cell Biochem* **69**: 233–243.

Buffoni F. 1966. Histaminase and related amine oxidases. *Pharmac Rev* **18**: 1163–1199.

Cogé F, Guénin SP, Audinot V, Renouard-Try A, Beauverger P, Macia C, Ouvry C, Nagel N, Rique H, Boutin JA, Galizzi JP. 2001a. Genomic organization and characterization of splice variants of the human histamine H3 receptor. *Biochem J* **355**: 279–288.

Cogé F, Guénin SP, Rique H, Boutin JA, Galizzi JP. 2001b. Structure and expression of the human histamine H4-receptor gene. *Biochem Biophys Res Commun* **284**: 301–309.

Collucci R, Fleming JV, Xavier R, Wang TC. 2001. L-Histidine decarboxylase decreases its own transcription through downregulation of ERK activity. *Am J Physiol Gastrointest Liver Physiol* **281**: G1081–1091.

De Backer M, Loonen I, Verhasselt P, Neefs JM, Luyten WHM. 1998. Structure of the human histamine H1 receptor gene. *Biochem J* **335**: 663–670.

Del Valle J, Gantz I. 1997. Novel insights in histamine H2 receptor biology. *Am J Physiol* **273**: G987–996.

Fitzpatrick LA, Buzas E, Gagne TJ, Nagy A, Horvath C, Ferencz V, Mester A, Kari B, Ruan M, Falus A, Barsony J. 2003. Targeted deletion of histidine decarboxylase gene in mice increases bone formation and protects against ovariectomy-induced bone loss. Proc Natl Acad Sci USA **100**: 6027–6032.

Fleming JV, Wang TC. Amino- and carboxy-terminal PEST domains mediate gastrin stabilization of rat L-histidine decarboxylase isoforms. *Mol Cell Biol* **20**: 4932–4947.

Fulop AK, Foldes A, Buzas E, Hegyi K, Miklos IH, Romics L, Kleiber M, Nagy A, Falus A, Kovacs KJ. 2003. Hyperleptinemia, visceral adiposity, and decreased glucose tolerance in mice with a targeted disruption of the histidine decarboxylase gene. *Endocrinology* **144**: 4306–4314.

Garaczi E, Szell M, Janossy T, Koreck A, Pivarcsi A, Buzas E, Pos Z, Falus A, Dobozy A, Kemeny L. 2004. Negative regulatory effect of histamine in DNFB-induced contact hypersensitivity. *Int Immunol* **16**: 1781–1788.

Hayashi H, Watanabe T. Histamine and its related enzymes. 1993. In *Methods in Neurotransmitter and Neuropeptide Research*, Parvez SH, Naoi M, Nagatsu T, Parvez S (eds); 111–137.

Hegyi K, Fülöp AK, Tóth S, Buzás E, Watanabe T, Ohtsu H, Ichikawa A, Nagy A, Falus A. 2001. Histamine deficiency suppresses murine haptoglobin production and modifies hepatic protein tyrosine phosphorylation. *Cell Mol Life Sci* **58**: 850–854.

Hill SJ, Ganellin CR, Timmerman H, Schwartz JC, Shankley NP, Young JM, Schunack W, Levi R, Haas LH. 1997. International union of pharmacology. XIII. Classification of histamine receptors. *Pharmac Rev* **49**: 253–278.

Hirasawa N, Murakami A, Ohuchi K. 2001. Expression of 74-kDa histidine decarboxylase protein in a macrophage-like cell line RAW 264.7 and inhibition by dexamethasone. *Eur J Pharmac* **418**: 23–28.

Höcker M, Rosenberg I, Xavier R, Henihan RJ, Wiedenmann B, Rosewicz S, Posolsky DK, Wang TC. 1998. Oxidative stress activates the human histidine decarboxylase promoter in AGS gastric cancer cells. *J Biol Chem* **273**: 23046–23054.

Igaz P, Novák I, Lázár E, Horváth B, Héninger E, Falus A. 2001. Bidirectional communication between histamine and cytokines. *Inflamm Res* **50**: 123–128.

Igaz P, Fitzimons CP, Szalai C, Falus A. 2002. Histamine genomics *in silico*: polymorphisms (sequence variants) of the human genes involved in the synthesis, action and degradation of histamine. *Am J Pharmacogenom* **2**: 67–72.

Inoue I, Yanai K, Kitamura D, Taniuchi I, Kobayashi T, Niimura K, Watanabe T, Watanabe T. 1996. Impaired locomotor activity and exploratory behavior in mice lacking histamine H1 receptors. *Proc Natl Acad Sci USA* **93**: 13316–13320.

Joseph DR, Sullivan PM, Wang YM, Kozak C, Fenstermacher A, Behrendsen ME, Zahnow CA. 1990. Characterization and expression of the complementary DNA encoding rat histidine decarboxylase. *Proc Natl Acad Sci USA* **87**: 733–737.

Jutel M, Watanabe T, Klunker S, Akdis M, Thomet OAR, Malolepszy J, Zak-Nejmark T, Koga R, Kobayashi T, Blaser K, Akdis CA. 2001a. Histamine regulates T-cell and antibody responses by differential expression of H1 and H2 receptors. *Nature* **413**: 420–425.

Jutel M, Klunker S, Akdis M, Malolepszy J, Thomet OAR, Zak-Nejmark T, Blaser K, Akdis CA. 2001b. Histamine upregulates Th1 and downregulates Th2 responses due to different patterns of surface histamine 1 and 2 receptor expression. *Int Arch Allergy Immunol* **124**: 190–192.

Kobayashi T, Tonai S, Ishihara Y, Koga R, Okabe S, Watanabe T. 2000. Abnormal functional and morphological regulation of the gastric mucosa in histamine H2 receptor-deficient mice. *J Clin Invest* **105**: 1741–1749.

Kozma GT, Losonczy G, Keszei M, Komlosi Z, Buzas E, Pallinger E, Appel J, Szabo T, Magyar P, Falus A, Szalai C. 2003. Histamine deficiency in gene-targeted mice strongly reduces antigen-induced airway hyper-responsiveness, eosinophilia and allergen-specific IgE. *Int Immunol* **15**: 963–973.

Kubota H, Hayashi H, Watanabe T, Taguchi Y, Wada H. 1984. Mechanism of inactivation of mammalian L-histidine decarboxylase by (*S*)-α-fluoromethylhistidine. *Biochem Pharmac* **33**: 983–990.

Kuramasu A, Satoi H, Suzuki S, Watanabe T, Ohtsu H. 1998. Mast cell/basophil specific transcriptional regulation of human L-histidine decarboxylase gene by CpG methylation in the promoter region. *J Biol Chem* **273**: 31607–31614.

Leurs R, Blandina P, Tedford C, Timmerman H. 1998. Therapeutic potential of histamine H3 receptor agonists and antagonists. *Trends Pharmac Sci* **19**: 177–183.

Lovenberg TW, Roland BL, Wilson SJ, Jiang X, Pyati J, Huvar A, Jackson MR, Erlander MG. 1999. Cloning and functional expression of the human histamine H3 receptor. *Mol Pharmacol* **55**: 1101–1107.

Malzac P, Mattei MG, Thibault J, Bruneau G. 1996. Chromosomal localization of the human and mouse histidine decarboxylase genes by *in situ* hybridization: exclusion of the hdc gene from the Prader–Willi syndrome region. *Hum Genet* **97**: 359–361.

Mamune-Sato R, Tanno Y, Maeyama K, Miura Y, Takishima T, Kishi K, Fukuda T, Watanabe T. 1990. Histidine decarboxylase in human basophilic leukemia (KU-812-F) cells. Characterization and induction by phorbol myristate acetate. *Biochem Pharmac* **40**: 1125–1129.

Mamune-Sato R, Yamauchi K, Tanno Y, Ohkawara Y, Ohtsu H, Katayose D, Maeyama K, Watanabe T, Shibahara S, Takashima T. 1992. Functional analysis of the human histidine decarboxylase gene and its expression in human tissues and basophilic leukemia cells. *Eur J Biochem* **209**: 533–539.

Maslinski C, Fogel WA. 1991. Catabolism of histamine. In *Handbook of Experimental Pharmacology. Vol 97. Histamine and Histamine Antagonists*, Uvnäs B (ed.). Berlin: Springer; 165–189.

Mazzoni A, Young HA, Spitzer JH, Visintin A, Segal DM. 2001. Histamine regulates cytokine production in maturing dendritic cells, resulting in altered T cell polarization. *J Clin Invest* **108**: 1865–1873.

Morse KL, Behan J, Laz TM, West RE Jr, Greenfeder SA, Anthes JC, Umland S, Wan Y, Hipkin RW, Gonsiorek W, Shin N, Gustafson EL, Qiao X, Wang S, Hedrick JA, Greene J, Bayne M, Monsma FJ Jr. 2001. Cloning and characterization of a novel human histamine receptor. *J Pharmac Exp Ther* **296**: 1058–1066.

Noszál B, Kraszni A, Rácz A. 2004. Histamine: fundamentals of biological chemistry. In *Histamine: Biology and Medical Aspects* Falus A, Grosman N, Darvas ZS (eds). Karger/SpringMed: Barel; 15–28.

Ohgoh M, Yamamoto J, Kawata M, Yamamura I, Fukui T, Ichikawa A. 1993. Enhanced expression of the mouse L-histidine decarboxylase gene with combination of dexamethasone and 12-*O*-tetradecanoylphorbol-13-acetate. *Biochem Biophys Res Commun* **196**: 1113–1119.

Ohmori E, Fukui T, Imanishi N, Yatsunami K, Ichikawa A. Purification and characterization of L-histidine decarboxylase from mouse mastocytoma P-815. *J Biochem* **107**: 834–839.

Ohtsu H, Tanaka S, Terui T, Hori Y, Makabe-Kobayashi Y, Pejler G, Tchougounova E, Hellman L, Gertsenstein M, Hirasawa N, Sakurai E, Buzas E, Kovacs P, Csaba G, Kittel A, Okada M, Hara M, Mar L, Numayama-Tsuruta K, Ishigaki-Suzuki S, Ohuchi K, Ichikawa A, Falus A, Watanabe T, Nagy A. 2001. Mice lacking histidine decarboxylase exhibit abnormal mast cells *FEBS Lett* **502**: 53–56.

Olmo MT, Urdiales JL, Pegg AE, Medina MA, Sánchez-Jimenez F. 2000. *In vitro* study of proteolytic degradation of rat histidine decarboxylase. *Eur J Biochem* **267**: 1527–1531.

Orange PR, Heath PR, Wright SR, Pearson RC. 1996. Allelic variations of the human histamine H2 receptor gene. *Neuroreport* **7**: 1293–1296.

Pár G, Szekeres-Bartho J, Buzas E, Pap E, Falus A. 2003. Impaired reproduction of histamine deficient (histidine-decarboxylase knockout) mice is caused predominantly by a decreased male mating behavior. *Am J Reprod Immunol* **50**: 152–158.

Pos Z, Hegyesi H, Rivera ES. 2004. Histamine and cell proliferation. In *Histamine: Biology and Medical Aspects*. Falus A, Grosman N, Darvas ZS (eds). Karger/SpringMed: Basel; 199–217.

Pos Z, Safrany G, Muller K, Toth S, Falus A, Hegyesi H. 2005. Phenotypic profiling of engineered mouse melanomas with manipulated histamine production identifies histamine H2 receptor and rho-C as histamine-regulated melanoma progression markers. *Cancer Res* **65**: 4458–4466.

Quintana Quintana FJ, Buzas E, Prohaszka Z, Biro A, Kocsis J, Fust G, Falus A, Cohen IR. Knock-out of the histidine decarboxylase gene modifies the repertoire of natural autoantibodies. *J Autoimmun* **22**: 297–305.

Raychowdhury R, Fleming J, McLaughlin J, Bulitta C, Wang TC. 2002. Identification and characterization of a third gastrin response element (GAS-RE3) in the human histidine decarboxylase gene promoter. *Biochem Biophys Res Commun* **297**: 1089.

Rodriguez-Agudo D, Olmo MT, Sanchez-Jimenez F, Medina MA. 2000. Rat histidine decarboxylase is a substrate for *m*-calpain *in vitro*. *Biochem Biophys Res Commun* **19271**: 777–781.

Roskoski R. *Biochemistry*. 1996. Philadelphia, PA: WB Saunders; 412–413.

Sasaki Y, Ihara K, Ahmed S, Yamawaki K, Kushuhara K, Nakayama H, Nishima S, Hara T. 2000. Lack of association between atopic asthma and polymorphisms of the histamine H1 receptor, histamine H2 receptor and histamine *N*-methyltransferase genes. *Immunogenetics* **51**: 238–240.

Savany A, Cronenberger L. 1982. Isolation and properties of multiple forms of histidine decarboxylase from rat gastric mucosa. *Biochem J* **205**: 405–412.

Schneider E, Rolli-Derkinderen M, Arock M, Dy M. 2002. Trends in histamine research: new functions during imune response and hematopoiesis. *Trends Immunol* **23**: 255–263.

Smit MJ, Hoffmann M, Timmerman H, Leurs R. 1999. Molecular properties and signalling pathways of the histamine H1 receptor. *Clin Exp Allergy* **29**(suppl 3): 19–28.

Sullivan PM, Petrusz P, Szpirer C, Joseph DR. 1991. Alternative processing of androgen-binding protien RNA transcripts in fetal rat liver. Identification of a transcript formed by *trans* splicing. *J Biol Chem* **266**: 143–154.

Suzuki S, Tanaka S, Nemoto K, Ichikawa A. 1998. Membrane targeting and binding of the 74-kDa form of mouse L-histidine decarboxylas via its carboxyl-terminal sequence. *FEBS Lett* **437**: 44–48.

Suzuki-Ishigaki S, Numayama-Tsuruta K, Kuramasu A, Sakurai E, Makabe Y, Shimura S, Shirato K, Igarashi K, Watanabe T, Ohtsu H. 2000. The mouse L-histidine decarboxylase gene: structure and transcriptional regulation in the promoter region. *Nucl Acids Res* **28**: 2627–2633.

Tanaka S, Nemoto K, Yamamura E, Ohmura S, Ichikawa A. 1997. Degradation of the 74 kDa form of L-histidine decarboxylase ubiquitin–proteasome pathway in a rat basophilic/mast cell (RBL-2H3). *FEBS Lett* **417**: 203–207.

Tanaka S, Nemoto K, Yamamura E, Ichikawa A. 1998. Intracellular localization of the 74- and 53-kDa forms of L-histidine decarboxylase in a rat basophilic/mast cell line, RBL-2H3. *J Biol Chem* **273**: 8177–8182.

Toyota H, Dugovic C, Koehl M, Laposky AD, Weber C, Ngo K, Wu Y, Lee DH, Yanai K, Sakurai E, Watanabe T, Liu C, Chen C, Barbier AJ, Turek FW, Fung-Leung WP, Lovenberg TW. 2002. Behavioral characterization of mice lacking histamine H(3) receptors. *Mol Pharmac* **62**: 389–397.

Watanabe T, Nakamura H, Liang LY, Yamatodani A, Wada H. 1979. Partial purification and characterization of L-histidine decarboxylases from fetal rats. *Biochem Pharmac* **28**: 1149–1155.

Watson F, Kiernan RS, Dimaline R. 2002. GATA proteins are potential negative regulators of HDC expression in the gastric epithelium. *Biochim Biophys Acta* **1576**: 198–202.

Wiener Z, Andrásfalvy M, Pállinger É, Kovács P, Szalai Cs, Erdei A, Tóth S, Nagy A, Falus A. 2002. Bone marrow-derived mast cell differentiation strongly reduced in histidine decarboxylase knockout, histamine-free mice. *Int. Immunol.* **14**: 381–387.

Yamada M, Watanabe T, Fukui H, Taguchi Y, Wada H. 1984. Comparison of histidine decarboxylases from rat stomach and brain with that from whole bodies of rat fetus. *Agents Actions* **14**: 143–152.

Yamamoto J, Yatsunami K, Ohmori E, Sugimoto Y, Fukui T, Katayama T, Ichikawa A. 1990. cDNA-derived amino acid sequence of L-histidine decarboxylase from mouse mastocytoma P-815 cells. *FEBS Lett* **276**: 214–218.

Yamamoto J, Fukui T, Suzuki K, Tanaka S, Yatsunami K Ichikawa A. 1993. Expression and characterization of recombinant mouse mastocytoma histidine decarboxylase. *Biochim Biophys Acta* **1216**: 431–440.

Yamauchi K, Sato R, Tanno Y, Ohkawara Y, Maeyama K, Watanabe T, Satoh K, Yoshizawa M, Shibara S, Takishima T. 1990. Nucleotide sequence of the cDNA encoding L-histidine decarboxylase derived from human basophilic leukemia cell line, KU-812-F. *Nucl Acids Res* **18**: 5891.

Yamauchi K, Mamune-Sato R, Tanno Y, Ohtsa H, Takishima T, Shibahara S, Maeyama K, Watanabe T. 1993. Molecular biological aspects of human L-histidine decarboxylase. *Adv Biosci* **89**: 177–196.

Yatsunami K, Öhtsu H, Tsuchikawa M, Higuchi T, Ishibashi K, Shida A, Shima Y, Nakagawa S, Yamauchi K, Yamamoto M, Hayashi N, Watanabe T, Ichikawa A. 1994. Structure of the L-histidine decarboxylase gene. *J Biol Chem* **269**: 1554–1559.

Yatsunami K, Tsuchikawa M, Kamada M, Hori K, Higuchi T. 1995. Comparative studies of human recombinant 74- and 54-kDa L-histidine ecarboxylases. *J Biol Chem* **270**: 30813–30817.

Zahnow CA, Yi HF, McBride OW, Joseph DR. 1991. Cloning of the cDNA encoding human histidine decarboxylase from an erythroleukemia cell line and mapping of the gene locus to chromosome 15. *DNA Seq* **1**: 395–400.

Zahnow CA, Panula P, Yamatodani A, Millhorn DE. 1998. Glucocorticoid hormones downregulate histidine decarboxylase mRNA and enzyme activity in rat lung. *Am J Physiol* **275**: L407–413.

Zhang Z, Höcker M, Koh TJ, Wang TC. The Human histidine decarboxylase rromoter is regulated by gastrin and phorbol 12-myristate 13-acetate through a downstream *c*is-acting element. *J Biol Chem* **271**: 14188–14197.

17

The Histamine H_4 Receptor: Drug Discovery in the Post-genomic Era

Niall O'Donnell, Paul J. Dunford and **Robin L. Thurmond**

Abstract

This chapter describes a post-genomic pharmacology approach used to identify a new histamine receptor, the histamine H_4 receptor. The discovery of the histamine H_4 receptor is indicative of the new frontier of genomics and reverse pharmacology: the identification of an interesting sequence and the elucidation of that sequence's function despite a paucity of previously developed pharmacological knowledge. The availability of genomic information has accelerated the identification of potential drug targets. However, in many cases the function of the protein is not known and therefore selective pharmacological agents are essential tools for uncovering the biological role of the protein. This chapter discusses this process in relation to the histamine H_4 receptor – from the cloning of the receptor to the identification of selective ligands and the discovery of the biological role for the receptor in inflammation and immune responses. The data suggest that ligands to the histamine H_4 receptor could have a variety of applications, including the treatment of allergic and inflammatory conditions.

17.1 Introduction

The modulation of histamine receptors has proven therapeutically successful and financially profitable for pharmaceutical companies. Histamine is a ubiquitous mediator that is involved in many different physiological processes. The best characterized of these are inflammation and gastric acid secretion. During inflammation histamine is released from preformed stores in mast cells and basophils, leading to vasodilation and local swelling, typified by a wheal and flare response. This response is blocked by histamine H_1 receptor (H_1R) antagonists such as diphenhydramine, leading to their use in the treatment of inflammatory disorders and allergy (Ash and Schild, 1966). Enterochromaffin-like cells in the gut control gastric acid release and also release histamine. This effect of histamine is mediated by the

Immunogenomics and Human Disease Edited by András Falus

histamine H_2 receptor (H_2R) and antagonists of this receptor are used for the treatment of gastric acid disorders (Black *et al.*, 1972).

Given the fact that histamine is such an important physiological mediator and that together H_1R and H_2R antagonists have generated several blockbuster drugs, the discovery of two new histamine receptors, H_3 receptor (H_3R) (Lovenberg *et al.*, 1999) and H_4 receptor (H_4R) (Oda *et al.*, 2000; Zhu *et al.*, 2001; Nguyen *et al.*, 2001; Morse *et al.*, 2001; Liu *et al.*, 2001; O'Reilly *et al.*, 2002; Coge *et al.*, 2001), in a short period of time beginning in the late 1990s generated enormous excitement in the scientific community. Of note was the difficulty in cloning these histamine-specific G-protein-coupled receptors (GPCR). The pharmacology of H_3R had been partially elucidated previously (reviewed Leurs *et al.*, 2005). However, the molecular identification and cloning, an essential step in the modern paradigm of small-molecule inhibitor generation, proved elusive. The reasons for these difficulties are now clear; traditional homology searches using H_1R and H_2R, as 'backbones' were unlikely to work in the identification of these novel histamine receptors due lack of sequence homology between H_3R/H_4R and H_1R/H_2R.

The discovery of H_3R is representative of the traditional approach to drug discovery where pharmacology led to the development of ligands before the sequence of the receptor was known. However, the H_4R is representative of a post-genomic paradigm for drug discovery. That is, an interesting DNA sequence, bearing the hallmarks of a drug target, is discovered that eventually leads to the development of ligands as potential drugs. One drawback to such a genomics approach is that in most cases the function of the receptor is not known and, therefore, selective ligands become an essential tool for uncovering the biology.

This chapter describes a post-genomic pharmacology approach that has led to the identification of selective ligands for H_4R and the indication that they may be useful for the treatment of inflammatory diseases. The chapter will give the background to the cloning of the H_3R and H_4R, before focusing on the development of H_4R-selective ligands and the characterization of its role in inflammation. Preliminary studies on H_4R and specific antagonists suggest that another rich vein of histamine receptor therapeutics, likely to generate more blockbusters and medical breakthroughs, could be on the horizon.

17.2 Cloning of H_3R and H_4R

The identification of the molecular structure of H_4R owes much to the earlier discovery of H_3R. The existence of H_3R had been shown pharmacologically in 1983, and was identified as a presynaptic autoreceptor (Arrang, Garbarg and Schwartz, 1983; Leurs *et al.*, 2005). However, the cloning of the gene for H_3R remained frustratingly slow, leading to suggestions that the receptor was not a GPCR. The advent of reverse pharmacology added new vigor to the search (Mertens *et al.*, 2004, Wise, Jupe and Rees, 2004).

Rather than using degenerate PCR strategies based on known sequences in an attempt to clone the desired receptor, the idea behind reverse pharmacology was to obtain putative GPCR based on general sequence homologies (e.g. the conical seven-transmembrane domains). The putative GPCR could be expressed in cell lines and a panel of ligands screened. Transfected cells that responded were likely to contain a receptor specific for a given ligand, e.g. histamine. Thus, using a ligand as a 'hook', the responding GCPR could be 'caught' and identified by their function, not just by sequence homologies. The human genome project generated several hundred novel, putative GPCR that could be deorphanized using this approach. Armed with sequence information from H_1R and H_2R, and cognizant of the conserved bioamine sequence motifs, Tim Lovenberg's group at J&J PRD performed a large-scale search for putative GPCRs. An expressed sequence tag (EST) from the Incyte EST library was identified using α_2-adrenergic receptor as a sequence query. This sequence shared 35 per cent sequence homology with the seventh transmembrane (TM) domain of α_2-adrenergic receptor. Using this EST as a probe, a full-length 2.7 kb clone was identified. Translation of this clone, designated GPCR97, indicated a protein of 445 amino acids with approximately 25 per cent homology to the biogenic amine receptor superfamily. Most telling was the conserved biogenic amine receptor residues, e.g. an aspartate in the third TM domain, discussed later. GPCR97 was expressed and deorphanized using a cell-based assay system which indicated that the receptor responded specifically to histamine, with a pharmacological profile matching that previous identified for H_3R (Lovenberg *et al.*, 1999).

Examination of the sequence for H_3R indicated why previous, exhaustive cloning attempts based on H_1R and H_2R sequences had ended in frustration and failure: H_3R receptor exhibited only 22 per cent and 20 per cent sequence homology to H_1R and H_2R, respectively. H_3R was much more closely related to the α_{2a}, α_{2c} and muscarinic acetylcholine receptors.

Armed with the H_3R cloning breakthrough, Lovenberg then used the H_3R sequence as a basis for further homology searches. Searches using the H_3R sequence as a query of a human genomic database identified an unordered draft contiguous sequence. The sequences were reassembled in the correct orientation, analysed using Genewise (the Sanger Centre) and shown to share intron–exon junctions with H_3R (Liu *et al.*, 2001). H_3R is encoded by three exons, contrasting with the single exon of H_1R and H_2R. The clone, designated GPCR105, was approximately 35 per cent identical to H_3R, possessed seven putative transmembrane domains, and exhibited all the conserved residues seen in the class A GPCR, e.g. asparagine in TM I, aspartic acid in TM II and three conserved prolines in TM V, VI and VII. In addition, GPCR105 possessed two key conserved amino acids found in bioamine receptors: an aspartic acid at position 114 in TM III and a tryptophan at position 341 in TM VII. Transfection of the full-length cloned putative receptor conferred the ability to bind histamine to transfected cell lines. GPCR105 was redesignated as the histamine H_4 receptor (H4R), a new member of the histamine receptor family. Phylogenetic analysis (together with the exon–intron organization) strongly suggests that H_3R and H_4R are not closely related to H_1R or H_2R, and thus may have evolved from different 'ancestor' receptors (Figure 17.1). Overall H_4R gene, located on human chromosome 18, shared 40 per cent sequence identity with H_3R. The binding of pharmacological compounds to the transfected H_4R indicated a profile distinct from H_3R and the two 'classic'

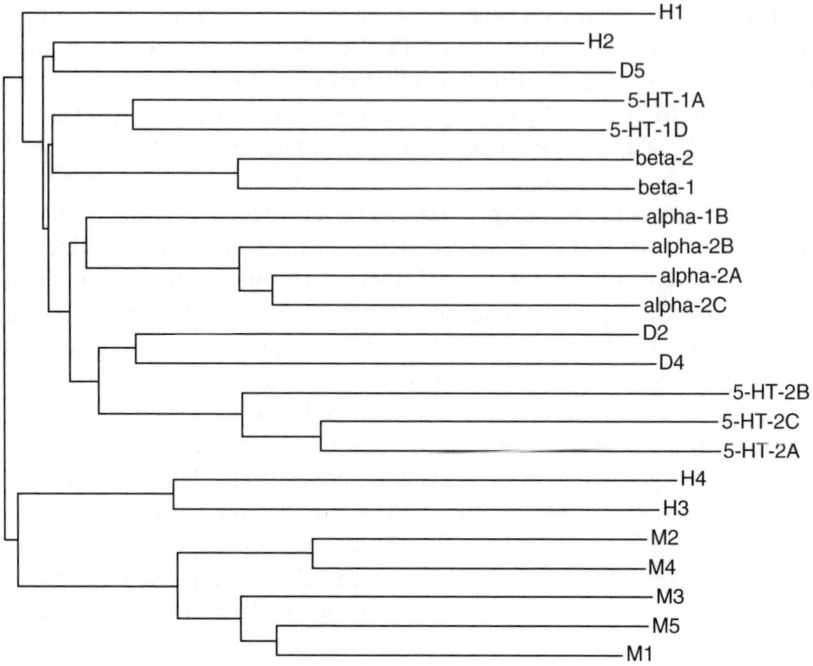

Figure 17.1 Phylogenetic tree showing the degree of homology between selected members of the bioamine GPCR families. An initial pairwise and group alignments, and N-J tree were performed using the CLUSTALX program. The tree was visualized and refined using NJPLOT.

histamine receptors. Perhaps the most striking feature of this new histamine receptor was its expression profile: H_4R was abundantly expressed in bone marrow and in haematopoietic cells such as eosinophil, basophils and mast cells, and possibly dendritic cells and leukocytes (Hofstra *et al.*, 2003; Desai *et al.*, 2004; Oda *et al.*, 2000; Zhu *et al.*, 2001; Nguyen *et al.*, 2001; Morse *et al.*, 2001; Liu *et al.*, 2001; O'Reilly *et al.*, 2002; Coge *et al.*, 2001). Given the prominent role of histamine in allergic responses and the central role of these cells types in allergic ailments, this expression pattern, along with the stellar track record of previous anti-histamines in the drug market, excited the interest of immunologists and drug companies.

17.3 Generation of H_4R-Specific Antagonists

H_4R bound a ligand of known importance in inflammatory responses, and had an expression pattern suggesting a role in inflammation, but what was the actual function of the receptor? In modern drug discovery the use of mice with deletions of targeted genes has been central in validating targets. While valuable, this approach has drawbacks: there is no guarantee that the mice will be viable, there may be functional compensation or redundancy, and the insights obtained may only apply to mice.

Another way to validate targets is to use more traditional pharmacological approaches. The increasing complexity of histamine biology necessitates the generation of antagonists and agonists that are exquisitely specific for each of the four known receptors. Without these tools the function of H_4R would be extremely difficult to distinguish from those of other receptors. For example, without specific pharmacological tools, elucidation of a given receptor's role in a signalling process could only be inferred on cell types expressing two or more histamine receptors.

17.4 High-Throughput Screening

In order to aid understanding of the biology of H_4R and to provide a pharmacological lead for drug design, it was decided to screen a large library of compounds against the expressed receptor (Jablonowski *et al.*, 2003). Cell pellets from SK-N-MC cells stably transfected with the human H_4R were used in receptor binding assays using radiolabelled histamine. This approach identified certain indolypiperazine compounds that were shown to be specific, high-affinity H_4R binders. A medicinal chemistry effort led to the discovery of JNJ7777120 (Jablonowski *et al.*, 2003). This compound is a potent ligand for H_4R and exhibits at least 1000-fold selectivity over other the histamine receptors as well as having no cross-reactivity against 50 other targets (Thurmond *et al.*, 2004).

Receptor binding data only give an indication that a particular compound is a ligand for the receptor. The creation of cell lines stably expressing H_4R allows for further elucidation of the functional aspects of compound binding, i.e. whether the ligand is an agonist or antagonist. These assays require the identification of signalling pathways activated by the receptor. Activation of H_1R leads to changes in intracellular calcium levels via $G\alpha_q$-proteins, H_2R signal through $G\alpha_s$-proteins, leading to cAMP increases, and H_3R activate $G\alpha_{i/o}$ proteins, resulting in inhibition of cAMP production (Table 17.1). As with H_3R, histamine binding to H_4R transfected into

Table 17.1 Comparison of the known histamine receptors

Receptor	H_1	H_2	H_3	H_4
Protein homology to H_4R (%)	17	16	35	100
Exons	One	One	Three	Three
$G\alpha$ protein and signalling mediator	$G\alpha_q$ $\uparrow Ca^{2+}$	$G\alpha_s$ $\uparrow$cAMP	$G\alpha_{i/o}$ $\downarrow$cAMP	$G\alpha_{i/o}$ $\uparrow Ca^{2+}$
Tissue expression	Ubiquitous	Ubiquitous	Neuronal cells	Mast cells, eosinophils (dendritic and T cells)
Function	Bronchoconstriction and vasodilation	Gastric acid secretion	Neurotransmitter release	Mast cell and eosinophil chemotaxis

SK-N-MC cells resulted in an inhibition of forskolin-induced cAMP increases. However, the effect was significantly less than that the H_3R. The effect on cAMP can be blocked by pretreating the cells with pertussis toxin (PTX), suggesting a role for $G\alpha_{i/o}$ proteins. Ligands such as thioperamide inhibit this response and therefore are classified as antagonists or inverse agonists. Using such a system it was determined that JNJ7777120 acts as an antagonist for H_4R.

One note should be made regarding the use of transfected cell lines. The fact that H_4R can couple to a $G\alpha_{i/o}$ receptor in this cell type and leads to inhibition of cAMP is no indication of the actual coupling in the natural state. Indeed, an increase in intracellular calcium can be observed when H_4R is cotransfected with promiscuous or chimeric G proteins like $G\alpha_{qi5}$, $G\alpha_{qi1/2}$, $G\alpha_{qi3}$, $G\alpha_{15}$ or $G\alpha_{16}$. However, subsequent studies using primary cells, discussed later, suggest that H_4R does signal through $G\alpha_{i/o}$ family members to increase intracellular calcium levels (Hofstra *et al.*, 2003).

17.5 Functional Studies

Cells transfected with the receptor of interest are great tools for screening potential ligands and for their initial characterization. However, in the end, one must resort to primary cells or *in vivo* systems in order to probe the physiological role of the receptor in question. This is doubly true for histamine receptors, where an important consideration in determining the biological function of the H_4R is the potential 'cross signalling' events generated by the other histamine receptors responding to the shared ligand. Conflicting results might result from the presence of multiple histamine receptors on the same cell type, each potentially with different biological functions. Judicious use of specific histamine receptor antagonists (Table 17.2) and, where available, histamine receptor knockout mice can ensure that the distinct roles of histamine receptors can be elucidated. The use of pharmacological approaches alone always leaves open the question of selectivity, whereas with knockout approaches changes in development may mislead or mask the function of a target in adults. Therefore, a dual genetic and pharmacological ablation of target function avoids

Table 17.2 Selected histamine receptor ligands and histamine receptor affinities (adapted from Fung-Leung *et al.*, 2004)

		K_i (nM)			
Compound	Mechanism of action	H_1R	H_2R	H_3R	H_4R
Histamine	Agonist	12 500	2 000	5.4	8.1
Diphenhydramine	H_1R antagonist	15	×	>10 000	>10 000
Ranitidine	H_2R antagonist	>10 000	85	>10 000	>10 000
R-α-methylhistamine	H_3R/H_4R agonist	>10 000	>10 000	0.7	146
JNJ5207852	H_3R antagonist	>10 000	>10 000	0.57	>10 000
Thioperamide	H_3R/H_4R antagonist	>10 000	>10 000	25	27
JNJ7777120	H_4R antagonist	>10 000	>10 000	5 000	4.1

these pitfalls and is an extremely powerful approach, aiding the definitive characterization of gene function. Furthermore, the viability and phenotypic analysis of mice that lack H_4R receptors can yield important information, e.g. the survival and normal lifespan of mice lacking H_4R are strongly indicative that the receptor is not essential for murine and possibly human survival. Also, the administration of H_4R agonists and antagonists to mice lacking H_4R is an important means of determining compound specificity and elucidating the difference between toxicity induced by the compounds versus toxicity induced by H_4R blockage. Such information is useful in the drug development pathway. With both H_4R-selective antagonist and knockout mice in hand, we set out to elucidate the function of the receptor in primary cells and *in vivo*.

In Vitro

Eosinophils are important effector cells in the late phase allergic response. Resident generally in tissues, such as lung, intestinal and uteral systems, eosinpohils can rapidly migrate towards inflamed tissues. Degranulation of eosinophils releases toxic granule proteins and is believed to be an important factor in airway epithelial damage and bronchial hyperreactivity in asthma. Early data indicated that there was a histamine receptor in eosinophils that mediated histamine-induced chemotaxis and calcium responses which was not H_1R, H_2R or H_3R (Clark, Gallin and Kaplan, 1975; Clark *et al.*, 1977; Raible *et al.*, 1994). The use of dual H_3R–H_4R ligands and JNJ7777120, the specific H_4R antagonist, provided definitive evidence that this receptor was H_4R (Ling *et al.*, 2004; Buckland, Williams and Conroy, 2003). Histamine induces chemotaxis of human eosinophils and antagonists specific for the H_4R are able to block this effect. Histamine signalling through H_4R also leads to actin polymerization, a key step in the polarization process required for chemotaxis, as well as the upregulation of cell surface adhesion molecules, such as CD11b and CD54 (Ling *et al.*, 2004; Buckland, Williams and Conroy, 2003).

H_4R is also expressed in mast cells. Mast cells are important early effector cells in allergic disease. Binding of antigens to specific IgE on the surface of mast cells causes the release of inflammatory mediators such as histamine, serotonin, prostaglandins and leukotrienes. The correlative relationships between mast cells, histamine and allergic diseases made this cell type an attractive starting point for additional function studies. Purified mouse mast cells were shown to be chemotactic towards histamine, with an associated calcium mobilization (Hofstra *et al.*, 2003). The inhibition of this migration by PTX was consistent with histamine receptor signalling in these primary cells being mediated via the $G\alpha_{i/o}$ G proteins. From Table 17.1, it can be seen that both H_1R and H_4R ligation results in calcium mobilization. However, the calcium response in mast cells can be blocked by PTX. H_1R calcium mobilization is not affected by PTX, suggesting different signalling mechanisms and outcomes.

The histamine-induced chemotaxis and calcium mobilization in mast cells is mediated by the H_4R (Hofstra *et al.*, 2003). Chemotactic migration in response to

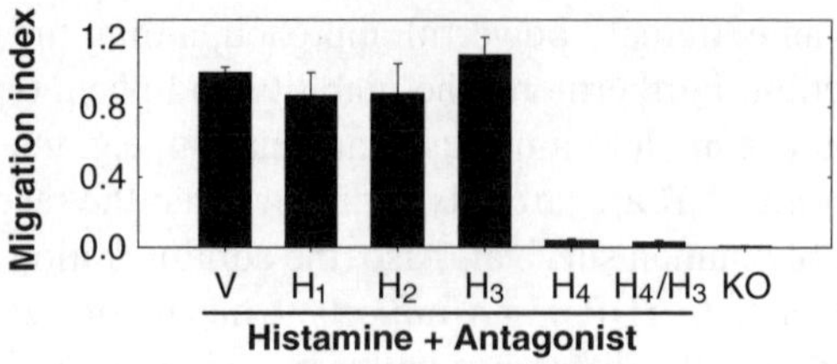

Figure 17.2 Elucidation of the role of H_4R-mediated mast cell chemotaxis in response to histamine.

histamine (and calcium mobilization) was unaffected by specific H_1R, H_2R and H_3R antagonists, diphenhydramine, ranitidine and JNJ5207852, respectively, but could be blocked by the dual H_3R/H_4R antagonist thioperamide and a H_4R specific antagonists (Figure 17.2). Strikingly, mast cells from mice lacking H_4R did not show histamine chemotactic responses and did not mobilize calcium in response to histamine. This clearly illustrates the power of the combined pharmacological and genetic approaches for gene function analysis The use of a specific H_4R antagonist such as JNJ7777120 in combination with cells from H_4R knockout mice clearly demonstrates a central role for H_4R in mast cell chemotaxis.

It should be noted that, compared with 'traditional' eosinophil chemotactic signalling molecules, such as the chemokines eotaxin-2 and MCP-3, histamine is a less potent chemotactic agent, although it appears to be just as efficacious. This is also true for mast cell chemotaxis. Given the short half-life of histamine in circulation, it seems likely that this histamine chemotactic event is local in nature, with tissue concentrations of histamine rising rapidly in response to mast cell degranulation. This triggers eosinophil homing to the site of mast cell degranulation. Further, histamine appears to have an adjuvant effect, enhancing the chemotactic power of the 'traditional' chemokines mentioned.

In Vivo

The availability of a selective H_4R antagonist allowed for the exploration of the *in vivo* functions of the receptor. *In vivo* studies focusing on mast cell kinetics in the lung reinforced the *in vitro* chemotactic findings (Thurmond *et al.*, 2004). Immature mast cells are known to migrate from the bone marrow to connective and mucosal tissues, where they mature into separate phenotypes. During disease states such as asthma and rhinitis local increases in mast cells numbers are apparent in the affected airway tissues. Stem cell factor is likely the main driver of mast cell chemotaxis into inflamed tissues, however, based on our *in vitro* findings, we were interested in investigating a modulatory role for histamine, in this process. Mice were subjected to aerosolized histamine, leading to increases in the total number of tracheal mast cells and an increase in the number of mast cells in the epithelial cell layer (Thurmond *et al.*, 2004). Administration of JNJ7777120 decreased this mast cell migration toward the chemotactic gradient produced by histamine inhalation, thus suggesting a chemotactic role for histamine *in vivo*. We postulate that antigen inhalation in rhinitis or asthma sufferers leads to mucosal mast cell activation and release of a variety of chemotactic signals, including histamine. This provides an amplification cascade with an influx of more mast cells, increasing the allergic response. The ability of JNJ7777120 to block this phenomenon suggests a potential role for H_4R antagonism in modulating allergic disease.

The fact that H_4R antagonists had effects *in vivo* led us to investigate more general roles in pathological responses. Given the inflammatory 'guilt-by-association' of H_4R expression, our early *in vivo* studies focused on general models of inflammation. Peritonitis models in mice allowed us to probe for general anti-inflammatory properties of H_4R antagonists. Results showed that mice treated with JNJ7777120 or thioperimide had a reduction in infiltrating neutrophils after intra-peritoneal zymosan or urate crystal injection (Thurmond *et al.*, 2004; Desai *et al.*, 2004). These peritonitis models are mast cell-dependent and therefore the effects seen may be mediated by H_4R expressed on this cell type. Consistent with this, the antagonist has no effect in inhibiting thioglycollate-induced peritonitis, which has been reported to be mast cell independent (Ajuebor *et al.*, 1999). Since it does not appear that the H_4R is expressed on neutrophils, it is probable that the decreased infiltration is a secondary effect of H_4R antagonism. In support of this, other workers have used thioperamide in a zymosan pleurisy model to suggest a role for H_4R in the release of LTB_4, a potent neutrophil chemoattractant, in this model (Takeshita, Bacon and Gantner, 2004).

The role of the H_4R in inflammation models, as well as its function in eosinophils and mast cells, suggests a potential role in allergic disease. This hypothesis was explored in a mouse model of asthma. Mice were sensitized to ovalbumin before being challenged repeatedly over 4 days with inhaled ovalbumin. Cells counts of the bronchoalveolar lavage fluid from the mice showed a 50 per cent reduction in eosinophils in the lung after H_4R antagonist administration (Figure 17.3; Desai *et al.*, 2004). Thus, it appears that H_4R antagonists can attenuate inflammatory cell influx into the lung, suggesting that they may have beneficial effects in treating allergic rhinitis or asthma in humans.

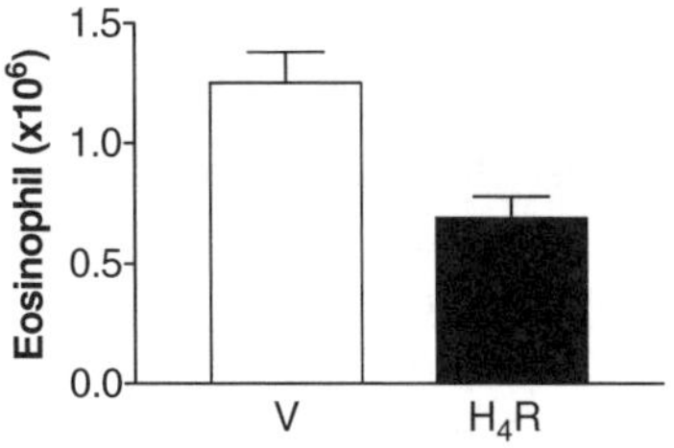

Figure 17.3 H_4R antagonist reduces eosinophil influx into the lung in a mouse model of asthma (V, vehicle-treated animals; H_4R, H_4R antagonist-treated).

Allergic dermatological diseases display similar aetiology and immunology to allergic airway diseases in that skin mast cells, stimulated via allergen-specific IgE, liberate histamine and other inflammatory mediators to cause acute and chronic inflammatory sequalae. A main symptom of atopic dermal inflammation is itch. Recently it has been reported that thioperamide can block histamine-mediated itch in mice (Bell, McQueen and Rees, 2004). The involvement of H_3R, however, in these responses could not be fully disproved with the pharmacological tools available. These findings need to be expanded with the use of specific agonists and antagonists, and in knockout mice, but could have important implications for the future treatment of dermatitis and itch, which are poorly managed by non-sedating H_1R antagonists.

17.6 Future Prospects

The work with the H_4R clearly illustrates the strategies and hurdles of drug discovery in the post-genomic age. Classical pharmacology had correctly identified and characterized H_3R. However, the cloning proved intractable until the combination of *in silico* sequence searches of the human genome, available in the genomic age, and the advent of reverse pharmacology. The discovery of H_4R is indicative of the new frontier of genomics and reverse pharmacology: the identification of an interesting sequence and the elucidation of that sequence's function despite a paucity of previously developed pharmacological knowledge.

Identification of interesting sequences is relatively easy; determining the function of the sequences is more difficult. Elucidation of function requires information from genomic tools, i.e. when and where the protein is expressed, but also pharmacological information, i.e. selective ligands. The unique expression pattern of H_4R, the receptor's high affinity for histamine relative to H_1R and H_2R, and the very different route of calcium mobilization with H_4R ligation compared with H_1R, suggests a distinct and central role for H_4R in inflammatory responses. The initial correlation-based postulates, linking H_4R to a role in allergy and inflammation, based largely on the nature of the ligand and the expression pattern of the receptor, have been proven largely correct. Blocking chemotactic responses in mast cells, eosinophils and possibly basophils in response to the inflammatory release of histamine at the site of insult by H_4R antagonists may be important in relieving many disease pathologies. In an allergic response large amounts of histamine can be generated locally by mast cells in response to local antigen exposure. The trafficking of mast cells, eosinophils and basophils, cells with robust H_4R expression, to the sites of inflammation often correlates with disease severity. It appears that histamine may play an important role in this recruitment. Blocking this process could be of great therapeutic benefit to allergic disease sufferers. In addition, H_4R antagonists have been shown to be efficacious in a variety of mouse models of human disease, including inflammation and asthma. Thus the potential for H_4R antagonists to bring relief to disease sufferers seems promising.

References

Ajuebor MN, Das AM, Virag L, Flower RJ, Szabo C, Perretti M. 1999. Role of resident peritoneal macrophages and must cells in chemokine production and neutrophil migration in acute inflammation: evidence for an inhibitory loop involving endogeneous IL-10 *J Immunol* **162**: 1685–1691.

Arrang JM, Garbarg M, Schwartz JC. 1983. Auto-inhibition of brain histamine release mediated by a novel class (H3R) of histamine receptor. *Nature* **302**: 832–837.

Ash ASF, Schild HO. 1966. Receptors mediating some actions of histamine. *Br J Pharmac* **27**: 427–439.

Bell JK, McQueen DS, Rees JL. 2004. Involvement of histamine H4 and H1 receptors in scratching induced by histamine receptor agonists in Balb/C mice. *Br J Pharmac* **142**: 374–380.

Black JW, Duncan WAM, Durant CJ, Ganelin CR, Parsons EM. 1972. Definition and antagonism of histamine H2 receptors. *Nature* **236**: 385–390.

Buckland KF, Williams TJ, Conroy DM. 2003. Histamine induces cytoskeletal changes in human eosinophils via the H4 receptor. *Br J Pharmac* **140**: 1117–1127.

Clark RA, Gallin JI, Kaplan AP. 1975. The selective eosinophil chemotactic activity of histamine. *J Exp Med* **142**: 1462–1476.

Clark RA, Sandler JA, Gallin JI, Kaplan AP. 1977. Histamine modulation of eosinophil migration. *J Immunol* **118**: 137–145.

Coge F, Guenin SP, Rique H, Boutin JA, Galizzi JP. 2001. Structure and expression of the human histamine H4-receptor gene. *Biochem Biophys Res Commun* **284**: 301–309.

Desai PJ, Dunford PJ, Hofstra CL, Karlsson L, Fung-Leung W-P, Ling P, Thurmond RL. 2004. Use of Histamine H4 Receptor modulators for the treatment of allergy and asthma. Janssen Pharmaceutica, NV.

Fung-Leung WP, Thurmond RL, Ling P, Karlsson L. 2004. Histamine H4 receptor antagonists: the new antihistamines? *Curr Opin Invest Drugs* **5**: 1174–1183.

Hofstra CL, Desai PJ, Thurmond RL, Fung-Leung W-P. 2003. Histamine H4 receptor mediates chemotaxis and calcium mobilization of mast cells. *J Pharmac Exp Ther* **305**: 1212–1221.

Jablonowski JA, Grice CA, Chai W, Dvorak CA, Venable JD, Kwok AK, Ly KS, Wei J, Baker SM, Desai PJ. *et al.* 2003. The first potent and selective non-imidazole human histamine H4 receptor antagonists. *J Med Chem* **46**: 3957–3960.

Leurs R, Bakker RA, Timmerman H, de Esch IJ. 2005. The histamine H3 receptor: from gene cloning to H3 receptor drugs. *Nat Rev Drug Discov* **4**: 107–120.

Ling P, Ngo K, Nguyen S, Thurmond RL, Edwards JP, Karlsson L, Fung-Leung W-P. 2004. Histamine H4 receptor mediates eosinophil chemotaxis with cell shape change and adhesion molecule upregulation. *Br J Pharmac* **142**: 161–171.

Liu C, Ma X, Jiang X, Wilson SJ, Hofstra CL, Blevitt J, Pyati J, Li X, Chai W, Carruthers N, Lovenberg TW. 2001. Cloning and pharmacological characterization of a fourth histamine receptor H4 expressed in bone marrow. *Mol Pharmac* **59**: 420–426.

Lovenberg TW, Roland BL, Wilson SJ, Jiang X, Pyati J, Huvar A, Jackson MR, Erlander MG. 1999. Cloning and functional expression of the human histamine H3 receptor. *Mol Pharmac* **55**: 1101–1107.

Mertens I, Vandingenen A, Meeusen T, De Loof A, Schoofs, L. 2004. Postgenomic characterization of G-protein-coupled receptors. *Pharmacogenomics* **5**: 657–672.

Morse KL, Behan J, Laz TM, West RE. Jr, Greenfeder SA, Anthes JC, Umland S, Wan Y, Hipkin RW, Gonsiorek W. *et al.* 2001. Cloning and characterization of a novel human histamine receptor. *J Pharmac Exp Ther* **296**: 1058–1066.

Nguyen T, Shapiro DA, George SR, Setola V, Lee DK, Cheng R, Rauser L, Lee SP, Lynch KR, Roth BL, O'Dowd BF. 2001. Discovery of a novel member of the histamine receptor family. *Mol Pharmac* **59**: 427–433.

Oda T, Morikawa N, Saito Y, Masuho Y, Matsumoto S-I. 2000. Molecular cloning and characterization of a novel type of histamine receptor preferentially expressed in leukocytes. *J Biol Chem* **275**: 36781–36786.

O'Reilly M, Alpert R, Jenkinson S, Gladue RP, Foo S, Trim S, Peter B, Trevethick M, Fidock M. 2002. Identification of a histamine H4 receptor on human eosinophils-role in eosinophil chemotaxis. *J Recept Signal Transduct Res* **22**: 431–448.

Raible DG, Lenahan T, Fayvilevich Y, Kosinski R, Schulman ES. 1994. Pharmacologic characterization of a novel histamine receptor on human eosinophils. *Am J Respir Crit Care Med* **149**: 1506–1511.

Takeshita K, Bacon KB, Gantner F. 2004. Critical role of L-selectin and histamine H4 receptor in zymosan-induced neutrophil recruitment from the bone marrow: comparison with carrageenan. *J Pharmac Exp Ther* **310**: 272–280.

Thurmond RL, Desai PJ, Dunford PJ, Fung-Leung W-P, Hofstra CL, Jiang W, Nguyen S, Riley JP, Sun S, Williams KN. *et al.* 2004. A potent and selective histamine H4 receptor antagonist with anti-inflammatory properties. *J Pharmac Exp Ther* **309**: 404–413.

Wise A, Jupe SC, Rees S. 2004. The identification of ligands at orphan G-protein coupled receptors. *A Rev Pharmac Toxicol* **44**: 43–66.

Zhu Y, Michalovich D, Wu H-L, Tan KB, Dytko GM, Mannan IJ, Boyce R, Alston J, Tierney LA, Li X. *et al.* 2001. Cloning, expression, and pharmacological characterization of a novel human histamine receptor. *Mol Pharmac* **59**: 434–441.

18

Application of Microarray Technology to Bronchial Asthma

Kenji Izuhara, Kazuhiko Arima, Sachiko Kanaji, Kiyonari Masumoto and **Taisuke Kanaji**

Abstract

Bronchial asthma is a disorder affected by genetic and environmental factors. It is widely accepted that it is a Th2-type inflammation originating in the lung and caused by inhalation of ubiquitous allergens. However, the complicated and diverse pathogenesis of this disease is yet to be clarified. Microarray technology is a powerful tool to analyse whole gene expression profiling under given conditions. Several attempts to clarify the pathogenesis of bronchial asthma have been carried out using microarray technology, providing us with some novel pathogenic mechanisms of bronchial asthma as well as information about gene expression profiling. In this article, we review the outcomes of these analyses by the microarray approach as applied to bronchial asthma.

18.1 Introduction

The human genome project has been completed, and we are now in the postgenomic era. At present, a lot of attention is being paid to the 'transcriptome', meaning the complete set of transcribed genes expressed as mRNAs, because only 5 per cent of genes are active in a particular cell at any given point in time and gene expression forms part of our knowledge of the role of genes in human diseases (Russo, Zegar and Giordano, 2003; Sevenet and Cussenot, 2003). Microarray technology is now the most powerful tool for generating massive parallel gene expression profiling (Butte, 2002; Churchill, 2002; Xiang *et al.*, 2003). Other methods, such as differential display or serial analysis of gene expression (SAGE), are technically complex (Sevenet and Cussenot, 2003). In contrast, the microarrays are easier to use, do not require large-scale DNA sequencing, and allow parallel quantification of thousands of genes from multiple samples (Russo, Zegar and Giordano, 2003; Izuhara *et al.*, 2004).

Immunogenomics and Human Disease Edited by András Falus

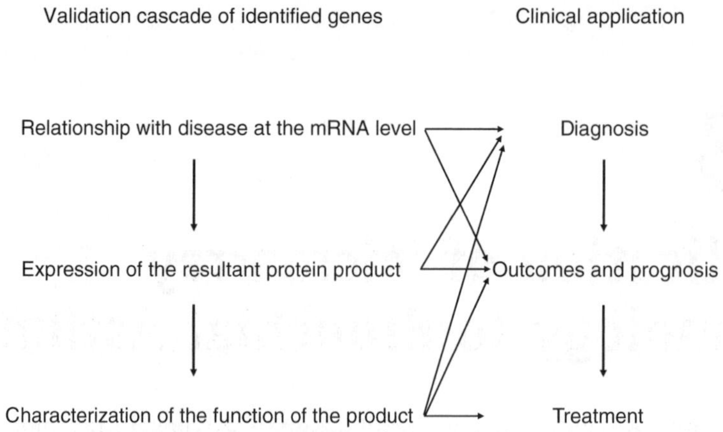

Figure 18.1 Validation cascade and clinical relevance of the identified genes.

It is hoped that application of microarray technology will help us to refine and redefine disease diagnosis (prediction of diseases severity, discovery of disease class), disease outcomes and prognosis (prediction of treatment response or prognosis in advance of the initiation of therapy), and disease treatment (discovery of biological pathways of disease or new targets for therapy; Dudda-Subramanya *et al.*, 2003; Sevenet and Cussenot, 2003).

Validation of the pathological roles of the identified genes is simple (Williams, 2003; Figure 18.1). The first step is the investigation of the relationship with a particular disease at the mRNA level. The next step is the confirmation of expression of the resultant protein product in the disease. The last is characterization of the function of the product in the disease. Information from each step can be used at each stage of clinical application: diagnosis, outcomes and prognosis, and treatment (Figure 18.1). However, the most difficult step is the last, characterizing the function of the product, because most human diseases are pathologically complex and heterogeneous. Bronchial asthma is particularly so. Multiple genetic and environmental factors predispose people to it, and various cells and mediators are involved (Holgate, 1999). To deal with such a complexity, the design of microarray experiments is critical.

Application of microarray technology to the clarification of the pathogenesis of bronchial asthma is classified into four strategies (Figure 18.2). 'Study condition' to analyse the expression genes may include comparison between samples derived from asthmatic and normal subjects as a broad condition, or comparison between samples in the presence or absence of a particular stimulus as a specific condition. In addition, 'sample source' may be lung tissues as a broad condition or particular cells as a specific condition. The various combinations of 'study condition' and 'sample source' yield four strategies. The comparison of asthma patients vs normal donors as 'study condition', or the analysis of lung tissues as 'sample source', has the advantage of covering broader phenotypic changes of bronchial asthma. In contrast, the compar-

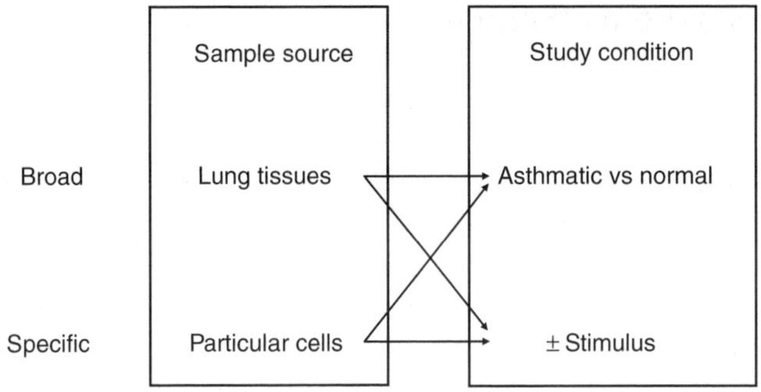

Figure 18.2 Strategies for applying microarray technology to bronchial asthma.

ison of the effects of a particular stimulus as 'study condition', or the analysis of particular cells as 'sample source', has the advantage of focusing on identification of the genes correlated with a particular pathological aspect of bronchial asthma. Several trials have been performed based on each strategy, leading to novel pathological mechanisms of bronchial asthma. It is important to perform minute investigations for the involved cells or stimuli, when either broader 'study condition' or 'sample source' is selected. Conversely, it is important to study the role of the identified gene product in the whole pathogenesis of bronchial asthma, in the case of a particular 'study condition' or 'sample source'. We next describe results based on these strategies (Table 18.1).

Table 18.1 Summary of the applications of microarray technology to bronchial asthma

Source	Species/cell	Condition	Reference
Lung tissue	Mouse	Ovalubumin-inducible asthma	Zimmermann *et al.* (2003)
		Aspergillus-inducible asthma	
		A/J vs C3H/HeJ	Karp *et al.* (2000)
		ADA-deficient	Banerjee *et al.* (2002)
	Monkey	*Ascaris*-inducible asthma	Zou *et al.* (2002)
		IL-4-inducible asthma	
	Human	Asthmatic vs healthy	Laprise *et al.* (2004)
Particular cell	PBMC	Asthmatic vs healthy	Brutsche *et al.* (2001)
	T cell	Th1 vs Th2	Li *et al.* (2001)
	Mast cell	± PMA/ionophore	Cho *et al.* (2000, 2003)
	Eosinophil	± IL-5	Temple *et al.* (2001)
	Macrophage	± IL-4	Welch *et al.* (2002)
		± DEP	Verheyen *et al.* (2004)
	BEC	± IL-13	Lee *et al.* (2001)
	BSMC		
	Fibroblast		
	BEC	± IL-4/IL-13	Yuyama *et al.* (2002)
	BSMC	± IL-13	Syed *et al.* (2005)

18.2 Lung Tissue as 'Source'

Mouse

Ovalubumin sensitization is a typical way of generating asthma-model mice. In addition, *Aspergillus* is a ubiquitous and common aeroallergen. Zimmermann *et al.* (2003) tried to identify inducible genes in lung tissue derived from ovalubumin- or *Aspergillus*-inducible asthmatic mice, finding that 291 genes overlapped among the identified genes. Among the overlapping genes, expression of arginase I, arginase II and cationic amino acid transporter 2 (CAT) was significantly augmented. These three molecules are involved in the uptake and metabolism of arginine, but their precise role in the pathogenesis of bronchial asthma is unclear. However, because arginine is metabolized to nitric oxide (NO) by NO synthase, induction of arginase I, arginase II and CAT may affect NO synthesis. Alternatively, induction of these molecules may enhance collagen synthesis. Furthermore, the cells predominantly expressing arginase turn out to be macrophages, and expression of arginase I and arginase II is enhanced by IL-4 and IL-13, abundantly expressed in the lung tissues of asthmatic mice.

It is known that A/J mice and C3H/HeJ mice are highly susceptible to and highly resistant to allergen-inducing airway hyper-responsiveness (AHR), respectively. Karp *et al.* (2000) designed a unique experiment using microarray technology to try to identify a susceptibility factor to explain this difference. They found that, among 7350 genes, 227 exhibited great change in expression, and paid attention to a complement C5 among the listed genes, because C5 was situated near a locus correlated with allergen-induced bronchial hyper-responsiveness. It turned out that A/J mice, but not C3H/HeJ mice, have a 2 bp deletion in the 5′ exon of the C5 gene that renders them deficient in C5 mRNA and protein production. Furthermore, they found that C5 induced IL-12 production in monocytes. These results indicated that C5 is involved in determining susceptibility to bronchial asthma by inducing IL-12 production, which counterbalances Th2-type immune responses.

Adenosine deaminase (ADA)-deficient mice show a combined immunodeficiency that has been linked to profound disturbances in purine metabolism. ADA-deficient mice also develop asthmatic features such as lung eosinophilia, IgE elevation and mucus hypersecretion. To clarify the mechanism of the link between ADA deficiency and eosinophilia, Banerjee *et al.* (2002) investigated genes expressed in lung tissues in ADA-deficient mice, and found that several genes mediating eosinophil trafficking – such as MCP-3, CD32 and osteopontin – were included in the list.

Monkey

IL-4 is a Th2 cytokine, which has an important role in the pathogenesis of bronchial asthma. IL-4 acts on B cells, inducing IgE synthesis, and also acts on various cells including resident cells in the lung tissue (Izuhara *et al.*, 2002). The allergic monkey

(*Macaca fascicularis*) has a natural hypersensitivity to the antigen of the nematode *Ascaris*, developing IgE-mediated allergic reactions. Zou *et al.* (2002) tried to identify *Ascaris*- or IL-4-inducible genes in lung tissues derived from these asthmatic monkeys by the microarray approach. Expression of MCP-1, MCP-3, eotaxin (chemokine), VCAM-1 (adhesion molecule), TIMP-1, the plasminogen activator inhibitor type-1 (PAI-1), chitinase and α1-antichymotrypsin (proteinase/proteinase inhibitor) showed significant increases in both conditioned monkeys. MCP-1, MCP-3 and eotaxin are known to recruit Th2 cells, eosinophils, basophils, macrophages and dendritic cells. TIMP-1, PAI-1 and chitinase are thought to be involved in tissue remodelling.

Human

Laprise *et al.* (2004) performed bronchial biopsies against four healthy subjects and eight asthma patients and then applied the pooled samples to microarray analysis. Thus far, this is the only report directly comparing the gene expression profiles of bronchial tissues of asthma patients and of normal donors. They found that, among about 6000 genes, 79 showed significant differences in expression, and classified 74 of the 79 into 12 functional categories such as immune signalling molecules, extracellular proteins, proteolytic enzymes, and so on. These genes include nitric oxide synthase 2A, CD14 antigen, T cell receptor α locus, RANTES, mucin, tryptase and cathepsin C, and have all been shown in previous studies to be up-regulated and associated with asthma. Glutathione peroxidase 3, an important antioxidant enzyme, was also confirmed to be reduced in asthma, as previously described. The fractalkine receptor (CX3CR1), which binds to fractalkine, a pro-Th1 membrane-bound chemokine, was downregulated in asthma samples.

18.3 Particular Cell as 'Source'

Immune cell

Peripheral blood mononuclear cell

Brutsche *et al.* (2001) compared gene expression of peripheral blood mononuclear cells (PBMCs) derived from healthy donors, atopic asthma patients and atopic non-asthmatic patients, and found that 12 genes (including IL-6, IL-1 type I receptor and myc) were upregulated in asthmatic subjects. However, it remains unclear which cell in PBMCs expresses these genes.

T cell

It is widely accepted that bronchial asthma is a Th2-type inflammation originating in lung caused by inhalation of ubiquitous allergens (Wills-Karp, 2001; Umetsu, Akbari and Dekruyff, 2003). High expression of Th2 cytokines such as IL-4, IL-5, IL-9 and IL-13 in the lesions is a cardinal feature of bronchial asthma (Robinson *et al.*, 1992; Huang *et al.*, 1995; Kotsimbos, Ernst and Hamid, 1996; Humbert *et al.*, 1997; Bodey *et al.*, 1999). Therefore, it is of interest to investigate dominantly expressed genes in Th1 or Th2 cells. Li *et al.* (2001) compared gene expression in human Th1 and Th2 clones, and found that 14 genes are specifically expressed in Th2 clones. STAT6, a transcriptional factor critical for IL-4/IL-13 signals, was included in these genes, and its product was augmented in bronchoalveolar lavage cells derived from asthma patients.

Mast cell

Mast cells are multifunctional effector cells, playing an important role in many inflammatory and pathologic processes of bronchial asthma. Cho *et al.* (2000, 2003) tried to identify inducible genes in a human mast cell line, HMC-1, activated by phorbol ester (PMA) and calcium ionophore, finding that expression of PAI-1 and activin A was up-regulated. Expression of both PAI-1 and activin A was observed in lung mast cells derived from asthma patients. Induction of PAI-1 in asthmatic monkey lung was observed by Zou *et al.* (2002). PAI-1 inhibits the plasminogen activator converting plasminogen to plasmin, which enhances proteolytic degradation of the extracellular matrix. Activin A is a member of the TGF-β superfamily, involved in inflammatory and wound repairing processes. These results have indicated that activated mast cells have an important role in airway remodeling by secreting PAI-1 and/or activin A.

Eosinophil

Eosinophil infiltration is another cardinal feature of bronchial asthma. Eosinophils play a critical role in airway remodelling of bronchial asthma (Humbles *et al.*, 2004; Lee *et al.*, 2004). In contrast, IL-5 is a potent factor for survival of eosinophils. To clarify the anti-apoptotic mechanism of IL-5 on eosinophils, Temple *et al.* (2001) combined microarray analyses of IL-5-induced genes in primary eosinophils and downregulated genes in IL-5-withdrawn TF-1.8 cells. Four genes – Pim-1, DSP-5, CD24 and SLP-76 – were included in both groups. These four gene products are either anti-apoptotic factors or have a strong role in anti-apoptotic signalling pathways. These results suggested that IL-5 exerts anti-apoptotic actions on eosinophils by inducing these gene products.

Monocyte

Antigen-presenting cells such as dendritic cells and macrophages recognize various allergens invading the body, and present these to naive T cells, inducing T cell activation and differentiation. To clarify activation or differentiation mechanisms of macrophages, Welch *et al.* (2002) analysed expression genes in thioglycolate-elicited macrophages stimulated by IL-4. Expression of Ym1 and arginase showed the highest induction. Ym1 might act as a chitinase, protecting against chitin-containing microorganisms such as parasite, fungi and bacteria. This induction of a chitinase-like protein may coincide with induction of acidic mammalian chininase (AMCase) by IL-13 in epithelial cells and macrophages (Zhu *et al.*, 2004). Arginase is involved in NO synthesis, and this result is in accordance with the report of Zimmermann *et al.* (2003).

Diesel exhaust particles (DEP), constituting a fraction of an aerodynamic diameter of less than 10 μm, are coated with polycyclic aromatic hydrocarbons and nitropolyaromatic hydrocarbons. DEP is thought to be an important air pollutant, exacerbating allergic inflammation. To analyse the effects of DEP on macrophages, Verheyen *et al.* (2004) studied genes expressed in a human monocytic cell line, THP-1 cells, exposed to DEP. They found that several genes changed their expression and particularly that CYP1B1 correlated with metabolism of polycyclic aromatic hydrocarbons was induced.

Nonimmune cell

Bronchial epithelial cell and lung fibroblast

IL-13 is a member of the Th2-type cytokines, thought to play key roles in the pathogenesis of bronchial asthma (Corry, 1999; Izuhara, 2003; Wills-Karp and Chiaramonte, 2003). Administration of IL-13 in mice has shown that IL-13 induces asthma-like phenotypes independent of lymphocytes (Grünig *et al.*, 1998; Wills-Karp *et al.*, 1998). In particular, the action of IL-13 on bronchial epithelial cells (BECs) is crucial for generation of airway hyper-responsiveness and mucous production (Zhu *et al.*, 1999; Kuperman *et al.*, 2002). Furthermore, in humans, the *IL13* variant Gln110Arg is genetically associated with bronchial asthma (Heinzmann *et al.*, 2000; Arima *et al.*, 2002). Therefore, great attention has been paid to identifying IL-13-inducible genes in resident cells of the bronchial tissue, particularly BECs. Lee *et al.* (2001) analysed IL-13-inducible genes in BECs, bronchial smooth muscle cells (BSMCs) and lung fibroblasts. The induced genes in BECs included OTF-2 (transcription factor), bigycan, tenascin-C (extracellular matrix protein), SCCA2, cathepsin C (protease/protease inhibitor) and phosphotyrosine phasphatase MEG (signalling molecule). The induced genes in BSMCs included JNK1b2, Fgr (signalling molecule), CXCR2 (chemokine receptor), sarcolipin, dystroglycan-associated protein, smooth muscle myosin heavy chain, (contractile protein), KCNQ2 (ion channel) and bFGF (secreted protein). Furthermore, the induced genes in lung fibroblasts included MCP-1 and IL-6 (secreted protein). Very few genes overlapped in expression profiling of these three kinds of cells.

IL-4 is another Th2 cytokine sharing receptor and signalling pathways with IL-13 (Izuhara, Arima and Yasunaga, 2002). STAT6 is a critical transcription factor for both

IL-4 and IL-13, so biological activities of these cytokines are similar. Therefore, we identified the genes induced by either IL-13 or IL-4 in human BECs using the microarray approach and found that 12 genes overlapped between two conditioned analyses (Yuyama *et al.*, 2002). The overlapped genes were SCCA1, SCCA2, carboxypeptidase M, cathepsin C (protease/protease inhibitor), periostin, tenascin C (extracellular matrix protein), IL-13Rα2, endothelin-A receptor (receptor), CYP1B1, carbonic anhydrase II (metabolizing enzyme), KAL1 (secreted protein) and DD96 (unknown protein) (Yuyama *et al.*, 2002). Induction of SCCA2, cathepsin C and tenascin C was also observed in other microarray analyses (Lee *et al.*, 2001; Laprise *et al.*, 2004). We confirmed that expression of four genes among these 12 – SCCA1, SCCA2, KAL1 and DD96 – was upregulated or present in asthma samples taken by bronchial biopsies. Serum levels of SCCA were high in asthmatic children particularly at the attack phase, and correlated with the serum level of IL-13 (Yuyama *et al.*, 2002; Nishi *et al.*, 2005). These results suggested that SCCA is a novel biomarker for bronchial asthma. We recently found that SCCA is also a good biomarker for atopic dermatitis (Mitsuishi *et al.*, in press). It is known that SCCA1 and SCCA2 are serine/cysteine proteinase inhibitors; however, the roles of these molecules in the pathogenesis of bronchial asthma remain obscure. We found that SCCA2 inhibited the cysteine proteinase activity of a major mite allergen, Der p 1 (Sakata *et al.*, 2004), indicating the possibility that these molecules may have a protective role against an extrinsic proteinase activity derived from microorganisms. Furthermore, IL-13Rα2 acts as a decoy receptor, inhibiting the IL-13 signal (Bernard *et al.*, 2001; Kawakami *et al.*, 2001; Arima *et al.*, 2002; Wood *et al.*, 2003; Chiaramonte *et al.*, 2003; Yasunaga *et al.*, 2003). Together with the results of our microarray analysis, these findings suggest that induction of IL-13Rα2 is associated with negative feedback regulation for the IL-13 signal in BECs.

Bronchial smooth muscle cell

In addition to the report of Lee *et al.* (2001), Syed *et al.* (2005) analysed genes expressed in human BSMCs stimulated by IL-13. Expression of VCAM-1 (adhesion molecule), IL-13Rα2, tenascin C and histamine H1 receptor was augmented, although the fold changes were low. The identified genes did not overlap with those reported by Lee *et al.* (2001).

18.4 Conclusions

In summary, the microarray analyses that have been performed so far have provided us not only with information about gene expression profiling, but also some novel pathogenic mechanisms of bronchial asthma. Application of microarray technology, using either broad or specific ‘sample source’ or ‘study condition’, has both advantages and disadvantages in clarifying the pathogenesis of bronchial asthma. It is of great importance to combine each result to further characterize the function of the gene products in bronchial asthma.

Acknowledgements

We thank Dr Dovie R. Wylie for critical review of this manuscript. This work was supported in part by a Research Grant for Immunology, Allergy and Organ Transplant from the Ministry of Health and Welfare of Japan, a grant-in-aid for Scientific Research from the Ministry of Education, Science, Sports and Culture of Japan.

References

Arima K, Umeshita-Suyama R, Sakata Y, Akaiwa M, Mao XQ, Enomoto T, Dake Y, Shimazu S, Yamashita T, Sugawara N, Brodeur S, Geha R, Puri RK, Sayegh MH, Adra CN, Hamasaki N, Hopkin JM, Shirakawa T, Izuhara K. 2002. Upregulation of IL-13 concentration *in vivo* by the IL13 variant associated with bronchial asthma. *J Allergy Clin Immunol* **109**: 980–987.

Banerjee SK, Young HW, Volmer JB, Blackburn MR. 2002. Gene expression profiling in inflammatory airway disease associated with elevated adenosine. *Am J Physiol Lung Cell Mol Physiol* **282**: L169–182.

Bernard J, Treton D, Vermot-Desroches C, Boden C, Horellou P, Angevin E, Galanaud P, Wijdenes J, Richard Y. 2001. Expression of interleukin 13 receptor in glioma and renal cell carcinoma: IL13Rα2 as a decoy receptor for IL13. *Lab Invest* **81**: 1223–1231.

Bodey KJ, Semper AE, Redington AE, Madden J, Teran LM, Holgate ST, Frew AJ. 1999. Cytokine profiles of BAL T cells and T-cell clones obtained from human asthmatic airways after local allergen challenge. *Allergy* **54**: 1083–1093.

Brutsche MH, Brutsche IC, Wood P, Brass A, Morrison N, Rattay M, Mogulkoc N, Simler N, Craven M, Custovic A, Egan JJ, Woodcock A. 2001. Apoptosis signals in atopy and asthma measured with cDNA arrays. *Clin Exp Immunol* **123**: 181–187.

Butte A. 2002. The use and analysis of microarray data. *Nat Rev Drug Discov* **1**: 951–960.

Chiaramonte MG, Mentink-Kane M, Jacobson BA, Cheever AW, Whitters MJ, Goad ME, Wong A, Collins M, Donaldson DD, Grusby MJ, Wynn TA. 2003. Regulation and function of the interleukin 13 receptor α 2 during a T helper cell type 2-dominant immune response. *J Exp Med* **197**: 687–701.

Cho SH, Tam SW, Demissie-Sanders S, Filler SA, Oh CK. 2000. Production of plasminogen activator inhibitor-1 by human mast cells and its possible role in asthma. *J Immunol* **165**: 3154–3161.

Cho SH, Yao Z, Wang SW, Alban RF, Barbers RG, French SW, Oh CK. 2003. Regulation of activin A expression in mast cells and asthma: its effect on the proliferation of human airway smooth muscle cells. *J Immunol* **170**: 4045–4052.

Churchill GA. 2002. Fundamentals of experimental design for cDNA microarrays. *Nat Genet* **32**: 490–495.

Corry DB. 1999. IL-13 in allergy: home at last. *Curr Opin Immunol* **11**: 610–614.

Dudda-Subramanya R, Lucchese G, Kanduc D, Sinha AA. 2003. Clinical applications of DNA microarray analysis. *J Exp Ther Oncol* **3**: 297–304.

Grünig G, Warnock M, Wakil AE, Venkayya R, Brombacher F, Rennick DM, Sheppard D, Mohrs M, Donaldson DD, Locksley RM, Corry DB. 1998. Requirement for IL-13 independently of IL-4 in experimental asthma. *Science* **282**: 2261–2263.

Heinzmann A, Mao XQ, Akaiwa M, Kreomer RT, Gao PS, Ohshima K, Umeshita R, Abe Y, Braun S, Yamashita T, Roberts MH, Sugimoto R, Arima K, Arinobu Y, Yu B, Kruse S, Enomoto T, Dake Y, Kawai M, Shimazu S, Sasaki S, Adra CN, Kitaichi M, Inoue H, Yamauchi K, Tomichi N, Kurimoto F, Hamasaki N, Hopkin JM, Izuhara K, Shirakawa T, Deichmann KA. 2000. Genetic variants of IL-13 signalling and human asthma and atopy. *Hum Mol Genet* **9**: 549–559.

Holgate ST. 1999. The epidemic of allergy and asthma. *Nature* **402**: B2–4.

Huang SK, Xiao HQ, Kleine-Tebbe J, Paciotti G, Marsh DG, Lichtenstein LM, Liu MC. 1995. IL-13 expression at the sites of allergen challenge in patients with asthma. *J Immunol* **155**: 2688–2694.

Humbert M, Durham SR, Kimmitt P, Powell N, Assoufi B, Pfister R, Menz G, Kay AB, Corrigan CJ. 1997. Elevated expression of messenger ribonucleic acid encoding IL-13 in the bronchial mucosa of atopic and nonatopic subjects with asthma. *J Allergy Clin Immunol* **99**: 657–665.

Humbles AA, Lloyd CM, McMillan SJ, Friend DS, Xanthou G, McKenna EE, Ghiran S, Gerard NP, Yu C, Orkin SH, Gerard C. 2004. A critical role for eosinophils in allergic airways remodeling. *Science* **305**: 1776–1779.

Izuhara K. 2003. The role of interleukin-4 and interleukin-13 in the non-immunologic aspects of asthma pathogenesis. *Clin Chem Lab Med* **41**: 860–864.

Izuhara K, Arima K, Yasunaga S. 2002. IL-4 and IL-13: their pathological roles in allergic diseases and their potential in developing new therapies. *Curr Drug Targets Inflamm Allergy* **1**: 263–269.

Izuhara K, Arima K, Yuyama N, Sakata Y, Masumoto K. 2004. Application of functional genomics to bronchial asthma. *Curr Pharmacogenom* **2**: 351–356.

Karp CL, Grupe A, Schadt E, Ewart SL, Keane-Moore M, Cuomo PJ, Kohl J, Wahl L, Kuperman D, Germer S, Aud D, Peltz G, Wills-Karp M. 2000. Identification of complement factor 5 as a susceptibility locus for experimental allergic asthma. *Nat Immunol* **1**: 221–226.

Kawakami K, Taguchi J, Murata T, Puri RK. 2001. The interleukin-13 receptor $\alpha 2$ chain: an essential component for binding and internalization but not for interleukin-13-induced signal transduction through the STAT6 pathway. *Blood* **97**: 2673–2679.

Kotsimbos TC, Ernst P, Hamid QA. 1996. Interleukin-13 and interleukin-4 are coexpressed in atopic asthma. *Proc Assoc Am Physicians* **108**: 368–373.

Kuperman DA, Huang X, Koth LL, Chang GH, Dolganov GM, Zhu Z, Elias JA, Sheppard D, Erle DJ. 2002. Direct effects of interleukin-13 on epithelial cells cause airway hyperreactivity and mucus overproduction in asthma. *Nat Med* **8**: 885–889.

Laprise C, Sladek R, Ponton A, Bernier MC, Hudson TJ, Laviolette M. 2004. Functional classes of bronchial mucosa genes that are differentially expressed in asthma. *BMC Genom* **5**: 21–30.

Lee JH, Kaminski N, Dolganov G, Grunig G, Koth L, Solomon C, Erle DJ, Sheppard D. 2001. Interleukin-13 induces dramatically different transcriptional programs in three human airway cell types. *Am J Respir Cell Mol Biol* **25**: 474–485.

Lee JJ, Dimina D, Macias MP, Ochkur SI, McGarry MP, O'Neill KR, Protheroe C, Pero R, Nguyen T, Cormier SA, Lenkiewicz E, Colbert D, Rinaldi L, Ackerman SJ, Irvin CG, Lee NA. 2004. Defining a link with asthma in mice congenitally deficient in eosinophils. *Science* **305**: 1773–1776.

Li XD, Essayan DM, Liu MC, Beaty TH, Huang SK. 2001. Profiling of differential gene expression in activated, allergen-specific human Th2 cells. *Genes Immun* **2**: 88–98.

Mitsuishi K, Nakamura T, Sakata Y, Yuyama N, Arima K, Yugi S, Suto H, Izuhara K, Ogawa H. The squamous cell carcinoma antigens as relevant biomarkers of atopic dermatitis, *Clin Exp Allergy*, in press.

Nishi N, Miyazaki M, Tsuji K, Hitomi T, Muro E, Zaitsu M, Yamamoto S, Inada S, Kobayashi I, Ichimaru T, Izuhara K, Nagumo F, Yuyama N, Hamasaki Y. 2005. Squamous cell carcinoma-related antigen (SCCA) in children with acute asthma. *Ann Allergy Asthma Immunol* **94**: 391–397.

Robinson DS, Hamid Q, Ying S, Tsicopoulos A, Barkans J, Bentley AM, Corrigan C, Durham SR, Kay AB. 1992. Predominant T_H2-like bronchoalveolar T-lymphocyte population in atopic asthma. *New Eng J Med* **326**: 298–304.

Russo G, Zegar C, Giordano A. 2003. Advantages and limitations of microarray technology in human cancer. *Oncogene* **22**: 6497–6507.

Sakata Y, Arima K, Takai T, Sakurai W, Masumoto K, Yuyama N, Suminami Y, Kishi F, Yamashita T, Kato T, Ogawa H, Fujimoto K, Matsuo Y, Sugita Y, Izuhara K. 2004. The squamous cell carcinoma antigen 2 inhibits the cysteine proteinase activity of a major mite allergen, Der p 1. *J Biol Chem* **279**: 5081–5087.

Sevenet N, Cussenot O. 2003. DNA microarrays in clinical practice: past, present, and future. *Clin Exp Med* **3**: 1–3.

Syed F, Panettieri RA, Jr, Tliba O, Huang C, Li K, Bracht M, Amegadzie B, Griswold D, Li L, Amrani Y. 2005. The effect of IL-13 and IL-13R130Q, a naturally occurring IL-13 polymorphism, on the gene expression of human airway smooth muscle cells. *Respir Res* **1**: 9.

Temple R, Allen E, Fordham J, Phipps S, Schneider HC, Lindauer K, Hayes I, Lockey J, Pollock K, Jupp R. 2001. Microarray analysis of eosinophils reveals a number of candidate survival and apoptosis genes. *Am J Respi Cell Mol Biol* **25**: 425–433.

Umetsu DT, Akbari O, Dekruyff RH. 2003. Regulatory T cells control the development of allergic disease and asthma. *J Allergy Clin Immunol* **112**: 480–487.

Verheyen GR, Nuijten JM, Van Hummelen P, Schoeters GR. 2004. Microarray analysis of the effect of diesel exhaust particles on *in vitro* cultured macrophages. *Toxicol in Vitro* **3**: 377–391.

Welch JS, Escoubet-Lozach L, Sykes DB, Liddiard K, Greaves DR, Glass CK. 2002. TH2 cytokines and allergic challenge induce Ym1 expression in macrophages by a STAT6-dependent mechanism. *J Biol Chem* **277**: 42821–42829.

Williams M. 2003. Target validation. *Curr Opin Pharmacol* **3**: 571–577.

Wills-Karp M. 2001. IL-12/IL-13 axis in allergic asthma. *J Allergy Clin Immunol* **107**: 9–18.

Wills-Karp M, Chiaramonte M. 2003. Interleukin-13 in asthma. *Curr Opin Pulmou Med* **9**: 21–27.

Wills-Karp M, Luyimbazi J, Xu X, Schofield B, Neben TY, Karp CL, Donaldson DD. 1998. Interleukin-13: central mediator of allergic asthma. *Science* **282**: 2258–2261.

Wood N, Whitters MJ, Jacobson BA, Witek J, Sypek JP, Kasaian M, Eppihimer MJ, Unger M, Tanaka T, Goldman SJ, Collins M, Donaldson DD, Grusby MJ. 2003. Enhanced interleukin (IL)-13 responses in mice lacking IL-13 receptor α 2. *J Exp Med* **197**: 703–709.

Xiang Z, Yang Y, Ma X, Ding W. 2003. Microarray expression profiling: analysis and applications. *Curr Opin Drug Discov Devel* **6**: 384–395.

Yasunaga S, Yuyama N, Arima K, Tanaka H, Toda S, Maeda M, Matsui K, Goda C, Yang Q, Sugita Y, Nagai H, Izuhara K. 2003. The negative-feedback regulation of the IL-13 signal by the IL-13 receptor α2 chain in bronchial epithelial cells. *Cytokine* **24**: 293–303.

Yuyama N, Davies DE, Akaiwa M, Matsui K, Hamasaki Y, Suminami Y, Lu Yoshida N, Maeda M, Pandit A, Lordan JL, Kamogawa Y, Arima K, Nagumo F, Sugimachi M, Berger A, Richards I, Roberds SL, Yamashita T, Kishi F, Kato H, Arai K, Ohshima K, Tadano J, Hamasaki N, Miyatake S, Sugita Y, Holgate ST, Izuhara K. 2002. Analysis of novel disease-related genes in bronchial asthma. *Cytokine* **19**: 287–296.

Zhu Z, Homer RJ, Wang Z, Chen Q, Geba GP, Wang J, Zhang Y, Elias JA. 1999. Pulmonary expression of interleukin-13 causes inflammation, mucus hypersecretion, subepithelial fibrosis, physiologic abnormalities, and eotaxin production. *J Clin Invest* **103**: 779–788.

Zhu Z, Zheng T, Homer RJ, Kim YK, Chen NY, Cohn L, Hamid Q, Elias JA. 2004. Acidic mammalian chitinase in asthmatic Th2 inflammation and IL-13 pathway activation. *Science* **304**: 1678–1682.

Zimmermann N, King NE, Laporte J, Yang M, Mishra A, Pope SM, Muntel EE, Witte DP, Pegg AA, Foster PS, Hamid Q, Rothenberg ME. 2003. Dissection of experimental asthma with DNA microarray analysis identifies arginase in asthma pathogenesis. *J Clin Invest* **111**: 1863–1874.

Zou J, Young S, Zhu F, Gheyas F, Skeans S, Wan Y, Wang L, Ding W, Billah M, McClanahan T, Coffman RL, Egan R, Umland S. 2002. Microarray profile of differentially expressed genes in a monkey model of allergic asthma. *Genome Biol* **3**: research0020.0021–0020.0013.

19

Genomic Investigation of Asthma in Human and Animal Models

Csaba Szalai

Abstract

Asthma is a pulmonary disease characterized by increased bronchial responsiveness to a variety of stimuli. It is the most common chronic disease of childhood and the most frequent reason for paediatric hospital admission, and its incidence is on the rise. Several studies have suggested that asthma is a multifactorial disease influenced by genetic and environmental factors. In this review I will systematically summarize collective evidence from linkage and association studies that have consistently reported suggestive linkage or association of asthma or its associated phenotypes to polymorphic markers and single nucleotide polymorphisms in selected chromosomes. Next, I will show some results from animal models of asthma investigating the pathomechanism and the genomic background of the disease.

19.1 Introduction

Asthma is a pulmonary disease characterized by intermittent narrowing of the small airways of the lung with subsequent airflow obstruction, increased bronchial responsiveness to a variety of stimuli and symptoms of wheeze, cough and breathlessness. The majority of asthmatics are also atopic, with manifestation of allergic diathesis including clinical allergy to aeroallergens and foods, or subclinical allergy manifest by skin test reactivity to allergen or elevated serum IgE. Allergic asthma can present for the first time at any age, but the incidence is highest in children (Dodge and Burrows, 1980). It is the most common chronic disease of childhood and the most frequent reason for paediatric hospital admission, and its incidence is on the rise (Mannino *et al.*, 1998). A doctor-diagnosed asthma in populations with developed health services is typically reported by about 5 per cent (Janson *et al.*, 1997) of those aged 20–44 years and in more than 10 per cent of children (ISAAC Steering

Immunogenomics and Human Disease Edited by András Falus

Committee, 1998). Prevalence rates tend to be highest in economically developed countries with a temperate climate and low in rural subsistence and economically developing communities, and they increase with adoption of a more affluent lifestyle (Janson *et al.*, 1997). The increased prevalence of asthma over time in the developed world seems to be part of a generalized trend of increasing prevalence of allergic sensitization and allergic disease.

One increasingly popular explanation for the rise in disease prevalence is referred to as the 'hygiene hypothesis'. This hypothesis states that early childhood infections alter the TH1/TH2 balance or directly prevent TH2-mediated immune responses and so inhibit the tendency to develop allergic disease (Strachan, 1989). Regardless of the specific nature of the changing environmental influences, it is most probable that changes in environmental exposures have led to the expression of asthmatic phenotypes in previously unaffected, but genetically susceptible, individuals.

Previous studies suggest that asthma is a multifactorial disease influenced by genetic and environmental factors (CGSA, 1997). Studies of twins have shown generally that concordance rates for asthma are significantly higher in monozygotic twins than dizygotic twins, and that the heritability of asthma varies between 36 and 79 per cent (Weiss and Raby, 2004). Importantly, there is evidence that genetic liability for asthma, airway responsiveness and allergic traits are regulated through distinct loci, although there is probably some shared overlap as well (Barnes, 2000).

Given the likely presence of genes of strong effect, it is a reasonable expectation that understanding the genetics of asthma will lead to improvements in its diagnosis, prevention and treatment. As a result, programmes aimed at the discovery of genes that predispose individuals to this illness are being carried out worldwide. Studies on the genetics of asthma are hampered by the fact that there is no standard definition of asthma (Tattersfield *et al.*, 2002). Attempts to define asthma have generally resulted in descriptive statements invoking notions of variable airflow obstruction over short periods of time, sometimes in association with markers of airway hyperresponsiveness (AHR) and cellular pathology of the airway; they have not, however, provided validated quantitative criteria for these characteristics to enable diagnosis of asthma to be standardized for clinical, epidemiological or genetic purposes. For this reason, investigators have defined and commonly used objective quantitative traits, such as total and specific immunoglobulin E (IgE) levels, AHR and skin prick test, as surrogate markers of asthma. The danger in using these intermediate phenotypes is the assumption that their genetic basis is the same as that of the disease and that they represent the full range of disease states. For example, although atopy is one of the strongest risk factors for asthma, it alone is not sufficient to induce asthma, as many atopic individuals do not have asthmatic symptoms.

The other difficulty that hampers the efforts to identify the specific genes involved in asthma is the multigenic nature of the disease. This means that more than one gene in each individual might interact to produce the disease phenotype (polygenic inheritance), different disease alleles might exist in different individuals (genetic heterogeneity) and interaction with the environment might lead to incomplete penetrance. An individual might also develop disease owing to environmental factors alone (phenocopy). By contrast with single-gene disorders, genes that predispose to asthma will not usually contain mutations that lead to a gross aberration in function. Most often they will be variants of normal genes, the evolutionary advantages of which have become obscure (Anderson and Cookson, 1999).

19.2 Methods for Localization of Asthma Susceptibility Genes

Despite these obstacles, tremendous effort has been expended over the past decade to find the genes and map the chromosomal locations that are involved in susceptibility to asthma. There are two general methods possible for identifying genetic effects on a disease. The first is the 'candidate gene approach', the second is the 'genome-wide screen approach' also known as 'positional cloning'.

'Candidate genes' are selected because their biological function suggests that they could play a role in the pathogenesis of asthma. Association studies between variation in these candidate genes and asthma-related phenotypes are mostly conducted in unrelated case and unrelated control samples by comparing allele or genotype frequencies between samples. Controls for association studies are often derived from the population that shares ethnic or geographic similarities with the cases. Alternatively, family-based controls can be used, in which control subjects are selected from families of affected probands. The advantages of association studies include their independence from inheritance models, superior power to detect susceptibility genes and their applicability to populations as well as families. This approach is powerful if the susceptibility gene is indeed one of the candidates selected for study. However, because there is a multitude of potential candidate genes for a complex disease such as asthma, the work involved in a comprehensive candidate-gene approach might be overwhelming without prior support from genome-wide screens. In addition, as it is now recognized that the results of individual polymorphism studies might be misleading, the candidate-gene approach has evolved such that multiple variants are evaluated simultaneously (Wills-Karp and Ewart, 2004).

In contrast to the candidate gene studies, the genome-wide screens do not require prior knowledge of the pathomechanism of the disease. The genome-wide screen method investigates the whole genome using microsatellite markers [or in the future single-nucleotide polymorphisms (SNPs)] through linkage analysis. Genetic linkage analysis is performed by LOD (logarithm of odds) score analysis. This method assesses the likelihood that a trait cosegregates with a marker, which is expressed as an LOD score, i.e. the log of the ratio of the likelihood of linkage and the likelihood of no linkage. A value of $+3$ is traditionally taken as evidence for linkage and a value of -2 is considered evidence against linkage.

The hypothesis-independent nature of this approach means that it might more reliably identify susceptibility genes, particularly those in pathways that are not obviously implicated in the asthma phenotype. However, genome-wide screens are costly, labour-intensive and can suffer from lack of statistical power. In addition, when this approach identifies linkage regions, they are generally broad chromosomal regions that contain many genes that could potentially be the source of the effect. So the move from broad linkage regions to causal genes requires additional approaches, including fine mapping and positional cloning of genes within these narrowed regions (Wills-Karp and Ewart, 2004).

Statistical power has been an important problem in conducting genome-wide screens for complex genetic diseases such as asthma. Clearly, the best solution is to study large populations. However, this is not always feasible. Nonetheless, the few studies that have met the stringent requirements for genome-wide linkage have been those that have studied large, more homogenous populations. The lack of replication of some regions in independent studies might reflect true genetic heterogeneity, insufficient power to detect linkages in replicative studies, differences in the strategies for collecting subjects or differences in the phenotype definition between the individual studies. Indeed, a recent study that compared linkage data derived from differing ethnic backgrounds – in which the ascertainment schemes, phenotypic definitions and analytic approaches were held constant – underlines the importance of genetic heterogeneity in defining asthma genes. Thus, there is a possibility that genes identified in homogenous populations will not be applicable to other ethnic backgrounds. Nonetheless, as I discuss below, follow-up studies of several of these chromosomal regions have yielded exciting new asthma genes.

19.3 Results of the Association Studies and Genome-Wide Screens in Humans

Until the beginning of 2005 more than 300 gene association studies for asthma were published and they identified more than 70 genes as potential susceptibility loci (Hoffjan, Nicolae and Ober, 2003). Some genes were associated with asthma phenotypes rather consistently across studies and populations. In particular, variation in eight genes has been associated with asthma phenotypes in five or more studies: interleukin-4 (IL-4), interleukin-13 (IL-13), β2 adrenergic receptor (ADRB2), human leukocyte antigen (HLA) DRB1, tumour necrosis factor-α (TNF-α), lymphotoxin-α (LTA), high-affinity IgE receptor (FcεRI-β) and IL-4 receptor (IL-4R). These loci probably represent true asthma or atopy susceptibility loci or genes important for disease modification. Chromosomal localization and possible function of some candidate genes in asthma and related diseases are presented in Table 19.1.

The genetic complexity of asthma is underlined by the results of genome-wide screens, which have highlighted 20 genomic regions as being likely to contain asthma susceptibility genes. The size of these linked regions (10–20 million base pairs) means that they typically harbour hundreds of candidate genes. For the most part, each study has indicated that several loci are linked to asthma or related traits, supporting the multigenic model for the disease. However, few of the reported linkages have met the accepted criteria for significant genome-wide linkage. This lack of clear front-runners has made it difficult to set priorities for narrowing gene regions and gene discovery. Despite this, many of these linkages have been replicated in multiple screens. The consistency with which these regions have been detected in asthma scans indicates that they might indeed contain asthma-susceptibility genes.

Next I will systematically summarize collective evidence from linkage and association studies that have consistently reported suggestive linkage or association of asthma or its associated phenotypes to polymorphic markers and SNPs in selected chromosomes.

Chromosome 2

Evidence for linkage of asthma and related phenotypes to chromosome 2q arm has been reported in several studies (Wjst *et al.*, 1999; Allen *et al.*, 2003). Mouse genome screens have also linked AHR to the region homologous to 2q in the human (De Sanctis *et al.*, 1995).

This 2q14 region includes the IL-1 gene family. Single-marker, two-locus and three-locus haplotype analysis of SNPs yielded several significant results for asthma ($p < 0.05$–0.0021) for the human IL1RN gene encoding the IL-1 receptor antagonist protein, an anti-inflammatory cytokine that plays an important role in maintaining the balance between inflammatory and anti-inflammatory cytokines (Gohlke *et al.*, 2004). The study was carried out in a German population and was replicated and confirmed in an independent Italian family sample. Furthermore, a single G/T base exchange at +4845 in exon 5 of the *IL1A* gene results in an amino acid substitution of alanine for serine. This SNP and a haplotype of the *IL1A*, *IL1B* and *IL1RN* genes containing the same SNP as *IL1A* have recently been shown to be associated with atopy in nonasthmatic adults.

The 2q33 region harbours the candidate gene cytotoxic T-lymphocyte antigen 4 (CTLA-4), an important regulator of T-cell activation and differentiation. Transmission disequilibrium test analysis showed that several SNPs in the CTLA-4 gene were significantly associated with serum IgE levels, allergy, asthma and FEV1 per cent (forced expiratory volume 1 s) predicted below 80 per cent, but not with AHR, and CTLA-4 polymorphisms of potentially direct pathogenic significance in atopic disorders were identified. The particular SNP alleles found to be positively associated with these phenotypes were previously shown to be associated negatively with autoimmune disorders. Since autoimmune disorders are TH1-skewed diseases and asthma and atopic diseases are TH2 diseases, these data suggest a role for CTLA-4 polymorphisms in determining the TH1/TH2 balance (Munthe-Kaas *et al.*, 2004).

Allen and colleagues positionally cloned a novel asthma gene through an effort that was aimed at mining the candidate linkage region on 2q (Allen *et al.*, 2003). They found and replicated association between asthma and the D2S308 microsatellite, 800 kb distal to the IL1 cluster on 2q14. They sequenced the surrounding region and constructed a comprehensive, high-density, SNP linkage disequilibrium map. The strongest associations were with SNP WTC122, in close proximity to D2S308.

Table 19.1 Chromosomal localization and possible function of candidate genes in asthma, and related phenotypes

Chromosomal region	Candidate gene	Function	Phenotype
2q14	IL-1 gene family	Influencing inflammatory response	Asthma, atopy
2q33	CTLA-4	Regulator of T-cell activation and differentiation	Asthma, high IgE
5q31–q33	IL-4, IL-13, GM-CSF	IgE isotype switching, induction of Th2 response	High IgE, asthma, AHR
	IL-5	Eosinophil activation, maturation	Asthma
	IL-9	Role in T, B and mast cell functions	Asthma
	SPINK5	Possible epithelial differentition	Atopy, asthma
	CD14	Bacterial LPS binding receptor	High IgE, atopy
	TIM1, TIM3	Th1, Th2 differentiation	Asthma
	ADRB2	Influencing the effect of β2-agonists and smoking	Asthma
	Leukotriene C_4 synthase	Enzyme for leukotriene synthesis	Aspirin-intolerant asthma, asthma
6p21.3	HLA-D	Antigen presentation	Specific IgE
	TNFα	Proinflammatory cytokine	Asthma
	LTA	Induces the expression of cell adhesion molecules and cytokines	Asthma
7p	GPRA	Unknown	Asthma, atpoy
11q13	FcεRI-β	High-affinity IgE receptor	Atopy, asthma
	CC16	Regulation of airway inflammation	Asthma
11p	ETS-2, ETS-3	Transcription factors	Asthma
12q14.3–q24.31	INF-γ	Inhibition of IL-4 activity	Asthma, atopy, high IgE
	SCF	IL-4 production, mast cell maturation	
	STAT6	Cytokine regulated transcriptin factor	
	NFY-β	Elevation of IL-4 and HLA-D gene transcription	
	NNOS	NO: vasodilation, inflammatory regulation	Asthma
13q	PHF11, (SETDB2, RCBTB1(?))	Transcriptional regulation (?)	High IgE
14q11.2–q13	T cell receptor α and γ	Interaction with MHC–peptide complex	High specific IgE
14q22.1	DP2R	T cell chemotaxis	Asthma
16p21	IL-4R	α subunit is part of the receptor for IL-4 and IL-13	Atopy, asthma

(*Continued*)

Table 19.1 (*Continued*)

Chromosomal region	Candidate gene	Function	Phenotype
17q11.2	RANTES, MCP-1, eotaxin	Attracting and stimulating of leukocytes	Asthma
20p13	ADAM33	Unknown	Asthma, AHR

Abbreviations in alphabetical order: ADRB2, β2 adrenergic receptor; CC16, Clara cell protein 16; CTLA-4, cytotoxic T-lymphocyte antigen 4; DP2R, prostanoid D2 receptor; ETS, epithelium-specific transcription factor; FcεRI-β, high-affinity IgE receptor β subunit; GM-CSF, granulocyte–macrophage colony-stimulating factor; GPRA, G protein-coupled receptor for asthma susceptibility; HLA, human leukocyte antigen; IL, interleukin; INF, interferon; LTA, lymphotoxin-α; MCP-1, monocyte chemoattractant protein-1; MHC, major histocompatibility complex; NFY-β, β-subunit of nuclear factor Y; NNOS, neuronal nitric oxid synthase; PHF11, plant homeodomain finger protein-11; RANTES, regulated on activation normal T cell expressed and secreted; SCF, stem cell factor; STAT6, signal tranducer and activator of transcription 6; TIM, T-cell integrin mucin-like receptor; TNF, tumour necrosis factor.

Weaker associations with SNPs at the other end of this linkage disequilibrium (LD) block were found to be due to their association with WTC122 on a shared haplotype, which also included the asthma-associated D2S308*3 allele. After an extensive search for the gene that contains the associated SNPs, they identified DPP10. This gene encodes a homologue of dipeptidyl peptidases (DPPs), which are thought to cleave terminal dipeptides from various proteins. Based on homology of this gene to other members of this family, the authors speculate that DPP10 regulates the activity of various chemokine and cytokine genes by removing N-terminal dipeptides from them in a proline-specific manner. They suggest that DPP10 might cleave various pro-inflammatory and regulatory chemokines and cytokines. If this is the case, DPP10 might modulate inflammatory processes in the airways. As comparisons between DPP10 expression in asthmatic and normal tissues are yet to be done, it is still unclear whether differences in the quantity or pattern of expression of DPP10 will be associated with asthma.

Chromosome 5

After an original observation of genetic linkage of total IgE levels to the 5q31 region in extended Amish pedigrees and confirmation of linkage to the same region, chromosome 5q31–33 has become one of the most studied candidate asthma regions (Marsh *et al.*, 1994). It contains the cytokine gene cluster that plays an important role in the pathomechanism of asthma and atopic disorders.

IL-4 is important in IgE isotype switching and the regulation of allergic inflammation. The 3017 G/T variant of IL-4 or the haplotype it identifies was found significantly influence IL-4's ability to modulate total serum IgE levels. Large-scale association studies in 1120 German schoolchildren were conducted to determine the effect of all polymorphisms present in the IL-4 gene on the phenotypic expression of atopic diseases. A total of 16 polymorphisms were identified in the IL-4 gene. A significant association between a cluster of polymorphisms in strong linkage disequilibrium with each other and a physician's diagnosis of asthma and total serum IgE levels was found (Kabesh *et al.*, 2003).

IL-13 is one of the major cytokines in asthma. It enhances mucus production, AHR and the production of the main eosinophil chemoattractant eotaxin. The receptors for IL-13 and IL-4 share a common α chain and the functions of the two cytokines overlap. Several polymorphisms were found in the IL-13 gene. The most significant associations were observed to asthma, AHR and skin-test responsiveness with the −1111 promoter polymorphism. The Q110 IL-13 variant displayed significantly increased binding capacity to its receptor compared with R110 IL-13 and was associated with elevated IgE level and asthma (Chen *et al.*, 2004).

CD14 is located on chromosome 5q31 and it is a receptor that has specificity for lipopolysaccharides (LPS) and other bacterial wall-derived components. Engagement of CD14 by these bacterial components is associated with strong IL-12 responses by antigen-presenting cells, and IL-12 is regarded as an obligatory signal for the maturation of naive T cells into Thl cells. A C→T SNP at position −159 in the promoter of the gene encoding CD14 was found to be associated with increased levels of soluble CD14 and decreased total serum IgE.

The dissection of the molecular mechanisms that underlie the transcriptional effects of CD14–159C/T highlights some of the complex ways in which genetic variation can affect the expression of a critical allergy gene. A genetically determined increase in CD14 expression could result in enhanced LPS responsiveness in early life, when the relative balance between TH1- and TH2-mediated immunity is finely adjusted. Robust CD14-mediated reactions to pathogens would elicit strong IL-12/IL-18 expression by innate immune pathways. TH1 differentiation, rather than TH2 differentiation, would thus be favoured, decreasing the likelihood of vigorous IgE-dependent responses after allergen exposure. This role of CD14 in the TH1/TH2 balance can be one of the mechanisms of how infection in early life may be protective against allergic disease, and a possible explanation for the hygiene hypothesis.

The gene underlying Netherton disease (SPINK5) encodes a 15-domain serine proteinase inhibitor (LEKTI) which is expressed in epithelial and mucosal surfaces and in the thymus.

SPINK5 is at the distal end of the cytokine cluster on 5q31. A Glu420→Lys variant was found to be significantly associated with atopic dermatitis and atopy with weaker correlation with asthma in two independent panels of families (Walley *et al.*, 2001).

The 5q31–33 is an important pharmacogenomic region for asthma. β2–Agonists are used widely via inhalation for the relief of airway obstruction. These drugs act via binding to the β2 adrenergic receptor (ADRB2), a cell surface G protein-coupled receptor located on 5q32. Responses to this drug are currently the most investigated pharmacogenomic pathway in asthma. Two coding variants (at positions 16 and 27) within the ADRB2 gene have been shown *in vitro* to be functionally important. The Gly-16 receptor exhibits enhanced downregulation *in vitro* after agonist exposure. In contrast, Arg-16 receptors are more resistant to downregulation. Because of linkage disequilibrium, individuals who are Arg/Arg at position 16 are much more likely to be Glu/Glu at position 27; individuals who are Gly/Gly at position 16 are much more likely to be Gln/Gln at position 27. The position 27 genotypes influence but do not abolish the effect of the position 16 polymorphisms with regard to downregulation of phenotypes *in vitro*. Retrospective studies and prospective clinical trials have suggested that adverse effects occur in patients homozygous for arginine (Arg/Arg), rather than glycine (Gly/Gly), at position 16. Bronchodilator treatments avoiding β2-agonist may be appropriate for patients with the Arg/Arg genotype (Palmer *et al.*, 2002). Additionally, smoking subjects homozygous for Arg16 had an almost 8-times risk for developing asthma compared with non-smoking subjects with Gly/Gly genotype at position 16.

Leukotrienes, released by eosinophils, mast cells and alveolar macrophages, are among the main mediators in asthma, inducing airway obstruction, migration of eosinophils and proliferation of smooth muscle. Leukotriene C_4 (LTC4) synthase is a membrane-bound glutathione transferase expressed only by cells of haematopoietic origin and is a key enzyme in the synthesis of cys-LTs, converting LTA4 to LTC4. The gene encoding LTC4 synthase is located on 5q35. An adenine to cytosine transversion has been found 444 bp upstream (−444) of the translation start site of the LTC4 synthase gene and it has been reported that the polymorphic C −444 allele occurred more commonly in patients with aspirin-intolerant asthma (AIA). A 5-fold greater expression of LTC4 synthase has been demonstrated in individuals with AIA when compared with patients with aspirin-tolerant asthma; furthermore, the expression of LTC4 synthase mRNA has also been shown to be higher in blood eosinophils from asthmatic subjects compared with control subjects and was particularly increased in eosinophils from patients with AIA. In addition, it was found that, among subjects with asthma treated with zafirlukast (a leukotriene receptor antagonist), those homozygous for the A allele at the −444 locus had a lower FEV1 response than those with the C/C or C/A genotype (Palmer *et al.*, 2002).

Chromosome 6

The MHC region on chromosome 6p21.3 has shown consistent linkage to asthma-associated phenotypes in several studies and is considered to be a major locus influencing allergic diseases (Wjst *et al.*, 1999; Hakonarson and Wjst, 2001). This region contains many molecules involved in innate and specific immunity. The class II genes of the MHC have recognized influences on the ability to respond particular allergens. The strongest and most consistent association is between the minor component of ragweed antigen (Amb a V) and HLA-DR2. It was demonstrated that all but two of 80 white IgE responders to Amb a V carried HLA-DR2 and Dw2 (DR2.2). This was significantly higher than the frequency of this haplotype among nonresponders (approximately 22 per cent; Marsh *et al.*, 1989). Many other possible positive and negative associations of the MHC with allergen reactivity have been described. Stronger HLA effects may be seen when the antigen is small and contains a single or very few antigenic determinants. This may be the case with aspirin-induced asthma and DPB1*0301, and sensitivity to inhaled acid anhydrides and HLA-DR3 (Cookson, 1999).

Both class I and class III genes of the MHC as well as nonclassic MHC genes may also affect asthma through allergic or nonallergic pathways, respectively. The TNF-α gene is located on chromosome 6 between the class I and III clusters of the human MHC. It is a potent proinflammatory cytokine, which is found in excess in asthmatic airways. The −308A allele in the promoter region of the TNF-α gene is transcribed *in vitro* at seven times the rate of the −308G allele. Several reports found associations between the −308A allele and asthma.

The lymphotoxin-α (LTA) gene is located next to the TNF-α gene, and induces the expression of cell adhesion molecules and proinflammatory cytokines including E selectin, TNF-α and IL-1A and B. An intronic SNP (+252) in the LTA gene was found to be associated with increased expression of the gene and increased susceptibility to several inflammatory diseases, including asthma. These latter results emphasize the inflammatory nature of the asthmatic response, as distinct from its allergic basis.

Chromosome 7

The first published genome-wide scan in asthma suggested six tentative genetic loci, among them chromosome 7p, which was then implicated in a study of Finnish and Canadian families and confirmed in West Australian families (Daniels *et al.*, 1996; Laitinen *et al.*, 2001).

Using strategies of genetic mapping and positional cloning, Laitinen *et al.* (2004) identified new molecular players in asthma and allergy on chromosome 7p. They adopted a hierarchical genotyping design, leading to the identification of a 133 kb risk-conferring segment containing two genes. One of these coded for an orphan G protein-coupled receptor named GPRA (G protein-coupled receptor for asthma susceptibility), which showed distinct distribution of protein isoforms between bronchial biopsies from healthy and asthmatic individuals. In three cohorts from Finland and Canada, SNP-tagged haplotypes associated with high serum immunoglobulin E or asthma. The murine orthologue of GPRA was upregulated in a mouse model of ovalbumin-induced inflammation. The properties of GPRA make it a strong candidate for involvement in the pathogenesis of asthma. GPRA might act as a receptor for an unidentified ligand. The putative ligand, isoforms of GPRA and their putative downstream signalling molecules may define a new pathway that is critically altered in asthma.

Chromosome 11

Linkage of atopy to a genetic marker on chromosome 11q13 was first reported in 1989 (Cookson *et al.*, 1989). The β chain of the high-affinity receptor for IgE (FcεRI-β)

was subsequently localized to the region (Hill and Cookson, 1996). Affected sib-pair analysis showed that linkage of atopy to chromosome 11 markers was to maternal alleles in many families (Cookson *et al.*, 1992). Several coding and noncoding polymorphisms have been identified in the gene that encodes FcεRI-β. Ile/Leu181 and Val/Leu183 have been found in several populations. Maternal inheritance of both these variants was found to be associated with severe atopy. Ile/Leu181 has also been associated with levels of IgE in heavily parasitized Australian aborigines, implying a protective role for the gene in helminthiasis. Another polymorphism, E237G, was found to be associated with various measures of atopy, as well as bronchial reactivity to methacholine in the Australian population and with asthma in the Chinese population. No such associations were found in some other populations.

The 237G allele and a silent substitution in the nitric oxide synthase (NOS)2A gene were associated with reduced IL-13 responses in cord blood. A significant gene–gene interaction between FcεRI-β 237G and NOS2A D346D was detected, with individuals carrying the minor allele for both polymorphisms having the lowest cord blood IL-13 levels.

There has been much controversy over the linkage of atopy to chromosome 11q13. Several studies using a variety of genetic markers and phenotypes in populations with differing ethnic background have failed to reproduce the results of the earlier studies. In addition, groups (including ours, unpublished data) have been unable to identify the Ile/Leu181 and Val/Leu193 variants either by assay or by direct sequencing, leading to doubts over its existence. This might be due to duplicate sequences in a pseudogene or in a homologous gene that interfere with the assays, all of which rely on PCR amplification. Even in cases when associations were shown, the strength of the chromosome 11q13 linkage cannot be explained. This implies that there should be further functional polymorphisms in proximity to FcεRI-β.

One of the possible candidates is the Clara cell secretory protein (CC16), a 16- kDa protein, which is also located on 11q13 and primarily expressed in the respiratory tract by nonciliated bronchiolar secretory cells, accounting for 7 per cent of the total protein content in the bronchoalveolar lavage (BAL) fluid of healthy non-smokers. The immunomodulatory activity of CC16 has been well documented and CC16 mRNA levels have been proposed as markers of lung maturation and epithelial differentiation (Oshika *et al.*, 1998). The CC16 gene was screened for mutations and a polymorphism (A38G) was identified and associated with an increased risk of physician-diagnosed asthma in a population of Australian children, and increased AHR in a population of Australian infants. The 38A sequence was associated with reduced plasma CC16 levels and individuals with lower plasma CC16 levels were more likely to have asthma (Laing *et al.*, 2000), although studies on populations of Japanese and British adults and North American children did not replicate these associations.

A collaboration between Sequana Therapeutics, Boehringer-Ingelheim and the University of Toronto identified ETS-2 and ETS-3 genes, which are adjacent to each other on chromosome 11p and code for epitheliumspecific transcription factors, as being asthma-associated genes in a genome-wide screen of the population of Tristan da Cunha, a volcanic island in the South Atlantic Ocean (Brooks-Wilson, 1999). This small population of 290 inbred individuals, derived from a small set of founders who settled on the island in the nineteenth century, has a high incidence of asthma ($\sim$30 per cent) and shares a relatively homogeneous living environment. This association was replicated in an outbred population from Toronto and also in three population samples from the USA and Denmark; however, no such association was reported in a study of an outbred European-American population. ETS-2 and ETS-3 are expressed in airway epithelial cells and may function as transcriptional activators or repressors of genes expressed in these cells.

Chromosome 12

This chromosome has been linked to both atopy and asthma. Several asthma-associated genes are located on chromosome 12q21–24 including stem cell factor (SCF), interferon-γ (IFN-γ), signal transducer and activator of transcription-6 (STAT6).

According to gene association studies, IFN-γ does not seem to be responsible for the linkage.

STAT6 is a critical signalling molecule in the Th2 signalling pathway, and mice lacking STAT6 are protected from allergic pulmonary manifestations. The importance of STAT6 in asthma is also evident from studies showing that STAT6 gene expression is markedly upregulated in airway epithelial cells in asthma. A number of common polymorphisms have been identified, including a GT repeat in exon 1 and three common SNPs (G4219A, A4491G and A4671G) in the human STAT6 gene. All four of these polymorphisms and a haplotype have been shown to be associated with allergic phenotypes in various populations (Schedel *et al.*, 2004; Gao *et al.*, 2004).

Neurally derived nitric oxide (NO), produced by neuronal NO synthase (NOS1), is physiologically linked to asthma as it is a neurotransmitter for bronchodilator nonadrenergic noncholinergic nerves. Mice lacking a functional NOS1 gene were shown to be hyporesponsive to methacholine challenge compared with wild-type mice (De Sanctis *et al.*, 1999). The frequencies of the number of a CA repeat in exon 29 were significantly different between Caucasian asthmatic and nonasthmatic population (Grasemann *et al.*, 2000). Recently, the NOS1 intron 2 GT repeat and STAT6 exon 1 GT repeat were associated with childhood asthma in a Japanese population (Shao *et al.*, 2004).

Chromosome 13

Linkage of chromosome 13q to atopy, asthma and allergy to house dust mites in children with asthma was found in different studies and genome-wide scans. Recently, Zhang *et al.* (2003) progressed with monumental efforts from broad linkage to gene identification in this region. First, they confirmed linkage of atopy and total serum IgE concentrations to this region using standard linkage approaches. Next, they made a saturation map that indicated that the locus associated with atopy was within 7.5 cM of the linkage peak. Confining their analysis to those polymorphisms with a minor allele frequency >0.15, they identified 49 SNPs, four deletion-insertion polymorphisms and a GGGC repeat. To determine whether IgE levels were associated with any of these SNPs, they did LD analyses in Australian families and found significant associations of the natural log of IgE concentrations (lnIgE) within a 100 kb region on chromosome 13. They confirmed their findings using transmission tests of association. Subsequent haplotype analysis indicated that the region of association to lnIgE centred on one gene, PHF11 [plant homeodomain (PHD) finger protein-11], and extended to two flanking genes, SETDB2 and RCBTB1. By conducting an analysis in which they held each SNP in PHF11 constant in a serial fashion, they identified three SNPs in introns 5 and 9 and in the 3′-untranslated region as having independent effects. They then showed that the intron 5 and 3′-untranslated region variants were also associated with severe clinical asthma, an association that was confirmed in an unrelated British population.

The precise function of PHF11 gene has not been determined, but the presence of two zinc finger motifs in the translated protein suggests a role in transcriptional regulation. The gene is expressed in most tissues, but Zhang and colleagues observed consistent expression in many immune-related tissues. Moreover, they identified multiple transcript isoforms, including variants expressed exclusively in the lung and in peripheral blood leukocytes. Because variation in this gene was strongly associated with serum IgE levels and, as described by Zhang, with circulating IgM, and because the gene is expressed heavily in B-cells, the authors suggest that this locus may be an important regulator in immunoglobulin synthesis.

The two genes that colocalize with PHF11 are within the same LD blocks (SETDB2 and RCBTB1) may also be important, particularly SETDB2, which has a similar expression profile to PHF11 in immune-related cells. This possibility highlights one of the important limitations of LD mapping of disease genes: if strong linkage disequilibrium extends over a segment of genome that harbours multiple genes, it is difficult (if not impossible) to identify which gene within the cluster is the disease gene using LD mapping alone. Given that no functional data has been presented, the next major focus of investigation for this locus should include functional dissection of these three genes and the IgE-associated variants.

Chromosome 14

Using 175 extended Icelandic families that included 596 patients with asthma, Hakonarson *et al.* (2002) performed a genome-wide scan with 976 microsatellite markers. Linkage of asthma was detected to chromosome 14q24, with an allele-sharing LOD score of 2.66. After the marker density was increased within the locus to an average of one microsatellite every 0.2 cM, the LOD score rose to 4.00.

Prostanoid DP receptor2 (DP2R) is located on chromosome 14q22.1 and was found to be required for the expression of the asthma phenotype in mice (Matsuoka *et al.*, 2000). Prostanoid DP mediates the chemotaxis of T cells, which follows the degranulation of mast cells. Six SNPs in DP2R and its vicinity have been found. These define four common three-SNP haplotypes, which vary in their ability to support transcription of DP2R and have distinct DNA-binding-protein affinity profiles. Individual DP2R SNPs were significantly associated with asthma in white and black populations in the USA. Multivariate analysis of the haplotype combinations (diplotypes) demonstrated that both whites and blacks who had at least one copy of the haplotype with a low transcriptional efficiency had a lower risk of asthma than subjects with no copies of the haplotype. These functional and genetic findings identify DP2 as an asthma-susceptibility gene (Ogume *et al.*, 2004).

Chromosome 16

Several studies have shown linkage between region on chromosome 16p21 and atopic phenotypes of specific IgE. The strongest candidate gene in this region is the IL-4 receptor (IL-4R), which also serves as the α-chain of the IL-13R. At least three of the eight reported SNPs that result in amino acid substitutions in the IL-4R gene have been associated with the atopic phenotypes and less commonly with asthma (Hakonarson and Wjst, 2001), although the alleles or haplotypes showing the strongest evidence differed between the populations.

Chromosome 17

Linkage between asthma and chromosome 17 was detected in several ethnic groups, although no such linkage was shown to other atopic diseases (Barnes, 2000; Szalai *et al.*, 2001). There are several candidate genes for asthma in this region, but the most important of them are genes in the chemokine gene cluster. Chemotactic cytokines, or chemokines, are small signalling proteins which are deeply involved in the physiology and pathophysiology of acute and chronic inflammatory processes, by attracting and stimulating specific subsets of leukocytes. A number of chemokines have been identified in human asthma whose production appears to be related to the severity of asthmatic inflammation and reactive airway responses. Monocyte chemoattractant protein-1 (MCP-1) may play a significant role in the allergic responses because of its ability to induce mast cell activation and leukotriene C_4 release into the airway, which directly induces AHR. Neutralization of MCP-1 drastically reduces bronchial hyperreactivity, lymphocyte-derived inflammatory mediators, and T cell and eosinophil recruitment to the lung. A biallelic A/G polymorphism in the MCP-1 distal gene regulatory region at position −2518 has been found that affects the level of MCP-1 expression in response to an inflammatory stimulus. Associations were found between carrying G at −2518 of the MCP-1 gene regulatory region and the presence of childhood asthma and between asthma severity and homozygosity for the G allele. In asthmatic children the MCP-1 −2518G also correlated with increased eosinophil level (Szalai *et al.*, 2001).

RANTES (regulated on activation normal T cell expressed and secreted) is one of the most extensively studied chemokines in allergic and infectious diseases. RANTES is likely to be important in airway inflammation because blocking antibodies to RANTES inhibit airway inflammation in a murine model of allergic airway disease. Furthermore, eosinophil chemotactic activity that appears in the BAL fluid of asthma patients following allergen challenge was found to be due to RANTES. Two polymorphisms in the RANTES promoter region (−28 C/G and –403 G/A) have been found to affect the transcription of the RANTES gene. Both polymorphisms have been found to be associated with asthma, or a phenotypic variant of asthma (−28G: near-fatal asthma) in some populations, but not in others (Szalai *et al.*, 2001; Al-Abdulhadi *et al.*, 2004; Yao *et al.*, 2003).

Eotaxin is the main chemoattractant for eosinophils, the most important cellular mediator of AHR. The expression of eotaxin mRNA and protein was found to be increased in the bronchial epithelium and submucosal layer of the airways of chronic asthmatics. In a Korean population the 123 G/A polymorphism was related to total serum IgE in asthmatics (Shin *et al.*, 2003), although the effects of the different SNPs in the eotaxin gene were quite inconclusive.

Chromosome 20

The first report of positional cloning of an asthma gene in a human population was published in 2002 (Van Eerdewegh *et al.*, 2002). In this study, a multipoint linkage analysis for asthma was done in 460 affected sibling-pair Caucasian families from the USA and the UK. The strongest linkage signal was to 20p13 (LOD score 2.94). The linked region, including a 1-LOD-support interval around marker D20S482, spans 4.28 cM, with a corresponding physical distance of 2.5 Mb. More than 40 genes were localized to this region, and 135 SNPs in 23 genes were then selected for genotyping in a case–control association study composed of probands that contributed to the LOD score from the linkage study and 'hypernormal' controls. Using this approach, the investigators identified a cluster of SNPs in the ADAM33 gene that demonstrated significant associations with asthma. The ADAM33 gene is expressed ubiquitously in muscle of every type, including the smooth muscle of bronchioles. It is also expressed in fibroblasts, lymph nodes, thymus and liver, but not in leukocytes or bone marrow. ADAM proteins are zinc-dependent metalloproteinases that belong to a disintegrin- and metalloproteinase-containing family. The exact function of ADAM33 is unknown but its expression profile and the functions of related proteins suggest a role for ADAM33 in bronchial contractility. Alternatively, it has been suggested that its position in these tissues might allow it to modify the process of bronchial remodelling (scarring) that follows chronic asthmatic airway inflammation. A further possibility is that ADAM33 might activate other as-yet-unknown cytokines.

19.4 Animal Models of Asthma

Although the mouse model of asthma does not replicate the human disease exactly, much of what is known about the immunobiology of eosinophilic inflammation and AHR in allergic asthma are based on mouse models of the disease (Leong and Huston, 2001). The literature of the animal model of allergic airway inflammation is impressive. The numbers of different factors (cells, allergens, proteins, genes, gene regulation, drugs, or different protocols etc.) that are investigated far exceed 100. In this review I will show only a few examples of these studies.

A major difference between mouse and human asthma is that airway inflammation persists longer in the latter. The perivascular and peribronchial distribution of eosinophils is similar in mice and humans. Recent studies suggest, however, that eosinophilic degranulation is absent in the lungs of asthmatic mice, while both intact and degranulated eosinophils are found in humans but their proportion bears no relation to asthma severity. An additional difference between mice and humans is that mice do not cough, while this is one of the most characteristic symptoms in human.

There is much similarity between mouse and human asthma, however, including production of specific IgE antibody, mast cells, T-lymphocytic and eosinophilic tissue infiltrate, Th2 cytokine pattern and AHR to non-specific bronchoconstrictor. The main advantages of the mouse model of asthma that it is much easier to study the histopathology of any murine tissue, particularly the bronchoalveolar lavage (BAL), airways, lungs, regional lymph nodes, bone marrow, serum and blood cells. There are numerous antibodies and reagents specific to the mouse available. The use of animal models to study the genetic basis of disease confers several additional advantages, such as a reduction in genetic heterogeneity, greater control of the phenotype to be studied and the ability to control environmental exposures. These advantages combined with the ability to manipulate the murine genome through selective breeding strategies and direct gene-targetting approaches, afford considerable power to the study of complex genetic traits in murine models.

To produce the asthma model, the test animal is injected with the antigen parenterally to induce systemic sensitization, then the same antigen is administered through the airways to focus the inflammatory process in the bronchi and lungs. One of the most important facts that we have learnt from the mouse model is that T cells are the main regulatory cells of the asthmatic response. Depletion of Th2-cytokine-producing CD4+ or CD8+ T cells prior to allergen sensitisation prevented airway inflammation and development of AHR. Transfer of CD8+ or CD4+ T cells from naive mice into CD8+- or CD4+-depleted animals fully restored the ability to generate airway inflammatory responses and AHR upon airway sensitization with allergen. Cytoplasmic staining of these cells from sensitized animals for cytokine production revealed that these cells produced significant amounts of Th2 cytokine IL-5, so allergic asthma is considered a Th2-predominant disease. In fact, only 1200 of these cells per animal are sufficient to cause asthma. Nevertheless, the co-operation of CD4+ Th1 cells and CD8+ cells is needed to mount a vigorous allergic response (Hamelmann *et al.*, 1996).

The important role of Th2 cells was verified by the finding that mice with a targeted deletion of the T-bet gene and severe combined immunodeficient mice receiving CD4+ cells from T-bet knockout (KO) mice spontaneously demonstrated multiple physiological and inflammatory features characteristic of asthma. T-bet, a Th1-specific T-box transcription factor, transactivates the IFN-γ gene in TH1 cells and has the unique ability to redirect fully polarized Th2 cells into Th1 cells, as demonstrated by simultaneous induction of IFN-γ and repression of IL-4 and IL-5. Even mice heterozygous for the deletion (with one functioning T-bet gene) produced asthmatic symptoms. In these mice the total or relative absence of T-bet sustains the differentiation of naive Th0 cells towards the Th2 type, which is enough for spontaneous asthmatic symptoms (Finotto *et al.*, 2002). The crucial role of Th2 cells was also confirmed in another study, in which the absent of GATA-3 in mice prevented the development of allergic inflammation after allergic sensitization and challenge. GATA-3 is the counter transcription factor to T-bet and it is important for the differentiation of naive Th0 cells towards Th2 cells.

With the exception of T cells it is possible to elicit allergic airway inflammation, or AHR in the absence of any other cells (e.g. B cells, mast cells) or molecules (e.g. IgE, histamine or cytokines) in certain mouse strains.

Increased production of IgE is the hallmark of atopic diseases such as allergic asthma. IgE, bound to high-affinity IgE receptors (FcϵRI), triggers the activation of mast cells following crosslinking with specific antigen, resulting in the synthesis and release of a variety of pro-inflammatory mediators and cytokines. IgE-dependent mast cell activation is the central mechanism of immediate allergic reactions such as allergen-induced bronchoconstriction. Still, genetically manipulated IgE deficient mice and IL-4 deficient mice (therefore lacking IgE as well) can both develop asthma.

Mice harbouring the spontaneous W/W^{v} mutation are deficient in mast cells, as the W gene product is necessary for fibroblast-dependent mast cell growth. It was shown that these mice could develop airway inflammation and late-phase AHR when sensitized and challenged, although early-phase AHR should be absent in these animals.

B-cell-deficient mice of C57BL/6 background can develop asthma but their eosinophils do not become activated. CD40-deficient mice cannot produce allergen-specific IgE, IgA or IgG but they can develop allergic asthma.

Among the Th2 cytokines, IL-4 has a central function in the induction of allergic responses. It induces and sustains Th2 differentiation of naive Th0 cells, induces the upregulation of the vascular cell adhesion molecule on endothelial cells which is required for eosinophilic infiltration into tissue, directs isotype switch of B cells towards IgE production and serves as a co-factor for mast cell differentiation. Early studies showed that, in the IL-4 KO mice, airway challenge of allergen-sensitized mice did not lead to elevation in specific and total IgE levels, increased numbers of eosinophils in lung tissues or BAL fluid, or AHR. Passive sensitization with allergen-specific IgE resulted in measurable amounts of total and allergen-specific IgE in the serum of IL-4 KO mice, but failed to significantly increase eosinophil airway accumulation or development of AHR to aerosolized methacholine. In contrast, a single exposure of sensitized IL-4 KO mice to IL-5 expressing adenovirus 1 day prior to the airway challenges with allergen resulted in a significant increase in numbers of eosinophils, and fully restored the development of AHR in systemically sensitized and challenged mice independently of IL-4. This study was carried out with C57BL/6 strain mice, and demonstrated that allergic airway disease is dependent only on IL-5 and eosinophilic inflammation in this mouse strain. In contrast, in Balb/c mice, inhibition of IL-4 and/or IL-5 did not prevent AHR, despite the absence of pronounced airway inflammation, but was abolished by depletion of CD4+ T cells. These seemingly conflicting results may perhaps be explained by a possible role of the Th2 cytokine IL-13, inducing T-cell-dependent AHR independently of IL-4, IL-5 or eosinophils. This is supported by data showing that mice deficient in STAT6 (STAT6 KO), which mediates cellular actions of both IL-4 and IL-13, are unable to develop airway eosinophilia and AHR following sensitization and challenge. However, IL-5 can restore the ability to develop AHR in STAT6 KO mice as well.

The essential role of IL-5 in enhancing eosinophilic accumulation and activation during allergen-induced inflammation has been well documented (Hamelmann and Gelfand, 2001). IL-5 promotes the terminal differentiation of eosinophil bone-marrow precursors and prolongs the viability of mature eosinophils by preventing apoptosis. Mice constitutively expressing the IL-5 gene (IL-5-transgenic mice) have high numbers of eosinophils in the peripheral blood, and eosinophilic infiltration in many organs, including lungs and lymphoid tissues. These animals spontaneously demonstrate multiple physiological and inflammatory features characteristic of serious asthma. Following systemic sensitization and airway challenge IL-5-deficient (IL-5 KO) Balb/c mice showed normal IgE production and T-cell-responses, indicating that IL-5 is not required for many B- or T-cell responses. In contrast to normal mice, IL-5 KO mice did not show eosinophil accumulation in the lungs and in the BAL fluids. These mice failed to develop AHR to aerosolized methacholine. Cumulatively, these data indicated that IL-5 mediated eosinophil accumulation is a prerequisite for the development of AHR in this model.

The results in murine model of asthma suggest that up to three independent but interrelated pathways are postulated to be important for the development of AHR in mice. The first pathway is dependent on IL-4 and mast cells, the second on IL-5 and eosinophils, and the third is independent of IL-4 and IL-5, and is likely to be mediated by IL-13. It was also shown that IL-13 had an inhibitory effect on the action of IL-5.

The existence of multiple mechanisms helps to explain why reducing eosinophilic inflammation may not lead to reduction in AHR, why AHR can occur in the absence of eosinophils, and why eosinophils alone may not be sufficient to cause AHR, which are questions raised in human asthma research. Airway inflammation orchestrated by T-lymphocytes is ultimately the cause of AHR, but the intermediaries could be eosinophils, mast cells or, apparently, solely IL-13. Besides immune cells, IgE and cytokines, discussed above, the role of several other factors in asthma was also studied in murine models.

Humbles *et al.* (2000) showed in complement C3a KO mice that innate immune system and complement C3a may be involved in the pathogenesis of asthma.

Using microarray analysis of pulmonary gene expression and SNP-based genotyping, Karp *et al.* (2000) identified complement C5 as a susceptibility locus for allergen-induced AHR in a mouse model of asthma. Comparing two mouse strains, a mutation that was present in C5 in the high-responder mouse strain led to mRNA and protein deficiency. Further studies showed strong associations between C5 gene expression, the genotype and susceptibility to development of allergen-driven AHR. Together, these studies provide robust evidence for the importance of C5 in susceptibility to allergen-induced AHR in this experimental model of asthma.

Histamine is an important mediator released from activated mast cells provoked by allergen and has a substantial role in the pathophysiology of asthma. Ovalbumin-sensitized and challenged histamine-deficient mice had significantly altered chemokine and cytokine expression profile, reduced airway hyperresponsiveness, lung inflammation, BAL eosinophilia and OVA-specific IgE compared with congenic wild-type littermates treated in the same way. These findings demonstrate that histamine, besides its role in immediate hypersensitivity, has a major effect on basic immunological processes and influences the asthmatic processes (Kozma *et al.*, 2003).

McIntire *et al.* (2001) used congenic inbred mouse strains, which differ only in a small chromosomal region syntenic with human chromosome 5q, to map a monogenic trait that confers reduced TH2 responsiveness and protects against AHR. Positional cloning identified the T cell immunoglobulin domain and mucin domain (TIM) gene family in which major sequence variants show strong association with TH1–TH2 differentiation and AHR. Recent work suggests that TIM-3 directs TH1 differentiation, and TIM-1 directs TH2 cytokine production. The finding that the human homologue of TIM-1 is the hepatitis-A virus receptor might explain the inverse relationship between hepatitis-A virus infection and reduced asthma susceptibility.

Homologues of candidate genes on human chromosome 5q31–q33 are found in four regions in the mouse genome, two on chromosome 18 and one each on chromosomes 11 and 13. Bronchial responsiveness was assessed as a quantitative trait in mice and found it linked to chromosome 13. IL-9 is located in the linked region and was analysed as a gene candidate. The expression of IL-9 was markedly reduced in bronchial hyporesponsive mice, and the level of expression was determined by sequences within the qualitative trait locus. The important role of IL-9 was confirmed in transgenic mice in which expression of the murine IL-9 cDNA was regulated by the rat Clara cell 10 protein promoter expressed selectively and constitutively in the lung. Lung selective expression of IL-9 caused massive airway inflammation with eosinophils and lymphocytes as predominant infiltrating cell types. A striking finding was the presence of increased numbers of mast cells within the airway epithelium of IL-9-expressing mice. Other impressive pathologic changes in the airways were epithelial cell hypertrophy associated with accumulation of mucus-like material within nonciliated cells and increased subepithelial deposition of collagen. Physiologic evaluation of IL-9-expressing mice demonstrated normal baseline airway resistance and markedly increased airway hyper-responsiveness to inhaled methacholine. These data suggest a role for IL-9 in the complex pathogenesis of bronchial hyper-responsiveness as a risk factor for asthma (Nicolaides *et al.*, 1997; Temann *et al.*, 1998).

19.5 Concluding Remarks

In the last few years our knowledge about the structure and function of the human genome improved considerably. Still, we are very far from the perfect understanding of the genomic background of complex diseases like allergy or asthma. Regarding the multiple gene–gene and gene–environmental interactions, it is very likely that we will never forecast whether a newborn will have asthma in the future, or not, but with the available sequence information, the completion of a high-quality physical map of the human and mouse genome, the different animal models (KO and transgenic animals) and the advance of bioinformatics and different methods (microarray, DNA sequencing, high throughput screenings) will make it possible for several additional asthma susceptibility genes and gene regulatory networks to be identified in the coming years. Hopefully, this knowledge will be translated into improved diagnosis, prevention and therapeutic strategies for this chronic disease.

References

Al-Abdulhadi SA, Helms PJ, Main M, Smith O, Christie G. 2004. Preferential transmission and association of the −403 G $\rightarrow$A promoter RANTES polymorphism with atopic asthma. *Genes Immun* **6**: 24–30.

Allen M, Heinzmann A, Noguchi E, *et al.* 2003. Positional cloning of a novel gene influencing asthma from chromosome 2q14. *Nat Genet* **35**: 258–263.

Anderson GG, Cookson WOCM. 1999. Recent advances in the genetics of allergy and asthma. *Mol Med Today* **5**: 264–273.

Barnes KC. 2000. Evidence for common genetic elements in allergic disease. *J Allergy Clin Immunol* **106**: S192–200.

Brooks-Wilson AR. 1999. Asthma related genes. International patent publication number W0 99/37809.

CGSA. 1997. A genome-wide search for asthma susceptibility loci in ethnically diverse populations. *Nat Genet* **15**: 389–392.

Chen W, Ericksen MB, Levin LS, Hershey GKK. 2004. Functional effect of the R110Q IL13 genetic variant alone and in combination with IL4RA genetic variants. *J Allergy Clin Immunol* **114**: 553–560.

Cookson W. 1999. The alliance of genes and environment in asthma and allergy. *Nature* **402**: B5–B11.

Cookson WO, Sharp PA, Faux JA, Hopkin JM. 1989. Linkage between immunoglobulin E responses underlying asthma and rhinitis and chromosome 11q. *Lancet* **1**: 1292–1295.

Cookson WO, Young RP, Sandford AJ *et al.* 1992. Maternal inheritance of atopic IgE responsiveness on chromosome 11q. *Lancet* **340**: 381–384.

Daniels SE, Bhattacharrya S, James A *et al.* 1996. A genome-wide search for quantitative trait loci underlying asthma. *Nature* **383**: 247–250.

De Sanctis GT, Merchant M, Beier DR *et al.* 1995. Quantitative locus analysis of airway hyperresponsiveness in A/J and C57BL/6J mice. *Nat Genet* **11**: 150–154.

De Sanctis GT, MacLean JA, Hamada K *et al.* 1999. Contribution of nitric oxide synthase 1,2, and 3 to airway hyperresponsiveness and inflammation in a murine model of asthma. *J Exp Med* **189**: 1621–1630.

Dodge RR, Burrows B. 1980. The prevalence and incidence of asthma and asthma-like symptoms in a general population sample. *Am Rev Respir Dis* **122**: 567–575.

Finotto S, Neurath MF, Glickman JN *et al.* 2002. Development of spontaneous airway changes consistent with human asthma in mice lacking T-bet. *Science* **295**: 336–338.

Gao PS, Heller NM, Walker W *et al.* 2004. Variation in dinucleotide (GT) repeat sequence in the first exon of the STAT6 gene is associated with atopic asthma and differentially regulates the promoter activity *in vitro*. *J Med Genet* **41**: 535–539.

Gohlke H, Illig T, Bahnweg M *et al.* 2004. Association of the interleukin-1 receptor antagonist gene with asthma. *Am J Respir Crit Care Med* **169**: 1217–1223.

Grasemann Yandava CN, H, Storm van's Gravesande K *et al.* 2000. A neuronal NO synthase (NOS1) gene polymprphism is associated with asthma. *Biochem Biopys Res Commun* **272**: 391–394.

Hakonarson H, Wjst M. 2001. Current concepts on the genetics of asthma. *Curr Opin Pediatr* **13**: 267–277.

Hakonarson H, Bjorndottir US, Halapi E *et al.* 2002. A major susceptibility gene for asthma maps to chromosome 14q24. *Am J Hum Genet* **71**: 483–491.

Hamelmann E, Gelfand E. 2001. IL-5-induced airway eosinophilia- the key to asthma? *Immunol Rev* **179**: 182–191.

Hamelmann E, Oshiba A, Paluh J *et al.* 1996. Requirement for CD8+ cells in the development of airway hyperresponsiveness in a murine model of airway sensitization. *J Exp Med* **183**: 1719–1730.

Hill MR, Cookson WOCM. 1996. A new variant of the b subunit of the high-affinity receptor for immunoglobin E (FεRI-β E237G): association with measures of atopy and bronchial hyperresponsiveness. *Hum Mol Genet* **5**: 959–962.

Hoffjan S, Nicolae D, Ober C. 2003. Association studies for asthma and atopic disease: a comprehensive review of the literature. *Respir Res* **4**: 14.

Humbles AA, Lu B, Nilsson CA *et al.* 2000. A role for the C3a anaphylatoxin receptor in the effector phase of asthma, *Nature* **406**: 998–1001.

ISAAC Steering Committee. 1998. Worldwide variations in the prevalence of asthma symptoms: the International Study of Asthma and Allergies in Childhood (ISAAC). *Eur Respir J* **12**: 315–335.

Janson C, Chinn S, Jarvis D, Burney P. 1997. Physician-diagnosed asthma and drug utilization in the European Community Respiratory Health Survey. *Eur Respir J* **10**: 1795–1802.

Kabesch M, Tzotcheva I, Carr D *et al.* 2003. A complete screening of the IL4 gene: novel polymorphisms and their association with asthma and IgE in childhood. *J Allergy Clin Immunol* **112**: 893–898.

Karp CL, Grupe A, Schadt E *et al.* 2000. Identification of complement factor 5 as a susceptibility locus for experimental allergic asthma. *Nat Immun* **1**: 221–226.

Kozma GT, Losonczy G, Keszei M *et al.* 2003. Histamine deficiency in gene-targeted mice strongly reduces antigen-induced airway hyperresponsiveness, eosinophilia and allergen specific IgE. *Int Immunol* **15**: 963–973.

Laing IA, Hermans C, Bernard A, Burton PR, Goldblatt J, Le Souef PN. 2000. Association between plasma CC16 levels, the A38G polymorphism, and asthma. *Am J Respir Crit Care Med* **161**: 124–127.

Laitinen T, Daly MJ, Rioux JD *et al.* 2001. A susceptibility locus for asthma-related traits on chromosome 7 revealed by genome-wide scan in a founder population. *Nat Genet* **28**: 87–91.

Laitinen T, Polvi A, Rydman P *et al.* 2004. Characterization of a common suscepitibiliy locus for asthma-related traits. *Science* **304**: 300–304.

Leong KP, Huston DP. 2001. Understanding the pathogenesis of allergic asthma using mouse models. *Ann Allergy Asthma Immunol* **87**: 96–110.

Mannino DM, Homa DM, Pertowski CA *et al.* 1998. Surveillance for asthma – United States, 1960–1995. *MMWR Morb Mortal Wkly Rep* **47**: 1–27.

Marsh DG, Zwollo P, Huang SK, Ghosh B, Ansari AA. 1989. Molecular studies of human response to allergens. *Cold Spring Harbor Symposia on Quantitative Biology*, Vol 320; 271–277.

Marsh DG, Neely JD, Breazeale DR *et al.* 1994. Linkage analysis of IL4 and other chromosome 5q31.1 markers and total serum immunoglobulin E concentrations. *Science* **264**: 1152–1155.

Matsuoka T, Hirata M, Tanaka H *et al.* 2000. Prostaglandin D2 as a mediator of allergic asthma. *Science* **287**: 2013–2017.

McIntire JJ, Umetsu SE, Akbari O *et al.* 2001. Identification of Tapr (an airway hyperreactivity regulatory locus) and the linked Tim gene family. *Nat Immunol* **2**: 1109–1116.

Munthe-Kaas MC, Carlsen KH, Helms PJ *et al.* 2004. CTLA-4 polymorphisms in allergy and asthma and the TH1/TH2 paradigm. *J Allergy Clin Immunol* **114**: 280–287.

Nicolaides NC, Holroyd KJ, Ewart SL *et al.* 1997. Interleukin 9: A candidate gene for asthma. *Proc Natl Acad Sci USA* **94**: 13175–13180.

Oguma T, Palmer LJ, Birben E, Sonna LA, Asano K, Lilly CM. 2004. Role of prostanoid DP receptor variants in susceptibility to asthma. *New Engl J Med* **351**: 1752–1763.

Oshika ES, Liu LP, Ung G *et al.* 1998. Glucocorticoid-induced effects on pattern formation and epithelial cell differentiation in early embryonic rat lungs. *Pediatr Res* **43**: 305–314.

Palmer LJ, Silverman ES, Weiss ST Drazen JM. 2002. Pharmacogenetics of asthma. *Am J Respir Crit Care Med* **165**: 861–866.

Schedel M, Carr D, Klopp N *et al.* 2004. A signal transducer and activator of transcription 6 haplotype influences the regulation of serum IgE levels. *J Allergy Clin Immunol* **114**: 1100–1105.

Shao C, Suzuki Y, Kamada F *et al.* 2004. Linkage and association of childhood asthma with the chromosome 12 genes. *J Hum Genet* **49**: 115–122.

Shin HD, Kim LH, Park BL *et al.* 2003. Association of eotaxin gene family with asthma and serum total IgE. *Hum Mol Genet* **12**: 1279–1285.

Strachan DP. 1989. Hay fever, hygiene, and household size. *Br Med J* **299**: 1259–1260.

Szalai C, Kozma GT, Nagy A *et al.* 2001. Polymorphism in the gene regulatory region ofMCP-1 is associated with asthma susceptibility and severity. *J Allergy Clin Immunol* **108**: 375–381.

Tattersfield AE, Knox AJ, Britton JR, Hall IP. 2002. Asthma. *Lancet* **360**: 1313–1322.

Temann U-A, Geba GP, Rankin JA, Flavell RA. 1998. Expression of interleukin 9 in the lungs of transgenic mice causes airway inflammation, mast cell hyperplasia, and bronchial hyperresponsiveness. *J Exp Med* **188**(7): 1307–1320.

Van Eerdewegh P *et al.* 2002. Association of the ADAM33 gene with asthma and bronchial hyperresponsiveness. *Nature* **418**: 426–430.

Walley AJ, Chavanas S, Moffatt MF *et al.* 2001. Gene polymorphism in Netherton and common atopic disease. *Nat Genet* **29**: 175–178.

Weiss ST, Raby BA. 2004. Asthma genetics 2003. *Hum Mol Genet* **13**(special issue no. 1): R83–89.

Wills-Karp M, Ewart SL. 2004. Time to draw breath: asthma-susceptibility genes are identified. *Nat Genet* **5**: 376–387.

Wjst M, Fischer G, Immervoll T *et al.* 1999. A genome-wide search for linkage to asthma. German Asthma Genetics Group. *Genomics* **58**: 1–8.

Yao TC, Kuo ML, See LC *et al.* 2003. The RANTES promoter polymorphism: a genetic risk factor for near-fatal asthma in Chinese children. *J Allergy Clin Immunol* **111**: 1285–1292.

Zhang Y, Leaves NI, Anderson GG *et al.* 2003. Positional cloning of a quantitative trait locus on chromosome 13q14 that influences immunoglobulin E levels and asthma. *Nat Genet* **34**: 181–186.

20 Primary Immunodeficiencies: Genotype–Phenotype Correlations

Mauno Vihinen and **Anne Durandy**

Abstract

Close to 150 different forms of immunodeficiencies are known. The affected gene has been identified in more than 100 disorders and a large number of disease-causing mutations are listed in dedicated databases. Strict genotype–phenotype correlations are not very common. However, a number of cases are known. In addition, somewhat weaker correlations are apparent in certain disorders. Here the genotype–phenotype correlations are discussed for some of the most prominent and clear-cut cases, namely adenosine deaminase deficiency, severe combined immune deficiency caused by mutations in recombinant activating gene 1 or 2 (RAG 1 and 2), *AICDA* gene defects in hyper-IgM deficiency and Wiskott–Aldrich syndrome (WAS), as well as X-linked agammaglolubulinaemia (XLA) as an example of weaker correlations. In addition, whenever possible, the protein structural consequences of the mutations are also described. The study is based on systematic analysis of all known mutations; altogether some 3500 cases are available.

20.1 Introduction

The immune system is always alert to recognize and neutralize invading microbes and foreign molecules. The highly sophisticated system can handle a very wide spectrum of substances and organisms. Innate immunity mounts rapidly, but the response is usually nonspecific, whereas adaptive immunity facilitates specific recognition. When components of the machinery are mutated, the affected individuals suffer from immunodeficiencies (IDs). These disorders vary greatly with regard to genotype, phenotype, symptoms, infection-causing organisms and severity of the disease, because many cells and molecules are required for both natural and adaptive immunity.

Close to 150 primary IDs are now known, and have been grouped according to the components of the immune system affected (Väliaho *et al.*, 2005a; http://bioinf.uta.fi/IDR;

Immunogenomics and Human Disease Edited by András Falus

Table 20.1 Classification of immunodeficiencies

Classification	Diseases	Genes	Patients	Unique mutations
Combined B and T cell IDs	26	24	388	345
Deficiencies predominantly affecting antibody production	22	17	1045	513
Defects in lymphocyte apoptosis	4	4	4	4
Other well-defined immunodeficiency syndromes	11	9	344	138
Defects of phagocyte function	25	24	1130	533
Defects of innate immune system, receptors and signalling components	10	10	127	90
DNA breakage-associated syndromes and DNA epigenetic modification syndromes	7	7	58	49
Defects of the classical complement cascade proteins	16	16	117	42
Defects of the alternative complement pathway	4	4	94	49
Defects of complement regulatory proteins	8	8	191	105
Total	133	123	3499	1868

Table 20.1). Most IDs are relatively rare disorders. Antibody deficiency disorders are defects in immunoglobulin-producing B cells. T-cell deficiencies affect the capability to kill infected cells or help other immune cells. Both T cells and antibody production are defective in combined immunodeficiencies. Life-threatening symptoms can arise within the first few days of life in patients with severe combined immune deficiency (SCID).

B-cell immunodeficiencies are antibody deficiencies restricted to antibody function. They are the most frequent IDs (around 70 per cent). Either the development or the function of B-lymphocytes is impaired. All or some selected subsets of immunoglobulins may be deficient. Treatment is achieved by regular immunoglobulin subsitution. In combined B- and T-cell immunodeficiencies, which are the most severe IDs, all adaptive immune functions are absent. The condition is fatal unless the immune system can be reconstituted either by transplants of immunocompetent tissue or by enzyme replacement, and possibly in the future by gene therapy. The immunological, genetic and enzymatic characteristics of these diseases are very diverse. SCIDs have an average frequency of approximately 1 in 75 000 births.

Other IDs affect, for example, the complement system or phagocytic cells, impairing antimicrobial immunity. Secondary immunodeficiencies may allow similar infections to primary IDs, but are associated with other factors such as malnutrition, drugs, age, tumours or infections, including human immunodeficiency virus (HIV).

The immune system is based on a large number of molecules and processes. A particular ID can originate from defects in any one of the molecules in sequential steps that are essential for a certain response. The incidence of IDs varies greatly from about 1:500 live births to only a few known cases of the most rare disorders. In the different ID classes the numbers of patients vary greatly (Table 20.1). The most common IDs are antibody production defects, different forms of SCIDs and defects in the phagocyte system. The largest numbers of identified cases with gene mutations are also in these categories. ESID registry of European ID patients contains close to 11 000 patients (www.esid.org). Immunodeficiency mutation databases (IDbases; Vihinen *et al.*, 2001) contain genetic and clinical information for about 3500 individuals.

ID-related genes are distributed throughout the human genome; only from chromosomes 3, 8, 18 and Y have no genes been identified. Some 86 per cent of the IDs are autosomal recessive (AR), although the best known cases are X-linked forms (10 per cent of diseases, but 52 per cent of cases investigated on a sequence level). In addition there are 4 per cent autosomal dominant IDs. The majority of the genes code for multidomain proteins. Consanguinity is common in families with AR forms of IDs. In X-linked disorders a single mutated gene causes the phenotype in males, because X-linked recessive diseases generally have full penetrance.

20.2 Immunodeficiency Data Services

From the billions of pages of information on the internet it is often difficult to find the specific knowledge one is looking for. This is especially true for rare disorders for which there may not be much available data. This is a key issue since the internet is already the primary source of information for scientists and medical doctors as well as for patients and the general public. ImmunoDeficiency Resource (IDR) aims to provide comprehensive integrated knowledge on immunodeficiencies online at http://bioinf.uta.fi/idr/ (Väliaho, Riikonen and Vihinen, 2000; Väliaho *et al.*, 2002, 2005a,b). This resource includes data for clinical, biochemical, genetic, structural and computational analyses. IDR also includes articles, instructional resources, analysis and visualization tools as well as advanced search routines. Extensive cross-referencing and links to other services are available. All information in IDR is validated by expert curators.

The disease- and gene-specific information is stored in fact files that are extensible markup language (XML) based. Specific inherited disease markup language (IDML) was developed to distribute and collect information (Väliaho *et al.*, 2005b). IDR is continuously updated and new features will be added in order to provide a comprehensive navigation point.

Diagnosis of immunodeficiencies can be very difficult, because several disorders can have similar symptoms. Numerous IDs are very rare. Early and reliable diagnosis is in many instances crucial for efficient treatment, because delayed diagnosis and management can cause severe and irreversible complications, even the death of the patient. The European and Pan American societies for IDs have released guidelines for the diagnosis of some common immunodeficiencies (Conley, Notarangelo and Etzioni, 1999).

The definitive diagnostics of IDs depends on genetic and clinical tests since the physical signs may be nonspecific, very discreet or even absent. Because of the rareness of IDs there are generally not many laboratories analysing a particular disease. The IDdiagnostics registry (http://bioinf.uta.fi/IDdiagnostics/) contains two databases, genetic and clinical, which provide a service for those looking for laboratories conducting ID testing (Samarghitean, Väliaho and Vihinen, 2004).

Immunodeficiency-causing mutations have been identified from several genes. The immunodeficiency mutation databases (IDbases) are available for more than 100 diseases (Table 20.1; http://bioinf.uta.fi/IDbases; Vihinen *et al.*, 2001). The databases maintained at the University of Tampere contain some 3500 entries (Table 20.1). Mutation data is distributed along with patient-related clinical information. The first immunodeficiency mutation database, the BTKbase, was founded in 1994 (Vihinen *et al.*, 1995a, 1999) for X-linked agammaglobulinaemia (XLA). The contents of IDbases are checked by curators. For the maintenance of the IDbases a computer program suite was developed, MUTbase (Riikonen and Vihinen, 1999), that automatically handles submission, generation of the distribution version and other routine tasks.

The IDbases have been used to retrieve retrospective and prospective information on clinical presentation, immunological phenotype, long-term prognosis and efficacy of available therapeutic options. The information may be essential in developing new treatments, including drug design. Here, the data compiled on IDbases have been used to address genotype–phenotype (GP) correlations in IDs.

Missense mutations are the single most common mutation type (Table 20.2). When calculated together, null mutations (nonsense, out-of-frame deletions and insertions) are the biggest group. The most frequent point mutations are C to T and G to A transitions (Figure 20.1). There are clear differences in the mutation spectra in different IDs for several reasons, including function and structure of the encoded protein, founder effects etc.

Table 20.2 Mutation types in immunodeficiencies

	ADA	RAG1	RAG2	AID	BTK
Missense	83	45	24	101	390
Nonsense	9	13	3	8	154
Deletion inframe	1	0	2	6	30
Deletion frameshift	17	19	0	4	141
Deletion undefined	6	0	1	3	15
Insertion inframe	0	0	0	0	2
Insertion frameshift	0	1	2	2	61
Insertion undefined	0	0	0	0	4
Splice site inframe	2	0	0	0	19
Splice site frameshift	11	0	0	0	29
Splice site undefined	4	0	0	0	101
Multiple	1	0	0	0	5
Unclassified	134	78	32	126	955

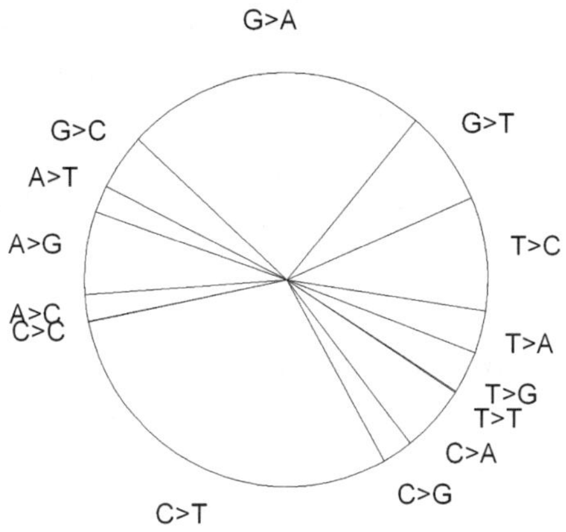

Figure 20.1 Mutation types in immunodoficiencies on nucleotide level.

20.3 Genotype–Phenotype Correlations

GP correlations have been in the focus of intense research, because the knowledge would help the proper treatment and diagnosis and understanding of clinical features, and provide the means for basic research into functionally and structurally important regions in genes and proteins. There seem to be very few clear-cut GP correlations in IDs, whereas weaker correlations are apparent in many disorders. IDs can be caused by mutations in numerous genes and proteins. The proteins in IDs are necessary for several cellular functions, including cell surface receptors, signal transduction, transcription factors, nucleotide metabolism, gene diversification and phagocytosis. Certain diseases can arise from defects in several genes, which further complicates the picture.

GP correlations are discussed here for some of the most prominent and clear-cut cases, namely adenosine deaminase (ADA) deficiency, SCID caused by mutations in recombinant activating gene 1 or 2 (RAG 1 and 2), *AICDA* gene defects in hyper-IgM deficiency (HIGM) (Revy *et al.*, 2000; Quartier *et al.*, 2004), and Wiskott–Aldrich syndrome (WAS), as well as for X-linked agammaglolubulinaemia as an example of weaker but still consistent correlations. When possible, protein structural consequences of the mutations are described. There are also some other IDs where GP correlations have been implicated, including ELA2 in cyclic and congenital neutropenia (Dale *et al.*, 2000), serpin (complement component 1) mutations in hereditary angioedema (HANE) (Verpy *et al.*, 1996) and tafazzin in Barth syndrome (Johnston *et al.*, 1997).

20.4 ADA Deficiency

ADA is a monomeric zinc enzyme in which mutations lead to SCID. ADA deficiency is found in about 15 per cent of SCIDs, and 85–90 per cent of ADA-deficient patients have SCID. They are usually diagnosed by one year of age. The remaining ~15 per cent have milder immunodeficiency, which is diagnosed later in the first decade ('delayed onset'), or during the second to fourth decades ('late/adult onset'). Further, some healthy individuals have ADA deficiency with normal immune functions (for a review see Hershfield, 2003). The red cell dATP level is increased in these patients 0–30-fold compared with over 300-fold in SCID patients. These persons have partial ADA deficiency.

The ADA deficiency patients have very profound lymphopaenia, involving B, T and NK (natural killer) cells. In addition to immunological defects, most patients with ADA deficiency also have skeletal abnormalities, central nervous system dysfunctions, and elevated hepatic transaminase levels appear occasionally. ADA deficiency can be treated with bone marrow transplantation or by enzyme replacement therapy with polyethylene glycol-modified bovine ADA (PEG-ADA). Gene therapy trials have been performed succesfully (Aiuti *et al.*, 2002).

ADA activity protects lymphoid cells, especially immature thymocytes, from the toxic effects of its substrates adenosine and deoxyadenosine, which are generated in large amounts in the thymus during T cell maturation by active cell turnover, as well as in lymph nodes during the response to the antigen. Already very low levels of ADA either prevent or reduce the adverse effects of the deficiency.

ADA protein consists of a single catalytic domain. The structure of highly related mouse ADA has been solved (Wilson, Rudolph and Quiocho, 1991). The protein has an α/β TIM barrel structure, where the essential zinc ion is located in the active site pocket [Figure 2(A)]. The catalytic centre is, as typical in the enzymes of this fold, in C terminal ends of the barrel-forming β-strands and in the loops connecting to the surrounding α-helices.

A total of 134 patients (83 missense, 23 deletions; Table 20.2) have been identified in the ADAbase (http://bioinf.uta.fi/ADAbase/). Roughly half of the mutations (69/135) are unique for single families. The majority of the patients are heteroallelic (70 per cent) the homozygotes being from consanguineous families. Missense mutations are more common than is generally the case in IDs. There are two common mutations, R211H and G216R, which account altogether for about a quarter of all the cases. The great majority of the missense mutations are known to result in unstable proteins (Aredondo-Vega *et al.*, 1998).

ADA mutations have good correlations between genotypes and both clinical and metabolic phenotypes. The phenotypes have been grouped based on the symptoms and age at diagnosis (Aredondo-Vega *et al.*, 1998). Hershfield and coworkers have investigated the mutant forms in an *Escherichia coli* strain from which the endogeneous *ADA* gene has been deleted (Aredondo-Vega *et al.*, 1998; Hershfield, 2003). They grouped the alleles and genotypes into four categories with increasing ADA activity in the *E. coli* expression system. The activity in group 0 is 0, in group I it is about 0.015 per cent of the wild type, in group II it is 0.1 per cent, in group III 0.42 per cent and in group IV 8.3 per cent of wild-type activity. In their cohort, 10 patients had genotype 0/0 (i.e. both alleles belong to group 0) or 0/I. All these patients had SCID, as did 18 out of 20 patients in group I/I, while only 1 out of 31 patients had

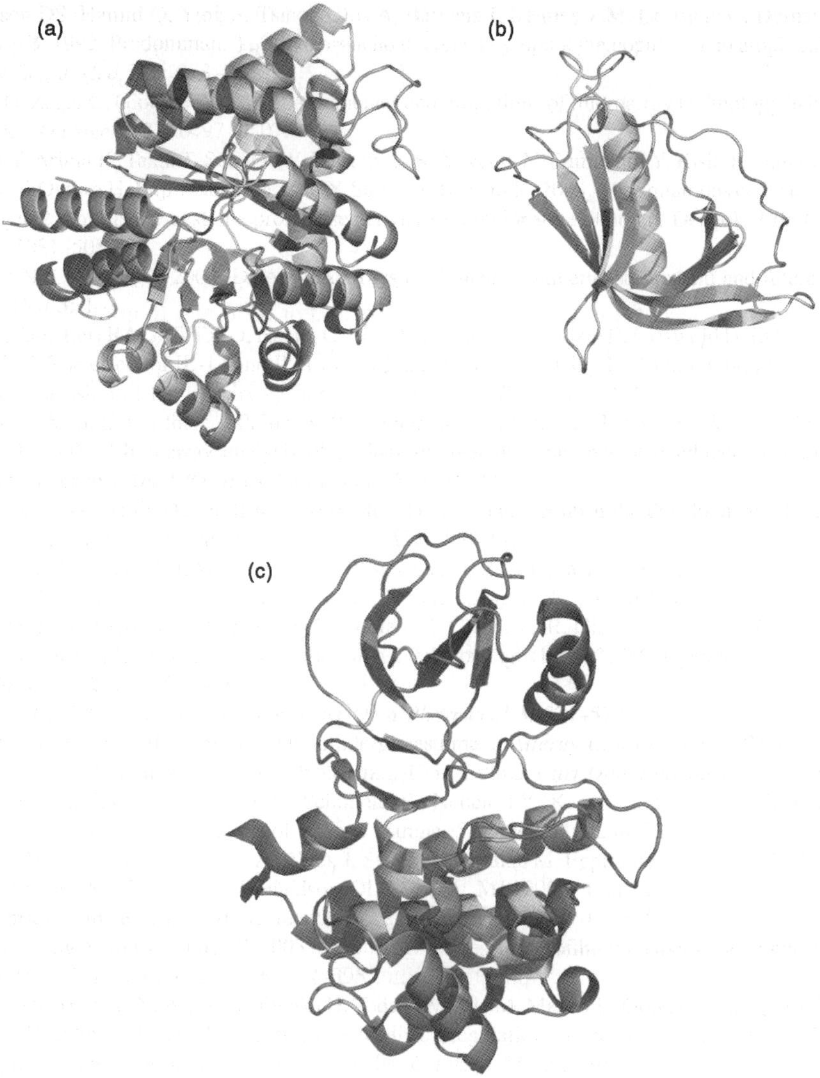

Figure 20.2 Three-dimensional structures of ID related proteins. (a) Missense mutations in ADA (PDB code 1a4m). Positions for group I mutations are in red, group II in yellow, group III in green and combination of groups in the same position in magenta. (b) Missense mutations in WASP PH/WH1 domain (lmke). Positions where mutations lead to WAS are in red, those leading to XLT in yellow, and mixed outcome in green. (c) Missense mutations in BTK (1k2p). Mutations in the upper lobe are in yellow and in the lower lobe in cyan. (A colour reproduction of this figure can be viewed in the colour plate section)

SCID with the other allele forms. Thus, there is a clear distinction between SCID and the milder forms of ADA deficiency.

When looking at the distribution and consequences of the mutations in the ADA protein [Figure 20.2(A)], it is possible to pinpoint correlation between functionally and structurally important residues and the severity of the disease. The SCID-causing alterations are clearly clustered on the protein core as well as the catalytic site residues, while those positions linked to milder forms are located either further from the active site or on the surface loops.

20.5 RAG1 and RAG2 Deficiency

The T−B− SCID patients have no mature T and B cells; in addition, lymph nodes and tonsils are absent. Infections start in the second or third month after birth. Omenn syndrome (OS) is a rare combined ID with the presence of a substantial number of oligoclonal, activated T cells, the lack of B lymphocytes and characteristic clinical features (Villa *et al.*, 1998). The patients have opportunistic infections, chronic persistent disease of the airways and local and systemic bacterial infections leading to a failure to thrive. Epstein–Barr virus (EBV) and Cytomegalovirus (CMV) infections can cause lethal complications.

SCID and OS patients are treated with antibiotics and supportive measures. Bone marrow transplantation is the only curative tool. T−B− SCID and OS are caused by mutations in either of the two recombination activating genes, RAG1 and RAG2 (Schwarz *et al.*, 1996; Villa *et al.*, 1998).

The process of V(D)J recombination, leading to the assembly of genes coding for immunoglobulins (Igs) and T cell receptors (TCRs), is central for the differentiation of B and T cells, and thus for the adaptive immune system. V(D)J recombination leads to the generation of variable domains of the recognition molecules through the assembly of one segment each from a set of variable (V), joining (J) and, in some cases, diversity (D) elements.

RAG1 and RAG2 are the first proteins in the V(D)J recombination process. V(D)J recombination is directed by recombination signal sequences (RSSs), which flank each receptor gene segment and consist of a conserved heptamer and a nonamer separated by a nonconserved spacer of either 12 or 23 nucleotides. RAG1 recognizes and binds the nonamer sequence. After the recruitment of RAG2, the complex RAG1–RAG2 cleaves the double strand between the coding element and the heptamer of aRSS. The RAG genes are also involved in the following steps of DNA-end processing. During end-joining, nucleotides can be removed or added. Nonhomologous DNA end-joining (NHEJ) proteins are involved in the joining. Artemis functions in a crucial hairpin opening as well as in end joining. Ligase IV is also required for end joining. Mutations in the recombination process proteins RAG1, RAG2, Artemis (Moshous *et al.*, 2001) and Ligase IV (O'Driscoll *et al.*, 2001) lead to IDs.

To date, 110 cases have been reported, 78 with mutations in RAG1 and 32 in RAG2. Some 58 and 75 per cent of the cases are amino acid substitutions in RAG1 and RAG2, respectively (http://bioinf.uta.fi/RAG1base/, http://bioinf.uta.fi/RAG2base/). RAG1 protein consists of several domains including ring finger domain, zinc finger domains, homeodomain and core domains. Less is known about RAG2.

Mutations are distributed widely within the proteins. Certain clustering is apparent in RAG1.

Analysis of 44 patients from 41 families with RAG defects revealed that the cases can be grouped into four categories (Villa *et al.*, 2001). Null mutations in both alleles cause the classical T−B− SCID. OS patients have missense mutations at least in one allele, but maintain partial RAG activity. A third group, called atypical SCID/OS, has RAG defects with some, but not all, of the clinical and immunological features of OS. Missense mutations are also over-represented in this group, as in OS patients. The engraftment of maternal T cells into foetuses with RAG deficiency, regardless of the type and site of mutations, may result in a clinical and immunological phenotype that mimicks OS.

SCID-causing mutations appear in both RAGs. Typically there are either nonsense or frameshift mutations and often the patients are homozygous for the defect. OS patients seem to have at least one missense mutation at least in one allele, which can explain the residual activity. The GP correlation is not complete. The analysis indicated that, in the mutational hotspot, R229 (mutated either to Q or W) in RAG2 appeared both in SCID and OS when both homo- and heterozygous patients were considered (Villa *et al.*, 2001). There is no overlap if only homozygous patients are taken into account. Owing to missing structural information, structure–function correlations and protein structural causes of the disease forms cannot be investigated.

20.6 AID Deficiency

AID (activation-induced cytidine deaminase) is a B cell-specific molecule involved in B cell terminal differentiation in secondary lymphoid organs. Mutations in *AICDA* gene (http://bioinf.uta.fi/AICDAbase) lead to an HIGM syndrome characterized by increased (or normal) serum IgM levels and complete absence of other isotypes (IgG, IgA and IgE), pinpointing a defect in the immunoglobulin class switch recombination (CSR) process. AID-deficient patients are therefore prone to bacterial infections from childhood, although a few patients are diagnosed later on. Impressive lymphadenopathies are common, owing to the presence of giant germinal centres. Auto-immunity is found in 20 per cent of patients. Immunoglobulin substitution, associated or not with prophylactic antibiotics, can easily control the infections.

AID is a 198 amino-acid protein, with a nuclear localization signal and a nuclear export signal located respectively in the N and C terminal domains, allowing the protein to shuttle from cytoplasm to nucleus (Figure 20.3). It also possesses a catalytic domain (cytidine deaminase domain) and an APOBEC-1-like domain, the role of which is unknown.

Forty-seven patients have been reported in the Paris group. Seventeen different missense mutations, six nonsense, two in-frame deletions, three frameshift deletions or insertions, three splice site defects and one deletion of the whole coding region have been observed. Mutations are scattered all along the gene, without any hotspot. Most of the mutations are found in consanguineous families and homozygous.

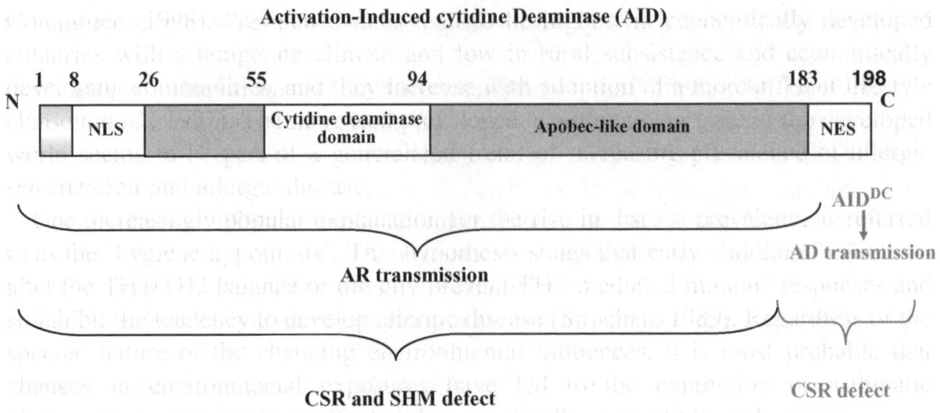

Figure 20.3 Penotype–genotype correlations in AID.

Although clinical phenotype (HIGM) is similar in all patients, it is striking to note that:

- Almost all *AICDA* mutations lead to a defect in CSR, but also in the other event of terminal B cell differentiation [the generation of the somatic hypermutation (SHM) in the variable region of the IgM; Revy *et al.*, 2000; Quartier *et al.*, 2004]. Mutations located in the C terminal part of AID result only in a CSR defect, without affecting the SHM process (Ta *et al.*, 2003). This observation suggests that AID, besides its cytidine deaminase activity, acts in CSR as a docking protein recruiting by its C terminal part CSR-specific AID cofactors.
- An autosomal recessive inheritance is observed in almost all cases, but a peculiar mutation (R190X) located in the nuclear export signal is responsible for a HIGM inherited in an autosomal dominant fashion (Imai *et al.*, 2005).

20.7 WAS

In WAS mutations appear in the gene encoding the Wiskott–Aldrich syndrome protein (WASP; Derry, Ochs and Francke, 1994). WAS is an X-linked recessive ID associated with eczema, haemorrhagic episodes and recurrent severe infections. Symptoms include thrombocytopenia with small platelets, lymphopenia, eczema and recurrent infections. WAS patients have increased risk of autoimmunity and malignancies. Treatment modalities include antibiotics and supportive measures. The only curative therapy is haematopoietic stem cell transplantation.

X-linked thrombocytopenia (XLT) differs from classical WAS since it is characterized by thrombocytopenia with small platelets, and usually the absence of or only mild and transient complications of WAS (Ochs, 1998).

WASP is involved in reorganization of actin cytoskeleton by activating Arp2/3-mediated polymerization, which is controlled by Cdc42, a small GTPase. WASP-related signalling via adaptor proteins and interaction with protein kinases is crucial for cell motility and trafficking as well as for immunological synapse formation. WASP interacts with a very large number of partners and is essential for numerous pathways and processes.

WASP is localized to cytoplasm and consists of several regions, including N-terminal pleckstrin homology/WASP1 homology domain 1 (PH/WH1) (Volkman *et al.*, 2002), GTPase-binding domain (GBD), proline-rich region (PRR), verprolin homology (VH) and C terminal cofilin homology (CH) domain.

Of 169 patients for which genotype and phenotype could be verified, 88 had XLT and 79 WAS (Imai, Nonoyama and Ochs, 2003). In another study 124 patients had WAS and 116 XLT (Notarangelo and Ochs, 2003). Note that some patients were included in both these studies. A recent report on Japanese patients also indicated GP correlations in 50 patients (Imai *et al.*, 2004).

The three-dimensional structure was initially modelled for the PH/WH1 domain (Rong and Vihinen, 2000), and subsequently the structure of the PH/WH1 (Volkman *et al.*, 2002), and GBD domain bound to Cdc42 (Abdul-Manan *et al.*, 1999) have been reported. The PH/WH1 domain has the pleckstrin homology domain fold, which is typical for phosphoinositol binding domains. The WASP PH/WH1 domain is responsible for binding with the polyproline region of WIP, which participates in fillopodium, ruffle and vesicle formation. WIP interacts via its PRR with WASP, to which it binds constitutively.

The WASP mutations affecting the coding region are unevenly distributed along the WASP gene. The most common mutational events in *WAS* gene are missense mutations followed by nonsense and splice site mutations. Missense mutations account for two-thirds in XLT patients but only about 20 per cent in WAS (Notarangelo and Ochs, 2003).

There is strong GP correlation for WASP mutations. In an analysis of 169 patients (Imai, Nonoyama and Ochs, 2003), it was reported that the majority of alterations affect the PH/WH1 domain (89 per cent). In addition, there are four missense mutations in the GBD, three in PRR, one in VH/WH2 and four in CH domain. All the mutations in PRR and VH/WH2 domain are from WAS patients. PH/WH1 and CH domains also contain a number of XLT-causing substitutions. There is very clear separation between the clinical phenotype and genotype. As a third variant for WASP mutations, X-linked neutropenia was noticed in two patients with mutations in GBD domain. These patients are without microcephaly, thrombopenia and eczema.

The WIP and WASP PH/WH1 domain form a tight complex where WIP wraps around the WASP domain based on the three-dimensional structure (Volkman *et al.*, 2002). Of the five direct polyproline binding residues, disease-causing mutations appear only in A56. Interestingly, all six cases lead to XLT. Almost all of the WASP missense mutations are in residues, which are in contact with the peptide by disrupting the WIP interaction or by affecting the interaction either by destabilizing the WASP structure or causing conformational changes that weaken or prevent the interction with WIP. The WASP–WIP complex structure (Volkman *et al.*, 2002) is not for the complete protein; therefore conclusions based on the ends of the WIP are somewhat unclear. The WIP has an exceptionally long contact area with WASP and therefore many mutations even outside the PRR binding site are deleterious. An example is R86, in which four kinds of substitutions are known. Imai, Nonoyama and Ochs (2003) report that 19 patients with mutations in this codon have XLT and 17 have WAS. The disease-related mutations are indicated in Figure 20.2(B) together with the WIP structure.

20.8 XLA

X-linked agammaglobulinaemia (XLA) is caused by mutations in the gene coding for Bruton tyrosine kinase (Btk; Tsukada *et al.*, 1993; Vetrie *et al.*, 1993). In XLA, B cell differentiation is blocked, resulting in severely decreased numbers of B lymphocytes and an almost complete lack of plasma cells, and negligible or very low immunoglobulin levels of all isotypes. B lymphocyte and plasma cell numbers are decreased, whereas T lymphocyte subsets are normal and may show a relative increase. The patients have increased susceptibility to mainly bacterial infections because of virtually absent humoural immune responses. The increased susceptibility most often begins during the first year of life when the transferred maternal Ig has been catabolized. The onset of symptoms varies extensively; most patients show an increased frequency of infections during their first year of life, whereas a few are asymptomatic until adults. Patients are treated with both antibiotics and immunoglobulin substitution therapy.

The first IDbase was established for XLA (Vihinen *et al.*, 1995a) (http://bioinf.uta.fi/BTKbase) and there are altogether 955 patients in the database. The patients represent 823 unrelated families (Table 20.3), and there are 554 unique mutations (58 per cent). XLA mutations are scattered all along the *BTK* gene. Btk belongs to the Tec family of related cytoplasmic protein tyrosine kinases. They consist of five distinct structural domains, from the N-terminus PH domain, Tec homology (TH) domain, Src homology 3 (SH3) domain, SH2 domain and the catalytic kinase domain of about 280 residues. The distribution of the mutations in the domains is approximately according to the length of the domains, except for the TH domain.

Traditionally it has been considered that there are no GP correlations in XLA. However, detailed analysis indicates that there are most likely certain GP-related features. The severity of XLA can be classified according to susceptibility to infections or based on decreased B lymphocyte numbers and/or immunoglobulin levels. Most of the data in the BTKbase is for severe (classical) XLA. However, a high frequency of mutations resulting in mild XLA causes classical XLA in other families. Certain cases have only mild symptoms, especially without severe infections, and with late disease onset. The highest age at diagnosis has been 51 years. The characteristic Ig and CD19 and CD20 levels are also higher for some XLA patients. Together these data indicate that there are also certain GP correlations in XLA. The picture, however, is not clear because the same mutation can even within certain kindreds cause mild or moderate disease in some and in others a classical, severe XLA.

The effects of mutations to BTK structure have been investigated with several methods. It has been possible to provide putative explanation for all the alterations.

Many of the mutations affect functionally significant, conserved residues and the most frequently affected sites are CpG dinucleotides (Ollila, Lappalainen and Vihinen, 1996; Figure 20.1). The majority of the missense mutations in the PH domain are in the inositol compound binding region. In the TH domain the missense mutations affect Zn^{2+} binding. No missense mutations are known to affect the SH3 domain. Most of the amino acid substitutions in the SH2 domain impair phosphotyrosine binding. In the catalytic kinase domain, the mutations are mainly on one face of the molecule, which is in charge of the ATP, Mg^{2+} and substrate binding. Although several missense mutations are expressed as protein, in many instances the mutated protein is unstable (Conley *et al.*, 1998; Futatani *et al.*, 1998).

The 33 CpG dinucleotides in the coding region comprise only 3.3 per cent of the BTK gene, although CG to TG or CA mutations constitute 24 per cent of all single base substitutions. Replacements to related amino acids have been proposed to result in a milder form of disease (Lindvall *et al.*, 2005) and a genotype–phenotype correlation would be apparent if these individuals were identified. Similarly, incomplete forms of splice site mutations are likely to result in a mild disease as well as splice defects resulting in only modestly reduced levels of Btk. All mutations in CpGs are not thought to cause XLA. Already in 1997 we reported, based on 368 entries in BTKbase, that only eight out of 18 CpG containing arginine codons were found to be mutated (Vihinen *et al.*, 1997). Now that there are 955 entries, the number remains still at eight.

Most PH domains bind either plasma membrane phosphoinositides or cytosolic inositol phosphates. The specific binding to inositol phosphates is important for signal-dependent membrane targetting. Although the average sequence identity of PH domains is only 17 per cent, all the determined three-dimensional structures have the same fold of seven β-strands in two antiparallel β-sheets and a C-terminal α-helix.

Amino acid conservation in PH domains was investigated on three levels using information theoretical methods and physicochemical properties (Shen and Vihinen, 2004). Type I conservation means invariance, which is apparent from sequence alignment. In type II, physicochemical properties are conserved. To identify type II conservation, information and entropies were calculated. Since the sequence identity of PH domains is very low, only a small number of residues with type I or II conservation were identified. Type III conservation indicates covariance, inferred by calculating mutual information. Mutations in any of the three types of conserved sites may cause structural destabilization or a loss of function.

Covariant residues may make contacts that maintain structural stability, form binding sites/catalytic centres, or may be otherwise structurally and/or functionally crucial. Regions with low entropy, i.e. high information and significant mutual information, generally contribute significantly to tertiary structure and/or function of a protein. Residues in areas of low entropy are typically conserved in protein families and may be structurally or functionally important.

Covariant residues can form networks of three or more amino acids. The more contacts a residue makes, the more constrained it is with respect to acquiring mutations. Among the 10 residues in the PH domain signature motif for 3-phosphoinositide binding, one is located at a conserved position and seven other residues are present in a network. Among the 11 residues involved in conserved residue networks, six contain XLA-causing mutations (Shen and Vihinen, 2004).

Another method, ranking of amino acid contacts with program RankViaContact (Shen and Vihinen, 2003), revealed that the contact energies of many Btk residues involved in XLA-causing mutations are ranked at the top among all residues.

The structural bases of XLA have been studied by modelling the individual domain structures for PH (Vihinen *et al.*, 1995b), SH3 domain (Zhu *et al.*, 1994), SH2 domain (Vihinen, Nilsson and Smith, 1994), and kinase domain (Vihinen *et al.*, 1994). Subsequently the experimental structures have been determined for PH domain (Hyvönen *et al.*, 1997), SH3 domain (Hansson *et al.*, 1998) and kinase domain (Mao, Zhao and Uckun, 2001). The models and structures have been used to provide reasons for the disease (Vihinen, Nilsson and Smith, 1994; Vihinen *et al.*, 1994, 1995a,b, 1997, 1999, 2001; Zhu *et al.*, 1994; Shen and Vihinen, 2003, 2004). The distribution of the mutations in kinase domain is indicated in Figure 2(C).

Our earlier study indicates that missense mutations will only cause XLA if they affect certain residues, while other locations are tolerant of such alterations. The mutation tolerance in many sites is probably sequence and residue sensitive, such that a particular location in BTK may allow certain amino acid substitution(s). Analysis of PH domain mutations revealed that there are differently conserved positions in protein domains (Shen and Vihinen, 2004). Many of the mutations affect either mutually conserved residues, in the case of PH domain mainly involved in phosphatidyl inositol ligand binding or those forming strong contacts.

There also seem to be GP correlations in XLA. It is apparent that such correlations are very common in many hereditary diseases, the widely reported clear correlations representing part of the continuum. The actual phenotype is in many instances dependent on the type of mutation; some alterations in certain positions may be tolerated, some causing mild and others possibly severe disease.

20.9 Why GP Correlations are not More Common

During the last few years GP correlations have been reported in increasing numbers from several diseases. Despite of this flurry of data, only in a small fraction of diseases can a clear correlation be noticed. There are apparently several reasons for this phenomenon. Disease-causing mutations are not simple on/off switches. There is a spectrum of symptoms in each disorder. The actual symptoms vary for several reasons including genotype, phenotype, environmental factors etc. In the case of IDs, the history of prior infections is very important because of memory cells. There may be large number of modifiers for genes/proteins, the functions of which, owing to natural variations (such as SNPs), may be different.

Many signalling pathways are to a certain extent redundant, meaning that, even a drastic mutation does not necessarily cut off a certain pathway. In the case of *BTK*, signalling, related Tec can substitute it to certain extent (Ellmeier *et al.*, 2000). A small amount of active protein may be enough to prevent at least the most adverse effects of a mutation such as in ADA deficiency (Hershfield, 2003). Another view is provided by RAGs, where nonsense mutations in one allele can lead to milder OS (Villa *et al.*, 2001).

In the case of IDs, only a small fraction of patients and diseases are diagnosed. It has been estimated that some 85 per cent of ID patients remain undiagnosed. Among these cases are many mild disease-causing mutations, which when identified will add to our knowledge about protein function, reactions, interactions and eventually of phenotype–genotype correlations. When put together, these data will allow better understanding of clinical parameters of IDs and eventually better and more personalized treatment.

References

Abdul-Manan N, Aghazadeh B, Liu GA, Majumdar A, Ouerfelli O, Siminovitch KA, Rosen MK. 1999. Structure of Cdc42 in complex with the GTPase-binding domain of the 'Wiskott-Aldrich syndrome' protein. *Nature* **399**: 379–383.

Arredondo-Vega FX, Santisteban I, Daniels S, Toutain S, Hershfield MS. 1998. Adenosine deaminase deficiency: genotype–phenotype correlations based on expressed activity of 29 mutant alleles. *Am J Hum Genet* **63**: 1049–1059.

Aiuti A, Slavin S, Aker M, Ficara F, Deola S, Mortellaro A, Morecki S, Andolfi G, Tabucchi A, Carlucci F, Marinello E, Cattaneo F, Vai S, Servida P, Miniero R, Roncarolo MG, Bordignon C. 2002. *Science* 2410–2413.

Conley ME, Mathias D, Treadaway J, Minegishi Y, Rohrer J. 1998. Mutations in btk in patients with presumed X-linked agammaglobulinemia. *Am J Hum Genet* **62**: 1034–1043.

Conley ME, Notarangelo LD, Etzioni A. 1999. Diagnostic criteria for primary immunodeficiencies. Representing PAGID (Pan-American Group for Immunodeficiency) and ESID (European Society for Immunodeficiencies). *Clin Immunol* **93**: 190–197.

Dale DC, Person RE, Bolyard AA, Aprikyan AG, Bos C, Bonilla MA, Boxer LA, Kannourakis G, Zeidler C, Welte K, Benson KF, Horwitz M. 2000. Mutations in the gene encoding neutrophil elastase in congenital and cyclic neutropenia. *Blood* **96**: 2317–2322.

Derry JM, Ochs HD, Francke U. 1994. Isolation of a novel gene mutated in Wiskott–Aldrich syndrome. *Cell* **78**: 635–644.

Ellmeier W, Jung S, Sunshine MJ, Hatam F, Xu Y, Baltimore D, Mano H, Littman DR. 2000. Severe B cell deficiency in mice lacking the tec kinase family members Tec and Btk. *J Exp Med* **192**: 1611–1624.

Futatani T, Miyawaki T, Tsukada S, Hashimoto S, Kunikata T, Arai S, Kurimoto M, Niida Y, Matsuoka H, Sakiyama Y, Iwata T, Tsuchiya S, Tatsuzawa O, Yoshizaki K, Kishimoto T. 1998. Deficient expression of Bruton's tyrosine kinase in monocytes from X-linked agammaglobulinemia as evaluated by a flow cytometric analysis and its clinical application to carrier detection. *Blood* **91**: 595–602.

Hansson H, Mattsson PT, Allard P, Haapaniemi P, Vihinen M, Smith CI, Hard T. 1998. Solution structure of the SH3 domain from Bruton's tyrosine kinase. *Biochemistry* **37**: 2912–2924.

Hershfield MS. 2003. Genotype is an important determinant of phenotype in adenosine deaminase deficiency. *Curr Opin Immunol* **15**: 571–577.

Imai K, Nonoyama S, Ochs HD. 2003. WASP (Wiskott–Aldrich syndrome protein) gene mutations and phenotype. *Curr Opin Allergy Clin Immunol* **3**: 427–436.

Imai K, Morio T, Zhu Y, Jin Y, Itoh S, Kajiwara M, Yata J, Mizutani S, Ochs HD, Nonoyama S. 2004. Clinical course of patients with WASP gene mutations. *Blood* **103**: 456–464.

Imai K, Zhu Y, Revy P, Morio T, Mizutani S, Fischer A, Nonoyama S, Durandy A. 2005. Analysis of class switch recombination and somatic hypermutations in patients affected with autosomal dominant hyper-IgM type 2. *Clin Immunol* **115**: 277–285.

Hyvönen M, Saraste M. 1997. Structure of the PH domain and Btk motif from Bruton's tyrosine kinase: molecular explanations for X-linked agammaglobulinaemia. *EMBO J* **16**: 3396–2304.

Johnston J, Kelley RI, Feigenbaum A, Cox GF, Iyer GS, Funanage VL, Proujansky R. 1997. *Am J Hum Genet* **61**(5): 1053–1058.

Lindvall JM, Blomberg KE, Valiaho J, Vargas L, Heinonen JE, Berglof A, Mohamed AJ, Nore BF, Vihinen M, Smith CI. 2005. Bruton's tyrosine kinase: cell biology, sequence conservation, mutation spectrum, siRNA modifications, and expression profiling. *Immunol Rev* **203**: 200–215.

Mao C, Zhou M, Uckun FM. 2001. Crystal structure of Bruton's tyrosine kinase domain suggests a novel pathway for activation and provides insights into the molecular basis of X-linked agammaglobulinemia. *J Biol Chem* **276**: 41435–41443.

Moshous D, Callebaut I, de Chasseval R, Corneo B, Cavazzana-Calvo M, Le Deist F, Tezcan I, Sanal O, Bertrand Y, Philippe N, Fischer A, de Villartay JP. 2001. Artemis, a novel DNA double-strand break repair/V(D)J recombination protein, is mutated in human severe combined immune deficiency. *Cell* **105**: 177–186.

Notarangelo LD, Ochs HD. 2003. Wiskott–Aldrich Syndrome: a model for defective actin reorganization, cell trafficking and synapse formation. *Curr Opin Immunol* **15**: 585–591.

Ochs HD. 1998. The Wiskott–Aldrich syndrome. *Semin Hematol* **35**: 332–345.

O'Driscoll M, Cerosaletti KM, Girard PM, Dai Y, Stumm M, Kysela B, Hirsch B, Gennery A, Palmer SE, Seidel J, Gatti RA, Varon R, Oettinger MA, Neitzel H, Jeggo PA, Concannon P. 2001. DNA ligase IV mutations identified in patients exhibiting developmental delay and immunodeficiency. *Mol Cell* **8**: 1175–1185.

Ollila J, Lappalainen I, Vihinen M. 1996. Sequence specificity in CpG mutation hotspots. *FEBS Lett* **396**: 119–122.

Quartier P, Bustamante J, Sanal O, Plebani A, Debre M, Deville A, Litzman J, Levy J, Fermand JP, Lane P, Horneff G, Aksu G, Yalcin I, Davies G, Tezcan I, Ersoy F, Catalan N, Imai K, Fischer A, Durandy A. 2004. Clinical, immunologic and genetic analysis of 29 patients with autosomal recessive hyper-IgM syndrome due to activation-induced cytidine deaminase deficiency. *Clin Immunol* **110**: 22–29.

Revy P, Muto T, Levy Y, Geissmann F, Plebani A, Sanal O, Catalan N, Forveille M, Dufourcq-Labelouse R, Gennery A, Tezcan I, Ersoy F, Kayserili H, Ugazio AG, Brousse N, Muramatsu M, Notarangelo LD, Kinoshita K, Honjo T, Fischer A, Durandy A. 2000. Activation-induced cytidine deaminase (AID) deficiency causes the autosomal recessive form of the Hyper-IgM syndrome (HIGM2). *Cell* **102**: 565–575.

Riikonen P, Vihinen M. 1999. MUTbase: maintenance and analysis of distributed mutation databases. *Bioinformatics* **15**: 852–859.

Rong SB, Vihinen M. 2000. Structural basis of Wiskott–Aldrich syndrome causing mutations in the WH1 domain. *J Mol Med* **78**: 530–537.

Samarghitean C, Valiaho J, Vihinen M. 2004. Online registry of genetic and clinical immunodeficiency diagnostic laboratories, IDdiagnostics. *J Clin Immunol* **24**: 53–61.

Schwarz K, Gauss GH, Ludwig L, Pannicke U, Li Z, Lindner D, Friedrich W, Seger RA, Hansen-Hagge TE, Desiderio S, Lieber MR, Bartram CR. 1996. RAG mutations in human B cell-negative SCID. *Science* **274**: 97–99.

Shen B, Vihinen M. 2003. RankViaContact: ranking and visualization of amino acid contacts. *Bioinformatics* **19**: 2161–2162.

Shen B, Vihinen M. 2004. Conservation and covariance in PH domain sequences: physicochemical profile and information theoretical analysis of XLA-causing mutations in the Btk PH domain. *Protein Eng Des Select* **17**: 267–276.

Ta VT, Nagaoka H, Catalan N, Durandy A, Fischer A, Imai K, Nonoyama S, Tashiro J, Ikegawa M, Ito S, Kinoshita K, Muramatsu M, Honjo T. 2003. AID mutant analyses indicate requirement for class-switch-specific cofactors. *Nat Immunol* **4**(9): 843–848.

Tsukada S, Saffran DC, Rawlings DJ, Parolini O, Allen RC, Klisak I, Sparkes RS, Kubagawa H, Mohandas T, Quan S, Belmont JW, Cooper MD, Conley ME, Witte ON. 1993. Deficient expression of a B cell cytoplasmic tyrosine kinase in human X-linked agammaglobulinemia. *Cell* **72**: 279–290.

Väliaho J, Riikonen P, Vihinen M. 2000. Novel immunodeficiency data servers. *Immunol Rev* **178**: 177–185.

Väliaho J, Pusa M, Ylinen T, Vihinen M. 2002. IDR: the ImmunoDeficiency Resource. *Nucl Acids Res* **30**: 232–234.

Väliaho J, Samarghitean C, Piirilä H, Pusa M, Vihinen M. 2005a. *Primary Immunodeficiency Diseases*, Ochs HD, Smith CIE, Puck JM (eds).

Väliaho J, Riikonen P, Vihinen M. 2005b. Biomedical data description with XML. Distribution of immunodeficiency fact files – from Web to WAP. *BMC Med Inform Dec Making*, **5**: 21.

Verpy E, Biasotto M, Brai M, Misiano G, Meo T, Tosi M. 1996. Exhaustive mutation scanning by fluorescence-assisted mismatch analysis discloses new genotype-phenotype correlations in angiodema. *Am J Hum Genet* **59**: 308–319.

Vetrie D, Vorechovsky I, Sideras P, Holland J, Davies A, Flinter F, Hammarstrom L, Kinnon C, Levinsky R, Bobrow M, Smith CIE, Bentley DR. 1993. The gene involved in X-linked agammaglobulinaemia is a member of the src family of protein–tyrosine kinases. *Nature* **361**: 226–233.

Vihinen M, Nilsson L, Smith CIE. 1994a. Structural basis of SH2 domain mutations in X-linked agammaglobulinemia. *Biochem Biophys Res Commun* **205**: 1270–1277.

Vihinen M, Vetrie D, Maniar HS, Ochs HD, Zhu Q, Vorechovský I, Webster ADA, Notarangelo LD, Nilsson L, Sowadski JM, Smith CIE. 1994b. Structural basis for chromosome X-linked agammaglobulinemia: a tyrosine kinase disease. *Proc Natl Acad Sci USA* **91**: 12803–12807.

Vihinen M, Cooper MD, de Saint Basile G, Fischer A, Good RA, Hendriks RW, Kinnon C, Kwan S-P, Litman GW, Notarangelo LD, Ochs HD, Rosen FS, Vetrie D, Webster ADB, Zegers BJM, Smith CIE. 1995a. BTKbase: a database of XLA-causing mutations. International Study Group. *Immunol Today* **16**: 460–465.

Vihinen M, Zvelebil MJ, Zhu Q, Brooimans RA, Ochs HD, Zegers BJ, Nilsson L, Waterfield MD, Smith CI. 1995b. Structural basis for pleckstrin homology domain mutations in X-linked agammaglobulinemia. *Biochemistry* **34**: 1475–1481.

Vihinen M, Belohradsky BH, Haire RN, Holinski-Feder E, Kwan S-P, Lappalainen I, Lehväslaiho H, Lester T, Meindl A, Ochs HD, Ollila J, Vorechovský I, Weiss M, Smith CIE. 1997. BTKbase, mutation database for X-linked agammaglobulinemia (XLA). *Nucl Acids Res* **25**: 166–171.

Vihinen M, Kwan SP, Lester T, Ochs HD, Resnick I, Valiaho J, Conley ME, Smith CI. 1999. Mutations of the human BTK gene coding for bruton tyrosine kinase in X-linked agammaglobulinemia. *Hum Mutat* **13**: 280–285.

Vihinen M, Arredondo-Vega FX, Casanova JL, Etzioni A, Giliani S, Hammarstrom L, Hershfield MS, Heyworth PG, Hsu AP, Lahdesmaki A, Lappalainen I, Notarangelo LD, Puck JM, Reith W, Roos D, Schumacher RF, Schwarz K, Vezzoni P, Villa A, Valiaho J, Smith CI. 2001. Primary immunodeficiency mutation databases. *Adv Genet* **43**: 103–188.

Villa A, Santagata S, Bozzi F, Giliani S, Frattini A, Imberti L, Gatta LB, Ochs HD, Schwarz K, Notarangelo LD, Vezzoni P, Spanopoulou E. 1998. Partial V(D)J recombination activity leads to Omenn syndrome. *Cell* **93**: 885–896.

Villa A, Sobacchi C, Notarangelo LD, Bozzi F, Abinun M, Abrahamsen TG, Arkwright PD, Baniyash M, Brooks EG, Conley ME, Cortes P, Duse M, Fasth A, Filipovich AM, Infante AJ, Jones A, Mazzolari E, Muller SM, Pasic S, Rechavi G, Sacco MG, Santagata S, Schroeder ML, Seger R, Strina D, Ugazio A, Valiaho J, Vihinen M, Vogler LB, Ochs H, Vezzoni P, Friedrich W, Schwarz K. 2001. V(D)J recombination defects in lymphocytes due to RAG mutations: severe immunodeficiency with a spectrum of clinical presentations. *Blood* **97**: 81–88.

Volkman BF, Prehoda KE, Scott JA, Peterson FC, Lim WA. 2002. Structure of the N-WASP EVH1 domain-WIP complex: insight into the molecular basis of Wiskott–Aldrich Syndrome. *Cell* **111**: 565–576.

Wilson DK, Rudolph FB, Quiocho FA. 1991. Atomic structure of adenosine deaminase complexed with a transition-state analog: understanding catalysis and immunodeficiency mutations. *Science* **252**: 1278–1284.

Zhu Q, Zhang M, Rawlings DJ, Vihinen M, Hageman T, Saffran DC, Kwan S-P, Nilsson L, Smith CIE, Witte ON, Chen S-H, Ochs HD. 1994. Deletion within the Src homology domain 3 of Bruton's tyrosine kinase resulting in X-linked agammaglobulinemia (XLA). *J Exp Med* **180**: 461–470.

21

Transcriptional Profiling of Dendritic Cells in Response to Pathogens

Maria Foti, Francesca Granucci, Mattia Pelizzola, Norman Pavelka, Ottavio Beretta, Caterina Vizzardelli, Matteo Urbano, Ivan Zanoni, Giusy Capuano, Francesca Mingozzi and **Paola Ricciardi-Castagnoli**

Abstract

The immune system has developed mechanisms to detect and initiate responses to a continual barrage of immunological challenges. Dendritic cells (DC) play a major role as immune surveillance agents. To accomplish this function, DC are equipped with highly efficient mechanisms to detect pathogens, to capture, process and present antigens, and to initiate T-cell responses. The recognition of molecular signatures of potential pathogens is accomplished by membrane receptor of the Toll-like family, which activates DC, leading to the initiation of adaptive immunity. High-density DNA microarray analysis of host gene expression provides a powerful method of examining microbial pathogens from a novel perspective. The ability to survey the responses of a large subset of the host genome and to find patterns among the profiles from many different microorganisms and hosts allow fundamental questions to be addressed about the basis of pathogen recognition, the features of the interaction between host and pathogen and the mechanisms of host defence and microbial virulence. The biological insights thus gained are likely to lead to major shifts in our approach to the diagnosis, treatment, assessment of prognosis and prevention in many types of infectious diseases within a decade.

Keywords

dendritic cells; microarray; transcription analysis; pathogen interaction; microbial stimuli; expression profiling; innate immunity

Immunogenomics and Human Disease Edited by András Falus

21.1 Transcriptional Profiling to Study the Complexity of the Immune System

The immune response is extraordinarily complex. It involves dynamic interaction of a wide array of tissues, cells and molecules. Traditional approaches are based on one-by-one gene analysis, shying away from complexity, but providing detailed knowledge of a particular molecular entity. The completion of draft sequences of the human and mouse genomes offers many opportunities for gene discovery in the field of immunology through the application of the methods of computational genomics. In concert with emerging genomic and proteomic technologies, it permits the definition of the biology of the immune system. The initiation and regulation of the immune response is complicated and occurs on many levels. Multicellular organisms have been obliged to develop multifaceted innate and adaptive immune systems to cope with the challenges to survival originating from microorganisms and their products. The diversity of innate immune mechanisms is in large part conserved in all multicellular organisms (Mushegian and Medzhitov, 2001). Some basic principles of microbial recognition and response are emerging, and recently the application of computational genomics has played an important role in extending such observations from model organisms, such as *Drosophila*, to higher vertebrates, including humans.

The analysis of gene expression in tissues, cells and biological systems has evolved in the last decade from the analysis of a selected set of genes to an efficient high-throughput whole-genome screening approach of potentially all genes expressed in a tissue or cell sample. Development of sophisticated methodologies such as microarray technology allows an open-ended survey to identify comprehensively the fraction of genes that are differentially expressed between samples and define the sample's unique biology. This discovery-based research provides the opportunity to characterize either new genes with unknown function or genes not previously known to be involved in a biological process.

Microarrays were developed in 1995 (Schena *et al.*, 1995) and have now been widely applied in the field of immunology. Two types of microarrays are commonly used, two-colour microarrays and oligonucleotide microarrays. In a two-colour microarray, collections of DNA samples [i.e. expressed sequence tag (ESTs) or other clones] are deposited onto a glass slide using robotics. These microarrays are highly flexible as they may be constructed from anonymous clones found in genomic, subtractive, differentially displayed or normalized libraries or from commercially synthesized long ($n = 50-70$) oligonucleotides (Duggan *et al.*, 1999). Oligonucleotide arrays are constructed from 25-mer oligonucleotides synthesized *in situ* on a solid substrate (Lipshultz *et al.*, 1999). This type of microarray requires exact sequence information and bioinformatic design prior to the construction of the microarray. To date, oligonucleotide microarrays cannot be produced in-house and must be purchased from commercial sources. They are still expensive enough to limit the number and scale of experiments that can be performed by a typical laboratory.

However, oligonucleotide microarrays are highly consistent and offer sequence-specific detection of gene expression, which is especially important in the study of gene families. With both types of microarray analysis, data aggregation from multiple experiments is possible, allowing higher order analyses of transcript profiles.

Large-scale gene expression analysis is of great relevance in the field of immunology to generate a global view of how the immune system attacks invading microorganisms, maintains tolerance or creates a memory for past infections. Besides the availability of large-scale or full-genome microarrays, specialized microarrays that contain a tailored set of DNA sequences related to immunology are generated and used. Fundamental questions in immunology address how the immune system distinguishes between self and nonself, and how immune cell differentiation and growth are regulated. The exciting part of microarray studies is that the many data points that are generated cause unpredictable and unexpected results, which may lead to new insights in immunology.

The study of host–pathogen interactions is instrumental for the control of infectious diseases. Host eukaryotes are constantly exposed to attacks by microbes seeking to colonize and propagate in host cells. To counteract them, host cells utilize a whole battery of defence systems to combat microbes. However, in turn, successful microbes evolve sophisticated systems to evade host defence. As such, interactions between hosts and pathogens are perceived as evolutionary arms races between genes of the respective organisms (Bergelson *et al.*, 2001; Kahn *et al.*, 2002; Woolhouse *et al.*, 2002). Any interaction between a host and its pathogen involves alterations in cell signalling cascades in both partners that may be mediated by transcriptional or post-translational changes. The basic challenge is how to select target genes to be studied in detail from among thousands of genes encoded in the genome. Transcriptomics is one of the methodologies to serve this purpose. Analytical techniques for transcriptomics include differential display (DD, Liang and Pardee, 1992), cDNA-amplified restriction fragment length polymorphism (AFLP) (Bachem *et al.*, 1996), random EST sequencing (Kamoun *et al.*, 1999), microarray (Schena *et al.*, 1995), serial analysis of gene expression (SAGE, Velculescu *et al.*, 1997) and massively parallel signature sequencing (MPSS, Brenner *et al.*, 2000). Among them, microarray has recently been used more frequently than other platforms. Most of the gene expression studies addressing host–pathogen interactions in reality have examined either host or pathogen. However, the simultaneous monitoring of gene expression of both host and pathogen, preferably during the infection process and *in situ*, has already been investigated (Birch and Kamoun, 2000). This approach is necessary to elucidate the host–pathogen interplay in molecular detail.

21.2 DC Subsets and Functional Studies

Dendritic cells (DC) are professional antigen-presenting cells, which play a crucial role in initiating immune responses. DC are continuously produced from haematopoietic stem cells within the bone marrow and are subdivided into subsets characterized by their tissue distribution, morphology, surface markers and functions. DC are present in most tissues. We will refer to them as resident DC (R-DC). In particular,

R-DC are found in those tissues that interface with the external environment, where microorganisms can enter.

The R-DC are present in an immature state, they are highly phagocytic and continuously internalize soluble and particulate antigens that are processed and presented to T cells. The interaction of immature DC with T cells induces an abortive T cell activation with the induction of T cell anergy (Hawiger *et al.*, 2001; Sotomayor *et al.*, 2001) or the differentiation of regulatory T cells (Jonuleit *et al.*, 2001). In contrast, microbial stimuli that are recognized through a complex DC innate receptor repertoire induce DC maturation and migration that is completed after 24 h (Granucci *et al.*, 2003a; Figure 21.1). Mature and migratory DC (M-DC) express high levels of stable peptide±MHC complexes and costimulatory molecules at the cell surface and efficiently prime naive T cells. The extent and the type of innate and adaptive responses induced by DC are related to the type of signal they have received. Indeed, DC are able to distinguish different pathogens through the expression of pattern-recognition receptors (PRRs) that interact with specific microorganism molecular structures, called microbe-associated molecular patterns (PAMPs). These constitutive and conserved microbial structures are absent in host mammalian cells and represent the signature of microorganisms (Medzhitov, 2001). Well-defined PRRs are Toll-like

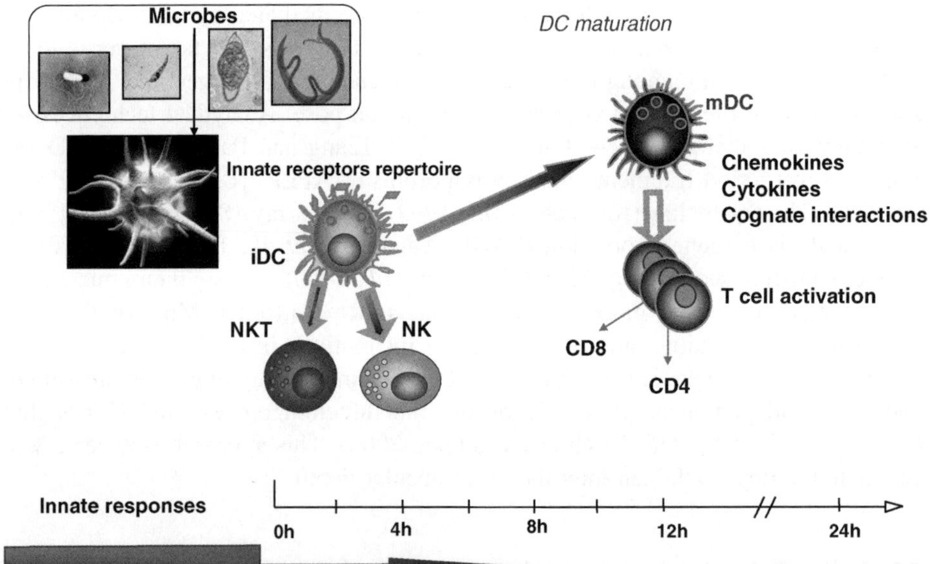

Figure 21.1 Coordinated process of DC maturation. After interaction with appropriate stimuli, DC respond to danger signals by stimulating a controlled and effective immune response. Dendritic cells first regulate leukocyte recruitment at the site of inflammation, through the production of chemokines and inflammatory cytokines, and then they acquire migratory properties and undergo a rapid switch in chemokine receptors. This allows them to leave the inflamed tissue and to reach the lymph node T cell area.

receptors (TLRs). The stimulation of different TLRs at the DC surface results in the activation of different signalling pathways and the induction of diverse maturation processes that influence the outcome of adaptive immunity.

DC are a heterogeneous group of cells that display differences in anatomic localization, cell surface phenotype and function. However, DC have several features in common (Banchereau and Steinman, 1998; Hart, 1997).

Murine DC have been classified into two main lineages: myeloid DC and lymphoid DC. The definition of DC subset phenotypes and the attribution of specific functions to defined DC stages has been a very difficult task; DC are characterized by a very high functional plasticity and can adapt their responses upon antigen encounter; they are able to segregate in time different functions, which will dictate the outcome of the immune response.

At least five major populations of DC have been described in the central and peripheral lymphoid organs of mice. In murine spleen, three DC subtypes are delineated, namely $CD4^{-}CD8\alpha^{+}DEC205^{+}CD11b^{-}$, $CD4^{+}CD8\alpha^{-}DEC205^{-}CD11b^{+}$ and $CD4^{-}CD8\alpha^{-}DEC205^{-}CD11b^{+}$ (Kamath *et al.*, 2000; Shortman, 2000; Vremec *et al.*, 2000). In lymph nodes, these three subtypes are present together with a fourth population, $CD4^{-}CD8\alpha^{low}DEC205^{+}$ with various levels of CD11b (Anjuere *et al.*, 1999; Shortman, 2000). The mouse thymus appears to contain two DC types, one that overlaps with a lymph node subtype and one that may be unique, $CD4^{-}CD8^{\alpha-/low}DEC205^{+}CD11b^{-}$ and $CD4^{-}CD8\alpha^{-}DEC205^{+}CD11b^{-}$, respectively. All three subtypes were classed as mature, because they expressed CD80, CD86 and CD40 and efficiently activated allogeneic T cells (Vremec *et al.*, 2000).

In the last decade, understanding of DC physiology and function improved, but many aspects of the mouse DC subset biology remain unclear, including their ontogeny, function, cytokine production potential and Ag presentation capacity. Functional comparisons of DC subsets have been controversial. Given this, it would be useful to define DC subsets at the molecular level in an attempt to understand more about their relationship to each other, provide better insights into their biology, and find alternative markers for their isolation. Toward this end, different studies have been conducted to study DC biology at the molecular level both in human and in the mouse systems. Here, we review microarray research to characterize gene expression in the immunology of DC to broaden our understanding of the biology of immunologic processes.

In this regard, microarray analysis has been conducted using mRNA purified from $CD11c^{high}$ splenic DC subsets immediately after cell isolation (Edwards *et al.*, 2003). The data presented in this study suggested that all three subsets constitute unique populations and that the $CD4^{+}$ and the double negative DC subsets are more similar to each other than to the $CD8\alpha^{+}$ DC subset. Selective expression of some previously reported genes was confirmed and CD5, CD72 and CD22 identified as novel markers for the $CD8\alpha^{-}$ DC subsets (Edwards *et al.*, 2003).

The lists of differentially expressed genes included many immunological relevant molecules not previously described as differentially expressed in DC, such as some chemokines and their receptors (e.g. CXCR1, CXCL9 and CCL22), signalling components (e.g. RyR3, IFN regulatory factor 4 and STAT-4), MHC molecules (e.g. H-2DMb2), proteins involved in resistance to CTL lysis (e.g. SPI6), and many others. Differentially expressed genes have been grouped according to Gene Ontology biological process and they showed a significant association with responses to biotic stimuli, signal transduction from cell surface receptors and development. Such an association was not seen when total DC-expressed genes were similarly grouped. Therefore, this suggests that $CD4^+$ and $CD8\alpha^+$ DC are likely to show significant differences in the biological process they participate in.

We have conducted a similar study applying global transcriptional analysis on DC to better understand DC biology at the molecular level. To investigate the effects of different stimuli on DC function, we have used the Affymetrix GeneChip® technology (Lockhart *et al.*, 1996), which permits the simultaneous analysis of the expression of thousands of genes. These analyses require homogeneous cell populations to avoid dilution and contamination of information. Bone marrow-derived mouse DC are extremely unstable and it is not possible to obtain homogeneous immature DC without contamination with mature and intermediate DC. Cell lines that closely parallel fresh DC functions are a valid alternative. Thus we took advantage of the previously described mouse DC line, D1 (Winzler *et al.*, 1997). D1 cells are a splenic, myeloid and growth factor-dependent DC line that can be maintained indefinitely in culture in the immature state. This cell line can be driven to full maturation using different stimuli. In particular, D1 cells reach a mature state 18 h after lipopolysaccharide (LPS) or bacterial stimulation, as assessed by phenotypical (upregulation of class II and costimulatory molecules) and functional characteristics, such as antigen presentation, inhibition of migration, blocking of antigen uptake, cytoskeleton rearrangements (Winzler *et al.*, 1997; Rescigno *et al.*, 1998).

The transcription profile is a major determinant of cellular phenotype and function. Differences in gene expression are indicative of morphological, phenotypical and functional changes induced in a cell by environmental factors and perturbations. The most common stimuli used to activate DC are tumour necrosis factor (TNF)-α and LPS. Thus, to evaluate the differences in DC response to LPS and TNF-α we performed a genome-wide transcriptional analysis of activated DC and compared it with the expression analysis performed on immature DC (Granucci *et al.*, 2001b; Table 21.1). We observed that only LPS was able to induce the transcription of genes responsible for DC growth arrest and it was much more effective than TNF-α in activating the expression of genes involved in antigen processing and T cell stimulation. Moreover LPS- but not TNF-α-stimulated DC expressed genes able to control the inflammation during the immune response. The transcriptional program analysis suggests that TNF-α is an ineffective stimulus for terminal DC differentiation. The observation that the expression of several genes found with the GeneChip® utilization corresponds to a number of functional characteristics of DC validates the applicability of the oligonucleotide microarray technology for monitoring gene expression in mouse cells.

Similar transcriptional studies have also been conducted on human DC subsets. In humans, DC are also found as precursor populations in bone marrow and blood

Table 21.1 Differential gene expression analysis in LPS- and TNFα-stimulated vs unstimulated D1 cells

ID[a]	Gene name	LPS 6 h	LPS 18 h	TNFα 6 h	TNFα 18 h
		Cell surface and membrane proteins			
L09754	CD30L	NC	NC	I 36[*]	I 42[b]
U12763	OX40L	NC	I 70	NC	NC
M83312	CD40	I 345[b]	I 201[b]	NC	I 4.3[b]
M34510	CD14	U 11.6	U 3.2	NC	NC
Y08026	IAP38	S 73[c]	S 73[c]	NC	S 73[c]
U10484	Jaw1	S 242[c]	S 242[c]	NC	D 2.6
X93328	F4-80	D 2.6	D 2.4	NC	NC
Z16078	CD53	D 3.9	D 7.3	D 2	D 2.3
X68273	Macrosialin	D 2.3	D 2.1	NC	NC
U47737	TSA1	D 4.1	D 9.9	NC	NC
U18372	CD37	D 3.8	D 2.9	NC	NC
U05265	gp49	NC	D 4.9	D 2.9	D 3.7
L08115	CD9	NC	D 4.3	D 2.9	D 2.4
X72910	HSA-C	NC	D 5.5	D 2.8	D 2.5
U25633	TMP	S 165[c]	S 165[c]	NC	D 7.4
		Cell cycle and apoptosis			
D86344	TIS	NC	I 273[b]	NC	NC
L49433	c-IAP-1	I 71[b]	I 68[b]	NC	I 77[*]
L16846	BTG1	U 3.2	U 2.4	NC	NC
M83749	Cyclin D2	U 3.7	U 3.7	NC	NC
U19860	GAS	NC	U 2.8	NC	NC
D50494	RCK	U 3	U 2.2	NC	NC
M64403	CYL-1	D 2.4	D 3.8	NC	NC
U70210	TR2L	NC	NC	D 2.4	D 3.5
U58633	p34CDC2	NC	D 5.5	D 2.2	D 3.2
X82786	Ki-67	NC	D 6.4	NC	D 2.6
Z26580	Cyclin A	NC	S 147[c]	NC	NC
X66032	Cyclin B2	NC	S 108[c]	NC	S 108[c]
X64713	Cyclin B1	NC	S 47[c]	NC	S 47[c]
D86725	mMCM2	NC	S 125[c]	D 2.7	NC
Z72000	BTG3	S 46[c]	S 46[c]	NC	S 46[c]
		Antigen processing and presentation			
U60329	PA28	U 3.2	U 3.2	NC	NC
X97042	UBcM4	U 2.3	U 2.6	U 2.2	U 2.6
M55637	TAP-1cas	NC	U 2.8	NC	NC
U35323	H-2Mβ2	D 12.5	D 9.2	D 2.2	D 2.2
U35323	H-2Mα	S 304[c]	S 304[c]	NC	NC
U35323	H-2Mβ1	D 6.9	S 149[c]	NC	NC
D83585	Proteasome Z subunit	NC	NC	D 2.5	D 2.2
K01923	I-Aα	D 2.8	D 3.5	NC	NC
V01527	I-Aβ	NC	D 3.7	NC	NC
		Secreted molecules			
J03783	IL-6	I 45[b]	I 78[b]	NC	NC

(Continued)

Table 21.1 (*Continued*)

ID[a]	Gene name	LPS 6 h	LPS 18 h	TNFα 6 h	TNFα 18 h
M86671	IL-12p40	I 820[b]	I 759[b]	NC	NC
M64404	IL-1RA	U 13.4	U 9	NC	NC
X03505	Serum amyloid	NC	I 256[b]	NC	I 67[b]
M15131	IL-1β	U 82	U 34.8	U 7.4	U 11.8
M73061	MIP-1α	U 3.9	U 5.5	NC	NC
X53798	MIP-2	U 20.2	U 5.7	NC	NC
U02298	RANTES	U 3.7	U 28	U 3	U 3.4
X58861	C1qα	S 628[c]	S 628[c]	NC	NC
X66295	C1qC	D 15	S 656[c]	D 3.7	D 3.5
X16151	ETA1	S 485[c]	S 485[c]	NC	D 7.9
M19681	JE	D 5.9	S 348	D 6	D 4.6
X06086	MEP	D 2.6	D 2.9	D 2	S 126[c]
U50712	MCP5	S 83[c]	S 83[c]	S 83[c]	S 83[c]
X83601	PTX3	D 3.8	S 211[c]	D 3.6	S 211[c]
M22531	C1qB	D 8.7	D 42.8	D 2.4	D 4.2
M58004	C10	D 3.1	D 6.5	D 2.6	D 5.2
L19932	βig-h3	NC	D 5.1	NC	NC
X12905	Properdin	D 4.7	D 16.2	D 2.2	NC

[a]Accession number. NC, no change in the level of expression; I, induced (detected only in stimulated cells); U, upregulated; D, downregulated; S, suppressed (detected only in unstimulated cells). [b]AvgDiff reached after stimulation. [c]AvgDiff in the baseline, values without superscript letters represent the fold change in stimulated vs unstimulated cells.

and as more mature forms in lymphoid and nonlymphoid tissues. Three distinct subtypes of human DC have been delineated based on studies of skin DC (Cerio *et al.*, 1989), DC generated *in vitro* from CD34$^+$ haematopoietic progenitors (Caux *et al.*, 1996), and blood DC precursors (Romani *et al.*, 1994). Human skin contains two of the three DC subtypes in immature form: Langerhans cells (LC) and interstitial DC. Both subtypes emerge in cultures from CD34$^+$ bone marrow and CD11c$^+$ blood precursors in the presence of GM-CSF and either IL-4 or TNF-α (Caux *et al.*, 1999; Romani *et al.*, 1994; Sallusto and Lanzavecchia, 1994). The CD11c$^+$ DC precursor expresses myeloid markers, including CD13 and CD33. Upon activation by CD40L, immature myeloid DC undergo maturation and produce IL-12 (Cella *et al.*, 1996).

Human DC have also been characterized at molecular level using global strategies. The only concern until now about these studies has been the source and the purification issue in obtaining these cell types. Until approximately 10 years ago, it was challenging to perform studies on human DC because of the difficulty of isolating them. This changed with the development of systems for the differentiation of DC from precursor cells that are easier to isolate. One of the most common methods for preparing DC is by the differentiation of monocytes with IL-4 and GM-CSF to immature DC. Although less commonly, CD34$^+$ stem cells have also been used as a source for precursor cells in DC profiling studies.

Human DC transcriptional profiling has been determined during the maturation process; maturation of DC occurs through their interaction with infectious agents or cytokines, or contact with other cell types. Infectious agents or components from such agents used for profiling studies included LPS (Hashimoto *et al.*, 2000; Baltathakis *et al.*, 2001; Matsunga *et al.*, 2002; Messmer *et al.*, 2003) poly I:C (Huang *et al.*, 2001), bacteria (e.g. *Escherichia coli*; Granucci *et al.*, 2001b; Huang *et al.*, 2001), yeast (e.g. *Candida albicans*; Huang *et al.*, 2001), parasites (e.g. *Leishmania major*; Chaussabel *et al.*, 2003), and viruses (e.g. influenza virus, HIV-1; Huang *et al.*, 2001; Izmailova *et al.*, 2003). Cytokines and molecules reflecting cell–cell contact that have been used to mature DC for profiling studies include TNF-α (Dietz *et al.*, 2000; Lapteva *et al.*, 2001; Le Naour *et al.*, 2001; Moschella *et al.*, 2001; Ahn *et al.*, 2002) and CD40L (Moschella *et al.*, 2001; Bleharski *et al.*, 2001).

Other transcriptional profile studies are performed on freshly isolated cells. Two types of human DC have been identified from blood, a $CD11c^{-}CD123^{+}$ plasmacytoid and a $CD11c^{+}CD123^{-}$ myeloid dendritic cell type (Kohrgruber *et al.*, 1999). The plasmacytoid DC are potent antigen-presenting cells (Brière *et al.*, 2003). They are best known for their capacity to produce high levels of interferon (IFN)α upon stimulation (Brière *et al.*, 2003), unlike myeloid DC. The myeloid class of DC more closely represents the DC that are obtained by the differentiation of precursor cells with GM-CSF/IL-4. They have a strong ability to take up, process, and present antigen (Kohrgruber *et al.*, 1999). Thus myeloid DC, unlike plasmacytoid DC, are capable of being stimulated by agents such as LPS and zymosan.

21.3 DC at the Intersection Between Innate and Adaptive Immunity

When higher organisms are exposed to pathogenic microorganisms, innate immune responses occur immediately, both in terms of cell activation and inflammation. The initial response is characterized by phagocytosis or endocytosis and subsequent destruction or degradation of pathogens.

At the initial stage of primary infection, DC constitute an integral part of the innate immune response, supported by the recruitment activity of bone-marrow-derived immune cells and various resident tissue cells.

DC and macrophages are acutely activated; during innate responses they produce pro-inflammatory cytokines and chemokines such as TNF-α and interleukin (IL)-1β, and effector cytokines such as IL-12 (p40 subunit) and type I IFNs (Manger and Relman, 2000). This cytokine and chemokine production occurs in waves at a precise time point during the process of DC maturation, as shown by the MIP-1γ and MIP-1β gene expression profiling (Figure 21.2). During this phase, DC enhance presentation of the products of pathogen degradation (antigenic peptides) via the MHC class I or II presentation pathway to antigen reactive T cells, and they produce bactericidal effector substances such as nitric oxide. Thus, innate immune cells and in particular DC represent not only a first line of defence towards infections but also play an

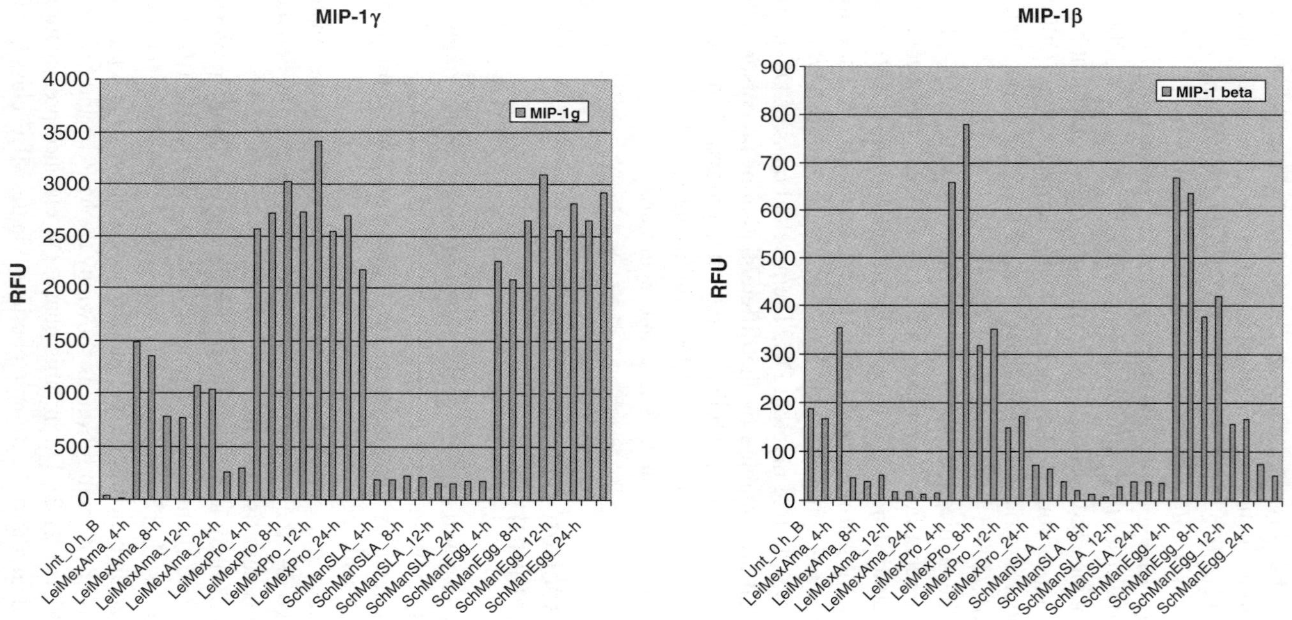

Figure 21.2 Examples of Chemokine gene expression profiling as measured by microarray analysis. MIP-1γ mRNA production is upregulated between 4 and 24 h of parasite stimulation whereas MIP-β shows maximum level of expression at 4 h. Unt, untreated cell; LeiMexPro, *Leishmania mexicana* promastigote; SchManEgg, *Schistosoma mansoni* eggs. (A colour reproduction of this figure can be viewed in the colour plate section)

instructive role in shaping the adaptive immune responses (Fearon and Locksley, 1996).

Another interesting feature of DC is their ability to delay the processing of the internalized antigens by antigen retention in a storage compartment that we have identified with its mildly acidic pH content (Lutz *et al.*, 1997). In these vesicles the internalized antigens are not immediately degraded and the fusion with the lysosomes is delayed; this mechanism is apparently coordinated with the generation of newly synthesized MHC class I molecules that occurs 12–18 h following DC activation (Rescigno *et al.*, 1998). How degraded antigens derived from the exogenous pathway can access the MHC class I loading compartment has been recently reported with the discovery of endoplasmic reticulum (ER)–phagosome fusions (Guermonprez *et al.*, 2003). Therefore, DC can efficiently present on class I molecules peptides generated by the exogenous pathway. Adaptive immunity is controlled by the generation of MHC-restricted effector T cells and production of cytokines (Moser and Murphy, 2000). DC are able to stimulate naive T helper (Th) cells, which in turn they may differentiate into Th1- vs Th2-polarized subsets; Th1 cells secrete primarily interferon IFN-γ, whereas Th2 cells produce IL-4, IL-5, IL-10 and IL-13. Upon activation, DC upregulate the expression of costimulatory molecules, such as CD80 and CD86, thereby increasing immunogenicity of peptide antigens presented. Finally, DC activation triggers the production of cytokines, such as IL-12, IL-18, IL-4 or IL-10, which are able to polarize emerging T cell responses. Added to the complexity, it is to date not clear whether all forms of activation of DC necessarily result in increased immunogenicity. Furthermore, DC can produce different cytokines in response to different activating stimuli (Moser and Murphy, 2000). An example is shown by the observation that murine DC interacting either with yeast or hyphae of *Candida albicans* produce IL-12 or IL-4 respectively, and *in vivo* drive either Th1 or Th2 differentiation, respectively (D'Ostiani *et al.*, 2000). Therefore, DC and macrophages are important at the interface in bridging the innate and adaptive immune system (Kahn *et al.*, 2002).

21.4 DC and Infectious Diseases

DC are the first immune cells that come into contact with foreign microorganisms. Not surprisingly, DC play an important role in the generation of protective immunity towards intracellular parasites (Ludewig *et al.*, 1998, Flohe *et al.*, 1998) However, DC function may be subverted as part of the life cycle of a pathogen. A number of viruses use molecules expressed by DC as receptors; examples include CD4, CCR5 and CXCR4 (HIV; Baluvelt, 1997) CD13 (coronavirus and cytomegalovirus; Yeager *et al.*, 1992, Soderberg *et al.*, 1993) and CD46 (measles virus, MV; Schnorr *et al.*, 1997) The most studied example of DC involvement in infection is HIV. This lentivirus can remain latent in DC and exploits the trafficking of DC towards lymphoid tissue as a strategy to enhance the infection of permissive $CD4^+$ lymphocytes (Masurier *et al.*, 1998; Granelli-Piperno *et al.*, 1999).

HIV appears to be activated in DC by CD40 ligation or by the presence of Th cells. The activation status of the DC themselves is thought to have an impact on viral replication, with immature and cutaneous DC supporting productive infection of macrophage tropic virus, while mature DC are able to transport HIV but appear unable to replicate both T cell and macrophage tropic strains of virus (Granelli-Piperno *et al.*, 1998).

Therefore, microbes have learned to directly invade DC in peripheral tissues and replicate intracellularly. By either a productive nonlytic infection and/or by killing DC, the agent can then be spread locally, or infected DC can carry the agent to draining lymph nodes. Furthermore, infectious agents can interfere with MHC class I and class II antigen processing and presentation pathways or activate T cells indiscriminately by presenting bacterial superantigens diverting an effective immune response (reviewed in Austin, 2000; Banchereau *et al.*, 2000; Reis e Sousa *et al.*, 1999). Among organisms that have developed the ability to subvert DC function are the viruses *human immunodeficiency virus I* (HIV), *Epstein–Barr virus* (EBV), *human choriomeningitis virus* (HCMV), murine *lymphocytic choriomeningitis virus* (LCMV), *human cytomegalovirus* (HCMV), *Herpes simplex virus* (HSV) and *measles virus*, the bacteria *M. tuberculosis*, *Yersinia enterocolitica*, *Salmonella* sp. and *Listeria monocytogenes*, and the parasites *Leishmania major*, *L. donovani* and *Plasmodium falciparum*.

21.5 DC and Bacteria Interaction

The interactions between a host and microbial pathogens are diverse and regulated. The molecular mechanisms of microbial pathogenesis show common themes that involve families of structurally and functionally related proteins such as adherence factors, secretion systems, toxins and regulators of microbial pathogens. Microarray expression analysis of pathogen-infected cells and tissues can identify, simultaneously and in the same sample, host and pathogen genes that are regulated during the infectious process.

Both bacterial and mammalian (mouse, human) genome sequences can be used in microarray technology to define the expression profile of pathogens and the host cells. The global transcription effects on host cells of the innate immunity by various bacterial pathogens, including *Listeria monocytogenes*, *Salmonella*, *Pseudomonas aeruginosa* and *Bordetella pertussis*, have been analysed using microarray technology (Rappuoli, 2000). The infection of macrophages with *S. typhimurium* identified novel genes whose expression levels are altered (Rosenberger *et al.*, 2000). Similarly, *L. monocytogenes*-infected human promyelocytic THP1 cells identified 74 upregulated RNAs and 23 down-regulated host RNAs (Cohen *et al.*, 2000). Many of the upregulated genes encode proinflammatory cytokines (e.g. IL-8, IL-6 and growth-related oncogene-1) and many of the downregulated genes encode transcription factors and cellular adhesion molecules.

Understanding the molecular basis of the host response to bacterial infections is critical for preventing disease and tissue damage resulting from the host response. Furthermore, an understanding of host transcription changes induced by the microbes can be used to identify specific protein targets for drug development.

During the initial phase of an infection, the invading microbes can avoid various innate immune defences, but there are also mechanisms to counteract the development of an adaptive immune response (Hornef *et al.*, 2002). As a principal component of the innate immunity able to prime adaptive responses, the DC play a pivotal role in immunity and consequently are a particularly interesting target for pathogens (Palucka and Banchereau, 2002).

To study DC and bacteria interaction, we carried out a kinetic analysis of gene expression in immature mouse DC stimulated at different time-points with live Gram-negative bacteria (Granucci *et al.*, 2001a). For this study we used the well-characterized DC line, D1, which shows similar maturation to that seen with fresh splenic or bone marrow-derived DC (Rodriguez *et al.*, 1999; Rescigno *et al.*, 2000; Singh-Jasuja *et al.*, 2000). We activated D1 cells with Gram-negative *Escherichia coli* and transcriptionally analysed immature cells as well as mature cells that had been stimulated for 4, 6, 12, 18, 24 or 48 h with high-density oligonucleotide arrays that displayed probes for 11 000 genes and ESTs. At each time-point after stimulation, D1 cells were phenotypically characterized for their state of developmental synchronization by analysing surface expression of major histocompatibility complex (MHC) class II, B7-2 and CD40.

The main finding from this study was the discovery that DC are able to produce IL-2 upon bacterial encounter (Granucci *et al.*, 2001a). As well as the other inflammatory cytokines produced by DC during the maturation process, IL-2 is also expressed with a strictly defined kinetic, between 2 and 8 h after bacterial uptake. IL-2 is a cytokine able to sustain T, B and natural killer (NK) cell growth and, during the late phases of antigen-specific T cell responses, it contributes to the maintenance of T cell homeostasis by promoting activation-induced cell death (AICD) of effector T lymphocytes (Sporri and Reis e Sousa, 2005). Given the important regulatory role exerted by IL-2 in the immune system, IL-2-deficient mice show a generalized immune system deregulation (Schimpl *et al.*, 2002). The observation that DC, other than T cells, can also produce IL-2 opens new possibilities in understanding the mechanisms by which DC control innate and adaptive immunity (Lebecque, 2001). *In vivo*, both $CD8\alpha^{+}$ and $CD8\alpha^{-}$ splenic DC can produce IL-2 following microbial activation. Interestingly, only microbial stimuli and not inflammatory cytokines are able to induce IL-2 secretion by DC (Granucci *et al.*, 2003b), indicating that DC can distinguish between the actual presence of an infection and a cytokine-mediated inflammatory process (Sporri and Reis e Sousa, 2005).

Recent studies have focused on the function of DC during the early phases of the immune response, and a predominant role for DC in activation of NK cells has been described (Ferlazzo *et al.*, 2002; Piccioli *et al.*, 2002; Gerosa *et al.*, 2002; Fernandez *et al.*, 1999).

21.6 DC and Virus Interaction

Studies on virus infections in humans and animals revealed that the infected host responds to virus infection by induction of the two arms of the immune system: the first line of defence is the appearance of antiviral cytotoxic T cells ($CD8^+$ and $CD4^+$) and the second is the synthesis of antiviral antibodies.

Few studies have been reported about the modulation of DC gene expression in response to virus infections. One such describes the DC transcriptome analysis upon *influenza virus* infection (Huang *et al.*, 2001). The immature DC were exposed to different pathogens and the reprogramming of DC gene expression during the maturation process was studied using oligonucleotide microarrays. Some 166 genes were reported to be activated as core response genes to all the pathogens tested; additionally, 58 genes were modulated in response only to influenza virus. The IFN-α and -β genes were markedly induced together with a subset of genes linked to the inhibition of the immune response. These genes included pro-apoptotic genes that may induce early death of infected cells, the gene that encodes mcp-1, which is capable of blocking IL-12 production in macrophages, genes that inhibit NO synthesis and genes that code for proteins which inhibit T cell activation. Among the IFN genes, the infected DC induced the expression of the genes Mx1 and Mx2 transiently. The expression of human Mx in transgenic mice enhanced resistance to *influenza virus* infection (Pavlovic *et al.*, 1995).

DC and HIV interaction have been studied recently by analysing the pattern of gene expression induced by this virus (Izmailova *et al.*, 2003). The authors used DNA microarrays to identify DC genes whose expression is modified by HIV-1 infection and by expression of Tat alone. HIV-1 and Tat induced expression of chemokines that recruit activated T cells and macrophages, the ultimate cellular targets of HIV-1 infection. A new role for Tat in facilitating expansion of the viral infection is proposed. Both HIV-1 infection and Tat expression caused increased expression of the IFN-inducible genes. The products of these genes are responsible for the diverse effects of IFNs, including antiviral growth, immune modulation and antitumour activity (Sen, 2000). Other viruses, such as *polyoma*, *human papilloma type 31* and mumps, down-regulate expression of these genes (Chang and Laimins, 2000 Fujii *et al.*, 1999; Weihua *et al.*, 1998). Therefore, the gene expression program of immature DC is modified by HIV-1 infection and Tat expression. The observation that Tat regulates chemokine gene expression in DC suggests that therapies designed to affect Tat function may produce the combined benefit of limiting viral transcription and reducing the interactions between infected DC and T cells that contribute to the expansion of viral infection.

21.7 DC and Parasite Interaction

Shistosoma mansoni molecular signature

Expression analyses have shown that, after microbial interaction, DC undergo a multistep maturation process (Granucci *et al.*, 2001b) and acquire specific immune functions, depending on the type of microbe they have encountered. We have defined in detail the transcriptome induced in murine DC by different pathogens such as *Shistosoma mansoni* (Trottein *et al.*, 2004) and *Leishmania mexicana*. The data clearly demostrate that individual parasites induce both common and individual regulatory networks within the cell. This suggests a mechanism whereby host–pathogen interaction is translated into an appropriate host inflammatory response.

Shistosoma mansoni is a helminth parasite and has a complex life cycle that is initiated by the transcutaneous penetration of the larvae followed by its rapid transformation into schistosomula (SLA; Pearce and MacDonald, 2002). Once in the skin, SLA closely interact with immunocompetent cells, including DC, to manipulate the host immune response (Ramaswamy *et al.*, 2000; Angeli *et al.*, 2001). SLA then begins a long vascular journey to reach the intrahepatic venous system, where they mature into adult male and egg-producing female worms. Eggs that accumulate in the liver, spleen and lungs induce inflammation and an intense granulomatous hypersensitivity reaction (Rumbley and Phillips, 1999).

We have investigated DC–schistosome interactions using a genome-wide expression study. We have used a near-homogeneous source of mouse DC, the well-defined, long-term D1 splenic population (Winzler *et al.*, 1997). The kinetic global gene expression analysis of mouse DC stimulated with eggs or SLA indicated that genes encoding inflammatory cytokines, chemokines and IFN-inducible proteins were oppositely regulated by the two stimuli (Figure 21.3). Interestingly, eggs, but not SLA, induced the expression of IFN-β that efficiently triggered the type I IFN receptor (IFNAR) expressed on DC, causing phosphorylation of STAT-1 with consequent upregulation of IFN-induced inflammatory products.

Clustering techniques applied to 283 differentially expressed genes distinguished the two stimuli from different points of view (Figure 21.4). The egg time-course experiment was compatible with a progressive cell differentiation process, such as maturation, whereas observations from SLA-stimulated DC samples suggested the occurrence of a stable blocking event within the first 4 h. Moreover, eggs modulated different amounts and subsets of genes in comparison with SLA, indicating that the two developmental stages of *S. mansoni* affected distinct intracellular pathways in DC, possibly by triggering specific receptors. The egg stage sustains the maximization of Ag presentation efficiency in DC by inducing the upregulation of H-2M, which plays a crucial role in the peptide loading of MHC class II molecules (Kovats *et al.*, 1998) and of the costimulatory molecules CD40 and ICAM-1. Cathepsins D and L, which are believed to remove the invariant chain from its complex with MHC class II molecules (Villadangos *et al.*, 1999), are downregulated by SLA, but are not modulated by eggs, suggesting a reduction in the Ag processing capacity exerted by the larval stage on DC. Moreover, the egg stage induced the expression of proinflammatory cytokine transcripts, such as TNF-α, and chemokines, such as IP-10 (CXCL10), monocyte chemoattractant protein-5 (CCL12), MIP-1α (CCL3), MIP-1β (CCL4), MIP-1γ (CCL9) and MIP-2 (CXCL2), that are known to collectively attract granulocytes, immature DC, NK cells and activated T cells (Greaves and Schall, 2000). *S. mansoni* eggs, but not SLA, induced the production of high amounts of

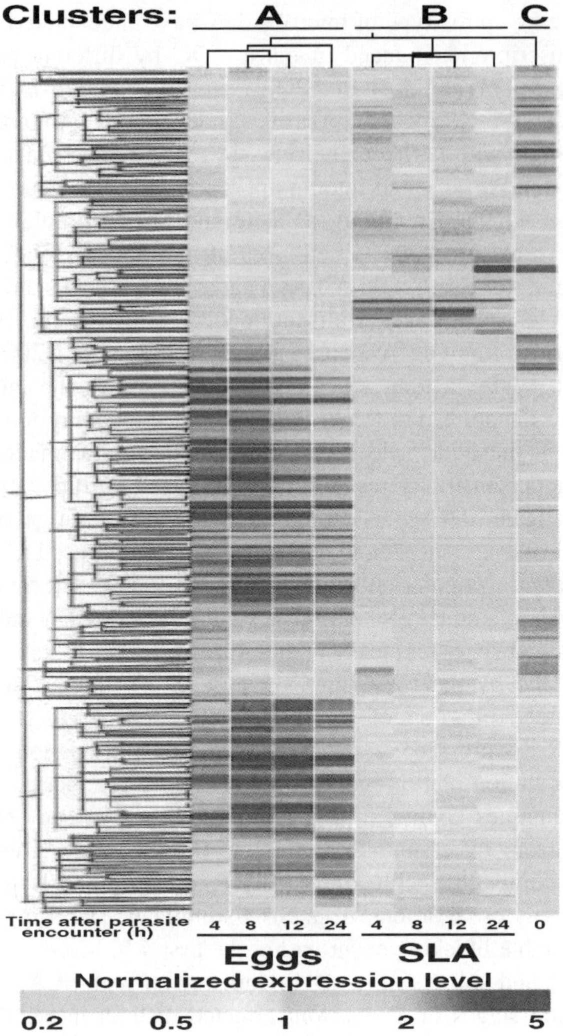

Figure 21.3 Expression profile clustering of 283 genes differentially expressed during DC–schistosome interaction. Two-way hierarchical clustering of gene expression profiles measured in time-course experiments; normalized expression levels relative to median are displayed in yellow (median expression), red (increased expression), or cyan (decreased expression) according to the colour bar. (A colour reproduction of this figure can be viewed in the colour plate section)

IL-2, which could be important for DC-mediated activation of NK cells or NKT cells (Fujii *et al.*, 2002) as well as for priming naive T cells (Granucci *et al.*, 2003b).

Mouse myeloid DC, in response to helminth eggs, activate a strong interferon response compared with SLA. We have observed that the DC-derived IFN-β molecule efficiently triggered the IFNAR expressed on DC, thus providing an autocrine and/or paracrine stimulation mechanism. Therefore, our data indicate

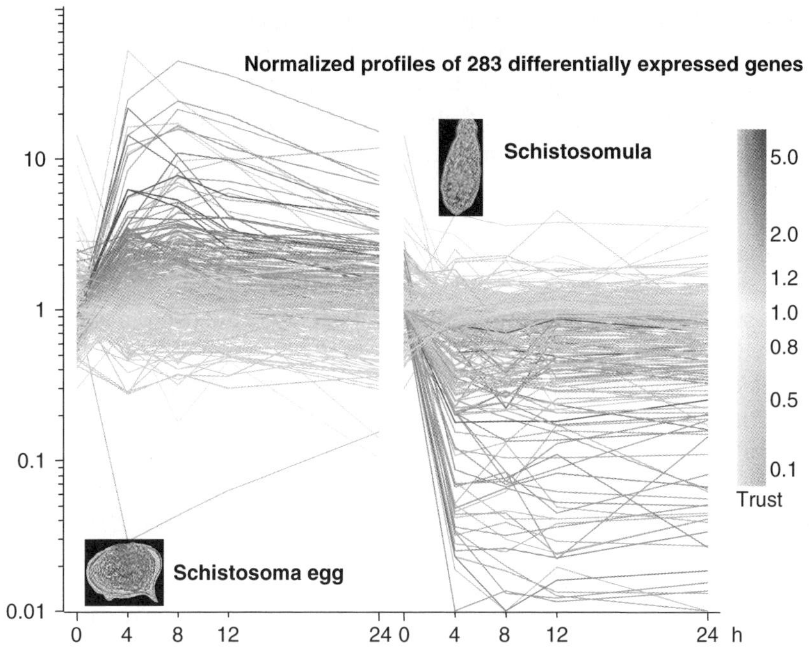

Figure 21.4 Comparative kinetic analysis of DC transcriptomes generated by *S. mansoni*. Expression profile clustering of 283 genes differentially expressed during DC–schistosome interaction. Supervised clustering of kinetic gene expression profiles. Each panel groups genes that share same transcriptional response (enhanced, silenced or unchanged) relative to the two developmental stages of the parasite. Each line represents the expression profile of a particular gene and is coloured according to its normalized expression level 4 h after encountering *S. mansoni* eggs. (A colour reproduction of this figure can be viewed in the colour plate section)

myeloid DC as one possible mediator of type I IFN signalling as well as one plausible source of IP-10 and MIP-1α production, also in response to helminth infections. The comparative gene expression analysis revealed two different DC global transcriptional modifications induced by either *Schistosoma* eggs or SLA, consistent with the different responses induced *in vivo* by these two parasite stages. Taken as a whole, our observations have provided new molecular insights into the host–parasite interaction established in the course of schistosomiasis, leading to the identification of a type I IFN-dependent mechanism by which DC may amplify inflammatory reactions in response to helminth infection.

21.8 *Leishmania Mexicana* Molecular Signature

A molecular signature was generated by DC after interaction with the protozoan parasite *Leishmania mexicana* (Aebischer *et al.*, 2005). Parasite infection is initiated by the transfer of insect vector-borne promastigotes. These are taken up by phagocytic cells and transformed into the obligate intracellular amastigotes, which dwell within vacuoles with endosomal–early-lysosomal characteristics (Antoine *et al.*, 1998). Uncontrolled replication of the parasites causes disease; to resolve the infection the host generates leishmanicidal mechanisms consisting primarily of the activation of phagocytes through cytokines, e.g. IFN-γ and TNF-α (McConville and Blackwell, 1991).

DC activation was shown upon interaction with *L. mexicana* promastigotes but not with *L. mexicana* amastigotes, as indicated by MHC II, CD86, CD54 expression and IL-12p40 synthesis. The transformation from the extracellular promastigote stage to the intracellular amastigote stage of *Leishmania* is associated with a dramatic change in the molecular composition of the parasite surface. The surface of promastigotes is almost entirely coated by the lipid-anchored glycan, lipophosphoglycan (LPG). In contrast, in amastigotes of most species, LPG expression is severely downregulated (Bahr *et al.*, 1993) or completely undetectable (Winter *et al.*, 1994). Instead, the cell surface of amastigotes is dominated by small glycolipids, mostly glycoinositolphospholipids, and is poor in proteins.

We have investigated the nature of the *L. mexicana* promastigote-derived DC activating signals; again we employed comparative global transcriptional analysis to study the response of DC cell line D1 to *L. mexicana* promastigotes and amastigotes and to lpg1−/− promastigotes (Figure 21.5). Only LPG expressing *L. mexicana* promastigotes induced a pro-inflammatory programme in DC and this suggests that the amastigotes have evolved to downregulate LPG expression in the mammalian host in order to achieve immune escape.

We have shown that DC mount a proinflammatory transcriptional response to *L. mexicana* parasites that depends on LPG expression. Both hierarchical clustering and principal component analysis allowed visualizion of the distinct impact of proinflammatory genes within the global effects of *L. mexicana* infection on DC. The significant role of LPG expression for this response was further supported by the finding that infection with mutant parasites, where LPG-expression was restored, reproduced the expression pattern of these signature genes. Hierarchical clustering of genes classified according to Venn diagram allowed the identification of additional genes differentially modulated by LPG-positive and LPG-negative parasite forms. Overall, the response to LPG-expressing parasites is consistent with the biological role of DC, i.e. their migration to local lymphoid centres, increase in Ag presentation capacity and control of immunity in the early phase via the production of key cytokines following the recognition of a pathogen.

LPG downregulation in the amastigote stage may be the result of evolutionary adaptation. However, it also became clear during this study that infection with *L. mexicana* amastigotes is not a silent process but results in a complex response in DC. Apart from not triggering a proinflammatory response, this may have additional functional consequences which need to be addressed in future studies.

(a)

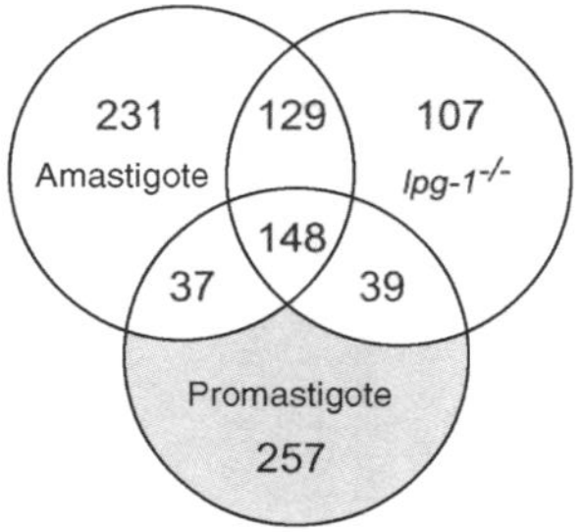

(b)

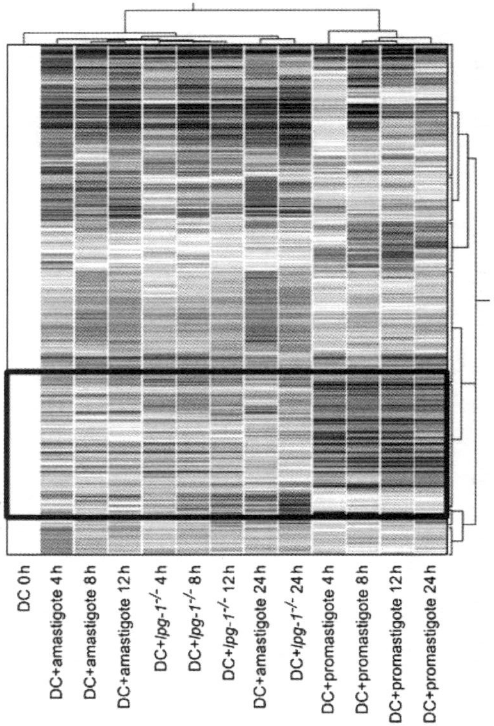

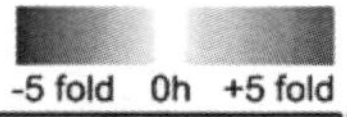

(c)

Interferon induced genes 17/260				Inflammatory cytokines 18/260			
name	probeset	name	probeset	name	probeset	name	probeset
Ifit2	103639_at	Irf1	102401_at	Il2	AFFX-MurIL2_at	Ccl6	92849_at
Ifi47	104750_at	Igtp	160933_at	Ccl4-Mip1b	94146_at	Ccl9-Mip1g	104388_at
Irf7	104669_at	Gtpi-p	98410_at	Ccl1-Tca3	94166_g_at	Il1b	103486_at
Isg15	98822_at	Ifrd1	160092_at	Cxcl10-IP10	93858_at	Ccl6	162198_f_at
Ifi204	98465_f_at	Isgf3	103634_at	Tnfa	102629_at	Ccl5-RANTES	98406_at
Ifit1	100981_at	Icsbp1	98002_at	Il12b-p40	100779_at	Ccl22	102310_at
Irf7	162202_f_at	Ifi30	97444_at	BC017607	96515_at	Cxcl2-Mip2a	101160_at
Ifi1	97409_at			Cxcl1	95349_g_at	Il1a	94755_at
Ifrg15	93997_at			Cxcl1	95348_at		
Ifi203	93321_at			Ccl3	102424_at		

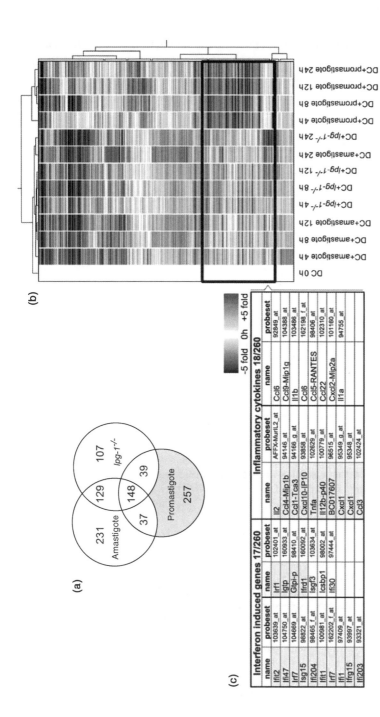

Figure 21.5 *Leishmania mexicana* promastigotes expressing LPG trigger a proinflammatory transcriptional response in D1 cells. (a) A total of 948 differentially expressed mRNAwere organized according to common or specific modulation in D1 cells infected with *L. mexicana* amastigotes or wild-type or lpg1−/− mutant promastigotes (lpg1−/−). (b) Hierarchical clustering according to samples (vertical) and gene expression profiles (horizontal). Mean gene expression values are shown, and gene signals are divided by their 0 h value. Red and blue indicate up- and downregulation, respectively, in comparison to 0 h time point values (white). A cluster of 260 genes that contain differentially expressed genes belonging to interferon-induced genes ($n = 17$) and inflammatory cytokine genes ($n = 18$) is boxed. (c) These genes, which belong to the promastigote-specific group of the Venn diagram (A), are colour-coded in yellow. (A colour reproduction of this figure can be viewed in the colour plate section)

21.9 Conclusions

Microarray technology can provide insights into the interaction between the pathogen and host by revealing global host expression responses to a range of pathogenic stimuli. Pathogens may manipulate host–cell gene expression, for example by causing upregulation of cellular support for pathogen replication and downregulation of MHC expression to allow pathogens to evade the immune system.

In conclusion, the amount of data generated by microarray experiments is enormous. A simple experiment comparing stimulation of immune cells by two different bacteria in two individuals at three different time points requires at least 12 microarrays.

With up to 20 000 genes on an array, the number of data points leaps to 240 000. Such quantities of data require specialized statistical expertise and software to decipher patterns from the entire expression repertoire. The bottleneck in genetic analysis has therefore moved from the speed at which an experiment can be done to the speed at which the resulting data can be analysed.

References

Aebischer T, Bennett CL, Pelizzola M, Vizzardelli C, Pavelka N, Urbano M, Capozzoli M, Luchini A, Granucci F, Blackburn CC, Ricciardi-Castagnoli P. 2005. A critical role for lipophosphoglycan in proinflammatory responses of dendritic cells to Leishmania mexicana. *Eur J Immunol* **35**: 476–478.

Ahn JH, Lee Y, Jeon CJ, Lee SJ, Lee BH, Choi KD, Bae YS. 2002. Identification of the genes differentially expressed in human dendritic cell subsets by cDNA subtraction and microarray analysis. *Blood* **100**: 1742–1754.

Angeli V, Faveeuw C, Roye O, Fontaine J, Teissier E, Capron A, Wolowczuk I, Capron M, Trottein F. 2001. Role of the parasite-derived prostaglandin D2 in the inhibition of epidermal Langerhans cell migration during schistosomiasis infection. *J Exp Med* **193**: 1135.

Anjuere F, Martin P, Ferrero I, Fraga ML, Del Hoyo GM, Wright N, Ardavin C. 1999. Definition of dendritic cell subpopulations present in the spleen, Peyer's patches, lymph nodes, and skin of the mouse. *Blood* **93**: 590–598.

Antoine JC, Prina E, Lang T, Courret N. 1998. The biogenesis and properties of the parasitophorous vacuoles that harbour Leishmania in murine macrophages. *Trends Microbiol* **6**: 392–401.

Austin JM. 2000. Antigen-presenting cells. Experimental and clinical studies of dendritic cells. *Am J Respir Crit Care Med* **162**: S146–S150.

Bachem CW, Van der Hoeven RS, De Bruijn SM, Vreugdenhil D, Zabeau M, Visser RG. 1996. Visualization of differential gene expression using a novel method of RNA fingerprinting based on AFLP: analysis of gene expression during potato tuber development. *Plant J* **9**: 745–753.

Bahr V, Stierhof YD, Ilg T, Demar M, Quinten M, Overath P. 1993. Expression of lipophosphoglycan, high-molecular weight phosphoglycan and glycoprotein 63 in promastigotes and amastigotes of *Leishmania mexicana. Mol Biochem Parasitol* **58**: 107–121.

Baltathakis J, Alcantara O, Boldt DH. 2001. Expression of different NF-kB pathway genes in dendritic cells (DCs) or macrophages assessed by gene expression profiling. *J Cell Biochem* **83**: 281–290.

Baluvelt A. 1997. The role of skin dendritic cells in the initiation of human immunodeficiency virus infection. *Am J Med* **102**: 16–20.

Banchereau J, Steinman RM. 1998. Dendritic cells and the control of immunity. *Nature* **392**: 245–252.

Banchereau J, Briere F, Caux C, Davoust J, Lebecque S, Liu YJ, Pulendran B, Palucka K. 2000. Immunobiology of dendritic cells. *A Rev Immunol* **18**: 767–811.

Bergelson J, Kreitman M, Stahl EA, Tian D. 2001. Evolutionary dynamics of plant R-genes. *Science* **292**: 2281–2284.

Birch PRJ, Kamoun S. 2000. Studying interaction transcriptomes: coordinated analyses of gene expression during plant–microorganism interactions. In *New Technologies for Life Sciences: a Trends Guide* (supplement, Elsevier Trends Journals, December). London: Elsevier; 77–82.

Bleharski JR, Niazi KR, Sieling PA, Cheng G, Modlin RL. 2001. Signaling lymphocytic activation molecule is expressed on CD40 ligand- activated dendritic cells and directly augments production of inflammatory cytokines. *J Immunol* **167**: 3174–3181.

Brenner S, Johnson M, Bridgham J, Golda G, Lloyd DH, Johnson D, Luo S, McCurdy S, Foy M, Ewan M, Roth R, George D, Eletr S, Albrecht G, Vermaas E, Williams SR, Moon K, Burcham T, Pallas M, DuBridge RB, Kirchner J, Fearon K, Mao J, Corcoran K. 2000. Gene expression analysis by massively parallel signature sequencing (MPSS) on microbead arrays. *Nat Biotechnol* **18**: 630–634.

Brière F, Bendriss-Vermare N, Delale T, Burg S, Corbet C, Rissoan MC, Chaperot L, Plumas J, Jacob MC, Trinchieri G, Bates EE. 2003. Origin and filiation of human plasmacytoid dendritic cells. *Hum Immunol* **63**: 1081–1092.

Caux C, Vanbervliet B, Massacrier C, Dezutter-Dambuyant C, De Saint-Vis B, Jacquet C, Yoneda K, Imamura S, Schmitt D, Banchereau J. 1996. $CD34^+$ hematopoietic progenitors from human cord blood differentiate along two independent dendritic cell pathways in response to GM-CSF1TNF alpha. *J Exp Med* **184**: 695–706.

Caux C, Massacrier C, Dubois B, Valladeau J, Dezutter-Dambuyant C, Durand I, Schmitt D, Saeland S. 1999. Respective involvement of TGF-beta and IL-4 in the development of Langerhans cells and non-Langerhans dendritic cells from $CD34^+$ progenitors. *J Leukoc Biol* **66**: 781–791.

Cella M, Schiedegger D, Palmer-Lehmann K, Lane P, Lanzavecchia A, Alber G. 1996. Ligation of CD40 on dendritic cells triggers production of high levels of interleukin-12 and enhances T cell stimulatory capacity: T-T help via APC activation. *J Exp Med* **184**: 747–752.

Cerio R, Griffiths CE, Cooper KD, Nickoloff BJ, Headington JT. 1989. Characterization of factor XIIIa positive dermal dendritic cells in normal and inflamed skin. *Br J Dermatol* **121**: 421–431.

Chang YE, Laimins LA. 2000. Microarray analysis identifies interferon-inducible genes and Stat-1 as major transcriptional targets of human papillomavirus type 31. *J. Virol* **74**: 4174–4182.

Chaussabel D, Semnani RT, McDowell MA, Sacks D, Sher A, Nutman TB. 2003. Unique gene expression profiles of human macrophages and dendritic cells to phylogenetically distinct parasites. *Blood* **102**: 672–681.

Cohen P, Bouaboula M, Bellis M, Baron V, Jbilo O, Poinot-Chazel C, Galiegue S, Hadibi EH, Casellas P. 2000. Monitoring cellular responses to Listeria monocytogenes with oligonucleotide arrays. *J Biol Chem* **275**: 11181–11190.

Dietz AB, Bulur PA, Knutson GJ, Matasíc, R. and Vuc-Pavlovíc S. 2000. Maturation of human monocyte-derived dendritic cells studied by microarray hybridization. *Biochem Biophys Res Commun* **275**: 731–738.

D'Ostiani CF, Del Sero G, Bacci A, Montagnoli C, Spreca A, Mencacci A, Ricciardi-Castagnoli P, Romani L. 2000. Dendritic cells discriminate between yeasts and hyphae of the fungus Candida albicans. Implications for initiation of T helper cell immunity *in vitro* and *in vivo*. *J Exp Med* **191**: 1661–1674.

Duggan DJ, Bittner M, Chen Y, Meltzer P, Trent JM. 1999. Expression profiling using cDNA microarrays. *Nat Genet* **21**: 10–14.

Edwards AD, Chaussabel D, Tomlinson S, Schulz O, Sher A, Reis e Sousa C. 2003. Relationships among murine CD11c(high) dendritic cell subsets as revealed by baseline gene expression patterns. *J Immunol* **1**(171): 47–60.

Fearon DT, Locksley RM. 1996. The instructive role of innate immunity in the acquired immune response. *Science* **272**: 50–53.

Ferlazzo G, Tsang ML, Moretta L, Melioli G, Steinman RM, Munz C. 2002. Human dendritic cells activate resting natural killer (NK) cells and are recognized via the NKp30 receptor by activated NK cells. *J Exp Med* **195**: 343.

Fernandez NC, Lozier A, Flament C, Ricciardi-Castagnoli P, Bellet D, Suter M, Perricaudet M, Tursz T, Maraskovsky E, Zitvogel L. 1999. Dendritic cells directly trigger NK cell functions: cross-talk relevant in innate anti-tumor immune responses *in vivo*. *Nat Med* **5**: 405.

Flohe SB, Bauer C, Flohe S, Moll H. 1998. Antigen-pulsed epidermal Langerhans cells protect susceptible mice from infection with the intracellular parasite Leishmania major. *Eur J Immunol* **28**: 3800–3811.

Fujii N, Yokosawa N, Shirakawa S. 1999. Suppression of interferon response gene expression in cells persistently infected with mumps virus, and restoration from its suppression by treatment with ribavirin. *Virus Res* **65**: 175–185.

Fujii S, Shimizu K, Kronenberg M, Steinman RM. 2002. Prolonged IFN-gamma producing NKT response induced with α-galactosylceramide-loaded DCs. *Nat Immunol* **3**: 867.

Gerosa F, Baldani-Guerra B, Nisii C, Marchesini V, Carra G, Trinchieri G. 2002. Reciprocal activating interaction between natural killer cells and dendritic cells. *J Exp Med* **195**: 327.

Granelli-Piperno A, Delgado E, Finkel V, Paxton W, Steinman RM. 1998. Immature dendritic cells selectively replicate macrophage tropic (M-tropic) human immunodeficiency virus type 1, while mature cells efficiently transmit both M- and T-tropic virus to T cells. *J. Virol* **72**: 2733–2737.

Granelli-Piperno A, Finkel V, Delgado E, Steinman RM. 1999. Virus replication begins in dendritic cells during the transmission of HIV-1 from mature dendritic cells to T cells. *Curr Biol* **14**: 21–29.

Granucci F, Vizzardelli C, Pavelka N, Feau S, Persico M, Virzi E, Rescigno M, Moro G, Ricciardi-Castagnoli P. 2001a. Inducible IL-2 production by dendritic cells revealed by global gene expression analysis. *Nat Immunol* **2**: 882.

Granucci F, Vizzardelli C, Virzi E, Rescigno M, Ricciardi-Castagnoli P. 2001b. Transcriptional reprogramming of dendritic cells by differentiation stimuli. *Eur J. Immunol* **31**: 2539–2546.

Granucci F, Feau S, Zanoni I, Pavelka N, Vizzardelli C, Raimondi G, Ricciardi-Castagnoli P. 2003a. The immune response is initiated by dendritic cells via interaction with microorganisms and interleukin-2 production. *J Infect Dis* **15**(187): S346–350.

Granucci F, Feau S, Angeli V, Trottein F, Ricciardi-Castagnoli P. 2003b. Early IL-2 production by mouse dendritic cells is the result of microbial-induced priming. *J Immunol* **170**: 5075.

Greaves DR, Schall TJ. 2000. Chemokines and myeloid cell recruitment. *Microb Infect* **2**: 331.

Guermonprez P, Saveanu L, Kleijmeer M, Davoust J, Van Endert P, Amigorena S. 2003. ER-phagosome fusion defines an MHC class I cross-presentation compartment in dendritic cells. *Nature* **425**: 397–402.

Hart DN. 1997. Dendritic cells: unique leukocyte populations which control the primary immune response. *Blood* **90**: 3245–3287.

Hashimoto SI, Suzuki T, Nagai S, Yamashita T, Toyoda N, Matsushima K. 2000. Identification of genes specifically expressed in human activated and mature dendritic cells through serial analysis of gene expression. *Blood* **96**: 2206–2214.

Hawiger D, Inaba K, Dorsett Y, Guo M, Mahnke K, Rivera M, Ravetch JV, Steinman RM, Nussenzweig MC. 2001. Dendritic cells induce peripheral T cell unresponsiveness under steady state conditions *in vivo*. *J Exp Med* **194**: 769–779.

Hornef MW, Wick MJ, Rhen M, Normark S. 2002. Bacterial strategies for overcoming host innate and adaptive immune responses. *Nat Immunol* **3**: 1033.

Huang Q, Liu D, Majewski P, Schulte LC, Korn JM, Young RA, Lander ES, Hacohen N. 2001. The plasticity of dendritic cell responses to pathogens and their components. *Science* **294**: 870–875.

Izmailova E, Bertley FMN, Huang Q, Makori N, Miller CJ, Young RA, Aldovini A. 2003. HIV-1 tat reprograms immature dendritic cells to express chemoattractants for activated T cells and macrophages. *Nat Med* **9**: 191–197.

Jonuleit H, Schmitt E, Steinbrink K, Enk AH. 2001. Dendritic cells as a tool to induce anergic and regulatory T cells. *Trends Immunol* **22**: 394–400.

Kahn RA, Fu H, Roy CR. 2002. Cellular hijacking: a common strategy for microbial infection. *Trends Biochem Sci* **27**: 308–314.

Kamath AT, Pooley J, O'keeffe MA, Vremec D, Zhan Y, Lew AM, D'amico A, Wu L, Tough DF, Shortman K. 2000. The development, maturation, and turnover rate of mouse spleen dendritic cell populations. *J Immunol* **165**: 6762–6770.

Kamoun S, Hraber P, Sobral B, Nuss D, Govers F. 1999. Initial assessment of gene diversity for the oomycete pathogen *Phytophthora infestans* based on expressed sequences. *Fungal Genet Biol* **28**: 94–106.

Kohrgruber N, Halanek N, Gröger M, Winter D, Rappersberger K, Schmitt-Egenolf M, Stingl G, Maurer D. 1999. Survival, maturation, and function of CD11c- and CD11c+ peripheral blood dendritic cells are differentially regulated by cytokines. *J Immunol* **163**: 3250–3259.

Kovats S, Grubin CE, Eastman S, deRoos P, Dongre A, Van Kaer L, Rudensky AY. 1998. Invariant chain-independent function of H-2M in the formation of endogenous peptide-major histocompatibility complex class II complexes *in vivo*. *J Exp. Med* **187**: 245.

Lapteva N, Nieda M, Ando Y, Ide K, Hatta-Ohashi Y, Dymshits G, Ishikawa Y, Juji T, Tokunaga K. 2001. Expression of renin-angiotensin system genes in immature and mature dendritic cells identified using human cDNA microarray. *Biochem Biophys Res Commun* **285**: 1059–1065.

Lebecque S. 2001. A new job for dendritic cells. *Nat Immunol* **2**: 830.

Le Naour F, Hohenkirk L, Grolleau A, Misek DE, Lescure P, Geiger JD, Hanash S, Beretta L. 2001. Profiling changes in gene expression during differentiation and maturation of monocyte-derived dendritic cells using both oligonucleotide microarrays and proteomics. *Proc Natl Acad Sci USA* **276**: 17920–17931.

Liang P, Pardee AB. 1992. Differential display of eukaryotic messenger RNA by means of the polymerase chain reaction. *Science* **257**: 967–971.

Lipshultz RJ, Fodor SP, Gingeras TR, Lockhart DJ. 1999. High density synthetic oligonucleotide arrays. *Nat Genet* **21**: 20–24.

Lockhart DJ, Dong H, Byrne MC, Follettie MT, Gallo MV, Chee MS, Mittmann M, Wang C, Kobayashi M, Horton H, Brown EL. 1996. Expression monitoring by hybridization to high-density oligonucleotide arrays. *Nat. Biotechnol.* **14**: 1675–1680.

Ludewig B, Ehl S, Karrer U, Odermatt B, Hengartner H, Zinkernagel RM. 1998. Dendritic cells efficiently induce protective antiviral immunity. *Virology* **72**: 3812–3818.

Lutz MB, Rovere P, Kleijmeer MJ, Rescigno M, Assmann CU, Oorschot VM, Geuze HJ, Trucy J, Demandolx D, Davoust J, Ricciardi-Castagnoli P. 1997. Intracellular routes and selective retention of antigens in mildly acidic cathepsin D/lysosome-associated membrane protein-1/MHC class II-positive vesicles in immature dendritic cells. *J Immunol* **159**: 3707–3716.

Manger ID, Relman DA. 2000. How the host 'sees' pathogens: global gene expression responses to infection. *Curr Opin Immunol* **12**: 215–218.

Masurier C, Salomon B, Guettari N, Pioche C, Lachapelle F, Guigon M, Klatzmann D. 1998. Dendritic cells route human immunodeficiency virus to lymph nodes after vaginal or intravenous administration to mice. *J Virol* **72**: 7822–7829.

Matsunga T, Ishida T, Takekawa M, Nishimura S, Adachi M, Imai K. 2002. Analysis of gene expression during maturation of immature dendritic cells derived from peripheral blood monocytes. *Scand J Immunol* **56**: 593–601.

McConville MJ, Blackwell JM. 1991. Developmental changes in the glycosylated phosphatidylinositols of Leishmania donovani. Characterization of the promastigote and amastigote glycolipids. *J Biol Chem* **266**: 15170–15179.

Medzhitov R. 2001. Toll-like receptors and innate immunity. *Nat Rev Immunol* **1**: 135–145.

Messmer D, Messmer B, Chiorazzi N. 2003. The global transcriptional maturation program and stimuli-specific gene expression profiles of human myeloid dendritic cells. *Int Immunol* **15**: 491–503.

Moschella F, Maffei A, Cantanzaro RP, Papadopoulos KP, Skerrett D, Hesdorffer CS, Harris PE. 2001. Transcript profiling of human dendritic cells maturation-induced under defined culture conditions: comparison of the effects of tumour necrosis factor alpha, soluble CD40 ligand trimer and interferon gamma. *Br J Haematol* **114**: 444–457.

Moser M, Murphy KM. 2000. Dendritic cell regulation of TH1–TH2 development. *Nat Immunol* **1**: 199–205.

Mushegian A, Medzhitov R. 2001. Evolutionary perspective on innate immune recognition. *J Cell Biol* **26**(155): 705–710.

Palucka K, Banchereau J. 2002. How dendritic cells and microbes interact to elicit or subvert protective immune responses. *Curr Opin Immunol* **14**: 420.

Pavlovic J, Arrzet HA, Hefti HP, Frese M, Rost D, Ernst B, Kolb E, Stahelli P, Haller O. 1995. Enhanced virus resistance of transgenic mice expressing the human Mx protein. *J Virol* **69**: 4506–4510.

Pearce EJ, MacDonald AS. 2002. The immunobiology of schistosomiasis. *Nat Rev Immunol* **2**: 499.

Piccioli D, Sbrana S, Melandri E, Valiante NM. 2002. Contact-dependent stimulation and inhibition of dendritic cells by natural killer cells. *J Exp Med* **195**: 335.

Ramaswamy K, Kumar P, He YX. 2000. A role for parasite-induced PGE2 in IL-10-mediated host immunoregulation by skin stage schistosomula of *Schistosoma mansoni*. *J Immunol* **165**: 4567.

Rappuoli R. 2000. Pushing the limits of cellular microbiology: microarrays to study bacteria-host cell intimate contacts. *Proc Natl Acad Sci USA* **97**: 13467–13469.

Reis e Sousa C, Sher A, Kaye P. 1999. The role of dendritic cells in the induction and regulation of immunity to microbial infection. *Curr Opin Immunol* **11**: 392–399.

Rescigno M, Citterio S, Thery C, Rittig M, Medaglini D, Pozzi G, Amigorena S, Ricciardi-Castagnoli P. 1998. Bacteria-induced neo-biosynthesis, stabilization, and surface expression of functional class I molecules in mouse dendritic cells. *Proc Natl Acad Sci USA* **95**: 5229–5234.

Rescigno M, Piguet V, Valzasina B, Lens S, Zubler R, French L, Kindler V, Tschopp J, Ricciardi-Castagnoli P. 2000. Fas engagement induces the maturation of dendritic cells (DCs), the release of interleukin (IL)-1b, and the production of interferon gamma in the absence of IL-12 during DC-T cell cognate interaction. A new role for fas ligand in inflammatory responses. *J Exp Med* **192**: 1661–1668.

Rodriguez A, Regnault A, Kleijmeer M, Ricciardi-Castagnoli P, Amigorena S. 1999. Selective transport of internalized antigens to the cytosol for MHC class I presentation in dendritic cells. *Nat Cell Biol* **1**: 362–368.

Romani N, Gruner S, Brang D, Kampgen E, Lenz A, Trockenbacher B, Konwlinka G, Fritsch PO, Steinman RM, Schuler G. 1994. Proliferating dendritic cell progenitors in human blood. *J Exp Med* **180**: 83–93.

Rosenberger CM, Scott MG, Gold MR, Hancock RE, Finlay BB. 2000. *Salmonella typhimurium* infection and lipopolysaccharide stimulation induce similar changes in macrophage gene expression. *J Immunol* **164**: 5894–5904.

Rumbley CA, Phillips SM. 1999. The schistosome granuloma: an immunoregulatory organelle. *Microb Infect* **1**: 499.

Sallusto F, Lanzavecchia A. 1994. Efficient presentation of soluble antigen by cultured human dendritic cells is maintained by granulocyte/macrophage colony-stimulating factor plus interleukin 4 and downregulated by tumor necrosis factor alpha. *J Exp Med* **179**: 1109–1118.

Schena M, Shalon D, Davis RW, Brown PO. 1995. Quantitative monitoring of gene expression patterns with a complementary DNA microarray. *Science* **270**: 467–470.

Schimpl A, Berberich I, Kneitz B, Kramer S, Santner-Nanan B, Wagner S, Wolf M, Hunig T. 2002. IL-2 and autoimmune disease. *Cytokine Growth Factor Rev* **13**: 369.

Schnorr JJ, Xanthakos S, Keikavoussi P, Kampgen E, Ter Meulen V, Schneider-Schaulies S. 1997. Induction of maturation of human blood dendritic cell precursor by measles virus is associated with immunosuppression. *Proc Natl Acad Sci USA* **94**: 5326–5331.

Sen GC. 2000. Novel functions of interferon-induced proteins. *Sem Cancer Biol* **10**: 93–101.

Shortman K. 2000. Burnet oration: dendritic cells: multiple subtypes, multiple origins, multiple functions. *Immunol Cell Biol* **78**: 161–165.

Singh-Jasuja H, Toes RE, Spee P, Munz C, Hilf N, Schoenberger SP, Ricciardi-Castagnoli P, Neefjes J, Rammensee HG, Arnold-Schild D, Schild H. 2000. Cross-presentation of glycoprotein 96-associated antigens on major histocompatibility complex class I molecules requires receptor-mediated endocytosis. *J Exp Med* **191**: 1965–1974.

Soderberg C, Giugni TD, Zaia JA, Larsson S, Wahlberg JM, Moller E. 1993. CD13 (human aminopeptidase N) mediates human cytomegalovirus infection. *J Virol* **67**: 6576–6585.

Sotomayor EM, Borrello IM, Rattis F, Cuenca AG, Abrams J, Staveley-O'Carroll K, Levitsky HI. 2001. Cross-presentation of tumor antigens by bone marrow-derived antigen-presenting cells is the dominant mechanism in the induction of T-cell tolerance during Bcell lymphoma progression. *Blood* **98**: 1070–1077.

Sporri R, Reis e Sousa C. 2005. Inflammatory mediators are insufficient for full dendritic cell activation and promote expansion of CD4+ T cell populations lacking helper function. *Nat Immunol* **6**: 163.

Trottein F, Pavelka N, Vizzardelli C, Angeli V, Zouain CS, Pelizzola M, Capozzoli M, Urbano M, Capron M, Belardelli F, Granucci F, Ricciardi-Castagnoli P. 2004. A type I IFN-dependent pathway induced by Schistosoma mansoni eggs in mouse myeloid dendritic cells generates an inflammatory signature. *J Immunol* **172**: 3011–3017.

Velculescu V, Zhang L, Zhou W, Vogelstein B, Basrai MA, Bassett DE, Hieter P, Vogelstein B, Kinzler KW. 1997. Characterization of the yeast transcriptome. *Cell* **88**: 243–251.

Villadangos JA, Bryant RA, Deussing J, Driessen C, Lennon-Dumenil AM, Riese RJ, Roth W, Saftig P, Shi G, Chapman HA, Peters C, Ploegh HL. 1999. Proteases involved in MHC class II antigen presentation. *Immunol Rev* **172**: 109.

Vremec D, Pooley J, Hochrein H, Wu L, Shortman K. 2000. CD4 and CD8 expression by dendritic cell subtypes in mouse thymus and spleen. J Immunol **164**: 2978–2986.

Weihua X, Ramanujam S, Lindner DJ, Kudaravalli RD, Freund R, Kalvakolanu DV. 1998. The polyoma virus T antigen interferes with interferon-inducible gene expression. *Proc Natl Acad Sci USA* **95**: 1085–1090.

Winter G, Fuchs M, McConville MJ, Stierhof YD, Overath P. 1994. Surface antigens of *Leishmania mexicana* amastigotes: characterization of glycoinositol phospholipids and a macrophage-derived glycosphingolipid. *J Cell Sci* **107**: 2471–2482.

Winzler C, Rovere P, Rescigno M, Granucci F, Penna G, Adorini L, Zimmermann VS, Davoust J, Ricciardi-Castagnoli P. 1997. Maturation stages of mouse dendritic cells in growth factor-dependent long-term cultures. *J Exp Med* **185**: 317–321.

Woolhouse MEJ, Webster JP, Domingo E, Charlesworth B, Levin BR. 2002. Biological and biomedical implications of the co-evolution of pathogens and their host. *Nat Genet* **32**: 569–577.

Yeager CL, Ashmun RA, Williams RK, Cardellichio CB, Shapiro LH, Look AT, Holmes KV. 1992. Human aminopeptidase N is a receptor for human coronavirus 229E. *Nature* **357**: 420–422.

22

Parallel Biology: a Systematic Approach to Drug Target and Biomarker Discovery in Chronic Obstructive Pulmonary Disease

Laszlo Takacs

Abstract

Genome research is a paradigm-free fact-gathering approach to complex biological problems, like a disease. We chose a global gene expression profiling, proteomics and genetics approach to find the best drug targets and biomarkers for chronic obstructive pulmonary disease. Here we describe the general genomics and data integration aspects of a comprehensive clinical study.

22.1 Introduction

Molecular biologists working busy on cycles of hypothesis-driven experimental research are rarely willing to acknowledge the existence of another major source of knowledge, one that derives from encyclopaedic approaches. History provides much evidence to support the significance of observation-based, descriptive research and, in reality, science evolves by consecutive but simultaneous cycles of multiple lines of observation and experimentation (Figure 22.1).

As pre-existing knowledge has little influence, all observable changes in a given organism or experiment should be considered equally important (Kuhn, 1962). There are two important factors: (i) the depth or 'globality' of the analysis; and (ii) the sensitivity. Under optimal circumstances, all elements should be measured at a sensitivity level that is in the range of random change in a steady state. Requirements are optimally fulfilled for parallel molecular observation of a system or an organism if 'globality' is 100 per cent, and the sensitivity is orders of magnitude higher than is

Immunogenomics and Human Disease Edited by András Falus

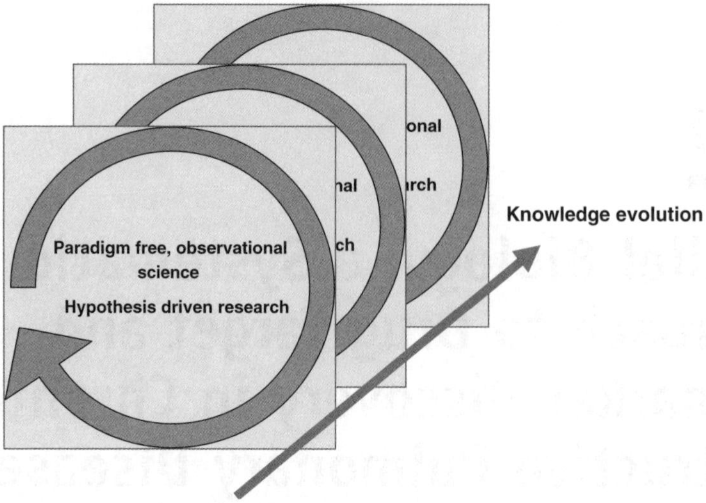

Figure 22.1 Progress of scientific research via cycles of observation and hypothesis-driven experiments.

required for reliable measurement of biologically relevant changes for all classes of molecular entities. As the observation of changes is done in parallel, the best descriptor of the process is parallel biology. When applied for the analysis of experimental systems, it is often called 'systems biology'.

Duke Fredrico Cesi, an early observationalist, founded the first academy of sciences in Italy in 1603. The organization was driven to understand the universe via observation, thus he named the academy after the lynx (a feline with very precise vision), 'Academia dei Lincei' (www.scholarly-societies.org/history/1603al.html). Simple visual depiction and accurate recording fuelled the interest of many natural scientists from times before and during the renaissance, up to today, and resulted among other effects, in the creation of 'binominal nomenclature', the inventory and the natural selection theory of species by Linnaeus and Darwin, respectively.

Figure 22.2 Synthesis of microscopic observations required about 150 years.

An excellent encyclopaedic example in the medical biology field is the discovery of the microscope that resulted in a leap forward into a new and better pathophysiology paradigm when Virchow summarized his observations in the first histopathological textbook entitled '*Cellular Pathology*'. It took surprisingly long, about 150 years, from Antonie van Leeuwenhoek's first biologically relevant observations until Virchow published the comprehensive synthesis and data set, which provided a deeper level of understanding to all pathological conditions (Figure 22.2).

22.2 Genome Research is a Specific Application of Parallel Biology Often Regarded as Systems Biology

Genome research is parallel biology, hypothesis-free, and global molecular interrogation of all elements of the entire organism or observed experiment. Today, there are only a limited number of technologies that are sufficiently specific and sensitive enough to provide accurate and global molecular-level information, which once obtained does not have to be frequently double-checked and is ready to feed into standardized databases (Collins, Morgan and Patrinos, 2003). Genome sequence, measurement of mRNA levels approximate to the requirements for 'globality' and sufficient sensitivity, protein and metabolite level measurements lag behind in specific and sensitive areas.

Genome sequence, steady-state transcript levels and genome-wide genetic association satisfy the criteria. Proteome information is close to qualifying in bacteria and yeast. Genome-scale mutation, knock-out and knock-down information are becoming available for some bacteria, yeast, *Drosophila* and *C. elegans*, while other technologies (metabonomics, single-cell transcriptomics and proteomics, individual human genome re-sequencing) are expected to evolve in the near future. Chronic obstructive pulmonary disease (COPD) was selected to test the encyclopaedic genome research approach and apply it for drug and biomarker discovery.

22.3 Chronic Obstructive Pulmonary Disease

COPD is a lethal disease commonly caused by smoking and it is the fifth leading cause of death in the Western world. The incidence of COPD is increasing, via a mechanism that is unlikely to be associated with elevated pollution or other environmental factors damaging the lung (Molfino, 2005). More importantly, the overall improvement in life expectancy, a process that is due directly to better treatment of cardiovascular disease, brings among other factors, chronic organ or system diseases as life-limiting factors into the foreground.

COPD is assumed to be driven by a single pathological process and lead to three main symptom groups: (i) chronic and progressive emphysema; (ii) so-called 'non-reversible' airway obstruction that does not respond to beta adrenergic stimulants or cholinergic inhibitors in acute tests; and (iii) recurrent exacerbations of chronic bronchitis. Disease diagnosis is built on symptoms of recurrent bronchitis, and low levels of lung function. Nonresponsiveness to adrenergic stimulants is used to discriminate COPD from asthma (Bolton *et al.*, 2005). Standardized criteria have been built based on these most informative clinical tests. Because many patients first look for treatment when the lung damage is already extensive (stage II COPD), there is a need for earlier disease detection and effective treatment.

Today, COPD treatment is symptomatic (Cooper and Tashkin, 2005). Current thinking suggests that an acute initial inflammatory process leads to progressive and chronic airway destruction in the susceptible population of smokers that could be manifested in loss of active alveolar surface and chronic reactive obstruction of mid-way bronchi (Fujimoto *et al.*, 2005). However, neither the initial inflammatory 'noxa' nor the mechanisms that drive it in the genetically susceptible individuals to an irreversible life-threatening disease are understood. Genetic or environmental factors could be responsible for observed geographical differences which show that the highest incidence of death directly caused by COPD to be in males in Hungary (Figure 22.3; Kleeburger and Peden, 2005).

Cells of the innate host defence and immune systems, the granulocyte and the macrophage have been suggested to play central roles in the disease. Recent evidence puts the macrophage in the centre, because the first histopathological picture is evidenced by higher number of sub-mucosal bronchial macrophages in early stage I COPD patients who have only minimal symptoms (Bolton *et al.*, 2005). Animal models, namely the 'smoking mouse', seem to support this idea. Surprisingly it has been shown that homozygous deletion of the transforming growth factor beta (TGF-β) processing enzyme protects from cigarette smoke-caused COPD in a susceptible

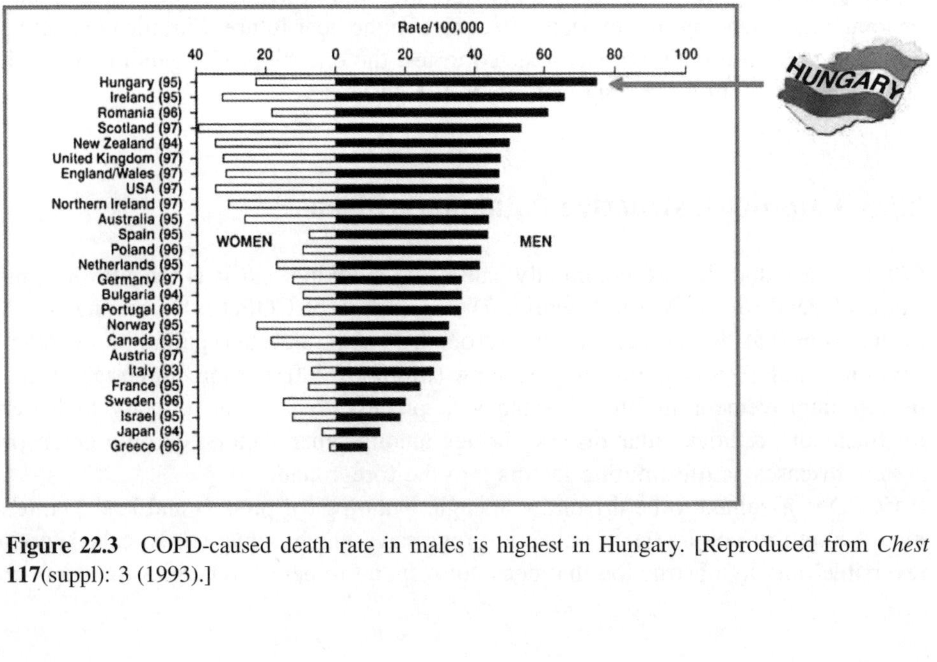

Figure 22.3 COPD-caused death rate in males is highest in Hungary. [Reproduced from *Chest* **117**(suppl): 3 (1993).]

inbred mouse strain (Groneberg and Chung, 2004; Morris *et al.*, 2003). TGF-beta is abundant in tissue macrophages and has been detected in human alveolar macrophages.

Our approach, described below, indicates that without a particular hypothesis in mind genomescale interrogation of the disease provides accessory data to strengthen already existing hypotheses, like TGF-beta, and provide sufficient information to generate a number of new hypotheses.

22.4 Goals of the Study

As the mechanism of the disease is not known, only drugs to treat symptoms are available, which address the phenomenology of the disease pathology, and include bronchodilators, antibiotics and steroids. The primary goal of the present approach is to discover the most relevant disease mechanism and those genes that can be altered via novel drugs. As COPD progresses over a time period that is likely to be longer than 5 years, it is unlikely that mechanism-based drugs will improve the disease in a measurable manner over the time that is feasibly available for phase II and phase III clinical trials. The secondary goal of this project is thus the discovery of biomarkers that will report disease progression before this becomes apparent via current clinical testing.

22.5 Methods

Models and data sources

Macrophage, a disease process central cell type, was chosen, and bronhoalveolar macrophages were isolated from 40 well-defined stage II COPD patients and 40 controls. For genetic analysis, DNA was collected from 250 COPD patients and 250 control individuals. Simultaneous studies were performed in mice and bronchoalveolar macrophages were tested from cigarette smoke, LPS or protease-induced murine COPD disease models. Human macrophage cell lines were stimulated with LPS in the presence and absence of drugs and were also tested (Figure 22.4).

Genome-scale technologies

Global and steady-state mRNA expression level analysis

Affymetrix gene-chip experiments were done on human cell models and human and mouse bronchoalveolar (BAL) macrophages using the widest genome-covering chips. Printed microarrays were prepared by printing 50-mer human and mouse transcript-specific oligonucleotides onto glass slides. The transcripts were selected on the basis of their likelihood to provide relevant information for drug discovery and biomarker research. The selection analysis was done via a need-driven but global informatics

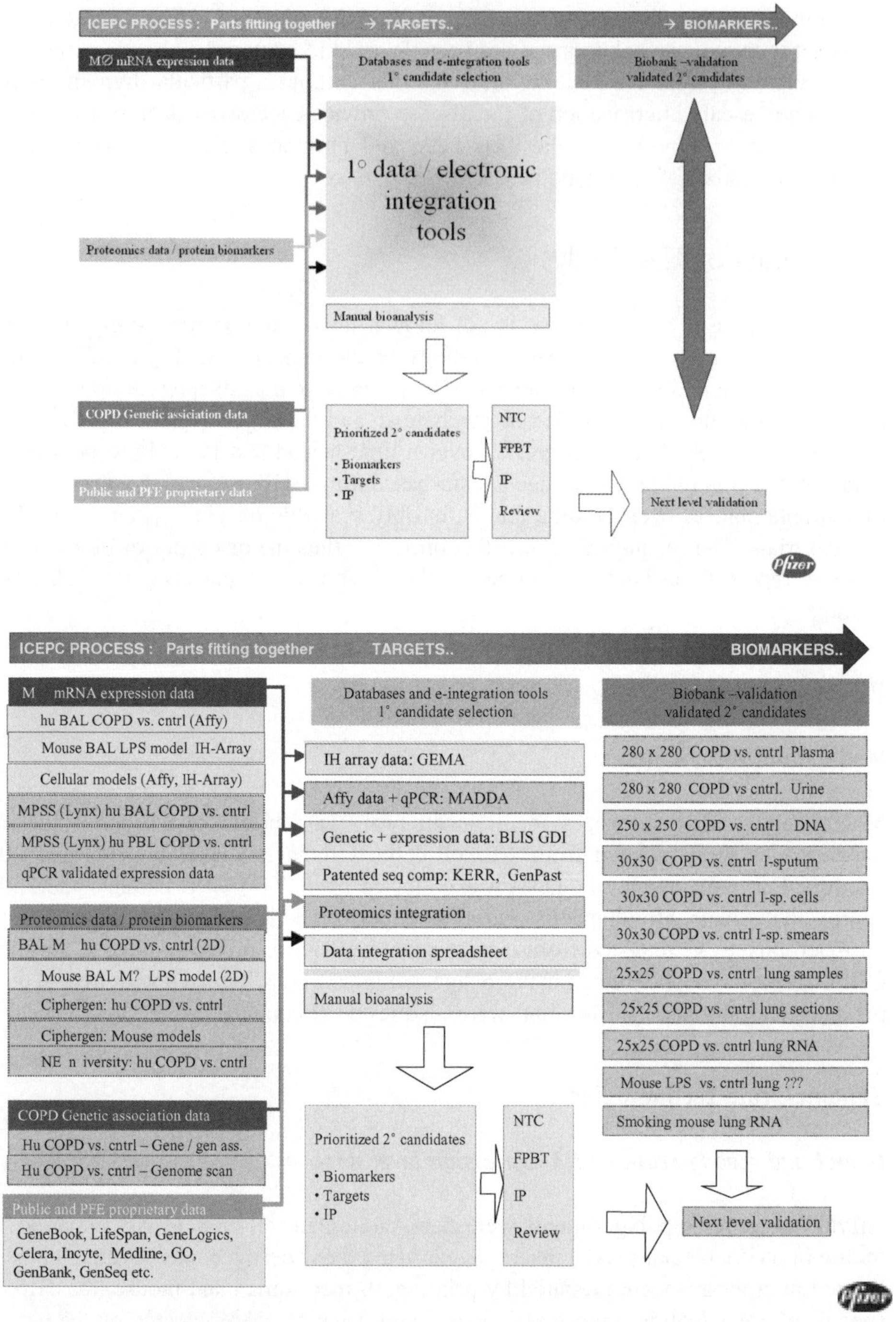

Figure 22.4 COPD target and biomarker discovery process.

integration and prioritization process of the COPD literature and in-house databases. Finally, transcript sequences were selected and three specific olgo sequences were chosen for each selected gene. As hybridization-driven expression analysis is less sensitive than deep sequencing-based expression profiling technologies, we chose massive paralallel sequencing signature analysis (MPSS), provided by Lynx Inc. (Stolovitzky *et al.*, 2005). Human BAL macrophages, peripherial blood mononuclear cell-derived monocytes and *in vitro* differentiated macrophages from these cells were tested via sequencing 1 million tags from a representative set of patients and controls. Important candidates were validated via an automated quantitative PCR assay using multiple robots and high-throughput ABI qPCR machines.

Proteomics technologies

Although not global and not sufficiently sensitive, proteomics analysis is important as disease-specific changes in steady-state mRNA expression levels might reflect pathology less efficiently than protein levels. To this end we applied two-dimensional gel-electrophoresis and subsequent protein identification via mass spectrometry on human and mouse BAL macrophages and on some of the human cellular models. Plasma and urine samples were analysed from 30 COPD patients and 30 controls via MALDI-TOF technology, provided by Ciphergen Inc.

Genetic technologies

The data collection for this technology was not complete at the time of the writing of this manuscript. A full or partial genomescan as well as gene-specific approaches are planned to find disease-, response- and symptom-specific genetic associations via single-nudeotide polymorphism screening.

Additional information gathering

Public databases, including medline, Gene Bank and NCBI, were interrogated for relevant information via automatic and manual processes. Private databases, such as GeneBook, Celera, Incyte and Gene-Logics, were interrogated via similar technologies.

Priamary data analysis and integration

For data treatment, a specific MIAMI-compliant data warehouse was constructed and global and local normalization processes were applied to expression data first. Subsequently, both inferential (parametric and nonparametric tests) and descriptive statistics (various clustering techniques based on Euclidean distances of Pearson coefficients as well as *K*-means distances) were used for data analysis. The results were various priority lists that showed no more that 5–10 per cent overlap, even if the broadest selection criteria were applied, suggesting that the analysis process and/or the analyser had a greater effect on the outcome than the biological difference or the experiment itself. Nevertheless, lists were used as input for the secondary data analysis process.

Secondary data analysis

List were generated from all different approaches based on simply 'give me the genes that differ most' by comparing across two to five expression/patient descriptors, e.g. the gene expression difference between those COPD patients who quit smoking and those who did not (three descriptors). The lists were compiled in a matrix of genes vs best attributes (e.g. smoking cessation in the previous example). Both binary and linear scales were applied as needed. Attributes were weighted on the basis of perception as to which would contribute more to a final selection concept; factors were used to multiply the binary or linear values, which after a normalization step were simply computed. The sum was then used to re-sort the matrix. Empiric tests revealed that weighting factors can be applied in such a manner that the currently best gene targets will be on the top of the list.

Biobanking

To validate the findings, a set of relevant tissues and cells was collected in order to test the hypotheses derived from the global approach via qPCR, histology, ELISA and other techniques.

22.6 Results

Secondary data integration revealed that, if weighting factors are applied in such a manner that the highest scoring candidate is the current best target of the disease, a set of the next candidates contains both the known second-best candidates and novel gene targets or biomarkers. Many of these were successfully validated on the bio-banked samples. However, as the real disease mechanism is not known, the assumption of putting the highest weight on the currently best target might be wrong. Weights could be set to prioritize genes that are the currently best literature-derived discovery candidates (e.g. TGF-beta) but derive entirely from mouse experiments. While arranging prioritization weights using the results of hypothesis-driven experiments can be useful, generic weighting factors that derive from multiple-case testing of the specific value of the particular attribute types is expected to provide superior final results, because it eliminates the hypothesis from the analysis process.

Appendix

Individuals and various teams participating in the study (PGRD-Fresnes, Pfizer Global Research and Development Fresnes Laboratories, France; U of Debrecen, University of Debrecen, Hungary; North-Eastern U, Northeastern University, Boston, USA; PGRD-Groton, Pfizer Global Research, Groton Laboratories, Groton, USA). Team leaders are underlined.

PGRD-FRESNES
Myriam Artola
Patrick Berna
Karine Boudjoulian
Estelle Carlos-Diaz
Clothilde Dantier
Isabelle De Mendez
Frederic Dubois
Manuel Duval
Thomas Eichholtz
Guido Grentzmann
Stephane Guerif
William Hempel
Carole Malderez-Bloes
Malika Ouagued
Jean-Michel Planquois
Marie-Pierre Pruniaux
Zoltan Rozsnyay
Laszlo Takacs
Sophie Trouilhet
Christine Vanhee-Brossollet

U of DEBRECEN
Istvan Andrejkovics
Julia Buslig
Eszter Csanky
Laszlo Fesus
Pal Gergely
Erika Marhas
Laszlo Nagy
Zsuzsa Reveszne
Beata Scholtz
Titanilla Tolgyesi
Attila Vasko

PFIZER HUNGARY
Klara Der
Peter Horovitz

NORTH-EASTERN U
Barry Karger
William Hancock
Andras Guttman

CIPHERGEN
Davies Huw
Isabelle Buckle

LYNX
Michael Kramer
Thomas Vasicek
Maria Schramke

PGRD-Groton
Tom turi

References

Bolton CE, Ionescu AA, Edwards PH, Faulkner TA, Edwards SM, Shale DJ. 2005. Attaining a correct diagnosis of COPD in general practice. *Respir Med* **99**(4): 493–500.

Collins FS, Morgan M, Patrinos A. 2003. The Human Genome Project: lessons from large-scale biology. *Science* **300**(5617): 286–290.

Cooper CB, Tashkin DP. 2005. Recent developments in inhaled therapy in stable chronic obstructive pulmonary disease. *Br Med J* **330**(7492): 640–644.

Fujimoto K, Yasuo M, Urushibata K, Hanaoka M, Koizumi T, Kubo K. 2005. Airway inflammation during stable and acutely exacerbated chronic obstructive pulmonary disease. *Eur Respir J* **25**(4): 640–646.

Groneberg DA, Chung KF. 2004. Models of chronic obstructive pulmonary disease. *Respir Res* **5**(1): 18.

Kleeberger SR, Peden D. 2005. Gene–environment interactions in asthma and other respiratory diseases. *Rev Med* **56**: 383–400.

Kuhn T. 1962. *The Structure of Scientific Revolutions*. University of Chicago Press: Chicago, IL.

Molfino NA. 2005. Drugs in clinical development for chronic obstructive pulmonary disease. *Respiration* **72**(1): 105–112.

Morris DG, Huang X, Kaminski N, Wang Y, Shapiro SD, Dolganov G, Glick A, Sheppard D. 2003. Loss of integrin alpha(v)beta6-mediated TGF-beta activation causesMmp12-dependent emphysema. *Nature* **422**(6928): 169–173.

Stolovitzky GA, Kundaje A, Held GA, Duggar KH, Haudenschild CD, Zhou D, Vasicek TJ, Smith KD, Aderem A, Roach JC. 2005. Statistical analysis of MPSS measurements: application to the study ofLPS-activated macrophage gene expression. *Proc Natl Acad Sci USA* **102**(5): 1402–1407.

23
Mycobacterial Granulomas: a Genomic Approach

Laura H. Hogan, Dominic O. Co and **Matyas Sandor**

Abstract

Mycobacterium-induced chronic granulomatous inflammation is an ecosystem in which mycobacterial species seek survival in the surrounding environment of a sequestered inflammatory granuloma. Granulomas restrict dissemination of pathogens and protect the host, yet fail to achieve sterilizing immunity. Mycobacteria parasitize macrophage using various immunosuppressive mechanisms, but are ultimately forced into a quiescent state during latency. The ability of the pathogen to achieve this state is probably an advantage allowing its long-term survival in the host environment. The complex changes in gene expression both in macrophages and in mycobacteria have been successfully studied using genomic approaches typified by DNA-based microarray. It is clear that, alongside advances in genetics, proteomics, and informatics tools, studying global gene expression changes during co-accommodation of host and pathogen will significantly contribute to our understanding of the local inflammatory reactions that represent the direct interface between the infectious agents and the host.

23.1 Introduction

Granulomatous inflammation represents a localized delayed-type hypersensitivity induced by CD4 T-cells. Many different agents can induce granulomatous responses, including macrophage infected by intracellular microbes (Co *et al.*, 2004a, b), parasitic worm eggs, and particulate material including antigen absorbing silica or metals (Saltini, 1995). Granuloma-inducing pathogens include *Leishmania*, *Schistosoma* eggs, mycobacteria, *Listeria*, and *Histoplasma* among many other intracellular pathogens. While granulomas contain a variety of cell types depending upon the stimulating event, the presence of macrophage and CD4 T lymphocytes responsible for directing the granulomatous response is universal. The consequence of granuloma formation can be protective to the host by containing the pathogen to prevent

Immunogenomics and Human Disease Edited by András Falus

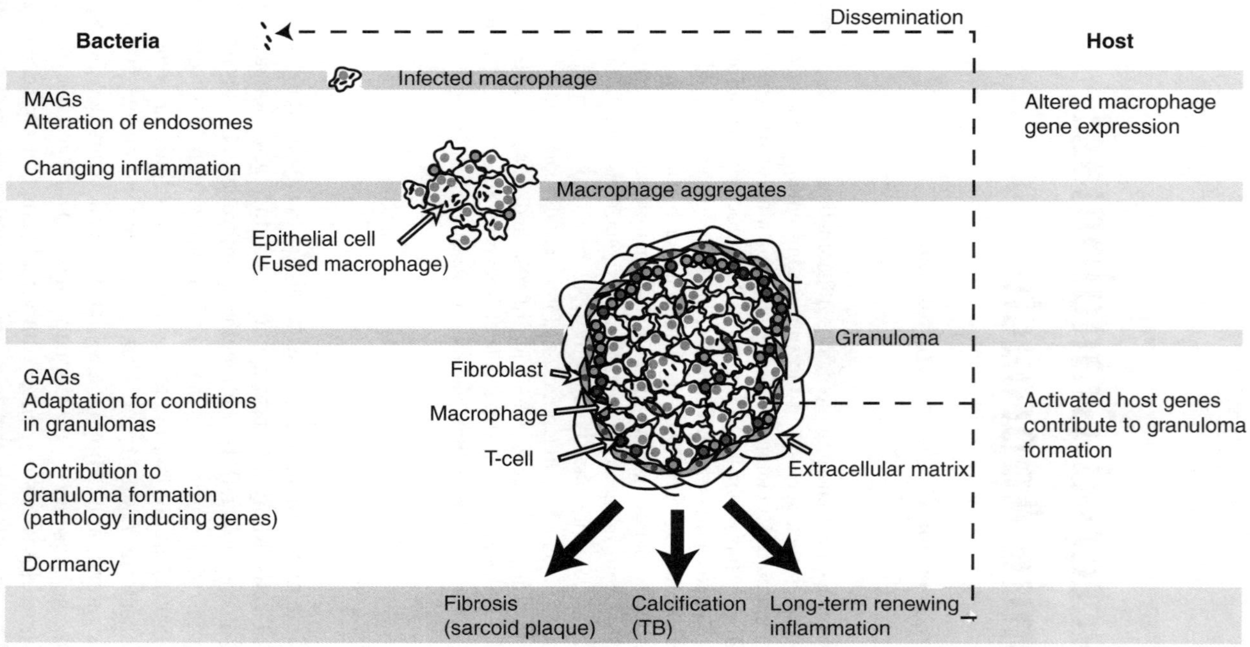

Figure 23.1 Continuous adjustment of host and pathogen gene expression during the course of granuloma formation in response to mycobacterial infection. Mycobacterial genes: MAG, macrophage-activated genes; GAG, granuloma-activated genes.

systemic dissemination and insulating the surrounding tissue from damaging inflammation, but they are always pathologic since the long-term fibrosis and progressive scarring lead to chronic organ damage, including the lungs, liver, gut, and vasculature.

In the case of infection by pathogenic mycobacteria, granuloma formation around infected macrophage is an essential component of the initial successful control of the bacterium (Sandor, Weinstock and Wynn, 2003). Macrophage are the primary host cells for mycobacteria, and mycobacterial infection begins with exponential growth within the macrophage phagolysosome (Hingley-Wilson, Sambandamurthy and Jacobs, 2003; Figure 23.1). Granuloma formation begins with aggregation of infected and noninfected macrophages. Bacterial loads begin to fall at 2–4 weeks after infection as cell-mediated immunity controls the infection; however, mycobacteria remain present at a very low levels for extended periods in a condition variously described as dormancy or nonreplicating persistence (reviewed in Wayne and Sohaskey, 2001). CD4 T-cells are essential in the granuloma as they regulate granuloma function, allow recruitment of other cell types including lymphocytes, macrophages, and fibroblasts, and promote maturation and resolution of the lesion. A physical barrier to bacterial dissemination is provided by an extracellular matrix shell secreted by fibroblasts in the lesion. Progressive fibrosis can eliminate granulomas and lead to scarring of the tissue.

The calcification and fibrosis associated with granuloma formation in the human lung is diagnostic of infection even in the absence of apparent symptoms and may persist for many decades. Yet at the same time, protective immunity conferred by granulomatous inflammation is rarely sterilizing immunity. Reactivation and disease arising from the failure of the adaptive immune system many years or decades beyond the initial infection is a significant cause of mycobacteria-associated morbidity. While the primary reservoir of infected macrophage in the body is unknown, it is not unreasonable to suppose that some reactivation may result from mycobacteria disseminating from failing granulomas. The granuloma is an advantageous adaptation for both the host and for the bacterium. The isolation of dormant bacteria in the granuloma prevents host disease until immunity weakens. Conversely, mycobacteria can survive in the host for an extended period, and the granulomatous barrier that prevents dissemination also protects the pathogen from sterilizing immunity. This inflammatory interface is the point at which the host genome and the mycobacterial genome negotiate changes in gene expression that lead to long-term coexistence. This review will focus upon the coexistence of mycobacteria and infected host in granulomas that are the primary site of interaction.

23.2 Initial Infection of Macrophage

Mycobacteria can infect dendritic cells (DC), alveolar macrophage, epithelial cells, and, rarely, endothelial cells. However, the major targeted cell is the macrophage. There are a growing number of cell surface receptors that have been identified as participating in the binding and uptake of mycobacteria. The interaction of mycobacteria and mycobacterial products with macrophage cell surface receptors utilizes multiple redundant pathways (Vergne *et al.*, 2004). The phagocytosis of mycobacteria via macrophage surface receptors can include interactions with complement receptors CR1, CR3, and CR4, mannose receptor, surfactant protein A (Sp-A) and its receptors, CD14, scavenger receptor class A, a variety of Toll-like receptors (TLR), or Fc receptor. DC-specific receptors such as DC-SIGN are involved in DC–mycobacteria interactions via mycobacterial mannose-capped lipoarabinomannan (ManLAM; Tailleux *et al.*, 2003). Mycobacterial products having known interactions with macrophage cell surface receptors include 19 kDa lipoprotein, ManLAM, and soluble tuberculosis factor (STF; Krutzik and Modlin, 2004). Those interactions are with extracellular TLR, notably TLR2 and TLR4 (Krutzik and Modlin, 2004), and CD14, in addition to multiple intracellular targets accessed after release of mycobacterial products into the infected cell cytoplasm. The specific receptors initiate different endocytic and signalling pathways, so the receptor set involved in binding has an important consequence for determining whether macrophage act as a host for bacterial growth, are able to effectively kill mycobacteria, or have an intermediate response.

Altered macrophage gene expression

After phagocytosis of mycobacteria, the macrophage initiates a complex response that involves altered expression levels of hundreds of genes. In recent years, DNA microarray analysis has frequently been used to examine global changes in macrophage gene expression specific to mycobacterial infection (Kendall *et al.*, 2004). Examination of a few of the numerous recent studies reveals some common themes.

The response of freshly cultured human monocytes to various bacterial pathogens or their derived components was compared using high-density DNA microarrays (Nau *et al.*, 2002). Some 977/6800 genes changed on exposure to one or more of the bacteria examined. This study defined a shared transcriptional response of macrophage to infection by laboratory and virulent *E. coli* strains, *Salmonella typhi*, *S. typhimurium*, *Staphylococcus aureus*, *Listeria monocytogenes* and mycobacteria relative to responses measured after exposure to latex beads. Regulated genes include proinflammatory genes, cytokines, chemokines, signalling molecules, transcription factors, adhesion molecules, tissue remodelling factors, enzymes, and anti-apoptotic molecules. The strength of this study was in the examination of gene expression induced by multiple pathogens. This breadth allowed cross-comparisons of individual gene significance in combination with detailed kinetic analysis of a few of the pathogens. *M. tuberculosis*-specific gene expression changes were identified as a unique pattern containing a subset of the shared transcriptional responses as well as some individually regulated genes. Some of the mycobacteria-specific changes were dominant in mixed co-culture over opposing changes induced by other bacteria. Many of the *M. tuberculosis*-induced gene expression patterns were specific and suggestive of specialized signalling pathways. Some of the observed changes have been documented in the literature, such as declining IL-12 and IL-15 production (Nau *et al.*, 2002). The specificity of the macrophage response to mycobacteria was also observed in a study comparing gene expression profiles of human primary blood monocytes in response to *M. tuberculosis* or to various parasitic pathogens. Each pathogen elicited a characteristic functional gene cluster as in the bacterial pathogen study (Chaussabel *et al.*, 2003).

These studies can be compared with a similar series of experiments (Ehrt *et al.*, 2001) in which murine bone marrow-derived macrophage were assayed for global changes in gene expression upon exposure to IFN-γ and/or *M. tuberculosis*. The transcriptional signatures of *M. tuberculosis* infection and IFN-γ exposure are similar, but not totally overlapping, and include $\sim$25 per cent of the monitored genome over 24 h of stimulation by the pathogen. The expression of more than half of these genes is modulated by inducible nitric oxide synthase and phagocyte NADPH oxidase, as demonstrated using genetically deficient mice as bone marrow donors.

Wahl and co-workers used *in vitro* co-culture of human monocyte-derived macrophage cultures and *M. avium* to show innate proinflammatory responses within 2 h of infection (IL-1, TNF-α) along with enhanced expression of numerous diverse transcription factors and signalling molecules (Greenwell-Wild *et al.*, 2002). They examined a time course of acute changes in gene expression *in vitro*, but the evolving patterns of gene expression (sustained expression of diverse chemokines and adhesion molecules) and the formation of multinucleated giant cells in co-culture suggests that at least some aspects of chronic host–pathogen interactions were modelled. Their data reinforces the observation that mycobacterial species can interact with numerous diverse host signalling pathways. The transient expression of many macrophage gene products probably reflects significant mycobacterial interference with host signalling pathways observed in more direct studies.

Others have studied the expression of immunoregulatory gene subsets using microarrays of 375 selected genes and *M. tuberculosis* infection of the human macrophage cell line THP-1 (Ragno *et al.*, 2001). At 6 and 12 h, inflammatory genes including IL-1β, IL-2, and TNF-α; chemokines IL-8, osteopontin, MCP-1, MIP-1α, RANTES, MIP-1β, MIP-3α, MPIF-1, PARC, and eotaxin; and cell migration factors MMP-9, VEGF, and CCR3 exhibited changed expression.

These studies are largely consistent with the known capacity of mycobacterial species to induce specific pro-inflammatory changes in infected macrophage (Monack, Mueller and Falkow, 2004). Typically, each study detects at least a subset of genes for the known cytokines, chemokines, adhesion molecules, and signal transduction components responding to phagocytosis of pathogenic mycobacteria. The contribution of microarray analysis has been in delineating the specificity of the response to mycobacteria in comparison to other bacterial pathogens and other intracellular parasites. Despite this, much remains to be understood about the nature of the macrophage response and the consequences of particular changes in gene expression. What is understood is that mycobacteria change the macrophage to accommodate their long-term survival at that site by stopping phagolysosomal fusion, preventing phagosomal acidification, blocking apoptotic pathways to keep the macrophage alive as a host, and suppressing the immune response.

Many of these changes are accomplished by the direct modification of macrophage signal transduction (Koul *et al.*, 2004). A defining characteristic of pathogenic mycobacterial infection is blockage of the normal sequences of phagosomal fusion with lysosomal and late endosomal compartments in the infected cells, resulting in the survival of the bacterium within this organelle. Other events arising from infection include downregulation of major histocompatibility complex (MHC) class II accompanied by decreased antigen processing, decreased bactericidal responses, and decreased apoptosis (Vergne *et al.*, 2004). Many of these events are mediated by the ability of mycobacteria and secreted virulence factors to directly interfere with macrophage signal transduction pathways. The direct mechanisms by which mycobacteria are able to redirect macrophage signal transduction have recently been extensively reviewed (Koul *et al.*, 2004).

The mechanistic data reviewed by Koul *et al.* suggests direct mycobacterial inhibition of the typical macrophage responses in infected host cells. A growing body of evidence suggests that secretion of virulence molecules such as Man-LAM and other lipids and glycolipids by pathogenic mycobacteria is responsible for extensive disruption of host cell intracellular signalling via various signal transduction pathways including MAPKs, IFN-γ, Ca^{2+}/calmodulin, TLRs, and JAK/STATs. Many of the alterations reflect blockade of pre-existing response pathways of macrophage to a variety of pathogens, rather than specialized adaptive responses of the macrophage to infection by mycobacteria. Mycobacteria secrete classes of kinase and phosphatase genes capable of direct interference with signalling pathways. Mycobacterial mutants defective in serine/threonine protein kinase gene *pknG* have reduced survival in both immunocompetent and immunodeficient mice. This kinase gene and others are thought to be responsible for altering host signalling pathways and contributing to blocked phagosomal maturation to allow intracellular survival of mycobacteria. Likewise, mycobacterial mutants defective in secreted protein tyrosine phosphatase B (MptpB) are attenuated after infection and their survival in IFN-γ activated macrophage is much lower. These mutants support a model of active mycobacterial alteration of host signalling responses to IFN-γ, resulting in enhanced intracellular survival.

Overall, changes in macrophage gene expression in response to mycobacterial infection encompass the range of macrophage-specific gene expression including signalling, transcription factor levels and metabolic pathways. There is good data now about the consequence of those gene changes for mycobacterial gene expression.

23.3 Mycobacterial Gene Expression in the Host

In the macrophage, mycobacteria are confronted by a new environment containing bactericidal factors, stress factors, and altered food sources. The altered mycobacterial gene regulation starts during the earliest stages of phagocytosis, is adjusted for bacterial expansion within the phagolysosome and culminates with expression of a dormancy programme. Very interestingly, mycobacterial factors regulate host responses by altering macrophage functions and promoting granulomatous reactions. Knowledge of the complete genome of *M. tuberculosis* and the sequence of all ORFs (Cole *et al.*, 1998) greatly facilitated identification of gene sequences involved in mycobacterial virulence. There are 4000 mycobacterial genes, and early gene array studies established that hundreds of genes have altered expression levels in macrophages (macrophage-activated genes, MAGs; Chan *et al.*, 2002). When infected macrophage fuse and recruit further inflammatory cells, granulomas are established. A few dozen additional mycobacterial genes are activated within the granuloma (granuloma activated genes, GAGs).

The identification of many mycobacterial genes regulated by infection of macrophage or exposure to the phagolysosome historically followed a 'promoter trap' approach. A classic demonstration of the usefulness of this approach is outlined in a study using LacZ reporter gene expression as a measure of induction (Hobson *et al.*, 2002). Responses to acidified sodium nitrate as a surrogate for stimuli received during intracellular macrophage growth were used for screening, followed by validation of promoter activity during infection of THP-1 macrophage cell lines and of corresponding gene expression during *M. tuberculosis* infection of macrophage. As is typical, this approach yielded a relatively small number of genes, some of which were of unknown function. Certain clones failed to exhibit equivalent regulation in various screens, which can be attributed with equal plausibility to either different assay conditions in different screens or to differences in gene expression between *M. bovis* strain bacille Calmette Guérin (BCG) and *M. tuberculosis*. Upregulation of the DNA binding protein *IdeR* in this study may represent either a direct response to iron limitation, resulting in the regulation of iron uptake and storage genes, or alternatively, a more complex pleiotropic regulation of multiple gene networks including oxidative-stress responses (Gold *et al.*, 2001).

Another study applying the promoter trap approach used *M. marinum* infection and green fluorescent protein (GFP) reporter expression coupled with fluorescent activated cell sorting (Chan *et al.*, 2002). This study concluded that most genes expressed in granulomas *in vivo* are also expressed during log-phase growth *in vitro* and favoured the hypothesis that a large fraction of granuloma-residing bacteria are metabolically active. These investigators distinguished granuloma-, macrophage-, and constitutively-activated promoters (gaps, maps, and caps, respectively) for the regulation of GAG, MAG, and CAG gene groups. Many of the gene sets do not overlap so that few, if any, GAGs are also activated in cultured macrophage, suggesting a distinction between the immediate response to the phagolysosome and the long-term accommodation to granuloma-dwelling macrophage. Although *M. marinum* reside in activated macrophage within granuloma, activation of infected macrophage with IFN-γ or exposure of *in vitro* cultures to stationary phase or hypoxia fails to activate a majority of GAG promoters. Many gene families, including stress-response genes and PE-PGRS genes, have members characterized individually as GAGs, MAGs, or CAGs. In addition, many mycobacterial genes expressed during *in vivo* granuloma residence could not be upregulated during *in vitro* culture in response to various stimuli known to induce the dormancy regulon (see below). Thus, while many genes expressed in granuloma are from constitutive promoters, many of them are specialized adaptations to the inflammatory microenvironment and cannot be modelled *in vitro*.

Early reviews of individual mycobacterial genes upregulated upon macrophage internalization covered an array of sophisticated technical approaches including promoter trap strategies, signature tagged mutagenesis, selective capture of transcribed sequences (SCOTS), and DNA microarray (Domenech, Barry and Cole, 2001; Triccas and Gicquel, 2000). Macrophage-regulated genes revealed in those screens include those predicted to be involved in stress responses, transcriptional regulation, and cell wall metabolism. The discrete set of genes known to be upregulated by macrophage internalization included *sigE, sigH, pks2, aceA (icl), mce1B, uvrA, acr (hspX), mtrA*, and *prrA*, while the set of genes upregulated in response to isoniazid included FASII system components, *fbpC, fadE23, fadE24, ahpC, efpA*, and *iniAB*.

Just a few years later, Kendall *et al.* (2004) revisited the topic in a review of DNA microarray studies applied to gene expression in *M. tuberculosis* covering a total of 51 studies involving mycobacteria, greatly expanding the list of identified regulated genes. Their presentation made the strong overall point that microarray studies are primarily best at generating hypotheses for testing by other means. The authors also discussed conceptual issues relating to comparing and reconciling data across disparate studies. Perhaps their most provocative point was to questioned the underlying assumption common to all microarray experiments, that changes in gene expression are uniformly biologically important. In their view, a primary focus upon gene expression changes ignores the contribution of constitutively expressed genes, de-emphasizes the absolute level of gene expression, does not take into account that the significance of changing gene expression may be different for different proteins, and finally, does not consider that some gene changes represent co-regulation in regulons irrespective of individual significance to the phenomenon being tested. Despite these valid points, DNA microarray analysis has contributed very significantly to our understanding of mycobacterial interactions with macrophage in granulomas, and is likely to continue to do so.

Schnappinger *et al.* (2003) examined the adaptation of *M. tuberculosis* to the macrophage environment by comparing their transcriptional response to *in vitro* culture and to murine bone marrow macrophage with and without activation by IFN-γ. Their study found that $\sim$15 per cent of the *M. tuberculosis* ORFs (601 genes) are regulated by parasitism of macrophage. They considered broad classes of *M. tuberculosis* gene expression as indicators of the biochemical environment of the phagolysosome. The enhanced expression of fatty acid-degrading enzymes, DNA repair proteins, secreted siderophores, lipid metabolism enzymes, and transcriptional regulators is consistent with specific metabolic adaptations of the pathogen to the phagolysosome. Changes in expression of a subset of the regulated genes was validated by qrtPCR of transcript levels in RNA isolated from infected mouse lungs. Regulated genes include *fadD26, mbtABCDEFGH, icl (aceA), furA, umaA, alkB*, and *fadD33*.

Pathology inducing genes of mycobacteria contribute to granuloma formation

So-called mycobacterial pathology mutants (*pat* mutants) demonstrate decreased or altered host pathology after infection. Many of these mutants have alterations in putative transcription factors and their phenotypes suggest the possibility that mycobacteria may actively contribute to granuloma formation by using trancription factors to modulate a potentially wide array of gene expression involved in host interactions. The *pat* mutants have phenotypes including less severe host pathology, altered cellular composition of granulomas, decreased bacterial control, and increases in host mean survival times. Mycobacterial transcription factors which may modulate host pathology include putative transcription factor WhiB3 (Steyn *et al.*, 2002), transcription factor *sigE* (Manganelli *et al.*, 2004; Ando *et al.*, 2003), and *sigH* (Kaushal *et al.*, 2002). The altered pathology induced by *sigH* mutants is either correlated with or dependent upon T cell-mediated immunity. Infection of SCID mice by a *sigH* mutant leads to the same survival time and pathology as seen in a wild-type infection. Given the centrality of CD4 T cell in granuloma function, it would not be unexpected if the virulence phenotype requires adaptive immunity.

Taken together, this data suggests the evolutionary selection for *M. tuberculosis* capable of promoting very specific types of granulomatous responses in the host. An independent line of evidence also points to the preference of mycobacteria for granulomas as a 'shelter' in which they can persist for extended periods of time and maintain a reservoir of infection without killing the host. *M. marinium* tagged with either GFP or dsRed can be used to establish a frog infection model in which fluorescently tagged mycobacteria are visible within granulomas. Reinfection of frogs infected with GFP-tagged *M. marinum* with dsRed-tagged *M. marinum* results in homing of the dsRed-tagged bacteria to established granulomas containing GFP-tagged bacteria (Cosma, Humbert and Ramakrishnan, 2004). Finally, a putative evolutionary preference of virulent mycobacteria for one particular type of granulomatous inflammation is consistent with the prevalent idea that caseating granulomas promote the contagiousness of *M. tuberculosis*.

Host factors which might influence mycobacterial gene expression and persistence are less clear. Expression of a group of mycobacterial genes referred to as a persistence programme correlates with mycobacterial entry into granulomas (Ramakrishnan, Federspiel and Falkow, 2000; Chan *et al.*, 2002; Schnappinger *et al.*, 2003). Host Th1 genes, such as IFN-γ and iNOS, may be required for expression of these genes (Shi *et al.*, 2003) since *M. tuberculosis* does not express this persistence programme after infection of IFN-γ-deficient mice. Overall, it is interesting to speculate that mycobacteria and T-cells are partners in the development of granulomatous responses.

Regulation of mycobacterial genes during dormancy in the granuloma

The extended period of clinically nonapparent disease in the host followed by reactivation disease has long suggested the existence of a prolonged state of latency by the mycobacteria in a host reservoir. The lack of symptoms is accompanied by only a few granuloma in the lung evident on X-ray (Balasubramanian *et al.*, 1994) and an inability to culture mycobacteria from tissue samples. Murine models indicate that mycobacterial DNA persists during this interval, suggesting the presence of dead bacilli, free DNA or dormant bacilli (de Wit *et al.*, 1995). Mycobacterial latency can be alternatively modelled as either a nonreplicative, metabolically and transcriptionally inactive state similar to sporulation in *Bacillus* species and controlled by specialized sigma factor SigF (DeMaio *et al.*, 1996; Geiman *et al.*, 2004), or as a process of dynamic long-term growth and replacement which is replicatively, transcriptionally, and metabolically active (Manabe and Bishai, 2000). The true character of mycobacterial adaptation to long-term survival in the host is likely to be a hybrid of the two concepts, given the strong evidence for both a dormancy, regulon (Voskuil, Visconti and Schoolnik, 2004; Zhang, 2004) and for dynamic traffic into the granuloma at a chronic stage of infection (Cosma, Humbert and Ramakrishna, 2004) accompanied by GAG expression (Chan *et al.*, 2002; Ramakrishnan, Federspiel and Falkow, 2000).

Initially, dormancy was modelled using *in vitro* culture under conditions of reduced oxygen tension to model the assumed oxygen environment of the granuloma (Wayne, 1994; Cunningham and Spreadbury, 1998). Transmission electron microscopy and protein profiling indicated that, under oxygen stress, the mycobacterial cell wall thickens and a 16 kDa α crystalline homologue (HspX/acr) is produced (Cunningham and Spreadbury, 1998). Different mycobacterial species including *M. smegmatis* (Dick, Lee and Murugasu-Oei, 1998; Smeulders *et al.*, 1999; Mayuri *et al.*, 2002) and BCG (Michele, Ko and Bishai, 1999, Lim *et al.*, 1999; Hutter and Dick, 1999, 2000; Boon *et al.*, 2001; Sander *et al.*, 2001) have been used to model dormancy, as well as *M. tuberculosis*. These studies described the expression of gene products including a 27 kDa homologue of histone-like protein H1p, whose disruption did not affect viability during a prolonged stationary phase (Lee, Murugasu-Oei and Dick, 1998), glycine dehydrogenase and α crystallin type protein (Lim *et al.*, 1999; Boon *et al.*, 2001), fused nitrite reductase (*narX*; Hutter and Dick, 1999), and nitrite extrusion protein (*narK2*; Hutter and Dick, 2000) among others.

Expression of a *sigF* reporter in BCG indicated that SigF is increased by antibiotic treatment and stress treatments including anaerobic metabolism, cold shock, and oxidative stress (Michele, Ko and Bishai, 1999). The 100-fold higher expression of SigF in stationary phase growth vs log phase growth suggests that it may control mycobacterial persistence genes during chemotherapy (see above). BCG *recA* mutants do not affect the *in vitro* dormancy response or lead to attenuation in a mouse infection model (Sander *et al.*, 2001).

Corresponding studies of *M. tuberculosis* dormancy indicated the regulated expression of alkylhydroxperoxide reductase (*ahpC*; Master *et al.*, 2002).

M. tuberculosis aphC mutants are more susceptible to peroxynitrite (a RNI and RNO product), but not to NO alone; and have decreased survival in resting macrophage, but not in activated macrophage. Other studies described regulation of genes, including hspX and DevR/S two-component response regulator (Mayuri *et al.*, 2002; Saini *et al.*, 2004; Saini, Malhotra and Tyagi, 2004) and regulated protein expression of α-crystallin homologue (HspX), elongation factor Tu (Tuf), GroEL2, ScoB, mycolic acid synthase (CmaA2), thioredoxin (TrxB2), β-ketoacyl-ACP synthase (KasB), I-alanine dehydrogenase (Ald), Rv2005c, Rv2629, Rv0560c, Rv2185c, and Rv3866 using the modified Wayne dormancy model (Starck *et al.*, 2004).

Ultimately, a response regulator for gene expression regulated in response to oxygen depletion was described (Park *et al.*, 2003; Boon and Dick, 2002). A gene knock-out of the BCG Rv3133c response regulator dies during oxygen depletion and blocks expression of 3 major proteins on two-dimensional gels (Boon and Dick, 2002). This gene product was named DosR for dormancy survival regulator. An *M. tuberculosis* sequence motif for DosR binding was described that is found upstream of regulated genes, notably *acr* (Park *et al.*, 2003), and it was demonstrated that DosR is a member of the two-component response regulator class responsible for regulation in response to hypoxia. Two genes adjacent to DosR encode two sensor kinases, DosS and DosT. These sensor proteins can activate DosR transcription factor function via phosphorylation (Roberts *et al.*, 2004).

Subsequently, gene expression in response to low concentrations of NO was used to model anaerobic metabolism. Voskuil *et al.*, (2003) modelled gene expression during bacterial persistence in the granuloma using DNA microarrays to assay changing patterns of gene expression in response to low concentrations of NO. Strikingly, this approach induced a 48-gene dormancy regulon via the response regulator DosR (Boon and Dick, 2002). The same genes are induced by *in vitro* culture in limiting oxygen, and genes previously described as being regulated by low oxygen are found in the dormancy regulon (Sherman *et al.*, 2001). Induction of this regulon is very rapid and mRNA accumulates within 5 min of exposure to the NO-producing pulse of diethylenetriamine/nitric oxide. These genes are not co-regulated by oxidative stress, low iron, starvation, or SDS, indicating that they are not part of the general stress response. In contrast, high concentrations of NO similar to those produced by antimicrobial activity induce ~400 genes including those involved in responding to oxidative stress (i.e. *katG, ahpC, trxC, trxB2, sigB*, and *sigH*). Currently the *M. tuberculosis* dormancy state appears to be initiated by gradual nutrient depletion and reduced oxygen tension, leading to nonreplication and upregulation of genes involved in β-oxidation of fatty acids and virulence (Hampshire *et al.*, 2004; Daniel *et al.*, 2004).

The classification of *M. tuberculosis* genes into a schema of virulence classes (Glickman and Jacobs, 2001) was recently covered in an extensive review (Smith, 2003). These classes include *sgiv* (severe growth *in vivo*), *giv* (growth *in vivo*), or *per* (persistence) mutants (Glickman and Jacobs, 2001). Unlike other bacterial pathogens, mycobacterial species do not elaborate classical virulence factors such as toxins. Putative virulence factors encompass cell secretion and envelope function products, enzymes involved in general cellular metabolism, and transcriptional regulators. In all, 41 gene products are described encompassing most of the genes covered by this review. The focus upon genes from the sole viewpoint of the pathogen allows a comprehensive description of how various gene products function within the life cycle of the bacteria, and how they may function as virulence factors in the context of regulation for either constitutive or inducible expression.

Rapid advances in the fields of mycobacterial genetics are bringing forth many new tools. The availability of custom mycobacterial libraries genetically deficient for specific genes and the emergence of improved protein interaction databases are only two examples of technologies which will facilitate a better understanding of the genetic program of host–mycobacteria co-existence.

23.4 Host Genes Important to Granuloma Formation

IFN-γ and TNF-α are crucial cytokines for protective granuloma formation against mycobacteria, and genes related to regulation, production and effects of these agents all play a role in granuloma function. Additionally, many genes important for recruitment of inflammatory cells (pro- and anti-inflammatory cytokines, chemokines, osteopontin M, etc.; Flynn and Chan, 2001) have been shown to be modifying factors for granuloma formation. A majority of people exposed to mycobacteria do not get infected, indicating the importance of innate response elements for resistance. However, in granuloma formation, CD4 T-cells play a central role and we will focus on these cells in the rest of this review.

Despite the importance of CD4 T-cells in granuloma formation, important questions remain about their role. Rag$^{-/-}$ immunodeficient mice or CD4 T-cell-deficient HIV-infected people do not form functional granulomas and mycobacterial infections are disseminated. The adoptive transfer of CD4 T-cells to Rag$^{-/-}$ mice restores granuloma formation and prevents dissemination, clearly showing the central role of these cells in granuloma formation. T-cell function begins very early on during the initial aggregation of infected macrophage as granuloma formation is initiated. Cytokines and effector molecules elaborated by T-cells are some of the most essential host genes contributing to protective granuloma function by recruiting inflammatory cells and regulating their interactions (Co *et al.*, 2004a). Granuloma resolution either by apoptosis or fibrosis is also regulated by CD4 T-cells via a series of cytokines involving TGF-β. So CD4 T-cells are important for initiation, upkeep and resolution of granulomas. Relatively little is known about the recruitment of T-cells to the granuloma and the importance of their T-cell receptor (TCR) repertoire for full protective function. Recent studies have shed some light upon the role of both specific and nonspecific T-cells. Data suggest that, while initial T-cell infiltration is commonly specific for local antigens, during chronic inflammation, such as within granulomas, recruitment of nonspecific T-cells occurs (Steinman, 1996; Hogan, Weinstock and Sandor, 1999).

In addition, the available data for mycobacteria-induced granulomas suggest that local antigen specificity may not be an absolute requirement for accumulation of T-cells (Hogan *et al.*, 2001). Nonspecific T-cells having an activated phenotype accumulate in BCG-induced granuloma, but antigen-specific T-cells preferentially accumulate. The TCR repertoire of the BCG-induced lesion is quite diverse even at the level of the single granuloma, similar to what is observed during *S. mansoni* infection (Hogan *et al.*, 2002). Furthermore, injection of a GFP-tagged CNS-autoantigen specific T-cell line into BCG-infected mice results in the localization of GFP-positive T-cells in the granuloma (Sewell *et al.*, 2003).

Protective granulomas can be induced after infection of TCR transgenic 5C.C7 Rag2$^{-/-}$ mice with rBCG expressing the recognized epitope (Hogan *et al.*, 2001) or with lowered effectiveness after infection with wtBCG where there is no specificity (Hogan *et al.*, manuscript in preparation). Our data suggest a model in which activated T-cells can move to local inflammatory sites more efficiently than naive cells. Increased T-cell accumulation and optimal protection are observed when infiltrating T-cells recognize their cognate antigen within the granuloma (Co and Sandor, manuscript in preparation). To examine any differences in macrophage function arising from T-cell activation by cognate antigen vs nonspecific activation, DNA microarrays were used to compare gene expression in granuloma Mac-1^{+} sorted cells at a chronic stage of infection of TCR transgenic mice when the inducing BCG did or did not express the T-cell epitope recognized by the monospecific T-cell population (rBCG-PCC vs wtBCG infection of 5C.C7 Rag2$^{-/-}$ mice at 6 weeks). Our results indicate that fewer than 50 macrophage genes are upregulated by the presence of antigen in the granuloma (Hogan and Sandor, unpublished data). More work will be required to understand the significance of the upregulated genes and to confirm their expression using other experimental approaches.

In summary, in mycobacteria-induced granulomas, mycobacteria-specific T-cells are present and important for the induction of granulomas that are protective to both the host and the pathogen. For ongoing infections, the role of nonmycobacteria-specific T-cells in the maintenance of on-going granuloma function becomes more and more important. Activated nonspecific T-cells are able to accumulate in chronic granulomas, bringing in protective cytokines and augmenting protective function as the frequency of mycobacteria-specific T-cells decreases. During immunosuppression, when the levels of both specific and nonspecific activated T-cells are diminished, granuloma function fails, and bacteria reactivate and disseminate.

23.5 Granulomatous Inflammation as an Ecological System

Within the containment of granulomas, the long-term survival of macrophage and persistent mycobacteria for the duration of a clinically latent infection is a challenging problem. Understanding this phenomenon is not only intellectually fascinating, but of profound importance to public health. Tuberculosis continues to be a widespread cause of morbidity and mortality worldwide. The unique synergism of HIV and *M. tuberculosis* co-infections extends beyond the reliance of protective granuloma formation upon adequate CD4 T-cell numbers and involves cooperative interactions of the infecting microorganisms at the cellular level (Deretic *et al.*, 2004). Even if HIV is omitted from the public health equation, gradual sustained improvements in healthcare worldwide will ultimately expose more and more of the Earth's population to the diseases of old age, including failures of immunity arising naturally and from chemotherapies. Chronic infections such as tuberculosis will be an ongoing challenge in this environment.

This review has approached chronic granuloma-inducing infections as inflammatory sites that comprise an ecosystem in which the mycobacterial species represent the organism seeking survival and the granulomatous inflammation represents the surrounding environment. This is similar to the hypothesis in which the host immune response is directly influenced by the replicating pathogen and involves a complex interaction between two competing genomes (Srinivasan and McSorley, 2004). For example, *Salmonella* is known to downregulate class II presented epitopes and express virulence factors directly influencing T-cell responses, while other organisms can directly initiate apoptosis of dendritic cells or T-cells. The complex changes in gene expression both in macrophages and in mycobacteria represent an interaction by which both components of the ecosystem accommodate each other to their mutual advantage. The infected macrophages restrict dissemination of the microorganism and protect the health of the host, while failing to achieve sterilizing immunity. Mycobacteria parasitize macrophage using a variety of immunosuppressive mechanisms, but are ultimately forced into a quiescent persistent or nonreplicative dormancy state in latently infected individuals. It is likely that the ability of the pathogen to achieve this state is an advantage, allowing its long-term survival in the face of a less-than-optimal host environment.

This review has attempted to delineate some of the more successful approaches to understanding co-accommodation at the level of gene expression using genomic approaches, significantly DNA microarrays. This technology can be enormously informative, but is ultimately only as useful as the underlying rigour of the experiment and the appropriateness of the model used. Under the best of circumstances, the volume of information generated can be overwhelming. It is clear that, alongside advances in genetics, proteomics and informatics tools, this approach will contribute significantly to our understanding of the local inflammatory reactions that represent the direct interface between the infectious agents and the host.

References

Ando M, Yoshimatsu T, Ko C, Converse PJ, Bishai WR. 2003. Deletion of Mycobacterium tuberculosis sigma factor E results in delayed time to death with bacterial persistence in the lungs of aerosol-infected mice. *Infect Immun* **71**: 7170–2.

Balasubramanian V, Wiegeshaus EH, Taylor BT, Smith DW. 1994. Pathogenesis of tuberculosis: pathway to apical localization. *Tuber Lung Dis* **75**: 168–78.

Boon C, Dick T. 2002. Mycobacterium bovis BCG response regulator essential for hypoxic dormancy. *J Bacteriol* **184**: 6760–7.

Boon C, Li R, Qi R, Dick T. 2001. Proteins of *Mycobacterium bovis* BCG induced in the Wayne dormancy model. *J Bacteriol* **183**: 2672–6.

Chan K, Knaak T, Satkamp L, Humbert O, Falkow S, Ramakrishnan L. 2002. Complex pattern of Mycobacterium marinum gene expression during long-term granulomatous infection. *Proc Natl Acad Sci USA* **99**: 3920–5.

Chaussabel D, Semnani RT, McDowell MA. Sacks D, Sher A, Nutman TB. 2003. Unique gene expression profiles of human macrophages and dendritic cells to phylogenetically distinct parasites. *Blood* **102**: 672–81.

Co D, Hogan LH, Kim SI, Sandor M. 2004a. T cell contributions to the different phases of granuloma formation. *Immunol Lett* **92**: 135–142.

Co DO, Hogan LH, Kim SI, Sandor M. 2004b. Mycobacterial granulomas: keys to a long-lasting host–pathogen relationship. *Clin Immunol* **113**: 130–6.

Cole ST, Brosch R, Parkhill J, Garnier T, Churcher C, Harris D, Gordon SV, Eiglmeier K, Gas S, Barry CE 3rd, Tekaia F, Badcock K, Basham D, Brown D, Chillingworth T, Connor R, Davies R, Devlin K, Feltwell T, Gentles S, Hamlin N, Holroyd S, Hornsby T, Jagels K, Barrell BG *et al.* 1998. Deciphering the biology of *Mycobacterium tuberculosis* from the complete genome sequence. *Nature* **393**: 537–44.

Cosma CL, Humbert O, Ramakrishnan L. 2004. Superinfecting mycobacteria home to established tuberculous granulomas. *Nat Immunol* **5**: 828–35.

Cunningham AF, Spreadbury CL. 1998. Mycobacterial stationary phase induced by low exygen tension: cell wall thickening and localization of the 16-kilodalton alpha-crystallin homolog. *J Bacteriol* **180**: 801–8.

Daniel J, Deb C, Dubey VS, Sirakova TD, Abomoelak B, Morbidoni HR, Kolattukudy PE. 2004. Induction of a novel class of diacylglycerol acyltransferases and triacylglycerol accumulation in *Mycobacterium tuberculosis* as it goes into a dormancy-like state in culture. *J Bacteriol* **186**: 5017–30.

de Wit D, Wootton M, Dhillon J, Mitchison DA. 1995. The bacterial DNA content of mouse organs in the Cornell model of dormant tuberculosis. *Tuber Lung Dis* **76**: 555–62.

DeMaio J, Zhang Y, Ko C, Young DB, Bishai WR. 1996. A stationary-phase stress-response sigma factor from *Mycobacterium tuberculosis. Proc Natl Acad Sci USA* **93**: 2790–4.

Deretic V, Vergne I, Chua J, Master S, Singh SB, Fazio JA, Kyei G. 2004. Endosomal membrane traffic: convergence point targeted by *Mycobacterium tuberculosis* and HIV. *Cell Microbiol* **6**: 999–1009.

Dick T, Lee BH, Murugasu–Oei B. 1998. Oxygen depletion induced dormancy in *Mycobacterium smegmatis. FEMS Microbiol Lett* **163**: 159–64.

Domenech P, Barry CE 3rd, Cole ST. 2001. *Mycobacterium tuberculosis* in the post-genomic age. *Curr Opin Microbiol* **4**: 28–34.

Ehrt S, Schnappinger D, Bekiranov S, Drenkow J, Shi S, Gingeras TR, Gaasterland T, Schoolnik G, Nathan C. 2001. Reprogramming of the macrophage transcriptome in response to interferon-gamma and Mycobacterium tuberculosis: signaling roles of nitric oxide synthase-2 and phagocyte oxidase. *J Exp Med* **194**: 1123–40.

Flynn JL, Chan J. 2001. Immunology of tuberculosis. *A Rev Immunol* **19**: 93–129.

Geiman DE, Kaushal D, Ko C, Tyagi S, Manabe YC, Schroeder BG, Fleischmann RD, Morrison NE, Converse PJ, Chen P, Bishai WR. 2004. Attenuation of late-stage disease in mice infected by the *Mycobacterium tuberculosis* mutant lacking the SigF alternate sigma factor and identification of SigF-dependent genes by microarray analysis. *Infect Immun* **72**: 1733–45.

Glickman MS, Jacobs WR, Jr. 2001. Microbial pathogenesis of *Mycobacterium tuberculosis*: down of a discipline. *Cell* **104**: 477–85.

Gold B, Rodriguez GM, Marras SA, Pentecost M, Smith I. 2001. The Mycobacterium tuberculosis IdeR is a dual functional regulator that controls transcription of genes involved in iron acquisition, iron storage and survival in macrophages. *Mol Microbiol* **42**: 851–65.

Greenwell-Wild T, Vazquez N, Sim D, Schito M, Chatterjee D, Orenstein J. M, Wahl SM. 2002. Mycobacterium avium infection and modulation of human macrophage gene expression. *J Immunol* **169**: 6286–97.

Hampshire T, Soneji S, Bacon J, James BW, Hinds J, Laing K, Stabler RA, Marsh PD, Butcher PD.v2004. Stationary phase gene expression of Mycobacterium tuberculosis following a progressive nutrient depletion: a model for persistent organisms? *Tuberculosis (Edinb)* **84**: 228–38.

Hingley-Wilson SM, Sambandamurthy VK, Jacobs WR, Jr. 2003. Survival perspectives from the world's most successful pathogen, *Mycobacterium tuberculosis*. *Nat Immunol* **4**: 949–955.

Hobson RJ, McBride AJ, Kempsell KE, Dale JW. 2002. Use of an arrayed promoter-probe library for the identification of macrophage-regulated genes in *Mycobacterium tuberculosis*. *Microbiology* **148**: 1571–9.

Hogan LH, Weinstock JV, Sandor M. 1999. TCR specificity in infection induced granulomas. *Immunol Lett* **68**: 115–20.

Hogan LH, Macvilay K, Barger B, Co D, Malkovska I, Fennelly G, Sandor M. 2001. Mycobacterium bovis strain bacillus Calmette–Guerin-induced liver granulomas contain a diverse TCR repertoire, but a monoclonal T cell population is sufficient for protective granuloma formation. *J Immunol* **166**: 6367–75.

Hogan LH, Wang M, Suresh M, Co DO, Weinstock JV, Sandor M. 2002. CD4+ TCR repertoire heterogeneity in *Schistosoma mansoni*-induced granulomas. *J Immunol* **169**: 6386–93.

Hutter B, Dick T. 1999. Up-regulation of narX, encoding a putative 'fused nitrate reductase' in anaerobic doemant *Mycobacterium bovis* BCG. *FEMS Microbiol Lett* **178**: 63–9.

Hutter B, Dick T. 2000. Analysis of the dormancy-inducible narK2 promoter in *Mycobacterium bovis* BCG. *FEMS Microbiol Lett* **188**: 141–6.

Kaushal D, Schroeder BG, Tyagi S, Yoshimatsu T, Scott C, Ko C, Carpenter L, Mehrotra J, Manabe YC, Fleischmann RD, Bishai WR. 2002. Reduced immunopathology and mortality despite tissue persistence in a *Mycobacterium tuberculosis* mutant lacking alternative sigma factor, SigH. *Proc Natl Acad Sci USA* **99**: 8330–5.

Kendall SL, Rison SC, Movahedzadeh F, Frita R, Stoker NG. 2004. What do microarrays really tell us about *M. tuberculosis*? *Trends Microbiol* **12**: 537–44.

Koul A, Herget T, Klebl B, and Ullrich A. 2004. Interplay between mycobacteria and host signalling pathways. *Nat Rev Microbiol* **2**: 189–202.

Krutzik SR, Modlin RL. 2004. The role of Toll-like receptors in combating mycobacteria. *Semin Immunol* **16**: 35–41.

Lee BH, Murugasu-Oei B, Dick T. 1998. Upregulation of a histone-like protein in dormant *Mybobacterium smegmatis*. *Mol Gen Genet* **260**: 475–9.

Lim A, Eleuterio M, Hutter B, Murugasu-Oei B, Dick T. 1999. Oxygen depletion-induced dormancy in *Mycobacterium bovis* BCG. *J Bacteriol* **181**: 2252–6.

Manabe YC, Bishai WR. 2000. Latent *Mycobacterium tuberculosis*-persistence, patience, and winning by waiting. *Nat Med* **6**: 1327–9.

Manganelli R, Fattorini L, Tan D, Iona E, Orefici G, Altavilla G, Cusatelli P, Smith I. 2004. The extra cytoplasmic function sigma factor sigma(E) is essential for *Mycobacterium tuberculosis* virulence in mice. *Infect Immun* **72**: 3038–41.

Master SS, Springer B, Sander P, Boettger EC, Deretic V, Timmins GS. 2002. Oxidative stress response genes in *Mycobacterium tuberculosis*: role of ahpC in resistance to peroxynitrite and stage-specific survival in macrophages. *Microbiology* **148**: 3139–44.

Mayuri Bagchi G, Das TK, Tyagi JS. 2002. Molecular analysis of the dormancy response in Mycobacterium smegmatis: expression analysis of genes encoding the DevR-DevS tow-component system, Rv3134c and chaperone alpha-crystallin homologues. *FEMS Microbiol Lett* **211**: 231–7.

Michele TM, Ko C, Bishai WR. 1999. Exposure to antibiotics induces expression of the Mycobacterium tuberculosis sigF gene: implications for chemotherapy against mycobacterial persistors. *Antimicrob Agents Chemother* **43**: 218–25.

Monack DM, Mueller A, Falkow S. 2004. Persistent bacterial infections: the interface of the pathogen and the host immune system. *Nat Rev Microbiol* **2**: 747–65.

Nau GJ, Richmond JF, Schlesinger A, Jennings EG, Lander ES, Young RA. 2002. Human macrophage activation programs induced by bacterial pathogens. *Proc Natl Acad Sci USA* **99**: 1503–8.

Park HD, Guinn KM, Harrell MI, Liao R, Voskuil MI, Tompa M, Schoolnik GK, Sherman DR. 2003. Rv3133c/dosR is a transcription factor that mediates the hypoxic response of *Mycobacterium tuberculosis*. *Mol Microbiol* **48**: 833–43.

Ragno S, Romano M, Howell S, Pappin DJ, Jenner PJ, Colston MJ. 2001. Changes in gene expression in macrophages infected with *Mycobacterium tuberculosis*: a combined transcriptomic and proteomic approach. *Immunology* **104**: 99–108.

Ramakrishnan L, Federspiel NA, Falkow S. 2000. Granuloma-specific expression of *Mycobacterium virulence* proteins from the glycine-rich PE-PGRS family. *Science* **288**: 1436–9.

Roberts DM, Liao RP, Wisedchaisri G, Hol WG, Sherman DR. 2004. Two sensor kinases contribute to the hypoxic response of *Mycobacterium tuberculosis*. *J Biol Chem* **279**: 23082–7.

Saini DK, Malhotra V, Tyagi JS. 2004. Cross talk between DevS sensor kinase homologue, Rv2027c, and DevR response regulator of *Mycobacterium tuberculosis*. *FEBS Lett* **565**: 75–80.

Saini DK, Malhotra V, Dey D, Pant N, Das TK, Tyagi JS. 2004. DevR-DevS is a bona fide two-component system of *Mycobacterium tuberculosis* that is hypoxia-responsive in the absence of the DNA-binding domain of DevR. *Microbiology* **150**: 865–75.

Saltini C. 1995. A genetic marker for chronic beryllium disease. In *Biomarkers and Occupational Health: Progress and Perspectives*. Joseph Henry Press: Washington DC; 293–303.

Sander P, Papavinasasundaram KG, Dick T, Stavropoulos E, Ellrott K, Springer B, Colston MJ, Bottger EC. 2001. *Mycobacterium bovis* BCG recA deletion mutant shows increased susceptibility to DNA-damaging agents but wild-type survival in a mouse infection model. *Infect Immun* **69**: 3562–8.

Sandor M, Weinstock JV, Wynn TA. 2003. Granulomas in schistosome and mycobacterial infections: a model of local immune responses. *Trends Immunol* **24**: 44–52.

Schnappinger D, Ehrt S, Voskuil MI, Liu Y, Mangan JA, Monahan IM, Dolganov G, Efron B, Butcher PD, Nathan C, Schoolnik GK. 2003. Transcriptional adaptation of *Mycobacterium tuberculosis* within macrophages: insights into the phagosomal environment. *J Exp Med* **198**: 693–704.

Sewell DL, Reinke EK, Co DO, Hogan LH, Fritz RB, Sandor M, Fabry Z. 2003. Infection with *Mycobacterium bovis* BCG diverts traffic of myelin oligodendroglial glycoprotein autoantigen-specific T cells away from the cental nervous system and ameliorates experimental autoimmune encephalomyelitis. *Clin Diagn Lab Immunol* **10**: 564–72.

Sherman DR, Voskuil M, Schnappinger D, Liao R, Harrell MI, Schoolnik GK. 2001. Regulation of the *Mycobacterium tuberculosis* hypoxic response gene encoding alphacrystallin. *Proc Natl Acad Sci USA* **98**: 7534–9.

Shi L, Jung YJ, Tyagi S, Gennaro ML, North RJ. 2003. Expression of Th1-mediated immunity in mouse lungs induces a *Mycobacterium tuberculosis* transcription pattern characteristic of non-replicating persistence. *Proc Natl Acad Sci USA* **100**: 241–6.

Smeulders MJ, Keer J, Speight RA, Williams HD. 1999. Adaptation of *Mycobacterium smegmatis* to stationary phase. *J Bacteriol* **181**: 270–83.

Smith I. 2003. *Mycobacterium tuberculosis* pathogenesis and molecular determinants of virulence. *Clin Microbiol Rev* **16**: 463–96.

Srinivasan A, McSorley SJ. 2004. Visualizing the immune response to pathogens. *Curr Opin Immunol* **16**: 494–8.

Starck J, Kallenius G, Marklund BI, Andersson DI, Akerlund T. 2004. Comparative proteome analysis of *Mycobacterium tuberculosis* grown under aerobic and anaerobic conditions. *Microbiology* **150**: 3821–9.

Steinman L. 1996. A few autoreactive cells in an autoimmune infiltrate control a vast population of nonspecific cells: a tale of smart bombs and the infantry. *Proc Natl Acad Sci USA* **93**: 2253–6.

Steyn AJ, Collins DM, Hondalus MK, Jacobs WR Jr, Kawakami RP, Bloom BR. 2002. *Mycobacterium tuberculosis* WhiB3 interacts with RpoV to affect host survival but is dispensable for *in vivo* growth. *Proc Natl Acad Sci USA* **99**: 3147–52.

Tailleux L, Schwartz O, Herrmann JL, Pivert E, Jackson M, Amara A, Legres L, Dreher D, Nicod LP, Gluckman JC, Lagrange PH, Gicquel B, Neyrolles O. 2003. DC-SIGN is the major *Mycobacterium tuberculosis* receptor on human dendritic cells. *J Exp Med* **197**: 121–7.

Triccas JA, Gicquel B. 2000. Life on the inside: probing mycobacterium tuberculosis gene expression during infection. *Immunol Cell Biol* **78**: 311–7.

Vergne I, Chua J. Singh SB, Deretic V. 2004. Cell biology of *Mycobacterium tuberculosis* phagosome. *A Rev Cell Dev Biol* **20**: 367–94.

Voskuil MI, Visconti KC, Schoolnik GK. 2004. *Mycobacterium tuberculosis* gene expression during adaptation to stationary phase and low-oxygen dormancy. *Tuberculosis (Edinb)* **84**: 218–27.

Voskuil MI, Schnappinger D, Visconti KC, Harrell MI, Dolganov GM, Sherman DR, Schoolnik GK. 2003. Inhibition of respiration by nitric oxide induces a *Mycobacterium tuberculosis* dormancy program. *J Exp Med* **198**: 705–13.

Wayne LG. 1994. Dormancy of *Mycobacterium tuberculosis* and latency of disease. *Eur J Clin Microbiol Infect Dis* **13**: 908–14.

Wayne LG, Sohaskey CD. 2001. Nonreplicating persistence of *Mycobacterium tuberculosis. A Rev Microbiol* **55**: 139–63.

Zhang Y. 2004. Persistent and dormant tubercle bacilli and latent tuberculosis. *Front Biosci* **9**: 1136–56.

Index

Page numbers in italic, e.g. *101*, refer to figures. Page numbers in bold, e.g. **355**, denote entries in tables.

Immunogenomics and Human Disease Edited by András Falus

Index compiled by John Holmes